Mycobacterium tuberculosis: Molecular Infection Biology, Pathogenesis, Diagnostics and New Interventions

Seyed Ehtesham Hasnain
Nasreen Z. Ehtesham • Sonam Grover
Editors

Mycobacterium tuberculosis: Molecular Infection Biology, Pathogenesis, Diagnostics and New Interventions

Springer

Editors
Seyed Ehtesham Hasnain
JH Institute of Molecular Medicine,
Jamia Hamdard
New Delhi, Delhi, India

Dr Reddy's Institute of Life Sciences
University of Hyderabad Campus
Hyderabad, India

Nasreen Z. Ehtesham
Inflammation Biology and Cell Signaling
Laboratory
ICMR-National Institute of Pathology,
Safdarjung Hospital Campus
New Delhi, Delhi, India

Sonam Grover
JH Institute of Molecular Medicine,
Jamia Hamdard
New Delhi, Delhi, India

ISBN 978-981-32-9412-7 ISBN 978-981-32-9413-4 (eBook)
https://doi.org/10.1007/978-981-32-9413-4

This Springer imprint is published by the registered company Springer Nature Singapore Pte Ltd.
The registered company address is: 152 Beach Road, #21-01/04 Gateway East, Singapore 189721, Singapore

Foreword

I am delighted to write the foreword for *Mycobacterium tuberculosis: Molecular Infection Biology, Pathogenesis, Diagnostics and New Interventions*, edited by Prof. Seyed Hasnain and colleagues. Prof. Hasnain has brought together an impressive set of experts and authors to contribute to this seminal work.

Tuberculosis (TB) remains the leading infectious killer disease of our times, with ten million new cases every year. A large portion of the burden is borne by countries in Asia and sub-Saharan Africa. Recent years have also seen growing numbers of drug and multidrug-resistant TB, leading to higher mortality and morbidity. While TB has been documented in Egyptian mummies, it has learnt to adapt and survive with mankind and continues to pose significant challenges for its control. The major risk factors include those associated with poverty like undernutrition, overcrowding, tobacco and alcohol use, low awareness, and lack of access to quality healthcare services. While HIV infection has been a major driver of TB in sub-Saharan Africa, the growing burden of diabetes and medication-induced immunosuppression are significant risk factors elsewhere.

While the End TB targets of the WHO as well as the SDG global targets envision an end to this global epidemic, many gaps in understanding its pathogenesis, transmission, and factors that enhance or reduce the risk of virulence or the development of resistance mutations remain. Furthermore, there is presently no rapid diagnostic test that can diagnose all forms of TB simply and affordably, and as a consequence, millions of patients face delayed or wrong diagnoses. While a few new drugs have been developed in the last few years, patients with TB still face months or years of treatment with drugs, which could have potentially severe side effects, especially in the case of MDRTB treatment. Attempts have also been made to develop a vaccine that would be more effective than BCG, but the lack of surrogate markers of protection makes clinical trials of new vaccines time-consuming and expensive.

This book addresses fundamental biology of tuberculosis as well as non-tuberculous mycobacteria, an increasingly recognized cause of chronic lung disease in many parts of the world. I am sure that this book will serve as a comprehensive knowledge base for *M. tb*, the pathogen that causes TB, as well as non-tuberculous mycobacteria (NTM). It highlights developments in understanding the disease at the molecular level, identification and validation of molecular drug targets, molecular epidemiology, and novel diagnostics. A special chapter on TB vaccines describing present and future perspectives is also one of the highlights of this book. Posttranslational and epigenetic modifications in *Mycobacterium* with emphasis on methyltransferases are another key chapter. Functional biology of the unique group of proteins called PE/PPE, present exclusively in the genus *Mycobacterium* and nowhere else in the living kingdom, is also discussed in the book.

The book has a total of 25 chapters written by eminent scientists drawn from abroad and India, such as from Robert Koch Institute, Berlin; Trinity College, Dublin; Harvard T.H. Chan School of Public Health, Boston, MA, USA; Yale School of Medicine, New Haven, USA; Emory University, Atlanta, USA; John Hopkins University, Baltimore, USA; Max Planck Institute for Infection Biology, Berlin, Germany; Indian Institute of Technology, Delhi; Indian Institute of Science, Bangalore; All India Institute of Medical Sciences (AIIMS), Delhi; National Institute of Pathology, New Delhi; National Institute of Immunology, New Delhi; Delhi University; University of Hyderabad; Jamia Hamdard; etc. A blend of basic scientists, drug discovery researchers, epidemiologists, clinicians, and vaccinologists has collectively focused on various facets of the disease and the causal organisms. With state-of-the-art contributions from experts in the respective domains, I am sure that this volume will be an informative resource for current practitioners as well as students and researchers.

Chief Scientist, WHO
Geneva, Switzerland

Soumya Swaminathan, MD

Preface

Tuberculosis (TB), the disease, and *Mycobacterium tuberculosis* (*M.tb*), the causative pathogen, need no introduction. This book on TB will serve as a comprehensive knowledge base for *M.tb,* as well as nontubercular bacteria (NTM).

Presenting a wealth of information on the molecular infection biology of *M.tb*, as well as nontuberculous mycobacteria (NTM), this book provides an overview of the functional role of the PE/PPE group of proteins, which is exclusive to the genus *Mycobacterium*, of host-pathogen interactions, and of virulence. It also explores the pathogenesis of the infection, pathology, epidemiology, and diagnosis of NTM. Last but not least, it discusses current and novel approaches in vaccine development against tuberculosis, including the role of nanotechnology. This book highlights developments in understanding the disease at the molecular levels and identification and validation of molecular drug targets, molecular epidemiology, and novel diagnostic techniques. With state-of-the-art contributions from experts in the respective domains, this volume will be an informative resource for current practitioners as well as medical students, postgraduates, and researchers and is expected to be much in demand. The book is mainly divided into three sections covering all major aspects of TB.

Pathogenesis and Epidemiology of Tuberculosis

This section starts with a chapter on the history of tuberculosis with a description of fascinating events starting from the discovery of the elusive TB bacteria. A chapter on TB comorbidities describes the explicit involvement of TB in promoting other lung diseases such as bronchiectasis and chronic obstructive pulmonary disease during childhood. To reduce the global burden of chronic lung diseases, the management of TB in the early stages is imperative. A chapter on the role of host factors in susceptibility to TB serves as a primer to understand the efforts mounted by the host which can be useful in designing host-directed therapies and also reduce tissue damage due to chronic TB infections. A chapter on extrapulmonary tuberculosis (EPTB) provides an in-depth overview of the clinical diagnosis and treatment of the different forms of EPTB. This section also covers nontubercular mycobacteria (NTM), a high priority due to its misdiagnosis and association with cystic fibrosis, COPD, and nosocomial infections. The last chapter in this section discusses the spread of *M.tb* strains circulating in New Zealand and emphasizes the need for early detection and treatment of active cases together with contact tracing, if one has to minimize the emergence of epidemiological clusters of the disease. This chapter also illustrates the importance of a political will and decisive legislation in bringing down the incidence of disease from 150 cases per 100,000 to less than 10 cases per 100,000 and steps a country need to take to avoid new cases of TB.

Molecular and Infection Biology of *M.tb*

M.tb is highly adapted for survival in the extremely hostile intracellular environment in host macrophages. In order to cope up with stresses encountered by the bacterium in macrophages, it modulates its gene expression in various ways. This section is a combination of chapters where molecular aspects of *M.tb* have been highlighted to understand bacterial survival strategy. This section begins with a chapter on methyltransferases, present in large number in *M.tb,* modifying DNA, RNA, proteins, lipids, and other biomolecules of both pathogen and host and illustrates how these epigenetic modifiers impact virulence and pathogenesis of *M.tb*. Three consecutive chapters provide a comprehensive overview on functional biology of the unique group of proteins called PE/PPE, which is present exclusively in the genus *Mycobacterium* and nowhere else in the living kingdom. The discussions include evolutionary developments, characterization, and transcriptional regulation of the PE/PPE family along with their involvement in host immunomodulation and also showcase the functional role of intrinsically disordered regions concentrated within this protein family. A chapter on the importance of cell wall-associated poly-α-L-glutamine in the biology of pathogenic mycobacteria is also presented in this book which dissects the unique cell wall architecture of *M.tb* contributing to the unrestricted success of the bacterium during infection. Cellular stress responses during *M.tb* infection have been highlighted in the next chapter including the strategies adopted by *M.tb* to influence the host innate and adaptive immune responses. The role of heat shock proteins in the mycobacterial pathogenesis has also been discussed in a separate chapter. The importance of endoplasmic reticulum (ER) stress in *M.tb.* pathogenesis including the ER stress pathways evoked by *M.tb* proteins such as ESAT-6, HBHA, 38-kDa antigen, and PE_PGRS5 is nicely presented in the next chapter. This is followed by a discussion on the role of toxin-antitoxin (TA) systems in bacterial pathogenesis and the abundance of such TA systems in *M.tb*, which in turn helps the pathogen fight against various cellular stresses including acidic, hypoxic, and oxidative and also challenges posed by the immune system. A detailed description of the nucleotide excision repair (NER) pathway in mycobacteria is in a separate chapter where in-depth studies of the structure-function relationship of DNA repair throw light on the molecular mechanisms employed by *M.tb* for self-defense. Such studies will enable the development of novel therapeutic interventions against TB.

The chapter on mesenchymal stem cells (MSC) describes its potential in *M.tb* persistence, resuscitation, and disease reactivation and highlights studies on MSCs as a reservoir of *M.tb,* which could be a probable target for developing new therapeutics against non-replicating bacteria and viable but non-culturable bacteria. The last chapter in this section emphasizes biofilm and its importance in drug resistance and how biofilm disruption could be an important strategy not only to contain the TB bacteria but also to reduce IC_{50} of antitubercular drugs.

Vaccines, Diagnostics, and New Interventions

This section begins with a detailed description of the best practices for mycobacteria research laboratories, given the deadly nature of this airborne pathogen. It also covers the occupational risks of working with *M.tb* and identifies the primary and secondary barriers to *M.tb* exposure, controls that reduce the risk of exposure and also the good practices and guidelines to be followed during experimental manipulation so as to avoid getting infected with the bacterium. Of particular interest are the deployment of biosafety cabinets, good waste management practices, spills management, and of course occupational health programs and facility decontamination. This is followed by comprehensive review on TB vaccines. The presently used whole cell vaccine such as BCG is inefficient in TB-endemic population. Thus, there is an urgent need for novel vaccines capable of providing complete protection in all age groups as well as in different stages of infection. This is followed by a chapter describing the immunotherapeutic properties of *Mycobacterium indicus pranii (MIP)* and its likely use as an alternative vaccine candidate. *MIP*, a nonpathogenic saprophytic mycobacterium, shares many cross-reactive antigens with *M.tb* and proved to be effective in category II TB as an adjunct to MDT.

History as well as present status of TB diagnostics has also been covered in a chapter entitled "TB Diagnostics: Journey from Smear Microscopy to Whole Genome Sequencing." Developing point of care diagnostic with fast, reliable, and affordable features is vital to curb and control TB and HIV-TB burden in resource-poor countries. This is discussed in the chapter that also focusses on the journey of such diagnostic techniques starting from older strategies such as smear microscopy to modern-day techniques like whole genome sequencing.

An interesting chapter describes how to break the transmission of TB by bridging the gap in controlling TB in endemic settings. In order to control the spread of MDR-TB, there is a need to use better diagnostic tool and health measures to interrupt pathogen transmission. Challenges and advances in TB drug discovery are also discussed in detail including the advancement of methodologies for enhancing bioavailability of drugs by different means such as nanoparticle-/liposome-mediated delivery systems and devices for sustained release.

The last chapter in this section is based on the experience of actual drug discovery gained by the researchers at Johns Hopkins University. While describing the processes involved in TB drug discovery, and a review of existing candidate drugs, the authors illustrate their efforts in using drug repurposing strategy for development of new drugs against MDR and XDR strains of *M.tb*. The very powerful carbapenem antibiotic, belonging to the beta-lactam group, has been used as a model new drug. In addition, the use of antisense RNA-based therapeutics and CRISPER-CAS system-based new interventions are also discussed.

New Delhi, Delhi, India

Seyed Ehtesham Hasnain
Nasreen Z. Ehtesham
Sonam Grover

Acknowledgments

This book is made possible, thanks to the unqualified support, hard work and adherence to the deadlines by scientists, clinicians, and academicians from around the globe who took their time off from their other responsibilities to present new knowledge about TB, a disease that largely affects the underprivileged in underdeveloped countries. We are indeed honored to be associated with renowned personalities, working on different aspects of TB, and are humbled by their commitment, kindness, and endurance.

We do hope that this book will add new knowledge about the unique attributes of this super smart bacteria so as to enable the development of new interventions including drugs and vaccines and also point-of-care, rapid, and cost-effective diagnosis of TB.

The contributions of our reviewers who generously donated their time to review every chapter are gratefully acknowledged. We would like to put on record the efforts of Dr. Soumya Swaminathan, former Director General, Indian Council of Medical Research and Secretary, Department of Health Research, presently Chief Scientist, WHO Geneva, for writing the foreword which elegantly and succinctly summarizes the salient features of the mammoth effort that have gone into compiling this book.

We would like to duly acknowledge the editorial support provided by Ms. Ruby Palta (PS to SEH) and Dr. Salma Jamal and Dr. Rishabh Gangwar, Postdocs in our lab for their technical help. Springer Nature deserves full credit for their persistence, professionalism, and enthusiasm during the publication of the book.

Finally, our respective families deserve all the credit for allowing us to bypass family and social commitments during the completion of this project.

Contents

About the Editors

Prof. Seyed Ehtesham Hasnain has contributed significantly to tuberculosis (TB) research, focusing on infection biology and functional epidemiology. He is Fellow of Robert Koch Institute (Berlin), American Academy of Microbiology (USA), TWAS, and all major Indian Science Academies (FNA/FASc/FNASc) and member of German National Academy of Sciences Leopoldina. He has received numerous recognitions and honors, including Alexander von Humboldt Research Award (Germany), Shanti Swarup Bhatnagar Prize, GD Birla Award, JC Bose Fellowship Award, and many others. A former member of the Science Advisory Council to the Prime Minister of India for two terms (2004–2014), he was the first director of the Centre for DNA Fingerprinting and Diagnostics, Hyderabad, and later vice-chancellor of the prestigious University of Hyderabad (2005–2011) and served as invited professor at the IIT Delhi (2011–2019). Currently, he is the Vice-Chancellor of Jamia Hamdard and also professor at JH Institute of Molecular Medicine, where he leads a very large, active, and generously funded TB research group.

Dr. Nasreen Z. Ehtesham received her Master's degree from Aligarh Muslim University, Aligarh, and graduate training from the University of Alberta, Edmonton, Canada. She pursued her Ph.D. at the National Institute of Immunology, New Delhi (1991), and joined the ICGEB as a Rockefeller postdoc fellow. She was appointed as Deputy Director, National Institute of Nutrition, Hyderabad, where her group focused on nutrition and inflammation biology. She established the link between pathogenesis, inflammation, and stress response using the *Mycobacterium tuberculosis* infection model. An elected Fellow of the National Academy of Sciences, she has served as a member of several decision-making bodies of the Health and S&T ministries. She is the recipient of many awards, including the ICMR Kshanika Oration Award for her contributions to tuberculosis biology. She has published over 70 research papers in prestigious journals such as *PNAS* and serves on the editorial boards of a number of journals. Currently, she is Director-in-Charge of the ICMR National Institute of Pathology, New Delhi.

Dr. Sonam Grover is a UGC Assistant Professor at JH Institute of Molecular Medicine, Jamia Hamdard, New Delhi. She completed her Ph.D. at JNU (2014) and joined Kusuma School of Biological Sciences, IIT Delhi, as a postdoctoral fellow in the same year. During this tenure, she received a Women Scientist Grant from the Department of Health Research, Government of India. Her main areas of research include *Mycobacterium tuberculosis* host-pathogen interactions, drug repurposing, and identifying novel drug targets against tuberculosis. She has published over 25 research papers in various international peer-reviewed journals including *mBio* and *Molecular Neurobiology*. She has delivered talks and presented her work at several national and international conferences and symposia. She is PI/co-PI in research projects funded by agencies such as DBT and DHR. She was also awarded the prestigious "ASCB Travel Award for Graduate Students" for presenting her research at international conferences.

Contributors

Fahad Abdullah Department of Respiratory Medicine, Deccan College of Medical Sciences, Hyderabad, India

Javed N. Agrewala Immunology Laboratory, Institute of Microbial Technology, Chandigarh, India
Centre for Biomedical Engineering, Indian Institute of Technology-Ropar, Rupnagar, India

Javeed Ahmad Inflammation Biology and Cell Signaling Laboratory, ICMR-National Institute of Pathology, Safdarjung Hospital Campus, New Delhi, Delhi, India

Anwar Alam JH Institute of Molecular Medicine, Jamia Hamdard, New Delhi, Delhi, India
Inflammation Biology and Cell Signaling Laboratory, ICMR-National Institute of Pathology, Safdarjung Hospital Campus, New Delhi, Delhi, India

Mohammad Aqdas Immunology Laboratory, Institute of Microbial Technology, Chandigarh, India

Mohd Arish JH Institute of Molecular Medicine, Jamia Hamdard, New Delhi, Delhi, India

Nirupama Banerjee Molecular Biology and Genetic Engineering Laboratory, School of Biotechnology, Jawaharlal Nehru University, New Delhi, Delhi, India

Hilal Bashir Immunology Laboratory, Institute of Microbial Technology, Chandigarh, India

Janendra K. Batra National Institute of Immunology, New Delhi, Delhi, India
Department of Biochemistry, School of Chemical and Life Sciences, Jamia Hamdard, New Delhi, Delhi, India

Pranami Bharadwaj Department of Microbiology, Tripura University (A Central University), Agartala, Tripura, India

Sangeeta Bhaskar Product Development Cell-I, National Institute of Immunology, New Delhi, Delhi, India

Rakesh Bhatnagar Molecular Biology and Genetic Engineering Laboratory, School of Biotechnology, Jawaharlal Nehru University, New Delhi, Delhi, India
Banaras Hindu University, Varanasi, Uttar Pradesh, India

Harish Chandra Department of Microbiology, School of Biomedical & Pharmaceutical Sciences, Babasaheb Bhimrao Ambedkar University, Lucknow, India

K. K. Chopra New Delhi Tuberculosis Center, New Delhi, Delhi, India

Bikul Das Department of Stem Cell and Infectious Diseases, Kavikrishna Laboratory, Guwahati Biotech Park, Indian Institute of Technology, Guwahati, India
Department of Stem Cells and Infectious Diseases, Thoreau Laboratory for Global Health, M2D2, University of Massachusetts, Lowell, MA, USA

Deepjyoti K. Das Immunology Laboratory, Institute of Microbial Technology, Chandigarh, India

Nasreen Z. Ehtesham Inflammation Biology and Cell Signaling Laboratory, ICMR-National Institute of Pathology, Safdarjung Hospital Campus, New Delhi, Delhi, India

Ravikrishnan Elangovan Department of Biochemical Engineering and Biotechnology, Indian Institute of Technology-Delhi, New Delhi, Delhi, India

Rishabh Gangwar JH Institute of Molecular Medicine, Jamia Hamdard, New Delhi, Delhi, India

Rajni Garg Jawaharlal Nehru Centre for Advanced Scientific Research, Bangalore, Karnataka, India
Department of Microbiology and Cell Biology, Indian Institute of Science, Bangalore, Karnataka, India

Jaishree Garhyan Department of Microbiology, University of Massachusetts, Amherst, MA, USA
Molecular Biology and Genetic Engineering Laboratory, School of Biotechnology, Jawaharlal Nehru University, New Delhi, Delhi, India

Uthaman Gowthaman Department of Laboratory Medicine, Immunobiology and Orthopedics, Yale University School of Medicine, New Haven, CT, USA

Sonam Grover JH Institute of Molecular Medicine, Jamia Hamdard, New Delhi, Delhi, India

Manish Gupta Molecular Biology and Genetic Engineering Laboratory, School of Biotechnology, Jawaharlal Nehru University, New Delhi, Delhi, India

Ashfaq Hasan Department of Respiratory Medicine, Deccan College of Medical Sciences, Hyderabad, India

Seyed Ehtesham Hasnain JH Institute of Molecular Medicine, Jamia Hamdard, New Delhi, Delhi, India
Dr Reddy's Institute of Life Sciences, University of Hyderabad Campus, Hyderabad, India

Salma Jamal JH Institute of Molecular Medicine, Jamia Hamdard, New Delhi, Delhi, India

K. Kala jyothi Department of Biotechnology and Bioinformatics, School of Life Sciences, University of Hyderabad, Hyderabad, Telangana, India

Gurpreet Kaur Immunology Laboratory, Institute of Microbial Technology, Chandigarh, India
Centre for Biomedical Engineering, Indian Institute of Technology-Ropar, Rupnagar, India

Junaid Khan Immunology Laboratory, Institute of Microbial Technology, Chandigarh, India

Nooruddin Khan Department of Biotechnology and Bioinformatics, School of Life Sciences, University of Hyderabad, Hyderabad, Telangana, India

Garima Khare Department of Biochemistry, University of Delhi South Campus, New Delhi, Delhi, India

Ashutosh Kumar Department of Microbiology, Tripura University (A Central University), Agartala, Tripura, India

Pankaj Kumar Department of Biochemistry, School of Chemical and Life Sciences, Jamia Hamdard, Delhi, India

Gyanu Lamichhane Division of Infectious Diseases School of Medicine, Johns Hopkins University, Baltimore, MD, USA

Astrid Lewin Robert Koch Institute, Unit 16 Mycotic and Parasitic Agents and Mycobacteria, Berlin, Germany

Rajesh Mani Molecular Biology and Genetic Engineering Laboratory, School of Biotechnology, Jawaharlal Nehru University, New Delhi, Delhi, India

Gillipsie Minhas Department of Biotechnology and Bioinformatics, School of Life Sciences, University of Hyderabad, Hyderabad, Telangana, India

Alladi Mohan Department of Medicine, Sri Venkateswara Institute of Medical Sciences, Tirupati, India

Sangita Mukhopadhyay Laboratory of Molecular Cell Biology, Centre for DNA Fingerprinting and Diagnostics (CDFD), Hyderabad, Telangana, India

K. Muniyappa Department of Biochemistry, Indian Institute of Science, Bangalore, India

Prachi Nangpal Department of Biochemistry, University of Delhi South Campus, New Delhi, Delhi, India

Faiza Nazar Laboratory of Molecular Cell Biology, Centre for DNA Fingerprinting and Diagnostics (CDFD), Hyderabad, Telangana, India

Farha Naz Inflammation Biology and Cell Signaling Laboratory, ICMR-National Institute of Pathology, Safdarjung Hospital Campus, New Delhi, Delhi, India

Shikha Negi Immunology Laboratory, Institute of Microbial Technology, Chandigarh, India

Khairun Nisaa Department of Life Sciences, Ben Gurion University of the Negev, Beersheba, Israel

Ronan F. O'Toole School of Molecular Sciences, College of Science, Health and Engineering, La Trobe University, Melbourne, VIC, Australia
Department of Clinical Microbiology, School of Medicine, Trinity College Dublin, Dublin, Ireland
School of Medicine, College of Health and Medicine, University of Tasmania, Hobart, Tasmania, Australia

Ravi Pal Laboratory of Molecular Cell Biology, Centre for DNA Fingerprinting and Diagnostics (CDFD), Hyderabad, Telangana, India
Graduate Studies, Manipal Academy of Higher Education, Manipal, Karnataka, India

Sai Haranath Praveen Department of Pulmonary and Critical Care Medicine, Apollo Hospitals, Hyderabad, India

Mamta Rani Plant Microbe Interactions Laboratory, National Institute of Plant Genome Research, New Delhi, Delhi, India

Sabeeha Ali Kusuma School of Biological Sciences, Indian Institute of Technology, New Delhi, Delhi, India

Ruxana T. Sadikot Section of Pulmonary, Critical Care Medicine, Emory University, Atlanta, GA, USA

Mohammad Sajid Immunology Laboratory, Institute of Microbial Technology, Chandigarh, India

Jasmine Samal Inflammation Biology and Cell Signaling Laboratory, ICMR-National Institute of Pathology, Safdarjung Hospital Campus, New Delhi, Delhi, India

Hubert Schäfer Robert Koch Institute, Unit 16 Mycotic and Parasitic Agents and Mycobacteria, Berlin, Germany

Mohd Shariq Inflammation Biology and Cell Signaling Laboratory, ICMR-National Institute of Pathology, Safdarjung Hospital Campus, New Delhi, Delhi, India

Jyoti Sharma Department of Biotechnology and Bioinformatics, School of Life Sciences, University of Hyderabad, Hyderabad, Telangana, India

Surendra K. Sharma JH Institute of Molecular Medicine, Jamia Hamdard, New Delhi, Delhi, India
Departments of General Medicine & Pulmonary Medicine, JNMC, Datta Meghe Institute of Medical Sciences (DMIMS), Wardha, Maharashtra, India
Department of Internal Medicine (WHO Collaborating Centre for Research & Training in Tuberculosis, Centre of Excellence for EPTB, MoH & FW, GoI), All India Institute of Medical Sciences, New Delhi, Delhi, India

Tarina Sharma Molecular Infection and Functional Biology Lab, Kusuma School of Biological Sciences, Indian Institute of Technology-Delhi, New Delhi, Delhi, India

Javaid A. Sheikh Department of Biotechnology, School of Chemical and Life Sciences, Jamia Hamdard, New Delhi, Delhi, India

Noman Siddiqi Harvard TH Chan School of Public Health, Boston, MA, USA

Bindu Singh Product Development Cell-I, National Institute of Immunology, New Delhi, Delhi, India

Pooja Singh JH Institute of Molecular Medicine, Jamia Hamdard, New Delhi, Delhi, India

Sanpreet Singh Immunology Laboratory, Institute of Microbial Technology, Chandigarh, India

Urvashi B. Singh Tuberculosis Section, Department of Microbiology, A.I.I.M.S., New Delhi, Delhi, India

Yadvir Singh Department Molecular Biology, Max Planck Institute for Infection Biology, Berlin, Germany

Elizabeth Story-Roller Division of Infectious Diseases, School of Medicine, Johns Hopkins University, Baltimore, MD, USA

Sharmistha Tapadar Department of Microbiology, Tripura University (A Central University), Agartala, Tripura, India

Chandrakant Tarke Department of Pulmonary and Sleep Medicine, Apollo Hospitals, Hyderabad, India

Manoj Thakur Department of Biochemistry, Indian Institute of Science, Bangalore, India

Deeksha Tripathi Department of Microbiology, School of Life Sciences, Central University of Rajasthan, Ajmer, Rajasthan, India

Prajna Tripathi National Institute of Immunology, New Delhi, Delhi, India
JH Institute of Molecular Medicine, Jamia Hamdard, New Delhi, Delhi, India

Anil K. Tyagi Department of Biochemistry, University of Delhi South Campus, New Delhi, Delhi, India
Guru Gobind Singh Indraprastha University, New Delhi, Delhi, India

Himanshu Vashistha Viral Hepatitis Division, National Centre for Disease Control, New Delhi, Delhi, India

Sheeba Zarin Inflammation Biology and Cell Signaling Laboratory, ICMR-National Institute of Pathology, Safdarjung Hospital Campus, New Delhi, Delhi, India

Part I

Pathogenesis and Epidemiology of Tuberculosis

History of TB: Robert Koch and Beyond

1

Ashfaq Hasan

> *That which is not expressed will be forgotten.*
> *And what is forgotten will happen again.*
> Yevgeny Yevtushenko
>
> *The only thing we learn from history is that we learn nothing from history.*
> Georg Wilhelm Friedrich Hegel

Abstract

Tuberculosis is a disease like no other. It has claimed more lives down the ages than perhaps any other infectious illness and has resisted man's efforts to isolate and conquer it. A disease once thought to be on the verge of extinction has rebounded in its most dangerous avatar yet. This chapter describes the culmination of the hunt for the elusive microbe of tuberculosis and the fascinating sequence of events that followed.

Keywords

Infection · Tuberculosis · World Health Organization

Abbreviations

BCG	Bacillus Calmette-Guerin
DOTS	Directly observed treatment
INH	Isonicotinylhydrazide

A. Hasan (✉)
Department of Respiratory Medicine, Deccan College of Medical Sciences, Hyderabad, India
e-mail: principal@deccancollegeofmedicalsciences.com

S. E. Hasnain et al. (eds.), Mycobacterium tuberculosis: *Molecular Infection Biology, Pathogenesis, Diagnostics and New Interventions*,
https://doi.org/10.1007/978-981-32-9413-4_1

MDR-TB	Multidrug-resistant tuberculosis
MTB	*Mycobacterium tuberculosis*
PAS	Para-aminosalicylic acid
PPD	Purified protein derivative
SCCT	Short-course chemotherapy
TB	Tuberculosis
WHO	World Health Organization
XDR-TB	Extensively drug-resistant tuberculosis

Anton van Leeuwenhoek, the Dutch microscopist, had seen protozoa in 1674 and bacteria in 1676. Since Leeuwenhoek's specimens were obtained from healthy individuals, he did not associate microorganisms with disease. The first time tuberculosis (TB) was conceptualized as an infection was in 1720. Benjamin Marten in his *A New Theory of Consumption* proposed that the etiologic agents of TB were "wonderfully minute living creatures" which were transmissible by the airborne route and that "it may be therefore very likely that by an habitual lying in the same bed with a consumptive patient, constantly eating and drinking with him, or by very frequently conversing so nearly as to draw in part of the breath he emits from the lungs, a consumption may be caught by a sound person." But a century and a half later, not much progress upon the matter had been made.

Louis Pasteur, the great French researcher—biochemist, microbiologist, and immunologist *par excellence*—was instrumental in destroying the *Theory of Miasma* and thus paved the way for the incrimination of microbes as the causative agents of infection. By his time, microscopes were achieving much higher magnifications than even Leeuwenhoek's superb lenses (that had been capable of ×250 to ×300 magnification).

Much later, Ernst Abbe, Professor at the University of Jena, working for Carl Zeiss took optics to a different level. One of Zeiss' microscopes found its way into the hands of a District Medical Officer in the pastoral German district of Wollstein.

Robert Koch, once an army doctor, now settled into the tedium of peacetime medical practice, became drawn to a mysterious illness rampant among the livestock of Wollstein—anthrax. At that time, in 1871, after the recently concluded Franco-German War, some of the bitterness between the two warring nations had spilled over into medicine itself. Louis Pasteur was already a colossus and a household name. Yet, despite his successes, he and other notable French scientists had failed to clarify the life cycle of what was to be called the *Bacillus anthracis*.

By day, Heinrich Hermann Robert Koch was a typical country doctor. But at night, with hardly any equipment to speak of (though he did have a microscope presented to him by his wife on his 29th birthday) and using slivers of wood that served as lancets, he worked ceaselessly to unravel the life cycle of anthrax.

Professor Ferdinand Cohn was Director of the Botanical Institute at the University of Breslau and widely regarded as the founder of modern bacteriology. Cohn had described, for the first time, endospores of *Bacillus subtilis*, a close relation of what today is known as *Bacillus anthracis* (Cohn 1872). It was natural that Koch turned to Cohn. "After many failures," Koch wrote to him, "I have finally succeeded in completely elucidating the life cycle of Bacillus anthracis" (Sakula 1982). Cohn, not a little surprised that a country practitioner using his rudimentary tools was claiming to have unraveled a mystery that had so far eluded the elite researchers of France and Germany, promptly summoned Koch, who then, before Cohn and pathologists Karl Weigert and Julius Cohnheim, proceeded to give a demonstration of "meticulous methodical research."

"We thus see that anthrax tissues, regardless of whether they are relatively fresh, putrefying, dried or many years old can produce anthrax when these substances contain bacilli capable of developing spores of Bacillus anthracis"—and by stating this, Koch brought into focus once again the role of germs at the heart of transmissible disease (Koch 2010).

Cohnheim went on to state "I consider this the greatest discovery in the field of bacteriology" and prophetically added that "...Koch will again astonish and shame us with still further discoveries" (Wagner 1885).

Unfortunately, Koch's views found no favor with Rudolph Virchow the highly influential chief pathologist of Berlin's Charite Hospital, who also served as leader of the opposition in the German Parliament. Virchow held that it was an intrinsic derangement in the function of the body's cells that caused all disease and dismissed out of hand the role of "infection" from "external agents." He considered Koch's theory "improbable."

Koch, disillusioned but determined, returned to Wollstein with a new Carl Zeiss microscope—with its special new oil immersion lenses crafted by the Ernst Abbe, the finest lens maker in Germany—that magnified images more than a thousandfold. In 1879, he published another brilliant paper, "Investigations of the Etiology of Wound Infections" (Koch 2010), in which he proposed that different types of infections existed.

When Koch moved to Breslau and then eventually to Berlin—mainly through the efforts of Cohn and Cohnheim—he took charge of his new laboratory with some of the most brilliant names to adorn medical textbooks: Friedrich Loeffler (discoverer of the diphtheria bacillus) and Georg Gaffky (discover of the germ of typhoid and co-discoverer—with Koch—of the germ of cholera) were assigned to him.

Koch's new team initially concentrated on refining culture media. Liquid media had many limitations, not the least of which was that separation of the pathogen from contaminants was difficult; Koch admitted that it was "truly depressing to attempt pure cultivation" (Sakula 1982). In 1881, Koch tried garden potatoes then gelatin, and finally—on an inspired suggestion of his assistant Walter Hess' wife and assistant Fannie—a seaweed extract, agar. Success was immediate. Gratified, Koch began to use solid media in flat-bottomed glass dishes devised by another of his illustrious assistants, Julius Richard Petri.

Armed with his novel culture media, Koch finally turned his attention—secretly—to something close to his heart: tuberculosis!

But despite all the innovations that went behind his culture media and staining methods (Koch was provided a variety of stains created by yet another brilliant colleague, Paul Ehrlich), the TB bacillus refused to show itself under the microscope.

Slightly differential versions of the following story exist: but essentially, it appears that Koch forgot to discard an old slide of a 10-day-old culture lying near a warm stove, covered with dye, in the ammonia-heavy air of his laboratory. The next morning an astonished Ehrlich showed Koch the vivid blue clumps of mycobacteria (Sakula 1982). The ammonia fumes had permeated into the cell walls of mycobacteria, and the alkalinity and the heat had carried the methylene blue dye into the cell. The bacillus lay stark and exposed.

In the spring of 1882, armed with hundreds of slides, test tubes, flasks, culture plates, human and animal tissue specimens, and the records of nearly 500 experiments, Koch proceeded—not to the Berlin Pathological Society, where Virchow was influential—but to the Berlin Physiological Society, where, in the words of Loeffler, he "...spoke slowly and haltingly, but what he said was clear, simple, logically stated—in short, pure unadulterated gold."

"On the basis of my recent observations" Koch stated, "I consider it as proved that in all tuberculous conditions of man and animals there exists a characteristic bacterium which I have designated as the Tubercle Bacillus, which has specific properties which allow it to be distinguished from all other microorganisms" ([The etiology of tuberculosis by Dr. Robert Koch. From the Berliner Klinische Wochenschrift, Volume 19 (1882)] 1982).

The date was 24th of March, a date celebrated today as the World Tuberculosis Day.

What Robert Koch did on that date was not only to unearth mankind's most deadly infectious killer but also to firmly entrench the concept of microbes as agents of infection, a concept that Pasteur, his old rival, had led credence to. However Koch was not done with Pasteur yet. In an unprovoked assault on Pasteur at the Fourth International Congress for Hygiene and Demography in Geneva in September (Carter 1988), Koch challenged both Pasteur's research methods and his scientific temper. Pasteur demanded gratification—the great rivalry between two of medicine's colossuses sank to the level of a petty quarrel.

By the end of the decade, with the tenth International Medical Conference in Berlin approaching, letters personally signed by the Kaiser Wilhelm II went out to over 6000 delegates, and the pressure on Koch to do something special to showcase Germany's medical prowess mounted steadily (Burke 1993).

On the third day of the conference, Koch finally unveiled his much anticipated vaccine against TB. "I have at last hit upon a substance that has the power of preventing the growth of the tubercle bacilli not only in the test tube but in the body of an animal," read out Koch. But both his emphasis on the word "animal" and his statement that tuberculin was effective only if the disease were "not too advanced" were completely drowned by the ensuing pandemonium.

With the world's expectations raised sky-high, Koch began testing this vaccine in earnest, beginning with himself. Initially the vaccine seemed to work, and a relieved Koch pronounced the first human trial successful. Both the *British Medical Journal* and the *Lancet* carried a complete translation of Koch's paper, and the *Lancet* described Koch's discovery as "glad tidings of great joy" (Dubos 1952).

"The eagerness with which patients seize hold of everything under the name of remedy borders on madness," Benjamin Rush had written in 1783 in *Thoughts upon the Causes and Cure of Pulmonary Tuberculosis*. The floodgates literally opened. Patients poured into Berlin to receive the miracle drug. As to its composition, Koch only said: "As regards the origin and preparation of the remedy, I am unable to make any statement as my research is not yet concluded; I reserve this for a future communication. The remedy is a brownish, transparent liquid" (Koch 1890b).

Arthur Conan Doyle, already well known for his detective columns featuring Sherlock Holmes, undertook a pilgrimage to Berlin to meet the new prophet, Koch. Though he did not get to meet Koch, he did have a chance to look around and see what the vaccine could possibly do. Conan Doyle was disappointed in the "cure." Yet he foresaw that Koch's lymph would prove "an admirable aid to diagnosis" and a "single injection" would help determine if the pathology were "in any way tubercular." In all fairness, Koch himself had himself said: "I think I am justified in saying that the remedy will therefore in the future form an indispensable aid to diagnosis. By its aid we shall be able to diagnose doubtful cases of phthisis…" (Koch 1890a).

Within the space of a few short months, it was obvious that tuberculin did not work well. Many patients worsened; several died. The *British Medical Journal* in an editorial cautioned, "Already warning voices are being raised in Vienna and in Berlin in the attempt to stem the tide of misconception which is rolling in a veritable flood of consumptives to the Prussian capital…" (Campbell and Bah-Sow 2006). As to the formula which Koch still withheld, the editors urged that "…our patience is not to be severely taxed, for it is much to be hoped that Dr Koch….will give at an early date the promised account of the mode of preparation of his remedy."

A heavily decorated Koch became the object of censure—and ridicule, even. It was in 1905—nearly quarter of a century later—that he finally received the award that long ago should rightfully have been his—the Nobel Prize in Medicine.

Koch's vaunted vaccine, tuberculin ("old tuberculin"—OT), was after all just a glycerinated beef-broth culture of heat-killed bacilli. True to the deductions of Sherlock Holmes' creator, tuberculin became an important tool for the diagnosis of tubercular infection—a role it still plays after over 140 years of its discovery.

Austrian theologist and philosopher-turned-pediatrician, Baron Clemens Peter von Pirquet, found that by superficially scratching the skin through two drops of diluted OT, a skin reaction could be produced and used to effectively diagnose TB infection in Viennese children (Pirquet 1909). In 1907, Alfred Wolff-Eisner and Albert Calmette introduced a wafer of solidified tuberculin between the eyelids of bovine animals (Paine and Martinaglia 1929). Other methods of inoculating the subject with tuberculin—some bordering on the repugnant—were explored: Tuberculin in various forms was inserted into the nose, rectum, urethra, or vagina. Fortunately, these methods were soon abandoned. In 1908 in Munich, Ernst Moro prepared a paste of tuberculin and lanolin (Munchener Medizinische Wochenschrift/ 27 February 1912: On transplantation immunity by Georg Schone, Greifswald, 1978) and R Lautier a wad of tuberculin soaked in cotton, the same year, to apply to the skin (Grozin 1943).

In 1908 Charles Mantoux's method of injecting tuberculin directly into the epidermis (Mantoux 1910) replaced Pirquet's, and with a slight modification by Felix Mendel, the test took its current form as the popular Mendel-Mantoux—or simply the "Mantoux"—test.

In 1934, Florence Seibert's group at the University of Pennsylvania came up with an improved purified protein derivative (PPD) of tuberculin (Richards et al. 1979), and the WHO later further improved it by decreasing its adherence to containers by adding the detergent Tween 80 (Guld et al. 1958). By the late 1990s, a new lot, PPD-S2, had become the new standard (Villarino et al. 2000).

Today, in spite of its relative unsophistication, the tuberculin skin test—a cocktail of over 200 mycobacterial antigens—has managed to stand the test of time. In 1997, Ajit Lalvani and his colleagues at the Imperial College in London developed the ELISpot assay, an interferon-gamma assay based on a genomic segment unique to the MTB called the "region of difference-1" (Lalvani et al. 1997).

Yet in spite of everything, the footprint of TB was difficult to see. What was needed was a way to image the body and see the effect of disease on its tissues.

On November 8, 1895, at the University of Würzburg, 50-year-old Karl Wilhelm Conrad Roentgen noticed unusual fluorescence in crystals of barium platinocyanide that were kept in the proximity of a covered Crookes' tube. Roentgen arrived at the deduction that a new kind of invisible ray (he called it "X" for want of a better name) could penetrate and cast images of hidden structures (Tubiana 1996). That very day, Roentgen photographed a lead pipe and then his wife's hand. Within just a couple of years, X-rays had become a new diagnostic tool in medicine. Soon after, Italian physicist Enrico Salvioni unveiled fluoroscopy, which medicine also "appropriated."

With these devices, the shadows of tuberculosis were seen everywhere, but mostly in the lungs. In the 1930s, Brazilian pulmonologist Manuel Dias de Abreu found a way to take miniature X-rays (Ravindra 2002) and the contraption could be fitted into mobile units and transported to people's very doorsteps for community screening.

In 1901, Röntgen was awarded the first Nobel Prize in Physics, but being painfully stage-shy, he demurred when asked to give his Nobel acceptance speech. He never patented his discovery and refused to consider any financial gain from his invention, donating to his university the money that went with the Nobel.

However, although diagnosis was now a relatively simple, the treatment of TB was another matter altogether. People relied on rest cures, usually under the sun, for there was little else.

Pliny the Elder had said, "Sol est remediorum maximum—'The sun is the best remedy' (of all)" (Pliny 1938). Ancient Egyptians and the Babylonians had long known this; ancient Greeks had Helioses in which they sunbathed (Research 1996) and the Romans had their Solaria.

Over time, sunbathing, fresh air, and gentle exercise gained popularity as general measures to treat the ill. Thomas Sydenham the English physician (The "English Hippocrates") recommended horse riding (Sydenham and Dewhurst 1966), and George Bodington, practicing in Coldfield, Sutton, used the combination—minus the horse—to good effect (Keers 1980). That very year, Otto Walther started his

Nordrach Sanatorium in the Black Forest (McCarthy 2001), and in 1855, Austrian Arnold Rikli set up another one in Bled, Slovenia, writing no less than seven books detailing his methods.

Meanwhile, America had already opened its doors to settlers—and to consumptives in search of a natural cure. Iowa, Minnesota, and Wisconsin in the late 1840s gained popularity for the restorative qualities of their air and made this fact public through advertisements targeted at patients with TB.

Just about this time, a freshly graduated doctor's struggles of over four decades with TB were just beginning. In 1872, Edward Livingston Trudeau developed consumption. "I think I knew something of the feelings of the man at the bar who is told he is to be hanged on a given date," he wrote, "...it seemed to me that the world had grown suddenly dark...my dreams were all shattered now, and only exile and inevitable death remained" (Trudeau 1916).

Trudeau undertook a long and arduous journey to his beloved Adirondack mountains where he hoped to die in peace. Instead, he recovered and returned to his practice at New York City—whereupon he promptly suffered a relapse. He finally returned to the Adirondacks and made a home there. In 1882, Trudeau read about the travels of another consumptive doctor, Silesian Hermann Brehmer, who had regained his health at the foothills of the Himalayas (McCarthy 2001) and gone on to open a sanatorium-type "Kurhans" in Gorbersdorf. In 1884, in a quaint little wooden cottage called "Little Red," Trudeau established the Adirondack Cottage Sanatorium. By 1950, a total of 100,000 patients could be housed in sanatoria across the USA.

In 1956, funded by the Indian Council of Medical Research (ICMR) and the Government of Tamil Nadu in India and supported by the WHO, a landmark study was performed by Wallace Fox (Watts 2010) at the Medical Research Council at the newly founded Tuberculosis Chemotherapy Centre at Madras to see if patients could be effectively treated outside of Sanatoria. Wrote S Radhakrishna: "...the outcome of this trial was critical for the management of tuberculosis in India, as well as other developing countries with a high disease burden and inadequate resources" (Wallace 2010). After 5 years, the Madras trial proved that the vast majority of patients could be treated equally effectively at their homes.

In 1888, at about the same time that Trudeau was building "Little Red," Carlo Forlanini of Milan described the first successful artificial pneumothorax, intended to "rest" the lung (Sakula 1983), and in 1911, the invention of the water manometer made it possible to let in a precise amount of air into the chest—through a special needle invented by Saugmann. Later, collapse therapy began to be carried out with Lillington and Pearson's artificial pneumothorax apparatus under radiologic monitoring (Rakovich 2010). Over two-thirds of all patients admitted to sanatoria with all but the most minor forms of TB had collapse therapy performed on them. Andrew Banyai soon introduced a related procedure—the pneumoperitoneum—to push up patients' diaphragms and further immobilize the lung. In some patients, the phrenic nerve was divided—either in isolation or in combination with pneumoperitoneum—to achieve the same end.

The introduction of air into the pleural cavity was frequently impossible on account of the pleural adhesion to the thoracic wall that many patients had. Ferdinand Sauerbruch, the German thoracic surgeon—author of the first textbook on thoracic surgery—devised extrapleural thoracoplasty in which excision of the upper ribs allowed the cavity-cavity-containing lung apices to collapse inwards. In 1904, Sauerbruch performed surgery with the patient's torso enclosed within his pressure chamber (Cherian et al. 2001) which was capable of ventilating the patient by generating both negative and positive pressure.

In 1910, Swede Hans Christian Jacobaeus, while yet an intern, adapted a cystoscope to serve as a laparoscope, and later as a thoracoscope in 1922, describing 40 cases of pleural adhesiolysis (Jacobaeus 1922).

In 1947, to make the after-effects of thoracic surgery less disabling, Nagaishi and Wilson began inserting hollow spheres of polymethylmethacrylate—"lucite spheres"—into the apical pleural space. The procedure went by the unedifying name of "plombage" (Vigneswaran 1995). Heavy oil was sometimes used, "oleothorax," but it sometimes produced unpleasant inflammatory reactions and on occasion tracked into the bronchial tree with disastrous consequences.

In spite of everything, the 5-year mortality was close to a 100%. What was needed was an antibiotic that killed mycobacteriae.

The wonder drug penicillin was completely ineffective against the mycobacteriae whose highly complex cell walls appeared to have no vulnerability. Aspirin for some reason had quite the opposite effect: it stimulated a burst of activity. Bernheim remarked on this in a letter to Danish biochemist Jorgen Lehmann who was working on perfecting an anticoagulant, dicoumarol (Lehmann 1964). Later, Lehmann used the same strategy that he had used to construct dicoumarol. He swapped an amino group from the "ortho" to the "para" position, and the new variant of aspirin, para-aminosalicylic acid (PAS), showed effects opposite to that aspirin had—it inhibited bacterial synthesis.

In spite of disappointments with their earlier anti-TB drug ventures, Ferrosan, a Danish pharma company, decided to hedge on PAS. Chief chemist Karl Rosdahl churned out a meagre 13 grams for Lehmann, which Lehman routed through to Gylfe Vallentin, Head of Renstrom's Sanatorium (Lehmann 1964).

On October 1944, a full month before Selman Waksman used streptomycin in the USA, 24-year-old Sigrid was commenced on PAS—she recovered. In April 1947, the Swedish National Board for Tuberculosis began a major trial on PAS, which was later, on the evidence of a historic trial by the British Medical Council (Treatment of pulmonary tuberculosis with streptomycin and para-aminosalicylic acid; a Medical Research Council investigation 1950), incorporated into a combination regimen for TB. Over the next 20 years, the annual production of PAS rocketed to 3000 tons.

Meanwhile in America, Ukraine-born Selman Waksman, a widely regarded expert on soil microbes, was convinced that "The cure of tuberculosis must come not from enzymes, but from antibiotics." Yet, though his team had already processed over 10,000 specimens, it was to no avail.

In 1943, the tide turned. Twenty-three-year-old Albert Schatz joined Waksman's team. Less than 2 months later, one of Waksman's students, Doris Jones, presented the team with organisms cultured from the croup of a chicken from which Schatz—and another recent addition to the team, undergraduate student, Elizabeth Bugie—succeeded in isolating streptomycin. It was a while before the drug was put to use. "The discovery had consequently been made, but was not discovered by the discoverers themselves!" observed Birach in the Scandinavian *Journal of Respiratory Disease*.

At the Mayo Clinic, William H Feldman and physician Corwin Hinshaw finally undertook to test streptomycin. After several successful experiments on guinea pigs (Lehmann 1964), on November 20, 1944, the first round of injections was given to a seriously ill 21-year-old in Mineral Springs Sanatorium in Cannon Falls, Minnesota. Patricia weighing all of 75 pounds took a series of injections. She survived, going on to get married and have three children.

Although patients treated with either streptomycin or PAS appeared to do well at first, they soon relapsed. Yet, when combined, the drugs went on to become the keystone for anti-TB therapy. Waksman went on to win the Nobel Prize for the year 1952; Albert Schatz, who had always considered himself the true discoverer of streptomycin, and Jorgen Lehmann—arguably the discoverer of the first effective anti-tubercular agent—did not.

Then, in 1952, INH or, properly, isonicotinic-acid hydrazide was simultaneously discovered by three separate teams at three separate labs—teams led by Harry Yale at the Squibb Institute in New Brunswick, by Robert Schnitzer at Hoffmann-La Roche in Nutley, and by Gerhard Domagk in Germany. Domagk had discovered the anti-tubercular drug thiacetazone shortly before (McDermott 1969).

Then, Lepetit Research Labs in Milan brought forth the best drug for TB yet. Franco Parenti, holidaying in France in 1959, returned home with unusual gifts for his colleague Piero Sensei—clods of earth from a pine forest near Nice. From the *Nocardia mediterranei* he found within, Sensei extracted a new TB drug (Sensi et al. 1959) which he called rifomycin (after Rififi—*The Struggle*—a movie Parenti had fancied) (Dubos 1952).

In 1968, rifampin was launched in Europe and 6 years later in the USA. Ninety-five percent of all patients who took a rifampicin-INH combination-based regimen were pronounced cured, and anti-tubercular therapy was shortened to 9 months. In 1961 came ethambutol from Lederle Labs; and when pyrazinamide arrived, the anti-tubercular armory was complete; the extent of any TB regimen that initially incorporated four drugs was further reduced to a "mere" 6 months.

Meanwhile, radiology was undergoing a sea change. In 1973 the first CT scanners were unveiled at EMI and almost immediately thereafter at the Tufts University at Massachusetts by Allen McLeod Cormack, working independently. By 1970, Paul Lauterbur, a chemical engineer, was working on the first magnetic resonance imaging (MRI) scanner at the University of Illinois. Peter Mansfield a physicist at the University of Nottingham, UK, improved upon the techniques.

Although both Lauterbur and Mansfield got the 2003 Nobel Prize (Macchia et al. 2007), Raymond Damadian, who had constructed an MRI body scanner in 1969 (Harold 1972)—and had performed what was probably the first human MRI scan in 1977—did not (Damadian 1971).

By 1914, the world's first maritime SONAR machine had become fully functional in the USA largely through the efforts of Canadian Reginald Fessenden (Woo) and soon was adapted to medicine's needs.

With both diagnostic radiology and TB therapeutics up to speed, all that the world needed now was a preventive vaccine.

The twentieth century saw a violent beginning—World War I. Under the almost continuous bombing that went on overhead, at Lille, veterinarian and immunologist Camille Guérin along with the Director of the Pasteur Institute at Lille, Léon Charles Albert Calmette, began to wage a daily battle against the tubercle bacillus. In 1919, after incredible 11 years and 231 subcultures, they succeeded in weakening the *Mycobacterium* enough to make a vaccine of it (Weir et al. 2008).

In 1921, newborns at the Charité Hospital in Paris were successfully injected the vaccine of Bacillus Calmette-Guérin—now called the BCG. Over the next 9 years—between 1924 and 1925—the vaccine was exported to over 30 nations, and in 1928 it gained the seal of approval of the Health Committee of the League of Nations. But then, disaster struck! In 1930, in the German town of Lübeck, nearly all of the 249 infants who were administered the BCG vaccine developed TB, and 76 died over the following months. Overnight, almost, the remedy had turned into a poison.

It was eventually found that there was nothing wrong with the vaccine itself: a batch of BCG had become inadvertently contaminated with a virulent strain of tubercle bacillus, but by then, a devastated Calmette had passed on. BCG made its expected resurgence, and post-World War II—between 1945 and 1948—eight million babies in Eastern Europe received the vaccine.

Although the safety of the vaccine was proven beyond doubt, questions about its efficacy remained. Complexities in the vaccination programs run by different countries, racial and regional factors of the local population, and the technical skills of the operators—and, indeed, the differences in the vaccine strains used—impacted the efficacy of the vaccine (Weir et al. 2008).

In 1968, an ICMR-WHO-US Public Health Services-funded joint committee was appointed to study the vaccine, and a vaccine incorporating the Pasteur and Danish strains was administered to 360,000 people in the district of Chingleput 40 km from Chennai in India. Twelve and a half years later, the reports showed that protection rate in children was just 17%; that in adults was zero. A follow-up paper from the ICMR concluded: "...BCG offers no protection against adult type bacillary tuberculosis. Consequently, BCG cannot be expected to reduce the transmission of tuberculosis" (Tripathy 1987).

Nevertheless, in 1974, the BCG was included in the Expanded Program on Immunization, and by 2003, in most areas the use of BCG grew to achieve 95% coverage. Today BCG has been made mandatory in 64 countries and officially recommended in almost twice as many more (Colditz et al. 1994), and over four billion individuals have so far received the vaccine (Nuttall and Eley 2011).

In 1998, the Sanger Centre in Cambridge, Britain, and Paris' Pasteur Institute in France cracked the genetic code for the old H37Rv strain of tubercle bacillus, and then the genome of highly virulent CDC1551 (the "Oshkosh" strain) was mapped at the TIGR (Fine 1989).

By the 1980s, in the developed nations at least, TB seemed to be in decline. By the mid-1980s, this decline was no longer as manifest, and in some developed countries, the graph of new cases had begun to climb again.

A new disease had arrived.

The HIV epidemic pushed up the epidemiological curve of TB, and by December 1997, 11.7 million individuals had perished, with 16,000 new infections occurring each day. By the turn of the millennium, as the developed world inevitably came to terms with the HIV and gained a measure of control on it, the disease spiraled out of control in Asia and Africa. By 2015—the year the WHO had predicted would usher in a catastrophe—the situation had actually improved (UNAIDS 2015). Still, nearly 40 million people globally were living with HIV worldwide (74% of them in sub-Saharan Africa), but new HIV infections had fallen by 35% and the global response to HIV had saved millions from contracting the disease—even tuberculosis-related deaths had fallen by 32% since 2004 (UNAIDS 2015).

Footnote: Down the ages, a double armed heraldic cross had served as the emblem of the kings of Sumeria, flown upon the pennants of Joan of Arc, fueled the hermetic revival of René d'Anjou, and was revered as the Patriarchal Cross of Jerusalem. Later, it became the mark of the infamous Godfrey, Duke of Lorraine.

By 1967, four frontline drugs were available for treating TB: INH, pyrazinamide, ethambutol, and rifampicin. Wallace Fox and other researchers in British Medical Research Council units in East Africa, India, Hong Kong, and Singapore combined these drugs together into an effective regimen that shortened the treatment to 6 months. Short-course chemotherapy (SCCT) became the standard treatment for TB.

On the basis of a paper from the Tuberculosis Chemotherapy Centre, Madras, that showed the effectiveness of a twice-weekly administration of TB drugs, it was realized that treatment could now be dispensed under direct supervision. This new biweekly regimen came to be called the DOTS (World Health Organization 1994; Morse 1996).

Thus, after peaking in 2004, the incidence rates of TB began to decline, and by 2014, all six of WHO's regions had achieved the modest the Millennium Development Goal. TB was still at large, killing 1.5 million persons, and new unsettling words were increasingly being used—multidrug-resistant TB (MDR-TB) and extensively drug-resistant TB (XDR-TB). That year in 2014, 190,000 individuals died of MDR-TB.

According to WHO figures, nearly 10% of all MDR-TB have already metamorphosed into the deadlier XDR already—in over 100 countries (World Health Organization 2015b). The problem was that the world has run out of drugs to treat these new forms of TB (Ventola 2015).

Two new anti-TB drugs—bedaquiline and delamanid—have finally broken through the deadlock, but drug resistance to these virtually unused drugs is already being reported (Bloemberg et al. 2015).

As well, improving health technology has begun to drive earlier TB diagnosis (Sulaiman et al. 2013), and "rapid culture" tests—MODS, Griess, MGIT, thin-layer agar, colorimetric assays, and others—are competing for preference, along with the now well-accepted CBNAAT and line-probe assays (Medecins Sans Frontiers 2015).

Despite everything, given the new 2030 targets, TB control has been lagging nearly 150 years behind schedule (Hasan 2017). The WHO's End-TB Strategy 2016–2035—a costed, scalable blueprint for a world free of TB, zero deaths and zero disease and suffering due to TB—reflects the world's growing desperation to be done with a disease that for too long has remained unchallenged as mankind's most destructive infection. The new goals are by far the most ambitious yet (World Health Organization 2015a). As things stand today, things are far from perfect (Hasan 2017). TB accounts for a quarter of all avoidable deaths in developing nations.

Robert Koch's in his Nobel acceptance speech had said, "The fight has been ignited fully and the enthusiasm for this goal is so broad that I am not afraid that it will cease again...If we continue to work in such a powerful way, victory will be achieved."

Now, if ever, is the moment of truth.

∗*Footnote: The 2025 milestones for the End-TB Strategy are a 75% reduction in TB deaths (compared with 2015), a 50% reduction in TB incidence rate (less than 55 TB cases per 100,000 population), and no affected families facing catastrophic costs. The 2035 milestones are 95% reduction in TB deaths (compared with 2015), 90% reduction in TB incidence rate (less than 10 TB cases per 100,000 population), and no affected families facing catastrophic costs due to TB.*

References

Bloemberg GV, Keller PM, Stucki D, Trauner A, Borrell S, Latshang T, Coscolla M, Rothe T, Homke R, Ritter C, Feldmann J, Schulthess B, Gagneux S, Bottger EC (2015) Acquired resistance to bedaquiline and delamanid in therapy for tuberculosis. N Engl J Med 373 (20):1986–1988. https://doi.org/10.1056/NEJMc1505196

Burke DS (1993) Of postulates and peccadilloes: Robert Koch and vaccine (tuberculin) therapy for tuberculosis. Vaccine 11(8):795–804

Campbell IA, Bah-Sow O (2006) Pulmonary tuberculosis: diagnosis and treatment. BMJ (Clin Res Ed) 332(7551):1194–1197. https://doi.org/10.1136/bmj.332.7551.1194

Carter KC (1988) The Koch-Pasteur dispute on establishing the cause of anthrax. Bull Hist Med 62 (1):42–57

Cherian SM, Nicks R, Lord RS (2001) Ernst Ferdinand Sauerbruch: rise and fall of the pioneer of thoracic surgery. World J Surg 25(8):1012–1020

Cohn F (1872) Untersuchungen über Bacterien. Beiträge zur Biologie der Pflanzen 1(2):127–224

Colditz GA, Brewer TF, Berkey CS, Wilson ME, Burdick E, Fineberg HV, Mosteller F (1994) Efficacy of BCG vaccine in the prevention of tuberculosis. meta-analysis of the published literature. JAMA 271(9):698–702

Damadian R (1971) Tumor detection by nuclear magnetic resonance. Science (New York, NY) 171 (3976):1151–1153
Dubos R DJ (1952) The white plague: tuberculosis, man, and society. Little, Brown, Boston
Fine PE (1989) The BCG story: lessons from the past and implications for the future. Rev Infect Dis 11(Suppl 2):S353–S359
Grozin M (1943) The "visible tuberculin patch test": a new improvement. Am J Dis Child 66 (2):126–131. https://doi.org/10.1001/archpedi.1943.02010200026004
Guld J, Bentzon MW, Bleiker MA, Griep WA, Magnusson M, Waaler H (1958) Standardization of a new batch of purified tuberculin (PPD) intended for international use. Bul World Health Organ 19(5):845–951
Harold FM (1972) Conservation and transformation of energy by bacterial membranes. Bacteriol Rev 36(2):172–230
Hasan A (2017) Noughts and crosses (editorial). J Med Allied Sci 7(1):01–02
Jacobaeus H (1922) The practical importance of thoracoscopy in surgery of the chest. Surg Gynecol Obstet 34:296–298
Keers RY (1980) Two forgotten pioneers. James Carson and George Bodington. Thorax 35 (7):483–489
Koch R (1890a) A further communication on a remedy for tuberculosis. Br Med J 2 (1560):1193–1199
Koch R (1890b) Weitere Mittheilungen über ein Heilmittel gegen Tuberculose. Deutsche Medizinische Wochenschrift
Koch R (2010) Verfahren zur Untersuchung, zum Konservieren und Photographieren der Bakterien. In: Robert Koch-Institut. https://doi.org/10.25646/5065
Lalvani A, Brookes R, Hambleton S, Britton WJ, Hill AV, McMichael AJ (1997) Rapid effector function in CD8+ memory T cells. J Exp Med 186(6):859–865
Lehmann J (1964) Twenty years afterward historical notes on the discovery of the antituberculosis effect of paraaminosalicylic acid (PAS) and the first clinical trials. Am Rev Respir Dis 90:953–956. https://doi.org/10.1164/arrd.1964.90.6.953
Macchia RJ, Termine JE, Buchen CD (2007) Raymond V. Damadian, M.D.: magnetic resonance imaging and the controversy of the 2003 Nobel Prize in physiology or Medicine. J Urol 178(3 Pt 1):783–785. https://doi.org/10.1016/j.juro.2007.05.019
Mantoux C (1910) L'intradermo-reaction a la Tuberculin et son interpretation clinique. Presse Med 18:10–13
McCarthy OR (2001) The key to the sanatoria. J R Soc Med 94(8):413–417. https://doi.org/10.1177/014107680109400813
McDermott W (1969) The story of INH. J Infect Dis 119:678–683
Medecins Sans Frontiers STP (2015) Out of step 2015 TB policies in 24 countries: a survey of diagnostic and treatment practices. http://www.stoptb.org/assets/documents/news/report_out_of_step_2015_11_pdf_with_interactive_links.pdf
Morse D (1996) Directly observed therapy for tuberculosis. BMJ (Clin Res Ed) 312:719–720
[Munchener Medizinische Wochenschrift/27 February 1912: On transplantation immunity by Georg Schone, Greifswald] (1978) MMW, Munchener medizinische Wochenschrift 120 (14):480
Nuttall JJC, Eley BS (2011) BCG vaccination in HIV-infected children. Tuberc Res Treat 2011:712736-712736. doi:https://doi.org/10.1155/2011/712736
Paine R, Martinaglia G (1929) Tuberculosis in wild buck living under natural conditions. J Comp Pathol Ther 42:1–8. https://doi.org/10.1016/S0368-1742(29)80001-X
Pirquet CV (1909) Frequency of tuberculosis in childhood. JAMA LII(9):675–678
Pliny tER, H. (Harris) (1938) Natural history. W. Heinemann, London
Rakovich G (2010) Artificial pneumothorax: tapping into a small bit of history. CMAJ 182 (2):179–179. https://doi.org/10.1503/cmaj.090024
Ravindra S (2002) General information on MMR X-ray units & its present status. NTI Bulletin, pp 4–7

Research H (1996) Sun, air, and light baths. Health Research Books

Richards NM, Nelson KE, Batt MD, Hackbarth D, Heidenreich JG (1979) Tuberculin test conversion during repeated skin testing, associated with sensitivity to nontuberculous mycobacteria. Am Rev Respir Dis 120(1):59–65. https://doi.org/10.1164/arrd.1979.120.1.59

Sakula A (1982) Robert Koch: centenary of the discovery of the tubercle bacillus, 1882. Thorax 37 (4):246–251

Sakula A (1983) Carlo Forlanini, inventor of artificial pneumothorax for treatment of pulmonary tuberculosis. Thorax 38(5):326–332

Sensi P, Greco AM, Ballotta R (1959) Rifomycin. I. Isolation and properties of rifomycin B and rifomycin complex. Antibiot Annu 7:262–270

Sulaiman SAS, Khan AH, Ahmad N, Iqubal MS, Muttalif AR, Hassali MA (2013) Impact of diabetes mellitus on treatment outcomes of tuberculosis patients in tertiary care setup. Am J Med Sci 345(4):321–325. https://doi.org/10.1097/MAJ.0b013e318288f8f3

Sydenham T, Dewhurst K (1966) Dr. Thomas Sydenham (1624–1689); his life and original writings. [Ed. by] Kenneth Dewhurst. vol vii, 191 p. University of California Press, Berkeley

[The etiology of tuberculosis by Dr. Robert Koch. From the Berliner Klinische Wochenschrift, Volume 19 (1882)] (1982) Zentralblatt fur Bakteriologie, Mikrobiologie und Hygiene 1 Abt Originale A, Medizinische Mikrobiologie, Infektionskrankheiten und Parasitologie = International journal of microbiology and hygiene A, Medical micro 251(3):287–296

TREATMENT of pulmonary tuberculosis with streptomycin and para-aminosalicylic acid; a Medical Research Council investigation (1950) Br Med J 2(4688):1073–1085

Tripathy S (1987) Fifteen-year follow-up of the Indian BCG prevention trial. Bull Int Union Tuberc Lung Dis 62:69–72

Trudeau E (1916) An autobiography. Doubleday, New York

Tubiana M (1996) Wilhelm Conrad Rontgen and the discovery of X-rays. Bull Acad Natl Med 180 (1):97–108

UNAIDS (2015) How AIDS changed everything — MDG6: 15 years, 15 lessons of hope from the AIDS response. http://www.unaids.org/en/resources/documents/2015/MDG6_15years-15lessonsfromtheAIDSresponse

Ventola CL (2015) The antibiotic resistance crisis: part 1: causes and threats. P & T 40(4):277–283

Vigneswaran WT RS (1995) Paraffin plombage of the chest revisited. In: Presented at the sixty-first annual meeting of the American College of Chest Physicians, New York

Villarino ME, Brennan MJ, Nolan CM, Catanzaro A, Lundergan LL, Bock NN, Jones CL, Wang YC, Burman WJ (2000) Comparison testing of current (PPD-S1) and proposed (PPD-S2) reference tuberculin standards. Am J Respir Crit Care Med 161(4 Pt 1):1167–1171. https://doi.org/10.1164/ajrccm.161.4.9906050

Wagner E (1885) Cohenheim Gesammelte Abhandlungen. A. Hirschwald. Verlag, Berlin

Wallace F (2010) Obituary. Indian J Med Res:839–841

Watts G (2010) Obituary of Wallace Fox: clinician researcher who helped to revolutionise the treatment of tuberculosis. Lancet 375:1076

Weir RE, Gorak-Stolinska P, Floyd S, Lalor MK, Stenson S, Branson K, Blitz R, Ben-Smith A, Fine PEM, Dockrell HM (2008) Persistence of the immune response induced by BCG vaccination. BMC Infect Dis 8(1):9. https://doi.org/10.1186/1471-2334-8-9

Woo JA short History of the development of Ultrasound in Obstetrics and Gynecology. http://www.ob-ultrasound.net/site_index.html. Accessed 5 Apr 2016

World Health Organization (1994) Framework for effective tuberculosis control

World Health Organization (2015a) WHO end TB strategy: global strategy and targets for tuberculosis prevention, care and control after 2015. http://www.who.int/tb/post2015_strategy/en

World Health Organization (2015b) World Health Organization Global Tuberculosis Report. http://www.who.int/tb/publications/global_report/en

2 Tuberculosis as an Underlying Etiological Factor for Other Human Respiratory Diseases

Ronan F. O'Toole

Abstract

Tuberculosis (TB) does not occur in isolation from other human illnesses. There are multiple examples where TB combines with one of more comorbidities to amplify its prevalence. Noncommunicable diseases such as diabetes, or lifestyle behaviors including smoking and alcohol misuse, place people at a greater risk of presenting with active TB. But the epidemiological associations between TB and other human conditions are not confined to increasing susceptibility to TB disease. TB, in itself, is an underlying risk factor for the development of downstream respiratory illnesses later in life. This indicates that injury to the host resulted from an episode of TB persists beyond successful eradication of *Mycobacterium tuberculosis* infection by antimicrobial drug therapy. In this chapter, the specific role of TB in promoting other lung diseases is examined. In particular, TB during childhood increases the risk of development of progressive and poorly reversible airway diseases that include bronchiectasis and chronic obstructive pulmonary disease. It is apparent from the literature that prevention of TB disease offers a potential pathway for reducing the global burden of downstream chronic lung diseases.

Keywords

Tuberculosis · *Mycobacterium tuberculosis* · Bronchiectasis · Chronic obstructive pulmonary disease

R. F. O'Toole (✉)
School of Molecular Sciences, College of Science, Health and Engineering, La Trobe University, Melbourne, VIC, Australia

Department of Clinical Microbiology, School of Medicine, Trinity College Dublin, Dublin, Ireland

School of Medicine, College of Health and Medicine, University of Tasmania, Hobart, Tasmania, Australia
e-mail: otooler3@tcd.ie

S. E. Hasnain et al. (eds.), Mycobacterium tuberculosis: *Molecular Infection Biology, Pathogenesis, Diagnostics and New Interventions*,
https://doi.org/10.1007/978-981-32-9413-4_2

Abbreviations

AIDS	acquired immunodeficiency syndrome
BOLD	Burden of Obstructive Lung Disease
CI	confidence interval
COPD	chronic obstructive pulmonary disease
FEV_1	forced expired volume of air in the first second of expiration
FVC	forced vital capacity
HIV	human immunodeficiency virus
MDR	multidrug resistant
OR	odds ratio
TB	tuberculosis
UI	uncertainty interval
UK	United Kingdom
WHO	World Health Organization

2.1 Background

Tuberculosis (TB) is a major cause of infectious disease mortality worldwide, killing approximately 1.67 million people in 2016 alone (World Health Organization 2017). There were an estimated 10.4 million new cases of the disease in 2016 corresponding to an average annual incidence rate of approximately 140 per 100,000 persons globally (World Health Organization 2017). This included 490,000 cases that were multidrug resistant (MDR) and a further 110,000 cases that were resistant to rifampicin but susceptible to isoniazid in 2016. The involvement of HIV/AIDS in increasing the risk of TB development is well established (Chaisson and Martinson 2008; Geldmacher et al. 2008, 2010; Hwang et al. 2013; Swaminathan et al. 2000). Of the new cases of TB in 2016, 1.03 million occurred in HIV-positive individuals (World Health Organization 2017).

Noncommunicable diseases also contribute to the burden of TB. A systematic review by Jeon and Murray of 13 observational studies, comprising 17,698 TB patients, found that diabetes was associated with an approximate threefold heightened risk of developing active TB (relative risk of 3.11 [95% CI, 2.27–4.26]) (Jeon and Murray 2008). Diabetes is also a risk factor for adverse treatment outcomes in TB patients (Lonnroth et al. 2014).

A population-based study of 115,867 patients in Sweden, aged ≥40 years and with a hospital discharge diagnosis of chronic obstructive pulmonary disease (COPD), found that the relative risk of TB development in COPD patients was threefold higher than in controls (hazard ratio, HR, of 3.0 [95% confidence interval, 2.4–4.0]) (Inghammar et al. 2010). Furthermore, COPD patients who developed active TB exhibited a twofold greater risk of death from all causes in their first year post TB diagnosis with respect to general population controls with TB (odds ratio,

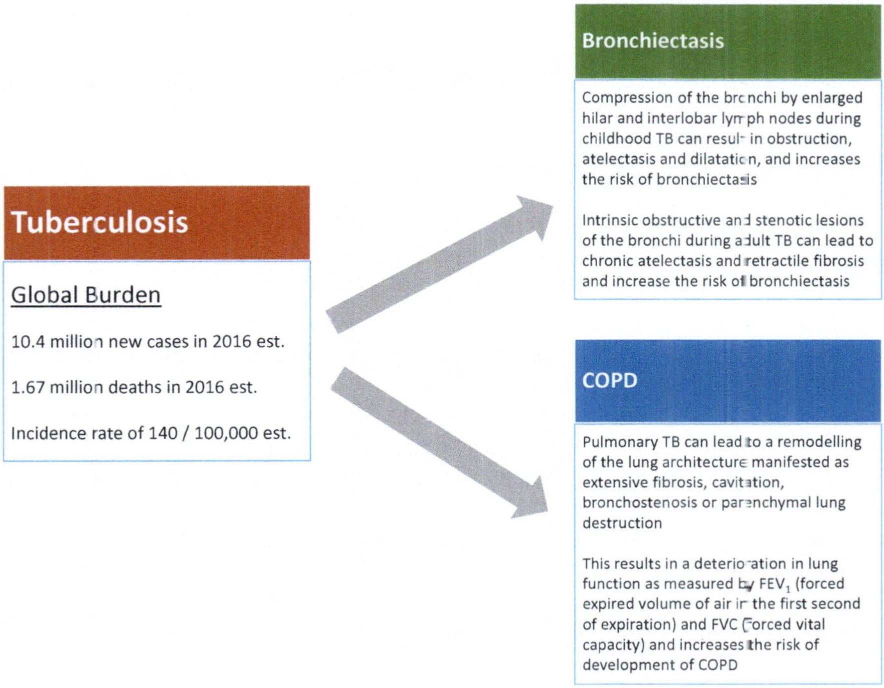

Fig. 2.1 Relationship between tuberculosis and the subsequent increased risk of development of airway diseases bronchiectasis and chronic obstructive pulmonary disease (COPD)

OR, of 2.2 [95% CI, 1.2–4.1]) (Inghammar et al. 2010). A study conducted on 23,594 COPD patients in Taiwan, and 47,188 non-COPD control subjects matched for age and gender, also identified COPD as an independent risk factor for TB development with a hazard ratio of 2.468 [95% CI, 2.205–2.762] (Lee et al. 2013) (Fig. 2.1).

Tobacco smoking has been well established as a dominant causative factor for the development of COPD (Lokke et al. 2006; Lundback et al. 2003). The percentage of COPD mortality attributable to smoking has been estimated to be in the region of 54% for men aged 30–69 years and 52% for men older than 70 years of age (Eisner et al. 2010). Smoking is also a risk factor for the development of TB. Following a meta-analysis of 24 studies, Bates and colleagues reported that smoking represents a significant risk factor for *Mycobacterium tuberculosis* infection as well as TB disease with relative risks estimated at 1.73 [95% CI, 1.46–2.04] and 2.33 [95% CI, 2.15–3.28], respectively (Bates et al. 2007). Other studies have corroborated the link between smoking and TB (Slama et al. 2007; Lin et al. 2007). Alcohol misuse is also a risk factor for TB (Lonnroth et al. 2008; Fok et al. 2008) as it is for ischemic heart disease (Laatikainen et al. 2003; Klatsky 2009).

Therefore, it is evident that TB shares risk factors with a range of human illnesses. As we will see in the proceeding sections, the relationship between TB and other conditions extends to its own role as an underlying risk factor for long-term airflow limitation of the respiratory system. The objective of this chapter is to review the epidemiological data that link TB and subsequent airway disease.

2.2 TB and Bronchiectasis

One of the first chronic airway diseases to be associated with post-tuberculosis was bronchiectasis which is defined as an "irreversible localized or diffuse dilatation, usually resulting from chronic infection, proximal airway obstruction, or congenital bronchial abnormality" (Cantin et al. 2009). In conjunction with enlargement of the bronchi is an impaired ability to maintain effective mucociliary clearance, an important innate defense mechanism in the bronchial tubes. This immune deficit permits microbes and other particles to accumulate in the mucus layer leading to inflammation and further damage of the airways.

In terms of the burden of bronchiectasis, global prevalence rates are not available; however, they have been estimated in a number of countries. The prevalence of bronchiectasis in the USA was previously calculated at 25 per 100,000 in the general population and 272 per 100,000 in persons aged 75 years and above (O'Donnell 2008). But the growing use of high-resolution computed tomography has contributed to an increase in the number of people diagnosed with bronchiectasis in the USA. Using healthcare claims data from 2009 to 2013, Weycker and co-workers determined the prevalence of bronchiectasis in the general US population to be 139 cases per 100,000 based on narrow case-finding criteria (Weycker et al. 2017). Age variability was evident with a prevalence rate of 7 per 100,000 observed in persons aged from 18 to 34 years and 812 per 100,000 in persons aged 75 years and above (Weycker et al. 2017). The authors concluded that the burden of bronchiectasis was higher than previously reported. In Germany, the total number of individuals with bronchiectasis in 2013 was estimated to be 53,807, corresponding to a prevalence rate of 67 [95% CI, 64–69] per 100,000 general population (Ringshausen et al. 2015).

Granger is attributed with recognizing in 1878 that bronchiectasis occurred as a consequence of TB (Grancher 1878). Jones and Cournand referred in 1933 to an enlargement of lymph nodes in primary tuberculosis as a possible causative factor in the development of bronchiectasis (Jones and Cournand 1933). In 1935, Wallgren cited the frequency of bronchiectasis in children after primary tuberculosis (Wallgren 1935). The etiological link between tuberculosis and bronchiectasis was established in a number of follow-up studies. These included the work by Valledor and colleagues published in 1952 (Valledor and Navarrete 1952). They examined 250 cases of primary tuberculosis in children using serial bronchoscopy and bronchography over a period of 15 years. They reported bronchiectasis of varying degrees in approximately 70% of cases (Valledor and Navarrete 1952). They attributed the cause of bronchiectasis to compression of the bronchi by enlarged hilar and interlobar lymph nodes during childhood TB leading to obstruction, atelectasis, and dilatation. They also examined TB in adults and reported that bronchiectasis in adult TB results from intrinsic obstructive and stenotic lesions of the bronchi culminating in chronic atelectasis and retractile fibrosis (Valledor and Navarrete 1952). Based on their work, they concluded that "bronchiectasis is a common sequel of chronic pulmonary tuberculosis, generally affecting the upper lobes."

While TB has been a major source of bronchiectasis, the introduction of anti-tubercular chemotherapeutics and other interventions to control the disease is believed to have reduced the incidence of TB-related bronchiectasis (Bilton and Jones 2011). This is evident in studies performed in low TB incidence countries such as the UK. In 2000, Shoemark and co-workers reported that among 165 patients with confirmed bronchiectasis at the Royal Brompton Hospital, London, only 2 patients had bronchiectasis associated with post-TB (Shoemark et al. 2007). Similarly, in a study conducted on 150 bronchiectasis patients at Papworth Hospital, Cambridge, UK, only 3 patients had a history of TB (Pasteur et al. 2000).

Studies have also been conducted in countries which have a high incidence of TB to examine the role of TB in contributing to the burden of bronchiectasis. In India, which had an estimated 2.79 million cases of TB in 2016, the highest number of TB cases in the world (World Health Organization 2017), one study involving a small number of bronchiectasis patients found that pulmonary TB was the most common etiology (10/19 or 52.63% of patients) (Nag et al. 2015). In earlier work by Sahoo and colleagues, it was reported that among 100 pulmonary cases of TB with detectable residual lung lesions in chest x-rays after completion of chemotherapy, 62 had bronchiectasis (Sahoo et al. 1988). In China, which had an estimated 895,000 cases of TB in 2016 (World Health Organization 2017), a study of non-cystic fibrosis bronchiectasis patients found that pulmonary TB was a major cause of bronchiectasis (2175/6977 or 31.17% of patients) ahead of pertussis and other bacterial infections (Xu et al. 2013). Therefore, it is apparent that reducing the burden of TB could assist in decreasing the prevalence of chronic lung diseases such as bronchiectasis, particularly in high TB incidence countries.

2.3 TB and COPD

Chronic obstructive pulmonary disease (COPD) refers to chronic airflow obstruction that is progressive and poorly reversible and includes chronic bronchitis and emphysema (Pauwels et al. 2001). The disease originates with structural damage to the small airways and neighboring alveoli and progresses over a number of years culminating in a reduction of elastic recoil and an increase in airway resistance (Petty 2006). These pathologies cause defects in lung function, measurable with spirometry, that become more pronounced with advancing age (Petty 2006). In terms of burden, COPD is emerging as the third highest cause of human mortality globally after heart disease and stroke (World Health Organization 2014). In 2015, 3.2 million people [95% uncertainty interval, UI, 3.1–3.3 million] died from COPD worldwide (Collaborators 2017). The World Health Organization (WHO) has estimated that over 65 million people are afflicted with moderate to severe COPD

(World Health Organization 2015). Recent data have found that the prevalence of COPD grew by 44.2% between 1990 and 2015 to 174.5 million individuals [95% UI, 160.2–189.0 million] (Collaborators 2017).

Tuberculosis is increasingly been recognized as a cause of obstructive pulmonary disease. A meta-analysis published in 2015 found a significant correlation between a past history of TB and the presence of COPD in individuals aged 40 years or more when adjusted for known COPD risk factors such as cigarette smoking and age (Byrne et al. 2015). Importantly, when the national annual TB incidence rate exceeded 100 cases per 100,000 population, the odds ratio (OR) of COPD development in patients who had a previous history of TB was more than 3 times that of patients with history of TB. For example, the adjusted OR for adults in the Philippines was 6.31 [95% CI, 2.67–15.0], and for males and females in South Africa were 4.9 [95% CI, 2.6–9.1] and 6.6 [95% CI, 3.7–11.7], respectively (Byrne et al. 2015).

In their review of nonsmoking-related COPD, Salvi and Barnes reported that almost half of fully treated TB patients later presented with obstructive airway disease when examined during follow-up (Salvi and Barnes 2009). In a study conducted on miners in South Africa, it was found that both FEV_1 (forced expired volume of air in the first second of expiration) and FVC (forced vital capacity) were reduced in parallel with an increasing number of episodes of TB (Jordan et al. 2010). In addition, the Korean National Health and Nutrition Examination Survey 2008–2012 reported that a previous history of pulmonary TB (OR of 2.314 [95% CI, 1.922–2.785]) and an inactive pulmonary TB lesion (OR of 2.300 [95% CI, 1.606–3.294]) represented predictors for subsequent obstructive airway disease [25]. In an analysis of patients hospitalized with COPD exacerbation ($n = 598$), Yakar and colleagues reported that COPD diagnosis and death was observed 5 years earlier in patients who had a past history of TB (Yakar et al. 2017). Mattila and co-workers, from a study of 6701 adults in Finland, determined the adjusted OR for airway obstruction ranged from 2.21 [95% CI, 1.52–3.21] in patients who had a TB scar recorded by a radiologist to 4.59 [95% CI, 2.86–7.37] in patients who had a known history of TB disease (Mattila et al. 2017).

A large, international, population-based Burden of Obstructive Lung Disease (BOLD) study examined the association of airflow obstruction and spirometric restriction with a previous history of TB. Examining a study population of 14,050 participants in 18 different countries, the study established a self-reported history of TB to be significantly associated with airflow obstruction (adjusted OR of 2.51 [95% CI, 1.83–3.42]) and also spirometric restriction (adjusted OR of 2.13 [95% CI, 1.42–3.19]) (Amaral et al. 2015). Based on the findings of the BOLD study, the authors concluded that a history of tuberculosis "should be considered as a potentially important cause of obstructive disease and low lung function, particularly where tuberculosis is common" (Amaral et al. 2015). Therefore, multiple strands of evidence now support the role that TB plays in increasing the likelihood that patients may go on to develop COPD in later life.

A number of studies have examined the specific aspects of TB pathology that affect lung structure and function. Pulmonary TB can result in extensive fibrosis, cavitation, traction bronchiectasis, bronchostenosis, or parenchymal lung destruction (Jordan et al. 2010; Dheda et al. 2005). In a South African-based study, Wilcox and Ferguson determined that in subjects with a previous history of TB treatment who were undergoing pulmonary function assessment, airway obstruction was prevalent in 68% of cases (Willcox and Ferguson 1989). They concluded that "treated pulmonary tuberculosis is a cause of significant chronic obstructive airways disease" (Willcox and Ferguson 1989). While antimicrobial chemotherapy may lead to an improvement in lung function in patients with pulmonary TB, Plit et al. (Plit et al. 1998) reported the presence of residual airflow limitation and a restrictive pattern in a sizeable proportion of patients, i.e., 28% and 24%, respectively. In a study conducted in India, 78% of obliterative bronchiolitis cases were identified as being post-TB (Gothi et al. 2007). Chronic deficits in lung function as estimated by spirometry were found by Hnizdo and colleagues to correlate with the number of TB episodes in which functional decreases of 18.4%, 27.1%, and 35.2% were associated with one, two, and three or more episodes of TB, respectively (Hnizdo et al. 2000). Therefore, airflow obstruction and a loss of pulmonary function can manifest in patients who have completed their TB treatment (Ehrlich et al. 2011). Pasipanodya and colleagues have shown that years lived with disability due to TB extend well beyond the acute TB disease due to long-term pulmonary impairment (Pasipanodya et al. 2010). Furthermore, significantly reduced lung function, as measured by FVC (43.58 $\pm$ 16.03% versus 72.06 $\pm$ 14.95% predicted) and FEV_1 (33.08 $\pm$ 15.64% versus 66.13 $\pm$ 19.87% predicted), was detected in patients who had the multidrug-resistant (MDR) form of TB with respect to patients who had non-MDR-TB (Di Naso 2011). It is therefore apparent that episodes of active TB cause structural and functional changes that heighten the risk of subsequent progression to COPD even after successful completion of TB treatment.

2.4 Conclusions and Future Perspectives

Epidemiological associations are emerging between communicable and noncommunicable respiratory diseases. TB in itself worsens the condition of the lungs such that they become more prone to chronic airway illnesses including bronchiectasis and chronic obstructive pulmonary disease. A previous history of TB increases significantly the propensity for subsequent bronchiectasis or COPD development even after the patient has successfully adhered to and completed their prescribed TB chemotherapy. These findings heighten the concern that the immense global incidence of TB and MDR-TB could be sustaining an underdiagnosed level of chronic airway disease, particularly, in high TB burden countries. The WHO predicts that mortalities due to COPD are destined to increase in the decades to come (Mathers and Loncar 2006); therefore, reducing the burden of TB may be an important strategy in reversing that trend. Further research is now required to define the level of contribution that TB makes to the incidence of bronchiectasis and COPD. In addition, investigations are required to establish the specific mechanistic relationships that exist between TB and chronic airway diseases at the cellular and molecular levels. An improved level of understanding of the interactions between TB and noncommunicable lung illnesses may translate into better interventions for the prevention, detection, and management of respiratory disease.

Funding Not applicable.

Competing Interests The author declares that he has no financial or other conflicts of interest.

Ethical Approval Not required.

References

Amaral AF, Coton S, Kato B, Tan WC, Studnicka M, Janson C, Gislason T, Mannino D, Bateman ED, Buist S, Burney PG, Group BCR (2015) Tuberculosis associates with both airflow obstruction and low lung function: BOLD results. Eur Respir J 46:1104–1112. https://doi.org/10.1183/13993003.02325-2014

Bates MN, Khalakdina A, Pai M, Chang L, Lessa F, Smith KR (2007) Risk of tuberculosis from exposure to tobacco smoke: a systematic review and meta-analysis. Arch Intern Med 167 (4):335–342. https://doi.org/10.1001/archinte.167.4.335

Bilton D, Jones AL (2011) Bronchiectasis: epidemiology and causes. Eur Respir Monogr 52:1–10

Byrne AL, Marais BJ, Mitnick CD, Lecca L, Marks GB (2015) Tuberculosis and chronic respiratory disease: a systematic review. Int J Infect Dis 32:138–146. https://doi.org/10.1016/j.ijid.2014.12.016

Cantin L, Bankier AA, Eisenberg RL (2009) Bronchiectasis. AJR Am J Roentgenol 193(3):W158–W171. https://doi.org/10.2214/AJR.09.3053

Chaisson RE, Martinson NA (2008) Tuberculosis in Africa – combating an HIV-driven crisis. N Engl J Med 358(11):1089–1092. https://doi.org/10.1056/NEJMp0800809

Collaborators GBDCRD (2017) Global, regional, and national deaths, prevalence, disability-adjusted life years, and years lived with disability for chronic obstructive pulmonary disease and asthma, 1990–2015: a systematic analysis for the global burden of disease study 2015. Lancet Respir Med 5(9):691–706. https://doi.org/10.1016/S2213-2600(17)30293-X

Dheda K, Booth H, Huggett JF, Johnson MA, Zumla A, Rook GA (2005) Lung remodeling in pulmonary tuberculosis. J Infect Dis 192(7):1201–1209. https://doi.org/10.1086/444545

Di Naso FPJ, Schuh SJ, Unis G (2011) Functional evaluation in patients with pulmonary tuberculosis sequelae. Rev Port Pneumol 17(5):216–221

Ehrlich RI, Adams S, Baatjies R, Jeebhay MF (2011) Chronic airflow obstruction and respiratory symptoms following tuberculosis: a review of South African studies [review article]. Int J Tuberc Lung Dis 15(7):886–891. https://doi.org/10.5588/ijtld.10.0526

Eisner MD, Anthonisen N, Coultas D, Kuenzli N, Perez-Padilla R, Postma D, Romieu I, Silverman EK, Balmes JR, Committee on Nonsmoking COPD Environmental and Occupational Health Assembly (2010) An official American Thoracic Society public policy statement: novel risk factors and the global burden of chronic obstructive pulmonary disease. Am J Respir Crit Care Med 182(5):693–718. https://doi.org/10.1164/rccm.200811-1757ST

Fok A, Numata Y, Schulzer M, FitzGerald MJ (2008) Risk factors for clustering of tuberculosis cases: a systematic review of population-based molecular epidemiology studies. Int J Tuberc Lung Dis 12(5):480–492

Geldmacher C, Schuetz A, Ngwenyama N, Casazza JP, Sanga E, Saathoff E, Boehme C, Geis S, Maboko L, Singh M, Minja F, Meyerhans A, Koup RA, Hoelscher M (2008) Early depletion of Mycobacterium tuberculosis-specific T helper 1 cell responses after HIV-1 infection. J Infect Dis 198(11):1590–1598. https://doi.org/10.1086/593017

Geldmacher C, Ngwenyama N, Schuetz A, Petrovas C, Reither K, Heeregrave EJ, Casazza JP, Ambrozak DR, Louder M, Ampofo W, Pollakis G, Hill B, Sanga E, Saathoff E, Maboko L, Roederer M, Paxton WA, Hoelscher M, Koup RA (2010) Preferential infection and depletion of Mycobacterium tuberculosis-specific CD4 T cells after HIV-1 infection. J Exp Med 207 (13):2869–2881. https://doi.org/10.1084/jem.20100090

Gothi D, Shah DV, Joshi JM (2007) Clinical profile of diseases causing chronic airflow limitation in a tertiary care Centre in India. J Assoc Physicians India 55:551–555

Grancher J (1878) La dilatation des bronches chez les tuberculeux. Gazette Medicale de Paris 146

Hnizdo E, Singh T, Churchyard G (2000) Chronic pulmonary function impairment caused by initial and recurrent pulmonary tuberculosis following treatment. Thorax 55(1):32–38

Hwang JH, Choe PG, Kim NH, Bang JH, Song KH, Park WB, Kim ES, Park SW, Kim HB, Kim NJ, Oh MD, Choe KW (2013) Incidence and risk factors of tuberculosis in patients with human immunodeficiency virus infection. J Korean Med Sci 28(3):374–377. https://doi.org/10.3346/jkms.2013.28.3.374

Inghammar M, Ekbom A, Engstrom G, Ljungberg B, Romanus V, Lofdahl CG, Egesten A (2010) COPD and the risk of tuberculosis – a population-based cohort study. PLoS One 5(4):e10138. https://doi.org/10.1371/journal.pone.0010138

Jeon CY, Murray MB (2008) Diabetes mellitus increases the risk of active tuberculosis: a systematic review of 13 observational studies. PLoS Med 5(7):e152. https://doi.org/10.1371/journal.pmed.0050152

Jones OR, Cournand A (1933) Shrunken pulmonary lobe with chronic bronchiectasis. Am Rev Tuberc 28:293

Jordan TS, Spencer EM, Davies P (2010) Tuberculosis, bronchiectasis and chronic airflow obstruction. Respirology 15(4):623–628. https://doi.org/10.1111/j.1440-1843.2010.01749.x

Klatsky AL (2009) Alcohol and cardiovascular diseases. Expert Rev Cardiovasc Ther 7 (5):499–506. https://doi.org/10.1586/erc.09.22

Laatikainen T, Manninen L, Poikolainen K, Vartiainen E (2003) Increased mortality related to heavy alcohol intake pattern. J Epidemiol Community Health 57(5):379–384

Lee CH, Lee MC, Shu CC, Lim CS, Wang JY, Lee LN, Chao KM (2013) Risk factors for pulmonary tuberculosis in patients with chronic obstructive airway disease in Taiwan: a nationwide cohort study. BMC Infect Dis 13:194. https://doi.org/10.1186/1471-2334-13-194

Lin HH, Ezzati M, Murray M (2007) Tobacco smoke, indoor air pollution and tuberculosis: a systematic review and meta-analysis. PLoS Med 4(1):e20. https://doi.org/10.1371/journal.pmed.0040020

Lokke A, Lange P, Scharling H, Fabricius P, Vestbo J (2006) Developing COPD: a 25 year follow up study of the general population. Thorax 61(11):935–939. https://doi.org/10.1136/thx.2006.062802

Lonnroth K, Williams BG, Stadlin S, Jaramillo E, Dye C (2008) Alcohol use as a risk factor for tuberculosis – a systematic review. BMC Public Health 8:289. https://doi.org/10.1186/1471-2458-8-289

Lonnroth K, Roglic G, Harries AD (2014) Improving tuberculosis prevention and care through addressing the global diabetes epidemic: from evidence to policy and practice. Lancet Diabetes Endocrinol 2(9):730–739. https://doi.org/10.1016/S2213-8587(14)70109-3

Lundback B, Lindberg A, Lindstrom M, Ronmark E, Jonsson AC, Jonsson E, Larsson LG, Andersson S, Sandstrom T, Larsson K, Obstructive Lung Disease in Northern Sweden S (2003) Not 15 but 50% of smokers develop COPD? – report from the obstructive lung disease in Northern Sweden studies. Respir Med 97(2):115–122

Mathers CD, Loncar D (2006) Projections of global mortality and burden of disease from 2002 to 2030. PLoS Med 3(11):e442. https://doi.org/10.1371/journal.pmed.0030442

Mattila T, Heliovaara M, Rissanen H, Knekt P, Puukka P, Vasankari T (2017) Tuberculosis, airway obstruction and mortality in a Finnish population. COPD 14(2):143–149. https://doi.org/10.1080/15412555.2016.1250253

Nag V, Nandan D, Singh S, Arora S, Tsanglao W, Arya N (2015) Pulmonary tuberculosis: a predominant cause of childhood bronchiectasis in India. J Commun Dis 47(1):14–18
O'Donnell AE (2008) Bronchiectasis. Chest 134(4):815–823. https://doi.org/10.1378/chest.08-0776
Pasipanodya JG, McNabb SJ, Hilsenrath P, Bae S, Lykens K, Vecino E, Munguia G, Miller TL, Drewyer G, Weis SE (2010) Pulmonary impairment after tuberculosis and its contribution to TB burden. BMC Public Health 10:259. https://doi.org/10.1186/1471-2458-10-259
Pasteur MC, Helliwell SM, Houghton SJ, Webb SC, Foweraker JE, Coulden RA, Flower CD, Bilton D, Keogan MT (2000) An investigation into causative factors in patients with bronchiectasis. Am J Respir Crit Care Med 162(4 Pt 1):1277–1284. https://doi.org/10.1164/ajrccm.162.4.9906120
Pauwels RA, Buist AS, Calverley PM, Jenkins CR, Hurd SS, Committee GS (2001) Global strategy for the diagnosis, management, and prevention of chronic obstructive pulmonary disease. NHLBI/WHO Global Initiative for Chronic Obstructive Lung Disease (GOLD) workshop summary. Am J Respir Crit Care Med 163(5):1256–1276. https://doi.org/10.1164/ajrccm.163.5.2101039
Petty TL (2006) The history of COPD. Int J Chron Obstruct Pulmon Dis 1(1):3–14
Plit ML, Anderson R, Van Rensburg CE, Page-Shipp L, Blott JA, Fresen JL, Feldman C (1998) Influence of antimicrobial chemotherapy on spirometric parameters and pro-inflammatory indices in severe pulmonary tuberculosis. Eur Respir J 12(2):351–356
Ringshausen FC, de Roux A, Diel R, Hohmann D, Welte T, Rademacher J (2015) Bronchiectasis in Germany: a population-based estimation of disease prevalence. Eur Respir J 46(6):1805–1807. https://doi.org/10.1183/13993003.00954-2015
Sahoo K, Panda BN, Singh VP, Kiran KG, Raha B, Bhalla JS (1988) Bronchographic evaluation of radiologic sequelae of pulmonary tuberculosis after short term chemotherapy. Indian J Tuberc 35(3):128–132
Salvi SS, Barnes PJ (2009) Chronic obstructive pulmonary disease in non-smokers. Lancet 374 (9691):733–743. https://doi.org/10.1016/S0140-6736(09)61303-9. pii
Shoemark A, Ozerovitch L, Wilson R (2007) Aetiology in adult patients with bronchiectasis. Respir Med 101(6):1163–1170. https://doi.org/10.1016/j.rmed.2006.11.008
Slama K, Chiang CY, Enarson DA, Hassmiller K, Fanning A, Gupta P, Ray C (2007) Tobacco and tuberculosis: a qualitative systematic review and meta-analysis. Int J Tuberc Lung Dis 11 (10):1049–1061
Swaminathan S, Ramachandran R, Baskaran G, Paramasivan CN, Ramanathan U, Venkatesan P, Prabhakar R, Datta M (2000) Risk of development of tuberculosis in HIV-infected patients. Int J Tuberc Lung Dis 4(9):839–844
Valledor T, Navarrete A (1952) Bronchiectasis in childhood and adult pulmonary tuberculosis. Dis Chest 22(1):46–54
Wallgren A (1935) Pulmonary tuberculosis in childhood. Am J Dis Child 49:1105
Weycker D, Hansen GL, Seifer FD (2017) Prevalence and incidence of noncystic fibrosis bronchiectasis among US adults in 2013. Chron Respir Dis 14:377–384. https://doi.org/10.1177/1479972317709649
Willcox PA, Ferguson AD (1989) Chronic obstructive airways disease following treated pulmonary tuberculosis. Respir Med 83(3):195–198
World Health Organization (2014) The top 10 causes of death. Fact Sheet No 310
World Health Organization (2015) Burden of COPD
World Health Organization (2017) Global tuberculosis report
Xu J-F, Ji X-B, Li H-P, Lu H-W (2013) Bronchiectasis caused by pulmonary tuberculosis: the clinical presentations and the differences. In: American thoracic society international conference, 2013, vol C58
Yakar HI, Gunen H, Pehlivan E, Aydogan S (2017) The role of tuberculosis in COPD. Int J Chron Obstruct Pulmon Dis 12:323–329. https://doi.org/10.2147/COPD.S116086

Host Factors in Tuberculosis

3

Ruxana T. Sadikot

Abstract

Tuberculosis (TB) caused by *Mycobacterium tuberculosis* (MTB) continues to remain one of the most deadly infectious diseases worldwide especially in the developing countries. The lack of new drugs for TB, vaccine availability, with increasing global antimicrobial resistance, has prompted more research into defining host factors and developing immunomodulatory strategies. Some of the known risk factors that increase the susceptibility to TB include immunocompromised states such as human immunodeficiency virus (HIV) infection, smoking, socioeconomic deprivation, malnutrition, diabetes mellitus, air pollution, occupational exposures, and alcohol consumption. In this chapter we review the available information about host factors that help understand susceptibility to TB. Defining the host factors is critical to develop host-directed therapies that may enhance host defenses and perhaps reduce tissue damage from chronic TB infections.

Keywords

Mycobacterium tuberculosis · Human immunodeficiency virus · Diabetes mellitus · Tumor necrosis factor · Interferon

Abbreviations

AIDS	Acquired immunodeficiency syndrome
DM	Diabetes mellitus
EPTP	Extrapulmonary tuberculosis

R. T. Sadikot (✉)
Section of Pulmonary, Critical Care Medicine, Emory University, Atlanta, GA, USA
e-mail: ruxana.sadikot@emory.edu

S. E. Hasnain et al. (eds.), Mycobacterium tuberculosis: *Molecular Infection Biology, Pathogenesis, Diagnostics and New Interventions*,
https://doi.org/10.1007/978-981-32-9413-4_3

HIV	Human immunodeficiency virus
IFN	Interferon
MBL2	Mannose-binding lectin gene
MDR-TB	Multidrug-resistant tuberculosis
MTB	*Mycobacterium tuberculosis*
TB	Tuberculosis
TNF	Tumor necrosis factor
WHO	World Health Organization

3.1 Introduction

Tuberculosis is the predominant infectious disease in mankind and the ninth leading cause of death globally, ranking on top of HIV/AIDS. *Mycobacterium tuberculosis*, the infectious organism, is carried by an estimated 2000 million people worldwide, and data from WHO show that it takes around 1.5 million lives every year (WHO 2017; Kyu et al. 2018). A variety of newer circulating strains are being identified, which have geographic variation, which are increasing the complexity of the disease. With advances in the diagnostic techniques and molecular genotyping, these variations are becoming apparent and being extensively studied. Similarly, a better understanding of host factors that increase the propensity for infection needs to be better defined.

It is increasingly recognized that there are several host factors that increase the propensity to develop TB such as demographics, genetics, socioeconomic variations, smoking, alcoholism (Simou et al. 2018), and immune status of the individual (Narasimhan et al. 2013). Some of the host factors that are related to genetics which increase predilection for TB occurrence and progression are being better defined with advances in human genome (Chapman and Hill 2012; Khor and Hibberd 2012). Ecological factors that are associated with distribution and variation of the disease constitute important targets for larger interventions to reduce TB infections. In this chapter host factors that have been associated with TB occurrence are discussed.

3.2 Gender Differences

In general, most studies have reported a significantly higher occurrence of TB in men than in women. By employing epidemiological surveys, Horton et al. have identified that this disease affects men population more than the women population, with a ratio close to 2 as reported by worldwide case notification rates (Horton et al. 2016). This gender-biased frequency of TB represents disease epidemiological patterns, but the reasons for this biasness are still unclear and need to be better understood. This phenomenon is quite well known and has been reported widely in different settings and geographic areas over the last century (Dolin et al. 1998; Diwan and Thorson

1999; Begum et al. 2001; Nhamoyebonde and Leslie 2014; Ting et al. 2014). It is becoming evident that the reasons for the observed gender bias may be far more complex and have been controversial in literature. It is likely that both gender- and sex-related factors may contribute to higher tuberculosis rates in men.

Biological differences may explain some of the differences in sex-biased rates of infection and disease. These include factors such as differences in immunity and variations in exposure to MTB associated with different social mixing patterns. Of note there is also a difference in health-seeking pattern among women especially in developing countries where resources are limited and cultural inhibitions of women. These are some of the epidemiological data that have been associated with poorer detection of disease among females (Bothamley et al. 1998). However, there is a scarcity of conclusive epidemiological evidence to support some of these theories. However, gender inequalities in accessing healthcare should be investigated more rigorously as this is a potentially correctable factor and may have significant impact on disease prevalence and spread as women in developing world also care for children who may get infected. Another important factor that may contribute to these differences includes lack of women reporting in passive-case finding contexts. Additionally, other factors such as misuse of alcohol and excessive smoking, which are generally more prevalent in men, may also contribute to the increased incidence of TB in men. Two separate reports have proposed that smoking is one of the key parameters for the observed differences in the incidence of TB in males and females (Maurya et al. 2002; Gajalakshmi et al. 2013).

While socioeconomic and cultural factors could be a cause in creating barriers to accessing healthcare leading to undernotification in women, in recent years, emerging data have also shown that biological mechanisms may contribute to the differences between men and women in susceptibility to TB infection. Interestingly, studies have shown that women have a better ability to clear pathogens, which has been linked to sex steroid-induced immunomodulation. A study by Nhamoyebonde showed that there is a difference in the effects of sex hormones such as the influence of androgens on gut microbiota, which may provoke host susceptibility to TB (Nhamoyebonde and Leslie 2014; Khan et al. 2016). Other studies have shown that sex hormones have different effects on immune cell function. For example, estrogen induces and testosterone represses T-helper cells and can impact macrophage stimulation and humoral immune responses, which play an important role in tuberculosis immune responses (Grossman 1985). Despite of these observations, the scientific evidences to define the immune differences are scant. Further studies are needed to investigate these biological gender differences to better define how sex differences predispose to TB.

3.2.1 Socioeconomic Factors

In developing countries, socioeconomic risk factors may play a key determinant role in the development and progression of TB disease and may impact the recurrence and development of resistance (Mohidem et al. 2018). Several studies have identified these determinants and include lack of social protection, low income, lack of availability of healthcare services, poor living conditions such as ventilation of residences, a high average number of inhabitants per household, and overall lack of hygienic environment (Souza et al. 2008; Hargreaves et al. 2011; Sales et al. 2018). Household income, educational level, and occupation leading to hazardous exposures have been evaluated more extensively in studies as socioeconomic factors. Data from several studies point toward the impact of poverty on treatment compliance as a particularly important factor. Other socioeconomic factors such as poor nutrition and hygiene are a result of poor living conditions. Malnutrition can particularly impact the innate and adaptive responses and may result in increased predilection for the disease and occurrence of TB (Sinha et al. 2018a, b).

In a recent study from Indonesia, Renggannis Wardani et al. showed that education, housing density index, and internal house transmission are the foremost parameters related to latent TB. Thus, TB control program should be integrated with improvement in education, hygiene, and a more rigorous examination of internal house contacts (Renggannis and Wahono 2018). Furthermore, it should be emphasized that these programs should be supported by public and health institutions. Such measures will likely advance TB control programs, particularly in low- and middle-income nations where with high-socioeconomic disparity is observed.

Worldwide, prisoners are also considered as one of the vulnerable groups to developing TB. Screening for incarcerated individuals should be more rigorous and systematic, as they may also be responsible for infecting other inmates. Understanding of socioeconomic differences has several practical implications for TB control programs. There is a requirement to strengthen screening for precise subgroups within the prisoners, in individuals with deprived socioeconomic class and comorbidities, along with consideration of probable role of systematic screening for latent TB infection. These findings reinforce the need for a community-centered approach for control of TB in areas of high endemicity and the need to deploy financial resources to these sectors which will have a broader impact on TB control programs.

3.3 Smoking and Alcohol

Smoking has been found to be a modifiable risk factor for TB infection and disease although the evidence is not yet considered to be conclusive (Kolappan and Gopi 2002). In general, smoking impairs innate and adaptive immune response (Mehta et al. 2008). A case-control study of smoking and mortality from India provides some evidence of this link although the study was limited to examining mortality among males as low smoking frequency among females. This study reported that smoking is attributing to the 61% of urban adult male mortality due to tuberculosis. The study also showed that smoking is responsible for around four times more deaths caused by tuberculosis as compared to nonsmokers. These limited data provide some evidence although larger epidemiologic and scientific studies to confirm the effect on smoking and TB are much needed.

A meta-analysis of 24 studies by Bates et al. showed that smoking is a risk factor for TB infection and disease (Bates et al. 2007). However, whether smoking contributes to increased mortality risk in persons who already have active TB is not established. The data from study by Bates et al. suggest that tuberculosis control policies should incorporate tobacco control including chewing and smoking cessation as a preventive intervention. A pooled analysis further reported hazard ratios were increased by threefold for developing TB in relation to alcohol consumption during follow-up. Study using exposure-response analysis reported that alcohol intake (10–20 g every day) leads to 12% increase in TB risk (Bates et al. 2007). Studies have revealed that smoking is connected with impairment of immune function and may be a risk factor to development of infections (Mehta et al. 2008). Together, these data suggest that smoking cessation should be considered as an integral component of TB control programs.

3.4 Diabetes Mellitus

Many studies suggest an increased risk of pulmonary tuberculosis among those with diabetes mellitus (DM) which poses a significant independent risk for the development of active tuberculosis. DM can also worsen disease outcome and complicate treatment. Tuberculosis and diabetes mellitus are both common and independently have immense public health significance globally. The association and consequences of the two diseases have been well-established in countries where the incidence of both the diseases is high. DM might impact the disease presentation and response to treatment. This is particularly true for patients who have a poor glycemic control. Recent studies suggest that this may be related to alterations in both innate and adaptive immune responses related to poor glycemic control. It is also interesting that diabetic patients who are smear-positive for TB have a higher possibility of failure for smear conversion compared to patients without DM.

Recent studies have recognized that TB can stimulate glucose intolerance and exacerbate dysglycemia in patients with DM. However, evidence on whether the association is bidirectional is yet to be established. In a recent study from the United Kingdom, Pearson et al. (2018) found that DM risk was substantially raised among individuals with a history of TB disease. This finding has implications for follow-up and screening for patients with TB, who may be at high risk of developing DM or related complications. In countries such as India where the incidence of TB is high and that of DM is increasing, this association has major health implications.

There is lack of clear evidence of whether TB-DM comorbidity is co-related with increased risk of developing multidrug-resistant tuberculosis (MDR-TB). In a recent meta-analysis, Tegegne et al. reviewed 24 observational studies from 15 different countries. Their analysis discovered that DM is highly associated with MDR-TB and DM can significantly enhance the probabilities of developing MDR-TB. Therefore, TB-DM patients must be provided with a more robust TB treatment and follow-up strategies. Additionally, efforts to control DM can have a considerable beneficial effect on TB consequences, specifically in patients who have MDR-TB (Tegegne et al. 2018). Other immunodeficiencies such as HIV can exacerbate the impact of DM and TB. By employing a retrospective database, in a community healthcare setting in KwaZulu-Natal Province of South Africa, Sinha et al. showed that DM and HIV synergistically increased the odds of TB symptoms (Sinha et al. 2018a, b). However, further studies are needed to define the impact of glycemic control on development of MDR-TB. With increased understanding with metabolomics, more scientific studies are needed to better define the host defense effects of hyperglycemia and TB propensity.

3.5 Genetics

The genes that impact immune responses are diverse as infectious diseases exert significant selective genetic pressure (Abel et al. 2017). Thus, the impact of host genetic variability on the susceptibility to exogenous pathogens is well recognized. Studies related to genomic analysis have revealed connection between polymorphisms in multiple histocompatibility complex and human leukocyte antigen genes that induces susceptibility to TB infection and disease. Modern studies that have employed genome-wide association approach have yielded several novel findings to define the susceptibility to TB. Some of these have been confirmed in experimental settings. The evidences also suggested the significant role of human genetic polymorphisms in susceptibility to latent TB infection and advancement of active infection such as mannose-binding lectin gene.

The host-pathogen relationship in TB has been described as sympatric (Fenner et al. 2013). This would mean that the pathogen and host may share some common ancestral geographic origin. When patients are infected with ana strain that originates from a different geographic origin than the patient, then the host may have a risk for greater lung pathology and impairment (Pasipandoya et al. 2013). Similarly, there is also evidence for associations between susceptibility to TB and human leukocyte antigen which depends on the strain of infection (Toyo-Oka et al. 2017). There is data that suggests an association of mannose-binding lectin gene (MBL2) polymorphisms with susceptibility to tuberculosis; however the data is not conclusive. Cao et al. performed a meta-analysis of 22 case-control studies to assess the effect of MBL2 polymorphisms and the risk of developing tuberculosis. They found that MBL2 rs1800451 polymorphism is linked with lower TB risk in the general population, and A/O, rs7096206, rs1800450, and rs1800451 were connected with the risk for some explicit ethnic groups (Cao et al. 2018). These data highlight the significance of genetic factors that can predispose to TB. As the genomic approaches to identify single nucleotide polymorphisms become more sophisticated, these associations will be better defined. This will help identify patients who are at high risk of developing TB or may be protected from TB infection and disease.

3.6 Immunocompromised Patients

It is well recognized that patients who have immunosuppressed states such as HIV, patients with solid or bone marrow transplant, and those who are on biologics are predisposed to developing TB. The prevalence of TB among people living with HIV/AIDS (PLWHA) is high. The incidence of both pulmonary and extrapulmonary TB is increased in PLWHA. Many studies from different countries have estimated the burden of combined HIV and TB. The prevalence is particularly high in sub-Saharan Africa and India.

Recent data suggest that the incidence of extrapulmonary TB is also considerably high. Mohammed et al. analyzed 31 studies with a total population of 28,659 from 1990 to 2017. They found that the prevalence of extrapulmonary TB (EPTB) among PLWHA ranged from 6.4% (Mohammed and Assefa 2008). Studies have shown that TB- and HIV-coinfected patients have a high mortality. In particular, a high mortality rate was observed during the first 3 months in those patients who did not take ART (Tshitenge et al. 2018). Thus, it is recommended that patients with HIV infection should be screened for TB early in their care, to minimize delays in diagnosis and treatment. This would allow for immediate treatment for both active and latent infection. This is particularly important in developing countries in Africa and Asia where there is a rise in HIV infection. It is also important to test TB cases for HIV. These rigorous screening programs will contribute to reduced morbidity and mortality among people living with AIDS. To reduce the burden of TB, the WHO's policy on collaborative TB/HIV activities recommend several measures for HIV-infected individuals. These include intensified case finding, isoniazid preventive therapy, infection control, and antiretroviral therapy for patients with HIV.

3.7 Immune Therapeutics and Susceptibility to TB

The recent advent of monoclonal antibodies and soluble receptors that target various molecules such as tumor necrosis factor (TNF)-α has modernized the therapeutic approach for many autoimmune and inflammatory diseases. The connection between anti-TNF therapy and reactivation of latent TB or de novo TB is well recognized. Due to the pleiotropic functions performed by TNF-alpha, blockade of TNF-α may intensify the incidence of serious infections such as TB. Physiological TNF-mediated signaling could be impaired by anti-TNF therapy and may lead to reactivation and dissemination of latent TB infection (Garziera et al. 2017). The maintenance of the granuloma is important for the containment of TB. Immune signaling through TNF and interferon IFN-γ is compulsory for host defenses against *M.tb* infections. TNF is important for multiple processes such as macrophage activation and recruitment of other immune cells including natural killer cells and lymphocytes, which either allows the formation of granuloma or kills the bacterium. TNF helps in the maintenance of the granuloma in patients with latent TB infection. TNF-mediated immune-inflammatory responses are attenuated by anti-TNF therapy, causing disruption of well-formed granulomas. This can cause a release of viable mycobacteria which in turn can lead to reactivation of TB (Cantini et al. 2018). Hence, it is recommended that careful screening is essential for patients who are to commence anti-TNF therapy.

3.8 Summary

Despite the significant efforts made by the World Health Organization (WHO) and partners to control TB, it is still a major public health issue. Recognition of risk factors that predispose to TB can help identify at-risk patients. Specialized interventions should be implemented to tackle the risk factors associated with this dreadful disease specifically in a high-risk population. Socioeconomic factors play an important role in the progression of TB disease and will need to be considered in TB control programs; these include smoking cessation and misuse of drugs and alcohol. Similarly, screening programs for patients with immunocompromised states especially those with HIV and those who receive biologics are recommended. Other correctable factors such as alcoholism and malnutrition may also play a critical role in TB control. Early diagnosis and rigorous treatment can prevent emergence of resistance and limit functional impairment of lungs. Studies to establish the gender differences are also much needed. Further research into the role of genetic factors and indoor and outdoor physical environments to better understand the association between the physical environment and the social environment with TB infection are much needed.

References

Abel L, Fellay J, Haas DW, Schurr E, Srikrishna G, Urbanowski M et al (2017) Genetics of human susceptibility to active and latent tuberculosis: present knowledge and future perspectives. Lancet Infect Dis 18(3):e64–e75

Bates MN, Khalakdina A, Pai M, Chang L, Lessa F, Smith KR (2007) Risk of tuberculosis from exposure to tobacco smoke: a systematic review and meta-analysis. Arch Intern Med 167:335–342

Begum V, de Colombani P, Das Gupta S et al (2001) Tuberculosis and patient gender in Bangladesh: sex differences in diagnosis and treatment outcome. Int J Tuberc Lung Dis 5:604–610

Bothamley G, Diwan VK, Thorson A, Winkvist A (1998) Gender and tuberculosis. Nordic School of Public Health; Sex and gender in the pathogenesis of infectious tuberculosis: a perspective from immunology, microbiology and human genetics, Stockholm

Cantini F, Nannini C, Niccoli L, Petrone L, Ippolito G, Goletti D (2018) Risk of tuberculosis reactivation in patients with rheumatoid arthritis, ankylosing spondylitis, and psoriatic arthritis receiving non-anti-TNF-targeted biologics. Mediat Inflamm 16:8909834

Cao Y, Wang X, Cao Z, Wu C, Wu D, Cheng X (2018) Genetic polymorphisms of MBL2 and tuberculosis susceptibility: a meta-analysis of 22 case-control studies. Arch Med Sci 14 (6):1212–1232

Chapman SJ, Hill AV (2012) Human genetic susceptibility to infectious disease. Nat Rev Genet 13:175–188

Diwan VK, Thorson A (1999) Sex, gender, and tuberculosis. Lancet 353:1000–1001

Dolin P, Diwan VK, Thorson A, Winkvist A (1998) Gender and tuberculosis. Nordic School of Public Health; Tuberculosis epidemiology from a gender perspective, Stockholm

Fenner L, Egger M, Bodmer T, Furrer H, Ballif M, Battegay M et al (2013) HIV infection disrupts the sympatric host-pathogen relationship in human tuberculosis. PLoS Genet 9:e1003318

Gajalakshmi V, Peto R, Kanaka TS, Jha P (2013) Smoking and mortality from tuberculosis and other diseases in India: retrospective study of 43 000 adult male deaths and 35 000 controls. Lancet 362:507–515

Garziera G, Morsch ALB, Otesbelgue F, Staub FL, Palominos PE, Brenol CV, Silva DR (2017) Latent tuberculosis infection and tuberculosis in patients with rheumatic diseases treated with anti-tumor necrosis factor agents. Clin Rheumatol 36(8):1891–1896

Grossman CJ (1985) Interactions between the gonadal steroids and the immune system. Science 227:257–261

Hargreaves JR, Boccia D, Evans CA, Adato M, Petticrew M, Porter J (2011) The social determinants of tuberculosis: from evidence to action. Am J Public Health 101(4):654–662

Horton KC, MacPherson P, Houben RM, White RG, Corbett EL (2016) Sex differences in tuberculosis burden and notifications in low- and middle-income countries: a systematic review and meta-analysis. PLoS Med 13:e1002119

Khan N, Vidyarthi A, Nadeem S, Negi S, Nair G, Agrewala JN (2016) Alteration in the gut microbiota provokes susceptibility to tuberculosis. Front Immunol 7:529

Khor CC, Hibberd ML (2012) Host-pathogen interactions revealed by human genome-wide surveys. Khor CC, Hibberd ML. Trends Genet 28:233–243

Kolappan C, Gopi PG (2002) Tobacco smoking and pulmonary tuberculosis. Thorax 57:964–966

Kyu HH, Maddison ER, Henry NJ, Mumford JE, Barber R, Shields C et al (2018) The global burden of tuberculosis: results from the global burden of disease (GBD) study 2015. Lancet Infect Dis 18:261–284

Maurya V, Vijayan VK, Shah A (2002) Smoking and tuberculosis: an association overlooked. Int J Tuberc Lung Dis 6:942–951

Mehta H, Nazzal K, Sadikot RT (2008) Cigarette smoking and innate immunity. Inflamm Res 57 (11):497–503

Mohammed H, Assefa N (2008) Mengistie B. prevalence of extrapulmonary tuberculosis among people living with HIV/AIDS in sub-Saharan Africa: a systemic review and meta-analysis. HIV AIDS (Auckl) 10:225–237

Mohidem NA, Hashim Z, Osman M, Shaharudin R, Muharam FM, Makeswaran P (2018) Demographic, socio-economic and behavior as risk factors of tuberculosis in Malaysia: a systematic review of the literature. Rev Environ Health 33:407–421

Narasimhan P, Wood J, Macintyre CR, Mathai D (2013) Risk factors for tuberculosis. Pulm Med 2013:828939

Nhamoyebonde S, Leslie A (2014) Biological differences between the sexes and susceptibility to tuberculosis. J Infect Dis 209(Suppl 3):S100–S106

Pasipanodya JG, Moonan PK, Vecino E, Miller TL, Fernandez M, Slocum P et al (2013) Allopatric tuberculosis host–pathogen relationships are associated with greater pulmonary impairment. Infect Genet Evol 16:433–440

Pearson F, Huangfu P, McNally R, Pearce M, Unwin N, Critchley JA (2018) Tuberculosis and diabetes: bidirectional association in a UK primary care data set. J Epidemiol Community Health 73:142–147

Rengganis Wardani DWS, Wahono EP (2018) Prediction model of tuberculosis transmission based on its risk factors and socioeconomic position in Indonesia. Indian J Community Med 43 (3):204–208

Sales CMM, Sanchez MN, Ramalho W, Bertolde AI, Maciel ELN (2018) Social determinants of tuberculosis via a zero-inflated model in small areas of a city in southeastern Brazil. Rev Soc Bras Med Trop 51(5):638–643

Simou E, Britton J, Leonardi-Bee J (2018) Alcohol consumption and risk of tuberculosis: a systematic review and meta-analysis. Int J Tuberc Lung Dis 22(11):1277–1285

Sinha P, Davis J, Saag L, Wanke C, Salgame P, Mesick J, Horsburgh CR Jr, Hochberg NS (2018a) Undernutrition and tuberculosis: public health implications. J Infect Dis 219:1356–1363
Sinha P, Moll AP, Brooks RP, Deng YH, Shenoi SV (2018b) Synergism between diabetes and human immunodeficiency virus in increasing the risk of tuberculosis. Int J Tuberc Lung Dis 22 (7):793–799
Souza WV, Ximenes R, Albuquerque MFM, Lapa TM, Portugal JL, Lima MLC et al (2008) The use of socioeconomic factors in mapping tuberculosis risk areas in a city of northeastern Brazil. Rev Panam Salud Publica 8(6):403–410
Tegegne BS, Mengesha MM, Teferra AA, Awoke MA, Habtewold TD (2018) Association between diabetes mellitus and multi-drug-resistant tuberculosis: evidence from a systematic review and meta-analysis. Syst Rev 7(1):161
Ting WY, Huang SF, Lee MC, Lin YY, Lee YC, Feng JY, Su WJ (2014) Gender disparities in latent tuberculosis infection in high-risk individuals: a cross-sectional study. PLoS One 9:e110104
Toyo-oka L, Mahasirimongkol S, Yanai H, Mushiroda T, Wattanapokayakit S, Wichukchinda N et al (2017) Strain-based HLA association analysis identified HLA-DRB1*09:01 associated with modern strain tuberculosis. HLA 90:149–156
Tshitenge S, Ogunbanjo GA, Citeya A (2018) A mortality review of tuberculosis and HIV co-infected patients in Mahalapye, Botswana: does cotrimoxazole preventive therapy and/or antiretroviral therapy protect against death? Afr J Prim Health Care Fam Med 10(1):e1–e5
WHO (2017) Global tuberculosis report 2017. World Health Organization, Geneva

Extrapulmonary Tuberculosis

4

Surendra K. Sharma and Alladi Mohan

Abstract

Extra-pulmonary tuberculosis (EPTB) refers to an isolated occurrence of tuberculosis (TB) at body sites other than the lung. The advent of human immunodeficiency virus (HIV) and acquired immunodeficiency syndrome (AIDS) pandemic and widespread use of immunosuppressive drugs and biologicals have changed the epidemiology of EPTB. In immunocompetent individuals, EPTB constitutes 15–20% of all TB; in immunosuppressed individuals, EPTB accounts for more than 50% of the cases. Because of paucibacillary nature of the disease and limited accessibility to the disease site, it is difficult to establish the diagnosis. However, advent of newer imaging methods like 18F-labelled 2-deoxy-D-glucose positron emission tomography (18FDG-PET)-computed tomography (CT) and PET-magnetic resonance imaging (MRI) has helped in better localisation and defining the extent of EPTB. With the availability of new molecular tools such as nucleic acid amplification tests (NAAT) and line probe assays (LPA), liquid culture method, the rapid diagnosis of EPTB, drug and drug susceptibility testing (DST)-guided treatment are possible. However, EPTB still remains a diagnostic and therapeutic challenge. Further, its treatment is cumbersome since endpoints are not well defined. For better adherence and

S. K. Sharma (✉)
JH Institute of Molecular Medicine, Jamia Hamdard, New Delhi, Delhi, India

Departments of General Medicine & Respiratory Medicine, JNMC, Datta Meghe Institute of Medical Sciences (DMIMS), Wardha, Maharashtra, India

Department of Internal Medicine (WHO Collaborating Centre for Research & Training in Tuberculosis, Centre of Excellence for EPTB, MoH & FW, GoI), All India Institute of Medical Sciences, New Delhi, Delhi, India
e-mail: sksharma@aiims.ac.in

A. Mohan
Department of Medicine, Sri Venkateswara Institute of Medical Sciences, Tirupati, India

S. E. Hasnain et al. (eds.), Mycobacterium tuberculosis: *Molecular Infection Biology, Pathogenesis, Diagnostics and New Interventions*,
https://doi.org/10.1007/978-981-32-9413-4_4

successful treatment outcome, active TB Drug-Safety Monitoring and Management (aDSM) is recommended.

Keywords
Tuberculosis, extrapulmonary · Diagnosis · Cartridge-based nucleic acid amplification tests · Treatment · Human immunodeficiency virus · Multidrug-resistant tuberculosis

Abbreviations

ATS	American Thoracic Society
CBNAAT	Cartridge-based nucleic acid amplification tests
CDC	Centers for Disease Control and Prevention
CT	Computed tomography
DR-TB	Drug-resistant TB
E	Ethambutol
EPTB	Extrapulmonary TB
^{18}FDG-PET-CT	^{18}F labelled 2-deoxy-D-glucose positron emission tomography
H	Isoniazid
HIV	Human immunodeficiency virus
IDSA	Infectious Disease Society of America
INDEX TB guidelines	Indian Extra-pulmonary TB guidelines
IRIS	Immune reconstitution inflammatory syndrome
MDR-TB	Multidrug-resistant TB
MRI	Magnetic resonance imaging
R	Rifampicin
RR-TB	Rifampicin-resistant TB
TB	Tuberculosis
TBM	TB meningitis
WHO	World Health Organization
Z	Pyrazinamide

4.1 Introduction

Tuberculosis (TB) is one of the top 10 causes of death among adults in the world. As per the Global TB Report 2018 (Global tuberculosis report 2018), in 2017, ten million people developed TB disease; TB caused an estimated 1.3 million deaths among human immunodeficiency virus (HIV)-seronegative people and an additional 300,000 deaths among HIV-seropositive individuals (Global tuberculosis report 2018). Drug-resistant TB (DR-TB) continues to remain a major public health problem world over. Multidrug-resistant TB (MDR-TB, defined as isolates of *Mycobacterium tuberculosis* (*M. tb*) resistant to isoniazid and rifampicin, with or without resistance to other anti-TB drugs)/rifampicin-resistant TB (RR-TB) has been documented in 3.6% of new TB cases and 17% of previously treated cases in 2017 (Global tuberculosis report 2018). Almost one-fourth of the world's population (23%, 1.7 billion) are estimated to have a latent TB infection (Global tuberculosis report 2018). With use of invasive sampling and widespread use of molecular diagnostic tests, extrapulmonary TB (EPTB) is increasingly being diagnosed and notified. In this chapter, we have attempted to provide a critical overview regarding the clinical detection, diagnosis and treatment of the more commonly encountered forms of EPTB.

4.2 Terminology

Pulmonary TB is the most common presentation of TB. The term "EPTB" refers to specific presence of TB in different body parts other than the lung. When an extrapulmonary focus is obvious in a patient with pulmonary TB, such patients are categorised as pulmonary TB (Sharma and Mohan 2004, 2013).

By this categorisation, intrathoracic mediastinal and/or hilar lymph node TB or TB pleural effusion, without radiographic abnormalities in the lungs, is categorised as EPTB; miliary TB is categorised as pulmonary TB.

The term "disseminated TB" refers to active TB disease characterised by coexisting participation of at least two noncontiguous body sites or evidence of *M. tuberculosis* in the blood or bone marrow. "Miliary TB" is a kind of disseminated TB as a consequence of a massive haematogenous dissemination of tubercle bacilli which results in minute discrete foci typically the size of millet seeds (1–2 mm) distributed consistently in the lungs and the other viscera (Sharma et al. 2005, 2012).

4.3 Epidemiology

In the pre-HIV era and in immunocompetent adults, EPTB constituted about 15–20% of all cases of TB (Sharma and Mohan 2004). In HIV-seropositive people with late HIV infection, the risk of TB increases as immunosuppression progresses. In immunosuppressed individuals, EPTB accounts for more than 50% of all cases of TB (Fig. 4.1) (Sharma and Mohan 2004). As per the Global TB Report 2018 (Global tuberculosis report 2018), EPTB constituted 14% of the 6.4 million incident TB cases (Table 4.1). EPTB can develop during primary or postprimary (reactivation, reinfection) TB. Peripheral lymph node TB and TB pleural effusion are two common forms of EPTB; DTB and MTB are more frequently encountered in HIV-seropositive and immunosuppressed individuals (Sharma and Mohan 2004, 2013).

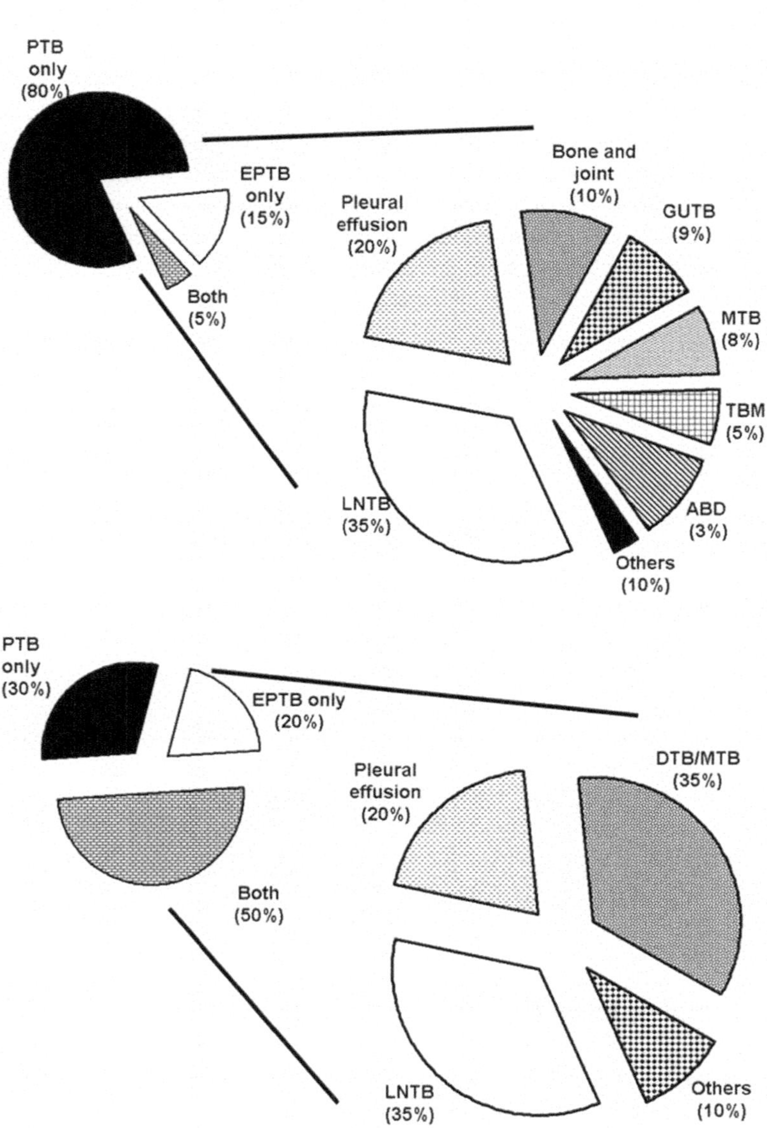

Fig. 4.1 Distribution of tuberculosis cases by anatomical site in immunocompetent (**a**) and immunosuppressed (**b**) adults

PTB pulmonary tuberculosis, *EPTB* extrapulmonary tuberculosis, *GUTB* genitourinary tuberculosis, *DTB* disseminated tuberculosis, *MTB* miliary tuberculosis, *TBM* tuberculosis meningitis, *ABD* abdominal tuberculosis, *LNTB* lymph node tuberculosis

Reproduced with permission from Sharma and Mohan (2004)

Table 4.1 Proportion of EPTB cases from among notified TB cases in 2017

WHO region	Total cases notified	EPTB[a] (%)
Africa	1,375,550	16
The Americas	228,943	15
Eastern Mediterranean	523,494	24
Europe	220,832	16
Southeast Asia	2,826,486	15
Western Pacific	1,350,202	08
Global	6,444,278	14

Data source: Global tuberculosis report (2018)

EPTB extrapulmonary TB, *TB* tuberculosis, *WHO* World Health Organization

[a]Includes new and relapse EPTB cases for which the "treatment history is unknown". It excludes cases that have been reregistered as *treatment after failure*, as *treatment after lost to follow-up* or as *other previously treated* (whose outcome after the most recent course of treatment is unknown or undocumented)

4.4 Clinical Presentation

4.4.1 Constitutional Symptoms

Patients with EPTB often (though not always) manifest general constitutional symptoms such as fever, anorexia, weight loss, malaise and fatigue. For example, fever is seldom present in peripheral lymph node TB; but, when the disease is positioned at an obscure site, EPTB may present with pyrexia of unknown origin (PUO), and this may be the only diagnostic hint in these patients (Sharma and Mohan 2004).

4.4.2 Symptoms and Signs Related to Site of Involvement

In addition to constitutional symptoms, patients with EPTB manifest symptoms of the organ system involved, and these are described under the respective anatomical sites (Sharma and Mohan 2004).

4.4.3 Lymph Node TB

Peripheral lymph nodes are most often affected site of EPTB. Lymph node TB is considered to be the local manifestation of a systemic disease and may develop at the time of primary infection or may occur in later stages of life as postprimary TB. Lymph node TB often affects children and young adults; a female preference has been reported in several studies. Peripheral lymph node TB has been classified into (i) stage 1, enlarged, firm mobile discrete nodes; (ii) stage 2, enlarged nodes fixed to surrounding tissue (periadenitis); (iii) stage 3, central softening (abscess formation); (iv) stage 4, collar-stud abscess; and (v) stage 5, sinus tract formation (Jones 1962).

Patients usually present with slowly enlarging lymphadenopathy and may otherwise remain asymptomatic. Though not common, fever and other constitutional symptoms may sometimes be evident in some patients. Cervical lymph nodes are usually affected followed by axillary and inguinal lymph nodes. Multifocal involvement, intrathoracic and intra-abdominal lymphadenopathy and associated pulmonary disease are more common in HIV-seropositive and immunosuppressed individuals (Sharma and Mohan 2004, 2013).

Physical examination findings are dependent on the stage of the disease. The enlarged lymph nodes could be of variable size, are typically firm and may be discrete or matted. Cystic consistency may be seen if necrosis and abscess formation have occurred. The lymph nodes are not tender in the absence of secondary bacterial infection. Lymph node abscess can burst causing chronic non-healing sinus and ulcer formation. TB sinuses have thin, bluish, undermined edges with scanty watery discharge (Sharma and Mohan 2004, 2013).

4.4.4 Uncommon Manifestations

Uncommon manifestations include dysphagia, oesophago-mediastinal/tracheo-oesophageal fistula, chylothorax, cardiac tamponade (mediastinal lymph node involvement), chylous ascites and chyluria (abdominal lymphadenopathy). Biliary obstruction due to enlarged lymph nodes can cause obstructive jaundice (Sharma and Mohan 2004).

4.4.5 Pleural Effusion and Empyema Thoracis

TB pleural effusion develops when a small subpleural focus ruptures in the pleural space and leads to the interaction between the tubercle bacilli or their particular components prompting a delayed-type hypersensitivity reaction. Other mechanisms such as rupture of a cavity having caseous material in the pleural space, rupture of caseous paratracheal lymph nodes, paravertebral abscess or osteomyelitis of the ribs can also result in empyema thoracis.

TB pleural effusion typically presents as an acute illness. The duration of symptoms may vary from a few days to few weeks. Pleuritic chest pain, dyspnoea and nonproductive cough are the most common symptoms. Fever is often present, with few exceptions. Rarely the onset could be subacute, with only mild chest pain, cough, low-grade pyrexia, anorexia and weight loss. Patients with TB empyema present with chest pain, dyspnoea, productive cough, fever and symptoms of TB toxaemia. Anaemia and hypoproteinaemia are often present.

Physical examination may reveal digital clubbing (empyema), findings suggestive of pleural effusion and intercostal tenderness. Sometimes, TB empyema may exist as a chest wall mass with an expansile cough impulse (*empyema necessitatis*) or a draining sinus tract.

4.4.6 Abdominal TB

The term "abdominal TB" in clinical usage encompasses TB of the gastrointestinal tract, peritoneum, omentum, mesentery and its nodes and other solid intra-abdominal organs such as the liver, biliary tract, spleen and pancreas (Sharma and Mohan 2004).

4.4.6.1 Gastrointestinal TB

Gastrointestinal TB can develop primarily within the intestinal tract or secondary to a focus elsewhere in the body. The most common site of involvement in gastrointestinal TB includes the ileum, ileocecal valve and caecum. Other sites of involvement include the colon, jejunum, stomach, duodenum and oesophagus.

Intestinal TB can develop at any age but is commonly seen in persons aged between 20 and 40 years; both genders are equally affected. Constitutional symptoms like fever, weakness, weight loss and night sweats are often present. Common presenting symptoms include colicky abdominal pain, distension, nausea, vomiting, constipation and rarely gastrointestinal bleeding (ulcero-constrictive disease). Diarrhoea may be seen in one-fourth to one-third of the patients. Patients with intestinal TB may also present with recurrent episodes of intestinal colic, partial or complete intestinal obstruction. Sometimes, patients may present with weight loss and pyrexia of unknown origin or features of malabsorption syndrome (Sharma and Mohan 2004).

4.4.7 Peritoneal TB

Peritoneal TB presents as a subacute disease and presents in three forms, namely, ascitic (wet type), encysted (loculated) and dry (plastic) forms. Abdominal pain that is often diffuse and non-localising is the most common presenting symptom. Abdominal distension is usually present except in the dry form. Tenderness on palpation is present in almost half of the patients. The classical "doughy" abdomen is seen in dry or plastic type of peritoneal TB (Sharma and Mohan 2004).

4.4.7.1 Hepatobiliary, Pancreatic and Splenic Tuberculosis

Hepatobiliary and pancreatic TB are rare and are encountered in immunosuppressed individuals and in miliary TB. The clinical presentation is non-specific. Constitutional symptoms, such as low-grade fever, anorexia, malaise, weight loss and night sweats, are often present. Other manifestations include malaena, pancreatic mass or abscess or obstructive jaundice. Clinical presentation of pancreatic TB may mimic acute or chronic pancreatitis or malignancy (Sharma and Mohan 2004).

While splenomegaly is common in patients with disseminated/miliary TB, isolated splenic TB is rare in immunocompetent individuals. Splenic TB presents as fever of unknown origin; imaging may reveal splenic abscess which may be multiple in immunosuppressed individuals (Sharma and Mohan 2004, 2013).

4.4.8 Neurological TB

Neurological TB is five times more frequent in immunosuppressed individuals compared with immunocompetent individuals.

4.4.8.1 TB Meningitis

TB meningitis (TBM) accounts for 7–80% of cases and is secondary to TB elsewhere in the body (Sharma and Mohan 2004). In developing countries, TBM is a childhood disease but is frequently being seen in adults as well. Onset is subacute over 2–6 weeks; nevertheless, acute onset has also been observed. The prodromal phase (2–3 weeks) is characterised by a history of vague ill-health, irritability, apathy, anorexia and behavioural changes. With the onset of meningitis, fever, headache and vomiting, focal neurological deficits and features of raised intracranial tension may become evident. Focal or generalised seizures and cranial nerve palsies (the sixth cranial nerve involvement being the most common) are often seen. Complete or partial loss of vision is a main complication of TBM. Progressive deterioration in the level of consciousness, pupillary abnormalities and pyramidal signs may indicate increasing hydrocephalus and tentorial herniation. Deep coma and decerebrate or decorticate posturing ensues. Untreated, death usually occurs in 5–8 weeks. Uncommon presentations include acute meningitic syndrome, progressive dementia, status epilepticus, psychosis, stroke syndrome, locked-in state, trigeminal neuralgia, infantile spasm and movement disorders (Sharma and Mohan 2004, 2013).

Patients with TBM are divided into three clinical stages: stage 1, patients are conscious and oriented with or without signs of meningeal irritation, but there is no focal neurological deficit; stage 2, altered sensorium or focal deficits are evident; and stage 3, coma and dense deficits are present. Clinical staging is helpful in optimising therapy and to predict the prognosis (Kennedy and Fallon 1979).

4.4.8.2 Other Forms of Neurological TB

Other forms of neurological TB include intracranial tuberculomas and single, small, enhancing brain lesions presenting with epilepsy and arachnoiditis.

4.4.9 TB and the Heart

Pericardial TB is the most common manifestation of TB of the heart. Pericardial involvement usually results from direct extension of infection from adjacent mediastinal lymph nodes or through lymphohaematogenous route from a focus elsewhere. The following stages of TB pericarditis have been described: (*i*) dry stage, (*ii*) effusive stage, (*iii*) absorptive stage and (*iv*) constrictive stage. However, the disease may progress sequentially from first to fourth stage or could be present as any stages.

TB pericarditis occurs most frequently in the third to fifth decade of life. The onset is insidious onset and patients present with fever, malaise and weakness and vague chest pain. Dyspnoea, cough and weight loss are commonly present. Pericardial TB can present as acute pericarditis, cardiac tamponade, pericardial effusion, effusive constrictive pericarditis or chronic constrictive pericarditis. *Pulsus paradoxus*, raised jugular venous pressure, ascites and dependant oedema are present; pericardial rub may be evident (Sharma and Mohan 2004).

Myocardial TB is increasingly being recognised. Patients with myocardial TB present with idiopathic ventricular tachycardia or unexplained heart failure (Mohan A et al. 2015).

4.4.10 Bone and Joint TB

Skeletal TB results from haematogenous dissemination and affects almost all bones. TB of the spine and hip joint is common; other sites include the knee, foot bones, elbow, hand bones and rarely shoulder joint. Granular and exudative (caseous) forms are described; and one form may predominate.

Spinal TB is the most frequently encountered form of skeletal TB. Mostly the affected individuals are under the age of 30 years. Constitutional symptoms, such as fever, weakness, anorexia, weight loss and night sweats, are usually present and may manifest before the symptoms related to the spine manifest. Lower thoracic and lumbar vertebrae are the mostly affected locations of spinal TB followed by middle thoracic and cervical vertebrae. Classically, two adjoining vertebrae are affected but several vertebrae may be involved and skip lesions also occur.

The disease process begins commonly in the cancellous area of vertebral body in epiphyseal location and rarely in central or anterior area of vertebral body. It spreads and destroys the epiphyseal cortex, the intervertebral disc and the adjacent vertebrae. The vertebral body becomes soft and gets easily compressed resulting in either wedging or total collapse. The exudate penetrates the ligaments and follows the path of least resistance to distant sites from the original bony lesion resulting in cold abscess formation. The most serious complication of spinal TB is Pott's paraplegia and may develop in almost one-third of the patients with spinal TB. Early-onset (during the active phase of disease) and late-onset (several years after the disease has become quiescent) paraplegia have been described.

Clinical presentation of TB of the hip and knee joints depends on the clinico-pathological stage and is characterised by definite patterns of clinical deformity. Pain, arthritis and "night cries" may be evident. TB osteomyelitis may resemble chronic osteomyelitis due to other causes (Sharma and Mohan 2004).

4.4.11 Genitourinary TB

Genitourinary TB develops due to haematogenous dissemination from an active site of TB; descending spread of infection, inflammation and scarring occur. Patients present with dysuria, haematuria which may be painless, flank pain, renal mass, sterile pyuria and recurrent urinary tract infection. Rarely, an acute presentation mimicking pyelonephritis has also been described. Other uncommon presentations include non-healing wounds, sinuses or fistulae and haemospermia among others (Sharma and Mohan 2004).

4.4.12 Female Genital TB

Female genital TB develops secondary to TB elsewhere in the body; haematogenous or lymphatic spread is the most common method of spread. Female genital TB is an important cause of infertility. Patients may also present with chronic lower abdominal or pelvic pain or menstrual abnormalities. Symptoms of TB toxaemia may not be apparent. Physical examination may be unremarkable (Sharma and Mohan 2004).

4.4.13 Cutaneous TB

Several clinical types of cutaneous TB have been reported. These include lupus vulgaris, scrofuloderma and *tuberculosis verrucosa cutis*; other lesions seen are tuberculids (*lichen scrofulosorum*, papulonecrotic tuberculid, erythema induratum and erythema nodosum). Miliary TB of the skin and TB chancre have also been described. Further, localised and generalised skin complications due to bacille Calmette-Guerin (BCG) vaccination have also been described (Sharma and Mohan 2004).

4.4.14 Ocular Tuberculosis

Ocular TB usually develops as a result of haematogenous dissemination. TB can affect all the parts of the eye; choroid is the most commonly affected site. Conjunctival TB and lupus vulgaris are the common manifestations of primary TB, while tuberculids and phlyctenulosis occur in postprimary TB (Sharma and Mohan 2004).

4.4.15 Tuberculosis in Otorhinolaryngology

Before the advent of anti-TB treatment, patients with active pulmonary TB often developed laryngeal, otological, nasal and paranasal sinus TB and deteriorated progressively. These forms have become rare with the availability of effective anti-TB treatment and are rarely seen presently in immunocompetent persons (Sharma and Mohan 2004).

4.4.16 TB of the Breast

TB of the breast can occur as primary disease or can be secondary to TB elsewhere in the body. Secondary involvement of the breast is more common and occurs as a result of lymphatic or haematogenous spread or by contiguous spread from the ribs, pleural space. Clinical presentation is atypical, and often, histopathological evidence suggests the diagnosis (Sharma and Mohan 2004).

4.4.17 Disseminated/Miliary Tuberculosis

Dissemination can occur during primary infection or later during reactivation of latent infection. Clinical manifestations of disseminated/miliary TB are non-specific. Clinical presentation is often subacute. Fever and inanition are relatively common. Cough and dyspnoea are frequently present. Chills and rigors can be present. Physical examination may reveal peripheral lymphadenopathy and cutaneous involvement. Choroidal tubercles (bilateral pale greyish white oblong patches), if present, can be a valuable diagnostic hint to miliary TB. Signs of hepatic involvement (icterus, hepatomegaly) and neurological involvement (meningitis or tuberculomas) are common. Clinically significant cardiac or renal involvement is rare. Overt adrenal insufficiency may be evident at initial presentation or may develop during treatment. Rarely, acute respiratory distress syndrome and acute kidney injury can develop in patients with miliary TB (Sharma and Mohan 2013; Sharma et al. 2012).

4.5 Diagnosis

In patients clinically suspected to have EPTB, attempt should be made to procure sputum; most accessible tissue/body fluids for histopathological, cytopathological, microbiological and molecular diagnosis. With the availability of ultrasonography, computed tomography (CT) and magnetic resonance imaging (MRI), ^{18}F labelled 2-deoxy-D-glucose positron emission tomography (^{18}FDG-PET)-CT, PET-MRI, bronchoscopy, mediastinoscopy, thoracoscopy, upper gastrointestinal endoscopy, colonoscopy, laparoscopy, cystoscopy, colposcopy, among others, procurement of tissue and fine needle aspiration and cytopathology (FNAC) under visual/image guidance, precise anatomical localisation of the lesions of EPTB antemortem and establishing definitive diagnosis has become possible (Sharma and Mohan 2004; Sharma and Mohan 2013).

Case definitions for EPTB are shown in Table 4.2 (Sharma et al. 2017). With availability of "universal drug-susceptibility testing (DST)" at the time of initial

Table 4.2 Case definitions for EPTB

Case	Definition
Presumptive case	A patient with symptoms and signs of EPTB who needs to be investigated
Bacteriologically confirmed case	A patient who has a microbiological diagnosis of EPTB, based on positive microscopy, culture or a validated PCR-based test
Clinically diagnosed case	A patient with negative microbiological tests for TB (microscopy, culture and validated PCR-based tests), but with strong clinical suspicion and other evidence of EPTB, such as compatible imaging findings, histological findings, ancillary diagnostic tests or response to anti-TB treatment

Source: Sharma SK et al. (2017)
TB tuberculosis, *EPTB* extrapulmonary tuberculosis, *PCR* polymerase chain reaction

presentation (Technical and Operational Guidelines for TB Control in India 2016; Guidelines on programmatic management of drug-resistant tuberculosis in India 2017 2017) and molecular diagnostic methods for diagnosis of EPTB (Sharma et al. 2014, 2017; Kohli et al. 2018), especially cartridge-based nucleic acid amplification tests (CBNAAT) (Fig. 4.2), rapid diagnosis of EPTB is possible.

The utility of Xpert MTB/RIF in the diagnosis of some forms of EPTB as recommended in the recently published evidence-based Indian Extra-pulmonary TB (INDEX TB) guidelines (Sharma et al. 2017) is shown in Table 4.3.

4.6 Treatment

4.6.1 Drug-Susceptible EPTB

Treatment of EPTB is similar to treatment of pulmonary TB. Initial intensive phase consists of rifampicin (R), isoniazid (H), pyrazinamide (Z) and ethambutol (E) administered daily for 2 months (2RHZE). In TBM, ethambutol is replaced by streptomycin (S) and the intensive phase would consist of 2SHRZ. In countries/settings where the level of H resistance among new cases is high and H susceptibility testing cannot be performed or results are not provided before the continuation phase starts, continuation phase consists of daily RHE (Technical and Operational Guidelines for TB Control in India 2016; Treatment of tuberculosis: guidelines. 2010; Treatment of tuberculosis. Guidelines for the treatment of drug-susceptible tuberculosis and patient care, update 2017).

There is no universal consensus regarding the *optimal duration* of treatment in patients with EPTB. Recommendations in some forms of EPTB as per the current guidelines are as follows. The optimal duration of standard first-line treatment for EPTB as recommended in the INDEX TB guidelines (Sharma et al. 2017) is as follows: peripheral lymph node TB (6 months), abdominal TB (6 months) and TBM (at least 9 months). The recently published American Thoracic Society (ATS), the Centers for Disease Control and Prevention (CDC) and the Infectious Disease

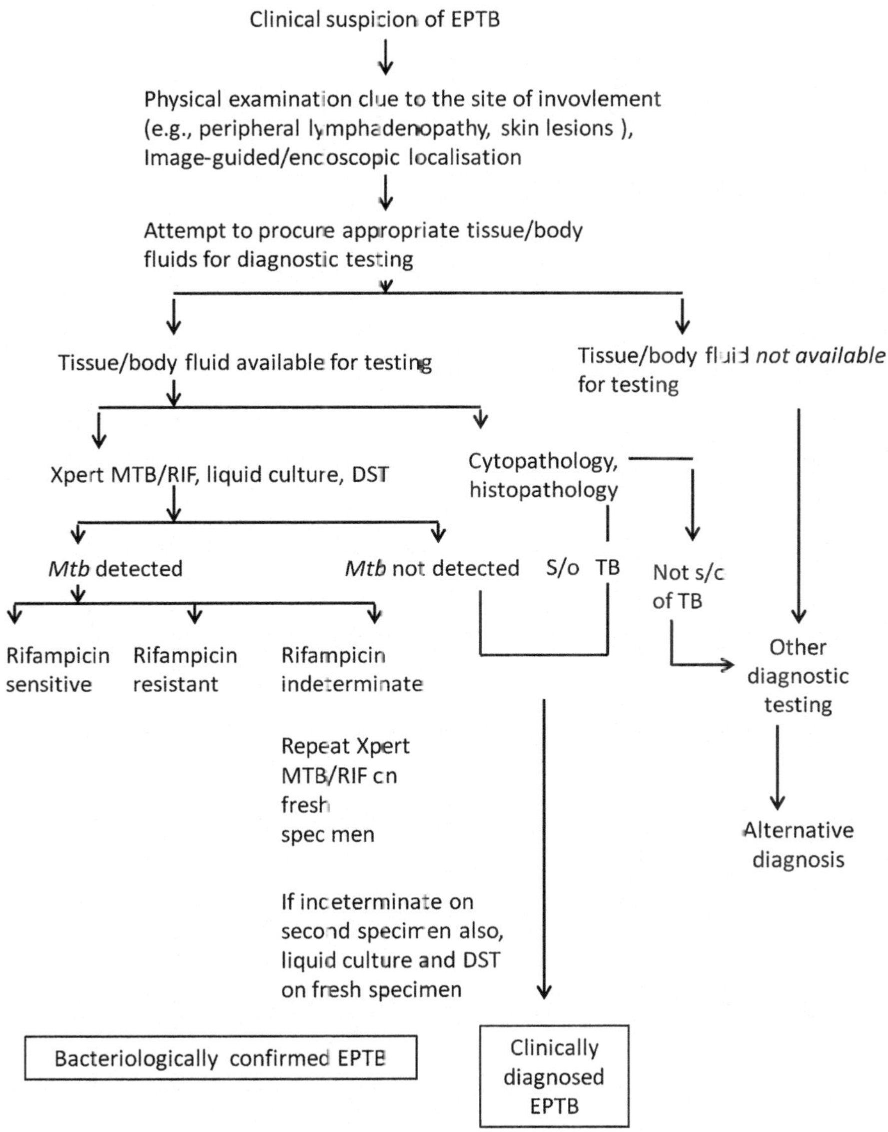

Fig. 4.2 Diagnostic algorithm for EPTB
EPTB extrapulmonary TB, *DST* drug-susceptibility testing, Mtb *Mycobacterium tuberculosis*, *S/o* suggestive of

Society of America (IDSA) guidelines (Nahid et al. 2016) state that some experts recommend 9–12 months of treatment for bone and joint TB; 6 months of standard treatment is considered adequate for pericardial TB and TB pleural effusion.

Table 4.3 Recommendations for use of Xpert MTB/RIF for diagnosis of EPTB as per the INDEX TB guidelines

Site of EPTB	Pooled sensitivity (%)	Pooled specificity (%)	Recommendation
Lymph node TB			Xpert MTB/RIF should be used as an additional test to conventional smear microscopy, culture and cytology in FNAC specimens (strong recommendation, low-quality evidence for sensitivity estimate, high-quality evidence for specificity estimate)
Against culture	83.1	93.6	
Against CRS	81.2	99.1	
TBM			Xpert may be used as an adjunctive test for TBM. A negative Xpert result on a CSF specimen does not rule out TBM. *The decision to give anti-TB treatment should be based on clinical features and CSF profile* (conditional recommendation, low-quality evidence for sensitivity estimate, high-quality evidence for specificity estimate)
Against culture	80.5	97.8	
Against CRS	62.8	98.8	
Pleural TB			Xpert MTB/RIF should not be used to diagnose pleural TB (strong recommendation, low-quality evidence for sensitivity estimate, high-quality evidence for specificity estimate)
Against culture	46.4	99.1	
Against CRS	21.4	100	

Data source: Sharma SK et al. (2017)

EPTB extrapulmonary tuberculosis, *TB* tuberculosis, *CRS* composite reference standard, *FNAC* fine needle aspiration and cytopathological examination, *TBM* tuberculosis meningitis, *CSF* cerebrospinal fluid

According to the World Health Organization (WHO) guidelines (Treatment of tuberculosis: guidelines 2010; Treatment of tuberculosis. Guidelines for the treatment of drug-susceptible tuberculosis and patient care, 2017 update 2017), 6 months of standard anti-TB treatment is recommended for miliary TB. The WHO guidelines (Treatment of tuberculosis: guidelines 2010) mention that some experts recommend 9–12 months of treatment for TBM and 9 months of treatment when bone and joint TB is also present. The recent ATS, CDC and IDSA guidelines (Nahid et al. 2016) state that 6 months of treatment is adequate in disseminated/miliary TB; these guidelines (Nahid et al. 2016) suggest that, when associated TBM is also present, treatment needs to be given for at least 12 months. The National Institute for Health and Clinical Excellence (NICE) (NICE guideline (NG33) 2016) guidelines from the United Kingdom recommend 6 months of treatment for miliary TB. In a given patient, the duration of continuation phase may be extended as per the clinical situation.

Patients with EPTB receiving anti-TB treatment should be carefully followed up and monitored for adverse drug reactions, especially anti-TB drug-induced

Table 4.4 Recommendations for corticosteroid usage in some forms of EPTB as per the INDEX TB guidelines

TB meningitis
Corticosteroids are recommended for TB meningitis in HIV-seronegative people. Duration of steroid treatment should be for at least 4 weeks with tapering as appropriate (strong recommendation)
Corticosteroids may be used for TB meningitis in HIV-seropositive people, where other life-threatening opportunistic infections are absent (conditional recommendation)
TB pericarditis
Corticosteroids are recommended for HIV-seronegative and HIV-seropositive patients with TB pericarditis with pericardial effusion (conditional recommendation)
TB pleural effusion
Corticosteroids are not routinely recommended in pleural TB (conditional recommendation)

Source: Sharma SK et al. (2017)
INDEX TB Indian Extra-pulmonary TB, *EPTB* extrapulmonary tuberculosis, *TB* tuberculosis, HIV human immunodeficiency virus

hepatotoxicity. Patients with HIV-EPTB coinfection should be monitored for immune reconstitution inflammatory syndrome (IRIS).

4.6.1.1 Corticosteroids

The usefulness of corticosteroids in the treatment of EPTB is not well established. When adrenal failure is present, corticosteroids are indicated. Use of corticosteroids can be life-saving in patients with some forms of EPTB. Recommendations for corticosteroid use as stated in the INDEX TB guidelines (Sharma et al. 2017) are shown in Table 4.4.

4.6.2 Drug-Resistant, Multidrug-Resistant, Extensively Drug-Resistant EPTB and HIV-EPTB Coinfection

Drug-resistant, multidrug-resistant, extensively drug-resistant EPTB (13) and HIV-EPTB coinfection (Nahid et al. 2016) should be treated as per standard guidelines (*Consolidated guidelines on the use of antiretroviral drugs for treating and preventing HIV infection: recommendations for a public health approach* 2016; Antiretroviral therapy for HIV infection in adults and adolescents. Recommendations for a public health approach 2010). A detailed description of this is beyond the scope of this chapter.

4.6.3 Surgery

In patients with EPTB, surgery is required to procure specimens for diagnostic testing and, in addition to anti-TB treatment, to ameliorate complications (e.g. intestinal perforation, spinal TB, hydrocephalus). Details regarding surgical management of EPTB are beyond the scope of this chapter.

4.7 Conclusions

A high index of clinical suspicion, use of invasive diagnostic methods for procuring tissue, body fluids for diagnostic testing, confirmation of diagnosis by use of molecular diagnostic tests and early institution of specific anti-TB treatment and close clinical monitoring are key to the successful outcomes in EPTB.

References

Global tuberculosis report (2018) Geneva: World Health Organization; 2018

Guidelines on programmatic management of drug-resistant tuberculosis in India 2017 (2017) Revised national tuberculosis control programme, Directorate General of Health Services, Ministry of Health and Family Welfare, Government of India

Jones PG CP (1962) Tuberculous lymphadenitis in childhood: the significance of anonymous mycobacteria

Kennedy DH, Fallon RJ (1979) Tuberculous meningitis. JAMA 241(3):264–268

Kohli M, Schiller I, Dendukuri N, Dheda K, Denkinger CM, Schumacher SG, Steingart KR (2018) Xpert((R)) MTB/RIF assay for extrapulmonary tuberculosis and rifampicin resistance. Cochrane Database Syst Rev 8:Cd012768. https://doi.org/10.1002/14651858.CD012768.pub2

Mohan A, Thachil A, Sundar G, Sastry BKS, Hasan A, Sridevi C, Narasimhan C (2015) Ventricular tachycardia and tuberculous lymphadenopathy. J Am Coll Cardiol 65(2):218–220

Nahid P, Dorman SE, Alipanah N, Barry PM, Brozek JL, Cattamanchi A, Chaisson LH, Chaisson RE, Daley CL, Grzemska M, Higashi JM, Ho CS, Hopewell PC, Keshavjee SA, Lienhardt C, Menzies R, Merrifield C, Narita M, O'Brien R, Peloquin CA, Raftery A, Saukkonen J, Schaaf HS, Sotgiu G, Starke JR, Migliori GB, Vernon A (2016) Official American thoracic society/ centers for disease control and prevention/infectious diseases society of America clinical practice guidelines: treatment of drug-susceptible tuberculosis. Clin Infect Dis 63(7):e147–e195. https://doi.org/10.1093/cid/ciw376

NICE guideline (NG33) (2016) National collaborating centre for chronic conditions, National Institute for Health and Clinical Excellence. https://www.nice.org.uk/guidance/ng33/resources/tuberculosis-1837390683589. Accessed 1 Mar 2019

Sharma SK, Mohan A (2004) Extrapulmonary tuberculosis. Indian J Med Res 120(4):316–353

Sharma SK, Mohan A (2013) Tuberculosis: from an incurable scourge to a curable disease - journey over a millennium. Indian J Med Res 137(3):455–493

Sharma SK, Mohan A, Sharma A, Mitra DK (2005) Miliary tuberculosis: new insights into an old disease. Lancet Infect Dis 5(7):415–430. https://doi.org/10.1016/s1473-3099(05)70163-8

Sharma SK, Mohan A, Sharma A (2012) Challenges in the diagnosis & treatment of miliary tuberculosis. Indian J Med Res 135(5):703–730

Sharma SK, Kohli M, Chaubey J, Yadav RN, Sharma R, Singh BK, Sreenivas V, Sharma A, Bhatia R, Jain D, Seenu V, Dhar A, Soneja M (2014) Evaluation of Xpert MTB/RIF assay performance in diagnosing extrapulmonary tuberculosis among adults in a tertiary care Centre in India. Eur Respir J 44(4):1090–1093. https://doi.org/10.1183/09031936.00059014

Sharma SK, Ryan H, Khaparde S, Sachdeva KS, Singh AD, Mohan A, Sarin R, Paramasivan CN, Kumar P, Nischal N, Khatiwada S, Garner P, Tharyan P (2017) Index-TB guidelines: guidelines on extrapulmonary tuberculosis for India. Indian J Med Res 145(4):448–463. https://doi.org/10.4103/ijmr.IJMR_1950_16

Technical and Operational Guidelines for TB Control in India 2016 (2016) Central TB Division, Ministry of Health and Family Welfare, Government of India. https://tbcindia.gov.in/index1.php?lang=1&level=2&sublinkid=4573&lid=3177. Accessed 1 Mar 2019

World Health Organization (2010) Treatment of tuberculosis: guidelines. WHO/HTM/TB/2009.420, 4th edn. World Health Organization, Geneva

World Health Organization (2017) Treatment of tuberculosis. Guidelines for the treatment of drug-susceptible tuberculosis and patient care, 2017 update. WHO/HTM/TB/2017.05. World Health Organization, Geneva

Infections with Nontuberculous Mycobacteria: Increased Awareness and Recent Developments

5

Astrid Lewin and Hubert Schäfer

Abstract

Nontuberculous mycobacteria represent a multispecies and extremely heterogeneous group of environmental bacteria and include many species that can cause severe disease. Still, they are given little attention as pathogenic agents, when compared to *Mycobacterium (M.) tuberculosis*. Individuals at risk to become infected by NTM usually suffer from immune defects or underlying diseases such as cystic fibrosis or chronic obstructive pulmonary disease. Therapeutic immunosuppression is an additional cause for increased susceptibility, but healthy, apparently immunocompetent individuals may also develop disease. NTM infections can manifest as lung infections, lymphadenitis, skin, bone, and soft tissue infections, or disseminated disease. Pulmonary disease is worldwide most frequently caused by different species of the *M. avium* complex. *M. abscessus* causes serious lung infections in patients afflicted by cystic fibrosis. Diagnosis of the infection is frequently delayed because there are few specific symptoms, and cultivation of the infective agent from clinical samples is not always successful. Isolation of NTM from a patient's lung, on the other hand, is no proof of infection, but might also result from colonization of the organ or from contamination with these ubiquitously occurring environmental bacteria. Therefore further diagnostic options have to be considered before initiating antibiotic intervention. NTM treatment is hampered by the high level of primary resistance against antibiotics, which culminates in the multidrug resistance of *M. abscessus*. New findings on infection rates, transmission routes, and nosocomial potential of NTM are currently changing our perception of the relevance of NTM infections. In the last two decades, an increase in NTM lung disease was reported in many regions of the world, which can be explained by demographic trends and medical progress prolonging life expectancy of persons vulnerable to NTM infections.

A. Lewin (✉) · H. Schäfer
Robert Koch Institute, Unit 16 Mycotic and Parasitic Agents and Mycobacteria, Berlin, Germany
e-mail: lewinA@rki.de

S. E. Hasnain et al. (eds.), Mycobacterium tuberculosis: *Molecular Infection Biology Pathogenesis, Diagnostics and New Interventions*,
https://doi.org/10.1007/978-981-32-9413-4_5

Transmission of *M. abscessus* between patients with cystic fibrosis has challenged the hitherto commonly accepted classification of NTM as environmental non-transmissible infectious agents. And finally the incidence of severe and fatal nosocomial infections has increased attention for NTM. Studies on the burden of disease considering infection numbers, mortality rates, and treatment costs have further substantiated the significance of NTM infections for public health. In view of these new perspectives, a higher priority for NTM disease both in basic research and when deciding on public health measures is indicated.

Keywords
Nontuberculous mycobacteria · Burden of disease · Public health impact · Nosocomial infections · Patient-to-patient transmission · Immunodiagnosis · Immunocompromised · Virulence

Abbreviations

NTM	Nontuberculous mycobacteria
MAC	*Mycobacterium avium* complex
CF	Cystic fibrosis
COPD	Chronic obstructive pulmonary disease
IFN	Interferon
IL	Interleukin
PCR	Polymerase chain reaction
RFLA	Restriction fragment length analysis
IS	Insertion sequences
MALDI-ToF MS	Matrix-assisted laser desorption ionization-time of flight mass spectroscopy
TST	Tuberculin skin test
IGRA	Interferon-gamma release assay
ELISA	Enzyme-linked immunosorbent assay
PPD	Purified protein derivative
BCG	Bacillus Calmette-Guerin
WHO	World Health Organization
TB	Tuberculosis
ATS	American Thoracic Society
BTS	British Thoracic Society
DST	Drug susceptibility testing
MIC	Minimal inhibitory concentration
rRNA	Ribosomal RNA
GPL	Glycopeptidolipids

5.1 The Public Health Attention to Nontuberculous Mycobacteria

The number of species in the genus *Mycobacterium* is continually expanding and currently comprises around 200 species (Parrish 2019), the vast majority of which belong to the group of nontuberculous mycobacteria (NTM). Although NTM have attracted increasing attention in the last two decades, they are still considered fairly insignificant pathogens. The dominance of *M. tuberculosis* within the genus *Mycobacterium* is also reflected by designating NTM as "mycobacteria other than tuberculosis" (MOTT) and "atypical mycobacteria". As a consequence research activities are focused on *M. tuberculosis.* Even among health professionals, NTM are rather taken as a marginal nuisance, than an infective threat. Undoubtedly, *M. tuberculosis* causing around 1.3 million deaths per year is the most important infectious bacterium worldwide. However, this should not lead to negligence toward other infectious mycobacteria. In this chapter we will point out some of the reasons for the small attention given to NTM infections and challenge the perception that NTM are of low significance for public health.

5.2 Environmental Habitat and Intracellular Lifestyle

Evolutionary early forms of mycobacteria probably inhabited a number of environmental niches as saprophytes. Different species of the genus *Mycobacterium* have adapted to various degrees of host invasion and even host dependence.

M. tuberculosis is a completely host-dependent pathogen, which has no environmental niche. Obviously this species has traded in most of the skills needed during environmental persistence and replication for capabilities that confer host adaption, intracellular lifestyle, immune evasion and human-to-human transmission (Banuls et al. 2015; Behr 2013). Similarly, *M. avium* subsp. *paratuberculosis*, a frequent pathogen in ruminants, is considered an obligate parasite of animals, meaning that its survival outside the host animal is finite (Eppleston et al. 2014). The obligate intracellular pathogen *M. leprae* has undergone extensive evolutionary genome reduction hinging these bacteria on host cell metabolism (Cole et al. 2001). These species represent the completely host-adapted end of mycobacterial lifestyle spectrum. The evolutionary changes leading to this adoption can be readily observed in their genomes (Rahman et al. 2014).

Basically all of the NTM species addressed in this chapter can both be considered as environmental mycobacteria and opportunistic pathogens. In some cases the distinction between mainly environmental and predominantly host dependent is quite narrow. Strains of the bird-pathogen *M. avium* subsp. *avium*, for example, are frequently found in surface waters, but it is not known whether all strains actively multiply in this environment or if some pathogenic strains get there only by excretion from birds. Similarly *M. avium* subsp. *hominissuis* is frequently found in household

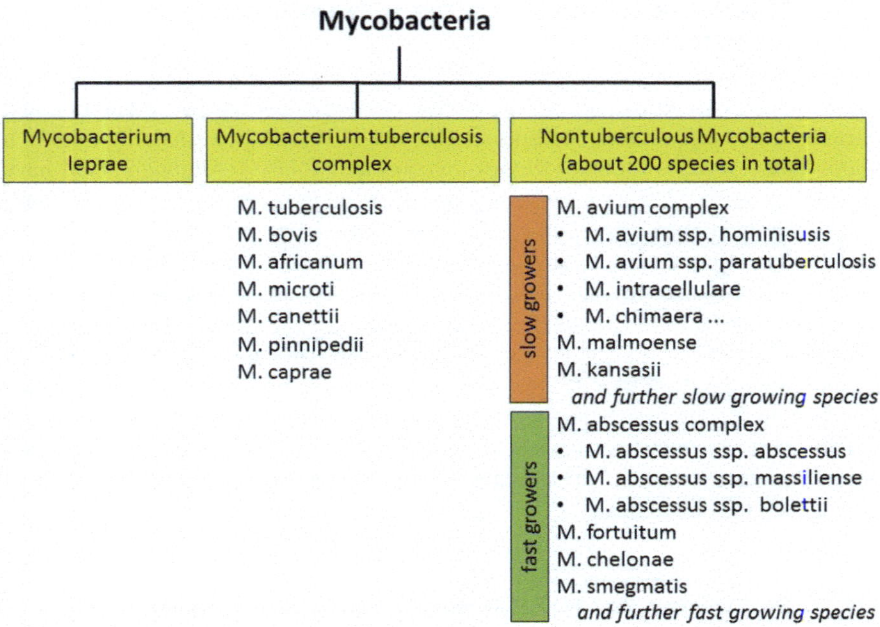

Fig. 5.1 Simplified overview of the taxonomy of the genus *Mycobacterium*

dust (Lahiri et al. 2014), but it is very unlikely that it grows there; other frequent NTM sources are shower heads, water pipes, and tap water (Falkinham 2016), but again it is not certain that those niches represent the main habitat of these NTM. Although the distinction between tuberculous and nontuberculous mycobacteria is quite clear (Fig. 5.1), the taxonomy of the latter is still open to debate.

Several attempts have been made to subdivide NTM into different groups according to systemic relatedness or physiological features such as growth speed or pigmentation (Runyon 1959). Most of these subgroups however show only little accordance of their individual members regarding pathogenicity or preference for the site of infection and certain predispositions in the hosts. Many of the NTM species so far identified as pathogens can cause opportunistic infections in various tissues of immunocompromised or otherwise predisposed hosts, although there is a certain preference for both aspects in most of the species (Table 5.1).

On the far end of the pathogenicity scale, we find species that have never been described as pathogens and thus live purely environmental, as commensals or as saprophytes (Primm et al. 2004). Some species that invade protozoans may be regarded as symbionts (Yu et al. 2007), but it has not been shown what the advantage for the protozoan partner would be.

Table 5.1 Examples for NTM species that cause disease in humans

Species	Disease(s)	Risk group (affected individuals)	Growth
M. avium complex	(a) Lymphadenitis	(a) Young children	Slow
	(b, c and d) Lung disease	(b) Immunocompromised adults	
	(e) Disseminated disease	(c) COPD patients	
		(d) CF patients	
		(e) AIDS patients	
M. abscessus complex	(a) Lung disease	(a) CF patients	Rapid
	(b) Skin and soft tissue infection	(b) Patients exposed to contaminated material after traumatic injury or surgery	
M. marinum	Skin and soft tissue infection	Recreational and professional handling of fish	Slow
M. kansasii	Lung disease	Patients suffering from silicosis (mine workers)	Slow
M. scrofulaceum	Lymphadenitis	Young children	Slow
M. malmoense	(a) Lung disease	(a) Immunocompromised adults	Slow
	(b) Lymphadenitis	(b) Young children	
M. fortuitum	Skin and soft tissue infection	Patients exposed intra- or postoperatively to contaminated instruments, water, soil, or air	Rapid
	Wound infection		
M. xenopi	Lung disease	Immunosuppressed individuals	Slow
	Bone infection		
M. chimaera	Invasive infection	Patients exposed to *M. chimaera* aerosols under open heart surgery	Slow

5.3 NTM Infections, Diagnosis, and Treatment

5.3.1 Predisposing Conditions and Risk Factors

Corresponding to the mainly opportunistic nature of NTM infections, most affected patients can be grouped into three categories of predisposition or risk of infection:

- Immunodeficiency, primarily of the cellular immune response (Casanova 2001, 2015; Casanova and Abel 2002; Doffinger et al. 2000; Doffinger et al. 2002; Reichenbach et al. 2001; Serour et al. 2007), and in particular the IFN-gamma/IL12 axis (Jouanguy et al. 1999). Immunosuppression in transplant patients (Knoll 2014) or treatment with anti-TNF-alpha to reduce inflammation in rheumatoid arthritis (van Ingen et al. 2008). People suffering from acquired immunodeficiency, such as HIV-infected and AIDS patients, were commonly affected by NTM before the widespread practice of HAART (El-Solh et al. 1998; Liao et al. 2004).

- Patients suffering from underlying diseases that affect the lung, mainly chronic obstructive pulmonary disease (COPD) (Char et al. 2014) or cystic fibrosis (CF), are often diagnosed with pulmonary NTM infections (Adjemian et al. 2018). Patients treated for lung cancer sometimes exhibit comorbidity with MAC lung infection (Daley and Iseman 2012).
- Occupational exposure to dust and working in mines can favor NTM infection especially with *M. kansasii* (Chobot et al. 1997; Policard et al. 1967). Professional handling of fish or leisure activities leading to frequent contact with fish or contaminated water may result in skin infections with *M. marinum* (Gray et al. 1990).

Persons without known immune defect or no obvious underlying disease may acquire NTM infections too. In the USA an increasing and large group of NTM patients are postmenopausal women above 50 years of age, with a low BMI and above-average height and frequently with pectus excavatum, scoliosis, and mitral valve prolapse (Chan and Iseman 2010). The concordant anatomical phenotype of these patients however indicates a shared, but so far unrecognized, hereditary predisposition.

Extrapulmonary infections in subjects with no known predisposing factors include lymphadenitis in small children, but closer scrutiny and additional studies in infection epidemiology might reveal hereditary or exposure-related risks (Garcia-Marcos et al. 2017; Serour et al. 2007).

5.3.2 Diagnosis of NTM Infections

The diagnosis of NTM infections is largely dependent on the organ system involved. The "gold standard" is represented by cultivation and identification of the infectious agent in samples taken from the site of infection.

Molecular tests are available now to identify NTM to species and subspecies levels. They involve hybridization techniques, PCR (where appropriate combined with RFLA), and gene-sequencing techniques as well as whole genome sequencing (reviewed in Koh 2017; O'Connor et al. 2015; Ryu et al. 2016; Slany and Pavlik 2012). Additionally, MALDI-TOF offers a promising approach for species and even subspecies determination (O'Connor et al. 2015; Alcaide et al. 2018), although this method is not yet established for use in routine diagnosis of NTM and still needs further development.

The hybridization techniques such as the line probe assays detect sequence polymorphisms in different parts of the rRNA operons. A variety of genus-, species-, and subspecies-specific PCRs have been developed, and some of these make use of the presence of different insertion sequences (IS elements) in different species/subspecies. PCR can be complemented by RFLA or sequencing of the product. Sequencing of the 16S rRNA gene is the reference method for identification of NTM to species or complex level, has a broad species coverage, and also identifies rarely encountered species. If further differentiation is needed, multigene sequencing (e.g., of 16S-23S rRNA internal spacer ITS, *rpoB*, *hsp65*) is recommended. Whole genome sequencing (WGS) followed by comparison with mycobacterial sequences present in public databases is the most accurate method and in addition to species identification provides valuable information such as presence of antibiotic resistance genes, virulence genes, or phylogenetic relations. An overview on molecular identification techniques is presented in Table 5.2.

In contrast to *M. tuberculosis* infections however, where a positive culture is readily accepted as proof of infection and reason for treatment, the situation is

Table 5.2 Techniques for molecular identification of NTM (reviewed in 32–35)

Technique	Commercial kits/ analyzed genes/ components	Main advantages	Main disadvantages
Hybridization	16 rRNA gene (AccuProbe, Hologic); 16S-23S ITS (INNO-LiPA, Innogenetics); 23S rRNA gene (GenoType Mycobacterium CM/AS, Hain Lifescience); 16S-23S ITS (Speed-Oligo Mycobacteria, Vircell);	Speed; ease	Cross-reactivity; for a limited number of NTM only
PCR	Different genus-, species-, and subspecies-specific DNA regions (e.g., IS elements)	Specificity; sensitivity; ease	Different primers for different species/ subspecies required
PCR followed by sequencing of the PCR product	16S rRNA gene; 16S-23S ITS; rpoB gene; hsp65 gene; others (dnaJ, secA1, recA, sodA)	16S rRNA gene sequencing is reference standard; good discrimination level (depending on gene selection to subspecies level)	Lengthy
Whole genome sequencing	Entire DNA	Most accurate and discriminatory method; provides additional information (e.g., on antibiotic resistance or virulence)	Costly; lengthy; bioinformatics expertise required
MALDI-TOF-MS	Whole cell preparations	Speed; nonlaborious; cost-effective (if instrument available)	Databases not yet complete and fully available; lack of discrimination for certain species; lack of standardization and interlaboratory reproducibility

complicated by the environmental habitat and ubiquitous occurrence of many NTM species. Especially the microscopic and culture confirmation of NTM from the lung is not necessarily indicative of an ongoing infection, but might result from mere colonization of certain areas of the organ. Therefore different or additional diagnostic criteria have to be considered to reach a substantiated decision. These additional measures largely depend on the site of infection.

5.3.2.1 Pulmonary Infections

As outlined earlier, patient groups mostly affected by NTM infection of the lung suffer either from CF, COPD, bronchiectasis, immunosuppression, immunodeficiency, or reduced pulmonary function for various reasons. Although the elevated risk for NTM infection in these patients is widely known, diagnosis is frequently delayed, because the presentation of symptoms is not disease-specific, and the isolation of NTM from clinical samples, although a prime demand for the diagnosis, does not distinguish between infection, colonization, and mere contamination of the sample (Cowman and Loebinger 2018; Catanzaro 2002). The situation is further complicated by the different mycobacterial species that are able to infect the lung. The first challenge is the differentiation between tuberculosis and NTM disease or even mixed infections (Ellis 2004; Grubek-Jaworska et al. 2009; Kendall et al. 2011; Kim et al. 2018; Kwon and Koh 2014; Pang et al. 2017; Thanachartwet et al. 2014; Yuan et al. 2014). If *M. tuberculosis* infection can be excluded, it is still critical to correctly identify the pathogen at least to the species level (Adelman and Addrizzo-Harris 2018). Certain patient groups are more susceptible to specific species, such as *M. abscessus* frequently infecting CF patients (Park and Olivier 2015), but there is no one-to-one relation between infective agent and underlying disease, and predominant pathogens differ between countries (Adelman and Addrizzo-Harris 2018). Furthermore, even experts rarely agree on the best scheme of diagnosis and clinical management of infection (Marras et al. 2015), and guidelines are not always followed by physicians (El-Zeenni et al. 2018). Proof of infection does not automatically indicate a need for treatment (Griffith et al. 2002); delayed treatment on the other hand might result in severe complications and disease progression. Chest imaging alone is not always sufficient because radiographic manifestations cannot reliably be attributed to NTM infection and in some cases it is even hard to distinguish this illness from lung cancer (Kobashi et al. 2004; Mohamad et al. 2012; Noguchi et al. 2016; Umehara et al. 2015). Therefore the decision to treat is usually based on the clinical presentation, radiological features, and microbiological findings (Aksamit et al. 2014; Kwon and Koh 2016; Ryu et al. 2016). Genetic testing and PCR-based recognition of the pathogen is increasingly performed in specialized laboratories (Chae et al. 2017; Hsiao et al. 2010; Jung et al. 2016; Kim et al. 2014; Kim et al. 2018; Wang et al. 2015; Wang et al. 2017). More recently an assay has been developed and marketed in Japan, which allows the serodiagnosis of *M. avium* lung infections with sufficient sensitivity (Shu et al. 2013; Kobayashi 2014) and the potential to distinguish between colonization and infection (Kitada et al. 2002).

Subjects without predisposing conditions are even more likely to stay undiagnosed for prolonged periods of time if they come up with NTM infections of the lung. Due to the unspecific symptoms these patients present with, they frequently have a long history of misdiagnosis and non-satisfactory treatment results, and only experienced pulmonologists or specialists for NTM lung infections come to a fast and reliable diagnosis. For such cases serological assays or PCR-based methods covering a broader spectrum of potential pathogens affecting the lung should proof very valuable.

5.3.2.2 Infections of Lymph Nodes of the Head and Neck

Cervicofacial lymphadenitis due to NTM infection is a rare but not uncommon condition in small children. Similar to other NTM infections, the symptoms are little specific, and differential diagnostics to exclude several infective and noninfective etiologies have to be considered. It has been observed that needle biopsy led only in about half of the cases to a culture-confirmed diagnosis (Willemse et al. 2018) and this procedure was therefore not considered to sufficiently prove or rule out NTM infection. Marks et al. (Marks et al. 1977) and Magdorf et al. (Magdorf et al. 2008) suggested skin tests with tuberculin and sensitin, to diagnose a mycobacterial infection and to distinguish between *M. tuberculosis* and NTM infection at the same time. A recent systemic review of diagnostic studies (Willemse et al. 2018) concluded that skin tests with sensitin-purified protein derivative (s-PPD) are the best diagnostic option so far available, but that they should be confirmed by PCR or culture if possible. An interferon-gamma release assay (IGRA) might provide a good preoperative alternative, since it is noninvasive, but has not been studied in sufficient numbers to allow reliable recommendations.

5.3.2.3 Skin, Soft Tissue, and Wound Infections

In skin and soft tissue infections, the infected area is usually readily accessible and samples can be taken for analysis. Molecular methods such as PCR, gene sequencing, and mass spectrometry allow the rapid and detailed analysis of the infecting agent, sometimes even without culture amplification of the bacteria (Misch et al. 2018). The main culprit leading to misdiagnosis is that NTM infections of the skin are relatively rare and may be misjudged as a manifestation or symptom of a predisposing condition (open wound, surgical treatment, autoimmune disease requiring immunosuppression), instead of a consequent infection (Touma et al. 2013). Skin infections with *M. marinum* in association with fish tanks or professional handling of fish or marine shells have been observed at least since the 1970s (Barrow and Hewitt 1971), and the disease is frequent enough to bear its own designation as "fish tank infection" or "fish tank granuloma." Most of these infections involve the fingers and hand (Aubry et al. 2002), and therefore the experienced dermatologist will suspect the cause of disease by the localization and mode of infection. The microscopic or culture confirmation however is difficult, because *M. marinum* requires temperatures below 30°C and several weeks to form colonies when cultivated from patient samples (Aubry et al. 2017).

Cutaneous infections with *M. abscessus massiliense* have also been reported in increasing frequency after tattooing of immunocompetent individuals. The source is probably contaminated ink or dilution of ink with non-sterile water. Surgery-related infections (see below) frequently involve skin and soft tissue and thus might be superficial or at least in the early stage restricted to the area affected by surgery. Diagnostic options therefore are isolation and culture of the pathogen.

5.3.2.4 Ocular Infections

In ocular NTM infections, early diagnosis and recognition of the infections agent is crucial for the outcome of the treatment (Kheir et al. 2015). Besides *M. tuberculosis* (Albert and Raven 2016), a number of additional mycobacterial species have been reported to infect the eye (Chu et al. 2015). As in many other NTM infections, diagnosis is often delayed, and hence treatment and management of the disease are in many instances far from optimal (Patel et al. 2013).

5.3.2.5 Immunodiagnosis

Both cellular and humoral immune responses can be used as starting points for the diagnosis of mycobacterial infections.

Cellular immunodiagnosis, applied in vivo as the tuberculin skin test (TST), is a well-established routine procedure to reveal contact with *M. tuberculosis*, including recent and latent infections which have not led to symptoms of disease. The TST however does not discriminate between active disease and latent infection or even cured disease. Earlier vaccination with *Bacillus Calmette-Guerin* (BCG) might also lead to a positive result. Therefore confirmation of infection is required by either mycobacteria-positive sputum smears or radiographic evidence for lung disease or extrapulmonary manifestations. IGRAs can be considered as an advanced in vitro version of the TST and carry several improvements: firstly the antigens used are specific for *M. tuberculosis* and are not cross-reactive with BCG and most NTM species; secondly, the test does not include injection of mycobacterial antigen; and thirdly results are obtained in the laboratory, circumventing the need for a second visit with the physician to assess the result (Thillai et al. 2014). In parallel to the TST, the IGRA cannot distinguish between latent or cured infection and active disease, but at least in low-incidence countries, it is recommended for the diagnosis of latent TB (WHO 2018) and screening of contacts.

Serological assays for TB infection, although commercially available in various forms, are not recommended by the WHO (WHO 2010), due to highly variable values for sensitivity and specificity. These recommendations have probably affected the development and use of serological assays for detection of NTM infections.

As mentioned earlier, two different approaches for the immunodiagnosis of NTM infections are used to a certain extent. The cellular approach was to transfer the TST principle of tuberculosis to different mycobacterial species and mycobacterial infections in humans and animals (Magnusson 1961) and to address sensitivity and specificity in different species (Magnusson et al. 1961; Magnusson 1961). PPD preparations of NTM were commonly designated as “sensitins” (s-PPD) independently of the actual source. The use of *M. avium* sensitin turned out to be useful for detection of *M. avium* infections in humans, despite intense cross-reactions with tuberculin (Fordham von Reyn et al. 1995). Generally, the ubiquitous pervasiveness of environmental mycobacteria may lead to sensitization of healthy and diseased individuals, so independently from cross-reactions with *M. tuberculosis*, a positive test result for s-PPD might not reflect current or preceding infection. Although the debate on the use and the value of this diagnostic tool is still going on, a recent review of the published literature attested sufficient specificity and sensitivity to use the s-PPD skin test as a first assay, which, if positive, has to be confirmed by culture or PCR methods (Willemse et al. 2018). Either of the latter measures is indispensable, because it will also provide details of the pathogen which are important to decide on treatment options. More recently the s-PPD skin test has also been used in its in vitro equivalent IGRA (Steindor et al. 2015) for the detection of NTM infections, but published data are not sufficient to come to a conclusion if this is actually an improvement.

In the absence of an all-purpose gold standard for the diagnosis of NTM infections, serodiagnosis tailored to detect immune reactions against specific pathogens might provide a valuable tool to assist in the differential diagnosis of infections with nonspecific symptoms. Commercial kits for the diagnosis of bovine *M. avium* subsp. *paratuberculosis* are available, but upon closer evaluation none of the assays turned out to be fully satisfactory (Kohler et al. 2008).

Starting in 2002, Kobayashi and co-workers suggested and developed the use of glycopeptidolipids (GPL) from *M. avium* as a diagnostic antigen for *M. avium* infections in humans (Kitada et al. 2002). It had been well known that GPL represent one of the immunodominant antigens in several mycobacterial species (Ikawa et al. 1989; Schaefer 1967), but the expression of different variations, due to diverse glycosylation patterns, even between different strains of the same subspecies, is a challenge for the diagnosis of *M. avium* infections in particular. The advantage of using GPL as diagnostic antigens resides in the absence of these cell wall components in tuberculous mycobacteria, thus ruling out cross-reactions with TB infections. After several refinements, the test was brought to the market (Kobayashi 2014) and is further evaluated in different settings and patient populations (Kitada et al. 2010, 2015, 2017; Shu et al. 2013). The problem of the highly variable GPL glycosylation was solved by using the GPL core antigen as a diagnostic target in *M. avium* infection, but it will have to be shown whether the same concept can be applied to the specific diagnosis of infections with other NTM.

In conclusion, both IGRA and ELISA might provide valuable tools for noninvasive diagnosis of suspected mycobacterial infections, and a consequent development and use of such assays should help to shorten the lack time between appearance of symptoms and treatment start in various settings.

5.3.3 Treatment

Compared to *M. tuberculosis* disease, there is no room for empirical treatment when NTM infection is suspected. Before initiation of treatment, clinical, radiological, and microbiological features must be taken into consideration, and ATS and BTS guidelines should be followed as closely as possible. The management options of mycobacterial infections are basically limited to antibiotic treatment and surgical removal of infected tissue. Surgery is quite successful for the cure of lymphadenitis at the head and neck in children (Lindeboom et al. 2007), although not without risk of permanent damage to facial nerves and of ill-favored scars of affected areas (Gonzalez et al. 2016). Excision of infected tissue is also an option for refractory lung disease (Griffith and Aksamit 2012; Sakane et al. 2018), if other treatment options have failed and localization of the infection permits surgical intervention.

The treatment strategies that need to be applied to eliminate NTM or reduce NTM load depend on the interplay of NTM species, patient characteristics, and disease manifestations. Mycobacteria can display considerable resistance toward antibiotics (Nasiri et al. 2017), and especially the multidrug-resistant *M. abscessus* is considered an "antibiotic nightmare" (Nessar et al. 2012). Antibiotic therapy is complicated by the combination of a number of independent obstacles that frequently occur in NTM infections: (i) primary antibiotic resistance of environmental mycobacteria; (ii) additional acquired resistance under treatment; (iii) serious, sometimes fatal side effects of the remaining effective antibiotics; (iv) long (months to years) treatment time needed for conversion; (v) coinfections with other bacteria or fungal pathogens; and (vi) underlying disease of the patients, including immunodeficiency or need for immunosuppression.

The design of appropriate treatment strategies is also hampered by difficulties in drug susceptibility testing (DST). Broth microdilution assays such as the Sensititre System (Trek Diagnostics) are suitable for determination of minimal inhibitory concentration (MIC). The protocol from the Clinical and Laboratory Standards Institute (Clinical and Laboratory Standards Institute M24-A2, 2011) (Clinical and Laboratory Standards Institute M62, 2018) proposes interpretations of MIC values for certain antibiotics and currently is the gold standard for DST of nontuberculous mycobacteria. A major problem in treatment of NTM infections is the lack of correlation between results of DST and clinical outcome (van Ingen and Kuijper 2014). Exceptions to this are amikacin and macrolides, whose MIC values correlate well with the clinical outcome in treatment of lung infections by MAC (Van Ingen et al. 2012; Brown-Elliott et al. 2013; Bastian et al. 2011; Prammananan et al. 1998). Macrolides and amikacin furthermore stand out because their efficacy against certain NTM can be predicted by molecular analysis of their target genes. Thus macrolide susceptibility of MAC and *M. abscessus* can be assessed by sequencing of the 23S rRNA gene (Griffith et al. 2006; Bastian et al. 2011), while sequencing of the 16S rRNA gene provides information on amikacin susceptibility (Brown-Elliott et al. 2013; Prammananan et al. 1998). In addition *M. abscessus* may have complete or truncated versions of the *erm*(41) gene causing inducible macrolide resistance (Zheng et al. 2019). DST of *M. abscessus* strains by broth microdilution assays should include additional plate reading after up to 14 days in order to detect inducible macrolide resistance (Clinical and Laboratory Standards Institute

M24-A2, 2011) (Clinical and Laboratory Standards Institute M62, 2018). Analysis of the molecular structure of the *erm*(41) gene from *M. abscessus* can also be used for subspecies identification, because the subspecies *M. abscessus massiliense* is distinguished from the other subspecies by a truncated nonfunctional *erm*(41) gene (Adekambi et al. 2017). Next to gene or genome sequencing, the commercial line probe assay GenoType NTM-DR (Hain Lifescience) provides a molecular assay for detection of amikacin and macrolide resistance in MAC and *M. abscessus*.

Although guidelines exist for the management of NTM infections in different countries and patient groups (Haworth et al. 2017; Haworth and Floto 2017; McGrath et al. 2008; Floto et al. 2016; Griffith et al. 2007), the factors listed above may limit adherence to general recommendations. Experts however criticize the lax interpretation of relevant guidelines and suppose that a more complete implementation will result in better treatment outcome for most of the diseased (El-Zeenni et al. 2018).

5.4 Obstacles in Evaluating the Public Health Impact of Nontuberculous Mycobacteria

Why is it problematic to assess the role of NTM infections for public health? The main reason is the diversity of NTM species, of hereditary predispositions within patient groups concerned, and of disease manifestations. This diversity prevents us from viewing and evaluating NTM as a whole.

First of all, their growth properties differ substantially leading to division into slow and rapid growers according to generation time. Species generating visible colonies on agar within 7 days belong to the rapid growers; the remaining ones are slow growers. There is a correlation between growth rate and pathogenicity: slow-growing mycobacteria are frequently more pathogenic (Lewin and Sharbati-Tehrani 2005). An exception to this rule is *M. abscessus*, a rapid-growing *Mycobacterium* of high clinical significance particularly in CF patients (Adjemian et al. 2018). NTM diversity does not end at species level. Even within the same species, different colony morphologies can occur that impact host-pathogen interaction (Gutierrez et al. 2018) and disease severity. NTM inhabit different environments such as water, soil, dust, and amoeba (Samba-Louaka et al. 2018) and can form biofilms on tissues, transplants, and medical devices (Ghosh et al. 2017). Generally, they are opportunistic pathogens but some species can also infect immunocompetent persons (Piersimoni and Scarparo 2009). As described above NTM cause a variety of diseases.

Owing to this complexity, most studies focus on selected conditions of NTM disease limiting their validity for the assessment of the public health impact of NTM.

5.5 New Developments Demanding a Re-evaluation of NTM

5.5.1 NTM Infection Rates

There is much debate on the question if infections by NTM are indeed increasing in incidence or if reports on increasing frequencies of NTM isolation and disease rather reflect a rise in awareness and improved diagnostic techniques. It is difficult to answer this question, because NTM are non-reportable diseases in most countries of the world, hampering the availability of complete infection numbers (Henkle et al. 2015). As a consequence numerous regional studies are performed that follow region-specific aims and differ therefore with respect to variables such as (i) the quantified epidemiologic measure (incidence, prevalence, or isolation numbers), (ii) the investigated diseases, (iii) the included patient groups, (iv) the geographical regions, and (v) the time periods included. These limitations impede comparability and definition of generally accepted trends. The majority of these studies conclude that NTM infections are increasing, and we refer to recent reviews on this topic (Johnson and Odell 2014; Prevots et al. 2010; Strollo et al. 2015). For example, a study comprising the US Medicare Beneficiaries came to the conclusion that the annual prevalence of NTM pulmonary infections increased from 1997 to 2007 from 20 to 47 cases/100,000 population or 8.2% per year (Adjemian et al. 2012). Especially studies on NTM lung infections report increasing infection rates, while extrapulmonary NTM infections do not show this effect. Different trends for pulmonary vs. extrapulmonary infections were also supported by data from Oregon, USA. While the incidence of extrapulmonary diseases in Oregon remained stable at around 1.5/100,000 population from 2007 until 2012 (Henkle et al. 2017), lung infections increased from 4.8/100,000 to 5.6/100,000 in the same period (Henkle et al. 2015). The only geographic region with a notification system for all NTM infections is Queensland, Australia, where from 2012 to 2015 a yearly increase in notification rates of approximately 17% was observed resulting in a notification of 25.9 cases per 100,000 population in 2015 (Thomson et al. 2017).

Altogether, there is sufficient body of evidence to support the assumption that, despite regional differences, there is an increasing trend in pulmonary NTM infections. These increasing NTM infection rates can be explained mainly by demographic trends and medical progress prolonging life expectancy of persons vulnerable to NTM infections.

5.5.2 Correlation Between Virulence and Genome Properties

Clinically relevant NTM are opportunistic pathogens, while host susceptibility certainly is the predominating condition determining if colonization with NTM will end in disease or not (Henkle and Winthrop 2015). The contribution of genomic features of different NTM species to cause illness (i.e., virulence factors) is still under study. Increasing use of whole genome sequencing in combination with association of clinical data should provide answers to this question.

Comparative genome analysis of *M. avium* strains provides first indications for the existence of bacterial genomic determinants for disease progression. Most studies performing cluster analysis with NTM genomes were conducted with genomes from *M. avium*. *M. avium* subsp. *hominissuis* is the clinically most important and also the genetically most variable of the four *M. avium* subspecies (Rindi and Garzelli 2014). The main determinants of genomic variation in *M. avium* subsp. *hominissuis* are small nucleotide polymorphisms (SNP), along with large sequence polymorphisms (LSP), minisatellite sequence polymorphisms, and the distribution of insertion elements (Rindi and Garzelli 2014). *M. avium* subsp. *hominissuis* has an open pan genome (the entire gene set of all strains of a species) supporting the genomic diversity of this subspecies (Uchiya et al. 2017). When comparing the genomes of a collection of 41 *M. avium* subsp. *hominissuis* containing 20 clinical and 21 environmental isolates from Germany, Sanchini and colleagues identified a new hypervariable genomic island that could be further divided into 8 different genome island types comprising 253 different genes in total. However their analysis did not show any correlation between presence or absence of certain types of this hypervariable genomic island and habitat of the isolates (Sanchini et al. 2016). The same strain collection was used to investigate the capacity of the strains for conjugative plasmid uptake. Interestingly, it was found that environmental isolates are more proficient in conjugative DNA uptake than clinical isolates and also contained a higher proportion of plasmid-derived genes (Shoulah et al. 2018). The capacity to act as conjugative recipient was correlated with absence of or mutations in a genomic DNA region containing the gene *radC* and the genes for components of a type I restriction/modification system. This region was differently organized in environmental compared to clinical isolates (Shoulah et al. 2018) pointing to its potential as marker predicting pathogenic potential. The results obtained with the strain collection of *M. avium* subsp. *hominissuis* from Germany differed from those reached by the use of a strain collection from Japan (Uchiya et al. 2017). Phylogenetic analysis of genome sequences from 46 clinical *M. avium* subsp. *hominissuis* isolates from Japan, 17 of which were classified as causing progressive disease and 29 as causing stable disease, positioned the isolates belonging to the progressive disease group in one cluster. DNA regions classified as "hot spot regions for mutation and/or recombination" could be identified in the core genomes. The noncore sequences of the progressive-type strains contained cluster-specific genes including known virulence-related genes. The plasmid pTH135 belonged to the noncore sequences of the progressive strain cluster, suggesting a contribution of this plasmid to virulence (Uchiya et al. 2017). This result is opposed to the result from Shoulah et al. (Shoulah et al. 2018), who found less genes from plasmids related to pTH135 in clinical isolates from Germany compared to environmental ones. Geographic strain variation may explain diverging results obtained with the German and Japanese isolates.

Progress in differentiating isolates with varying virulence potential has also been achieved for *M. abscessus*. A population-level study involving whole genome sequencing revealed existence of global transmissible *M. abscessus* clusters displaying increased antibiotic resistance, rates of chronic infection, phagocytic uptake, and intracellular survival in macrophages as well as greater bacterial burden upon infection of severe combined immunodeficient (SCID) mice (Bryant et al. 2016). Identification of markers for these particularly virulent clusters in the future may assist in the decision on the need for treatment in order to prevent chronic *M. abscessus* infections.

In summary, these studies suggest that bacterial virulence factors noticeably add to the progression of colonization by NTM to disease. Once these factors are determined, their use in diagnostics should support prognosis of disease progression and decision on the need for therapy.

5.5.3 Nosocomial Infection and Patient-to-Patient Transmission of NTM

NTM are environmental bacteria and can be isolated from water, soil, dust, and biofilms (Halstrom et al. 2015). They also survive in amoeba (Sharbati-Tehrani et al. 2005; Ovrutsky et al. 2013) and infect or colonize animals (Biet and Boschiroli 2014; Keller et al. 2018). Although more rarely, food was also shown to be contaminated by NTM (Cerna-Cortes et al. 2009). Finally infected humans represent a niche for NTM survival and growth.

The aforementioned habitats of NTM represent possible reservoirs for human infections. Proof of transmission requires molecular typing of the isolates from the patients and the potential infection source. Typing methods involve, for example, variable number tandem repeat typing (VNTR) (Torres-Coy et al. 2016), random amplification of polymorphic DNA (RAPD-PCR) (Sax et al. 2015), pulse field gel electrophoresis (PFGE) (De Groote et al. 2006), or whole genome sequencing combined with single nucleotide polymorphism (SNP) analysis, which has the best discriminatory power (Trovato et al. 2017) and is indeed the only method that can prove transmission routes beyond doubt.

As NTM are most abundant in environmental habitats, the environment is probably the most important source of infection. NTM are present in natural water reservoirs as well as in drinking water (Makovcova et al. 2014). Especially tap water has been frequently associated with NTM infection (Falkinham 2016). Resistance toward water disinfection procedures supports the survival and spread of NTM in drinking and municipal water systems. *M. avium* bacteria, for example, display high resistance against chlorine, monochloramine, chlorine dioxide, and ozone, and their calculated disinfection values ($CT_{99.9\%}$) for chlorine were shown to be 580–2300 times greater than those of *Escherichia coli* (Taylor et al. 2000). The hydrophobicity of the mycobacterial cell wall supports their adherence to surfaces including household plumbing materials or medical devices followed by the formation of biofilms. Biofilm-grown *M. avium* can reach densities as high as 10^7 CFU/cm^2 (Mullis and Falkinham 2013). They withstand flushing and are further protected against disinfectant activity. Oriani and colleagues tested the tolerance of *M. gordonae* and *M. chubuense* to the disinfectant sodium hypochlorite. Both NTM species when grown in biofilms require treatment at a concentration of 10 ppm for 60 min to achieve significant reduction of CFU, while a similar reduction was reached with a concentration of 2 ppm (*M. gordonae*) or 5 ppm (*M. chubuense*) during the same time period when planktonic cells were treated (Oriani et al. 2018). Resistance to disinfectants and biofilm formation are key elements for the success of waterborne NTM as infectious agents. This is especially true for those NTM belonging to the *M. avium* complex (MAC), which was repeatedly associated with waterborne infections involving use of tap water, for example, in hot tubs (Kahana et al. 1997) or during showering (Falkinham et al. 2008) [reviewed in, e.g., Halstrom et al. 2015].

Ecological studies involving isolation of NTM from environmental samples underline the importance of habitats other than water as infection sources. For example, Lahiri et al. (Lahiri et al. 2014) analyzed 130 environmental samples for the presence of cultivable *M. avium* subsp. *hominissuis* and found these in 20% of soil samples and 33% of dust samples but in none of the water samples. Dust and aerosolized soil should therefore be taken into consideration when searching for NTM infection sources and designing prevention measures. De Groote and colleagues succeeded in recovering *M. avium* from a patient's potting soil and linking it to the patient's isolate by pulse field gel electrophoresis (PFGE) (De Groote et al. 2006).

The relevance of zoonotic NTM infections is still not clear. Zoonotic transmissions are well known to occur by handling of fish or aquarium water. The most important NTM in this respect is *M. marinum*, which can penetrate the skin via minor traumata (Vincenzi et al. 1992; Keller et al. 2018; Gray et al. 1990). The nine-banded armadillo is known to represent a reservoir for *M. leprae,* and there is evidence of transmission to humans in Florida and Brazil (da Silva et al. 2018; Domozych et al. 2016). Possums and mosquitoes are suggested as zoonotic reservoirs of *M. ulcerans* (causing Buruli ulcer) in southeast Australia (Huang and Johnson 2014). The potential role of *M. avium* subsp. *paratuberculosis*, which causes Johne's disease in ruminants, as a causative agent of Crohn's disease in humans is still under debate (Waddell et al. 2008; Garvey 2018; Atreya et al. 2014). The presence of NTM in a wide number of animal species (Biet and Boschiroli 2014) suggests more frequent zoonotic transmission; the final proof by molecular typing has, however, only rarely been attempted. More frequent use of next-generation sequencing will shed new light on this still unresolved question.

The presence of NTM in livestock and animal-derived products suggests the occurrence of foodborne infections. NTM or DNA from NTM were found in milk products (Leite et al. 2003) and in powdered infant formula, ground beef, chicken sausage, and mortadella cold cut (Sevilla et al. 2017). Also products from fruits and vegetables/salads were shown to contain NTM, such as orange juice (Cerna-Cortes et al. 2016), salad (Cerna-Cortes et al. 2015), or chili sauce (Cerna-Cortes et al. 2009). Among the NTM found in animal-derived food products, *M. avium* seems to be prevailing (Sevilla et al. 2017; Klanicova-Zalewska and Slana 2014). Molecular typing by VNTR and IS1245 RFLP had revealed similarities between *M. avium* subsp. *hominissuis* isolates of human and porcine origin (Leao et al. 2014; Tirkkonen et al. 2007) suggesting either transmission between humans and pigs or infection by the same environmental source.

In medical settings, NTM species which are capable to withstand water disinfection procedures and which can form biofilms on the one side and persons vulnerable to NTM on the other side confront each other. In a study comparing occurrence of NTM in hospitals and home water systems, higher isolation rates were found in hospitals (Peters et al. 1995). Specific NTM strains seem to have the capacity for long-term persistence in hospital water supplies as shown by Crago et al. (Crago et al. 2014), who monitored NTM in a hospital water distribution system in Alberta and found the same *M. avium* strain in 34/183 sampling sites over 2.5 years. An overview on waterborne nosocomial infections is presented by Li et al. (Li et al. 2017). The authors of this review calculated the median attack rate to be 12.1% with a low mortality rate (Li et al. 2017). Other sources of nosocomial NTM infections can be devices (e.g., heater-cooler units, extracorporeal membrane oxygenation devices, endoscopes), surgical sources (e.g., in cosmetic and eye surgery), and aqueous solutions (e.g., use of contaminated ultrasound gel), reviewed in Sood and Parrish (2017). The increasing popularity of cosmetic surgical procedures, including cutaneous surgery, mesotherapy, liposuction, and laser resurfacing, leads to a rising incidence of NTM infections in otherwise healthy people (Green et al. 2017; Avanzi et al. 2018; Berliner et al. 2015; Chang et al. 2017; Torres-Coy et al. 2016). The manifestation is mostly in skin and soft tissues and the options of diagnosis have been described above. Transplant recipients are at high risk of NTM infections because they receive immunosuppressive treatment after surgery (Daley 2009; Huang et al. 2011; Jankovic Makek et al. 2016; Pandian et al. 2008; Patel et al. 1994; Piersimoni 2012; Rao and Silveira 2018; Shah et al. 2016). Recipients of allogeneic stem cell transplants are even more susceptible to NTM infections, which in most cases affect the lung (Beswick et al. 2018; Doucette and Fishman 2004; Knoll 2014). The frequently delayed identification of the infectious agents contributes to enhanced morbidity and mortality in this highly vulnerable patient population.

Since 2013 a global outbreak (Europe, Australia, USA), involving more than 100 patients, caused by *M. chimaera* present in heater-cooler devices used during heart surgery has attracted much attention (Sax et al. 2015; van Ingen et al. 2017). Persistence of *M. chimera* in the heater-cooler devices due to their tolerance to the disinfection protocols applied (Walker et al. 2017) in connection with biofilm formation and aerosolization into the operation area caused this outbreak. The patients suffered from prosthetic valve endocarditis, sternal wound infection, aortic graft infection, and disseminated disease. An evaluation of the disease progression of the first 30 UK cases revealed a mortality rate of 60% (Scriven et al. 2018). The likely source of this outbreak is a point-source contamination of the heater-cooler devices at the factory (van Ingen et al. 2017). This outbreak very clearly emphasizes the importance of NTM as nosocomial pathogens.

Until recently NTM were assumed to be non-transmissible between humans. In 2012 however, a report on probable transmission of *M. abscessus* subsp. *massiliense* between patients in a CF center was published (Aitken et al. 2012). The methodologies applied to compare the strains were PFGE and repetitive unit-sequence-based polymerase chain reaction, which may not have the discriminatory power to prove inter-patient transmission. PFGE was also used in a study that compared *M. abscessus* isolates from an outbreak among CF patients in Hawaii in 2012 and suggested patient-to-patient transmission (Johnston et al. 2016). In 2013, Bryant and colleagues could prove transmissions between patients of *M. abscessus* subsp. *massiliense* in a CF center in the UK by whole genome sequencing (Bryant et al. 2013). Further reports on possible transmissions of *M. abscessus* between patients with CF followed (Kapnadak et al. 2016). The dimension of the occurrence of patient-to-patient transmission of *M. abscessus* was finally demonstrated by a study comparing whole genome sequences of 1080 clinical isolates from 517 CF patients from Europe, the USA, and Australia. Phylogenetic analysis illustrated the presence of three human-transmissible global *M. abscessus* clusters that displayed increased adaptation to the human host (Bryant et al. 2016). This study has changed our view on NTM from a purely environmental to a human-transmissible group of bacteria. However, it is intriguing that all human-to-human transmissions so far described only concern CF patients and *M. abscessus* bacteria. Further epidemiologic studies including whole genome sequencing and cluster analysis are needed to work out if transmission between humans is a peculiarity of *M. abscessus* infections in CF patients or if it indicates a more general transmission path of nontuberculous mycobacteria.

5.6 Burden of Disease

As aforesaid, NTM infections are not reportable in most countries in the world, making it problematic to achieve sufficient data to fully describe the burden of disease.

The prevalence of NTM pulmonary infections is increasing in many areas of the world, and in certain geographic areas, it even exceeds the prevalence of pulmonary tuberculosis. When comparing prevalences of pulmonary NTM and tuberculosis in the 3-year period from 2004 to 2006 in four integrated healthcare delivery systems in the USA, the prevalence of pulmonary NTM disease was 2–2.9-fold higher than the tuberculosis prevalence (Prevots et al. 2010). This study also demonstrated that this increase mainly resulted from infections in the age group 80 years and older and within this age group women contributed more than men. It can be expected that the current demographic trend toward an aging population will further increase the prevalence of pulmonary NTM infections. The second driving force behind increasing NTM prevalence is the medical progress enhancing the life expectancy of vulnerable patient groups such as those with CF or COPD. The life expectancy for persons with CF, for example, is steadily increasing, and a UK study has shown that predictions of a mean survival of >50 years of age for infants born in the UK since the year 2000 are realistic (Dodge et al. 2007). As the percentage of CF patients with NTM increases with age (Adjemian et al. 2018), the longer life expectancy of CF patients will lead to increasing NTM prevalence rates. The prevalence of COPD, another risk factor for acquiring NTM pulmonary disease, is also increasing. In a study comparing and predicting the ranking of certain diseases with respect to their contribution to disease burden [measured in disability-adjusted life years (DALYs)] in 1990 and 2020, COPD was predicted to move from place 12 in 1990 to place 5 in 2020 (Murray and Lopez 1996). A Danish population-based case-control study estimated that COPD was associated with a 16.5-fold increased risk of NTM pulmonary disease (Andrejak et al. 2013).

The assessment of NTM-related mortality rates suffers from application of different study approaches and interpretations. Various studies consider different NTM disease manifestations, different NTM species, patient groups displaying different comorbidities, varying geographic areas, and variable follow-up times and apply different criteria for control groups (Yeung et al. 2016). We therefore firstly focus on the results of a systematic review reporting the 5-year mortality in patients with MAC lung disease (Diel et al. 2018), which is the most frequent cause of pulmonary NTM infection (Johnson and Odell 2014). By reviewing the literature until 2017, Diel et al. found a 5-year all-cause mortality of 27% with a variation between 10% and 48%. In another study Diel et al. (Diel et al. 2017) compared the mortality of patients with NTM pulmonary disease from Germany diagnosed in 2010 and 2011 and followed for 39 months with a control group that was matched with respect to age, sex, and comorbidities (e.g., COPD). The mortality rate of the patient group with NTM pulmonary disease during the 39 months was 22.4% compared to 6% in the control group. This comparison very clearly illustrates the high impact on mortality of NTM pulmonary infections.

The study from Diel and colleagues (Diel et al. 2017) also analyzed the costs caused by pulmonary NTM infections in this patient cohort. The annual direct costs plus the indirect work-loss costs attributable to NTM pulmonary disease were € 10,314.25 per patient and were about 4 times higher than for a matched control cohort of patients with the same underlying diseases. The annual average costs in 2010 for treatment of pulmonary NTM infection in the USA were estimated to be $ 9,451 and were related to the average direct costs of $ 5,113 per patient for treatment of multidrug-resistant tuberculosis (Strollo et al. 2015). An assessment of the costs of treatment of pulmonary NTM disease by Ballarina and colleagues (Ballarino et al. 2009) enrolling patients in 2004 and 2005 estimated the median annual cost to be $ 5,772 with higher costs for *M. avium* compared to *M. abscessus* infections ($ 5,196 versus $ 11,796) and concluded that treatment of NTM pulmonary disease is similarly expensive as treatment of MDR-TB and HIV/AIDS.

5.7 Conclusions: Prioritization of NTM Infections in Public Health

NTM lung infections represent severe chronic and hard-to-treat diseases requiring long-term treatments with pronounced side effects. Due to increasing numbers of elderly and vulnerable people in the absence of prevention measures, sufficiently informative diagnostics, and treatment options, the prevalence rates are increasing. The mortality rates of NTM infections are substantial, and the treatment costs are comparable to the costs for treatment of tuberculosis or HIV/AIDS. In the last years,

NTM have been established as important nosocomial pathogens causing severe surgery-related outbreaks. Finally, patient-to-patient transmissions of *M. abscessus* between CF patients have been proven. In view of these facts, we recommend giving NTM infections a higher priority both in basic research and when deciding on public health measures.

Note
After preparation of the manuscript, Humana Press published the book *Nontuberculous Mycobacterial Disease*, edited by **David E. Griffith**, in the book series "Respiratory Medicine," which provides valuable information to specific aspects of NTM disease.

References

Adekambi T, Sassi M, van Ingen J, Drancourt M (2017) Reinstating Mycobacterium massiliense and Mycobacterium bolletii as species of the Mycobacterium abscessus complex. Int J Syst Evol Microbiol 67(8):2726–2730. https://doi.org/10.1099/ijsem.0.002011

Adelman MH, Addrizzo-Harris DJ (2018) Management of nontuberculous mycobacterial pulmonary disease. Curr Opin Pulm Med 24(3):212–219. https://doi.org/10.1097/mcp.0000000000000473

Adjemian J, Olivier KN, Seitz AE, Holland SM, Prevots DR (2012) Prevalence of nontuberculous mycobacterial lung disease in U.S. Medicare beneficiaries. Am J Respir Crit Care Med 185 (8):881–886. https://doi.org/10.1164/rccm.201111-2016OC

Adjemian J, Olivier KN, Prevots DR (2018) Epidemiology of pulmonary nontuberculous mycobacterial sputum positivity in patients with cystic fibrosis in the United States, 2010–2014. Ann Am Thoracic Soc 15(7):817–826. https://doi.org/10.1513/AnnalsATS.201709-727OC

Aitken ML, Limaye A, Pottinger P, Whimbey E, Goss CH, Tonelli MR, Cangelosi GA, Dirac MA, Olivier KN, Brown-Elliott BA, McNulty S, Wallace RJ Jr (2012) Respiratory outbreak of Mycobacterium abscessus subspecies massiliense in a lung transplant and cystic fibrosis center. Am J Respir Crit Care Med 185(2):231–232. https://doi.org/10.1164/ajrccm.185.2.231

Aksamit TR, Philley JV, Griffith DE (2014) Nontuberculous mycobacterial (NTM) lung disease: the top ten essentials. Respir Med 108(3):417–425. https://doi.org/10.1016/j.rmed.2013.09.014

Albert DM, Raven ML (2016) Ocular Tuberculosis. Microbiology Spectrum 4(6). https://doi.org/10.1128/microbiolspec.TNMI7-0001-2016

Alcaide F, Amlerova J, Bou G, Ceyssens PJ, Coll P, Corcoran D, Fangous MS, Gonzalez-Alvarez I, Gorton R, Greub G, Hery-Arnaud G, Hrabak J, Ingebretsen A, Lucey B, Marekovic I, Mediavilla-Gradolph C, Monte MR, O'Connor J, O'Mahony J, Opota O, O'Reilly B, Orth-Holler D, Oviano M, Palacios JJ, Palop B, Pranada AB, Quiroga L, Rodriguez-Temporal D, Ruiz-Serrano MJ, Tudo G, Van den Bossche A, van Ingen J, Rodriguez-Sanchez B (2018) How to: identify non-tuberculous Mycobacterium species using MALDI-TOF mass spectrometry. Clin Microbiol Infect 24(6):599–603. https://doi.org/10.1016/j.cmi.2017.11.012

Andrejak C, Nielsen R, Thomsen VO, Duhaut P, Sorensen HT, Thomsen RW (2013) Chronic respiratory disease, inhaled corticosteroids and risk of non-tuberculous mycobacteriosis. Thorax 68(3):256–262. https://doi.org/10.1136/thoraxjnl-2012-201772

Atreya R, Bulte M, Gerlach GF, Goethe R, Hornef MW, Kohler H, Meens J, Mobius P, Roeb E, Weiss S (2014) Facts, myths and hypotheses on the zoonotic nature of Mycobacterium avium subspecies paratuberculosis. Int J Med Microbiol 304(7):858–867. https://doi.org/10.1016/j.ijmm.2014.07.006

Aubry A, Chosidow O, Caumes E, Robert J, Cambau E (2002) Sixty-three cases of Mycobacterium marinum infection: clinical features, treatment, and antibiotic susceptibility of causative isolates. Arch Intern Med 162(15):1746–1752

Aubry A, Mougari F, Reibel F, Cambau E (2017) Mycobacterium marinum. Microbiol Spectr 5(2). https://doi.org/10.1128/microbiolspec.TNMI7-0038-2016

Avanzi A, Bierbauer K, Vales-Kennedy G, Covino J (2018) Nontuberculous mycobacteria infection risk in medical tourism. JAAPA 31(8):45–47. https://doi.org/10.1097/01.JAA.0000541484.62234.7f

Ballarino GJ, Olivier KN, Claypool RJ, Holland SM, Prevots DR (2009) Pulmonary nontuberculous mycobacterial infections: antibiotic treatment and associated costs. Respir Med 103(10):1448–1455. https://doi.org/10.1016/j.rmed.2009.04.026

Banuls AL, Sanou A, Anh NT, Godreuil S (2015) Mycobacterium tuberculosis: ecology and evolution of a human bacterium. J Med Microbiol 64(11):1261–1269. https://doi.org/10.1099/jmm.0.000171

Barrow GI, Hewitt M (1971) Skin infection with Mycobacterium marinum from a tropical fish tank. Br Med J 2(5760):505–506

Bastian S, Veziris N, Roux AL, Brossier F, Gaillard JL, Jarlier V, Cambau E (2011) Assessment of clarithromycin susceptibility in strains belonging to the Mycobacterium abscessus group by erm (41) and rrl sequencing. Antimicrob Agents Chemother 55(2):775–781. https://doi.org/10.1128/aac.00861-10

Behr MA (2013) Evolution of Mycobacterium tuberculosis. Adv Exp Med Biol 783:81–91. https://doi.org/10.1007/978-1-4614-6111-1_4

Berliner JG, Aldabagh B, Mully T, Yu SS, Schwartz BS, Berger TG (2015) Non-tuberculous mycobacterial infections following cosmetic laser procedures: a case report and review of the literature. J Drugs Dermatol 14(1):80–83

Beswick J, Shin E, Michelis FV, Thyagu S, Viswabandya A, Lipton JH, Messner H, Marras TK, Kim DDH (2018) Incidence and risk factors for nontuberculous mycobacterial infection after allogeneic hematopoietic cell transplantation. Biol Blood Marrow Transplant J Am Soc Blood Marrow Transplant 24(2):366–372. https://doi.org/10.1016/j.bbmt.2017.09.015

Biet F, Boschiroli ML (2014) Non-tuberculous mycobacterial infections of veterinary relevance. Res Vet Sci 97(Suppl):S69–S77. https://doi.org/10.1016/j.rvsc.2014.08.007

Brown-Elliott BA, Iakhiaeva E, Griffith DE, Woods GL, Stout JE, Wolfe CR, Turenne CY, Wallace RJ Jr (2013) In vitro activity of amikacin against isolates of Mycobacterium avium complex with proposed MIC breakpoints and finding of a 16S rRNA gene mutation in treated isolates. J Clin Microbiol 51(10):3389–3394. https://doi.org/10.1128/jcm.01612-13

Bryant JM, Grogono DM, Greaves D, Foweraker J, Roddick I, Inns T, Reacher M, Haworth CS, Curran MD, Harris SR, Peacock SJ, Parkhill J, Floto RA (2013) Whole-genome sequencing to identify transmission of Mycobacterium abscessus between patients with cystic fibrosis: a retrospective cohort study. Lancet (London, England) 381(9877):1551–1560. https://doi.org/10.1016/s0140-6736(13)60632-7

Bryant JM, Grogono DM, Rodriguez-Rincon D, Everall I, Brown KP, Moreno P, Verma D, Hill E, Drijkoningen J, Gilligan P, Esther CR, Noone PG, Giddings O, Bell SC, Thomson R, Wainwright CE, Coulter C, Pandey S, Wood ME, Stockwell RE, Ramsay KA, Sherrard LJ, Kidd TJ, Jabbour N, Johnson GR, Knibbs LD, Morawska L, Sly PD, Jones A, Bilton D, Laurenson I, Ruddy M, Bourke S, Bowler IC, Chapman SJ, Clayton A, Cullen M, Daniels T, Dempsey O, Denton M, Desai M, Drew RJ, Edenborough F, Evans J, Folb J, Humphrey H, Isalska B, Jensen-Fangel S, Jonsson B, Jones AM, Katzenstein TL, Lillebaek T, MacGregor G, Mayell S, Millar M, Modha D, Nash EF, O'Brien C, O'Brien D, Ohri C, Pao CS, Peckham D, Perrin F, Perry A, Pressler T, Prtak L, Qvist T, Robb A, Rodgers H, Schaffer K, Shafi N, van Ingen J, Walshaw M, Watson D, West N, Whitehouse J, Haworth CS, Harris SR, Ordway D, Parkhill J, Floto RA (2016) Emergence and spread of a human-transmissible multidrug-resistant nontuberculous mycobacterium. Science (New York, NY) 354(6313):751–757. https://doi.org/10.1126/science.aaf8156

Casanova JL (2001) Mendelian susceptibility to mycobacterial infection in man. Swiss Med Wkly 131(31–32):445–454. doi: 2001/31/smw-09763

Casanova JL (2015) Severe infectious diseases of childhood as monogenic inborn errors of immunity. Proc Natl Acad Sci U S A 112(51):E7128–E7137. https://doi.org/10.1073/pnas.1521651112

Casanova JL, Abel L (2002) Genetic dissection of immunity to mycobacteria: the human model. Annu Rev Immunol 20:581–620. https://doi.org/10.1146/annurev.immunol.20.081501.125851

Catanzaro A (2002) Diagnosis, differentiating colonization, infection, and disease. Clin Chest Med 23(3):599–601, vi

Cerna-Cortes JF, Estrada-Garcia T, Gonzalez-y-Merchand JA (2009) Isolation of Mycobacterium mucogenicum from street-vended chili sauces: a potential source of human infection. J Food Prot 72(1):182–184

Cerna-Cortes JF, Leon-Montes N, Cortes-Cueto AL, Salas-Rangel LP, Helguera-Repetto AC, Lopez-Hernandez D, Rivera-Gutierrez S, Fernandez-Rendon E, Gonzalez-y-Merchand JA (2015) Microbiological quality of ready-to-eat vegetables collected in Mexico City: occurrence of aerobic-mesophilic bacteria, fecal coliforms, and potentially pathogenic nontuberculous mycobacteria. Biomed Res Int 2015:789508. https://doi.org/10.1155/2015/789508

Cerna-Cortes JF, Cortes-Cueto AL, Cano-Gaona MR, Leon-Montes N, Helguera-Repetto AC, Rivera-Gutierrez S, Salas-Rangel LP, Castro-Rosas J, Gonzalez YMJA (2016) Microbiological quality and occurrence of nontuberculous mycobacteria in fresh-squeezed orange juice samples purchased from street vendors in Mexico City. J Food Prot 79(12):2190–2195. https://doi.org/10.4315/0362-028x.jfp-16-197

Chae H, Han SJ, Kim SY, Ki CS, Huh HJ, Yong D, Koh WJ, Shin SJ (2017) Development of a one-step multiplex PCR assay for differential detection of major Mycobacterium species. J Clin Microbiol 55(9):2736–2751. https://doi.org/10.1128/jcm.00549-17

Chan ED, Iseman MD (2010) Slender, older women appear to be more susceptible to nontuberculous mycobacterial lung disease. Gend Med 7(1):5–18. https://doi.org/10.1016/j.genm.2010.01.005

Chang CH, Chang YY, Lu PH (2017) Non-tuberculous mycobacteria infection following autologous fat grafting on the face. Aesthet Surg J 38(1):NP1–NP5. https://doi.org/10 1093/asj/sjx168

Char A, Hopkinson NS, Hansell DM, Nicholson AG, Shaw EC, Clark SJ, Sedgwick P, Wilson R, Jordan S, Loebinger MR (2014) Evidence of mycobacterial disease in COPD patients with lung volume reduction surgery; the importance of histological assessment of specimens: a cohort study. BMC Pulm Med 14:124. https://doi.org/10.1186/1471-2466-14-124

Chobot S, Malis J, Sebakova H, Pelikan M, Zatloukal O, Palicka P, Kocurova D (1997) Endemic incidence of infections caused by Mycobacterium kansasii in the Karvina district in 1968-1995 (analysis of epidemiological data--review). Cent Eur J Public Health 5(4):164–173

Chu HS, Chang SC, Shen EP, Hu FR (2015) Nontuberculous mycobacterial ocular infections--comparing the clinical and microbiological characteristics between Mycobacterium abscessus and Mycobacterium massiliense. PLoS One 10(1):e0116236. https://doi.org/10.1371/journal.pone.0116236

CLSI (2011) Susceptibility testing of Mycobacteria, Nocardiae, and other aerobic actinomycetes—second edition: CLSI document M24-A2 Wayne, PA: Clinical and Laboratory Standards Institute

CLSI (2018) Performance standards for susceptibility testing of Mycobacteria, Nocardia spp., and other aerobic actinomycetes-first edition: CLSI supplement M62. Clinical and Laboratory Standards Institute, Wayne

Cole ST, Eiglmeier K, Parkhill J, James KD, Thomson NR, Wheeler PR, Honore N, Garnier T, Churcher C, Harris D, Mungall K, Basham D, Brown D, Chillingworth T, Connor R, Davies RM, Devlin K, Duthoy S, Feltwell T, Fraser A, Hamlin N, Holroyd S, Hornsby T, Jagels K, Lacroix C, Maclean J, Moule S, Murphy L, Oliver K, Quail MA, Rajandream MA, Rutherford KM, Rutter S, Seeger K, Simon S, Simmonds M, Skelton J, Squares R, Squares S, Stevens K, Taylor K, Whitehead S, Woodward JR, Barrell BG (2001) Massive gene decay in the leprosy bacillus. Nature 409(6823):1007–1011. https://doi.org/10.1038/35059006

Cowman SA, Loebinger MR (2018) Diagnosis of nontuberculous mycobacteria lung disease. Semin Respir Crit Care Med 39(3):343–350. https://doi.org/10.1055/s-0038-1651493

Crago B, Ferrato C, Drews SJ, Louie T, Ceri H, Turner RJ, Roles A, Louie M (2014) Surveillance and molecular characterization of non-tuberculous mycobacteria in a hospital water distribution system over a three-year period. J Hosp Infect 87(1):59–62. https://doi.org/10.1016/j.jhin.2014.03.002

da Silva MB, Portela JM, Li W, Jackson M, Gonzalez-Juarrero M, Hidalgo AS, Belisle JT, Bouth RC, Gobbo AR, Barreto JG, Minervino AHH, Cole ST, Avanzi C, Busso P, Frade MAC, Geluk A, Salgado CG, Spencer JS (2018) Evidence of zoonotic leprosy in Para, Brazilian Amazon, and risks associated with human contact or consumption of armadillos. PLoS Negl Trop Dis 12(6):e0006532. https://doi.org/10.1371/journal.pntd.0006532

Daley CL (2009) Nontuberculous mycobacterial disease in transplant recipients: early diagnosis and treatment. Curr Opin Organ Transplant 14(6):619–624. https://doi.org/10.1097/MOT.0b013e3283327cd6

Daley CL, Iseman M (2012) Mycobacterium avium complex and lung cancer: chicken or egg? Both? J Thoracic Oncol 7(9):1329–1330. https://doi.org/10.1097/JTO.0b013e318265a7ef

De Groote MA, Pace NR, Fulton K, Falkinham JO 3rd (2006) Relationships between mycobacterium isolates from patients with pulmonary mycobacterial infection and potting soils. Appl Environ Microbiol 72(12):7602–7606. https://doi.org/10.1128/aem.00930-06

Diel R, Jacob J, Lampenius N, Loebinger M, Nienhaus A, Rabe KF, Ringshausen FC (2017) Burden of non-tuberculous mycobacterial pulmonary disease in Germany. Eur Respir J 49(4). https://doi.org/10.1183/13993003.02109-2016

Diel R, Lipman M, Hoefsloot W (2018) High mortality in patients with Mycobacterium avium complex lung disease: a systematic review. BMC Infect Dis 18(1):206. https://doi.org/10.1186/s12879-018-3113-x

Dodge JA, Lewis PA, Stanton M, Wilsher J (2007) Cystic fibrosis mortality and survival in the UK: 1947–2003. Eur Respir J 29(3):522–526. https://doi.org/10.1183/09031936.00099506

Doffinger R, Altare F, Casanova JL (2000) Genetic heterogeneity of mendelian susceptibility to mycobacterial infection. Microbes Infect 2(13):1553–1557

Doffinger R, Dupuis S, Picard C, Fieschi C, Feinberg J, Barcenas-Morales G, Casanova J (2002) Inherited disorders of IL-12- and IFNgamma-mediated immunity: a molecular genetics update. Mol Immunol 38(12–13):903–909

Domozych R, Kim E, Hart S, Greenwald J (2016) Increasing incidence of leprosy and transmission from armadillos in Central Florida: a case series. JAAD Case Rep 2(3):189–192. https://doi.org/10.1016/j.jdcr.2016.03.004

Doucette K, Fishman JA (2004) Nontuberculous mycobacterial infection in hematopoietic stem cell and solid organ transplant recipients. Clin Infect Dis 38(10):1428–1439. https://doi.org/10.1086/420746

Ellis SM (2004) The spectrum of tuberculosis and non-tuberculous mycobacterial infection. Eur Radiol 14(Suppl 3):E34–E42. https://doi.org/10.1007/s00330-003-2042-1

El-Solh AA, Nopper J, Abdul-Khoudoud MR, Sherif SM, Aquilina AT, Grant BJ (1998) Clinical and radiographic manifestations of uncommon pulmonary nontuberculous mycobacterial disease in AIDS patients. Chest 114(1):138–145

El-Zeenni N, Chanoine S, Recule C, Stahl JP, Maurin M, Camara B (2018) Are guidelines on the management of non-tuberculous mycobacteria lung infections respected and what are the consequences for patients? A French retrospective study from 2007 to 2014. Eur J Clin Microbiol Infect Dis 37(2):233–240. https://doi.org/10.1007/s10096-017-3120-7

Eppleston J, Begg DJ, Dhand NK, Watt B, Whittington RJ (2014) Environmental survival of Mycobacterium avium subsp. paratuberculosis in different climatic zones of eastern Australia. Appl Environ Microbiol 80(8):2337–2342. https://doi.org/10.1128/AEM.03630-13

Falkinham JO 3rd (2016) Current Epidemiologic Trends of the Nontuberculous Mycobacteria (NTM). Curr Environ Health Rep 3(2):161–167. https://doi.org/10.1007/s40572-016-0086-z

Falkinham JO 3rd, Iseman MD, de Haas P, van Soolingen D (2008) Mycobacterium avium in a shower linked to pulmonary disease. J Water Health 6(2):209–213. https://doi.org/10.2166/wh.2008.032

Floto RA, Olivier KN, Saiman L, Daley CL, Herrmann JL, Nick JA, Noone PG, Bilton D, Corris P, Gibson RL, Hempstead SE, Koetz K, Sabadosa KA, Sermet-Gaudelus I, Smyth AR, van Ingen J, Wallace RJ, Winthrop KL, Marshall BC, Haworth CS (2016) US Cystic Fibrosis Foundation and European Cystic Fibrosis Society consensus recommendations for the management of non-tuberculous mycobacteria in individuals with cystic fibrosis. Thorax 71(Suppl 1): i1–i22. https://doi.org/10.1136/thoraxjnl-2015-207360

Fordham von Reyn C, Magnusson M, Chaparas SD, Margileth AM (1995) Use and standardization of the Mycobacterium avium sensitin skin-test reagent. Clin Infect Dis 21(6):1527–1528, 1528, 1529

Garcia-Marcos PW, Plaza-Fornieles M, Menasalvas-Ruiz A, Ruiz-Pruneda R, Paredes-Reyes P, Miguelez SA (2017) Risk factors of non-tuberculous mycobacterial lymphadenitis in children: a case-control study. Eur J Pediatr 176(5):607–613. https://doi.org/10.1007/s00431-017-2882-3

Garvey M (2018) Mycobacterium avium subspecies paratuberculosis: a possible causative agent in human morbidity and risk to public health safety. Open Vet J 8(2):172–181. https://doi.org/10.4314/ovj.v8i2.10

Ghosh R, Das S, Kela H, De A, Halcar J, Maiti PK (2017) Biofilm colonization of Mycobacterium abscessus: new threat in hospital-acquired surgical site infection. Indian J Tuberc 64 (3):178–182. https://doi.org/10.1016/j.ijtb.2016.11.013

Gonzalez CD, Petersen MG, Miller M, Park AH, Wilson KF (2016) Complex nontuberculous mycobacterial cervicofacial lymphadenitis: what is the optimal approach? Laryngoscope 126 (7):1677–1680. https://doi.org/10.1002/lary.25603

Gray SF, Smith RS, Reynolds NJ, Williams EW (1990) Fish tank granuloma. BMJ (Clin Res ed) 300(6731):1069–1070

Green DA, Whittier S, Greendyke W, Win C, Chen X, Hamele-Bena D (2017) Outbreak of rapidly growing nontuberculous mycobacteria among patients undergoing cosmetic surgery in the Dominican Republic. Ann Plast Surg 78(1):17–21. https://doi.org/10.1097/sap.0000000000000746

Griffith DE, Aksamit TR (2012) Therapy of refractory nontuberculous mycobacterial lung disease. Curr Opin Infect Dis 25(2):218–227. https://doi.org/10.1097/QCO.0b013e3283511a64

Griffith DE, Brown-Elliott BA, Wallace RJ Jr (2002) Diagnosing nontuberculous mycobacterial lung disease. A process in evolution. Infect Dis Clin N Am 16(1):235–249

Griffith DE, Brown-Elliott BA, Langsjoen B, Zhang Y, Pan X, Girard W, Nelson K, Caccitolo J, Alvarez J, Shepherd S, Wilson R, Graviss EA, Wallace RJ Jr (2006) Clinical and molecular analysis of macrolide resistance in Mycobacterium avium complex lung disease. Am J Respir Crit Care Med 174(8):928–934. https://doi.org/10.1164/rccm.200603-450OC

Griffith DE, Aksamit T, Brown-Elliott BA, Catanzaro A, Daley C, Gordin F, Holland SM, Horsburgh R, Huitt G, Iademarco MF, Iseman M, Olivier K, Ruoss S, von Reyn CF, Wallace RJ Jr, Winthrop K (2007) An official ATS/IDSA statement: diagnosis, treatment, and prevention of nontuberculous mycobacterial diseases. Am J Respir Crit Care Med 175(4):367–416. https://doi.org/10.1164/rccm.200604-571ST

Grubek-Jaworska H, Walkiewicz R, Safianowska A, Nowacka-Mazurek M, Krenke R, Przybylowski T, Chazan R (2009) Nontuberculous mycobacterial infections among patients suspected of pulmonary tuberculosis. Eur J Clin Microbiol Infect Dis 28(7):739–744. https://doi.org/10.1007/s10096-008-0694-0

Gutierrez AV, Viljoen A, Ghigo E, Herrmann JL, Kremer L (2018) Glycopeptidolipids, a double-edged sword of the Mycobacterium abscessus complex. Front Microbiol 9:1145. https://doi.org/10.3389/fmicb.2018.01145

Halstrom S, Price P, Thomson R (2015) Review: environmental mycobacteria as a cause of human infection. Int J Mycobacteriol 4(2):81–91. https://doi.org/10.1016/j.ijmyco.2015.03.002

Haworth CS, Floto RA (2017) Introducing the new BTS guideline: management of non-tuberculous mycobacterial pulmonary disease (NTM-PD). Thorax 72(11):969–970. https://doi.org/10.1136/thoraxjnl-2017-210929

Haworth CS, Banks J, Capstick T, Fisher AJ, Gorsuch T, Laurenson IF, Leitch A, Loebinger MR, Milburn HJ, Nightingale M, Ormerod P, Shingadia D, Smith D, Whitehead N, Wilson R, Floto

RA (2017) British thoracic society guideline for the management of non-tuberculous. BMJ Open Respir Res 4(1):2017–000242
Henkle E, Winthrop KL (2015) Nontuberculous mycobacteria infections in immunosuppressed hosts. Clin Chest Med 36(1):91–99. https://doi.org/10.1016/j.ccm.2014.11.002
Henkle E, Hedberg K, Schafer S, Novosad S, Winthrop KL (2015) Population-based Incidence of pulmonary nontuberculous mycobacterial disease in Oregon 2007 to 2012. Ann Am Thorac Soc 12(5):642–647. https://doi.org/10.1513/AnnalsATS.201412-559OC
Henkle E, Hedberg K, Schafer SD, Winthrop KL (2017) Surveillance of extrapulmonary nontuberculous mycobacteria infections, Oregon, USA, 2007-2012. Emerg Infect Dis 23 (10):1627–1630. https://doi.org/10.3201/eid2310.170845
Hsiao CH, Lin YT, Lai CC, Chou CH, Hsueh PR (2010) Identification of nontuberculous mycobacterial infection by IS6110 and hsp65 gene analysis on lung tissues. Diagn Microbiol Infect Dis 68(3):241–246. https://doi.org/10.1016/j.diagmicrobio.2010.05.017
Huang GK, Johnson PD (2014) Epidemiology and management of Buruli ulcer. Expert Rev Anti-Infect Ther 12(7):855–865. https://doi.org/10.1586/14787210.2014.910113
Huang HC, Weigt SS, Derhovanessian A, Palchevskiy V, Ardehali A, Saggar R, Saggar R, Kubak B, Gregson A, Ross DJ, Lynch JP 3rd, Elashoff R, Belperio JA (2011) Non-tuberculous mycobacterium infection after lung transplantation is associated with increased mortality. J Heart Lung Transplant 30(7):790–798. https://doi.org/10.1016/j.healun.2011.02.007
Ikawa H, Oka S, Murakami H, Hayashhi A, Yano I (1989) Rapid identification of serotypes of Mycobacterium avium-M. intracellulare complex by using infected swine sera and reference antigenic glycolipids. J Clin Microbiol 27(11):2552–2558
Jankovic Makek M, Pavlisa G, Jakopovic M, Redzepi G, Zmak L, Vukic Dugac A, Hecimovic A, Mazuranic I, Jaksch P, Klepetko W, Samarzija M (2016) Early onset of nontuberculous mycobacterial pulmonary disease contributes to the lethal outcome in lung transplant recipients: report of two cases and review of the literature. Transpl Infect Dis 18(1):112–119. https://doi.org/10.1111/tid.12481
Johnson MM, Odell JA (2014) Nontuberculous mycobacterial pulmonary infections. J Thoracic Dis 6(3):210–220. https://doi.org/10.3978/j.issn.2072-1439.2013.12.24
Johnston DI, Chisty Z, Gross JE, Park SY (2016) Investigation of Mycobacterium abscessus outbreak among cystic fibrosis patients, Hawaii 2012. J Hosp infect 94(2):198–200. https://doi.org/10.1016/j.jhin.2016.04.015
Jouanguy E, Doffinger R, Dupuis S, Pallier A, Altare F, Casanova JL (1999) IL-12 and IFN-gamma in host defense against mycobacteria and salmonella in mice and men. Curr Opin Immunol 11 (3):346–351
Jung YJ, Kim JY, Song DJ, Koh WJ, Huh HJ, Ki CS, Lee NY (2016) Evaluation of three real-time PCR assays for differential identification of Mycobacterium tuberculosis complex and nontuberculous mycobacteria species in liquid culture media. Diagn Microbiol Infect Dis 85 (2):186–191. https://doi.org/10.1016/j.diagmicrobio.2016.03.014
Kahana LM, Kay JM, Yakrus MA, Waserman S (1997) Mycobacterium avium complex infection in an immunocompetent young adult related to hot tub exposure. Chest 111(1):242–245
Kapnadak SG, Hisert KB, Pottinger PS, Limaye AP, Aitken ML (2016) Infection control strategies that successfully controlled an outbreak of Mycobacterium abscessus at a cystic fibrosis center. Am J Infect Control 44(2):154–159. https://doi.org/10.1016/j.ajic.2015.08.023
Keller C, Wenker C, Jermann T, Hirschi R, Schildger B, Meier R, Schmidt-Posthaus H (2018) Piscine mycobacteriosis – Involvement of bacterial species and reflection in pathology. Schweizer Archiv fur Tierheilkunde 160(6):385–393. https://doi.org/10.17236/sat00165
Kendall BA, Varley CD, Choi D, Cassidy PM, Hedberg K, Ware MA, Winthrop KL (2011) Distinguishing tuberculosis from nontuberculous mycobacteria lung disease, Oregon, USA. Emerg Infect Dis 17(3):506–509. https://doi.org/10.3201/eid1703.101164
Kheir WJ, Sheheitli H, Abdul Fattah M, Hamam RN (2015) Nontuberculous mycobacterial ocular infections: a systematic review of the literature. Biomed Res Int 2015:164989. https://doi.org/10.1155/2015/164989

Kim HS, Lee Y, Lee S, Kim YA, Sun YK (2014) Recent trends in clinically significant nontuberculous mycobacteria isolates at a Korean general hospital. Ann Lab Med 34 (1):56–59. https://doi.org/10.3343/alm.2014.34.1.56

Kim JU, Ryu DS, Cha CH, Park SH (2018) Paradigm for diagnosing mycobacterial disease: direct detection and differentiation of Mycobacterium tuberculosis complex and non-tuberculous mycobacteria in clinical specimens using multiplex real-time PCR. J Clin Pathol 71 (9):774–780. https://doi.org/10.1136/jclinpath-2017-204945

Kitada S, Maekura R, Toyoshima N, Fujiwara N, Yano I, Ogura T, Ito M, Kobayashi K (2002) Serodiagnosis of pulmonary disease due to Mycobacterium avium complex with an enzyme immunoassay that uses a mixture of glycopeptidolipid antigens. Clin Infect Dis 35 (11):1328–1335. https://doi.org/10.1086/344277

Kitada S, Kobayashi K, Nishiuchi Y, Fushitani K, Yoshimura K, Tateishi Y, Miki K, Miki M, Hashimoto H, Motone M, Fujikawa T, Hiraga T, Maekura R (2010) Serodiagnosis of pulmonary disease due to Mycobacterium avium complex proven by bronchial wash culture. Chest 138(1):236–237. https://doi.org/10.1378/chest.10-0248

Kitada S, Yoshimura K, Miki K, Miki M, Hashimoto H, Matsui H, Kuroyama M, Ageshio F, Kagawa H, Mori M, Maekura R, Kobayashi K (2015) Validation of a commercial serodiagnostic kit for diagnosing pulmonary Mycobacterium avium complex disease. Int J Tuberc Lung Dis 19(1):97–103. https://doi.org/10.5588/ijtld.14.0564

Kitada S, Maekura R, Yoshimura K, Miki K, Miki M, Oshitani Y, Nishida K, Sawa N, Mori M, Kobayashi K (2017) Levels of antibody against Glycopeptidolipid core as a marker for monitoring treatment response in Mycobacterium avium complex pulmonary disease: a prospective cohort study. J Clin Microbiol 55(3):884–892. https://doi.org/10.1128/JCM.02010-16

Klanicova-Zalewska B, Slana I (2014) Presence and persistence of Mycobacterium avium and other nontuberculous mycobacteria in animal tissues and derived foods: a review. Meat Sci 98 (4):835–841. https://doi.org/10.1016/j.meatsci.2014.08.001

Knoll BM (2014) Update on nontuberculous mycobacterial infections in solid organ and hematopoietic stem cell transplant recipients. Curr Infect Dis Rep 16(9):421 https://doi.org/10.1007/s11908-014-0421-1

Kobashi Y, Yoshida K, Miyashita N, Niki Y, Matsushima T (2004) Pulmonary Mycobacterium avium disease with a solitary pulmonary nodule requiring differentiation from recurrence of pulmonary adenocarcinoma. Intern Med 43(9):855–860

Kobayashi K (2014) Serodiagnosis of Mycobacterium avium complex disease in humans: translational research from basic mycobacteriology to clinical medicine. Jpn J Infect Dis 67 (5):329–332

Koh WJ (2017) Nontuberculous mycobacteria-overview. Microbiol Spectr 5(1). https://doi.org/10.1128/microbiolspec.TNMI7-0024-2016

Kohler H, Burkert B, Pavlik I, Diller R, Geue L, Conraths FJ, Martin G (2008) Evaluation of five ELISA test kits for the measurement of antibodies against Mycobacterium avium subspecies paratuberculosis in bovine serum. Berl Munch Tierarztl Wochenschr 121(5–6):203–210

Kwon YS, Koh WJ (2014) Diagnosis of pulmonary tuberculosis and nontuberculous mycobacterial lung disease in Korea. Tuberc Respir Dis (Seoul) 77(1):1–5. https://doi.org/10.4046/trd.2014.77.1.1

Kwon YS, Koh WJ (2016) Diagnosis and treatment of nontuberculous mycobacterial lung disease. J Korean Med Sci 31(5):649–659. https://doi.org/10.3346/jkms.2016.31.5.649

Lahiri A, Kneisel J, Kloster I, Kamal E, Lewin A (2014) Abundance of Mycobacterium avium ssp. hominissuis in soil and dust in Germany – implications for the infection route. Lett Appl Microbiol 59(1):65–70

Leao C, Canto A, Machado D, Sanches IS, Couto I, Viveiros M, Inacio J, Botelho A (2014) Relatedness of Mycobacterium avium subspecies hominissuis clinical isolates of human and porcine origins assessed by MLVA. Vet Microbiol 173(1-2):92–100. https://doi.org/10.1016/j.vetmic.2014.06.027

Leite CQ, Anno IS, Leite SR, Roxo E, Morlock GP, Cooksey RC (2003) Isolation and identification of mycobacteria from livestock specimens and milk obtained in Brazil. Memorias do Instituto Oswaldo Cruz 98(3):319–323

Lewin A, Sharbati-Tehrani S (2005) Slow growth rate of mycobacteria. Possible reasons and significance for their pathogenicity. Bundesgesundheitsbl Gesundheitsforsch Gesundheitsschutz 48(12):1390–1399. https://doi.org/10.1007/s00103-005-1171-x

Li T, Abebe LS, Cronk R, Bartram J (2017) A systematic review of waterborne infections from nontuberculous mycobacteria in health care facility water systems. Int J Hyg Environ Health 220(3):611–620. https://doi.org/10.1016/j.ijheh.2016.12.002

Liao CH, Chen MY, Hsieh SM, Sheng WH, Hung CC, Chang SC (2004) Discontinuation of secondary prophylaxis in AIDS patients with disseminated non-tuberculous mycobacteria infection. J Microbiol Immunol Infect 37(1):50–56

Lindeboom JA, Kuijper EJ, Bruijnesteijn van Coppenraet ES, Lindeboom R, Prins JM (2007) Surgical excision versus antibiotic treatment for nontuberculous mycobacterial cervicofacial lymphadenitis in children: a multicenter, randomized, controlled trial. Clin Infect Dis 44 (8):1057–1064. https://doi.org/10.1086/512675

Magdorf K, Schuck SD, Leitner S, Wahn U, Kaufmann SH, Jacobsen M (2008) T-cell responses against tuberculin and sensitin in children with tuberculosis and non-tuberculosis mycobacterial lymphadenopathy. Clin Microbiol Infect 14(11):1079–1083. https://doi.org/10.1111/j.1469-0691.2008.02084.x

Magnusson M (1961) Specificity of mycobacterial sensitins. I. Studies in guinea pigs with purified "tuberculin" prepared from mammalian and avian tubercle bacilli, Mycobacterium balnei, and other acid-fast bacilli. Am Rev Respir Dis 83:57–68. https://doi.org/10.1164/arrd.1961.83.1.57

Magnusson M, Bentzon MW, Bleiker MA, Engbaek HC, Griep WA, Manten A (1961) Comparative intradermal reactions of tuberculosis patients to purified tuberculin (PPD) and purified sensitin prepared from Mycobacterium kansasii. Acta Tuberc Scand 40:85–102

Makovcova J, Slany M, Babak V, Slana I, Kralik P (2014) The water environment as a source of potentially pathogenic mycobacteria. J Water Health 12(2):254–263. https://doi.org/10.2166/wh.2013.102

Marks J, Palfreyman M, Yates MD, Schaefer WB (1977) Differential tuberculin test for mycobacterial infection in children. Tubercle 58(1):19–23

Marras TK, Prevots DR, Jamieson FB, Winthrop KL, Pulmonary MACOG (2015) Variable agreement among experts regarding Mycobacterium avium complex lung disease. Respirology (Carlton, Vic) 20(2):348–351. https://doi.org/10.1111/resp.12440

McGrath EE, McCabe J, Anderson PB, American Thoracic S, Infectious Diseases Society of A (2008) Guidelines on the diagnosis and treatment of pulmonary non-tuberculous mycobacteria infection. Int J Clin Pract 62(12):1947–1955. https://doi.org/10.1111/j.1742-1241.2008.01891

Misch EA, Saddler C, Davis JM (2018) Skin and soft tissue infections due to nontuberculous mycobacteria. Curr Infect Dis Rep 20(4):6. https://doi.org/10.1007/s11908-018-0611-3

Mohamad B, Iqbal MN, Gopal KV, Arshad S, Daw HA (2012) Mai infection simulating metastatic breast cancer. BMJ Case Rep 2012. https://doi.org/10.1136/bcr-01-2012-5640

Mullis SN, Falkinham JO 3rd (2013) Adherence and biofilm formation of Mycobacterium avium, Mycobacterium intracellulare and Mycobacterium abscessus to household plumbing materials. J Appl Microbiol 115(3):908–914. https://doi.org/10.1111/jam.12272

Murray CJLL, Lopez AD (1996) The global burden of disease:1–43, published by The Harvard School of Public Health on Behalf of The World Health Organization and The World Bank distributed by Harvard University Press, ISBN 0-9655466-0-8

Nasiri MJ, Haeili M, Ghazi M, Goudarzi H, Pormohammad A, Imani Fooladi AA, Feizabadi MM (2017) New insights in to the intrinsic and acquired drug resistance mechanisms in mycobacteria. Front Microbiol 8:681. https://doi.org/10.3389/fmicb.2017.00681

Nessar R, Cambau E, Reyrat JM, Murray A, Gicquel B (2012) Mycobacterium abscessus: a new antibiotic nightmare. J Antimicrob Chemother 67(4):810–818. https://doi.org/10.1093/jac/dkr578

Noguchi S, Nishimoto Y, Taguchi Y, Kobashi Y, Noma S (2016) A case of disseminated Mycobacterium avium complex infection mimicking systemic metastases from lung cancer. J Thorac Imaging 31(6):W73–w75. https://doi.org/10.1097/rti.0000000000000239
O'Connor JA, O'Reilly B, Corcoran GD, O'Mahony J, Lucey B (2015) Mycobacterium diagnostics: from the primitive to the promising. Br J Biomed Sci 72(1):32–41
Oriani AS, Sierra F, Baldini MD (2018) Effect of chlorine on Mycobacterium gordonae and Mycobacterium chubuense in planktonic and Biofilm state. In J Mycobacteriol 7(2):122–127. https://doi.org/10.4103/ijmy.ijmy_30_18
Ovrutsky AR, Chan ED, Kartalija M, Bai X, Jackson M, Gibbs S, Falkinham JO 3rd, Iseman MD, Reynolds PR, McDonnell G, Thomas V (2013) Cooccurrence of free-living amoebae and nontuberculous mycobacteria in hospital water networks, and preferential growth of Mycobacterium avium in Acanthamoeba lenticulata. Appl Environ Microbiol 79(10):3185–3192. https://doi.org/10.1128/aem.03823-12
Pandian TK, Deziel PJ, Otley CC, Eid AJ, Razonable RR (2008) Mycobacterium marinum infections in transplant recipients: case report and review of the literature. Transpl Infect Dis 10(5):358–363. https://doi.org/10.1111/j.1399-3062.2008.00317.x
Pang Y, Lu J, Su B, Zheng H, Zhao Y (2017) Misdiagnosis of tuberculosis associated with some species of nontuberculous mycobacteria by GeneXpert MTB/RIF assay. Infection 45(5):677–681. https://doi.org/10.1007/s15010-017-1044-x
Park IK, Olivier KN (2015) Nontuberculous mycobacteria in cystic fibrosis and non-cystic fibrosis bronchiectasis. Semin Respir Crit Care Med 36(2):217–224. https://doi.org/10.1055/s-0035-1546751
Parrish N (2019) An update on mycobacterial taxonomy: 2016–2017. J Clin Microbiol. https://doi.org/10.1128/jcm.01408-18
Patel R, Roberts GD, Keating MR, Paya CV (1994) Infections due to nontuberculous mycobacteria in kidney, heart, and liver transplant recipients. Clinical Infect Dis 19(2):263–273
Patel SS, Saraiya NV, Tessler HH, Goldstein DA (2013) Mycobacterial ocular inflammation: delay in diagnosis and other factors impacting morbidity. JAMA Ophthalmol 131(6):752–758. https://doi.org/10.1001/jamaophthalmol.2013.71
Peters M, Muller C, Rusch-Gerdes S, Seidel C, Gobel U, Pohle HD, Ruf B (1995) Isolation of atypical mycobacteria from tap water in hospitals and homes: is this a possible source of disseminated MAC infection in AIDS patients? J Infect 31(1):39–44
Piersimoni C (2012) Nontuberculous mycobacteria infection in solid organ transplant recipients. Eur J Clin Microbiol Infect Dis 31(4):397–403. https://doi.org/10.1007/s10096-011-1329-4
Piersimoni C, Scarparo C (2009) Extrapulmonary infections associated with nontuberculous mycobacteria in immunocompetent persons. Emerg Infect Dis 15(9):1351–1358.; quiz 1544. https://doi.org/10.3201/eid1509.081259
Policard A, Gernez-Rieux C, Tacquet A, Martin JC, Devulder B, Le Bouffant L (1967) Influence of pulmonary dust load on the development of experimental infection by Mycobacterium kansasii. Nature 216(5111):177–178
Prammananan T, Sander P, Brown BA, Frischkorn K, Onyi GO, Zhang Y, Bottger EC, Wallace RJ Jr (1998) A single 16S ribosomal RNA substitution is responsible for resistance to amikacin and other 2-deoxystreptamine aminoglycosides in Mycobacterium abscessus and Mycobacterium chelonae. J Infect Dis 177(6):1573–1581
Prevots DR, Shaw PA, Strickland D, Jackson LA, Raebel MA, Blosky MA, Montes de Oca R, Shea YR, Seitz AE, Holland SM, Olivier KN (2010) Nontuberculous mycobacterial lung disease prevalence at four integrated health care delivery systems. Am J Respir Crit Care Med 182(7):970–976. https://doi.org/10.1164/rccm.201002-0310OC
Primm TP, Lucero CA, Falkinham JO (2004) Health impacts of environmental mycobacteria. Clin Microbiol Rev 17(1):98–106. https://doi.org/10.1128/cmr.17.1.98-106.2004
Rahman SA, Singh Y, Kohli S, Ahmad J, Ehtesham NZ, Tyagi AK, Hasnain SE (2014) Comparative analyses of nonpathogenic, opportunistic, and totally pathogenic mycobacteria reveal genomic and biochemical variabilities and highlight the survival attributes of Mycobacterium tuberculosis. MBio 5(6):e02020. https://doi.org/10.1128/mBio.02020-14

Rao M, Silveira FP (2018) Non-tuberculous mycobacterial infections in thoracic transplant candidates and recipients. Curr Infect Dis Rep 20(6):14. https://doi.org/10.1007/s11908-018-0619-8

Reichenbach J, Rosenzweig S, Doffinger R, Dupuis S, Holland SM, Casanova JL (2001) Mycobacterial diseases in primary immunodeficiencies. Curr Opin Allergy Clin Immunol 1 (6):503–511

Rindi L, Garzelli C (2014) Genetic diversity and phylogeny of Mycobacterium avium. Infect Genet Evol 21:375–383. https://doi.org/10.1016/j.meegid.2013.12.007

Runyon EH (1959) Anonymous mycobacteria in pulmonary disease. Med Clin North Am 43 (1):273–290

Ryu YJ, Koh WJ, Daley CL (2016) Diagnosis and treatment of nontuberculous mycobacterial lung disease: clinicians' perspectives. Tuberc Respir Dis (Seoul) 79(2):74–84. https://doi.org/10.4046/trd.2016.79.2.74

Sakane T, Matsuoka K, Kumata S, Watanabe R, Yamada T, Matsuoka T, Nagai S, Ueda M, Miyamoto Y (2018) The outcomes of anatomical lung resection for nontuberculous mycobacterial lung disease. J Thoracic Dis 10(2):954–962. https://doi.org/10.21037/jtd.2018.01.60

Samba-Louaka A, Robino E, Cochard T, Branger M, Delafont V, Aucher W, Wambeke W, Bannantine JP, Biet F, Hechard Y (2018) Environmental Mycobacterium avium subsp. paratuberculosis hosted by free-living Amoebae. Front Cellular Infect Microbiol 8:28. https://doi.org/10.3389/fcimb.2018.00028

Sanchini A, Semmler T, Mao L, Kumar N, Dematheis F, Tandon K, Peddireddy V, Ahmed N, Lewin A (2016) A hypervariable genomic island identified in clinical and environmental Mycobacterium avium subsp. hominissuis isolates from Germany. Int J Med Microbiol 306 (7):495–503. https://doi.org/10.1016/j.ijmm.2016.07.001

Sax H, Bloemberg G, Hasse B, Sommerstein R, Kohler P, Achermann Y, Rossle M, Falk V, Kuster SP, Bottger EC, Weber R (2015) Prolonged outbreak of Mycobacterium chimaera infection after open-chest heart surgery. Clin Infect Dis 61(1):67–75. https://doi.org/10.1093/cid/civ198

Schaefer WB (1967) Type-specificity of atypical mycobacteria in agglutination and antibody absorption tests. Am Rev Respir Dis 96(6):1165–1168. https://doi.org/10.1164/arrd.1967.96.6.1165

Scriven JE, Scobie A, Verlander NQ, Houston A, Collyns T, Cajic V, Kon OM, Mitchell T, Rahama O, Robinson A, Withama S, Wilson P, Maxwell D, Agranoff D, Davies E, Llewelyn M, Soo SS, Sahota A, Cooper M, Hunter M, Tomlins J, Tiberi S, Kendall S, Dedicoat M, Alexander E, Fenech T, Zambon M, Lamagni T, Smith EG, Chand M (2018) Mycobacterium chimaera infection following cardiac surgery in the United Kingdom: clinical features and outcome of the first 30 cases. Clin Microbiol Infect. https://doi.org/10.1016/j.cmi.2018.04.027

Serour F, Mizrahi A, Somekh E, Feinberg J, Picard C, Casanova JL, Dalal I (2007) Analysis of the interleukin-12/interferon-gamma pathway in children with non-tuberculous mycobacterial cervical lymphadenitis. Eur J Pediatr 166(8):835–841. https://doi.org/10.1007/s00431-006-0338-2

Sevilla IA, Molina E, Tello M, Elguezabal N, Juste RA, Garrido JM (2017) Detection of mycobacteria by culture and DNA-based methods in animal-derived food products purchased at Spanish supermarkets. Front Microbiol 8:1030. https://doi.org/10.3389/fmicb.2017.01030

Shah SK, McAnally KJ, Seoane L, Lombard GA, LaPlace SG, Lick S, Dhillon GS, Valentine VG (2016) Analysis of pulmonary non-tuberculous mycobacterial infections after lung transplantation. Transpl Infect Dis 18(4):585–591. https://doi.org/10.1111/tid.12546

Sharbati-Tehrani S, Stephan J, Holland G, Appel B, Niederweis M, Lewin A (2005) Porins limit the intracellular persistence of Mycobacterium smegmatis. Microbiology 151(Pt 7):2403–2410. https://doi.org/10.1099/mic.0.27969-0

Shoulah SA, Oschmann AM, Selim A, Semmler T, Schwarz C, Kamal E, Hamouda F, Galila E, Bitter W, Lewin A (2018) Environmental Mycobacterium avium subsp. hominissuis have a higher probability to act as a recipient in conjugation than clinical strains. Plasmid 95:28–35. https://doi.org/10.1016/j.plasmid.2018.01.003

Shu CC, Ato M, Wang JT, Jou R, Wang JY, Kobayashi K, Lai HC, Yu CJ, Lee LN, Luh KT (2013) Sero-diagnosis of Mycobacterium avium complex lung disease using serum immunoglobulin a antibody against glycopeptidolipid antigen in Taiwan. PLoS One 8(11):e80473. https://doi.org/10.1371/journal.pone.0080473

Slany M, Pavlik I (2012) Molecular detection of nontuberculous mycobacteria: advantages and limits of a broad-range sequencing approach. J Mol Microbiol Biotechnol 22(4):268–276. https://doi.org/10.1159/000342517

Sood G, Parrish N (2017) Outbreaks of nontuberculous mycobacteria. Curr Opin Infect Dis 30 (4):404–409. https://doi.org/10.1097/qco.0000000000000386

Steindor M, Nkwouano V, Mayatepek E, Mackenzie CR, Schramm D, Jacobsen M (2015) Rapid detection and immune characterization of Mycobacterium abscessus infection in cystic fibrosis patients. PLoS One 10(3):e0119737. https://doi.org/10.1371/journal.pone.0119737

Strollo SE, Adjemian J, Adjemian MK, Prevots DR (2015) The burden of pulmonary nontuberculous mycobacterial disease in the United States. Ann Am Thoracic Soc 12 (10):1458–1464. https://doi.org/10.1513/AnnalsATS.201503-173OC

Taylor RH, Falkinham JO 3rd, Norton CD, LeChevallier MW (2000) Chlorine, chloramine, chlorine dioxide, and ozone susceptibility of Mycobacterium avium. Appl Environ Microbiol 66(4):1702–1705

Thanachartwet V, Desakorn V, Duangrithi D, Chunpongthong P, Phojanamongkolkij K, Jitruckthai P, Kasetjaroen Y, Pitisuttithum P (2014) Comparison of clinical and laboratory findings between those with pulmonary tuberculosis and those with nontuberculous mycobacterial lung disease. Southeast Asian J Trop Med Public health 45(1):85–94

Thillai M, Pollock K, Pareek M, Lalvani A (2014) Interferon-gamma release assays for tuberculosis: current and future applications. Expert Rev Respir Med 8(1):67–78. https://doi.org/10.1586/17476348.2014.852471

Thomson R, Donnan E, Konstantinos A (2017) Notification of nontuberculous mycobacteria: an Australian perspective. Ann Am Thoracic Soc 14(3):318–323. https://doi.org/10.1513/AnnalsATS.201612-994OI

Tirkkonen T, Pakarinen J, Moisander AM, Makinen J, Soini H, Ali-Vehmas T (2007) High genetic relatedness among Mycobacterium avium strains isolated from pigs and humans revealed by comparative IS1245 RFLP analysis. Vet Microbiol 125(1-2):175–181. https://doi.org/10.1016/j.vetmic.2007.05.005

Torres-Coy JA, Rodriguez-Castillo BA, Perez-Alfonzo R, Dew JH (2016) Source investigation of two outbreaks of skin and soft tissue infection by Mycobacterium abscessus subsp. abscessus in Venezuela. Epidemiol Infect 144(5):1117–1120. https://doi.org/10.1017/s0950268815002381

Touma Z, Haddad A, Gladman DD, Uleryk EM, Urowitz MB (2013) Skin nontuberculous mycobacterial infection in systemic lupus erythematosus: an unusual skin infection mimicking lupus vasculitis. Semin Arthritis Rheum 42(5):498–506. https://doi.org/10.1016/j.semarthrit.2012.08.002

Trovato A, Baldan R, Costa D, Simonetti TM, Cirillo DM, Tortoli E (2017) Molecular typing of Mycobacterium abscessus isolated from cystic fibrosis patients. Int J Mycobacteriol 6 (2):138–141. https://doi.org/10.4103/ijmy.ijmy_33_17

Uchiya KI, Tomida S, Nakagawa T, Asahi S, Nikai T, Ogawa K (2017) Comparative genome analyses of Mycobacterium avium reveal genomic features of its subspecies and strains that cause progression of pulmonary disease. Sci Rep 7:39750. https://doi.org/10.1038/srep39750

Umehara T, Aoki M, Harada A, Watanabe Y, Sato M (2015) A Case of Nontuberculous Mycobacteria Highly Suspected as Lung Cancer Invading the Aortic Arch. J Thoracic Oncol 10(7):1118–1119. https://doi.org/10.1097/jto.0000000000000413

van Ingen J, Kuijper EJ (2014) Drug susceptibility testing of nontuberculous mycobacteria. Future Microbiol 9(9):1095–1110. https://doi.org/10.2217/fmb.14.60

van Ingen J, Boeree MJ, Dekhuijzen PN, van Soolingen D (2008) Mycobacterial disease in patients with rheumatic disease. Nat Clin Pract Rheumatol 4(12):649–656. https://doi.org/10.1038/ncprheum0949

Van Ingen J, Boeree MJ, Van Soolingen D, Mouton JW (2012) Resistance mechanisms and drug susceptibility testing of nontuberculous mycobacteria. Drug Resist Updat 15(3):149–161 van Ingen J, Kohl TA, Kranzer K, Hasse B, Keller PM, Katarzyna Szafranska A, Hillemann D, Chand M, Schreiber PW, Sommerstein R, Berger C, Genoni M, Ruegg C, Troillet N, Widmer AF, Becker SL, Herrmann M, Eckmanns T, Haller S, Holler C, Debast SB, Wolfhagen MJ, Hopman J, Kluytmans J, Langelaar M, Notermans DW, Ten Oever J, van den Barselaar P, Vonk ABA, Vos MC, Ahmed N, Brown T, Crook D, Lamagni T, Phin N, Smith EG, Zambon M, Serr A, Gotting T, Ebner W, Thurmer A, Utpatel C, Sproer C, Bunk B, Nubel U, Bloemberg GV, Bottger EC, Niemann S, Wagner D, Sax H (2017) Global outbreak of severe Mycobacterium chimaera disease after cardiac surgery: a molecular epidemiological study. Lancet Infect Dis 17(10):1033–1041. https://doi.org/10.1016/s1473-3099(17)30324-9

Vincenzi C, Bardazzi F, Tosti A, Varotti C, Morganti L (1992) Fish tank granuloma: report of a case. Cutis 49(4):275–276

Waddell LA, Rajic A, Sargeant J, Harris J, Amezcua R, Downey L, Read S, McEwen SA (2008) The zoonotic potential of Mycobacterium avium spp. paratuberculosis: a systematic review. Can J Public Health = Revue canadienne de sante publique 99(2):145–155

Walker J, Moore G, Collins S, Parks S, Garvey MI, Lamagni T, Smith G, Dawkin L, Goldenberg S, Chand M (2017) Microbiological problems and biofilms associated with Mycobacterium chimaera in heater-cooler units used for cardiopulmonary bypass. J Hosp Infect 96 (3):209–220. https://doi.org/10.1016/j.jhin.2017.04.014

Wang HY, Kim H, Kim S, Kim DK, Cho SN, Lee H (2015) Performance of a real-time PCR assay for the rapid identification of Mycobacterium species. J Microbiol (Seoul, Korea) 53(1):38–46. https://doi.org/10.1007/s12275-015-4495-8

Wang HY, Uh Y, Kim S, Lee H (2017) Performance of the Quantamatrix multiplexed assay platform system for the differentiation and identification of Mycobacterium species. J Med Microbiol 66(6):777–787. https://doi.org/10.1099/jmm.0.000495

WHO (2010) Commercial serodiagnostic tests for diagnosis of tuberculosis WHO. http://www.who.int/tb/publications/tb-serodiagnostic-report/en/. 2018

WHO (2018) Latent TB infection: updated and consolidated guidelines for programmatic management, 2018th edn. WHO, Geneva

Willemse SH, Oomens M, De Lange J, Karssemakers LHE (2018) Diagnosing nontuberculous mycobacterial cervicofacial lymphadenitis in children: a systematic review. Int J Pediatr Otorhinolaryngol 112:48–54. https://doi.org/10.1016/j.ijporl.2018.06.034

Yeung MW, Khoo E, Brode SK, Jamieson FB, Kamiya H, Kwong JC, Macdonald L, Marras TK, Morimoto K, Sander B (2016) Health-related quality of life, comorbidities and mortality in pulmonary nontuberculous mycobacterial infections: a systematic review. Respirology (Carlton, Vic) 21(6):1015–1025. https://doi.org/10.1111/resp.12767

Yu H, Jeong H, Hong Y, Seol S, Chung D, Kong H (2007) Natural occurrence of mycobacterium as an endosymbiont of acanthamoeba isolated from a contact lens storage case. Korean J Parasitol 45(1):11–18

Yuan MK, Chang CY, Tsai PH, Lee YM, Huang JW, Chang SC (2014) Comparative chest computed tomography findings of non-tuberculous mycobacterial lung diseases and pulmonary tuberculosis in patients with acid fast bacilli smear-positive sputum. BMC Pulm Med 14:65. https://doi.org/10.1186/1471-2466-14-65

Zheng H, Liu D, Lu J, Song Y, Wang S, Zhao Y, Ni X (2019) Genetic Correlation of antibiotic susceptibility and resistance genotyping for the Mycobacterium abscessus group. Antimicrob Agents Chemother 63(1). https://doi.org/10.1128/aac.01523-18

Tuberculosis in New Zealand: Historical Overview to Modern Epidemiology

6

Ronan F. O'Toole

Abstract

The burden of tuberculosis in New Zealand peaked during the Second World War with 2603 cases recorded in 1943. Control measures legislated for in the Tuberculosis Act of 1948 played a significant role in bringing down the national incidence rate from over 150 cases per 100,000 to below 10 per 100,000 by the 1990s. Today, New Zealand is considered to be a low TB incidence country; however, this designation can mask disparities in the burden of TB within the population. The Asian ethnic group exhibits rates of TB that are approximately 50 times greater than in the European or other ethnic groups. Furthermore, among New Zealand-born individuals, Māori experience a higher burden of TB compared to their Pākehā compatriots. In this chapter, factors involved in contributing to the risk of TB in New Zealand are examined. Data from molecular typing studies are explored to gain an insight into the transmission of the illness in the population. In terms of further measures to reduce the incidence of TB in New Zealand, newer technologies that specifically detect latent TB infection are required for pre-immigration screening from high TB burden countries. Greater utilisation of prophylactic therapy of latent TB infection in individuals at risk of developing TB could assist in preventing reactivated cases. And early detection and treatment of active TB cases combined with enhanced contact tracing would help minimise the emergence of epidemiological clusters of the disease.

R. F. O'Toole (✉)
School of Molecular Sciences, College of Science, Health and Engineering, La Trobe University, Melbourne, VIC, Australia

Department of Clinical Microbiology, School of Medicine, Trinity College Dublin, Dublin, Ireland

School of Medicine, College of Health and Medicine, University of Tasmania, Hobart, Tasmania, Australia
e-mail: otooler3@tcd.ie

S. E. Hasnain et al. (eds.), Mycobacterium tuberculosis: *Molecular Infection Biology, Pathogenesis, Diagnostics and New Interventions*,
https://doi.org/10.1007/978-981-32-9413-4_6

Keywords
Mycobacterium tuberculosis complex · MIRU-VNTR · Latent tuberculosis infection

Abbreviations

BCG	bacillus Calmette-Guérin
DHB	District Health Board
EEA	European Economic Area
EU	European Union
ICER	incremental cost-effectiveness ratio
IGRA	interferon gamma release assay
LTBI	latent tuberculosis infection
MIRU-VNTR	mycobacterial interspersed repetitive unit-variable number tandem repeat
MTBC	*Mycobacterium tuberculosis* complex
NTAC	National Tuberculosis Advisory Committee
NZ	New Zealand
TB	tuberculosis
TST	tuberculin skin test
UK	United Kingdom
USA	United States of America
WHO	World Health Organization

6.1 Introduction to Tuberculosis

In 2016, there were an estimated 10.4 million new cases of tuberculosis (TB) globally corresponding to an annual incidence rate of approximately 140 per 100,000 persons (World Health Organization 2017). Highest incidence rates of TB were seen in South Africa (781/100,000), Lesotho (724/100,000) and the Philippines (554/100,000) in 2016 (World Health Organization 2017). Five countries recorded more than 0.5 million new cases nationally and accounted for >50% of the global total in 2016, i.e. India (2,790,000 cases), Indonesia (1,020,000 cases), China (895,000 cases), the Philippines (573,000 cases) and Pakistan (518,000 cases) (World Health Organization 2017).

The World Health Organization (WHO) has set End TB Strategy targets to reduce global TB mortalities by 95%, and the incidence of TB by 90% to less than 10 cases per 100,000 by 2035 with respect to 2015 levels (World Health Organization 2014). While industrialised countries such as New Zealand are considered to have a low TB burden, the WHO has also established specific targets for low TB incidence countries for the pre-elimination of TB by 2035 (corresponding to <10 TB cases per million population) and the elimination of TB by 2050 (<1 TB case per million population) (Parish et al. 2001). Hence, an 85% drop in TB cases by 2035 and a 98.5% drop in TB cases by 2050 would be needed for New Zealand to meet international targets. But as the incidence rate of TB in New Zealand has been relatively stable from 2007 to 2014 at between 6 and 8 cases per 100,000 persons, achieving this target will not be straightforward. Reaching targets for the elimination of TB will require major advances in our understanding of the biology of the disease as well as improvements in the effectiveness and accessibility of preventive, diagnostic and treatment interventions.

6.2 History of Tuberculosis in New Zealand

There has been debate stretching back many years with regard to the origin of TB in New Zealand. A number of scholars have advocated that TB was likely brought by Europeans to New Zealand and that there is a lack of evidence that it was present in the indigenous Māori population prior to European colonisation (Dunsford 2008; Brothwell and Sandison 1967; Littleton et al. 2010). For example, Littleton and colleagues stated that 'there is no direct evidence of its effects on Māori people until the arrival of Europeans in the eighteenth century' (Littleton et al. 2010). The earliest recorded case of pulmonary tuberculosis in New Zealand dates from 1808, while there is evidence of tuberculosis on the first voyages of the French and English to New Zealand (Hanham 2003). The dominance of *Mycobacterium tuberculosis* strains belonging to the Euro-American lineage (lineage 4) in the Māori population would appear to correlate with the hypothesis of TB introduction with colonisation as establishment of lineage 4 in New Zealand is unlikely to have preceded the arrival of Europeans (Yen et al. 2013). However, one cannot rule out the possibility that TB existed in New Zealand Māori earlier and that strain displacement took place following introduction of the Euro-American lineage – a scenario that has been suggested for TB in relation to the post Columbian migration of Europeans to the Americas (Gagneux 2012). In line with this, there are also scholars who have proposed that TB was present in New Zealand prior to colonisation. In particular, Te Rangi Hiroa in his medical thesis, published in 1910, stated 'That tuberculosis existed among Māoris before the advent of Europeans, I feel certain' (Te Rangi Hiroa 1910). More investigations and experimental data are needed to verify a precise timeline for the emergence of TB in New Zealand. But what is clear is that Māori were disproportionately affected by TB in the years following colonisation of New Zealand. As noted by Te Rangi Hiroa, in relation to phthisis (tuberculosis):

> The change in environment has allowed phthisis to make great headway. The crowding together in single rooms aids its spread. There is no doubt that the lapse in the sanitary regulations and conditions, with the change in clothing, food and work, brought about by the first contact with the Europeans, made the conditions more favourable for tuberculosis, which was kept in check by the open-air active life of the old-time Māori. The privations and starvation owing to insufficient food cultivated which followed the waikato war helped to

disseminate the seeds of disease in that district until it became very widespread (Te Rangi Hiroa 1910).

The burden of TB in New Zealand increased until reaching a peak in 1943 during the Second World War of 2603 cases corresponding to a national rate of 159.1 cases per 100,000 persons (Das et al. 2006). Māori continued to experience higher rates of TB than their European counterparts. At the height of TB incidence in New Zealand, TB mortality in Māori was 422 per 100,000 as opposed to 39 per 100,000 in non-Māori in 1942 (Grant 2011). The Tuberculosis Act, which was introduced in New Zealand in 1948, sanctioned a number of provisions for the 'treatment, care, and assistance of persons suffering or having suffered from tuberculosis, and for preventing the spread of tuberculosis' (New Zealand Parliament 1948). The measures included the duty of medical practitioners to notify TB cases; powers for medical officers of health to ensure that TB patients obtain treatment; direct precautions to prevent further spread such as patient isolation, vocational training and industrial rehabilitation courses for patients; and a compensation scheme for healthcare workers believed to have contracted the disease in the course of their employment (New Zealand Parliament 1948). From the late 1940s onwards, the TB incidence rate in New Zealand steadily declined until it began to plateau at approximately 10 cases per 100,000 in the 1995–2004 period (Das et al. 2006). The post-Second World War decline in tuberculosis incidence in New Zealand, and in other industrialised countries such as the UK, can be attributed not only to the strengthening and coordination of control measures but also to improvements in socioeconomic development accompanied by better nutrition and living and working conditions (Glaziou et al. 2018).

6.3 Current Epidemiology of Tuberculosis in New Zealand

Today, New Zealand is regarded as a low TB incidence country with 302 notifications including 290 new cases and 12 relapse/reactivation cases in 2014 (incidence rate of 6.7 cases per 100,000 population) (Institute of Environmental Science and Research Ltd. (ESR) 2015). However, the distribution of the disease is not uniform across the population with some at-risk groups exhibiting disparate TB rates. For example, the notification rate of TB in the Asian ethnic group was 56.8 times higher (34.1 per 100,000) than in the European or other ethnic groups (0.6 per 100,000) in 2014 (Institute of Environmental Science and Research Ltd. (ESR) 2015). Furthermore, higher incidence rates are reported for Māori (5.3 per 100,000) and Pacific peoples (16.9 per 100,000) than for European or others.

Geographically, disparities also exist in the distribution of TB. In 2014, the highest rates were reported in the District Health Board (DHB) regions of Auckland (14.6 per 100,000, 69 cases) followed by Capital and Coast (11.8 per 100,000, 35 cases) and Counties Manukau (9.4 per 100,000, 48 cases), with lowest rates in Waitemata DHB region (6.4 per 100,000, 36 cases) (Institute of Environmental

Science and Research Ltd. (ESR) 2015). Intermittent outbreaks of TB can significantly impact the regional rates from year to year.

New Zealand TB notifications reported to the national notifiable diseases database (EpiSurv) by the country's four reference mycobacteriology laboratories located in Auckland, Hamilton, Wellington and Christchurch have high levels of data completeness. This includes data on patient risk factors for the development of TB. The most commonly reported TB risk factors have been birth outside New Zealand (76.6% of cases with risk factor data) and current or recent residence with person(s) born outside New Zealand (76.3%) (Bissielo et al Institute of Environmental Science and Research Ltd.; Institute of Environmental Science and Research Ltd. (ESR) 2015). These two risk factors have dominated TB cases in New Zealand over several years. Notification data from 2005 to 2009 showed that 70.1% of new TB cases occurred in persons born overseas (Ministry of Health New Zealand 2010). Patient age is also associated with TB with highest rates observed in the 15–39 and $\geq$ 60 age groups. The latter $\geq$60 age group experiences the highest rate of hospitalisations (43/58, 74.1% of cases). Furthermore, all four recorded fatalities from TB in 2014 occurred in the $\geq$60 age group (Institute of Environmental Science and Research Ltd. (ESR) 2015).

Among New Zealand-born individuals, Māori make up 15% of New Zealand's population but 51.5% of its TB cases, compared to 74% of the population being of European descent and 16.2% of TB cases recorded in the European or other ethnic groups (Institute of Environmental Science and Research Ltd. (ESR) 2015; Stats NZ 2013). This is a common trend seen in other jurisdictions where significant disparity exists between the indigenous and nonindigenous populations with respect to the burden of TB. Among indigenous populations in the USA, a 5.4-fold higher rate of TB has been reported in American Indians and Alaskan Natives compared to non-Hispanic Caucasians (Tollefson et al. 2013). The TB incidence rate in the Aboriginal population has been recorded at 5.3- and 12.9-fold higher in Australia and Canada, respectively, relative to the domestic-born non-Aboriginal population (Barry et al. 2012; Public Health Agency of Canada 2012).

With regard to drug resistance, there were three (1.2%) cases of multidrug-resistant (MDR) TB, defined as resistance to at least isoniazid and rifampicin, and an additional two cases that were mono-resistant to rifampicin, in New Zealand in 2014 (Institute of Environmental Science and Research Ltd. (ESR) 2015). One of the MDR-TB cases is believed to have developed during antituberculous drug treatment undertaken in New Zealand (Institute of Environmental Science and Research Ltd. (ESR) 2015). Although this may appear to be a relatively small number of cases, MDR-TB is more difficult and costly to treat. Globally, the treatment completion success rate drops from 83% overall for TB to 54% for MDR-TB cases due to treatment failure, loss to follow-up and death of patients (World Health Organization 2017). In addition, the estimated costs for treating TB (USD $17,000 in the USA, €10,282 in 15 European Union (EU) countries, per drug-susceptible TB case) increase substantially with multidrug-resistant forms of TB (USD $134,000 in the USA, €57,213 in 15 EU countries, per case) (Diel et al. 2014; Bhakta et al. 2004).

Therefore, only a small number of MDR-TB cases are required to place an additional financial strain on healthcare resources.

Regarding site of infection, of the 290 new TB cases notified in New Zealand in 2014, 179 cases (61.7%) had pulmonary disease, including 55 cases (19.0%) which also had extrapulmonary involvement. An additional 111 cases (38.3%) were reported as being exclusively extrapulmonary TB (Institute of Environmental Science and Research Ltd. (ESR) 2015). The most common extrapulmonary TB site of infection recorded was a lymph node (excluding abdominal), followed by pleural and intra-abdominal sites (excluding renal) (Institute of Environmental Science and Research Ltd. (ESR) 2015). There are detectable differences in the clinical characteristics of cases born in New Zealand with respect to overseas-born cases. Pulmonary disease was reported in approximately 75% of New Zealand-born TB cases between 2010 and 2014. In contrast, the level of pulmonary disease was lower, approximately 53%–57.5% between 2010 and 2014, in new TB cases in patients who were born outside New Zealand (Institute of Environmental Science and Research Ltd. (ESR) 2015).

6.4 Molecular Epidemiology of Tuberculosis in New Zealand

To better understand the biology of TB in New Zealand, it is necessary to examine the genetic make-up of the strains circulating in the population. One of the first nationwide studies on the molecular typing of *M. tuberculosis* complex in New Zealand was published in 2008 (Sexton et al. 2007). Here, clinical isolates from 2003 to 2007 were typed using the previously described mycobacterial interspersed repetitive unit-variable number tandem repeat (MIRU-VNTR) typing method (Supply et al. 2006). Of 1736 notifications in New Zealand for TB disease caused by species *M. tuberculosis* during the above time frame, 1328 cases were matched to available MIRU-VNTR data, 94.4% of which were new cases and 5.6% were due to relapse or reactivation (Sexton et al. 2007). A total of 866 cases (65.2%) had a unique MIRU-VNTR genotype and, therefore, were classified as non-cluster cases. On the other hand, 462 cases (34.8%) had a non-unique MIRU-VNTR genotype, which was suggestive of potential clustering of these TB cases (Sexton et al. 2007). 140 different MIRU-VNTR genotypes made up the clusters. The so-called Rangipo genotype was associated with the largest cluster size of 57 cases and with the most number of DHBs, i.e. 13 out of 21 then DHBs compared to other clusters which were associated with a median of 2 DHBs (Sexton et al. 2007). This genotype had previously been linked to a large outbreak of TB involving 61 cases which occurred in New Zealand between November 1996 and May 2000 (De Zoysa et al. 2001). A subsequent outbreak due to the Rangipo genotype occurred in Hawke's Bay in 2002 and entailed 19 active cases (McElnay et al. 2004). Next-generation sequencing studies are beginning to shed light on factors that may contribute to the relatively high prevalence of the Rangipo strain among cluster and outbreak cases (Colangeli et al. 2014; Gautam et al. 2017; Mac Aogain et al. 2016).

The cluster cases from 2003 to 2007 were more likely to be less than 20 years of age, to be of Māori or Pacific ethnicity and to reside in the Northland, Counties Manukau, Taranaki, Hawke's Bay, Whanganui or Hutt Valley DHB regions (Sexton et al. 2007). With four clusters involving patient numbers >20 cases each, this was strongly indicative of transmission among TB contacts. Indeed, one of the main TB risk factors associated with cluster cases was contact with a confirmed case of TB, followed by birth in New Zealand or a Pacific Island country, and the pulmonary form of TB. Non-cluster cases in contrast were more likely to be aged between 30 and 39 years or > 70 years, to be of Asian ethnicity and to reside in the Auckland or Nelson Marlborough DHBs. This indicates that a distinct set of contributory factors exist for TB clusters in New Zealand compared with unique cases.

It is important to note that *M. tuberculosis* is not a single organism but instead is a complex of seven distinct phylogenetic lineages to date based on genomic polymorphisms (Brites and Gagneux 2017; Comas et al. 2013). A number of studies have found that specific lineages within the *M. tuberculosis* complex (MTBC) differ with respect to geographic distribution, transmissibility, site of infection, virulence, immunogenicity and prevalence of drug resistance (Reed et al. 2009; Gagneux and Small 2007; Gagneux et al. 2006; Click et al. 2012; Nahid et al. 2010; Lan et al. 2003; Parwati et al. 2010; Borrell and Gagneux 2009; Niemann et al. 2010; National Institute for Public Health and the Environment 2006). A study by Yen and colleagues examined the relative abundance of the major *M. tuberculosis* lineages in New Zealand (Yen et al. 2013). They found that the predominant lineage in New Zealand was lineage 4 (Euro-American) constituting 37.8% of isolates in 2010–2011, followed by lineage 1 (Indo-Oceanic, 22.6%), lineage 2 (East Asian, 19.5%) and lineage 3 (East African Indian, 17.7%) (Yen et al. 2013). Lineage 2 was associated with multidrug resistance in New Zealand, in agreement with studies in other jurisdictions (National Institute for Public Health and the Environment 2006; Niemann et al. 2010), and accounted for 58.1% of MDR-TB cases between 2002 and 2011.

The investigators also compared the epidemiology of the MTBC lineages to determine whether any patterns emerged with respect to the country of origin and ethnicity of TB patients. Here, they found that in the New Zealand-born population, lineage 4 (Euro-American) was the dominant lineage among both the NZ European and Māori populations (82.6% and 64.4%, respectively) (Yen et al. 2013). Lineage 3 was significantly more prevalent in Māori than in NZ Europeans with no cases in the latter group in 2010 and 2011. The reasons behind this observation are not currently known.

Among migrant TB cases, the predominant lineages corresponded to MTBC lineages that are prevalent in their respective countries of origin (Yen et al. 2013). The significance of this finding is discussed in the next section. No significant relationships with respect to patient variables age or gender were detected with regard to MTBC lineages 1–4 over the years of the study; however, males had a higher association with *M. bovis* infection than females for New Zealand isolates from 2006 to 2010 which may reflect differences in relation to exposure risk during the period analysed (Yen et al. 2013).

6.5 Measures to Reduce the Incidence of Tuberculosis in New Zealand

In New Zealand, >70% of cases occur in overseas-born individuals. A trend that has been seen in many low TB burden countries is decreasing TB in the native-born population and increasing TB in the migrant population as a proportion of total cases (Hanway et al. 2016). Therefore, reducing the incidence of TB in the migrant population is key to bringing down the national rate of TB in New Zealand. Internationally, it is believed that most TB in the foreign-born population in industrialised countries occurs due to reactivation of latent TB infection (LTBI) rather than manifesting as a continuation of an existing case of active TB (White and Houben 2014; Ricks et al. 2011). This includes New Zealand's nearest neighbour, Australia, whose National Tuberculosis Advisory Committee (NTAC) recently stated that 'In 2013, 88% of 1322 notifications in Australia were in the overseas-born population (incidence 19.5 per 100,000 v. 1.0 per 100,000), with this proportion rising over the course of the last decade. Combined with epidemiological evidence of low local transmission, this strongly implies that the vast majority resulted from reactivation of latent infection acquired prior to immigration' (Stock and National Tuberculosis Advisory Committee (NTAC) 2017). Molecular genotyping studies conducted in a number of countries, including New Zealand, have identified a strong association between the genetic lineage of the MTBC strain causing TB in a migrant patient and the predominant MTBC lineage in the patient's country of origin (Yen et al. 2013; Filliol et al. 2006; Reed et al. 2009). Therefore, for a large proportion of overseas-born cases, initial infection with *M. tuberculosis* occurs prior to entry into New Zealand (O'Toole and Freeman 2013).

In New Zealand, confirmation of TB status in pre-immigration screening is based on a chest x-ray (New Zealand Immigration 2018b). Citizens of countries that are not on New Zealand Immigration's list of low TB incidence countries are required to undergo a chest x-ray if applying for a visa to stay in New Zealand for longer than 6 months (New Zealand Immigration 2018a). However, the chest x-ray is primarily intended to detect active pulmonary TB and has limited sensitivity in the detection of LTBI. Therefore, many cases of LTBI in migrants will be undetected with the currently used TB diagnostic test in pre-immigration screening. This problem is not unique to New Zealand, but as recently stated by the Australian NTAC, 'Given the very low rates of transmission within Australia, it is clear that progressing toward TB elimination is largely contingent on the implementation of strategies to detect and treat LTBI in migrants from high incidence countries' (Stock and National Tuberculosis Advisory Committee (NTAC) 2017).

One of the limitations with screening for LTBI has been the cross-reaction between the tuberculin skin test (TST) and previous vaccination with BCG, leading to a high proportion of migrants from high TB burden countries testing positive in the TST (Wells et al. 1997; Orlando et al. 2010). However, newer diagnostic tests based on the interferon-gamma release assay (IGRA) target non-BCG antigens, ESAT-6 and CFP-10, and, therefore, deliver greater specificity for latent TB (Lalvani et al. 2001; Lalvani 2003). While screening all migrants for LTBI is not economical, one study of 213 immigrants found that screening of migrants from countries with a TB incidence rate of greater than 150/100,000 with the QuantiFERON-TB Gold In-tube IGRA test detected 49–71% of LTBI cases with an incremental cost-effectiveness ratio (ICER) of GBP £31,867.10 per case averted (Pareek et al. 2013). Another study of 1229 immigrants reported that screening of immigrants (from countries with a > 150/100,000 TB incidence rate) specifically for LTBI with IGRA testing identified 92% of infected immigrants and prevented 29 additional TB cases at an ICER of GBP £20,819 per additional case averted (Pareek et al. 2011). The investigators concluded that 'Screening for latent infection can be implemented cost-effectively at a level of incidence that identifies most immigrants with latent tuberculosis, thereby preventing substantial numbers of future cases of active tuberculosis' (Pareek et al. 2011).

While further consideration will need to be devoted to the prevention and management of TB in the migrant population in New Zealand, it is important to note that a number of studies have established that immigrants are not a major source of TB infection for the native-born population (Barniol et al. 2009; Kamper-Jørgensen et al. 2012; Gautam et al. 2018). In Europe, a systematic review concluded that 'TB in a foreign-born population does not have a significant influence on TB in the native population in EU/EEA' (Sandgren et al. 2014). Therefore, it is important to maintain the focus on New Zealand's obligations as a 'global citizen' to assist in reducing the burden of TB internationally and in helping migrants to New Zealand from high TB incidence countries by minimising their risk of developing TB.

In terms of addressing TB in the New Zealand-born population, as noted above, Māori experience an approximate eightfold higher incidence rate of TB compared to the European or other groups and make up more than 50% of TB cases among New Zealand-born individuals (Institute of Environmental Science and Research Ltd. (ESR) 2015). Recent initiatives to overcome this disparity in TB rates include the announcement in 2018 by the Health Research Council of New Zealand of funding for a feasibility study to characterise the reservoir of LTBI among Māori and enable strategic intervention with preventive treatment to be considered (Health Research Council of New Zealand 2018).

6.6 Conclusions

New Zealand is regarded as a low TB incidence country with approximately 300 cases each year. Nevertheless, particular populations in New Zealand experience disproportionately high levels of TB. Furthermore, the country's national incidence rate must be reduced substantially in order to meet TB eradication targets for 2035 and 2050 that have been set by the World Health Organization. As annual TB rates among New Zealanders of European descent have dropped below 1 case per 100,000, reducing the burden of TB in New Zealand will require investment of healthcare resources towards groups who are at a higher risk of developing the illness. Measures to improve the detection of active TB may reduce diagnostic and treatment delays and hence minimise transmission and epidemiological clusters of

TB. In addition, where appropriate, further screening and preventive therapy for latent TB infection may assist in decreasing the numbers of reactivated cases.

Competing Interests The author declares that he has no financial or other conflicts of interest.

Ethical Approval Not required.

Funding Not applicable.

References

Barniol J, Niemann S, Louis VR, Brodhun B, Dreweck C, Richter E, Becher H, Haas W, Junghanss T (2009) Transmission dynamics of pulmonary tuberculosis between autochthonous and immigrant sub-populations. BMC Infect Dis 9(1):197. https://doi.org/10.1186/1471-2334-9-197

Barry C, Waring J, Stapledon R, Konstantinos A (2012) Tuberculosis notifications in Australia, 2008 and 2009. Commun Dis Intell Q Rep 36(1):82–94

Bhakta S, Besra GS, Upton AM, Parish T, Sholto-Douglas-Vernon C, Gibson KJ, Knutton S, Gordon S, DaSilva RP, Anderton MC, Sim E (2004) Arylamine N-acetyltransferase is required for synthesis of mycolic acids and complex lipids in Mycobacterium bovis BCG and represents a novel drug target. J Exp Med 199(9):1191–1199. https://doi.org/10.1084/jem.20031956. jem.20031956 [pii]

Bissielo A, Lim E, Heffernan H (Institute of Environmental Science and Research Ltd) Tuberculosis in New Zealand: annual report 2010

Borrell S, Gagneux S (2009) Infectiousness, reproductive fitness and evolution of drug-resistant Mycobacterium tuberculosis. Int J Tuberc Lung Dis 13(12):1456–1466

Brites D, Gagneux S (2017) The nature and evolution of genomic diversity in the Mycobacterium tuberculosis complex. Adv Exp Med Biol 1019:1–26. https://doi.org/10.1007/978-3-319-64371-7_1

Brothwell DR, Sandison AT (1967) Diseases in antiquity: a survey of the diseases, injuries, and surgery of early populations. Charles C Thomas Pub Ltd, Springfield

Click ES, Moonan PK, Winston CA, Cowan LS, Oeltmann JE (2012) Relationship between Mycobacterium tuberculosis phylogenetic lineage and clinical site of tuberculosis. Clin Infect Dis 54(2):211–219. https://doi.org/10.1093/cid/cir788

Colangeli R, Arcus VL, Cursons RT, Ruthe A, Karalus N, Coley K, Manning SD, Kim S, Marchiano E, Alland D (2014) Whole genome sequencing of *Mycobacterium tuberculosis* reveals slow growth and low mutation rates during latent infections in humans. PLoS One 9(3):e91024. https://doi.org/10.1371/journal.pone.0091024

Comas I, Coscolla M, Luo T, Borrell S, Holt KE, Kato-Maeda M, Parkhill J, Malla B, Berg S, Thwaites G, Yeboah-Manu D, Bothamley G, Mei J, Wei L, Bentley S, Harris SR, Niemann S, Diel R, Aseffa A, Gao Q, Young D, Gagneux S (2013) Out-of-Africa migration and Neolithic coexpansion of Mycobacterium tuberculosis with modern humans. Nat Genet 45 (10):1176–1182. https://doi.org/10.1038/ng.2744

Das D, Baker M, Calder L (2006) Tuberculosis epidemiology in New Zealand: 1995-2004. N Z Med J 119(1243):U2249

De Zoysa R, Shoemack R, Vaughan R, Vaughan A (2001) A prolonged outbreak of tuberculosis in the North Island. N Z Pub Health Rep 8

Diel R, Vandeputte J, de Vries G, Stillo J, Wanlin M, Nienhaus A (2014) Costs of tuberculosis disease in the European Union: a systematic analysis and cost calculation. Eur Respir J 43 (2):554–565. https://doi.org/10.1183/09031936.00079413

Dunsford DA (2008) Seeking the prize of eradication: a social history of tuberculosis in New Zealand from World War Two to the 1970s. University of Auckland

Filliol I, Motiwala AS, Cavatore M, Qi W, Hazbon MH, Bobadilla del Valle M, Fyfe J, Garcia-Garcia L, Rastogi N, Sola C, Zozio T, Guerrero MI, Leon CI, Crabtree J, Angiuoli S, Eisenach KD, Durmaz R, Joloba ML, Rendon A, Sifuentes-Osornio J, Ponce de Leon A, Cave MD, Fleischmann R, Whittam TS, Alland D (2006) Global phylogeny of Mycobacterium tuberculosis based on single nucleotide polymorphism (SNP) analysis: insights into tuberculosis evolution, phylogenetic accuracy of other DNA fingerprinting systems, and recommendations for a minimal standard SNP set. J Bacteriol 188(2):759–772. https://doi.org/10.1128/JB.188.2.759-772.2006. 188/2/759 [pii]

Gagneux S (2012) Host-pathogen coevolution in human tuberculosis. Philos Trans R Soc Lond Ser B Biol Sci 367(1590):850–859. https://doi.org/10.1098/rstb.2011.0316

Gagneux S, Small PM (2007) Global phylogeography of Mycobacterium tuberculosis and implications for tuberculosis product development. Lancet Infect Dis 7(5):328–337. https://doi.org/10.1016/S1473-3099(07)70108-1

Gagneux S, DeRiemer K, Van T, Kato-Maeda M, de Jong BC, Narayanan S, Nicol M, Niemann S, Kremer K, Gutierrez MC, Hilty M, Hopewell PC, Small PM (2006) Variable host-pathogen compatibility in Mycobacterium tuberculosis. Proc Natl Acad Sci U S A 103(8):2869–2873. https://doi.org/10.1073/pnas.0511240103

Gautam SS, Mac Aogain M, Bower JE, Basu I, O'Toole RF (2017) Differential carriage of virulence-associated loci in the New Zealand Rangipo outbreak strain of Mycobacterium tuberculosis. Infect Dis (Lond) 49(9):680–688. https://doi.org/10.1080/23744235.2017.1330553

Gautam SS, Mac Aogain M, Cooley LA, Haug G, Fyfe JA, Globan M, O'Toole RF (2018) Molecular epidemiology of tuberculosis in Tasmania and genomic characterisation of its first known multi-drug resistant case. PLoS One 13(2):e0192351. https://doi.org/10.1371/journal.pone.0192351

Glaziou P, Floyd K, Raviglione M (2018) Trends in tuberculosis in the UK. Thorax 73(8):702–703. https://doi.org/10.1136/thoraxjnl-2018-211537

Grant JM (2011) The determinants of tuberculosis transmission in indigenous people in Canada and New Zealand. University of Saskatchewan

Hanham DE (2003) The impact of introduced diseases in the pre-treaty period 1790–1850. University of Canterbury

Hanway A, Comiskey CM, Tobin K, O'Toole RF (2016) Relating annual migration from high tuberculosis burden country of origin to changes in foreign-born tuberculosis notification rates in low-medium incidence European countries. Tuberculosis 101:67–74. https://doi.org/10.1016/j.tube.2016.07.019

Health Research Council of New Zealand (2018) Feasibility study grant: towards elimination of tuberculosis in Māori through preventative treatment

Institute of Environmental Science and Research Ltd (ESR) (2015) Tuberculosis in New Zealand: annual report 2014

Kamper-Jørgensen Z, Andersen AB, Kok-Jensen A, Kamper-Jørgensen M, Bygbjerg IC, Andersen PH, Thomsen VØ, Lillebaek T (2012) Migrant tuberculosis: the extent of transmission in a low burden country. BMC Infect Dis 12(1):60. https://doi.org/10.1186/1471-2334-12-60

Lalvani A (2003) Spotting latent infection: the path to better tuberculosis control. Thorax 58 (11):916–918

Lalvani A, Pathan AA, McShane H, Wilkinson RJ, Latif M, Conlon CP, Pasvol G, Hill AV (2001) Rapid detection of Mycobacterium tuberculosis infection by enumeration of antigen-specific T cells. Am J Respir Crit Care Med 163(4):824–828. https://doi.org/10.1164/ajrccm.163.4.2009100

Lan NT, Lien HT, Tung le B, Borgdorff MW, Kremer K, van Soolingen D (2003) Mycobacterium tuberculosis Beijing genotype and risk for treatment failure and relapse, Vietnam. Emerg Infect Dis 9(12):1633–1635

Littleton J, Park J, Bryder L (2010) The end of a plague? Tuberculosis in New Zealand. In: Herring DA, Swedlund AC (eds) Plagues and epidemics: infected spaces past and present. Berg, New York

Mac Aogain M, Gautam S, Bower JE, Basu I, O'Toole RF (2016) Draft genome sequence of a New Zealand Rangipo strain of Mycobacterium tuberculosis. Genome Announc 4(4):1–2. https://doi.org/10.1128/genomeA.00657-16

McElnay C, Thornley C, Armstrong R (2004) A community and workplace outbreak of tuberculosis in Hawke's Bay in 2002. N Z Med J 117(1200):U1019

Ministry of Health New Zealand (2010) Guidelines for tuberculosis control in New Zealand

Nahid P, Bliven EE, Kim EY, Mac Kenzie WR, Stout JE, Diem L, Johnson JL, Gagneux S, Hopewell PC, Kato-Maeda M (2010) Influence of M. tuberculosis lineage variability within a clinical trial for pulmonary tuberculosis. PLoS One 5(5):e10753. https://doi.org/10.1371/journal.pone.0010753

National Institute for Public Health and the Environment N (2006) Beijing/W genotype Mycobacterium tuberculosis and drug resistance. Emerg Infect Dis 12(5):736–743. https://doi.org/10.3201/eid1205.050400

New Zealand Immigration (2018a) Countries with a low incidence of tuberculosis

New Zealand Immigration (2018b) When you need an x-ray or medical examination

New Zealand Parliament (1948) Tuberculosis Act

Niemann S, Diel R, Khechinashvili G, Gegia M, Mdivani N, Tang YW (2010) Mycobacterium tuberculosis Beijing lineage favors the spread of multidrug-resistant tuberculosis in the Republic of Georgia. J Clin Microbiol 48(10):3544–3550. https://doi.org/10.1128/JCM.00715-10

Orlando G, Merli S, Cordier L, Mazza F, Casazza G, Villa AM, Codecasa L, Negri E, Cargnel A, Ferrarese M, Rizzardini G (2010) Interferon-gamma releasing assay versus tuberculin skin testing for latent tuberculosis infection in targeted screening programs for high risk immigrants. Infection 38(3):195–204. https://doi.org/10.1007/s15010-010-0015-2

O'Toole RF, Freeman JT (2013) Molecular epidemiology of human tuberculosis in New Zealand and potential insights for disease control. N Z Med J 126(1373):1–3

Pareek M, Watson JP, Ormerod LP, Kon OM, Woltmann G, White PJ, Abubakar I, Lalvani A (2011) Screening of immigrants in the UK for imported latent tuberculosis: a multicentre cohort study and cost-effectiveness analysis. Lancet Infect Dis 11(6):435–444. https://doi.org/10.1016/S1473-3099(11)70069-X

Pareek M, Bond M, Shorey J, Seneviratne S, Guy M, White P, Lalvani A, Kon OM (2013) Community-based evaluation of immigrant tuberculosis screening using interferon gamma release assays and tuberculin skin testing: observational study and economic analysis. Thorax 68(3):230–239. https://doi.org/10.1136/thoraxjnl-2011-201542

Parish T, Turner J, Stoker NG (2001) amiA is a negative regulator of acetamidase expression in Mycobacterium smegmatis. BMC Microbiol 1:19

Parwati I, Alisjahbana B, Apriani L, Soetikno RD, Ottenhoff TH, van der Zanden AG, van der Meer J, van Soolingen D, van Crevel R (2010) Mycobacterium tuberculosis Beijing genotype is an independent risk factor for tuberculosis treatment failure in Indonesia. J Infect Dis 201 (4):553–557. https://doi.org/10.1086/650311

Public Health Agency of Canada (2012) Tuberculosis in Canada 2008

Reed MB, Pichler VK, McIntosh F, Mattia A, Fallow A, Masala S, Domenech P, Zwerling A, Thibert L, Menzies D, Schwartzman K, Behr MA (2009) Major Mycobacterium tuberculosis lineages associate with patient country of origin. J Clin Microbiol 47(4):1119–1128. JCM.02142-08 [pii] 521128/JCM.02142–08

Ricks PM, Cain KP, Oeltmann JE, Kammerer JS, Moonan PK (2011) Estimating the burden of tuberculosis among foreign-born persons acquired prior to entering the U.S., 2005–2009. PLoS One 6(11):e27405. https://doi.org/10.1371/journal.pone.0027405

Sandgren A, Sañé Schepisi M, Sotgiu G, Huitric E, Migliori GB, Manissero D, van der Werf MJ, Girardi E (2014) Tuberculosis transmission between foreign- and native-born populations in the

EU/EEA: a systematic review. Eur Respir J 43(4):1159–1171. https://doi.org/10.1183/09031936.00117213
Sexton K, Perera S, Pandey S (2007) Five years of molecular typing of Mycobacterium tuberculosis isolates in New Zealand, 2003 to 2007. Institute of Environmental Science and Research Limited
Stats NZ (2013) Census
Stock D, National Tuberculosis Advisory Committee (NTAC) (2017) National position statement for the management of latent tuberculosis infection
Supply P, Allix C, Lesjean S, Cardoso-Oelemann M, Rusch-Gerdes S, Willery E, Savine E, de Haas P, van Deutekom H, Roring S, Bifani P, Kurepina N, Kreiswirth B, Sola C, Rastogi N, Vatin V, Gutierrez MC, Fauville M, Niemann S, Skuce R, Kremer K, Locht C, van Soolingen D (2006) Proposal for standardization of optimized mycobacterial interspersed repetitive unit-variable-number tandem repeat typing of Mycobacterium tuberculosis. J Clin Microbiol 44 (12):4498–4510. https://doi.org/10.1128/JCM.01392-06
Te Rangi Hiroa (1910) Medicine amongst the Maoris. In: Ancient and modern times
Tollefson D, Bloss E, Fanning A, Redd JT, Barker K, McCray E (2013) Burden of tuberculosis in indigenous peoples globally: a systematic review [Review article]. Int J Tuberc Lung Dis 17 (9):1139–1150. https://doi.org/10.5588/ijtld.12.0385
Wells CD, Zuber PL, Nolan CM, Binkin NJ, Goldberg SV (1997) Tuberculosis prevention among foreign-born persons in Seattle--King County, Washington. Am J Respir Crit Care Med 156 (2. Pt 1):573–577
White RG, Houben RMGJ (2014) Towards elimination in industrialised countries: expanding diagnosis and treatment of LTBI among immigrants [Editorial]. Int J Tuberc Lung Dis 18 (4):380–380. https://doi.org/10.5588/ijtld.14.0156
World Health Organization (2014) The end TB strategy: global strategy and targets for tuberculosis prevention, care and control after 2015
World Health Organization (2017) Global tuberculosis report
Yen S, Bower JE, Freeman JT, Basu I, O'Toole RF (2013) Phylogenetic lineages of tuberculosis isolates in New Zealand and their association with patient demographics. Int J Tuberc Lung Dis 17(7):892–897. https://doi.org/10.5588/ijtld.12.0795

Part II

Molecular and Infection Biology of *M. tuberculosis*

Mycobacterial Methyltransferases: Significance in Pathogenesis and Virulence

7

Sonam Grover, Rishabh Gangwar, Salma Jamal, Sabeeha Ali, Khairun Nisaa, Nasreen Z. Ehtesham, and Seyed Ehtesham Hasnain

Abstract

Mycobacterium tuberculosis (*M.tb*) is a pathogen of incredible international prominence owing to its persistence for long duration inside human host in both active and latent form, complex eradication methods and imposing long-term treatment procedures. The mechanisms employed by *M.tb* to adjust and survive inside extreme host environment and to evade the immune system of host need to be explored in greater depth in order to enable the rational design of novel treatment strategies. Methylation of biomolecules plays a significant role in almost every kingdom of life but has not been extensively addressed in the case of *M.tb*. The genome of *M.tb* codes for 121 methyltransferases (MTases) in spite of the reductive evolution of its genome. In the present chapter, we will discuss in detail about various MTases modifying DNA, RNA, protein, mycolic acid and

S. Grover · R. Gangwar · S. Jamal
JH Institute of Molecular Medicine, Jamia Hamdard, New Delhi, Delhi, India

S. Ali
Kusuma School of Biological Sciences, Indian Institute of Technology, New Delhi, Delhi, India

K. Nisaa
Department of Life Sciences, Ben Gurion University of the Negev, Beersheba, Israel

N. Z. Ehtesham
Inflammation Biology and Cell Signaling Laboratory, ICMR-National Institute of Pathology, Safdarjung Hospital Campus, New Delhi, Delhi, India
e-mail: nasreenzehtesham.nip@gov.in

S. E. Hasnain (✉)
JH Institute of Molecular Medicine, Jamia Hamdard, New Delhi, Delhi, India

Dr Reddy's Institute of Life Sciences, University of Hyderabad Campus, Hyderabad, India
e-mail: vc@jamiahamdard.ac.in

S. E. Hasnain et al. (eds.), Mycobacterium tuberculosis: *Molecular Infection Biology Pathogenesis, Diagnostics and New Interventions*,
https://doi.org/10.1007/978-981-32-9413-4_7

other biomolecules of *M.tb* along with the host. This will also shed light on how methylation is implicated in virulence and influences the mechanism of pathogenesis of *M.tb*.

Keywords
Methyltransferases · Mycolic acid · Epigenetic regulation · Drug resistance

7.1 Introduction

Mycobacterium tuberculosis (*M.tb*), responsible for causing tuberculosis (TB), is a micro-organism that is posing greatest threats to humanity. According to the World Health Organization (WHO) report, around one-third of the population is exposed worldwide, and the 5% amongst the exposed ultimately develop disease that translates into eight million new TB cases per year (WHO 2017). The process of *M.tb* infection can be divided into a number of stages (Cooper 2009). The disease starts when the microbes are breathed out by infected individual which are further transferred by air droplets. These bacteria are then inhaled by healthy individuals which is the primary stage of infection during which the bacteria inhabit macrophages. Macrophages are large phagocytes that serve at the forefront of host immune defence against pathogenic microbes. These are also the primary sites of *M. tb* habitation contrasting other bacteria that need to avoid the process of phagocytosis for survival (Warner and Mizrahi 2007). *M.tb* preferentially targets macrophage vacuoles where they are constantly exposed to the macrophage's antimicrobial effectors that include reactive oxygen plus nitrogen species as well as low pH that could compromise the primary infection, latency and reinfection or kill *M.tb* (Gorna Alina et al. 2010). Hence, in order to maintain its viability and virulence, it is compulsory for *M.tb* to have effective systems to counteract the impact of such assaults generated by the environment and the host (Mizrahi and Andersen 1998; Warner and Mizrahi 2007; Dos Vultos et al. 2009; Kurthkoti and Varshney 2012; Voskuil et al. 2011).

Biology relies upon the enlarged repertoire of interactions that occur when biomolecules such as proteins, DNA or RNA are modified. The process of methylation of nucleic acid and protein expands the range of functions presented to their modified substrates and plays different roles in cellular signalling and regulation of macromolecular function. From current scientific research, it has become evident that methylation stands beside phosphorylation as a major regulating factor in the functioning of proteins. Methylation of DNA causes epigenetic changes in the genome, thus affecting gene expression, replication and chromatin remodelling. In the similar way, RNA methylation provides integrity and structural stability to ribosomes at the time of translation. These modifications increase multitudes of chemical interactions that play roles in variety of cellular mechanisms (Clarke and Tamanoi 2006). Thus, methylation controls numerous cellular processes of the organism according to the metabolic status.

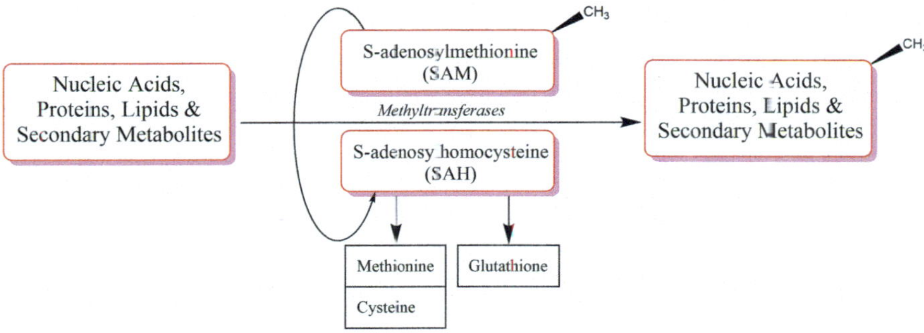

Fig. 7.1 Overview of methyltransferase reaction and fate of S-adenosylhomocysteine (SAH)

The process of methylation is catalysed by methyltransferases, during which a methyl group is transferred from S-adenosylmethionine (SAM, a methionine derivative) to nucleic acids, proteins, lipids and secondary metabolites. In methylation reaction, SAM is altered resulting in the formation of S-adenosylhomocysteine (SAH), which can then be broken down to methionine and cysteine which are sulphur-containing amino acids or can be further metabolized to the antioxidant glutathione (Fig. 7.1). Thus, methylation directly feeds into sulphur metabolism and oxidative balance. More than 1% of genes have been predicted to encode methyltransferases in the mammalian genome (Katz 2003), while in the case of *M. tb*, this number has increased to 3% in spite of genome reduction of *M.tb*. The large number of MTases is able to draw the attention of scientists to explore their role in pathogenicity and virulence.

7.2 Types of Methyltransferase

In *M.tb* proteome, about 121 methyltransferases (MTases) have been identified (Grover et al. 2016), which employ various substrates that include DNA, RNA, intermediates of mycolic acid biosynthesis, protein and other fatty acids. These MTases can be further divided into five main categories in case of *M.tb*: mycolic acid MTases, DNA MTases, RNA MTases, protein MTases and other MTases.

7.2.1 Mycolic Acid Methyltransferases

Mycolic acids (MAs) which are the 2-alkyl, 3-hydroxy long-chain fatty acids (FAs) form a distinctive feature of the thick lipid-rich membrane of *M.tb* and other related mycobacterium (Daffé and Draper 1998; Barry et al. 1998). MAs are very-long-chain (C_{60}–C_{90}) fatty acids, having an alkyl side chain at the α-position and a hydroxyl group at the β-position (Fig. 7.2a), either unbound or bound to the terminal penta-arabinofuranosyl units of arabinogalactan (AG) that, along with

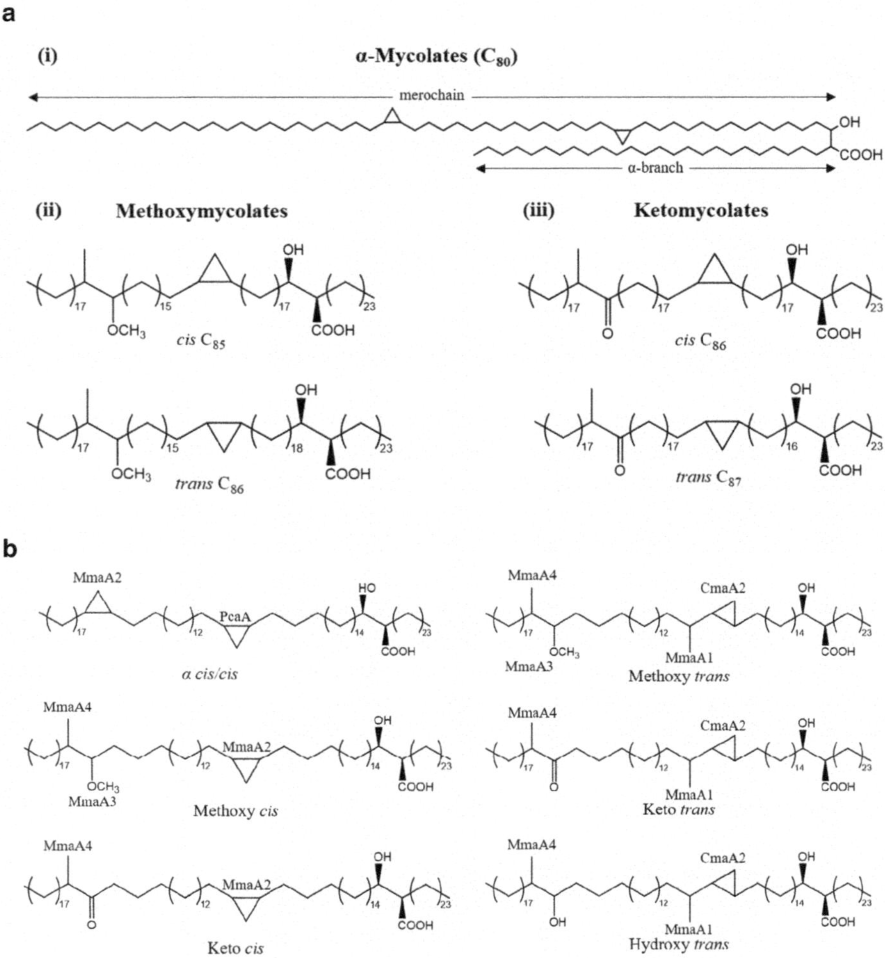

Fig. 7.2 **(a)** (i) Fundamental structure of a mycolic acid shown using *M.tb* α-mycolic acid as an example; (ii) structure of methoxy mycolic acid subclass from *M.tb*; (iii) structure of keto mycolic acid subclass from *M.tb* (Nataraj et al. 2015). **(b) Modifications of the mycolic acid mero-chains and involvement of MTases:** structures of the mero-chains of representative types of MAs from mycobacteria. The configuration of double bonds and cyclopropanes is either *cis* or *trans* (with an adjacent methyl branch). The established functions of the MTases are indicated

peptidoglycan, forms the insoluble cell wall skeleton (Daffé and Draper 1998; McNeil et al. 1991). Recent electron microscopy visualization revealed that the unbound and bound forms presumptively play a vital role within the exceptional architecture as well as the impermeable cell envelope, participating within the two leaflets of the outer membrane of the mycobacterium, additionally referred to as mycomembrane (Hoffmann et al. 2008; Sani et al. 2010; Zuber et al. 2008).

The mycomembrane serves as a physiological blockade between the bacteria and its environment and has a significant role in each persistence and virulence. MAs present in mycomembrane are responsible for not only the relative cell wall impermeability, specifically to antibacterial agents and drugs, but also mycobacterial acid fastness (Laval et al. 2008; Bhatt et al. 2007). It has been widely established in slow-growing mycobacteria responsible for pathogenesis that MAs are modified by addition of methyl branches, cyclopropane rings, methoxy groups and ketones to generate a series of three main mycolic acids: α-mycolates (two *cis*-cyclopropane rings), ketomycolates (a single *cis* or *trans* form of cyclopropane ring and a ketone group) and methoxymycolates (a single *cis*- or *trans*-cyclopropane ring and a methoxy group). Structural diversity in MAs is generated by the modification of unsaturated meromycolic acid, catalysed by a family of AdoMet-dependent MTases, to generate *cis* and *trans* forms of cyclopropanes and several other mycolates (Barry et al. 1998). So far, eight distinctive MTases have been characterized, namely, CmaA1 (Rv3392c), CmaA2 (Rv0503c), MmaA1 (Rv0645c), MmaA2 (Rv0644c), MmaA3 (Rv0643c), MmaA4 (Rv0642c), PcaA (formerly UmaA2; Rv0470c) and UmaA (Rv0469), which are responsible for certain modifications of mycolic acid moieties of the cell envelope (Fig. 7.2b).

The growth of *Mycobacterium*, in presence of the labelled methionine, points out that the methyl group of methionine can be directly integrated into MAs, and the bridging methylenes of the methyl branches adjacent to methoxy, *trans*-olefins and keto moieties, the carbon of the methoxy functionality and also the cyclopropane rings are all derivatives of methionine, presumptively via SAM (Daffé et al. 1991; Lacave et al. 1987; Danielson and Gray 1982).

The first direct evidence of the long proposed idea that the meromycolyl groups of cyclopropane were gained from the double bonds (Lederer 1969; Grogan and Cronan 1997) came with the identification of cyclopropane mycolic acid synthase (Cmas1), product of *cma1* gene from *M.tb* that enables *M. smegmatis* to produce cyclopropane-containing MAs in large quantities (Yuan et al. 1995). This gene has shown strong homology within the region known to bind SAM of other SAM-dependent methyltransferases and was also found to have 34% similarity with the cyclopropane fatty acid synthase (Cfas) of *E. coli* (Grogan and Cronan 1997; Wang et al. 1992).

Another gene from *M.tb*, *cma2*, was discovered on the basis of *cma1* sequence. Categorization of the purified chimeric mycolates is obtained as a consequence of heterologous expression of Cmas2, the gene product of *cma2* in *M. smegmatis*, shown to attach a cyclopropane ring at the proximal position (George et al. 1995). Interestingly, Cmas2 has the specificity for *cis* double bonds as it cyclopropanates the α1-mycolates and not the α2 from *M. smegmatis*. Although both *cis* and *trans* forms of cyclopropane-containing mycolates are produced by *M.tb* (Minnikin and Polgar 1967), only *cis*-cyclopropanes are produced by both *cma1* and *cma2*. The existence of *trans*-cyclopropanes suggests the existence of no less than one more enzyme in *M.tb*, capable of introducing such modification in MAs. The *trans*-cyclopropanes always go together with a nearby methyl branch, suggesting the involvement of two counterparts of SAM in their formation; the first forms a *trans*-olefin with a nearby methyl group and the second from the actual cyclopropane ring. Indeed, two additional homologs of *cma1* and *cma2* which might encode such activities have been revealed through the analysis of the complete *M.tb* genomic sequence.

Box 7.1

Bacterial methylation of an antibacterial agent is a unique mechanism of antimicrobial resistance.

Rv0560c of *Mycobacterium tuberculosis is* a putative methyltransferase and it has been shown to N-methylate the pyrido-benzimidazole *in vitro* and in *M.tb,* abrogating its mycobactericidal activity (Warrier et al. 2016). Rv0560, has been characterised as S-adenosyl-*L*-methionine-dependent methyltransferase

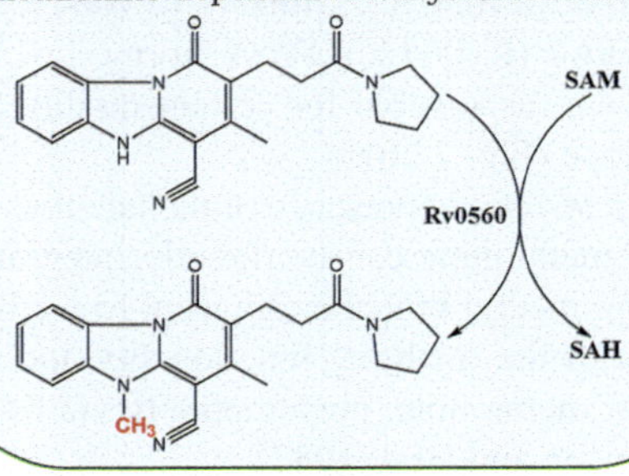

Box 7.2

Various novel MTases in *M.tb* complex were discovered on comparing the proteomes of *M.tb* H37Rv MTases with fifteen different *Mycobacterium* species.

Phylogenetic analysis of these novel MTases exhibited their relationship with extremophiles such as halophilic and acidophilic organisms. These results indicated an evolutionary relationship of *M.tb* with halotolerant organisms as well as the role of MTtases in pathogenesis, virulence and niche adaptation.

Remarkably, a group of four additional genes was identified by DNA hybridization to a *cma1* probe in *M.tb* (Yuan and Barry 1996). In addition to *cma1* and *cma2*, similarity was found amongst all the four genes. By analysing the resultant mycolates following the introduction of these genes, either alone or in combination, into *M. smegmatis*, it was shown that *mma4* gene product adds a hydroxyl group along with a nearby methyl branch at the distal position, whereas the *mma3* encodes an enzyme that methylates this hydroxyl resulting in the formation of a methyl ether. The new mycolate species were just like the ketomycolate and methoxymycolate series from *M.tb* (Yuan and Barry 1996). Interestingly, *mma2* encodes a cyclopropanation enzyme that is like *cma2* which alters the proximal position. The purpose for this obvious duplication of function is uncertain. One theory is that Cmas2 cyclopropanates α-series, whereas Mmas2 cyclopropanates the oxygenated series of mycolates. The heterologous expression of *mma1* seemed to have no consequence in *M. smegmatis*; however, when *mma1* was overexpressed in *M.tb*, a significant increase in the quantities of *trans* substituents has been observed at the proximal position (Yuan et al. 1997). Strain overexpressing *mma1*, showing the presence of *trans* form of cyclopropanes and *trans* double bonds, with nearby methyl branches, has led to the inference that Mmas1 transforms the proximal *cis* double bond to a *trans* double bond with an allylic methyl branch. The endogenous enzyme occurring in limiting quantities in *M.tb* would have catalysed *trans*-cyclopropane formation. A recent study reported these enzymes to be removed by homologous recombination, and a feeble immune response of the host as well as a substantial attenuation of *M.tb* was observed (Barkan et al. 2012).

While looking for the presence of these MTases in other mycobacterium species through BLAST analysis, it was found that UfaA (Rv0447), having cyclopropane-fatty-acyl-phospholipid synthase activity, is limited to true pathogens only and is lacking in all the non-pathogenic and opportunistic mycobacterium species. UfaA was revealed to catalyse tuberculostearic acid (10-methylstearic-acid, TSA) biosynthesis. TSA is a clinical marker of the disease which makes up a major lipid moiety of cell wall of the mycobacteria (Meena and Kolattukudy 2013). The transfer of the methyl group from SAM to the double bond of oleic acid in phosphatidylcholine or phosphatidylethanolamine to produce TSA is catalysed by UfaA (Meena and Kolattukudy 2013). The aforementioned results indicated that a specific kind of methylation is involved in immune response modulation and pathogenicity.

7.2.1.1 Mycolic Acid Methyltransferases' Relation to Virulence

The role of modification of MA in pathogenesis of *M.tb* is only moderately understood. Many of the genes responsible for MA modification are conserved amongst slow-growing pathogenic mycobacteria (such as *M. leprae*, *M. marinum* and other members of the *M.tb* complex) which indicates towards the role of these genes in pathogenesis. Overall, expression of these enzymes resulted to the synthesis of various subfamilies for which the studies delineated below have found remarkable and exact roles of these enzymes despite of very subtle differences in their structures. The studies on MAMTs which were lacking mutants in mice provided proof for the significance of modifications of MA in pathogenesis. It was demonstrated by

Dubnau et al. that inactivation of *hma* (known as *cmaA* also and recently as *mmaA4*) gene in *M.tb* leads to an extreme modification in the permeability of its envelope together with the loss of oxygenated MAs. When this mutant was tested in an infected mouse model by aerosolization, it exhibited an attenuated phenotype during the first 3 weeks of infection, indicating that the oxygenated MAs are essential in the infection process (Dubnau et al. 2000). Importantly, this study pointed towards the precise roles of each MAMTs in the biogenesis of MAs and their part in virulence. In the continuation of this line of research, Glickman et al. revealed the loss of cording by the deletion of *pcaA* gene, which encodes for an enzyme catalysing the proximal cyclopropanation of α-mycolates. This Δ*pcaA M.tb* strain (Glickman et al. 2000) has several phenotypes: defective persistence, defective growth in the first week of mouse infection (Rao et al. 2005) and unable to kill infected mice (Glickman et al. 2000). These observations indicated that the site-specific cyclopropanation of MAs could be an imperative determining factor of the *M.tb* and host interaction. Glickman et al. have demonstrated that an *M.tb* Δ*cmaA2* strain lacked the *trans* ring on the methoxy- and ketomycolates (Glickman et al. 2001). This mutant was also found to be overvirulent in mice, prompting an exaggerated immune response which leads to extreme tissue damage (Rao et al. 2006). Reconstituted methoxymycolate production was shown by *mmaA3*-complemented *M. bovis* BCG SSI as compared to wild type, which lacks functional MmaA3 protein. This mutant was also found to be attenuated (Belley et al. 2004).

The investigation of the vitality of each MAMT and deletion of the seven MAMTs existing in *M.tb* exhibited that none of them is as such essential and that all of them may be deleted in spite of the fact that a certain sequence in the deletion must be followed to permit cellular adjustment of permeability of the cell and consequently survival. The viable strain generated through this technically challenging sequential deletion of MAMTs was incapable of making cyclopropanated or oxygenated MAs, which resulted in massive changes in cell appearance such as susceptibility to detergents and total loss of acid fastness. Interestingly, it has been also shown that the total absence of cyclopropanation leads to severe reduction during the first week, whereas complete deletion of MAMTs confers attenuation in the second week of infection after aerosol infection in the mouse. This strain also generated a significant hyperinflammatory response from the host (Barkan et al. 2012). Thus, there are substantial evidences of immunomodulatory role of the MA modification, and MAMTs have many explicit functions in *M.tb* pathogenesis, which reinforces this enzyme class as an interesting target for drug development against mycobacteria.

7.2.2 DNA Methyltransferases

An interesting point to discuss is how *M.tb* intelligently sustains its gene expression profile. *M.tb* genome is rich in GC content (~ 65%). In the case of eukaryotes, a number of mechanisms are explored that heritably regulate expression patterns, but prokaryotes have DNA methylation as the only known mechanism to attain epigenetic inheritance. DNA could be methylated at adenine and cytosine residues, forming N^6-methyladenine as well as N^4-methylcytosine and 5-methylcytosine. In higher eukaryotes, cytosine methylation is validated as a significant regulatory mechanism of gene expression, but recent reports have proposed that methylation of cytosine also plays pivotal roles in prokaryotic gene expression (Militello et al. 2012; Kahramanoglou et al. 2012), although in prokaryotes N6-methylation of adenine is a well-established mechanism of gene expression via epigenetic regulation (Casadesus and Low 2006; Campbell and Kleckner 1990).

DNA MTases act by transferring the methyl group from SAM to specific residues in double-stranded DNA. Prokaryotic methylation can be regulated either as a part of host modification or in an independent manner such as by deoxyadenosine methylase (Dam) or deoxycytosine MTase (Dcm). In Dam-dependent methylation, GATC locus is recognized, and methylation of adenine occurs at N6 position (Marinus 1987), while in Dcm-dependent methylation, CCA/TGC locus is recognized and methylates the internal cytosine at the C5 position (Schlagman et al. 1976).

Total absence of both Dam and Dcm MTase activities in different strains of *M.tb* including H37Ra, H37Rv and *M. smegmatis* was reported (Hemavathy and Nagaraja 1995). However, analysis of genomic DNA from *M.tb* and *M. smegmatis* was shown to contain significant amount of 6-methyladenine and 5-methylcytosine indicating the presence of DNA MTase other than Dam or Dcm. Since then, a variety of MTases have been reported in *M.tb* (Srivastava et al. 1981). So far in *M.tb*, there are 121 characterized and hypothetical MTases reported in its genome (Grover et al. 2016). Out of 61 characterized MTases, 5 are characterized to be DNA MTases.

7.2.2.1 DNA Methylation Involved in Damage-Induced Repair

M.tb is an intracellular pathogen that persists and survives inside its host, i.e. human macrophages. Consequently, it is exposed to many unfavourable surroundings such as alkylating stress and exposure to oxygen as well as nitrogen radicals which can cause harmful effects comprising DNA damage. Cells have adopted numerous DNA repair pathways to counter the mutagenic/cytotoxic effect of methylation on DNA. A study by Yang et al. demonstrated two co-operonic DNA MTases in *ada* operon of *M.tb* which are involved in inducible repair of DNA alkylation damage. O^6-Methylguanine is one of the major mutagenic lesions removed by MTases under *ada* operon, a very well-characterized mechanism in *E. coli*. The *M.tb ada* operon encodes a fused protein made up of AdaA and AlkA (Rv1317c) in addition to a separate AdaB/OGT (Rv1316c) protein, both of which possess DNA MTase activity. The study suggested that AdaB neutralizes the mutagenic influence of O^6-methylation of guanine, whereas AdaA-AlkA by its MTase activity removes methyl groups from innocuous methylphosphotriesters in DNA (Yang et al. 2011).

Another detailed study on MtOGT (O^6-methylguanine MTase; Rv1316c) characterized its potential to ensure *M.tb* survival at the time of synthesis of potent DNA-alkylating metabolites by the host. It repairs double-stranded DNA at O^6-alkylguanine (Fig. 7.3a). MtOGT protects the *M.tb* DNA from G·C to A·T transition mutations associated with O^6-alkylated guanine in DNA (Miggiano et al. 2013). Gene encoded by MtOGT is part of the operon involved in adaptive response.

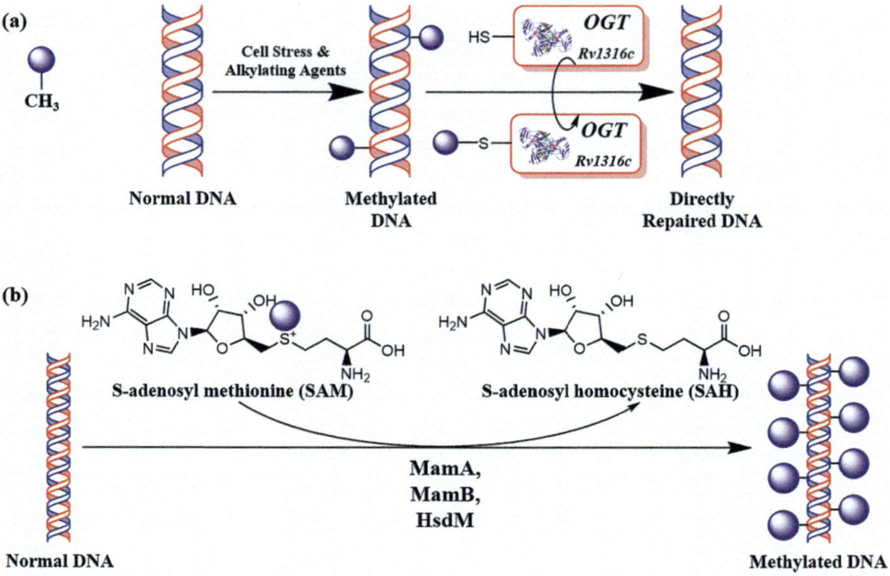

Fig. 7.3 **(a) Double-stranded DNA repair by MtOGT** (O^6-methylguanine MTase; Rv1316c), **(b) epigenetic modulation by *M.tb* MTases** (MamA, MamB and HsdM)

Structural features of MtOGT were shown by Miggiano et al. (2013) where they reported the structural conserved nature of this enzyme. Point mutations (T15S and R37L) in MtOGT were observed in geographically distributed isolates of *M.tb* and various drug-resistant strains. These mutational variations could be correlated with defect in alkylated DNA repair and equilibrium between genome stability and adaptability to the host in due course of evolution. In vitro activity analysis of MtOGT and the two-point mutations associated with it showed impaired activity of the mutated proteins in terms of alkylated DNA damage repair. MtOGT R37L mutant exhibited tenfold lesser affinity for methylated double-stranded DNA as compared to wild-type protein. Further crystal structure analysis of MtOGT and MtOGT-R37L showed huge structural conservation of members of this protein family (Miggiano et al. 2013).

In 2016, Miggiano and colleagues further revealed a peculiar structural assembly of this *Mt*OGT on the monoalkylated dsDNA molecule, shedding more light towards the mechanisms of protein clustering at sites of damaged DNA and disassemble of protein-DNA complexes at the time of repair (Miggiano et al. 2016).

7.2.2.2 DNA Methylation Leading to Epigenetic Regulation

DNA methylation performs regulatory function in a number of prokaryotic pathogens but has not been broadly deciphered in mycobacteria. The three predicted DNA MTases encoded by the genome of *M.tb* are MamA (Rv3263), MamB and HsdM (Rv2756c) (Fig. 7.3b). These are responsible for the methylation of the adenine at sixth position, out of which MamA has been very well characterized. Interestingly none of these are linked with an equivalent restriction endonuclease. A recent study, investigated the pattern of methylation in 12 different isolates of the *M.tb* complex using PacBio technology and described the presence of three unique methylation motifs which are associated with their respective MTases (Zhu et al. 2016). The study also identified a number of isolates where complete absence of methylation of specific motifs in the genome was observed. It was proposed that loss of function mutations in MTases led to the removal of methylation from the genome. Another recent study, highlighted the lineage-specific methylome of *M.tb* which is globally associated with kind of mutations present in the MTase genes (Phelan et al. 2018).

A comprehensive study by Shell et al. (2013) described one of the DNA MTases, Rv3263 gene encoding active DNA adenine MTase MamA (mycobacterial adenine methyltransferase). They reported that MamA methylates DNA at a six base-pair sequence (CTGGAG) in a strain-specific manner in *M.tb* genome affecting expression of several genes. The genes affected by MamA shared overlapping methylation site and sigma factor binding site as shown by mapping the transcriptional start site. This report highlighted that MamA deletion resulted in reduction of survival of *M.tb* in hypoxia and also that MamA is one of the predominant DNA MTases present in *M.tb* strain H37Rv. These findings proposed MamA as an important mediator of adaptation to different physiological stress (Shell et al. 2013).

Rv3204, one of another possible DNA MTase, encodes 101aa long protein and shows close homology with other bacterial MTases (Mycobrowser) (Kapopoulou et al. 2011). Rv3204 is also present in other *Mycobacterium* lineages like *M. bovis*, *M. leprae*, *M. marinum* and *M. smegmatis.*

Overall, DNA MTases in *M.tb* have been shown to play a significant role in tricking the host and ensuring its survival even in adverse conditions

7.2.3 RNA Methyltransferases

When it comes to RNA MTases in *M.tb* genome, a total of 18 are present in *M.tb* genome (Grover et al. 2016). RNA MTases are responsible for methylation of nucleotides of mRNA, rRNA and t-RNA at specific positions and affect the RNA metabolism in several ways such as pre-mRNA processing, post-transcriptional maturation, translation initiation and resistance to antibiotics (Shatkin and Manley 2000; Bussiere et al. 1998).

Modification of rRNAs after transcription is highly important and conserved phenomena and present in almost every kingdoms of life. In prokaryotes, rRNAs are usually modified by methylation by two different ways: (i) methylation of the hydroxyl group of pentose sugar and (ii) methylation of purine or pyrimidine bases. These kinds of methylation modifications increase the chemical and functional interaction of ribosomes and provide structural integrity to ribosomes by increasing RNA-protein interactions of 30S and 50S subunit assembly.

7.2.3.1 Functions of Different RNA MTases on the Basis of Structural Analysis

X-ray crystallographic structures of a number of SAM-MTases modifying protein, DNA, RNA, and small molecules have been reported. Regardless of very less sequence homology amongst several families of SAM-dependent MTases, majority of them have a structurally conserved catalytic SAM-binding motif.

One such SAM-dependent MTase of *M.tb*, RNA methyltransferase (Rv2118c), has signature sequence motifs which are conserved throughout the SAM-MTase family. Its crystal structure revealed a homotetrameric structure with the presence of N-terminal domain and C-terminal domain. The larger domain, C-terminal, binds the cofactor AdoMet, and the N-terminal domain mainly contains β-structure with an additional fold, absent in other MTases of known structures (Gupta et al. 2001). In this report, Rv2118c was suggested to be an RNA methyltransferase on the basis of structure analysis and homology to yeast Gcd14p. Later, Varshney et al. confirmed the function of Rv2118c as a t-RNA MTase forming N^1-methyladenosine of a highly conserved adenine at position 58 in the TΨC loop of t-RNA which helps in stabilization of t-RNA (Varshney et al. 2004). This kind of base modification of t-RNA acts as a virulence factor and growth determinant in pathogenic bacteria.

Rv2966c another RNA MTases of *M.tb* was identified as ortholog of ribosomal RNA small subunit methyltransferase D (RsmD) protein of *E. coli*. They showed that Rv2966c can complement RsmD-deleted *E. coli* cells, and its crystal structure revealed structural homology to RsmD. Recombinant Rv2966c uses 30S ribosomes as substrate and methylates guanidine of 16S rRNA at 966 position. Crystal structure of the protein suggested a two-domain structure with a short hairpin domain at the N-terminal and a SAM-MTase containing C-terminal domain. N-terminal domain is involved in target recognition and plays an important role in ensuring specific interactions with RNA near the P-site of the ribosome. The study further revealed that this domain is also able to bind to DNA; this property may perhaps be used under particular cellular growth stages (Kumar et al. 2011). However, latest report by Sharma et al. contradicts the function of Rv2966 by proving it as a DNA MTase. According to their study, Rv2966 is a 5-methylcytosine-specific DNA MTase, which secretes out from the *M.tb* after infection and localizes to the nucleus of the host cell. This MTase epigenetically modulates the host machinery by interacting not only with the host chromatin but also with the histones H3 and H4. DNA methylation activity of Rv2966 is predominantly in non-CpG context and dependent on its phosphorylation by mycobacterial kinases (Sharma et al. 2015).

Rv2372c is an RsmE-like MTase of *M.tb* which methylates N3 position of uridine 1498 in 16S rRNA. Methylated U1498 exists in decoding centre of ribosomes. A point mutation in U1498 inhibits formation of the first peptide bond at the time of translation. The crystal structure of Rv2372c revealed the presence of core fold of SPOUT superfamily MTases which is active in its dimeric form. The study proved that Rv2372c is orthologous to RsmE which is able to complement RsmE-deleted *E. coli* cells. A ternary model of Rv2372c with 16S rRNA fragment and cofactor SAM identified an RNA binding site at RNA interphase and SAM-binding pocket in the deep trefoil knot. The involvement of U1498 in translation fidelity as well as hygromycin resistance provides greater importance of its methylation (Kumar et al. 2014).

7.2.3.2 RNA MTases in Drug Resistance

Most of the antibiotics target ribosomes in the cell, and mutations at these sites result in antibiotic resistance. One such notorious candidate erm (Rv1988), probable rRNA MTase, has been shown to confer macrolide antibiotic resistance (Buriankova et al. 2004). In this report, through in silico studies, Rv1988 was suggested to be a macrolide-lincosamide-streptogramin (MLS) resistance gene, from the erythromycin resistance rRNA methylase (erm) family encoding 23S rRNA MTase. Further thorough experimental studies proved the gene to be conserved in the *M.tb* complex. Additionally, the gene was reported to show a high level of resistance through target modification.

Another example of involvement of RNA MTases in drug resistance is mutations in tlyA (Rv1694) gene encoding an O-methyltransferase which confers capreomycin resistance. Capreomycin is a cyclic aminoglycoside-like peptide antibiotic which binds the ribosome at the interface of large and small subunits (Johansen et al. 2006). Proper methylation by TlyA is prerequisite for the optimum binding of this drug and consequent inhibition of ribosomal activity. TlyA methylates ribose 2-hydroxy group of two cytidine residues, C1409 of 16S rRNA, present inside the 30S subunit of ribosome also known as decoding centre, and C1920 of 23S rRNA, which is found in a highly conserved region of the 50S subunit of ribosome close to the interface of both the subunits. This dual substrate activity (16S and 23S rRNA) of TlyA is conducted by two stable structural domains connected by a linker region, where N-terminal domain binds with the rRNA and C-terminal domain binds with the SAM, a methyl donor. By X-ray crystallographic studies, it was revealed that this inter-domain linker provides a structural plasticity which imparts TlyA to recognize its two structurally distinct rRNA substrates (Witek et al. 2017).

Reduced streptomycin resistance in clinical isolates of *M.tb* is associated with mutations found in gidB gene (Rv3919c) which is also an rRNA MTase. GidB methylates the N7 atom of guanine 518 in 16S rRNA. G518 methylation structurally stabilizes the 16S rRNA during protein translation by interacting with S12 protein of ribosome. Proline 44 of S12 interacts with mRNA at third wobble position making direct contact to the first two nucleotide of the codon anticodon helix (Ogle et al. 2001). Streptomycin, an aminoglycoside antibiotic, binds to a region in small ribosomal subunit known as accuracy centre and inhibits translation of proteins. Deletion of GidB gene from pathogenic strain of *M.tb* showed alteration in methylation of rRNA and subsequently interrupted the binding of streptomycin to its 16S rRNA substrate eventually generating the streptomycin resistance phenotype (Wong et al. 2013).

7.2.3.3 Dual Nature RNA MTases

The presence of multiple functional domains can lead to dual nature of proteins and can switch or multitask when required in specific environments. This interesting mechanism exists in *M.tb* MTases as well.

Rv1694 of *M.tb* also denoted as *tlyA* when expressed in *E.coli* showed haemolytic activity, processes of rupturing red blood cells (Wren et al. 1998). The product of this TlyA was annotated as a haemolysin. They also reported the presence of TlyA homologs in *Mycobacterium avium, Mycobacterium leprae and Mycobacterium bovis BCG* but lacking in non-pathogenic strain *M. smegmatis.* TlyA name comes from the *spirochaete Serpulina hyodysenteriae*, where it acts as main factor in swine dysentery pathogenesis (Hyatt et al. 1994).

Later Rv1694 was showed to function as ribosomal RNA MTase by Johansen et al. (2006). They showed this tlyA encoding 2′-O-methyltransferase that resulted in modification of C1409 nucleotide in helix 44 of 16S rRNA and C1920 nucleotide in helix 69 of 23S rRNA. This resulted in conferring resistance against the antibiotics capreomycin and viomycin.

In order to shed more light about dual nature of Rv1694, Rahman et al. reported the diverse functions of this MTase that included haemolytic activity and ribosomal RNA methylation. The study claimed Rv1694 is present inside the cell to methylate the ribosomal RNA and on the cell membrane to destabilize the target membranes (Rahman et al. 2010).

In general, the activity of MTases in *M.tb* has not been widely studied, but the data available in the literature points us to the potential roles of DNA and RNA MTases in enabling the *M.tb survival* in the range of extreme environment and helping the bacterium to develop multidrug resistance and maintain the diversity in strain-dependent manner.

7.2.4 Protein Methyltransferase

Methylation of protein is a kind of post-translational modification where addition of methyl group on the arginine or lysine amino acid residues of a protein sequence occurs. Protein methylation of histone proteins has been widely studied where addition of methyl group can epigenetically activate or repress the expression of certain genes.

Heparin-binding haemagglutinin (HBHA) and laminin-binding protein (LBP) are methylated surface-localized adhesins that facilitate the interaction of the *M.tb* with the non-phagocytic cells. C-terminal of HBHA, the active domain of the protein, shows a complex methylation pattern in the form of mono- and dimethyl lysines which protects it from proteolysis and affects the biochemical and immunological properties of this protein. Methylation of HBHA was absent when recombinant protein was expressed in *E.coli* cells (Pethe et al. 2000).

In a recent report, *M.tb* MTase Rv1988 has been found to be epigenetically modulating the host machinery by histone methylation. Rv1988 is part of the *M.tb* secretome and present only in pathogenic species of genus *Mycobacterium.* This MTase localizes to the nucleus of the host after infection and methylates histone H3 at arginine 42 position and subsequently downregulates the expression of the host genes involved in generation of immune response against pathogen. Deletion mutant of Rv1988 showed reduced survival inside the host proving it as an important virulence factor of *M.tb* which exploits the host epigenetic machinery in a noncanonical manner (Yaseen et al. 2015).

7.2.5 Methyltransferases Involved in Phenolglycolipid Biosynthesis

Phthiocerol dimycocerosates (DIM), glycosylated phenolphthiocerol and acid methanolysates are the three characteristic waxes of *M.tb*. Dimycocerosates of the phthiocerol detected in the free lipids along with glycosylated phenolphthiocerol, also known as phenolglycolipid (PGL), are two important groups of virulence factors in mycobacterial species, which are involved in pathogenesis in humans. Phthiotriol/phenolphthiotriol dimycocerosates are encoded by the gene Rv2952. Through insertional mutation analysis, it was found that the gene product is an MTase that catalyses the methyl groups' transfer onto the intermediate compounds such as phthiocerol and glycosylated phenolphthiotriol dimycocerosate to form phthiocerol dimycocerosates (DIM) and glycosylated phenolphthiocerol dimycocerosates (PGL) and other related p-hydroxybenzoic acid derivatives (p-HBAD). The gene has the similar sequence motif as the MTase which methylates the hydroxyl groups of fatty acids.

Phthiocerol dimycocerosates of *M.tb* play a major role in virulence primarily in the initial stages of infection when the mycobacterium encounters the host macrophages. Mutation of genes in the DIM biosynthesis pathway proved that DIM is involved in the receptor-dependent phagocytosis of *M.tb* and also inhibits the phagosomal acidification. DIM regulates *M.tb* invasion into host macrophages by altering organization of the lipid of the macrophage membrane, thus altering the biophysical properties of the membrane. The DIM induces changes in the lipid order and alters the efficiency of receptor-mediated phagocytosis by *M.tb*. These changes in lipid control the phagosomal pH further placing bacilli in a protecting environment.

The enzymes encoded by genes *Rv2955c*, *Rv2954c* and *Rv2956* are involved in the methylation process of phenolic glycolipids, produced by *M.tb*. They are necessary for O-methylation of the fucosyl residue of PGL-tb. These molecules impart pathogenicity by regulating the response of the immune system of the host during infection. The genes *Rv2954c*, *Rv2955c* and *Rv2956* were found to be MTases by the expression and purification analysis of phenolglycolipids produced by mutants of *M. tb* harbouring a deletion of these genes. It was revealed that the hydroxyl groups located at the positions 2, 3 and 4 of the fucosyl residue of PGL-tb were O-methylated. It was also established that the methylation process occurs in a sequential manner starting from the position 2, followed by 3 and 4 (Simeone et al. 2013).

An important variant of phenolic glycolipids in *M.tb* is PGL-tb which consists of an enormous-sized lipid core wrapped up by a glycosylated aromatic nucleus. The carbohydrate region playing a significant role in pathogenicity comprises three residues of sugar and two rhamnosyl units with a terminal fucosyl residue, which is per-O-methylated. This methylation imparts a significant pathogenic characteristic to the glycolipid component.

7.3 Conclusion

The huge number of MTases in *M.tb* has gained widespread attention from the TB researchers due to their significant roles in pathogenesis, virulence and adaptation inside the extreme environment of the host (Fig. 7.4). Interestingly, methylation of biomolecules is now acknowledged as a dynamic modification enhancing the complexity of interactions amongst DNA, RNA, protein and other components of cellular machinery. Methylation all together regulates biomolecular interactions, surface properties of the cell, structure and functional dynamics of chromatin and expression of genes. It has been estimated that 1% of total genes code for MTases in human genome where it is 3% in the case of *M.tb*, i.e. 121 in total number, and majority of them are uncharacterized so far. It is therefore important to understand the role of many of the uncharacterized MTases in pathomechanism of TB. In addition, understanding the potential of these MTases as drug targets would enable designing more effective treatment strategies to combat TB. New development on TB treatment should take into consideration the molecules that can uniquely inhibit *M.tb* MTases.

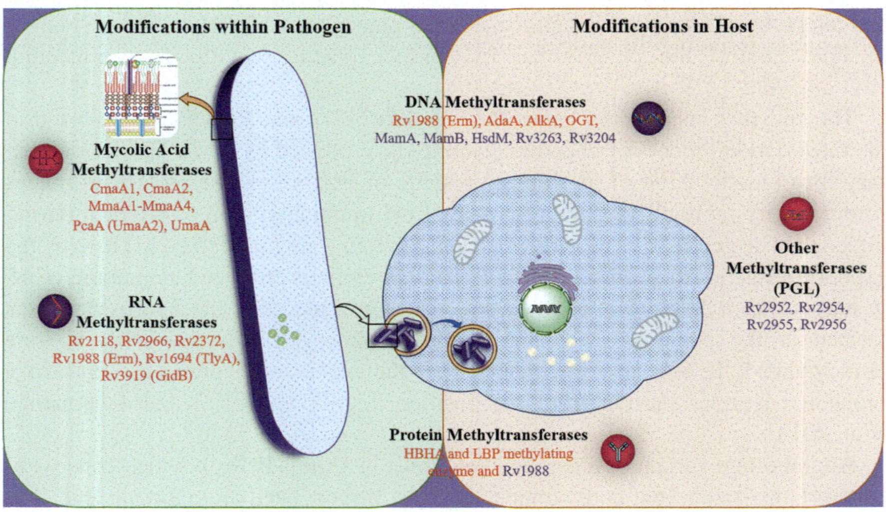

Fig. 7.4 Functional insights into the *Mycobacterial* methyltransferases and their site of action

References

Barkan D, Hedhli D, Yan H-G, Huygen K, Glickman MS (2012) Mycobacterium tuberculosis lacking all mycolic acid cyclopropanation is viable but highly attenuated and hyperinflammatory in mice. Infect Immun 80(6):1958–1968. https://doi.org/10.1128/iai.00021-12

Barry CE, Lee RE, Mdluli K, Sampson AE, Schroeder BG, Slayden RA, Yuan Y (1998) Mycolic acids: structure, biosynthesis and physiological functions. Prog Lipid Res 37(2):143–179. https://doi.org/10.1016/S0163-7827(98)00008-3

Belley A, Alexander D, Di Pietrantonio T, Girard M, Jones J, Schurr E, Liu J, Sherman DR, Behr MA (2004) Impact of methoxymycolic acid production by *Mycobacterium bovis* BCG vaccines. Infect Immun 72(5):2803–2809. https://doi.org/10.1128/iai.72.5.2803-2809.2004

Bhatt A, Fujiwara N, Bhatt K, Gurcha SS, Kremer L, Chen B, Chan J, Porcelli SA, Kobayashi K, Besra GS, Jacobs WR (2007) Deletion of *kasB* in *Mycobacterium tuberculosis* causes loss of acid-fastness and subclinical latent tuberculosis in immunocompetent mice. Proc Natl Acad Sci 104(12):5157–5162. https://doi.org/10.1073/pnas.0608654104

Buriankova K, Doucet-Populaire F, Dorson O, Gondran A, Ghnassia JC, Weiser J, Pernodet JL (2004) Molecular basis of intrinsic macrolide resistance in the Mycobacterium tuberculosis complex. Antimicrob Agents Chemother 48(1):143–150

Bussiere DE, Muchmore SW, Dealwis CG, Schluckebier G, Nienaber VL, Edalji RP, Walter KA, Ladror US, Holzman TF, Abad-Zapatero C (1998) Crystal structure of ErmC', an rRNA methyltransferase which mediates antibiotic resistance in bacteria. Biochemistry 37 (20):7103–7112. https://doi.org/10.1021/bi973113c

Campbell JL, Kleckner N (1990) E. coli oriC and the dnaA gene promoter are sequestered from dam methyltransferase following the passage of the chromosomal replication fork. Cell 62 (5):967–979

Casadesus J, Low D (2006) Epigenetic gene regulation in the bacterial world. Microbiol Mol Biol Rev 70(3):830–856. https://doi.org/10.1128/MMBR.00016-06

Cooper AM (2009) Cell-mediated immune responses in tuberculosis. Annu Rev Immunol 27 (1):393–422. https://doi.org/10.1146/annurev.immunol.021908.132703

Daffé M, Draper P (1998) The envelope layers of Mycobacteria with reference to their pathogenicity. Adv Microb Physiol 39:131–203

Daffé M, Lanéelle MA, Lacave C (1991) Structure and stereochemistry of mycolic acids of Mycobacterium marinum and Mycobacterium ulcerans. Res Microbiol 142(4):397–403. https://doi.org/10.1016/0923-2508(91)90109-N

Danielson SJ, Gray GR (1982) Structures of the two homologous series of dialkene mycolic acids from Mycobacterium smegmatis. J Biol Chem 257(20):12196–12203

Dos Vultos T, Mestre O, Tonjum T, Gicquel B (2009) DNA repair in *Mycobacterium tuberculosis* revisited. FEMS Microbiol Rev 33(3):471–487. https://doi.org/10.1111/j.1574-6976.2009.00170.x

Dubnau E, Chan J, Raynaud C, Mohan VP, Lanéelle MA, Yu K, Quémard A, Smith I, Daffé M (2000) Oxygenated mycolic acids are necessary for virulence of Mycobacterium tuberculosis in mice. Mol Microbiol 36(3):630–637. https://doi.org/10.1046/j.1365-2958.2000.01882.x

George KM, Yuan Y, Sherman DR, Barry CE (1995) The biosynthesis of cyclopropanated mycolic acids in *Mycobacterium tuberculosis*: identification and functional analysis of CMAS-2. J Biol Chem 270(45):27292–27298. https://doi.org/10.1074/jbc.270.45.27292

Glickman MS, Cox JS, Jacobs WR Jr (2000) A novel mycolic acid cyclopropane synthetase is required for cording, persistence, and virulence of *Mycobacterium tuberculosis*. Mol Cell 5 (4):717–727. https://doi.org/10.1016/S1097-2765(00)80250-6

Glickman MS, Cahill SM, Jacobs WR (2001) The *Mycobacterium tuberculosis* cmaA2 gene encodes a mycolic acid trans-cyclopropane synthetase. J Biol Chem 276(3):2228–2233. https://doi.org/10.1074/jbc.C000652200

Gorna Alina E, Bowater Richard P, Dziadek J (2010) DNA repair systems and the pathogenesis of *Mycobacterium tuberculosis*: varying activities at different stages of infection. Clin Sci 119 (5):187–202. https://doi.org/10.1042/cs20100041

Grogan DW, Cronan JE (1997) Cyclopropane ring formation in membrane lipids of bacteria. Microbiol Mol Biol Rev 61(4):429–441

Grover S, Gupta P, Kahlon PS, Goyal S, Grover A, Dalal K, Sabeeha, Ehtesham NZ, Hasnain SE (2016) Analyses of methyltransferases across the pathogenicity spectrum of different mycobacterial species point to an extremophile connection. Mol BioSyst 12(5):1615–1625. https://doi.org/10.1039/c5mb00810g

Gupta A, Kumar PH, Dineshkumar TK, Varshney U, Subramanya HS (2001) Crystal structure of Rv2118c: an AdoMet-dependent methyltransferase from Mycobacterium tuberculosis H37Rv. J Mol Biol 312(2):381–391. https://doi.org/10.1006/jmbi.2001.4935

Hemavathy KC, Nagaraja V (1995) DNA methylation in mycobacteria: absence of methylation at GATC (Dam) and CCA/TGG (Dcm) sequences. FEMS Immunol Med Microbiol 11 (4):291–296

Hoffmann C, Leis A, Niederweis M, Plitzko JM, Engelhardt H (2008) Disclosure of the mycobacterial outer membrane: Cryo-electron tomography and vitreous sections reveal the lipid bilayer structure. Proc Natl Acad Sci 105(10):3963–3967. https://doi.org/10.1073/pnas.0709530105

Hyatt DR, ter Huurne AA, van der Zeijst BA, Joens LA (1994) Reduced virulence of Serpulina hyodysenteriae hemolysin-negative mutants in pigs and their potential to protect pigs against challenge with a virulent strain. Infect Immun 62(6):2244–2248

Johansen SK, Maus CE, Plikaytis BB, Douthwaite S (2006) Capreomycin binds across the ribosomal subunit interface using tlyA-encoded 2'-O-methylations in 16S and 23S rRNAs. Mol Cell 23(2):173–182. https://doi.org/10.1016/j.molcel.2006.05.044

Kahramanoglou C, Prieto AI, Khedkar S, Haase B, Gupta A, Benes V, Fraser GM, Luscombe NM, Seshasayee AS (2012) Genomics of DNA cytosine methylation in Escherichia coli reveals its role in stationary phase transcription. Nat Commun 3:886. https://doi.org/10.1038/ncomms1878

Kapopoulou A, Lew JM, Cole ST (2011) The Mycobrowser portal: a comprehensive and manually annotated resource for mycobacterial genomes. Tuberculosis 91(1):8–13. https://doi.org/10.1016/j.tube.2010.09.006

Kumar A, Saigal K, Malhotra K, Sinha KM, Taneja B (2011) Structural and functional characterization of Rv2966c protein reveals an RsmD-like methyltransferase from Mycobacterium tuberculosis and the role of its N-terminal domain in target recognition. J Biol Chem 286 (22):19652–19661. https://doi.org/10.1074/jbc.M110.200428

Kumar A, Kumar S, Taneja B (2014) The structure of Rv2372c identifies an RsmE-like methyltransferase from Mycobacterium tuberculosis. Acta Crystallogr D Biol Crystallogr 70 (Pt 3):821–832. https://doi.org/10.1107/S1399004713033555

Kurthkoti K, Varshney U (2012) Distinct mechanisms of DNA repair in mycobacteria and their implications in attenuation of the pathogen growth. Mech Ageing Dev 133(4):138–146. https://doi.org/10.1016/j.mad.2011.09.003

Lacave C, Lanéelle MA, Daffé M, Montrozier H, Rols MP, Asselineau C (1987) Structural and metabolic study of the mycolic acids of *Mycobacterium fortuitum*. Eur J Biochem 163 (2):369–378. https://doi.org/10.1111/j.1432-1033.1987.tb10809.x

Laval F, Haites R, Movahedzadeh F, Lemassu A, Wong CY, Stoker N, Billman-Jacobe H, Daffé M (2008) Investigating the function of the putative mycolic acid methyltransferase UmaA: divergence between the *Mycobacterium smegmatis* and *Mycobacterium tuberculosis* proteins. J Biol Chem 283(3):1419–1427. https://doi.org/10.1074/jbc.M708859200

Lederer E (1969) Some problems concerning biological C-alkylation reactions and phytosterol biosynthesis. Q Rev Chem Soc 23(4):453–481. https://doi.org/10.1039/QR9692300453

McNeil M, Daffe M, Brennan PJ (1991) Location of the mycolyl ester substituents in the cell walls of mycobacteria. J Biol Chem 266(20):13217–13223

Meena LS, Kolattukudy PE (2013) Expression and characterization of Rv0447c product, potentially the methyltransferase involved in tuberculostearic acid biosynthesis in Mycobacterium tuberculosis. Biotechnol Appl Biochem 60(4):412–416. https://doi.org/10.1002/bab.1112

Miggiano R, Casazza V, Garavaglia S, Ciaramella M, Perugino G, Rizzi M, Rossi F (2013) Biochemical and structural studies of the Mycobacterium tuberculosis O6-methylguanine methyltransferase and mutated variants. J Bacteriol 195(12):2728–2736. https://doi.org/10.1128/JB.02298-12

Miggiano R, Perugino G, Ciaramella M, Serpe M, Rejman D, Pav O, Pohl R, Garavaglia S, Lahiri S, Rizzi M, Rossi F (2016) Crystal structure of Mycobacterium tuberculosis O6-methylguanine-DNA methyltransferase protein clusters assembled on to damaged DNA. Biochem J 473(2):123–133. https://doi.org/10.1042/BJ20150833

Militello KT, Simon RD, Qureshi M, Maines R, VanHorne ML, Hennick SM, Jayakar SK, Pounder S (2012) Conservation of Dcm-mediated cytosine DNA methylation in Escherichia coli. FEMS Microbiol Lett 328(1):78–85. https://doi.org/10.1111/j.1574-6968.2011.02482.x

Minnikin DE, Polgar N (1967) The methoxymycolic and ketomycolic acids from human tubercle bacilli. Chem Commun (22):1172–1174. https://doi.org/10.1039/C19670001172

Mizrahi V, Andersen SJ (1998) DNA repair in *Mycobacterium tuberculosis*. What have we learnt from the genome sequence? Mol Microbiol 29(6):1331–1339. https://doi.org/10.1046/j.1365-2958.1998.01038.x

Nataraj V, Varela C, Javid A, Singh A, Besra GS, Bhatt A (2015) Mycolic acids: deciphering and targeting the Achilles' heel of the tubercle bacillus. Mol Microbiol 98(1):7–16. https://doi.org/10.1111/mmi.13101

Ogle JM, Brodersen DE, Clemons WM Jr, Tarry MJ, Carter AP, Ramakrishnan V (2001) Recognition of cognate transfer RNA by the 30S ribosomal subunit. Science 292(5518):897–902. https://doi.org/10.1126/science.1060612

Pethe K, Aumercier M, Fort E, Gatot C, Locht C, Menozzi FD (2000) Characterization of the heparin-binding site of the mycobacterial heparin-binding hemagglutinin adhesin. J Biol Chem 275(19):14273–14280

Phelan J, de Sessions PF, Tientcheu L, Perdigao J, Machado D, Hasan R, Hasan Z, Bergval IL, Anthony R, McNerney R, Antonio M, Portugal I, Viveiros M, Campino S, Hibberd ML, Clark TG (2018) Methylation in Mycobacterium tuberculosis is lineage specific with associated mutations present globally. Sci Rep 8(1):160. https://doi.org/10.1038/s41598-017-18188-y

Rahman A, Srivastava SS, Sneh A, Ahmed N, Krishnasastry MV (2010) Molecular characterization of tlyA gene product, Rv1694 of Mycobacterium tuberculosis: a non-conventional hemolysin and a ribosomal RNA methyl transferase. BMC Biochem 11:35. https://doi.org/10.1186/1471-2091-11-35

Rao V, Fujiwara N, Porcelli SA, Glickman MS (2005) *Mycobacterium tuberculosis* controls host innate immune activation through cyclopropane modification of a glycolipid effector molecule. J Exp Med 201(4):535–543. https://doi.org/10.1084/jem.20041668

Rao V, Gao F, Chen B, Jacobs WR, Glickman MS (2006) Trans -cyclopropanation of mycolic acids on trehalose dimycolate suppresses *Mycobacterium tuberculosis* – induced inflammation and virulence. J Clin Invest 116(6):1660–1667. https://doi.org/10.1172/JCI27335

Sani M, Houben ENG, Geurtsen J, Pierson J, de Punder K, van Zon M, Wever B, Piersma SR, Jiménez CR, Daffé M, Appelmelk BJ, Bitter W, van der Wel N, Peters PJ (2010) Direct visualization by Cryo-EM of the mycobacterial capsular layer: a labile structure containing ESX-1-secreted proteins. PLoS Pathog 6(3):e1000794. https://doi.org/10.1371/journal.ppat.1000794

Sharma G, Upadhyay S, Srilalitha M, Nandicoori VK, Khosla S (2015) The interaction of mycobacterial protein Rv2966c with host chromatin is mediated through non-CpG methylation and histone H3/H4 binding. Nucleic Acids Res 43(8):3922–3937. https://doi.org/10.1093/nar/gkv261

Shatkin AJ, Manley JL (2000) The ends of the affair: capping and polyadenylation. Nat Struct Biol 7(10):838–842. https://doi.org/10.1038/79583

Shell SS, Prestwich EG, Baek SH, Shah RR, Sassetti CM, Dedon PC, Fortune SM (2013) DNA methylation impacts gene expression and ensures hypoxic survival of Mycobacterium tuberculosis. PLoS Pathog 9(7):e1003419. https://doi.org/10.1371/journal.ppat.1003419

Simeone R, Huet G, Constant P, Malaga W, Lemassu A, Laval F, Daffe M, Guilhot C, Chalut C (2013) Functional characterisation of three o-methyltransferases involved in the biosynthesis of phenolglycolipids in Mycobacterium tuberculosis. PLoS One 8(3):e58954. https://doi.org/10.1371/journal.pone.0058954

Srivastava R, Gopinathan KP, Ramakrishnan T (1981) Deoxyribonucleic acid methylation in mycobacteria. J Bacteriol 148(2):716–719

Varshney U, Ramesh V, Madabushi A, Gaur R, Subramanya HS, RajBhandary UL (2004) Mycobacterium tuberculosis Rv2118c codes for a single-component homotetrameric m1A58 tRNA methyltransferase. Nucleic Acids Res 32(3):1018–1027. https://doi.org/10.1093/nar/gkh207

Voskuil M, Bartek I, Visconti K, Schoolnik G (2011) The response of *Mycobacterium tuberculosis* to reactive oxygen and nitrogen species. Front Microbiol 2:105. https://doi.org/10.3389/fmicb.2011.00105

Wang AY, Grogan DW, Cronan JE (1992) Cyclopropane fatty acid synthase of Escherichia coli: deduced amino acid sequence, purification, and studies of the enzyme active site. Biochemistry 31(45):11020–11028. https://doi.org/10.1021/bi00160a011

Warner DF, Mizrahi V (2007) The survival kit of *Mycobacterium tuberculosis*. Nat Med 13:282. https://doi.org/10.1038/nm0307-282

WHO (2017) Global tuberculosis report 2017. WHO Publications. World Health Organization, Geneva

Witek MA, Kuiper EG, Minten E, Crispell EK, Conn GL (2017) A novel motif for S-Adenosyl-l-methionine binding by the ribosomal RNA Methyltransferase TlyA from Mycobacterium tuberculosis. J Biol Chem 292(5):1977–1987. https://doi.org/10.1074/jbc.M116.752659

Wong SY, Javid B, Addepalli B, Piszczek G, Strader MB, Limbach PA, Barry CE 3rd (2013) Functional role of methylation of G518 of the 16S rRNA 530 loop by GidB in Mycobacterium tuberculosis. Antimicrob Agents Chemother 57(12):6311–6318. https://doi.org/10.1128/AAC.00905-13

Wren BW, Stabler RA, Das SS, Butcher PD, Mangan JA, Clarke JD, Casali N, Parish T, Stoker NG (1998) Characterization of a haemolysin from Mycobacterium tuberculosis with homology to a virulence factor of Serpulina hyodysenteriae. Microbiology 144(Pt 5):1205–1211. https://doi.org/10.1099/00221287-144-5-1205

Yang M, Aamodt RM, Dalhus B, Balasingham S, Helle I, Andersen P, Tonjum T, Alseth I, Rognes T, Bjoras M (2011) The ada operon of Mycobacterium tuberculosis encodes two DNA methyltransferases for inducible repair of DNA alkylation damage. DNA Repair (Amst) 10(6):595–602. https://doi.org/10.1016/j.dnarep.2011.03.007

Yaseen I, Kaur P, Nandicoori VK, Khosla S (2015) Mycobacteria modulate host epigenetic machinery by Rv1988 methylation of a non-tail arginine of histone H3. Nat Commun 6:8922. https://doi.org/10.1038/ncomms9922

Yuan Y, Barry CE (1996) A common mechanism for the biosynthesis of methoxy and cyclopropyl mycolic acids in *Mycobacterium tuberculosis*. Proc Natl Acad Sci 93(23):12828–12833

Yuan Y, Lee RE, Besra GS, Belisle JT, Barry CE (1995) Identification of a gene involved in the biosynthesis of cyclopropanated mycolic acids in Mycobacterium tuberculosis. Proc Natl Acad Sci U S A 92(14):6630–6634

Yuan Y, Crane DC, Musser JM, Sreevatsan S, Barry CE (1997) MMAS-1, the branch point between cis- and trans-Cyclopropane-containing oxygenated Mycolates in Mycobacterium tuberculosis. J Biol Chem 272(15):10041–10049. https://doi.org/10.1074/jbc.272.15.10041

Zhu L, Zhong J, Jia X, Liu G, Kang Y, Dong M, Zhang X, Li Q, Yue L, Li C, Fu J, Xiao J, Yan J, Zhang B, Lei M, Chen S, Lv L, Zhu B, Huang H, Chen F (2016) Precision methylome characterization of Mycobacterium tuberculosis complex (MTBC) using PacBio single-molecule real-time (SMRT) technology. Nucleic Acids Res 44(2):730–743. https://doi.org/10.1093/nar/gkv1498

Zuber B, Chami M, Houssin C, Dubochet J, Griffiths G, Daffé M (2008) Direct visualization of the outer membrane of mycobacteria and Corynebacteria in their native state. J Bacteriol 190(16):5672–5680

The PE and PPE Family Proteins of *Mycobacterium tuberculosis*: What they Are Up To?

8

Ravi Pal, Faiza Nazar, and Sangita Mukhopadhyay

Abstract

An ebb and flow journey of over 125 years since the discovery of *Mycobacterium tuberculosis* (*M.tb*), the causative agent of tuberculosis (TB), clearly established it as an extremely resilient and elusive pathogen. Accurate diagnosis and effective treatment and full control of this dreaded pathogen are still beyond our reach due to its complex pathobiology. Other features like latency, emergence of multiple drug-resistant strains, and co-occurrence of TB with HIV further complicate development of effective vaccines and drugs to treat TB. Complete genome sequence of *M.tb* has identified a unique set of proteins belonging to PE and PPE family, which is characterized by the presence of highly conserved proline–glutamate (PE) and proline–proline–glutamate (PPE) residues at the N-terminal conserved domain. Ever since its discovery, the PE/PPE family has been the focus of attention in the field of TB research. The PE and PPE family of proteins appears to be critical for mycobacterial survival which modulates the protective innate and adaptive immune responses of the host. In this chapter, we summarize the findings on evolution, characterization, localization, and transcriptional regulation of the PE and PPE genes. We also describe the role of PE and PPE proteins in the modulation of host immune responses. A better understanding of the PE and PPE proteins may lead to the development of innovative and improved anti-TB therapeutics, vaccines, and diagnostics in the future.

R. Pal
Laboratory of Molecular Cell Biology, Centre for DNA Fingerprinting and Diagnostics (CDFD), Hyderabad, Telangana, India

Manipal Academy of Higher Education, Manipal, Karnataka, India

F. Nazar · S. Mukhopadhyay (✉)
Laboratory of Molecular Cell Biology, Centre for DNA Fingerprinting and Diagnostics (CDFD), Hyderabad, Telangana, India
e-mail: sangita@cdfd.org.in

S. E. Hasnain et al. (eds.), Mycobacterium tuberculosis: *Molecular Infection Biology, Pathogenesis, Diagnostics and New Interventions*,
https://doi.org/10.1007/978-981-32-9413-4_8

Keywords
Mycobacterial PE and PPE family of proteins · Evolution · Innate and adaptive immune responses · Vaccines · Virulence

Abbreviations

BCG	bacillus Calmette–Guérin
M.tb	*Mycobacterium tuberculosis*
PE	proline–glutamate
PPE	proline–proline–glutamate
TB	tuberculosis
TLR	Toll-like receptor

8.1 Introduction

Mycobacterium tuberculosis (*M.tb*) discovered by Robert Koch in 1882 is an obligate pathogenic bacterium. It is a part of family *Mycobacteriaceae*. Measured growth rate, hardy cell wall, and various other sneaky strategies adopted by the bacilli not only help it to survive inside the host but also to modulate the host immune response, thus creating a promising environment for further infection. Regardless of bacillus Calmette–Guérin (BCG) vaccination and directly observed treatment, short course (DOTS) therapy, tuberculosis still persists in the population of developing countries (WHO Global tuberculosis report 2017). Additionally, appearance of multidrug-resistant (MDR) and extensively drug-resistant (XDR) strains has intensified the illness caused by the bacillus. It causes deaths among millions of people every year worldwide. According to annual WHO report (2017), 10.4 million new TB cases were estimated, out of which 6.2 million were men, 3.2 million were women, and 1.0 million were children. In 2016, there were 1.7 million deaths due to TB (1.3 million HIV-negative people and 0.4 million HIV-positive people) (WHO Global tuberculosis report 2017). There is an indication that mortality of TB has fallen down to 37% since 2000.

With the annotation of complete *M.tb* H37Rv genome, around 4000 genes (with high GC content) were identified which led to the discovery of novel genes and gene families (Cole et al. 1998). PE/PPE family of genes constitutes approximately 10% of the *M.tb* genome with apparently no known biological functions. There are ~100 PE and ~70 PPE genes present in *M.tb*. Abundance of these genes has led to the understanding of their role in the context of host–pathogen interaction. PE and PPE proteins are named after the highly conserved proline–glutamate (PE) and proline–proline–glutamate (PPE) residues, respectively, near the start site of their translated proteins (Cole 1999). These genes code almost conserved N-terminal sequence of ~110 and ~180 amino acids in PE and PPE families (Fig. 8.1), respectively, (Cole

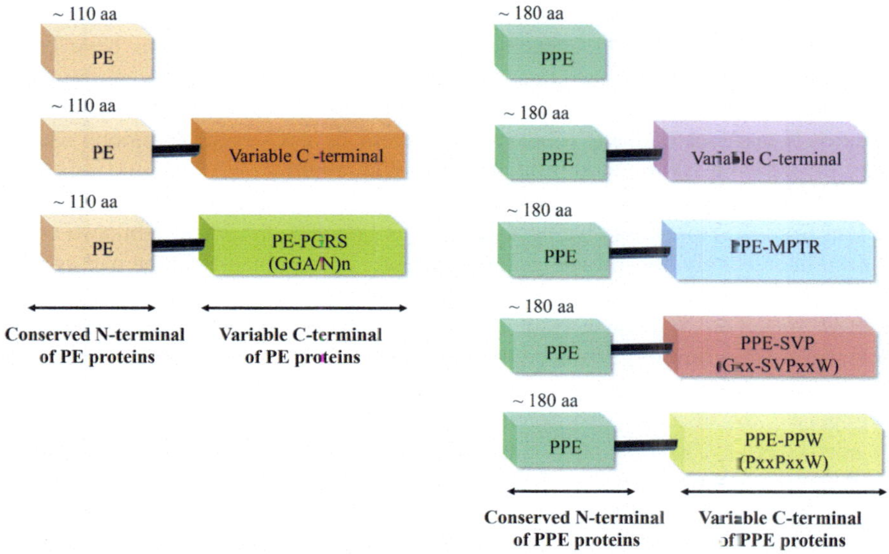

Fig. 8.1 Diagrammatic representation of PE and PPE gene families. PE and PPE family proteins possess a relatively conserved N-terminal of ~110 and ~180 amino acids (aa), respectively. These proteins are further divided into subgroups based on the variable C-terminal

et al. 1998) while the variable C-terminal is believed to be a cause of antigenic and genetic variations across the species. On the basis of variation in the length and sequence, these proteins are further classified into subgroups. PE family is subdivided mainly into two subgroups, i.e., PE_PGRS (polymorphic GC-rich sequences) and PE (with no such distinctive feature). The PPE family encompasses PPE_MPTR (major polymorphic tandem repeats), PPE_SVP (Gxx-SVPxxW motif), and PPE_PPW (PxxPxxW motif) subgroups (Fig. 8.1). Probably, diversity within these subgroups provides them the freedom to have different functions (Fishbein et al. 2015).

In the present manuscript, we have briefly summarized evolution, variation, and phylogeny of PE/PPE family of proteins and their roles in modulation of host immune responses and *M.tb* pathogenesis. Additionally, their potential roles as vaccines and use as diagnostic antigens in TB have been summed up.

8.2 Evolution of PE and PPE Genes

The PE/PPE family consists of around 100 and 70 members, respectively. The group can further be divided into subgroups comprising of members whose length and sequence features are varied. Multiple tandem repeats of Gly-Gly-Ala (or variant) are characteristics of PE_PGRS (polymorphic GC-rich sequence), while the polypeptides rich in repeats of Asn-X-Gly-X-Gly-Asn-X-Gly are present in

PPE_MPTR subgroup (Gey van Pittius et al. 2006). PE/PPE gene clusters are arranged in operons in *M.tb* genome. A total of 28 PE/PPE operons have been identified which are catalogued in a way that most PE members are placed upstream of PPE members. Only a few members of PE protein family contain relatively smaller N-terminal domain, while most of them have C-terminal regions which vary in size from 100 to 1400 residues (Cole et al. 1998).

PE/PPE family of genes is exclusively present in mycobacterial species and is absent from the closest relative bacteria such as *Nocardia farcinica* (Ishikawa et al. 2004). Interestingly, *pe* and *ppe* genes are present predominantly in the pathogenic species like *M. marinum*, *M. avium*, and *M.tb* but not in the nonpathogenic species like *M. smegmatis* (Abdallah et al. 2009), indicating toward a biased evolutionary selection of this family in pathogenic species mainly. Origin of PE/PPE family of genes is not clearly understood, but there are speculations that earliest form of *pe* and *ppe* genes was in close association with the ESX or type VII secretion system (Abdallah et al. 2007). In *M.tb*, there are five *esx* gene clusters (Gey Van Pittius et al. 2001), and phylogenetic studies indicate that *esx* gene clusters are expanded due to the multiple gene duplication. Therefore, it is possible that duplication of *esx* gene cluster has enabled duplication of closely associated *pe* and *ppe* genes (Fig. 8.2). Presence of frequent homologous regions among *pe* and *ppe* genes indicates that spontaneous homologous recombination might have played a crucial role in generating intragenic variations among them (Gey van Pittius et al. 2006; McEvoy et al. 2009, 2012). Insertions/deletions resulting in frame shift and transposon-mediated insertions have certain roles in creating genetic variation. Based on the phylogenetic study by Gey van Pittius et al. (2006), *pe* and *ppe* genes were further classified into sub-lineages, and it was observed that certain sub-lineages have more variations as compared to others. But the reason behind such biased selection is still unknown. In silico studies too have revealed that *pe* and *ppe* genes are more prone to mutations as compared to non-*pe and* non-*ppe* genes (McEvoy et al. 2012). There are reports stating that insertion elements (*IS* elements) have contributed to rearrangement of *ppe* genes (Sampson et al. 1999). *IS*6100, an insertion element, has been found to bring genetic changes by inserting into *pe* and *ppe* genes (Pérez-Lago et al. 2011).

8.3 Characterization, Localization, and Functional Studies of PE and PPE Proteins

Identification and characterization of PE and PPE proteins are important to understand their roles in *M.tb* metabolism and its pathophysiology. In fact, various studies have indicated that the PE and PPE proteins regulate growth and survival of pathogenic mycobacteria and modulate important host immune responses as well. PE11 protein contains GXSG motif and shows esterase activity (Singh et al. 2016). PE11, an essential protein of *M.tb*, is involved not only in regulating the cell wall architecture of mycobacteria but also in favoring mycobacterial replication in vivo (Singh et al. 2016; Rastogi et al. 2017). In silico studies by Sultana et al.

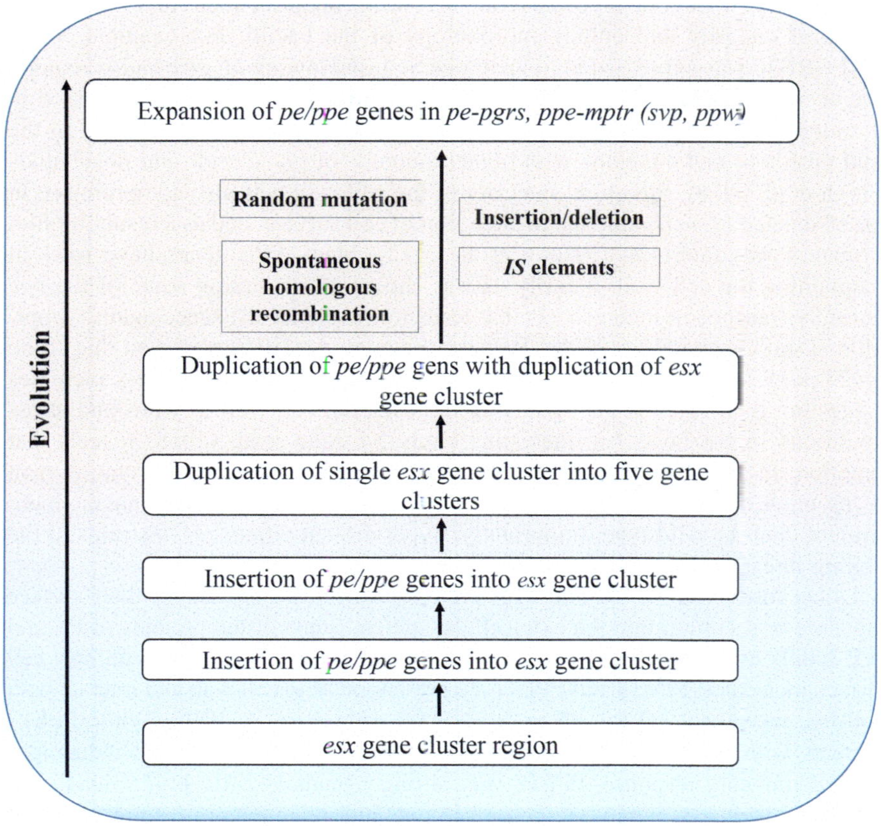

Fig. 8.2 Flow diagram predicting evolution and expansion of *pe* and *ppe* genes. During the course of evolution, *pe* and *ppe* genes were first inserted into the *esx* gene cluster. Duplication of *esx* cluster led to the duplication of *pe* and *ppe* genes. Spontaneous homologous recombination, random mutation, insertion/deletion, and *IS* elements contributed to the expansion and evolution of *pe* and *ppe* genes into different subgroups

(2016) reveal that certain PE and PPE proteins like PE1, PE2, PE3, PE4, PE16, PPE28, PPE42, and PPE63 have hydrolase activity domain. PE and PPE proteins have been detected in cell filtrate, mycobacterial membrane, and/or cell wall fraction. N- and C-terminals of the protein play an important role in the localization of the protein. In *Mycobacterium* (Cascioferro et al. 2007; Zumbo et al. 2013), the N-terminal PE domain of the well-defined PE_PGRS33 was found to be important for protein localization to the cell wall. Cascioferro et al. (2011) confirmed that first 30 amino acids of PE domain in PE_PGRS33 are crucial for its surface localization. Interestingly, in *Mycobacterium marinum*, processed PE domain of PE_PGRS33 was secreted into the culture medium when expressed alone. Chatrath et al. (2014) have demonstrated that the PE_PGRS domain of PE_PGRS30 is responsible for its localization at bacterial cell poles. Association of several PE and

PPE proteins with the mycobacterial cell wall suggests their involvement in structural integrity and colony morphology of the bacilli. For example, when PE_PGRS30 was expressed in nonpathogenic fast-growing *M. smegmatis*, colony size decreased (Chatrath et al. 2011). Another protein, PE11, when expressed in *M. smegmatis*, exhibited altered colony morphology and caused changes in the lipid profile of cell envelope with higher amounts of glycolipids and polar lipids (Singh et al. 2016). Structural integrity of the cell wall was also compromised in *ppe38*-deleted *M. marinum* due to alteration of cell surface, and as a result, biofilm formation was interrupted (Dong et al. 2012). Many PPE genes have roles in maintaining the cell wall integrity as was shown by disrupting some of the *ppe* genes by transposon mutagenesis that leads to increased resistance against ampicillin (Danilchanka et al. 2008). Ramakrishnan et al. (2016) observed that if the *ppe27-pe19 locus* is deleted, the sensitivity of *M.tb* cell wall to stress increases suggesting their roles in strengthening the bacterial cell wall to withstand stress conditions in the host. An interesting study by Ates et al. (2018) reveals that mutation in the PPE38 protein completely blocked secretion of more than 80 proteins of PPE_MPTR and PE_PGRS subsets and made the mutant more virulent than the wild type. Importantly, hypervirulent clinical *M.tb* strains of the Beijing lineage have such a mutation.

Direct interaction of PE and PPE proteins with host depends on their surface exposure or secretion into the extracellular milieu. Some of the proteins of PE and PPE family are cell wall localized and therefore can directly interact with host cell surface molecules. PPE18 and PPE17 are present on *M.tb* cell wall and interact with Toll-like receptor 2 (TLR2) (Nair et al. 2009; Udgata et al. 2016). While PPE17 interacts with the leucine-rich repeat (LRR) 16–20 domain of TLR2 inducing a pro-inflammatory response, PPE18 very strongly binds to LRR 11–15 domain of TLR2 and activates mainly the non-protective anti-inflammatory response. Detection of PPE41 in late endosomes (Abdallah et al. 2006) and PPE2 in macrophage nucleus (Bhat et al. 2017) suggests that some family members are secretory in nature. High-throughput proteomics data have also confirmed the localization of PE and PPE proteins on the *M.tb* cell wall and beyond. A detail knowledge of cellular localization and spatial and temporal regulation of PE/PPE proteins is likely to provide important clues on host–pathogen interaction and designing of novel anti-TB therapeutics.

The localization and functions of important PE and PPE proteins are listed in Table 8.1.

Table 8.1 Localization and functions of some PE and PPE proteins

Name	Location	Function
PE2 (Rv0152c)	Membrane (Mawuenyega et al. 2005)	Belongs to the esterase family; hydrolyzes short-to-medium-chain p-nitrophenyl esters (Bala et al. 2018)
PE3 (Rv0159c)	Membrane (Målen et al. 2011)	PE3 stimulates TNF-α, IL-6, and IL-2 cytokine and provides protection against *M.tb* in mice (Singh et al. 2013)
PE5 (Rv0285)	PE5 – culture filtrate (Målen et al. 2007)	Upregulates IL-10 and downregulates pro-inflammatory responses; supports enhanced bacterial survival in macrophages (Tiwari et al. 2012)
PE6 (Rv0335c)	Cell wall (Mawuenyega et al. 2005)	Essential gene. Function unknown
PE9 (Rv1088)	Membrane (Mawuenyega et al. 2005)	PE9 (Rv1088)–PE10 (Rv1089) protein pair forms heterodimers; induces macrophage apoptosis through TLR4 (Tiwari et al. 2015)
PE11 (lipX) (Rv1169)	Membrane (Målen et al. 2011; de Souza et al. 2011)	Cell wall remodeling and mycobacterial virulence (Singh et al. 2016)
PE12 (Rv1172c)	Membrane (Målen et al. 2011)	Induces high percentage of CD8+ T cells secreting IFN-γ and TNF-α (Stylianou et al. 2018)
PE13 (Rv1195)	Cell wall (Li et al. 2016)	Enhanced survival in THP-1 macrophages; upregulates IL-6 and IL-1β; downregulates SOCS3; modulates p38-ERK-NF-κB signaling; induces apoptosis in THP-1 cells (Li et al. 2016)
PE14 (Rv1214c)	Cell wall (Wolfe et al. 2010)	Nonessential gene. Function unknown
PE15 (Rv1386)	Membrane (Målen et al. 2011; de Souza et al. 2011)	Enhances survival of *Mycobacterium* in macrophages (Tiwari et al. 2012)
PE17 (Rv1646)	Whole cell lysate (de Souza et al. 2011)	Nonessential gene. Function unknown
PE20 (Rv1806)	Membrane (de Souza et al. 2011)	Nonessential gene. Function unknown
PE25 (Rv2431c)	Cell wall (Wolfe et al. 2010), membrane (Målen et al. 2011; de Souza et al. 2011)	PE25/PPE41 dimer induces necrosis in murine macrophages (Tundup et al. 2014), helps in activation and maturation of DCs; favors Th2 response; and induces strong humoral response in mice (Chen et al. 2016)
PE27 (Rv3018A)	Membrane (Målen et al. 2011)	Induces DC maturation by up-regulating CD80, CD86, MHC class I, and MHC class II; upregulates TNF-α, IL-1β, IL-6, and IL-12p70. Induces IFN-γ-producing memory T cell responses in *M.tb* infected mice; contributes to Th1-polarization (Kim et al. 2016)

(continued)

Table 8.1 (continued)

Name	Location	Function
PE31 (Rv3477)	Membrane; lysate not culture filtrate (de Souza et al. 2011)	Nonessential gene. Function unknown
PE35 (Rv3872)	PE35 – culture filtrate (Målen et al. 2007)	Co-operonic with PPE68; upregulates IL-10 by interaction with TLR2 via MAPK pathway; stimulates MCP-1 in THP-1 macrophages (Tiwari et al. 2014)
PE_PGRS11 (Rv0754)	Cell wall (Chaturvedi et al. 2010)	Interacts with TLR2 and imparts resistance to oxidative stress by upregulating COX-2 and Bcl2. Differential B cell responses during TB infection in human sera (Chaturvedi et al. 2010)
PE_PGRS12 (Rv0832)	Membrane (Mawuenyega et al. 2005)	Nonessential gene. Function unknown
PE_PGRS29 (Rv1468c)	Membrane (Målen et al. 2011; de Souza et al. 2011)	Nonessential gene. Function unknown
PE_PGRS30 (Rv1615c)	Bacterial cell pole (De Maio et al. 2014)	Colonizes lung tissue, causes tissue damage; inhibits phagosome-lysosome fusion in macrophage; and induces cell death in mice (Iantomasi et al. 2012)
PE_PGRS33 (Rv1818c)	Cell wall (Minerva et al. 2017)	Helps in entry into macrophages via TLR2 interaction (Palucci et al. 2016); induces TNF-α via TLR2 signaling (Zumbo et al. 2013); activates humoral response in mice and human (Cohen et al. 2014)
PE_PGRS38 (Rv2162c)	Membrane (Målen et al. 2011)	Nonessential gene. Function unknown
PE_PGRS39 (Rv2340c)	Membrane (Mawuenyega et al. 2005)	Nonessential gene. Function unknown
PE_PGRS56 (Rv3512)	Membrane (Mawuenyega et al. 2005)	Induces IFN-γ in (human peripheral blood mononuclear cells) PBMCs (Nebenzahl-Guimaraes et al. 2017)
PE_PGRS62 (Rv3812)	Cell wall (Thi et al. 2013)	Blocks phagosome maturation and inhibits iNOS expression (Thi et al. 2013); reduces phagolysosome maturation (Huang et al. 2012)
PPE2 (Rv0256c)	Culture Filtrate (Bhat et al. 2013)	Downregulates NO production by inhibiting *inos* gene transcription (Bhat et al. 2017)
PPE4 (Rv0286)	Membrane (Mawuenyega et al. 2005)	PPE4 is a part of the ESX-3 locus; role in iron acquisition and virulence (Tufariello et al. 2016)
PPE6 (Rv0305c)	Cytosol, cell wall, and membrane (Mawuenyega et al. 2005)	Nonessential gene. Function unknown

(continued)

Table 8.1 (continued)

Name	Location	Function
PPE7 (Rv0354c)	Membrane (Díaz et al. 2017)	Role in cell invasion in humans (Díaz et al. 2017)
PPE11 (Rv0453)	Cytosol (Mawuenyega KG et al. Mawuenyega et al. 2005), Culture filtrate (Målen et al. 2007)	Promotes cell death in THP1 macrophages; upregulates TNF-α, IL-6, IL-1β and downregulates IL-10; enhances bacterial survival in THP-1 and mice (Peng et al. 2018)
PPE17 (Rv1168c)	Cell wall (Donà et al. 2013)	Binds to TLR2 and induces TNF-α (Udgata et al. 2016)
PPE18 (Rv1196)	Cell wall (Wolfe et al. 2010)	Interacts with TLR2 and activates IL-10 cytokine (Nair et al. 2009). Inhibits proinflammatory cytokines by targeting the p38MAPK-SOCS3-Rel signaling; skews the anti-PPD T-cell response towards the Th2-type. (Nair et al. 2011). Imparts better survival of *M.tb* in mouse model of infection (Bhat et al. 2012)
PPE19 (Rv1361c)	Cell wall (Wolfe et al. 2010)	Nonessential gene. Function unknown
PPE20 (Rv1387)	Membrane (Målen et al. 2011)	Nonessential gene. Function unknown
PPE25 (Rv1787)	Cell wall (Mi et al. 2017)	Enhanced survival in murine macrophages and mice. Upregulates TNF-α, IL-1β and downregulates IL-6; activates NF-κB, p38MAPK, and ERK signaling pathway (Mi et al. 2017)
PPE26 (Rv1789)	Membrane (Målen et al. 2011; de Souza et al. 2011)	Interacts with TLR2 and activates MAPK and NF-κB signaling pathways; stimulates production of proinflammatory cytokines TNF-α, IL-6, and IL-12p40 in murine macrophages (Su et al. 2015)
PPE27 (Rv1790)	Cell wall (Yang et al. 2017)	Exhibits higher survival rate under several hostile conditions in vitro; longer persistence in mouse tissues. Upregulates TNF-α, IL-1β, nitric oxide (NO) and downregulates IL-6 in mice macrophages (Yang et al. 2017)
PPE29 (Rv1801)	Cell wall (Wolfe et al. 2010)	Nonessential gene. Function unknown
PPE31 (Rv1807)	Membrane (Målen et al. 2011; de Souza et al. 2011)	Confers the basal level of *M.tb* resistance to vancomycin (Provvedi et al. 2009)

(continued)

Table 8.1 (continued)

Name	Location	Function
PPE32 (Rv1808)	Membrane (Målen et al. 2011; de Souza et al. 2011)	Promotes proinflammatory cytokines and host cell apoptosis (Deng et al. 2016)
PPE33 (Rv1809)	Membrane (Målen et al. 2011; de Souza et al. 2011)	Nonessential gene. Function unknown
PPE34 (Rv1917c)	Surface exposed (Sampson et al. 2001)	Interacts with TLR2 and triggers functional maturation of human DCs (Bansal et al. 2010)
PPE36 (Rv2108)	Surface exposed (Mitra et al. 2017)	Role in nutrient acquisition (Mitra et al. 2017)
PPE37 (Rv2123)	Cell wall (Ahmad et al. 2018)	Downregulates TNF-α, IL-6, IL-12p70, and IL-1β. Lower levels of NF-κB, ERK, and p38 MAPK in murine macrophages (Daim et al. 2011)
PPE38 (Rv2352c)	Membrane (Målen et al. 2011)	Manipulates innate immune response of host and is shown to play a role in phagocytosis and mycobacterial (*M. marinum*) virulence (Dong et al. 2012)
PPE39 (Rv2353c)	Membrane (Målen et al. 2011)	Protective role against Beijing/K strain in mice (Kim et al. 2017)
PPE40 (Rv2356c)	Whole cell lysate (de Souza et al. 2011)	Essential gene. Function unknown
PPE41 (Rv2430c)	Culture filtrate (Målen et al. 2007)	PE25/PPE41 dimer induces necrosis in murine macrophages (Tundup et al. 2014); helps in activation and maturation of DCs; favors Th2 response and strong humoral response in mice (Chen et al. 2016)
PPE44 (Rv2770c)	Cell wall (Yu et al. 2017)	Induces apoptosis; upregulates IL-6 and IL-12p40 in THP-1 macrophages; and enhances intracellular survival (Yu et al. 2017)
PPE51 (Rv3136)	Cell wall (Wolfe et al. 2010); membrane (Målen et al. 2011)	Function unknown
PPE57 (Rv3425)	Cell wall (Xu et al. 2015)	Induces macrophage activation and drives Th1-type immune responses through TLR2 in mice (Xu et al. 2015)
PPE58 (Rv3426)	Cell membrane (Mawuenyega et al. 2005)	Function unknown
PPE60 (Rv3478)	Membrane (Målen et al. 2011)	Drives Th1/Th17 responses via Toll-like receptor 2-dependent maturation of dendritic cells in mice (Su et al. 2018)
PPE62 (Rv3533c)	Surface exposed (Mitra et al. 2017)	Role in nutrient acquisition (Mitra et al. 2017)
PPE63 (Rv3539)	Membrane (Mawuenyega et al. 2005)	Nonessential gene. Function unknown

(continued)

Table 8.1 (continued)

Name	Location	Function
PPE65 (Rv3621c)	Cell wall (Qureshi et al. 2019)	Binds to TLR2 and upregulates TNF-α and IL-6 in THP-1 macrophages (Qureshi et al. 2019)
PPE68 (Rv3873)	Membrane (Målen et al. 2011; de Souza et al. 2011)	Co-operonic with PE35 modulates MAPK signaling pathways; upregulates IL-10 in dose-dependent manner in THP-1 macrophages (Tiwari et al. 2014)

8.4 Effects of PE and PPE Proteins on Host Immune Responses

PE and PPE proteins modulate both innate and adaptive immune responses of the host (Fig. 8.3). Various immune effector functions that are modulated/manipulated by PE/PPE family of proteins are described below.

8.4.1 Innate Immune Responses

Innate immunity is the first component of the immune system which counteracts various microorganisms. Innate immune cells detect the presence of microorganism through pathogen-associated molecular patterns (PAMPs) and trigger downstream effector signaling cascades to eliminate the infection. These are followed by induction of another tier of the immune system, i.e., adaptive immune response.

M.tb infection is initiated by inhalation of aerosol droplets containing small numbers of bacilli (Smith 2003). Through the respiratory passage, bacilli reach to the lungs and are internalized by resident alveolar macrophages through phagocytosis. Macrophages are one of the important cells of the innate immune response which during *M.tb* infection encounter the bacilli and induce an array of antimicrobial mechanisms like secretion of antibacterial pro-inflammatory cytokines (TNF-α, IL-12, IL-1β), free radicals, nitric oxide (NO), and reactive oxygen intermediates (ROI) and induction of apoptosis and lysosomal degradation (Van Crevel et al. 2002; Hussain Bhat and Mukhopadhyay 2015). After being recognized by pattern recognition receptors (PRRs), bacilli trigger both innate and adaptive immune response (Qureshi and Medzhitov 2003). *M.tb* activates different families of pattern recognition receptors, majorly the TLRs and the nucleotide-binding oligomerization domain-like receptors (NODs) (Ferwerda et al. 2005).

Initial reports depicting the role of PE and PPE genes in modulating innate immune machinery come from *M. marinum* which has two PE_PGRS gene homologues to Rv1651c and Rv3812 of *M.tb*. These proteins are involved in granuloma formulation and enhanced survival of the bacilli inside macrophages

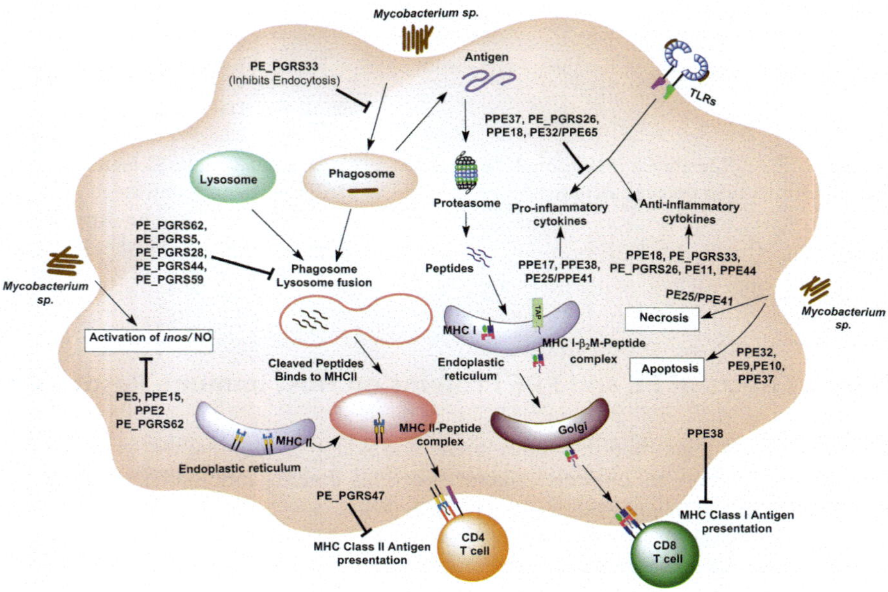

Fig. 8.3 Diagrammatic representation of macrophage effector functions modulated by PE and PPE proteins during *Mycobacterium* sp. infection. ROIs, reactive oxygen species; RNIs, reactive nitrogen species; TLR, Toll-like receptor; MHC, major histocompatibility complex

(Ramakrishnan et al. 2000). There are many PE and PPE proteins which act as virulence factors and affect various macrophage functions. For example, in BCG, a *pe_pgrs33* mutant has shown restricted entry into macrophages (Brennan et al. 2001) suggesting its role in promoting uptake of bacilli and its colonization inside macrophages during infection. Palucci et al. (2016) have shown that PE_PGRS33 contributes to *M.tb* entry in macrophages through TLR2-dependent manner. Similar observations were made when PE_PGRS33 was expressed in nonpathogenic *M. smegmatis* (Singh et al. 2008). An interesting study indicates that PE_PGRS30 is required for full virulence of *M.tb* as *pe_pgrs30* mutant was impaired in its ability to colonize lung tissue (Iantomasi et al. 2012). Camacho et al. (1999) have shown that attenuated PPE46 resulted in reduced *M.tb* growth in vitro. In macrophages, phagosome maturation and acidification is a crucial defense mechanism for killing of intracellular pathogens, and PE/PPE proteins can influence this process (Jha et al. 2010). For example, four *pe_pgrs* mutants and three *ppe_mptr* of BCG were found to be predominantly localized within acidified phagosomes which reflects their putative roles in inhibition of phagosome–lysosome fusion during infection (Stewart et al. 2005). There are several reports depicting role of PE/PPE proteins in cell viability. PE_PGRS5 localizes to endoplasmic reticulum and induces stress-mediated apoptosis through TL-4 (Grover et al. 2018).

PE_PGRS33 induces apoptosis in murine macrophages through TNF-α-dependent manner (Basu et al. 2007). Further, Cadieux et al. (2011) reported that the PGRS region of PE_PGRS33 is required for its localization into mitochondria but PE domain was crucial for induction of cell death. In *M. smegmatis*, PE9 and PE10 were found to form a heterodimer and induce apoptosis in THP-1 cells through TLR4 binding (Tiwari et al. 2015). In TB infection, necrotic cell death is known to be important for dissemination of bacilli. PE25/PPE41, a co-operonic protein complex, was found to induce necrosis in murine macrophages (Tundup et al. 2014). Inflammasome activation is one of the important events of *M.tb* infection. Inflammasome activation upregulates pro-inflammatory cytokines like IL-1β and IL-18 which leads to pyroptosis (Briken et al. 2013). PE8 and PPE15 are part of ESX-5a, and it was observed that *M.tb* lacking ESX-5a was not as efficient as the wild-type bacilli in activating inflammasome but did not affect host cell death (Shah et al. 2015).

There are various ways by which PE/PPE proteins modulate macrophage effector functions; few of them are reviewed and discussed here in detail.

8.4.1.1 Reactive Oxygen Intermediates (ROIs) and Reactive Nitrogen Intermediates (RNIs)

During *M.tb* infection, activated macrophages produce reactive oxygen intermediates (ROIs) via NADPH oxidase (NOX2/gp91phox) and reactive nitrogen intermediates (RNIs) through inducible nitric oxide synthase (iNOS) (Ehrt and Schnappinger 2009). Upon infection, superoxide ions (O^{2-}) are generated, which are acted upon by dismutase to produce H_2O_2 which finally generates toxic hydroxyl radicals ($OH^{\cdot}$) against the bacilli (Bedard and Krause 2007). RNIs are products of inducible nitric oxide synthase (iNOS), an enzyme that synthesizes nitric oxide (NO) from L-arginine in response to endogenous cytokines like IFN-γ and TNF-α or pathogen-derived molecules. NO and other chemical intermediates, i.e., NO^-, NO_2^-, $\bullet NO_2$, N_2O_3, and N_2O_4, are involved in imparting resistance during mycobacterial infection. In addition to activated macrophages, polymorpholeukocytes also generate ROIs and RNIs in response to *M.tb* infection (Hussain Bhat and Mukhopadhyay 2015).

Mycobacterium has evolved to surpass these effector responses by having scavenger compounds like phenolic glycolipid I (PGL-1) in the thick cell wall (Launois et al. 1989), enzyme catalase–peroxidase (KatG) neutralizing ROIs (Manca et al. 1999; Ng et al. 2004), superoxide dismutases SodA and SodC for neutralization of O^2 (Piddington et al. 2001), and the peroxidase and peroxynitrite reductase (Voskuil et al. 2011). An interesting study indicates that the PPE2 protein of *M.tb* can suppress NO production in murine macrophages by downregulating expression of *inos* gene resulting in increased survival inside macrophages (Bhat et al. 2013, 2017). PPE2 contains nuclear localization signal (NLS), which enables it to gain entry into the nucleus and a DNA-binding motif which facilitates the DNA binding at the promoter region of *inos* gene leading to its reduced transcription (Bhat et al. 2017). Furthermore, PPE2 reduces ROS levels in murine macrophages during infection (Srivastava et al. 2019). PE_PGRS62, a protein expected to alter the cell wall-related component, when expressed in *M. smegmatis*, has curtailed *inos* expression in infected macrophages without changing the transcriptional level by some unknown mechanism (Thi et al. 2013). PE11 (LipX or Rv1169c), which is upregulated during oxidative stress in macrophages (Schnappinger et al. 2003) as well as in patients with active TB (Rachman et al. 2006), modulates cell wall architecture of the bacilli to resist free radicals and various environmental and antibiotic stresses which is clearly reflected by the enhanced survival of *M. smegmatis* expressing PE11 in murine macrophages as well as in mouse model of infection (Singh et al. 2016). Other proteins, like PE5 and PPE15, suppress nitric oxide production by downregulating *inos* at the transcriptional level and provide protection to the pathogen against oxidative stresses (Tiwari et al. 2012).

8.4.1.2 Cytokine Signaling

Evidences from both experimental animal models and observation on patients have firmly established the role of cytokines and their signaling pathways in mycobacterial infection (Cooper and Khader 2008). During *M.tb* infection, macrophages secrete cytokines like TNF-α, IFN-γ, IL-10, and IL-12 which play an important role in the regulation of macrophage as well as other immune cell functions (Domingo-Gonzalez et al. 2016). Cytokines have an important role in dictating the T cell response phenotype. TNF-α, IL-12, and other pro-inflammatory cytokines enhance the antibacterial defense and activate induction of the protective T helper (Th) 1-type T cell response, whereas IL-10, the anti-inflammatory cytokine, is known to induce Th2-type T cell response which causes decreased protection of the host and enhanced mycobacterial survival. *M.tb* PPE17, a cell wall-localized protein, triggers a pro-inflammatory-type response in macrophages by TLR1/2 heterodimerization through PKCepsilon–IRAK3–p38MAPK signaling axis (Udgata et al. 2016), whereas PPE18, another cell wall-localized protein, binds to TLR2 but to a different domain (LRR 11–15) than PPE17 (LRR 16–20) and facilitates its homodimerization to induce IL-10 cytokine through activation of p38MAPK and, therefore, skews the T cell response toward the Th2 type (Nair et al. 2009; Udgata et al. 2016). PPE18 simultaneously inhibits IL-12 and TNF-α pro-inflammatory cytokines by SOCS3–p38MAPK–Rel signaling pathway (Nair et al. 2011). PPE37 reduces the level of pro-inflammatory cytokines, i.e., TNF-α and IL-6, which is due to reduced level of the phosphorylated p65 subunit of NF-κB, p38MAPK, and ERK1/2 (Daim et al. 2011). In an interesting study by Ahmad et al. (2018), it has been shown that N- and C-terminal of PPE37 performs different functions in THP-1 macrophages. N-terminal of recombinantly purified PPE37 causes proliferation and differentiation of monocytic THP-1 into DC-SIGN (dendritic cell-specific intercellular adhesion molecule-3-grabbing nonintegrin) which produces higher IL-10 than full length, whereas its C-terminal localizes to macrophage nucleus and induces caspase-3-dependent apoptosis. PE_PGRS33 and PE_PGRS26 were found to induce IL-10 in splenocytes after infection when expressed in *M. smegmatis* and provided the survival advantage to the bacteria (Singh et al. 2008). PE11 induces strong anti-inflammatory cytokines which help the bacterium in progression of disease and persistence in a mouse model of infection (Singh et al. 2016). Other than macrophages, dendritic cells (DCs) also show a change in the cytokine profiles during *M.tb* infection. PPE34 facilitates DC maturation via PI3K–MAPK and NF-κB signaling cascades and induces secretion of high levels of Th2 cytokines (IL-10) when co-cultured with the CD4+ T cells (Bansal et al. 2010). PPE38, when interacts with TLR2, induces both IL-6 and TNF-α in THP-1 macrophages (Dong et al. 2012) and downregulates class I antigen presentation in peritoneal macrophage, thereby reducing/delaying the protective CD8+ T cell activity (Meng et al. 2017). Khubaib et al. (2016) have shown that co-operonic PE32/PPE65 shows greater modulation of immune responses in combination rather than PE32 or PPE65 alone. Recombinant PE32 and PPE65 proteins together downregulate TNF-α and IL-6 whereas upregulate IL-10 at higher concentrations in vitro indicating a Th2-type response. In an in vivo experiment, they observed that recombinant PE32 + PPE65 suppress IFN-γ in splenocytes and there was a reduction in IFN-γ- and IL-2-positive $CD4^+$ and $CD8^+$ T cell populations.

8.4.1.3 Phagosome–Lysosome Fusion

Phagosome–lysosome membrane fusion is a highly regulated event that is essential for intracellular killing of microorganisms. Phagosome maturation and acidification is a process of internalizing particles and their trafficking into a series of increasingly acidified membrane-bound structures, ultimately leading to particle degradation. However, *M.tb* has evolved strategies to bypass this mechanism successfully. For example, PE_PGRS62, when expressed in *M. smegmatis*, could arrest phagosome–lysosome fusion by preventing the recruitment of Rab7 and LAMP1 to bacilli-containing phagosomes (Thi et al. 2013). Similarly, a role of PPE54 in the prevention of phagosome maturation was recorded (Brodin et al. 2010). BCG with four *pe_pgrs* mutants (*pe_pgrs5*, *pe_pgrs28*, *pe_pgrs44*, and *pe_pgrs59*) and three *ppe_mptr* mutants (*ppe_mptr10*, *ppe_mptr16*, and *ppe_mptr21*) were found to be localized more predominantly in acidified phagosomes as compared to the wild-type bacteria suggesting their roles in preventing phagosome maturation (Stewart et al. 2005). MAV-2928 (a homologue of PPE25) from *M. avium* was shown to prevent the fusion of the acidified phagosome with lysosomes inside macrophages (Jha et al. 2010). PE_PGRS30 and its homologue in *M. marinum*, MAG24–1, have also been found to interfere with phagosome maturation in macrophages via reduced recruitment of LAMP1 (Iantomasi et al. 2012).

8.4.2 Adaptive Immune Responses

8.4.2.1 Antigen Presentation

Antigen presentation is a vital process in which antigen-presenting cells (APCs) display processed antigens via major histocompatibility complex (MHC) which are further recognized by subsets of T cells depending upon the type of MHC molecules presenting the antigens. There are reports which state that PE_PGRS33 (Brennan and Delogu 2002) limits MHC class I antigen presentation. In a similar study, Koh et al. (2009a) have shown that PE_PGRS17 inhibits antigen processing by affecting ubiquitin–proteasome-dependent protein degradation and therefore can prevent the killing of infected host cells. In a genome-wide screening, Saini et al. (2016) have identified that PE_PGRS47 protein of *M.tb* is involved in inhibiting class II antigen presentation and autophagy in a nonredundant manner by an unknown mechanism. In an unpublished study from our laboratory, PPE18 has also been shown to inhibit class II antigen presentation in vitro. Dendritic cells (DCs) are one of the robust antigen-presenting cells. Some PE/PPE family proteins have been shown to manipulate maturation of dendritic cells. For example, Bansal et al. (2010) have shown that PPE34 possesses a SRC homology 3 (SH3) domain which interacts specifically with TLR2 and elicits functional maturation of human DCs. Similarly, PE25/PPE41 dimer is involved in maturation of mouse dendritic cells (Chen et al. 2016).

8.4.2.2 B Cell and T Cell Responses

PE and PPE family members are a potentially rich source of B cell epitopes and surface exposed, and a number of them are likely to generate humoral immune responses. The highly repetitive domains of both PE and PPE proteins are immunodominant. Ahmad et al. (2018) reveal that N- and C-terminal domains of PPE37 perform different functions during infection conditions. Full-length and N-terminal domain of PPE37 displayed stronger B cell response in TB patients and immunized mice. Full-length PPE37 showed a higher IgG response in pulmonary tuberculosis (PTB), extrapulmonary tuberculosis (EPTB), as well as patients with relapse cases of TB, whereas N-terminal of PPE37 showed higher IgG responses only in PTB cases (Ahmad et al. 2018). PPE57 has been shown to develop a significant IgG response compared to ESAT-6 in sera of active TB and EPTB patients as compared to healthy controls (Zhang et al. 2007). PE25 and PPE41 are coded by a single operon. In sera obtained from TB patients, it was observed that humoral response induced in TB patients against PE25/PPE41 complex was higher than PPE41 and PE25 alone (Tundup et al. 2008). PPE42 a member of PPE_MPTR class elicit humoral response primarily against domain rich in glycine and asparagine (Chakhaiyar et al. 2004). Both recombinantly purified PPE44 and BCG infection have shown antigen-specific IgG in mouse sera (Bonanni et al. 2005). The antibody responses in TB patients were predominantly mounted against the N-terminal domain of the PPE17 protein (Abraham et al. 2017). Co-operonic PE32 and PPE65 together generate higher IgG1/IgG2a ratio and are known to support enhanced survival of the bacilli (Khubaib et al. 2016).

T cell antigens have always been the center of attraction for development of therapeutics as well as for mediating immunity that play leading roles in the control of *M.tb* infections (Flynn et al. 1992; Caruso et al. 1999). *Mycobacterium* sp. have been reported to elicit $CD4^+$ and $CD8^+$ responses (Jouanguy et al. 1999). Anti-mycobacterial immunity is mediated by a Th1-type T cell. Th1 cells produce IFN-γ, interleukin (IL)-2, and tumor necrosis factor (TNF)-β, which activate macrophages and cell-mediated immunity. Studies have shown that PE/PPE proteins can be potent T cell antigens. Among the antigens identified by peptide library screening for $CD4^+$ T cell responses against *M.tb*, 45% were shown to be PE/PPE proteins. They constitute more than 5% of the open reading frames of *M.tb*. The epitopes from PE/PPE proteins constitute around 50% of the total pool of newly identified antigens suggesting the potential of these proteins in regulation of host immune responses

(Kunnath-Velayudhan and Porcelli 2013) and therefore can be used to develop TB vaccines. PPE44 lowers the IFN-γ in splenocytes after recombinant protein treatment as well as with the BCG infection. Additionally, PPE44 modulates IgG1/IgG2a ratio which skews the T cell response toward the Th2 type by producing higher IgG1and low IFN-γ (Bonanni et al. 2005). PPE18 has also been shown to generate a strong anti-inflammatory response by elevating IL-10 levels (Nair et al. 2009) skewing the T cell response toward the Th2 type. PPE60 activates Th1/Th17 responses via Toll-like receptor 2-dependent maturation of dendritic cells (Su et al. 2018). PPE57 causes macrophage activation and drives Th1-type immune responses through TLR2 (Xu et al. 2015). Balaji et al. (2007) have shown that PE_PGRS33 protein induces apoptosis in T cells in Smac-dependent activation of caspases. Choi et al. (2018) have shown that when PE/PPE peptides were fused with ESAT-6, it turned out to be a better T cell elicitor than ESAT-6 alone. In another study, it was shown that pool of peptides from PE/PPE family elicits a higher IFN-γ response in sensitized T cells obtained from both bovine and human hosts. It has been realized that a subset of PE/PPE proteins are major targets of the cellular immune response to tuberculosis and are recognized at multiple stages of infection and in different disease states (Vordermeier et al. 2012). Another report by Chaitra et al. (2008a, b) has shown that peptides derived from PE_PGRS33, PE_PGRS62, and PPE46 generate strong CTL response in PBMCs isolated from PPD-positive and TB patients. The group showed that immunization with PE_PGRS62 or PPE46 DNA construct induced CD8+ T cells and generated strong Th1-type response with high IFN-γ and low IL-4 (Chaitra et al. 2008a, b).

All these studies indicate that PE/PPE proteins can modulate both innate and adaptive immune responses of host and act as important virulence factors. Understanding individual and cumulative impact of PE/PPE proteins is an important aspect which could provide the basis for novel interventions. However, the pathophysiological functions of PE/PPE proteins are not fully understood. Therefore, further approaches are needed to unravel the biological roles of these proteins for identification of novel strategies to combat TB.

8.5 PE/PPE Proteins as Diagnostic Markers for Detecting TB Cases

Early and accurate diagnosis of TB is one of the primary challenges in curtailing the spread of TB. Current diagnosis of active pulmonary TB is mainly based on confirmation of acid fast bacilli by sputum smear microscopy and culture of mycobacteria. But sputum smear examination displays a low degree of sensitivity which is about 50–60%, and the sensitivity is further reduced especially in children and patients coinfected with HIV. Mycobacterial culture, a gold standard method of TB diagnosis, is time-consuming, and more importantly, TB bacilli fail to grow in 10–20% of positive cases (Andersen et al. 2000). Using conventional diagnostic method, it is cumbersome to diagnose both extrapulmonary and smear-negative TB. It is estimated that 13–17% of newer infections are caused by smear-negative

TB (Tostmann et al. 2008). WHO recommended rapid and fully automated Xpert MTB/RIF test for the diagnosis of TB in children and some forms of extrapulmonary TB (WHO 2014). Although molecular diagnostic tools like polymerase chain reaction (PCR) are rapid, highly sensitive, and specific for the diagnosis of different forms of TB, their application is limited due to higher cost, availability of reagents, requirement of technical expertise, and poor specificity under field conditions.

Serological assays that measure antibody responses to selective *M.tb* antigens in suspected cases of TB are still considered to be attractive tools due to simplicity and convenient detection of pathogens, low costs, easy operation, and rapid determination. According to recommendations of the WHO in order to replace the "gold standard" sputum culture, a serological test for active TB should have sensitivities of over 80% and test specificities of over 95% (WHO 2014). Recently WHO has estimated that more than 50 companies are involved in the development of rapid TB diagnostic tests (WHO 2014) and globally several attempts are being made to identify and test the serodiagnostic potential of various *M.tb* antigens (Baumann et al. 2014; Khurshid et al. 2014; Pukazhvanthen et al. 2014; Feng et al. 2013; Zhang et al. 2013).

Several groups have reported higher sensitivity of PPE proteins to detect active TB patients. Examples are PPE2 (Abraham et al. 2014), PPE17 (Abraham et al. 2016; Khan et al. 2008), PPE41 (Choudhary et al. 2003), PPE42 (Ireton et al. 2010; Chakhaiyar et al. 2004), PPE44 (Rindi et al. 2007), PPE57 (Zhang et al. 2007), PPE65 (He et al. 2011), and PPE68 (Xu et al. 2012) for diagnosis of active TB. Tundup et al. (2008) reported PPE41 was sensitive in detecting 45% of TB cases; however, its sensitivity was increased to 75% when the antigen was combined with a PE protein (PE25/PPE41 protein complex). Khan et al. (2008) showed that PPE17 could strongly discriminate active TB patients from BCG-vaccinated healthy controls compared to PPD, ESAT-6, and HSP60. PPE17 showed 81.3% sensitivity toward extrapulmonary TB. The serodiagnostic potential of PPE17 was also studied in smear-negative TB cases, and interestingly, PPE17 displayed higher sensitivity (75%) for this category of TB as well. In sera derived from TB patients, PPE2 protein too generates higher titer of IgG than PPD when compared with BCG-vaccinated healthy controls (Abraham et al. 2014). Though PPE2 could detect smear-negative and extrapulmonary TB patients, it is less sensitive than PPE17 (Abraham et al. 2016). PPE17 is surface exposed (Donà et al. 2013), and genes that are homologous to PPE17 are absent in the nontuberculous mycobacterial species. PPE17 is shown to be upregulated in conditions that mimic the macrophage environment features. Moreover, overexpression of PPE17 protein was observed in macrophages infected with various clinical isolates of *M.tb* (Homolka et al. 2010). All these features suggest that PPE17 could be a potential serodiagnostic marker for detecting active TB cases. Abraham et al. (2017) reported that the N-terminal domain of PPE17 protein plays a dominant role in inducing antibody responses in active TB patients including the smear-negative and extrapulmonary TB. Although the N-terminal domain of PPE proteins is conserved, PPE17 showed a unique amino acid sequence starting from 122 amino acid to 140 amino acid (IFGIHTPAIFALDALYAQY) in its N-terminal region and did not significantly cross-react with N-terminal domains of other PPE proteins such as PPE18, PPE44, and PPE65 (Abraham et al. 2017). Tundup et al. (2008) reported that PE25 (Rv2431c) has very low sensitivity as it could detect only 9% of TB cases (n = 32). Another group demonstrated that among the three PE antigens PE11, PE_PGRS17, and PE_PGRS33, PE_PGRS17 showed higher sensitivity toward newer and extrapulmonary TB patients (Narayana et al. 2007). Baumann et al. (2014) compared PE35 with other *M.tb* antigens and observed that PE35 could not display statistically significant IgG response, but showed higher IgA response (78.6%). Interestingly, Mukherjee et al. (2007) reported that PE35 displayed a higher sensitivity of 92.5% and 88.8% against pulmonary and extrapulmonary TB patients diagnosed in Kolkata, India.

PE/PPE proteins have been studied for serological diagnosis of latent TB as well. Koh et al. (2009b) studied the seroreactivity to PE-PGRS17 and PE-PGRS62 proteins and observed that PE-PGRS62 is associated with both latent and active TB. PPE44 showed a sensitivity of 37.5% in latent TB infection (Rindi et al. 2007). One of the highly immunogenic PPE protein, PPE55, was showed to be useful to differentiate between latent TB and incipient, subclinical TB (Singh et al. 2005). A recent study showed PPE68 could be used as a sensitive and specific biomarker for discriminating TB from LTBI (Pourakbari et al. 2015). An interesting study reveals that PPE17 could discriminate individuals with latent TB infection from the QFT-negative subjects and demonstrated a higher sensitivity (87%) for the detection of latent TB-positive individuals (Abraham et al. 2018).

Other than serological assays, Srivastava et al. (2006) employed PCR for amplification of 1291 bp fragment of *M.tb* that belongs to a PPE gene family, Rv0355 (PPE8). Interestingly, PCR was found to be highly sensitive for detection of *M.tb* in sputum, CSF, and pleural fluids (90.4%, 70%, and 77.7%, respectively) compared to the culture method (62.9%, 10%, and 24.4%, respectively). Recently Yuan and Xu (2017) screened genes of *M.tb* with comparative genomics approach and found that PPE66 (Rv3738c) can be amplified in clinical samples. However, this was a preliminary study with the sample size of only 15 and needs further detail testing.

Overall, with the limitations of conventional and existing diagnostic methods, it is cumbersome to diagnose active TB especially smear-negative, extrapulmonary, and latent TB infection. Thus, PE/PPE proteins can be used as potential markers for serodiagnosis of active TB especially smear-negative, extrapulmonary, and latent tuberculosis infection.

8.6 PE and PPE Proteins as Vaccine Candidates

Since 1940, BCG is the only vaccine against tuberculosis used around the globe. But partial effectiveness of BCG in demographically distinct populations and prevalence of MDR and XDR incidences demand designing of novel TB vaccines. There could be three strategies to develop new TB vaccines like i) improvement of existing BCG by adding immunogenic antigens, ii) creating more attenuated *M.tb* strains by genetic manipulation, and iii) use of DNA or protein subunits for boosting the immune responses. Being a potent T cell antigen, PE and PPE proteins can be promising candidates for TB vaccine. There are various preliminary reports stating the protective role of PE and PPE proteins in an animal model of infection. Delogu and Brennan (2001) have shown that PE_PGRS33 produces a prophylactic effect in the mouse as both DNA and protein subunit. In another study, PE4 was shown to act as a potent protective antigen (Singh et al. 2013a, b). Similarly, increased protection against *M.tb* was recorded during immunization with PPE14 (MTB41) in both mouse and guinea pig model (Skeiky et al. 2005). PPE57 protein is considered to be a potential antigen for the design of a vaccine against *M.tb* (Xu et al. 2015). Also, it has been noticed that immunization of mice with recombinant PE3 protein induced a strong protective immune response against challenge with live mycobacteria and can be used as a candidate to develop subunit vaccines (Singh et al. 2013a, b). Similarly, PPE15 was shown to induce significant protection when administered alone and as a boost to BCG vaccination in mice (Stylianou et al. 2018). There are studies which show improved efficacy of polyprotein subunit candidates, when fused with PE and PPE proteins. The fusion of MPT65 with PE domain of PE_PGRS33 (cell wall-localized protein) results in localization of MPT65 antigen on the surface of BCG and enhances the efficiency of BCG vaccine against *M.tb* infection in mice (Sali et al. 2010). Further, deletion of PE and PPE genes of ESX-5 caused the reduction in *M.tb* virulence in immunocompetent mice (Sayes et al. 2012). This *M.tb* mutant (*Δppe25-pe19*) strain was highly attenuated and was able to generate T cell response against the major ESX-1 virulence factors (ESAT-6, CFP-10) and other ESX-associated PE and PPE proteins (Sayes et al. 2012). Another study by Skeiky et al. (2004) revealed that immunization of C57BL/6 mice with Mtb72F (a 72 kDa polyprotein harboring $Mtb32_C$-Mtb39 (PPE18)-$Mtb32_N$ in linear order) resulted in protection against *M.tb* infection. However, there are clinical strains which have variations in PPE18 region, which can make Mtb72F less efficient. ID93, another vaccine candidate, contains PPE42 as one of the components of 93 kDa polyprotein complex (Rv3619-Rv1813-Rv3620-Rv2608/PPE42). This polyprotein subunit vaccine along with adjuvant has been shown to provide protection against drug-resistant *M.tb* in mice and guinea pig and therefore can be considered as the candidate molecule for improving the efficacy of the existing BCG vaccines (Bertholet et al. 2010).

Being a potent T cell antigen, PE/PPE proteins showcase themselves as a good candidate for vaccine development. Therefore, further research is needed to uncover the potential of PE/PPE proteins for improving the efficacy of existing vaccines.

8.7 Conclusion

M.tb has coevolved a long way along with the host and has adopted several strategies to evade host immune responses. MDR and XDR are posing new challenges in the field of TB vaccination and treatment. Therefore, detailed knowledge of "factors" that play crucial roles in the pathophysiology of the bacilli is needed to design better therapeutics to fight against this deadly disease. In a comparative genomic analysis of H37Rv and H37Ra, Kohli et al. (2012) have reported several key changes in the nucleotide sequences among PE and PPE homologues. These changes in the genes are often believed to be the cause of major or physiochemical changes in the encoded proteins, therefore affecting virulence. Further, physiological and biochemical assays are needed to understand the implication of these genetic variations. Moreover, presence of PE and PPE proteins only in the pathogenic species is a strong indicator of their roles in *M.tb* pathogenesis. Therefore, extensive studies are required to comprehend the role of PE and PPE proteins in host immunomodulation and then further expand it to understand the mechanisms of virulence, pathogenesis, and latency of TB. Detailed immunological studies of PE and PPE proteins might help to utilize them as therapeutics to treat other pathological conditions. For example, a novel role of PPE18 as an anti-sepsis molecule is established in mouse model of *E. coli*-induced and CLP-induced septicemia (Ahmed et al. 2018).

Acknowledgments The authors thank Dr. Philip Raj Abraham for his valuable suggestion in preparing the manuscript. The laboratory of SM is supported by the grants from the Department of Biotechnology (DBT), Govt. of India (BT/HRD/35/01/03/2018 and BT/PR12817/COE/34/23/2015), and Department of Science and Technology-Science and Engineering Research Board (DST), Govt. of India (EMR/2016/000644), and also core grant from CDFD by DBT. RP is supported by a fellowship from the Department of Biotechnology, Govt. of India.

References

Abdallah AM, Verboom T, Hannes F et al (2006) A specific secretion system mediates PPE41 transport in pathogenic mycobacteria. Mol Microbiol 62:667–679

Abdallah AM, Gey van Pittius NC, Champion PA et al (2007) Type VII secretion—mycobacteria show the way. Nat Rev Microbiol 5:883–891

Abdallah AM, Verboom T, Weerdenburg EM et al (2009) PPE and PE_PGRS proteins of *Mycobacterium marinum* are transported via the type VII secretion system ESX-5. Mol Microbiol 73:329–340

Abraham PR, Latha GS, Valluri VL et al (2014) *Mycobacterium tuberculosis* PPE protein Rv0256c induces strong B cell response in tuberculosis patients. Infect Genet Evol 22:244–249

Abraham PR, Udgata A, Sumanlata G et al (2016) The *Mycobacterium tuberculosis* PPE protein Rv1168c induces stronger B cell response than Rv0256c in active TB patients. Infect Genet Evol 40:339–345

Abraham PR, Pathak N, Pradhan G et al (2017) The N-terminal domain of *Mycobacterium tuberculosis* PPE17 (Rv1168c) protein plays a dominant role in inducing antibody responses in active TB patients. PLoS One 12:e0179965

Abraham PR, Devalraju KP, Jha V et al (2018) PPE17 (Rv1168c) protein of *Mycobacterium tuberculosis* detects individuals with latent TB infection. PLoS One 13:e0207787

Ahmad J, Farhana A, Pancsa R et al (2018) Contrasting function of structured N-terminal and unstructured C-terminal segments of *Mycobacterium tuberculosis* PPE37 protein. MBio 9:01712–01717

Ahmed A, Dolasia K, Mukhopadhyay S (2018) *Mycobacterium tuberculosis* PPE18 protein reduces inflammation and increases survival in animal model of sepsis. J Immunol 200:3587–3598

Andersen P, Munk ME, Pollock JM et al (2000) Specific immune-based diagnosis of tuberculosis. Lancet 356:1099–1104

Ates LS, Dippenaar A, Ummels R et al (2018) Mutations in ppe38 block PE_PGRS secretion and increase virulence of *Mycobacterium tuberculosis*. Nat Microbiol 3:181–188

Bala Divya M, Vemula M, Balakrishnan K et al (2018) *Mycobacterium tuberculosis* PE1 and PE2 proteins carrying conserved α/β-serine hydrolase domain are esterases hydrolyzing short to medium chain *p*-nitrophenyl esters. Prog Biophys Mol Biol 140:90–102

Balaji KN, Goyal G, Narayana Y et al (2007) Apoptosis triggered by Rv1818c, a PE family gene from *Mycobacterium tuberculosis* is regulated by mitochondrial intermediates in T cells. Microbes Infect 9:271–281

Bansal K, Sinha AY, Ghorpade DS et al (2010) Src homology 3-interacting domain of Rv1917c of *Mycobacterium tuberculosis* induces selective maturation of human dendritic cells by regulating PI3K-MAPK-NF-κB signaling and drives Th2 immune responses. J Biol Chem 285:36511–36522

Basu S, Pathak SK, Banerjee A et al (2007) Execution of macrophage apoptosis by PE_PGRS33 of *Mycobacterium tuberculosis* is mediated by Toll-like receptor 2-dependent release of tumor necrosis factor-alpha. J Biol Chem 282:1039–1050

Baumann R, Kaempfer S, Chegou NN et al (2014) Serologic diagnosis of tuberculosis by combining Ig classes against selected mycobacterial targets. J Infect 69:581–589

Bedard K, Krause KH (2007) The NOX family of ROS-generating NADPH oxidases: physiology and pathophysiology. Physiol Rev 87:245–313

Bertholet S, Ireton GC, Ordway DJ et al (2010) A defined tuberculosis vaccine candidate boosts BCG and protects against multidrug-resistant *Mycobacterium tuberculosis*. Sci Transl Med 2:53–74

Bhat KH, Ahmed A, Kumar S et al (2012) Role of PPE18 protein in intracellular survival and pathogenicity of *Mycobacterium tuberculosis* in mice. PLoS One 7:e52601

Bhat KH, Das A, Srikantam A et al (2013) PPE2 protein of *Mycobacterium tuberculosis* may inhibit nitric oxide in activated macrophages. Ann N Y Acad Sci 1283:97–101

Bhat KH, Srivastava S, Kotturu SK et al (2017) The PPE2 protein of *Mycobacterium tuberculosis* translocates to host nucleus and inhibits nitric oxide production. Sci Rep 7:397–406

Bonanni D, Rindi L, Lari N et al (2005) Immunogenicity of mycobacterial PPE44 (Rv2770c) in *Mycobacterium bovis* BCG-infected mice. J Med Microbiol 54:443–448

Brennan MJ, Delogu G (2002) The PE multigene family: a 'molecular mantra' for mycobacteria. Trends Microbiol 10:246–249

Brennan MJ, Delogu G, Chen Y et al (2001) Evidence that mycobacterial PE_PGRS proteins are cell surface constituents that influence interactions with other cells. Infect Immun 69:7326–7333

Briken V, Ahlbrand SE, Shah S (2013) *Mycobacterium tuberculosis* and the host cell inflammasome: a complex relationship. Front Cell Infect Microbiol 9:3–62

Brodin P, Poquet Y, Levillain F et al (2010) High content phenotypic cell-based visual screen identifies *Mycobacterium tuberculosis* acyltrehalose containing glycolipids involved in phagosome remodeling. PLoS Pathog 9:e1001100

Cadieux N, Parra M, Cohen H et al (2011) Induction of cell death after localization to the host cell mitochondria by the *Mycobacterium tuberculosis* PE_PGRS33 protein. Microbiology 157:793–804

Camacho LR, Ensergueix D, Perez E et al (1999) Identification of a virulence gene cluster of *Mycobacterium tuberculosis* by signature-tagged transposon mutagenesis. Mol Microbiol 34:257–267

Caruso AM, Serbina N, Klein E et al (1999) Mice deficient in CD4 T cells have only transiently diminished levels of IFN-γ, yet succumb to tuberculosis. J Immunol 162:5407–5416

Cascioferro A, Delogu G, Colone M et al (2007) PE is a functional domain responsible for protein translocation and localization on mycobacterial cell wall. Mol Microbiol 66:1536–1547

Cascioferro A, Daleke MH, Ventura M et al (2011) Functional dissection of the PE domain responsible for translocation of PE_PGRS33 across the mycobacterial cell wall. PLoS One 6: e27713

Chaitra MG, Shaila MS, Chandra NR et al (2008a) HLA-A∗0201-restricted cytotoxic T-cell epitopes in three PE/PPE family proteins of *Mycobacterium tuberculosis*. Scand J Immunol 67:411–417

Chaitra MG, Shaila MS, Nayak R (2008b) Characterization of T-cell immunogenicity of two PE/PPE proteins of *Mycobacterium tuberculosis*. J Med Microbiol 57:1079–1086

Chakhaiyar P, Nagalakshmi Y, Aruna B et al (2004) Regions of high antigenicity within the hypothetical PPE major polymorphic tandem repeat open-reading frame, Rv2608, show a differential humoral response and a low T cell response in various categories of patients with tuberculosis. J Infect Dis 190:1237–1244

Chatrath S, Gupta VK, Dixit A et al (2011) The Rv1651c-encoded PE-PGRS30 protein expressed in *Mycobacterium smegmatis* exhibits polar localization and modulates its growth profile. FEMS Microbiol Lett 322:194–199

Chatrath S, Gupta VK, Garg LC (2014) The PGRS domain is responsible for translocation of PE_PGRS30 to cell poles while the PE and the C-terminal domains localize it to the cell wall. FEBS Lett 588:990–994

Chaturvedi R, Bansal K, Narayana Y et al (2010) The multifunctional PE_PGRS11 protein from *Mycobacterium tuberculosis* plays a role in regulating resistance to oxidative stress. J Biol Chem 285:30389–30403

Chen W, Bao Y, Chen X et al (2016) *Mycobacterium tuberculosis* PE25/PPE41 protein complex induces activation and maturation of dendritic cells and drives Th2-biased immune responses. Med Microbiol Immunol 205:119–131

Choi SY, Kwon KW, Kim H et al (2018) Vaccine potential of ESAT-6 protein fused with consensus CD4(+) T-cell epitopes of PE/PPE proteins against highly pathogenic *Mycobacterium tuberculosis* strain HN878. Biochem Biophys Res Commun 503:2195–2201

Choudhary RK, Mukhopadhyay S, Chakhaiyar P et al (2003) PPE antigen Rv2430c of *Mycobacterium tuberculosis* induces a strong B-cell response. Infect Immun 71:6338–6343

Cohen I, Parada C, Acosta-Gío E et al (2014) The PGRS domain from PE_PGRS33 of *Mycobacterium tuberculosis* is target of humoral immune response in mice and humans. Front Immunol 5:236

Cole ST (1999) Learning from the genome sequence of *Mycobacterium tuberculosis* $H_{37}Rv$. FEBS Lett 452:7–10

Cole ST, Brosch R, Parkhill J et al (1998) Deciphering the biology of *Mycobacterium tuberculosis* from the complete genome sequence. Nature 393:537–544

Cooper AM, Khader SA (2008) The role of cytokines in the initiation, expansion, and control of cellular immunity to tuberculosis. Immunol Rev 226:191–204

Daim S, Kawamura I, Tsuchiya K et al (2011) Expression of the *Mycobacterium tuberculosis* PPE37 protein in *Mycobacterium smegmatis* induces low tumour necrosis factor alpha and interleukin 6 production in murine macrophages. J Med Microbiol 60:582–591

Danilchanka O, Mailaender C, Niederweis M (2008) Identification of a novel multidrug efflux pump of *Mycobacterium tuberculosis*. Antimicrob Agents Chemother 52:2503–2511

De Maio F, Maulucci G, Minerva M et al (2014) Impact of protein domains on PE_PGRS30 polar localization in mycobacteria. PLoS One 9:e112482

de Souza GA, Leversen NA, Målen H et al (2011) Bacterial proteins with cleaved or uncleaved signal peptides of the general secretory pathway. J Proteome 75:502–510

Delogu G, Brennan MJ (2001) Comparative immune response to PE and PE_PGRS antigens of *Mycobacterium tuberculosis*. Infect Immun 69:5606–5611

Deng W, Yang W, Zeng J et al (2016) *Mycobacterium tuberculosis* PPE32 promotes cytokines production and host cell apoptosis through caspase cascade accompanying with enhanced ER stress response. Oncotarget 7:67347–67359

Díaz DP, Ocampo M, Varela Y et al (2017) Identifying and characterizing PPE7 (Rv0354c) high activity binding peptides and their role in inhibiting cell invasion. Mol Cell Biochem 430:149–160

Domingo-Gonzalez R, Prince O, Cooper A et al (2016) Cytokines and chemokines in *Mycobacterium tuberculosis* infection. Microbiol Spectr 4:2–18

Donà V, Ventura M, Sali M et al (2013) The PPE domain of PPE17 is responsible for its surface localization and can be used to express heterologous proteins on the mycobacterial surface. PLoS One 8:e57517

Dong D, Wang D, Li M et al (2012) PPE38 modulates the innate immune response and is required for *Mycobacterium marinum* virulence. Infect Immun 80:43–54

Ehrt S, Schnappinger D (2009) Mycobacterial survival strategies in the phagosome: defence against host stresses. Cell Microbiol 11:1170–1178

Feng X, Xiu B, Chen K et al (2013) Enhanced serodiagnostic utility of novel *Mycobacterium tuberculosis* polyproteins. J Infect 66:366–375

Ferwerda G, Girardin SE, Kullberg BJ et al (2005) NOD2 and toll-like receptors are nonredundant recognition systems of *Mycobacterium tuberculosis*. PLoS Pathog 1:279–285

Fishbein S, van Wyk N, Warren RM et al (2015) Phylogeny to function: PE and PPE protein evolution and impact on *Mycobacterium tuberculosis* pathogenicity. Mol Microbiol 96:901–916

Flynn JL, Goldstein MM, Triebold KJ et al (1992) Major histocompatibility complex class I-restricted T cells are required for resistance to *Mycobacterium tuberculosis* infection. Proc Natl Acad Sci U S A 89:12013–12017

Gey Van Pittius NC, Gamieldien J, Hide W et al (2001) The ESAT-6 gene cluster of *Mycobacterium tuberculosis* and other high G+C Gram-positive bacteria. Genome Biol 2:44.1–44.8

Gey van Pittius NC, Sampson SL, Lee H et al (2006) Evolution and expansion of the *Mycobacterium tuberculosis* PE and PPE multigene families and their association with the duplication of the ESAT-6 (esx) gene cluster regions. BMC Evol Biol 15:6–95

Grover S, Sharma T, Singh Y et al (2018) The PGRS domain of *Mycobacterium tuberculosis* PE_PGRS protein Rv0297 is involved in endoplasmic reticulum stress-mediated apoptosis through toll-like receptor 4. MBio 9:e01017–e01018

He XY, Li J, Hao J et al (2011) Assessment of five antigens from *Mycobacterium tuberculosis* for serodiagnosis of tuberculosis. Clin Vaccine Immunol 18:565–570

Homolka S, Niemann S, Russell DG et al (2010) Functional genetic diversity among *Mycobacterium tuberculosis* complex clinical isolates: delineation of conserved core and lineage-specific transcriptomes during intracellular survival. PLoS Pathog 6:e1000988

Huang Y, Zhou X, Bai Y et al (2012) Phagolysosome maturation of macrophages was reduced by PE_PGRS 62 protein expressing in *Mycobacterium smegmatis* and induced in IFN-γ priming. Vet Microbiol 160:117–125

Hussain Bhat K, Mukhopadhyay S (2015) Macrophage takeover and the host-bacilli interplay during tuberculosis. Future Microbiol 10:853–872

Iantomasi R, Sali M, Cascioferro A et al (2012) PE_PGRS30 is required for the full virulence of *Mycobacterium tuberculosis*. Cell Microbiol 14:356–367

Ireton GC, Greenwald R, Liang H et al (2010) Identification of *Mycobacterium tuberculosis* antigens of high serodiagnostic value. Clin Vaccine Immunol 17:1539–1547

Ishikawa J, Yamashita A, Mikami Y et al (2004) The complete genomic sequence of *Nocardia farcinica* IFM 10152. Proc Natl Acad Sci U S A 101:14925–14930

Jha SS, Danelishvili L, Wagner D et al (2010) Virulence-related *Mycobacterium avium* subsp. *hominissuis* MAV_2928 gene is associated with vacuole remodeling in macrophages. BMC Microbiol 1:10–100

Jouanguy E, Döffinger R, Dupuis S et al (1999) IL-12 and IFN-γ in host defense against *mycobacteria* and *salmonella* in mice and men. Curr Opin Immunol 11:346–351

Khan N, Alam K, Nair S et al (2008) Association of strong immune responses to PPE protein Rv1168c with active tuberculosis. Clin Vaccine Immunol 15:974–980

Khubaib M, Sheikh JA, Pandey S et al (2016) *Mycobacterium tuberculosis* co-operonic PE32/PPE65 proteins alter host immune responses by hampering Th1 response. Front Microbiol 7:719

Khurshid S, Afzal M, Khalid R et al (2014) Improving sensitivity for serodiagnosis of tuberculosis using TB16.3-echA1 fusion protein. Tuberculosis 94:519–524

Kim WS, Kim JS, Cha SB et al (2016) *Mycobacterium tuberculosis* PE27 activates dendritic cells and contributes to Th1-polarized memory immune responses during in vivo infection. Immunobiology 221:440–453

Kim A, Hur YG, Gu S et al (2017) Protective vaccine efficacy of the complete form of PPE39 protein from *Mycobacterium tuberculosis* Beijing/K strain in mice. Clin Vaccine Immunol 24: e00219

Koh KW, Lehming N, Seah GT (2009a) Degradation-resistant protein domains limit host cell processing and immune detection of *mycobacteria*. Mol Immunol 46:1312–1318

Koh KW, Soh SE, Seah GT (2009b) Strong antibody responses to *Mycobacterium tuberculosis* PE-PGRS62 protein are associated with latent and active tuberculosis. Infect Immun 77:3337–3343

Kohli S, Singh Y, Sharma K et al (2012) Comparative genomic and proteomic analyses of PE and PPE multigene family of *Mycobacterium tuberculosis* H37Rv and H37Ra reveal novel and interesting differences with implications in virulence. Nucleic Acids Res 40:7113–7122

Kunnath-Velayudhan S, Porcelli SA (2013) Recent advances in defining the immunoproteome of *Mycobacterium tuberculosis*. Front Immunol 4:335

Launois P, Blum L, Dieye A et al (1989) Phenolic glycolipid-1 from *M. leprae* inhibits oxygen free radical production by human mononuclear cells. Res Immunol 140:847–855

Li H, Li Q, Yu Z et al (2016) *Mycobacterium tuberculosis* PE13 (Rv1195) manipulates the host cell fate via p38-ERK-NF-κB axis and apoptosis. Apoptosis 7:795–808

Målen H, Berven FS, Fladmark KE et al (2007) Comprehensive analysis of exported proteins from *Mycobacterium tuberculosis* H37Rv. Proteomics 7:1702–1718

Målen H, De Souza GA, Pathak S et al (2011) Comparison of membrane proteins of *Mycobacterium tuberculosis* H37Rv and H37Ra strains. BMC Microbiol 24:11–18

Manca C, Paul S, Barry CE 3rd et al (1999) *Mycobacterium tuberculosis* catalase and peroxidase activities and resistance to oxidative killing in human monocytes *in vitro*. Infect Immun 67:74–79

Mawuenyega KG, Forst CV, Dobos KM et al (2005) Mycobacterium tuberculosis functional network analysis by global subcellular protein profiling. Mol Biol Cell 16:396–404

McEvoy CR, Warren RM, van Helden PD et al (2009) Multiple, independent, identical IS6110 insertions in *Mycobacterium tuberculosis* PPE genes. Tuberculosis 89:439–442

McEvoy CR, Cloete R, Müller B et al (2012) Comparative analysis of *Mycobacterium tuberculosis pe* and *ppe* genes reveals high sequence variation and an apparent absence of selective constraints. PLoS One 7:e30593

Meng L, Tong J, Wang H et al (2017) PPE38 protein of *Mycobacterium tuberculosis* inhibits macrophage MHC class I expression and dampens CD8(+) T cell responses. Front Cell Infect Microbiol 7:68–79

Mi Y, Bao L, Gu D et al (2017) *Mycobacterium tuberculosis* PPE25 and PPE26 proteins expressed in *Mycobacterium smegmatis* modulate cytokine secretion in mouse macrophages and enhance mycobacterial survival. Res Microbiol 168:234–243

Minerva M, De Maio F, Camassa S et al (2017) Evaluation of PE_PGRS33 as a potential surface target for humoral responses against *Mycobacterium tuberculosis*. Pathog Dis 30:75

Mitra A, Speer A, Lin K et al (2017) PPE surface proteins are required for heme utilization by *Mycobacterium tuberculosis*. MBio 8:01720

Mukherjee P, Dutta M, Datta P et al (2007) The RD1-encoded antigen Rv3872 of *Mycobacterium tuberculosis* as a potential candidate for serodiagnosis of tuberculosis. Clin Microbiol Infect 13:146–152

Nair S, Ramaswamy PA, Ghosh S et al (2009) The PPE18 of *Mycobacterium tuberculosis* interacts with TLR2 and activates IL-10 induction in macrophage. J Immunol 183:6269–6281

Nair S, Pandey AD, Mukhopadhyay S (2011) The PPE18 protein of *Mycobacterium tuberculosis* inhibits NF-κB/rel-mediated proinflammatory cytokine production by upregulating and phosphorylating suppressor of cytokine signaling 3 protein. J Immunol 186:5413–5424

Narayana Y, Joshi B, Katoch VM et al (2007) Differential B-cell responses are induced by *Mycobacterium tuberculosis* PE antigens Rv1169c, Rv0978c, and Rv1818c. Clin Vaccine Immunol 14:1334–1341

Nebenzahl-Guimaraes H, van Laarhoven A, Farhat MR et al (2017) Transmissible *Mycobacterium tuberculosis* strains share genetic markers and immune phenotypes. Am J Respir Crit Care Med 195:1519–1527

Ng VH, Cox JS, Sousa AO et al (2004) Role of KatG catalase-peroxidase in mycobacterial pathogenesis: countering the phagocyte oxidative burst. Mol Microbiol 52:1291–1302

Palucci I, Camassa S, Cascioferro A et al (2016) PE_PGRS33 contributes to *Mycobacterium tuberculosis* entry in macrophages through interaction with TLR2. PLoS One 11:e0150800

Peng X, Luo T, Zhai X et al (2018) PPE11 of *Mycobacterium tuberculosis* can alter host inflammatory response and trigger cell death. Microb Pathog 126:45–55

Pérez-Lago L, Herranz M, Lirola MM et al (2011) Characterization of microevolution events in *Mycobacterium tuberculosis* strains involved in recent transmission clusters. J Clin Microbiol 49:3771–3776

Piddington DL, Fang FC, Laessig T et al (2001) Cu, Zn superoxide dismutase of *Mycobacterium tuberculosis* contributes to survival in activated macrophages that are generating an oxidative burst. Infect Immun 69:4980–4987

Pourakbari B, Mamishi S, Marjani M et al (2015) Novel T-cell assays for the discrimination of active and latent tuberculosis infection: the diagnostic value of PPE family. Mol Diagn Ther 19:309–316

Provvedi R, Boldrin F, Falciani F et al (2009) Global transcriptional response to vancomycin in *Mycobacterium tuberculosis*. Microbiology 155:1093–1102

Pukazhvanthen P, Anbarasu D, Basirudeen SA et al (2014) Assessing humoral immune response of 4 recombinant antigens for serodiagnosis of tuberculosis. Tuberculosis 94:622–633

Qureshi S, Medzhitov R (2003) Toll-like receptors and their role in experimental models of microbial infection. Genes Immun 4:87–94

Qureshi R, Rameshwaram NR, Battu MB et al (2019) PPE65 of M. tuberculosis regulate pro-inflammatory signalling through LRR domains of Toll like receptor-2. Biochem Biophys Res Commun 508:152–158

Rachman H, Strong M, Ulrichs T et al (2006) Unique transcriptome signature of *Mycobacterium tuberculosis* in pulmonary tuberculosis. Infect Immun 74:1233–1242

Ramakrishnan L, Federspiel NA, Falkow S (2000) Granuloma-specific expression of *Mycobacterium* virulence proteins from the glycine-rich PE_PGRS family. Science 288:1436–1439

Ramakrishnan P, Aagesen AM, McKinney JD et al (2016) Mycobacterium tuberculosis resists stress by regulating PE19 expression. Infect Immun 84:735–746

Rastogi S, Singh AK, Pant G et al (2017) Down-regulation of PE11, a cell wall associated esterase, enhances the biofilm growth of *Mycobacterium tuberculosis* and reduces cell wall virulence lipid levels. Microbiology 163:52–61

Rindi L, Peroni I, Lari N et al (2007) Variation of the expression of *Mycobacterium tuberculosis ppe44* gene among clinical isolates. FEMS Immunol Med Microbiol 51:381–387

Saini NK, Baena A, Ng TW et al (2016) Suppression of autophagy and antigen presentation by *Mycobacterium tuberculosis* PE_PGRS47. Nat Microbiol 1:16133

Sali M, Di Sante G, Cascioferro A et al (2010) Surface expression of MPT64 as a fusion with the PE domain of PE_PGRS33 enhances *Mycobacterium bovis* BCG protective activity against *Mycobacterium tuberculosis* in mice. Infect Immun 78:5202–5213

Sampson SL, Warren RM, Richardson M et al (1999) Disruption of coding regions by IS6110 insertion in *Mycobacterium tuberculosis*. Tuber Lung Dis 79:349–359

Sampson SL, Lukey P, Warren RM et al (2001) Expression, characterization and subcellular localization of the *Mycobacterium tuberculosis* PPE gene Rv1917c. Tuberculosis 81:305–317

Sayes F, Sun L, Di Luca M et al (2012) Strong immunogenicity and cross-reactivity of *Mycobacterium tuberculosis* ESX-5 type VII secretion: encoded PE-PPE proteins predicts vaccine potential. Cell Host Microbe 11:352–363

Schnappinger D, Ehrt S, Voskuil MI et al (2003) Transcriptional adaptation of *Mycobacterium tuberculosis* within macrophages: insights into the phagosomal environment. J Exp Med 198:693–704

Shah S, Cannon JR, Fenselau C et al (2015) A duplicated esat-6 region of esx-5 is involved in protein export and virulence of *mycobacteria*. Infect Immun 83:4349–4361

Singh KK, Dong Y, Patibandla SA et al (2005) Immunogenicity of the *Mycobacterium tuberculosis* PPE55 (Rv3347c) protein during incipient and clinical tuberculosis. Infect Immun 73:5004–5014

Singh PP, Parra M, Cadieux N et al (2008) A comparative study of host response to three *Mycobacterium tuberculosis* PE_PGRS proteins. Microbiology 154:3469–3479

Singh SK, Kumari R, Singh DK et al (2013a) Putative roles of a proline-glutamic acid-rich protein (PE3) in intracellular survival and as a candidate for subunit vaccine against *Mycobacterium tuberculosis*. Med Microbiol Immunol 202:365–377

Singh SK, Tripathi DK, Singh PK et al (2013b) Protective and survival efficacies of Rv0160c protein in murine model of *Mycobacterium tuberculosis*. Appl Microbiol Biotechnol 97:5825–5837

Singh P, Rao RN, Reddy JR et al (2016) PE11, a PE and PPE family protein of *Mycobacterium tuberculosis* is involved in cell wall remodeling and virulence. Sci Rep 6:21624

Skeiky YA, Alderson MR, Ovendale PJ et al (2004) Differential immune responses and protective efficacy induced by components of a tuberculosis polyprotein vaccine, Mtb72F, delivered as naked DNA or recombinant protein. J Immunol 172:7618–7628

Skeiky YA, Alderson MR, Ovendale PJ et al (2005) Protection of mice and guinea pigs against tuberculosis induced by immunization with a single *Mycobacterium tuberculosis* recombinant antigen, MTB41. Vaccine 23:3937–3945

Smith I (2003) *Mycobacterium tuberculosis* pathogenesis and molecular determinants of virulence. Clin Microbiol Rev 16:463–496

Srivastava R, Kumar D, Waskar MN et al (2006) Identification of a repetitive sequence belonging to a *PPE* gene of *Mycobacterium tuberculosis* and its use in diagnosis of tuberculosis. J Med Microbiol 55:1071–1077

Srivastava S, Battu MB, Khan MZ, Nandicoori VK, Mukhopadhyay S (2019) J Immunol 203 (5):1218–1229

Stewart GR, Patel J, Robertson BD et al (2005) Mycobacterial mutants with defective control of phagosomal acidification. PLoS Pathog 1:269–278

Stylianou E, Harrington-Kandt R, Beglov J et al (2018) Identification and evaluation of novel protective antigens for the development of a candidate tuberculosis subunit vaccine. Infect Immun 86:e00014–e00018

Su H, Kong C, Zhu L et al (2015) PPE26 induces TLR2-dependent activation of macrophages and drives Th1-type T-cell immunity by triggering the cross-talk of multiple pathways involved in the host response. Oncotarget 6:38517–38537

Su H, Zhang Z, Liu Z et al (2018) *Mycobacterium tuberculosis* PPE60 antigen drives Th1/Th17 responses via toll-like receptor 2-dependent maturation of dendritic cells. J Biol Chem 293:10287–10302

Sultana R, Tanneeru K, Kumar AB et al (2016) Prediction of certain well-characterized domains of known functions within the PE and PPE proteins of *mycobacteria*. PLoS One 11:e0146786

Thi EP, Hong CJ, Sanghera G et al (2013) Identification of the *Mycobacterium tuberculosis* protein PE_PGRS62 as a novel effector that functions to block phagosome maturation and inhibit *iNOS* expression. Cell Microbiol 15:795–808

Tiwari BM, Kannan N, Vemu L et al (2012) The *Mycobacterium tuberculosis* PE proteins Rv0285 and Rv1386 modulate innate immunity and mediate bacillary survival in macrophages. PLoS One 7:e51686

Tiwari B, Soory A, Raghunand TR (2014) An immunomodulatory role for the *Mycobacterium tuberculosis* region of difference 1 locus proteins PE35 (Rv3872) and PPE68 (Rv3873). FEBS J 281:1556–1570

Tiwari B, Ramakrishnan UM, Raghunand TR (2015) The *Mycobacterium tuberculosis* protein pair PE9 (Rv1088)-PE10 (Rv1089) forms heterodimers and induces macrophage apoptosis through toll-like receptor 4. Cell Microbiol 17:1653–1669

Tostmann A, Kik SV, Kalisvaart NA et al (2008) Tuberculosis transmission by patients with smear-negative pulmonary tuberculosis in a large cohort in the Netherlands. Clin Infect Dis 47:1135–1142

Tufariello JM, Chapman JR, Kerantzas CA et al (2016) Separable roles for *Mycobacterium tuberculosis* ESX-3 effectors in iron acquisition and virulence. Proc Natl Acad Sci U S A 113:348–357

Tundup S, Pathak N, Ramanadham M et al (2008) The co-operonic PE25/PPE41 protein complex of *Mycobacterium tuberculosis* elicits increased humoral and cell mediated immune response. PLoS One 3:e3586

Tundup S, Mohareer K, Hasnain SE (2014) *Mycobacterium tuberculosis* PE25/PPE41 protein complex induces necrosis in macrophages: role in virulence and disease reactivation? FEBS Open Bio 16:822–828

Udgata A, Qureshi R, Mukhopadhyay S (2016) Transduction of functionally contrasting signals by two mycobacterial PPE proteins downstream of TLR2 receptors. J Immunol 197:1776–1787

Van Crevel R, Ottenhoff TH, van der Meer JW (2002) Innate immunity to *Mycobacterium tuberculosis*. Clin Microbiol Rev 15:294–309

Vordermeier HM, Hewinson RG, Wilkinson RJ et al (2012) Conserved immune recognition hierarchy of mycobacterial PE/PPE proteins during infection in natural hosts. PLoS One 7: e40890

Voskuil MI, Bartek IL, Visconti K et al (2011) The response of *Mycobacterium tuberculosis* to reactive oxygen and nitrogen species. Front Microbiol 2:105

WHO report (2017) http://apps.who.int/iris/bitstream/handle/10665/259366/9789241565516-eng.pdf?sequence=1

Wolfe LM, Mahaffey SB, Kruh NA et al (2010) Proteomic definition of the cell wall of *Mycobacterium tuberculosis*. J Proteome Res 9:5816–5826

World Health Organization (2014) Latent TB infection fact sheet 2014. World Health Organization, Geneva

Xu JN, Chen JP, Chen DL (2012) Serodiagnosis efficacy and immunogenicity of the fusion protein of *Mycobacterium tuberculosis* composed of the 10-kilodalton culture filtrate protein, ESAT-6, and the extracellular domain fragment of PPE68. Clin Vaccine Immunol 19:536–544

Xu Y, Yang E, Huang Q et al (2015) PPE57 induces activation of macrophages and drives Th1-type immune responses through TLR2. J Mol Med (Berl) 93:645–662

Yang G, Luo T, Sun C et al (2017) PPE27 in *Mycobacterium smegmatis* enhances mycobacterial survival and manipulates cytokine secretion in mouse macrophages. J Interf Cytokine Res 37:421–431

Yu Z, Zhang C, Zhou M et al (2017) *Mycobacterium tuberculosis* PPE44 (Rv2770c) is involved in response to multiple stresses and promotes the macrophage expression of IL-12 p40 and IL-6 via the p38, ERK, and NF-κB signaling axis. Int Immunopharmacol 50:319–329

Yuan G, Xu G (2017) Highly unique and stable biomarkers for diagnosis of *Mycobacterium tuberculosis* pathogens. Biomed Res 28:9633–9637

Zhang H, Wang J, Lei J et al (2007) PPE protein (Rv3425) from DNA segment RD11 of *Mycobacterium tuberculosis*: a potential B-cell antigen used for serological diagnosis to distinguish vaccinated controls from tuberculosis patients. Clin Microbiol Infect 13:139–345

Zhang X, Su Z, Zhang X et al (2013) Generation of *Mycobacterium tuberculosis*-specific recombinant antigens and evaluation of the clinical value of antibody detection for serological diagnosis of pulmonary tuberculosis. Int J Mol Med 31:751–757

Zumbo A, Palucci I, Cascioferro A et al (2013) Functional dissection of protein domains involved in the immunomodulatory properties of PE_PGRS33 of *Mycobacterium tuberculosis*. Pathog Dis 69:232–239

Intrinsically Disordered Regions in PE/PPE Protein Family of *Mycobacterium tuberculosis*: Moonlighting Function

9

Farha Naz, Javeed Ahmad, Mohd Shariq, Mohd Arish, Javaid A. Sheikh, Seyed E. Hasnain, and Nasreen Z. Ehtesham

Abstract

Mycobacterium tuberculosis, the causative agent of tuberculosis, has the distinction of harboring an unusual set of plentiful antigens that are highly homologous. The abundance of similar proteins like PE/PPE that are restricted to pathogenic mycobacterium species implies their important role in either homeostasis or helping to adapt to the intracellular niche which is one of the major life cycle stages in pathogenesis of this bacterium. In spite of the various important functions delineated for this family, it is always challenging to study this extremely homologous family of proteins due to difficulties associated with expression and purification. Phylogenetic studies suggest that pe/ppe genes coevolved with esx loci, which encodes proteins that form secretion systems for the export of PE/PPE proteins. Despite the biochemical and immunological characterization of various PE/PPE proteins, there is dearth of structural data for PE/PPE complex. The structural details and thus the structure-function relation are essential to understand the utility of homologous protein repertoire in pathogenesis. Moreover, the atomic detail could unravel the novel targets to curb

F. Naz · J. Ahmad · M. Shariq · N. Z. Ehtesham (✉)
Inflammation Biology and Cell Signaling Laboratory, ICMR-National Institute of Pathology, Safdarjung Hospital Campus, New Delhi, Delhi, India
e-mail: nasreenzehtesham.nip@gov.in

M. Arish
JH Institute of Molecular Medicine, Jamia Hamdard, New Delhi, Delhi, India

J. A. Sheikh
Department of Biotechnology, School of Chemical and Life Sciences, Jamia Hamdard, New Delhi, Delhi, India

S. E. Hasnain
JH Institute of Molecular Medicine, Jamia Hamdard, New Delhi, Delhi, India

Dr Reddy's Institute of Life Sciences, University of Hyderabad Campus, Hyderabad, India
e-mail: vc@jamiahamdard.ac.in

S. E. Hasnain et al. (eds.), Mycobacterium tuberculosis: *Molecular Infection Biology, Pathogenesis, Diagnostics and New Interventions*,
https://doi.org/10.1007/978-981-32-9413-4_9

this obnoxious pathogen. This chapter will present in detail the structural organization as well as the disorder content of the PE/PPE family. This structural enigma of being highly homologous along with having highest disorder content will be explained with possible functional significance. The available crystal structure and predictions on the basis of those accompanied with functional aspect of their role in cell death pathways are also detailed herein.

Keywords
Structural disorder · PE/PPE proteins · Crystal structure · Programmed cell death · Apoptosis · Necrosis

Abbreviations

(M.tb)	*Mycobacterium tuberculosis*
LCR	Low complexity region
PCD	Programmed cell death
PE	Proline-glutamic acid
PPE	Proline-proline-glutamic acid
RMSD	Root-mean-square deviation
TB	Tuberculosis
TLR	Toll-like receptor

9.1 Introduction

Mycobacterium tuberculosis(M.tb) causes tuberculosis (TB), one of the most devastating diseases worldwide, and approximately kills two million people annually (Glaziou et al. 2013). *M.tb* has evolved from being nonpathogenic and attained pathogenicity during coevolution with humans. During the course of evolution, genome size has decreased (Rahman et al. 2014). The only exception to this prodigy is family of methyltransferase proteins and PE/PPE protein family (Gey van Pittius et al. 2006). The characteristic feature of the PE family is sequence of amino acids (proline-glutamic acid (PE) at eighth and ninth position) in a conserved N terminal domain of about 110 amino acids. PPE family also has a highly conserved N terminal but extended to about 180 amino acids and has the presence of proline-proline-glutamic acid (PPE) at 7th, 8th, and ninth position, respectively. Despite being highly conserved, there is no homology in this N terminus region between PE and PPE. The C terminals of the both PE/PPE protein families are of variable size with other distinctive features (Cole 1999). Although the precise function for this family of proteins is yet to be established, the fact that this family encodes about 4% proteins and has 10% of genome dedicated to it suggests some vital functions. The polymorphic nature of C terminal regions suggests their role in antigenic variation (Cole 1999). These proteins are involved in immune subversion as these proteins are known to interact and diminish the antigen processing and presentation (Delogu and Brennan 2001; Khubaib et al. 2016). Emerging evidence has implicated these proteins in a wide array of roles (Brennan 2017). Functional studies have validated that PE/PPE proteins localize to the membrane, thus forming an integral part of secretion systems. These proteins are associated with type VII mycobacterial secretion systems that are vital for pathogenesis of mycobacteria. This family also encodes lipolytic enzymes, and these enzymes form major virulence factors in mycobacterial metabolome. The evidence of their cell wall localization implicate that these proteins may directly interact with the host and be possibly involved in the modulation of innate as well as adaptive immune functions which makes this family of proteins highly desirable to be evaluated as prophylactic and therapeutic targets. Apart from immune modulatory strategies, proteins involved in the crucial pathways could be exploited as novel drug targets. However, exploring this promising family involves various challenges due to extensive homology and repetitive sequences. The functional redundancies due to the similarity do not allow ascertaining the functionality to a protein very precisely. Similar reasons impact the determining of crystal structure of these proteins that eventually affects elucidation of structure-function relationship. The limited crystal structures available limit the exploration of these ambiguous proteins, which is one of the major obstacles in determining the functional significance of the PE/PPE proteins. More verified crystal structures and stringent modeling approaches will help in determining an integrated view of the interactome of this protein family. This will also aid in the development of novel therapeutic approaches and interventions against this deadly pathogen (Fishbein et al. 2015).

The ability of these proteins to arm mycobacteria with sophisticated pathways to modulate host cell programmed death is very intriguing. This interplay of host-pathogen interaction is extremely vital for understanding the pathogenesis of tuberculosis. Unraveling the mechanism of modulating host cell death pathway could enable us to develop therapeutic interventions. Moreover, better understanding of pro-survival and pro-death factors could give insights into developing novel and efficacious vaccines which is the need of hour.

This chapter will familiarize the readers with structural features and some functional aspects of PE/PPE proteins. The structural disorder in this protein family and the information from the limited crystals are elucidated in detail. Moreover, the modulation of cell death pathways by these proteins will highlight the decoy mechanisms employed by this resilient pathogen.

9.2 Structural Organization of PE/PPE Operons: Functional Relevance

The PE proteins have conserved Pro-Glu (PE), and PPE have Pro-Pro-Glu (PPE) motifs at their N terminal end of the proteins. PE/PPE protein family accounts for about 10% of the coding capacity of the *M.tb* genome (Cole et al. 1998). The respective genes are systematized in a definite operonic pattern within the genome, where a pe gene is followed by a ppe gene (Tundup et al. 2006). The members of PE/PPE protein family are usually associated with members of the ESX family proteins that are important virulence factors, T cell antigens, and potential subunit vaccine candidates. Almost all of the PE/PPE family proteins probably evolved from the duplication of the ancestral esx-related pe/ppe genes, and they are significantly expanded in the slow-growing mycobacteria (Brennan and Delogu 2002; Cole et al. 1998). However, the functions of most of the PE/PPE proteins remain to be deciphered, and few of them have been shown to play important roles in mycobacterial virulence, required for growth of mycobacteria inside the macrophages or in the mouse infection model, or play important role in the inhibition of phagosome maturation of infected macrophages (Bottai and Brosch 2009; Bottai et al. 2012; Brennan et al. 2001; Brodin et al. 2010; Dong et al. 2012; Gey van Pittius et al. 2006; Goldstone et al. 2009; Iantomasi et al. 2012; Li et al. 2005; Sampson 2011; Sampson et al. 2001). PE/PPE proteins exhibit number of repetitive sequences and harbor abundant immunogenic regions, forming a rich source of B and T cell epitopes (Copin et al. 2014). The esx-5 region of *M.tb* (rv1782-rv1798) comprises two pe (pe18, pe19) and three ppe (ppe25, ppe26, ppe27) genes. These PE and PPE proteins are secreted through the transmembrane channel formed by the ESX-5 secretion apparatus involving the ESX conserved essential component EccD5 (Bottai et al. 2012). In addition, many of the other PE/PPE proteins not associated with esx-5 locus but having certain degree of sequence similarity with their esx-5 coded counterparts are also secreted via the ESX-5 secretion system (Abdallah et al. 2006, 2009; Majlessi et al. 2015). Earlier, it was assumed that PE, PPE, and EspG proteins that belong to a particular esx loci tend to interact with each other only and associate with other components of the same cluster, but recent evidence suggests that there may be some cross talk involved. This remains to be elucidated whether this cross talk has any functional relevance or it is just one esx loci compensating for the missing components of other esx loci.

There are ~99 PE and ~69 PPE proteins with ~100 and ~180 conserved N terminal residues, respectively, in the genome of *M.tb* (Brennan 2017). Among them, 61 falls in PE/PGRS (polymorphic GC-rich sequence) and the rest 38 belongs to PE only (Poulet and Cole 1995; Zumbo et al. 2013). Numbers can vary for different strains of *M.tb*. The PE/PGRS contributes significant role in pathogenesis of *M.tb*. The pe/pgrs genes contain significantly higher nucleotide diversity within the members of this family. The gene of PE/PGRS family proteins is mostly clustered in a region of *M.tb* genome as overlapping genes. To maintain the different functions in mycobacterial virulence, PPE proteins diversify to PPE/MPTR (major polymorphic tandem repeats), PPE/SVP (Gxx-SVPxxW motif), and PPE/PPW (PxxPxxW motif) subgroups (Fishbein et al. 2015). The PPE proteins play a key role in secretion and localization of different proteins (Daleke et al. 2011; Dona et al. 2013). All genes are scattered throughout the chromosome, but pe/ppe genes are found to be located in the vicinity of each other. There are certain pairs of pe/ppe genes which are found to be co-localized with each other, and these are pe1 and pe2, pe5 and ppe4, pe/pgrs12 and pe_pgrs13, pe7 and ppe14, pe9 and pe10, pe11 and ppe17, pe15 and ppe20, pe18 and ppe26, pe20 and ppe31, pe22 and ppe36, pe25 and ppe41, pe29, ppe47 and ppe48, ppe50 and ppe51, ppe54, pe/pgrs49 and pe/pgrs56, pe32 and ppe65, pe/pgrs60 and pe/pgrs61, ppe66 and ppe67, and pe35 and ppe68 (Table 9.1). There are approximately 22 predicted operonic clusters out of 69 PPE proteins (Akhter et al. 2012; Tundup et al. 2008). Many among these are transcribed and function together as there is less than 100 bp gap between them in 14 of the gene pairs. Some of these are secreted proteins or present on the cell surface together as heterodimers. pe5/ppe4 and pe15/ppe20 have only 2 bp and 4 bp intergenic distances, respectively, between their pe/ppe genes in *M.tb* and are co-operonic too, but they don't show any physical interaction between their PE/PPE proteins as the gene pair pe25/ppe41, which is co-operonic and shows interaction with each other to form stable protein complex (Tiwari et al. 2012).

9.3 Structural Disorder in PE/PPE: The Molecular Enigma

The molecular function of a protein is fundamentally related to its three-dimensional structure. This hypothesis laid down the foundation of sequence-structure-function paradigm. It is well-known datum that most of the proteins and polypeptide segments fold co-operatively into well-defined three-dimensional structure. However, earlier interpretation that proteins possess an absolute functional specificity with fixed structure disagrees with their marked ability to evolve and adapt new structures and functions. There is growing evidence from last few decades that suggests protein structural flexibility due to structural disorder plays an important role in many different biological systems. Many experimental evidences also suggest that disordered regions in proteins are involved in many types of molecular recognitions. One of the advantages of disordered binding sites is to recognize several targets with low affinity and high specificity allowed by their multiple meta-stable conformations (Dunker et al. 2001; Wright and Dyson 1999). Transitions between the native unfolded state and a globular structure induced by some type of interactions may also provide thermodynamic regulation of binding. These studies are now establishing the disorder-function paradigm. Such proteins are known as intrinsically disordered or unstructured proteins (Dunker et al. 2001). It has been shown in a number of studies that there are clear patterns that characterize disordered regions such as low sequence complexity regions and amino acid composition bias and, therefore, can be predicted successfully from amino acid sequence (Dunker et al. 2001; Iqbal and Hoque 2016).

Computational studies from our laboratory and other groups disclosed PE/PPE protein family is structurally highly disordered compared to the rest of the proteome of *M.tb* (unpublished data). The relatively high level of structural disorder signifies their important role and supports their anticipated functions in immune evasion and host-pathogen interactions, since many known pathogenic effector proteins are disordered. Amino acid composition analysis of PE/PPE protein family is given in Fig. 9.1. Moreover, the pathogenesis-related PE/PPE/PGRS family is highly enriched with glycine- and alanine-rich low complexity regions (Table 9.2).

Table 9.1 Co-operonic PE/PPE genes: PE and PPE genes generally exist in single operon, but some of them found to be present alone

Operon	GI	Synonym	Gene	Start	End	Strand	Length	Product
6924	57116694	Rv0151c	*pe1*	177,543	179,309	-	588	PE family protein
	57116695	Rv0152c	*pe2*	179,319	180,896	-	525	PE family protein
6950	57116714	Rv0280	*ppe3*	339,364	340,974	+	536	PPE family protein
	15607422	Rv0281	–	340,998	341,906	+	302	Hypothetical protein
6951	15607423	Rv0282	–	342,130	344,025	+	631	Hypothetical protein
	15607424	Rv0283	–	344,022	345,638	+	538	Hypothetical protein
	15607425	Rv0284	–	345,635	349,627	+	1330	Hypothetical protein
	57116715	Rv0285	*pe5*	349,624	349,932	+	102	PE family protein
	57116716	Rv0286	*ppe4*	349,935	351,476	+	513	PPE family protein
	15607428	Rv0287	*esxG*	351,525	351,818	+	97	Hypothetical protein
	15607429	Rv0288	*esxH*	351,848	352,138	+	96	Low-molecular-weight protein antigen 7 ESXH (10 kDa antigen) (CFP-7) (protein TB10.4)
	15607430	Rv0289	–	352,149	353,036	+	295	Hypothetical protein
	15607431	Rv0290	–	353,083	354,501	+	472	Transmembrane protein
	15607432	Rv0291	*mycP3*	354,498	355,883	+	461	Membrane-anchored mycosin
	15607433	Rv0292	–	355,880	356,875	+	331	Transmembrane protein
533901 Mycobacterium bovis BCG str. Tokyo 172	224988670	JTY_0292	–	343,390	345,285	+	631	Hypothetical protein
	224988671	JTY_0293	–	345,282	346,898	+	538	Hypothetical protein
	224988672	JTY_0294	–	346,895	350,887	+	1330	Hypothetical protein
	224988673	JTY_0295	*pe5*	350,884	351,192	+	102	PE family protein
	224988674	JTY_0296	*ppe4*	351,195	352,736	+	513	PPE family protein
	224988675	JTY_0297	*esxG*	352,785	353,078	+	97	Hypothetical protein
	224988676	JTY_0298	*esxH*	353,108	353,398	+	96	Low-molecular-weight protein antigen 7
	224988677	JTY_0299	–	353,409	354,296	+	295	Hypothetical protein
	224988678	JTY_0300	–	354,343	355,761	+	472	

	224988679	JTY_0301	–	355,758	357,143	+	461	Putative transmembrane protein
	224988680	JTY_0302	–	357,140	358,135	+	331	Putative protease precursor
								Putative transmembrane protein
260342	121636197	BCG_0322	–	373,058	374,953	+	631	Hypothetical protein
	121636198	BCG_0323	–	374,950	376,566	+	538	Hypothetical protein
	121636199	BCG_0324	–	376,563	380,555	+	1330	Hypothetical protein
	121636200	BCG_0325	*pe5*	380,552	380,860	+	102	PE family protein
	121636201	BCG_0326	*ppe4*	380,863	382,404	+	513	PPE family protein
	121636202	BCG_0327	*esxG*	382,453	382,746	+	97	Hypothetical protein
	121636203	BCG_0328	*esxH*	382,776	383,066	+	96	Low-molecular-weight protein antigen 7 CFP-7
	121636204	BCG_0329	–	383,077	383,964	+	295	Hypothetical protein
	121636205	BCG_0330	–	384,011	385,429	+	472	Putative transmembrane protein
	121636206	BCG 0331	–	385,426	386,811	+	461	Putative protease precursor
	121636207	BCG_0332	–	386,808	387,803	+	331	Putative transmembrane protein
6972	15607528	Rv0387c	–	466,672	467,406	–	244	Conserved hypothetical protein
	57116729	Rv0388c	*ppe9*	467,459	468,001	–	180	PPE family protein
6986	15607582	Rv0441c	–	530,296	530,724	–	142	Hypothetical protein
	57116731	Rv0442c	*ppe10*	530,751	532,214	–	487	PPE family protein
7057	15607885	Rv0745	–	835,154	835,681	+	175	Hypothetical protein
	57116773	Rv0746	*pe_pgrs9*	835,701	838,052	+	783	PE-PGRS family protein
7079	57116787	Rv0832	*pe_pgrs12*	924,951	925,364	+	137	PE-PGRS family protein
	57116788	Rv0833	*pe_pgrs13*	925,361	927,610	+	749	PE-PGRS family protein
7098	57116795	Rv0915c	*ppe14*	1,020,058	1,021,329	–	423	PPE family protein
	57116796	Rv0916c	*pe7*	1,021,344	1,021,643	–	99	PE family protein
7138	57116823	Rv1088	*pe9*	1,214,513	1,214,947	+	144	PE family protein
	57116824	Rv1089	*pe10*	1,214,769	1,215,131	+	120	PE family protein

(continued)

Table 9.1 (continued)

Operon	GI	Synonym	Gene	Start	End	Strand	Length	Product
7156	57116836	Rv1168c	*ppe17*	1,298,764	1,299,804	–	346	PPE family protein
	57116837	Rv1169c	*pe11*	1,299,822	1,300,124	–	100	PE family protein
7206	57116857	Rv1386	*pe15*	1,561,464	1,561,772	+	102	PE family protein
	57116858	Rv1387	*ppe20*	1,561,769	1,563,388	+	539	PPE family protein
7301	57116910	Rv1788	*pe18*	2,026,477	2,026,776	+	99	PE family protein
	57116911	Rv1789	*ppe26*	2,026,790	2,027,971	+	393	PPE family protein
7303	57116919	Rv1806	*pe20*	2,048,072	2,048,371	+	99	PE family protein
	57116920	Rv1807	*ppe31*	2,048,398	2,049,597	+	399	PPE family protein
7365	57116945	Rv2107	*pe22*	2,367,359	2,367,655	+	98	PE family protein
	57116946	Rv2108	*ppe36*	2,367,711	2,368,442	+	243	PPE family protein
7538	57117029	Rv2853	*pe_pgrs48*	3,162,268	3,164,115	+	615	PE-PGRS family protein
	15609991	Rv2854	–	3,164,152	3,165,192	+	346	Hypothetical protein
	57117030	Rv2855	*mtr*	3,165,20	3,166,5	+	459	Mycothione reductase
	15609993	Rv2856	*nict*	53,166,684	843,167,802	+	372	Nickel-transport integral membrane protein
7577	15610156	Rv3019c	*esxR*	3,378,711	3,379,001	–	96	Secreted ESAT-6 like protein ESXR (TB10.3) (ESAT-6 like protein 9)
	57117047	Rv3020c	*esxS*	3,379,036	3,379,329	–	97	Esat-6 like protein esxS
	57117048	Rv3021c	*ppe47*	3,379,376	3,380,452	–	358	PPE family protein
	57117049	Rv3022c	*ppe48*	3,380,440	3,380,682	–	80	PPE family protein
	57117050	Rv3022A	*pe29*	3,380,679	3,380,993	–	104	PE family protein
7599	57117064	Rv3135	*ppe50*	3,501,334	3,501,732	+	132	PPE family protein
	57117065	Rv3136	*ppe51*	3,501,794	3,502,936	+	380	PPE family protein
7657	57117092	Rv3343c	*ppe54*	3,729,364	3,736,935	–	2523	PPE family protein
	57117093	Rv3344c	*pe_pgrs49*	3,736,984	3,738,438	–	484	PE-PGRS family protein
	57117094	Rv3345c	*pe_pgrs50*	3,738,158	3,742,774	–	1538	PE-PGRS family protein

7698	57117116	Rv3511	*pe_pgrs55*	3,939,617	3,941,761	+	714	PE-PGRS family protein
	57117117	Rv3512	*pe_pgrs56*	3,941,724	3,944,963	+	1079	PE-PGRS family protein
7722	15610755	Rv3619c	*esxV*	4,059,984	4,060,268	–	94	ESAT-6 like protein ESXV (ESAT-6 like protein 1)
	15610756	Rv3620c	*esxW*	4,060,295	4,060,591	–	98	ESAT-6 like protein ESXW (ESAT-6 like protein 10)
	57117135	Rv3621c	*ppe65*	4,060,648	4,061,889	–	413	PPE family protein
	57117136	Rv3622c	*pe32*	4,061,899	4,062,198	–	99	PE family protein
7727	57117139	Rv3652	*pe_pgrs60*	4,093,632	4,093,946	+	104	PE-PGRS family-related protein
	57117140	Rv3653	*pe_pgrs61*	4,093,940	4,094,527	+	195	PE-PGRS family-related protein
7750	57117150	Rv3738c	*ppe66*	4,189,285	4,190,232	–	315	PPE family protein
	57117151	Rv3739c	*ppe67*	4,190,284	4,190,517	–	77	PPE family protein
7778	57117163	Rv3872	*pe35*	4,350,745	4,351,044	+	99	PE family-related protein
	57117164	Rv3873	*ppe68*	4,351,075	4,352,181	+	368	PPE family protein

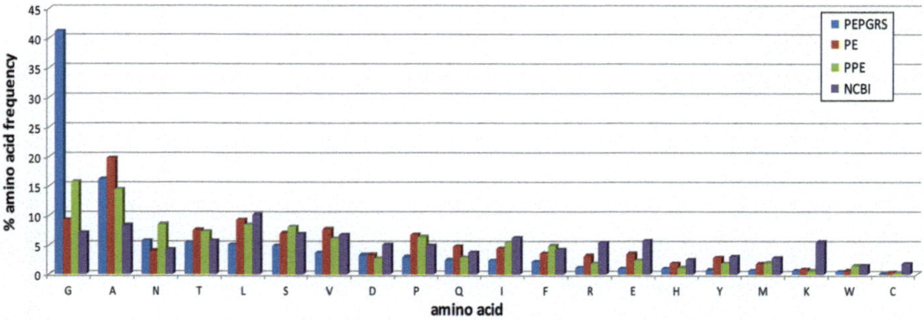

Fig. 9.1 Amino acid composition analysis of PE/PPE protein family: PE/PPE proteins are glycine and alanine rich. PE_PGRS has more number of glycine residues

Table 9.2 PE/PPE protein family is glycine and alanine rich and enriched with low complexity regions

	Alanine-rich LCRs		Glycine-rich LCRs	
Protein family	Terminal LCRs	Central LCRs	Terminal LCRs	Central LCRs
PE	26			1
PE/PGRS	57		4	57
PPE	58	11	2	22

The high structural disorder might also be the reason for difficult expression, purification, and crystallization of these proteins. Therefore, only few pairs have been crystallized till date (Chen et al. 2017; Ekiert and Cox 2014; Strong et al. 2006). The individual members of PE/PPE protein family are structurally more disordered and undergo mutual folding-upon-binding to attain higher structural order. Thus, the observed structural disorder in the free forms of PE and PPE domains and following a mutual folding-upon-binding mechanism of interaction could allow for faster evolutionary rates (unpublished data). These strategies might enable the pathogen to efficiently develop novel cognate pairs of PE/PPE proteins that might be crucial for its adaptive survival. Interestingly, the high structural disorder in PE/PPE protein family was observed mostly in variable C terminal regions which likely contain short linear interaction motifs embedded into disordered regions in some subfamily members (unpublished data) and may be more extended, domain-like interaction modules in some others (Ahmad et al. 2018).

The high structure disorder and expansion of PE/PPE protein family in pathogenic species implicate counterbalancing of reductive evolution by employing this specific PE/PPE family of proteins by *M.tb*. Many members of PE/PPE protein family accomplish structure ordered state through protein-protein interactions or retain partial disordered states even after the formation of the specific complex.

Numerous members of PE/PPE protein family are membrane localized and surface exposed showing high levels of functional specialization among family members through disordered and variable C terminal. PE/PPE protein complexes successfully impact the host immune system favoring pathogen survival through pro-pathogen immune response.

9.4 Crystal Structure of PE/PPE Proteins: The Way Forward

The expression of individual PE or PPE proteins is difficult, and they need each other to get folded (Strong et al. 2006). But, the expression and purification of individual PE5, PE15, PE32, and PPE65 proteins have been done (Khubaib et al. 2016; Tiwari et al. 2012). The structures of these PE/PPE proteins are not so clear because of the unavailability of crystal structures. There is no crystal known for individual PE/PPE protein till now. Only two PE/PPE pairs have been crystallized so far. The structures of the two known PE/PPE pairs are shown in Fig. 9.2.

The complex is heterodimeric and consists mainly of α-helix. It has one PE and one PPE protein. The PE protein consists of two-helix bundle and PPE protein has five helices. The two α-helices of PE interact with two α-helices of PPE protein to form four-helix bundle.

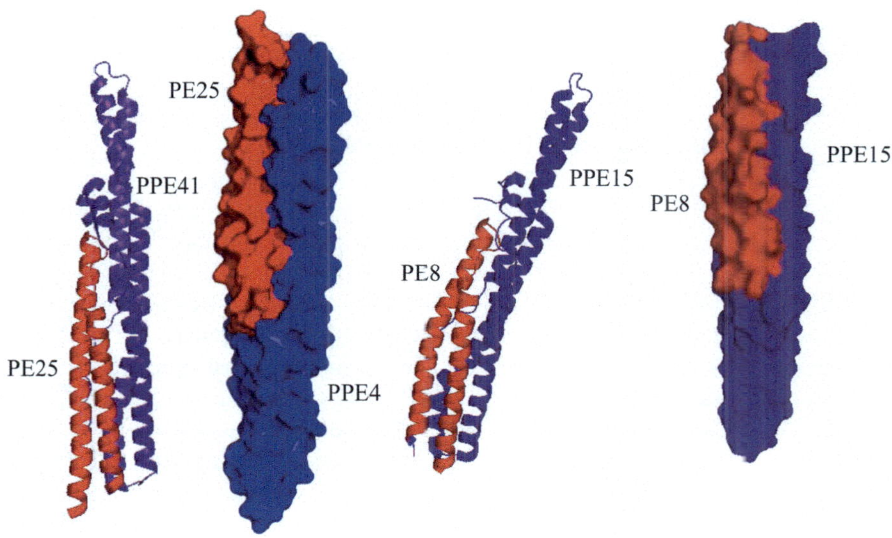

Fig. 9.2 3-D structure of PE/PPE proteins: PE proteins are represented in red; PPE proteins are shown in blue color. PE/PPE proteins are helix-rich proteins with number of interacting residues between them

9.5 Structural Organization of PE25/PPE41

The PE25 protein has 99 amino acids, along with two α-helices, and the PE motif, residues 8–9, is found to be present in the electron density map at N terminal of the protein. The residues 8–37 form one α-helix, and residues 45–84 form another α-helix of PE25. Both the helices are antiparallel with each other and connected by residues 38–44 forming loop. N and C termini of the protein are located at the top of the complex. Both the helices of PE protein interact with helices 2 and 3 of the PPE protein, and loop in between the helix of PE protein interacts with helices 2 and 5 of the PPE protein to make the PE protein stabilized. Due to the sequence similarities, nearly all the PE proteins of *M.tb* share similar structural organization.

The PPE41 protein has 194 amino acids, along with five α-helices, and the PPE motif, residues 7–9, is found to be present in the electron density map at N terminal "hook" of the protein. This hook protects the interacting PE protein. Extensive polar regions, residues 21–53, form helix 2, and residues 58–103 form helix 3 of PPE proteins which run antiparallel with each other and do interact with PE protein to make several hydrophobic and stearic interactions.

There are basically four regions of conserved residues, and the first high conservation region lies in the interior of four-helix bundle at the interface of PPE with PE protein and is conserved in many PE/PPE pairs due to sequence similarities. The second high conservation region lies in the PE loop that interacts with the PPE protein surface. The third high conservation region includes the PPE sequence motif, and the fourth high conservation region is in the C terminus of the PPE protein, a poly-proline-rich region. The tyrosine (Y139) of PPE and proline (P170, P171, P172, and P70) of the PE are one of the most conserved residues.

9.6 Structural Organization of PE8/PPE15

The purification of individual PE8 and PPE15 protein was difficult, so truncated $PE8_{1-99}$ and $PPE15_{1-194}$ were purified together. To get the crystal of these proteins, EspG5, a specific chaperone for PE/PPE proteins was included. All the three proteins exist in a stoichiometric ratio of 1:1:1 and form an elongated-shaped structure in solution (Chen et al. 2017). Residues 85–99 of PE8 (YxxxD/E secretion motif) and residues 174–194 of PPE15 are highly disordered. So the crystal structure which is known consists of 7–299 of EspG5, residues 7–84 of PE8, and residues 1–173 of PPE15. The structure of PE8/PPE15 is found to be similar to PE25/PPE41 with RMSD of 2.475 Å (PE) and 2.037 Å (PPE). The overall RMSD is calculated to be 1.9. Figure 9.3 EspG5 interacts only with helix 4 (α-4) and helix 5 (α-5) of PPE15 just at the opposite end of the PE/PPE heterodimer. The two helices of PE8 interact with the α-1, α-2, α-3, and α-5 of PPE15 to form a four-helix bundle.

The C terminal YxxxD/E secretion motif of PE and WxG motif of PPE in PE8/PPE15 and PE25/PPE41 are in different orientation. In PE25/PPE41, both the motifs are found to be in close proximity and form Van der Waals contacts between Y87 of PE25 and W56 of PPE41, whereas, in case of PE8/PPE15, there is no

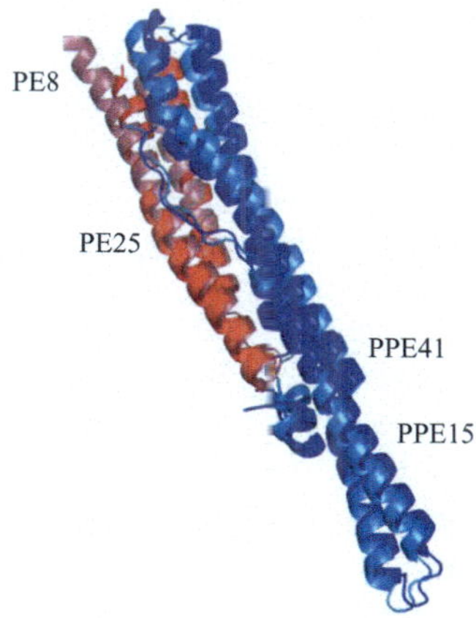

Fig. 9.3 Superimposed image of PE25/PPE41 and PE8/PPE15 proteins with 1.9 RMSD. Red, PE25; raspberry, PE8; blue, PPE41; cyan, PPE15

interaction found between the motifs because the side chain of WxG motif of PPE15 turns away from the PE/PPE binding interface.

The folding patterns are quite similar for both the known structures of PE25/PPE41 and $PE8_{1-99}$ and $PE15_{1-194.}$ Overlapping of both the pairs reveals the shifting in the helical direction of about 26°–29° and 20°–23° in α-1 and α-2 and α-2 and α-3 of PE8 and PPE15, respectively. These deviations may occur due to the presence of proline (Pro35 of PE8, Pro71 of PPE15) or glycine (Gly59 of PE8, Gly39 of PPE15). The α-2 helix of PPE15 that also contains two highly conserved glycine (Gly22 and Gly33) at the proximal to the kinks and various highly conserved alanine residues that are present in PE8 and PPE15. These helical bending, deviations, and presence of kinks on both PE and PPE to get same extent of bending are required for the specific PE/PPE pair formation and interaction. Both the electrostatic and hydrophobic interactions are present in the formation of the complex of PE25/PPE41 and $PE8_{1-99}$/$PPE15_{1-194}$. Ser48 in the α-2 helix of PE8 makes hydrogen bond (conserved hydrogen bond) with Tyr154 in the α-5 helix of PPE15. This hydrogen bond formation is critical for the minimal binding of the protein complex. All the four helices make several hydrophobic contacts, and at the upper and lower areas of the complex, distinct salt bridges and hydrogen bonds are present. One salt bridge (Glu46 of PE8 with Arg14 of PPE15) and three hydrogen bonds (Gln51 of PE8 with Ser93 of PPE15, His73 of PE8 with Tyr45 of PPE15, and Gln70 of PE8 with Tyr72 of PPE15) are also identified in PE8/PPE15 interactions. In PE25/PPE41, one salt bridge (Arg24 of PE25 with Glu37 of PPE41) and one hydrogen bond (Glu17 of PE25 with Thr48 of PPE41) are found at the interface of binding complex, but these two interactions are not found in PE8/PPE15 in spite that these residues (Ala24 of PE8 and Glu37 of PPE15 and Gln17 of PE8 and Val48 of PPE15) are present in the PE8/PPE15 complex. These interactions help in making the complex more stable and strong. Arg14, Tyr45, Ser93, and Tyr72 residues of PPE15 are essential for the PE8/PPE15 interaction (Chen et al. 2017). We can say that for them specificity and binding affinity of PE/PPE complex adopts unique sets of complementary residues.

9.7 PE/PPE Proteins and Host Cell Death Pathways: Effectors as Novel Targets

The PE/PPE family of proteins is commonly associated to the pathogenic mycobacterial species. Individual members of PE/PPE proteins have been demonstrated to involve in the disease process by providing antigenic variation, immune evasion, immune quorum sensing, and cell death (Choudhary et al. 2003; Deng et al. 2017; Nair et al. 2009, 2011; Tundup et al. 2008). The pathogenic strains of mycobacteria employ inhibition of apoptosis/programmed cell death/cell death as a means to survive within the infected host macrophages. In contrast, it has been shown that *M.tb* induces apoptosis for its successful dissemination and propagation (Davis and Ramakrishnan 2009). Therefore, inhibition of apoptosis by mycobacteria at an early stage in the infection process provides the pathogen a favorable niche for the replication and persistence in the host while at later phase in the infection process when nutrients are scare and bacterial burden is high, mycobacteria use apoptosis for its dissemination and successful propagation (Gupta and Gollapudi 2006; Haslinger-Loffler et al. 2005). In this regard, a recent study by Ahmed J et al. demonstrated that the *M.tb* multi-functional, intrinsically disordered protein PE/PPE family member PPE37 C terminal segment translocates to the nucleus and thereby induces caspase 3-dependent apoptosis of the host cells (Ahmad et al. 2018). The co-operonic PE/PPE proteins PE25/PPE41 which are secreted by the type VII secretion system of *M.tb* have been reported to elicit necrosis rather than apoptosis in RAW 264.7 macrophages (Tundup et al. 2014). It has been thought that induction of necrosis by *M.tb* is one of the important strategies deployed to enhance its multiplication, survival, and dissemination (Dobos et al. 2000; Saunders and Britton 2007). However, inhibition of apoptosis by virulent mycobacteria in infected macrophages is believed to help the bacteria escape the immune responses that would be generated when apoptotic bodies harboring bacterial antigens are taken up and presented by dendritic cells (Behar et al. 2010). The recombinant purified and *M. smegmatis* overexpressing PE/PGRS33 have been demonstrated to induce apoptosis in RAW 264.7 macrophages by signaling through TLR2 and the resultant production of TNF-α (Basu et al. 2007). Additionally, PGRS domain deletion of PE/PGRS33 resulted in its decreased ability to induce apoptosis (Basu et al. 2007). PE/PGRS33 has also been shown to localize in the Jurkat T cells mitochondria following overexpression to exhibit apoptotic effect (Balaji et al. 2007; Cadieux et al. 2011). Recombinant *M. smegmatis* overexpressing PE/PGRS33 demonstrated increased ability in inducing necrosis and had a better survival in cell and mouse infection models (Dheenadhayalan et al. 2006). However, wild-type *M. smegmatis* induces apoptosis in a PE domain-dependent manner (Cadieux et al. 2011). Therefore, it is believed that surface-localized PE/PGRS33 is involved in the modulation of host cell death pathways in several different ways for which the exact mechanism of action remains to be elucidated.

PE/PGRS18 is a cell envelope-associated protein of *M.tb*. PE/PGRS18 enhances the survival of recombinant overexpressing *M. smegmatis* within the macrophages possibly through the attenuation of apoptosis. These effects are largely attributed to the ability of recombinant *M. smegmatis* to alter the cytokine profile of the infected macrophages. PE/PGRS18 expressing *M. smegmatis* leads to the decreased production of IL-1β and IL-6 and increased production of IL-12 early in the infection process, demonstrating their role in physiology and pathogenesis of *M tb* (Yang et al. 2017). Another PE/PGRS family member, PE/PGRS41, has been shown to be associated with the cell envelope of recombinant overexpressing *M. smegmatis*. Overexpression of PE/PGRS41 leads to the enhanced survival of recombinant bacteria within the infected macrophages, and this effect was largely attributed due to the decreased apoptosis and autophagy, two of the important host innate defense mechanisms. PE/PGRS41 inhibits the production of pro-inflammatory cytokines IL-1β, TNF-α, and IL-6. It inhibits the apoptosis by modulating the activation of caspase 1, caspase 9, and caspase 3, whereas inhibition of autophagy by PE/PGRS41 is mediated through the inhibition of ATG-8. Interestingly, PE/PGRS41 helps in the dissemination of the overexpressing strain of *M. smegmatis* by inducing the necrosis (Deng et al. 2016). An important member of PE/PPE protein family PE13 has been shown to modulate the host-pathogen interactions through signaling via p38-ERK-NFκB pathways. Overexpression of PE13 in *M. smegmatis* demonstrated its cell wall association. Recombinant expression in *M. smegmatis* and infection studies showed that it enhances the survival of recombinant bacteria and also increased its ability of resistance to different stresses. Overexpressing strain of PE13 demonstrated increased ability to induce the host cell death by increasing apoptosis, but the underlying mechanism remains to be explored (Li et al. 2016).

PPE32 another member of PE/PPE protein family, which was found to be associated with the cell wall, modulates the cell death pathway, i.e., apoptosis, by the activation of caspases 1, 3, and 9. PPE32 induced the apoptosis of macrophages infected with recombinant PPE32 harboring *M. smegmatis*, possibly through the activation of endoplasmic reticulum stress-related pathways and subsequent activation of Map kinase, MAPK ERK1/2 (Deng et al. 2016). PE/PPE protein family member PPE44 has been demonstrated to be a virulence factor; its expression was found higher in the *M.tb*-infected guinea pig lungs (Yu et al. 2017). Yu Z et al. showed by heterologous expression of PE44 in *M. smegmatis* that it is a cell wall-associated protein. Infection studies with recombinant bacteria expressing PE44 showed that it leads to the enhanced secretion of IL-6 and IL-12 through NF-κB-P38-ERK1/2 axis. Infection of macrophages with recombinant *M. smegmatis* displayed increased secretion of LDH. PE44 induces enhanced apoptosis in infected macrophages, thus demonstrating its involvement in controlling cell death pathways, but the underlying mechanism is not known (Yu et al. 2017). PE11 a PE/PPE family member was earlier reported to be overexpressed after 24 h of starvation, or transient acid exposure suggested a role in pathogen persistence or dormancy, particularly its involvement in fatty acid metabolism (Fisher et al. 2002). Heterologous expression in *M. smegmatis* showed its cell wall association. Infection of macrophages with recombinant bacteria harboring PE11 leads to its enhanced survival. Its overexpression manipulates IL-6, IL-10, IL-4, and TNF-α secretion and also induces increased level of necrosis of infected macrophages as demonstrated by increased release of LDH, establishing its role in manipulating cell death (Deng et al. 2015; Singh et al. 2016). PE9 and PE10 are only of the two clusters that comprise two PE genes in tandem. It has been recently demonstrated that these two proteins physically interact with each other and form heterodimer. PE9/PE10 complex interacts with innate immune receptor TLR4 and leads to the enhanced secretion of IFN-β, IL-10, and IL-1β in treated macrophages. Interaction of PE9/PE10 to TLR4 elicits increased expression of pro-apoptotic genes bax, bam, and bid and activation of effector caspase, caspase 3, thus inducing increased apoptosis in the treated macrophages. This study therefore demonstrated the role of PE9/PE10 in manipulating the multiple hallmarks of apoptosis and thus importance of these genes in modulating the *M.tb* pathogenesis (Tiwari et al. 2015).

The PE/PPE family member, PE/PGRS5, expresses in the later stages of *M.tb* infection, i.e., in human granulomas. The PGRS domain of PE/PGRS5 harbors an intrinsically disordered endoplasmic reticulum-targeting sequence and targeted to endoplasmic reticulum. It induces TLR4-dependent endoplasmic reticulum stress response and generation of NO and oxidative stress responses. PGRS domain of PE/PGRS5 leads to the activation of caspase 8, thereby inducing apoptosis in treated macrophages. This demonstrates an earlier unknown function of PE/PGRS5 in endoplasmic reticulum stress-mediated induction of cell death (Grover et al. 2018). This topic is illustrated in Fig. 9.4.

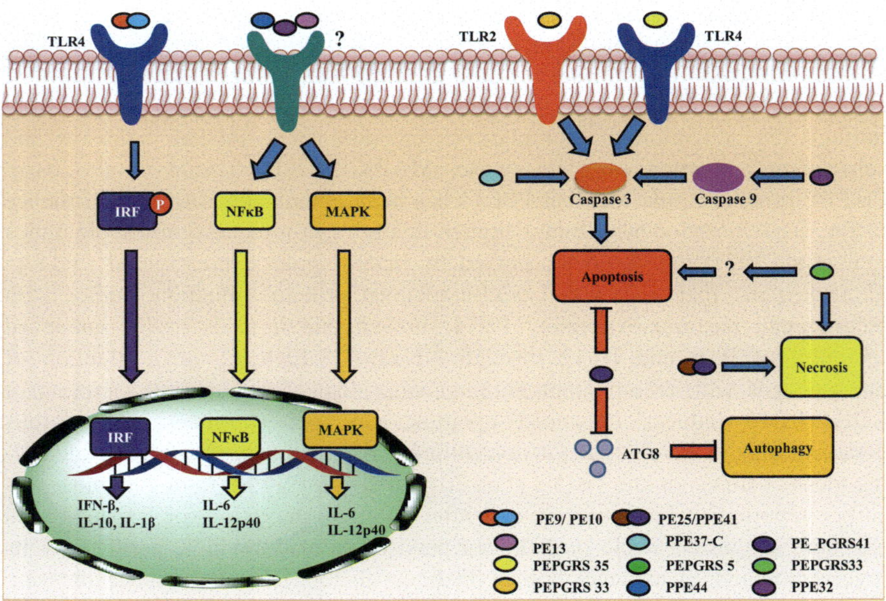

Fig. 9.4 Diagrammatic representation of role of PE and PPE proteins in cell death. TLR, Toll-like receptor

9.8 Conclusion

The abundance of PE/PPE protiens in the genome of slow-growing mycobacteria implies some important functions of this protein family in the life cycle of bacteria. Although a substantial progress has been made in the biochemical and molecular characterization since their discovery, their sybilline structural characterestics are still puzzling researchers. The presence of disordered regions in this family is additionally perplexing. The disorder regions in these proteins suggest their interactive potential and the ability of moonlighting functions. This higher level of intrinsic disorder suggests the positive evolutionaty selection. The study of these dynamic areas will complement the structural studies that are yet scarce in case of these PE/PPE proteins. The structural studies will aid in target development as well as biochemical characterization of hypothetical proteins. An extensive effort is required to reval the atomic coordinates of these proteins with their inhibitors to develop novel drug targets. Moreover, basic understanding of role of these proteins in manupilating host cell death pathways could provide insights into novel therapeutic approaches to curb infection.

References

Abdallah AM et al (2006) A specific secretion system mediates PPE41 transport in pathogenic mycobacteria. Mol Microbiol 62:667–679. https://doi.org/10.1111/j.1365-2958.2006.05409.x

Abdallah AM et al (2009) PPE and PE_PGRS proteins of Mycobacterium marinum are transported via the type VII secretion system ESX-5. Mol Microbiol 73:329–340. https://doi.org/10.1111/j.1365-2958.2009.06783.x

Ahmad J et al (2018) Contrasting function of structured N-terminal and unstructured C-terminal segments of Mycobacterium tuberculosis PPE37 protein. MBio 9. https://doi.org/10.1128/mBio.01712-17

Akhter Y, Ehebauer MT, Mukhopadhyay S, Hasnain SE (2012) The PE/PPE multigene family codes for virulence factors and is a possible source of mycobacterial antigenic variation: perhaps more? Biochimie 94:110–116. https://doi.org/10.1016/j.biochi.2011.09.026

Balaji KN, Goyal G, Narayana Y, Srinivas M, Chaturvedi R, Mohammad S (2007) Apoptosis triggered by Rv1818c, a PE family gene from Mycobacterium tuberculosis is regulated by mitochondrial intermediates in T cells. Microbes Infect 9:271–281. https://doi.org/10.1016/j.micinf.2006.11.013

Basu S et al (2007) Execution of macrophage apoptosis by PE_PGRS33 of Mycobacterium tuberculosis is mediated by toll-like receptor 2-dependent release of tumor necrosis factor-alpha. J Biol Chem 282:1039–1050. https://doi.org/10.1074/jbc.M604379200

Behar SM, Divangahi M, Remold HG (2010) Evasion of innate immunity by Mycobacterium tuberculosis: is death an exit strategy? Nat Rev Microbiol 8:668–674. https://doi.org/10.1038/nrmicro2387

Bottai D, Brosch R (2009) Mycobacterial PE, PPE and ESX clusters: novel insights into the secretion of these most unusual protein families. Mol Microbiol 73:325–328. https://doi.org/10.1111/j.1365-2958.2009.06784.x

Bottai D et al (2012) Disruption of the ESX-5 system of Mycobacterium tuberculosis causes loss of PPE protein secretion, reduction of cell wall integrity and strong attenuation. Mol Microbiol 83:1195–1209. https://doi.org/10.1111/j.1365-2958.2012.08001.x

Brennan MJ (2017) The enigmatic PE/PPE multigene family of mycobacteria and tuberculosis vaccination. Infect Immun 85. https://doi.org/10.1128/IAI.00969-16

Brennan MJ, Delogu G (2002) The PE multigene family: a 'molecular mantra' for mycobacteria. Trends Microbiol 10:246–249

Brennan MJ, Delogu G, Chen Y, Bardarov S, Kriakov J, Alavi M, Jacobs WR Jr (2001) Evidence that mycobacterial PE_PGRS proteins are cell surface constituents that influence interactions with other cells. Infect Immun 69:7326–7333. https://doi.org/10.1128/IAI.69.12.7326-7333.2001

Brodin P et al (2010) High content phenotypic cell-based visual screen identifies Mycobacterium tuberculosis acyltrehalose-containing glycolipids involved in phagosome remodeling. PLoS Pathog 6:e1001100. https://doi.org/10.1371/journal.ppat.1001100

Cadieux N, Parra M, Cohen H, Maric D, Morris SL, Brennan MJ (2011) Induction of cell death after localization to the host cell mitochondria by the Mycobacterium tuberculosis PE_PGRS33 protein. Microbiology 157:793–804. https://doi.org/10.1099/mic.0.041996-0

Chen X, Cheng HF, Zhou J, Chan CY, Lau KF, Tsui SK, Au SW (2017) Structural basis of the PE-PPE protein interaction in Mycobacterium tuberculosis. J Biol Chem 292:16880–16890. https://doi.org/10.1074/jbc.M117.802645

Choudhary RK, Mukhopadhyay S, Chakhaiyar P, Sharma N, Murthy KJ, Katoch VM, Hasnain SE (2003) PPE antigen Rv2430c of Mycobacterium tuberculosis induces a strong B-cell response. Infect Immun 71:6338–6343

Cole ST (1999) Learning from the genome sequence of Mycobacterium tuberculosis H37Rv. FEBS Lett 452:7–10

Cole ST et al (1998) Deciphering the biology of Mycobacterium tuberculosis from the complete genome sequence. Nature 393:537–544. https://doi.org/10.1038/31159

Copin R et al (2014) Sequence diversity in the pe_pgrs genes of Mycobacterium tuberculosis is independent of human T cell recognition. MBio 5:e00960–e00913. https://doi.org/10.1128/mBio.00960-13

Daleke MH et al (2011) Conserved Pro-Glu (PE) and Pro-Pro-Glu (PPE) protein domains target LipY lipases of pathogenic mycobacteria to the cell surface via the ESX-5 pathway. J Biol Chem 286:19024–19034. https://doi.org/10.1074/jbc.M110.204966

Davis JM, Ramakrishnan L (2009) The role of the granuloma in expansion and dissemination of early tuberculous infection. Cell 136:37–49. https://doi.org/10.1016/j.cell.2008.11.014

Delogu G, Brennan MJ (2001) Comparative immune response to PE and PE_PGRS antigens of Mycobacterium tuberculosis. Infect Immun 69:5606–5611

Deng W, Zeng J, Xiang X, Li P, Xie J (2015) PE11 (Rv1169c) selectively alters fatty acid components of Mycobacterium smegmatis and host cell interleukin-6 level accompanied with cell death. Front Microbiol 6:613. https://doi.org/10.3389/fmicb.2015.00613

Deng W, Yang W, Zeng J, Abdalla AE, Xie J (2016) Mycobacterium tuberculosis PPE32 promotes cytokines production and host cell apoptosis through caspase cascade accompanying with enhanced ER stress response. Oncotarget 7:67347–67359. https://doi.org/10.18632/oncotarget.12030

Deng W, Long Q, Zeng J, Li P, Yang W, Chen X, Xie J (2017) Mycobacterium tuberculosis PE_PGRS41 enhances the intracellular survival of M. smegmatis within macrophages via blocking innate immunity and inhibition of host defense. Sci Rep 7:46716. https://doi.org/10.1038/srep46716

Dheenadhayalan V, Delogu G, Brennan MJ (2006) Expression of the PE_PGRS 33 protein in Mycobacterium smegmatis triggers necrosis in macrophages and enhanced mycobacterial survival. Microbes Infect 8:262–272. https://doi.org/10.1016/j.micinf.2005.06.021

Dobos KM, Spotts EA, Quinn FD, King CH (2000) Necrosis of lung epithelial cells during infection with Mycobacterium tuberculosis is preceded by cell permeation. Infect Immun 68:6300–6310

Dona V et al (2013) The PPE domain of PPE17 is responsible for its surface localization and can be used to express heterologous proteins on the mycobacterial surface. PLoS One 8:e57517. https://doi.org/10.1371/journal.pone.0057517

Dong D et al (2012) PPE38 modulates the innate immune response and is required for Mycobacterium marinum virulence. Infect Immun 80:43–54. https://doi.org/10.1128/IAI.05249-11

Dunker AK et al (2001) Intrinsically disordered protein. J Mol Graph Model 19:26–59

Ekiert DC, Cox JS (2014) Structure of a PE-PPE-EspG complex from Mycobacterium tuberculosis reveals molecular specificity of ESX protein secretion. Proc Natl Acad Sci U S A 111:14758–14763. https://doi.org/10.1073/pnas.1409345111

Fishbein S, van Wyk N, Warren RM, Sampson SL (2015) Phylogeny to function: PE/PPE protein evolution and impact on Mycobacterium tuberculosis pathogenicity. Mol Microbiol 96:901–916. https://doi.org/10.1111/mmi.12981

Fisher MA, Plikaytis BB, Shinnick TM (2002) Microarray analysis of the Mycobacterium tuberculosis transcriptional response to the acidic conditions found in phagosomes. J Bacteriol 184:4025–4032

Gey van Pittius NC, Sampson SL, Lee H, Kim Y, van Helden PD, Warren RM (2006) Evolution and expansion of the Mycobacterium tuberculosis PE and PPE multigene families and their association with the duplication of the ESAT-6 (esx) gene cluster regions. BMC Evol Biol 6:95. https://doi.org/10.1186/1471-2148-6-95

Glaziou P, Falzon D, Floyd K, Raviglione M (2013) Global epidemiology of tuberculosis. Semin Respir Crit Care Med 34:3–16. https://doi.org/10.1055/s-0032-1333467

Goldstone RM, Goonesekera SD, Bloom BR, Sampson SL (2009) The transcriptional regulator Rv0485 modulates the expression of a pe and ppe gene pair and is required for Mycobacterium tuberculosis virulence. Infect Immun 77:4654–4667. https://doi.org/10.1128/IAI.01495-08

Grover S et al (2018) The PGRS domain of Mycobacterium tuberculosis PE_PGRS protein Rv0297 is involved in endoplasmic reticulum stress-mediated apoptosis through toll-like receptor 4. MBio 9. https://doi.org/10.1128/mBio.01017-18

Gupta S, Gollapudi S (2006) TNF-alpha-induced apoptosis in human naive and memory CD8+ T cells in aged humans. Exp Gerontol 41:69–77. https://doi.org/10.1016/j.exger.2005.10.001

Haslinger-Loffler B et al (2005) Multiple virulence factors are required for Staphylococcus aureus-induced apoptosis in endothelial cells. Cell Microbiol 7:1087–1097. https://doi.org/10.1111/j.1462-5822.2005.00533.x

Iantomasi R et al (2012) PE_PGRS30 is required for the full virulence of Mycobacterium tuberculosis. Cell Microbiol 14:356–367. https://doi.org/10.1111/j.1462-5822.2011.01721.x

Iqbal S, Hoque MT (2016) Estimation of position specific energy as a feature of protein residues from sequence alone for structural classification. PLoS One 11:e0161452. https://doi.org/10.1371/journal.pone.0161452

Khubaib M et al (2016) Mycobacterium tuberculosis co-operonic PE32/PPE65 proteins Alter host immune responses by hampering Th1 response. Front Microbiol 7:719. https://doi.org/10.3389/fmicb.2016.00719

Li Y, Miltner E, Wu M, Petrofsky M, Bermudez LE (2005) A Mycobacterium avium PPE gene is associated with the ability of the bacterium to grow in macrophages and virulence in mice. Cell Microbiol 7:539–548. https://doi.org/10.1111/j.1462-5822.2004.00484.x

Li H, Li Q, Yu Z, Zhou M, Xie J (2016) Mycobacterium tuberculosis PE13 (Rv1195) manipulates the host cell fate via p38-ERK-NF-kappaB axis and apoptosis. Apoptosis 21:795–808. https://doi.org/10.1007/s10495-016-1249-y

Majlessi L, Prados-Rosales R, Casadevall A, Brosch R (2015) Release of mycobacterial antigens. Immunol Rev 264:25–45. https://doi.org/10.1111/imr.12251

Nair S et al (2009) The PPE18 of Mycobacterium tuberculosis interacts with TLR2 and activates IL-10 induction in macrophage. J Immunol 183:6269–6281. https://doi.org/10.4049/jimmunol.0901367

Nair S, Pandey AD, Mukhopadhyay S (2011) The PPE18 protein of Mycobacterium tuberculosis inhibits NF-kappaB/rel-mediated proinflammatory cytokine production by upregulating and phosphorylating suppressor of cytokine signaling 3 protein. J Immunol 186:5413–5424. https://doi.org/10.4049/jimmunol.1000773

Poulet S, Cole ST (1995) Characterization of the highly abundant polymorphic GC-rich-repetitive sequence (PGRS) present in Mycobacterium tuberculosis. Arch Microbiol 163:87–95

Rahman SA, Singh Y, Kohli S, Ahmad J, Ehtesham NZ, Tyagi AK, Hasnain SE (2014) Comparative analyses of nonpathogenic, opportunistic, and totally pathogenic mycobacteria reveal genomic and biochemical variabilities and highlight the survival attributes of Mycobacterium tuberculosis. MBio 5:e02020. https://doi.org/10.1128/mBio.02020-14

Sampson SL (2011) Mycobacterial PE/PPE proteins at the host-pathogen interface. Clin Dev Immunol 2011:497203. https://doi.org/10.1155/2011/497203

Sampson SL, Lukey P, Warren RM, van Helden PD, Richardson M, Everett MJ (2001) Expression, characterization and subcellular localization of the Mycobacterium tuberculosis PPE gene Rv1917c. Tuberculosis 81:305–317. https://doi.org/10.1054/tube.2001.0304

Saunders BM, Britton WJ (2007) Life and death in the granuloma: immunopathology of tuberculosis. Immunol Cell Biol 85:103–111. https://doi.org/10.1038/sj.icb.7100027

Singh P, Rao RN, Reddy JR, Prasad RB, Kotturu SK, Ghosh S, Mukhopadhyay S (2016) PE11, a PE/PPE family protein of Mycobacterium tuberculosis is involved in cell wall remodeling and virulence. Sci Rep 6:21624. https://doi.org/10.1038/srep21624

Strong M, Sawaya MR, Wang S, Phillips M, Cascio D, Eisenberg D (2006) Toward the structural genomics of complexes: crystal structure of a PE/PPE protein complex from Mycobacterium tuberculosis. Proc Natl Acad Sci U S A 103:8060–8065. https://doi.org/10.1073/pnas.0602606103

Tiwari BM, Kannan N, Vemu L, Raghunand TR (2012) The Mycobacterium tuberculosis PE proteins Rv0285 and Rv1386 modulate innate immunity and mediate bacillary survival in macrophages. PLoS One 7:e51686. https://doi.org/10.1371/journal.pone.0051686

Tiwari B, Ramakrishnan UM, Raghunand TR (2015) The Mycobacterium tuberculosis protein pair PE9 (Rv1088)-PE10 (Rv1089) forms heterodimers and induces macrophage apoptosis through toll-like receptor 4. Cell Microbiol 17:1653–1669. https://doi.org/10.1111/cmi.12462

Tundup S, Akhter Y, Thiagarajan D, Hasnain SE (2006) Clusters of PE and PPE genes of Mycobacterium tuberculosis are organized in operons: evidence that PE Rv2431c is co-transcribed with PPE Rv2430c and their gene products interact with each other. FEBS Lett 580:1285–1293. https://doi.org/10.1016/j.febslet.2006.01.042

Tundup S, Pathak N, Ramanadham M, Mukhopadhyay S, Murthy KJ, Ehtesham NZ, Hasnain SE (2008) The co-operonic PE25/PPE41 protein complex of Mycobacterium tuberculosis elicits increased humoral and cell mediated immune response. PLoS One 3:e3586. https://doi.org/10.1371/journal.pone.0003586

Tundup S, Mohareer K, Hasnain SE (2014) Mycobacterium tuberculosis PE25/PPE41 protein complex induces necrosis in macrophages: role in virulence and disease reactivation? FEBS Open Bio 4:822–828. https://doi.org/10.1016/j.fob.2014.09.001

Wright PE, Dyson HJ (1999) Intrinsically unstructured proteins: re-assessing the protein structure-function paradigm. J Mol Biol 293:321–331. https://doi.org/10.1006/jmbi.1999.3110

Yang W et al (2017) Mycobacterium tuberculosis PE_PGRS18 enhances the intracellular survival of M. smegmatis via altering host macrophage cytokine profiling and attenuating the cell apoptosis. Apoptosis 22:502–509. https://doi.org/10.1007/s10495-016-1336-0

Yu Z et al (2017) Mycobacterium tuberculosis PPE44 (Rv2770c) is involved in response to multiple stresses and promotes the macrophage expression of IL-12 p40 and IL-6 via the p38, ERK, and NF-kappaB signaling axis. Int Immunopharmacol 50:319–329. https://doi.org/10.1016/j.intimp.2017.06.028

Zumbo A et al (2013) Functional dissection of protein domains involved in the immunomodulatory properties of PE_PGRS33 of Mycobacterium tuberculosis. Pathog Dis 69:232–239. https://doi.org/10.1111/2049-632X.12096

Comparative In Silico Analyses Reveal Crucial Factors for Virulence, Antigenicity, and Evolution in *M.tb*

10

Yadvir Singh

Abstract

The evolution of pathogenic mycobacteria includes not only loss of nonessential genes for survival in the host but also horizontal gene transfer-mediated gene acquisition critical for pathogenicity. Despite this pressure of reductive evolution, some multigene families have instead expanded down the evolution.

In silico comparative analyses of these multigene families (*mce*, *mmpL*, sigma, TCSS, and PE/PPE) in the context of virulence in *M.tb*, i.e., phenotypic difference between $H_{37}Rv$ and $H_{37}Ra$, have intensified the importance of PE/PPE gene family.

Although *Mycobacterium indicus pranii* (*MIP*), a nonpathogenic soil mycobacterium, is positioned much above *Mycobacterium tuberculosis* in evolutionary line, it shares ~75% of *M.tb* antigens. Upon comparing *MIP* with another vaccine candidate, *M. vaccae*, in terms of sharing antigens with *M.tb*, further highlighted the importance of PE/PPE proteins. This comparative antigenic analysis has also pointed to the PGRS domain acquisition by the PE proteins and further expansion through gene duplication events. These results have suggested gene co-option as another key player in mycobacterial evolution besides horizontal gene transfer and genomic reduction.

In conclusion, this chapter is an attempt to use in silico methodology to uncover the importance of PE and PE_PGRS proteins in virulence, antigenicity, and host immunomodulation in tuberculosis.

Keywords

PE/PPE gene family · Gene co-option · *MIP* · In silico analyses

Y. Singh (✉)
Department Molecular Biology, Max Planck Institute for Infection Biology, Berlin, Germany
e-mail: singh@mpiib-berlin.mpg.de

S. E. Hasnain et al. (eds.), Mycobacterium tuberculosis: *Molecular Infection Biology, Pathogenesis, Diagnostics and New Interventions*,
https://doi.org/10.1007/978-981-32-9413-4_10

Abbreviations

BLAST	Basic local alignment search tool
CNP	Carbon, nitrogen, and phosphate
ENA	European Nucleotide Archive
GRAVY	Grand average of hydropathicity
M.tb	Mycobacterium tuberculosis
MAC	*M. avium* complex
MCE	Mammalian cell entry
MIP	*Mycobacterium indicus pranii*
MMPL	Mycobacterial membrane protein large
NCBI	National Center for Biotechnology Information
PE	Pro-Glu
PGRS	Polymorphic GC-rich sequence
PPE	Pro-Pro-Glu
RND	Resistance, nodulation, and cell division proteins
RR	Response regulator
SK	Sensor kinase
SNVs	Single-nucleotide variations
TCSS	Two-component signal transduction systems

10.1 Introduction

10.1.1 Multigene Families in *M.tb*

In 1905, a parental strain of current laboratory strains H_{37}Rv and H_{37}Ra, H_{37}, was isolated from a human patient. Later, this strain evolved in two different strains H_{37}Rv and H_{37}Ra, exhibiting different phenotypes in terms of virulence (Steenken Jr. and Gardner 1946). H_{37}Rv exhibit a virulent phenotype in contrast to H_{37}Ra, which exhibits an avirulent phenotype. H_{37}Rv and H_{37}Ra are still used in laboratories as reference strains around the world to study *M.tb*.

In 1998, Stewart Cole and colleagues published the first genome sequence of *Mycobacterium tuberculosis* (H_{37}Rv, a laboratory strain) (Cole et al. 1998). This sequence data has provided a novel and significant understanding into the virulence of *M.tb* and documented the occurrence of a few highly duplicated gene families. For competent replication and transcription, huge pressure is always on pathogenic bacteria like *M.tb* to preserve their condensed genomes. Gene duplication facilitates the functional load on a single gene by dispensing it between the duplicate copies. Further, down the evolution, it also helps to explore new functions through acquisitions of variations. In *M.tb*, gene duplication events have contributed toward the emergence of multigene families encompassing ~ 50% of the proteome (Tekaia et al. 1999), i.e., mammalian cell entry (*mce*) gene family, *pe/ppe* gene family, and mycobacterial membrane protein large (*mmpL*) gene family (Mulder et al. 2009). Expansion of these gene families down the evolution highlights the significance of their use by the pathogenic bacteria (Sassetti and Rubin 2003). Some of the multigene families have been linked to pathogenesis of *M.tb*, i.e., mammalian cell entry (*mce*) gene family, mycobacterial membrane protein large (*mmpL*) gene family, sigma factor family, and two-component signal transduction systems.

10.1.2 Multigene Families Important in the Pathogenesis of Tuberculosis

10.1.2.1 Mammalian Cell Entry (mce) Gene Family

Mammalian cell entry (*mce*) gene family has been shown to play a role in entry and survival in host macrophages (Arruda et al. 1993). MCE proteins were first recognized as macrophage cell entry proteins upon expression in nonpathogenic *E. coli* (Arruda et al. 1993). Four *mce* operons are present in *M.tb* genome (mce1–4). Each *mce* operon comprises 8–13 genes. Overall, 36 members (Table 10.1) excluding Rv0590A (a probable continuation of *mce2B*) have been described in this family (Zhang and Xie 2011). The role in transmembrane transportation by MCE proteins is indicated by the presence of transmembrane helices (Casali and Riley 2007). Further, at the N-terminal anchoring in the membrane, MCE proteins also contain a hydrophobic stretch of amino acid residues (Chitale et al. 2001). These features correlate their cell surface location with proposed function in host-pathogen interactions (Harboe et al. 2002). Furthermore, a mutation study in *mce* operons had been shown to cause attenuation of *M.tb* virulence (Gioffre et al. 2005).

10.1.2.2 Mycobacterial Membrane Protein Large (*mmpL*) Gene Family

Mycobacterial membrane protein large (*mmpL*) gene family encodes for 13 RND proteins (Table 10.1) (resistance, nodulation, and cell division proteins) in mycobacteria and mainly consists of multidrug resistance pumps (Domenech et al. 2005). MMPL7 and MMPL8 proteins have been linked to complex lipid transport (Cox et al. 1999; Converse et al. 2003). Mutation studies have shown that MmpL4, MmpL5, MmpL7, MmpL8, MmpL10, and MmpL11 were associated with *M.tb* virulence in the mice model (Lamichhane et al. 2005).

10.1.2.3 Sigma Factors Gene Family

Alternative sigma factors control transcription in response to a particular stimulus in mycobacteria (Bashyam and Hasnain 2004). The ratio of alternative sigma factors to genome size is highest among the obligate pathogens (Rodrigue et al. 2006). The sigma factor family permits *M. tuberculosis* to modulate multiple phases of host-pathogen interactions like adhesion, entry, intracellular replication, and spreading to other sites by regulating the temporal expression of specific regulons (Rodrigue et al. 2006). This family has 13 members (Table 10.1), namely, sigma A to sigma M. SigA, SigC, SigE, SigF, SigH, and SigL proteins were linked to *M.tb* virulence (Smith 2003).

Table 10.1 List of members of *M.tb* $H_{37}Rv$ *mce* gene family, *mmpL* gene family, sigma factor gene family, TCSS, and PE/PPE (Kohli et al. 2012)

Mammalian cell entry (*mce*) gene family									
Rv0165c	Rv0166	Rv0167	Rv0168	Rv0169	Rv0170	Rv0171	Rv0172	Rv0173	Rv0174
Rv0586	Rv0587	Rv0588	Rv0589	Rv0590	Rv0590A	Rv0591	Rv0592	Rv0593	Rv0594
Rv1963c	Rv1964	Rv1965	Rv1966	Rv1967	Rv1968	Rv1969	Rv1970	Rv1971	Rv3494c
Rv3495c	Rv3496c	Rv3497c	Rv3498c	Rv3499c	Rv3500c	Rv3501c			
Mycobacterial membrane protein large (*mmpL*) gene family									
Rv0202c	Rv0206c	Rv0402c	Rv0450c	Rv0507	Rv0676c	Rv1145	Rv1146	Rv1183	Rv1522c
Rv1557	Rv2339	Rv2942	Rv3823c						
Sigma factors gene family									
Rv0182c	Rv0445c	Rv0735	Rv1189	Rv1221	Rv2069	Rv2703	Rv2710	Rv3223c	Rv3286c
Rv3328c	Rv3414c	Rv3911							
Two-component signal transduction systems (TCSS)									
Rv0260c	Rv0490	Rv0491	Rv0600c	Rv0601c	Rv0602c	Rv0757	Rv0758	Rv0818	Rv0844c
Rv0845	Rv0902c	Rv0903c	Rv0981	Rv0982	Rv1027c	Rv1028c	Rv1032c	Rv1033c	Rv1626
Rv2027c	Rv2884	Rv3132c	Rv3133c	Rv3143	Rv3220c	Rv3245c	Rv3246c	Rv3764c	Rv3765c

5. PE/PPE gene family									
Rv0151c	Rv0152c	Rv0159c	Rv0160c	Rv0285	Rv0335c	Rv0916c	Rv1040c	Rv1088	Rv1089
Rv1169c	Rv1172c	Rv1195	Rv1214c	Rv1386	Rv1430	Rv1646	Rv1788	Rv1791	Rv1806
Rv2099c	Rv2107	Rv2328	Rv2408	Rv2431c	Rv2519	Rv2769c	Rv3020c	Rv3477	Rv3622c
Rv3650	Rv3746c	Rv3872	Rv3893c	Rv0109	Rv0124	Rv0278c	Rv0279c	Rv0297	Rv0532
Rv0578c	Rv0742	Rv0746	Rv0747	Rv0754	Rv0832	Rv0833	Rv0834c	Rv0872c	Rv0977
Rv0978c	Rv0980c	Rv1067c	Rv1068c	Rv1087	Rv1091	Rv1243c	Rv1325c	Rv1396c	Rv1441c
Rv1450c	Rv1452c	Rv1468c	Rv1651c	Rv1759c	Rv1768	Rv1803c	Rv1818c	Rv1840c	Rv1983
Rv2098c	Rv2126c	Rv2162c	Rv2340c	Rv2371	Rv2396	Rv2487c	Rv2490c	Rv2591	Rv2615c
Rv2634c	Rv2741	Rv2853	Rv3097c	Rv3344c	Rv3345c	Rv3367	Rv3388	Rv3507	Rv3508
Rv3511	Rv3512	Rv3514	Rv3590c	Rv3595c	Rv3652	Rv3653	Rv3812	Rv0096	Rv0256c
Rv0280	Rv0286	Rv0304c	Rv0305c	Rv0354c	Rv0355c	Rv0388c	Rv0442c	Rv0453	Rv0755c
Rv0878c	Rv0915c	Rv1039c	Rv1135c	Rv1168c	Rv1196	Rv1361c	Rv1387	Rv1548c	Rv1705c
Rv1706c	Rv1753c	Rv1787	Rv1789	Rv1790	Rv1800	Rv1801	Rv1802	Rv1807	Rv1808
Rv1809	Rv1917c	Rv1918c	Rv2108	Rv2123	Rv2352c	Rv2353c	Rv2356c	Rv2430c	Rv2608
Rv2768c	Rv2770c	Rv2892c	Rv3018c	Rv3021c	Rv3022c	Rv3125c	Rv3135	Rv3136	Rv3144c
Rv3159c	Rv3343c	Rv3347c	Rv3350c	Rv3425	Rv3426	Rv3429	Rv3478	Rv3532	Rv3533c
Rv3539	Rv3558	Rv3621c	Rv3738c	Rv3739c	Rv3873	Rv3892c			

10.1.2.4 Two-Component Signal Transduction Systems Gene Family

Two-component signal transduction systems (TCSS) consist of a membrane-localized histidine sensor kinase (SK) involved in recognizing the external signal and a response regulator localized cytoplasmically (RR). *M.tb* uses TCSS for environmental adaptation within the host. TCSSs have also been shown to be regulating many physiological processes, i.e., competence, sporulation, drug resistance, CNP (carbon, nitrogen, and phosphate) use, and virulence (Krell et al. 2010). TCSSs consist of 11 core members, 5 orphaned RRs, and 2 orphaned SKs (Table 10.1), and most of them have been shown to be associated with virulence (Bretl et al. 2011). Compared to other bacterial species of similar genome size, *M. tuberculosis* has a smaller number of complete TCSSs, specifying the evolution of *M.tb* as a primarily intracellular human pathogen.

10.1.2.5 PE/PPE Gene Family

The PE/PPE gene family consists of proteins with conserved signature motif Pro-Glu (PE) and Pro-Pro-Glu (PPE) at the N-terminus, respectively. This gene family consists of ~107 PE and ~69 PPE members (Table 10.1). PE gene family has been further branched into *pe* and *pe_pgrs* gene family. In PE_PGRS, PE motif is followed by polymorphic GC-rich repetitive sequences (PGRS) at the C-terminal domain, which code for glycine and alanine repeats (Brennan and Delogu 2002; Cole et al. 1998). These glycine and alanine repeats further show variations in number, size, and sequence between different PE_PGRS proteins (Brennan and Delogu 2002).

10.1.3 Pathogenic *M.tb* vs Nonpathogenic Mycobacteria

The mycobacterium genus comprises more than 120 species, which can be easily grouped into non-pathogens like *M. smegmatis* opportunistic pathogens like *M. avium* and true pathogenic bacteria like *Mycobacterium tuberculosis* (*M.tb*). The genome size of nonpathogenic mycobacteria is larger than that of true pathogens suggesting that gene loss is a significant part of continuing evolution of slow-growing pathogenic mycobacteria (Brosch et al. 2002; Ahmed et al. 2008). There is also an acquisition of genes specific to intracellular survival in the host by *M.tb*. Comparative proteomic analysis of *M.tb* with nonpathogenic predecessor mycobacteria like *Mycobacterium indicus pranii* (*MIP*) has helped to elucidate the reasons of pathogenicity of *M.tb* (Singh et al. 2014).

10.1.3.1 Mycobacterium Indicus Pranii (MIP)

MIP is a saprophytic, nonpathogenic mycobacterium and taxonomically positioned in *M. avium* complex (*MAC*) of mycobacteria (opportunistic pathogens) (Ahmed et al. 2007; Saini et al. 2009). *MIP* has been used as an immunomodulator in various diseases (Ahmad et al. 2011; Katoch et al. 2008; Rakshit et al. 2012; Talwar 1999), i.e., an effective commercial immunotherapeutic vaccine "Immuvac" against leprosy (Nath 1998). Further, animal trials of *MIP* have shown that it can impart a protective response against *M.tb* infection and is under testing in phase III human clinical trials (Gupta et al. 2012; Gupta et al. 2009; Saini et al. 2012). This broad-spectrum antigenic potential of *MIP* is justified by an exceptional taxonomical position of *MIP* which leads to sharing more B- and T-cell antigens with other pathogenic mycobacteria including *M.tb* and *M. leprae* (Saini et al. 2012; Singh et al. 1992; Yadava et al. 1991).

10.1.4 Gene co-Option: Another Key Player in *M.tb* Evolution

Gene co-option includes modification of existing function or acquisition of a new function of a gene, possibly via gene duplication. In absence of gene duplication, genes can still gain a new function either by a change in coding sequences or gain of novel domains. While in presence of gene duplication, co-option can be acquired by either subfunctionalization (partial conservation of function) of the paralogous genes or neo-functionalization (gain of novel function) (True and Carroll 2002). Gene co-option has also been shown to elucidate the presence of virulence factors among environmental and pathogenic Rhodococci bacteria, which belong to the same taxonomical order like *M.tb* (Letek et al. 2010). Gene co-option could elucidate the emergence of pathogenic bacteria from nonpathogenic bacteria through adaptive changes in the niche factors.

10.2 Methods Used in In Silico Comparative Proteomics of *M.tb*

The first step for a comparative analysis is collection of gene sequences from various databases. Gene database of National Center for Biotechnology Information (www.ncbi.nlm.nih.gov/gene) and European Nucleotide Archive sequence database (www.ebi.ac.uk/ena) are valuable sources for listing and FASTA sequences of genes. With the help of these databases, FASTA sequences of these genes are searched in ENA database for corresponding genes in another strain/species after considering nucleotide position number (base range) as a reference point. Furthermore, BLASTn (nucleotide BLAST) tool from NCBI is helpful to explore differences between two strains/species in terms of insertions, deletions, and single-nucleotide variations (SNVs). Pipeline for in silico analyses is shown in Fig. 10.1.

Through various in silico analyses, the genetic and proteomic differences among these established gene families have been linked to the context of different virulence phenotypes of H_{37}Rv and H_{37}Ra. Various parameters can be calculated and compared using in silico tools. With the help of T-coffee align program, many proteins across these multigene families have been found to contain substitutions, insertions, and deletions in amino acid sequence between H_{37}Rv and H_{37}Ra. ProtParam tool from ExPASy portal (http://www.web.expasy.org/protparam/) is able to calculate various physicochemical properties like instability index, aliphatic index, and grand average of hydropathicity (GRAVY) which have been further explained in the following paragraphs.

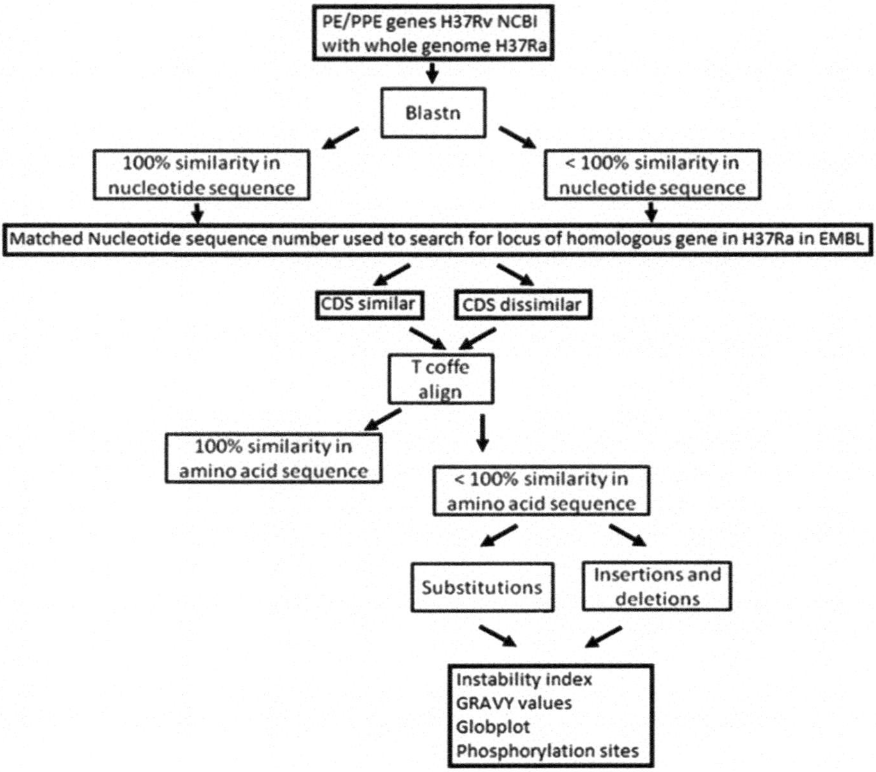

Fig. 10.1 Pipeline for in silico analyses for comparative proteomics of *M.tb*

The instability index correlates with the stability of the protein in vitro. Instability index >40 suggests unstable protein and < 40 suggests stable protein. Some dipeptides are present in different frequency in unstable vs stable proteins. The composition of such dipeptides can be correlated to the protein stability (Guruprasad et al. 1990).

The aliphatic index of a protein denotes the relative volume enveloped by aliphatic side chains. Increase in temperature can increase aliphatic hydrophobicity; therefore it has a positive correlation with the thermal stability of globular proteins (Ikai 1980).

Protein solubility is indicated by GRAVY, where a negative value correlates with hydrophilicity and positive with hydrophobicity. Hydrophilicity will increase the hydrogen bonding with water molecules and which in turn will increase solubility. ProtParam shows GRAVY score using Kyte-Doolittle's hydropathic indices of the amino acids (Kyte and Doolittle 1982).

GlobPlot (http://www.globplot.embl.de) tool indicates globularity in the proteins. GlobPlot analyzes the disordered and globular regions of a protein, using the Russell/Linding propensities (Linding et al. 2003). Globular domains in protein molecules might dispense special functions. Therefore, loss or gain of function in a protein can also be correlated to addition or deletion of globular domains.

Phosphorylation of amino acid residues like serine and threonine can play important role in protein-protein interactions. NetPhosBac (http://www.cbs.dtu.dk/services/NetPhosBac-1.0/) predicts serine and threonine phosphorylation sites in proteins using neural network algorithms on the dataset of bacterial proteins (Miller et al. 2009).

Besides physicochemical properties, antigenicity can also be calculated in silico, i.e., VaxiJen. Through VaxiJen v2 server (http://www.ddg-pharmfac.net/vaxijen/VaxiJen/VaxiJen.html), antigenicity index of proteins can be measured with an accuracy of 70–89%. This tool utilizes an alignment-free, autocross-covariance-based method to calculate the antigenicity of a protein sequence (Doytchinova and Flower 2007).

10.3 Results

10.3.1 Outcome of Comparative Proteomics of *M.tb Strains*

Nucleotide and amino acids sequence comparison of multigene families across the two strains of *M.tb* revealed the occurrence of a few differences between $H_{37}Rv$ and $H_{37}Ra$ genome in the perspective of established gene families except PE/PPE for virulence. Our in silico analyses (Kohli et al. 2012) revealed that substitutions only were not very substantial but insertions/deletions along with substitutions led to remarkable dissimilarities in the characteristics of these proteins.

MCE multigene family has a wider distribution among nonpathogenic and pathogenic mycobacterium; therefore *mce* operons in *M.tb* may not be directly involved in imparting virulence to *M.tb* (Haile et al. 2002). Interestingly our in silico analysis (Kohli et al. 2012) did not reveal any differences that could be associated with the attenuation of the $H_{37}Ra$ strain. The results were similar in MmpL gene family except for one member. The *mmpl13* gene of $H_{37}Rv$ and $H_{37}Ra$ was a case of gene splitting into *mmpL13a* and *mmpL13b* and another study predicted them to be functional open reading frames based on an another in silico analysis (Sandhu and Akhter 2015). A CDS mismatch in *mmpl13b* gene was detected which led to globular domain extension and phosphorylation site gain in $H_{37}Ra$ (Kohli et al. 2012). The significance of this observation is further subjected to the lack of any designated function of *mmpl13b* gene in virulence or pathogenesis of mycobacteria. Further in the same analyses, 1 out of 13 sigma factors, *sigM*, exhibited nucleotide and amino acid sequence change between $H_{37}Rv$ and $H_{37}Ra$. *sigM* (Rv3911) had a one nucleotide insertion causing a frameshift and subsequent truncation of the protein. This resulted in globular domain extension and phosphorylation site loss in $H_{37}Ra$. Another study has shown that *sigM* gene might be important for long-term in vivo adaptation (Agarwal et al. 2007).

In case of TCSS, a single-nucleotide change only in the case of *phoP* (response regulator of *phoP-phoR* TCSS) was observed, which led to an amino acid substitution from serine to leucine and subsequent phosphorylation site loss in $H_{37}Ra$ (Kohli et al. 2012). This substitution has been shown to decrease DNA-binding capacity of protein and further associated with $H_{37}Ra$ attenuation phenotype, although complementation partly restored the $H_{37}Ra$ persistence in a murine macrophage infection model (Lee et al. 2008).

10.3.2 Comparative Proteomics Highlights the Importance of PE/PPE Multigene Family in Virulence

PE/PPE gene family, which is the largest multigene family in *M.tb*, has also exhibited a number of differences in its member proteins among $H_{37}Ra$ and $H_{37}Rv$ than any other multigene family (Kohli et al. 2012). Another study regarding comparative genome analysis of H37Rv versus H37Ra has shown that *pe/ppe* multigene family exhibited the most variations between H37Rv and H37Ra (Zheng et al. 2008). Besides *phoP*, variation in PE/PPE proteins might be the most important factor involved in attenuation of H37Ra. Despite a large number of proteins in this family, only a few have been characterized so far (Akhter et al. 2012). The study of these proteins is very challenging as only one PE/PPE protein pair has been crystallized so far (Strong et al. 2006). The absence of proteolytic sites in these proteins further makes proteomic studies such as mass spectroscopy extremely difficult (Abdallah et al. 2009).

Nucleotide sequence comparison and protein annotation have revealed that some proteins were absent in $H_{37}Ra$, i.e., PE10, PE_PGRS49, PE_PGRS56, and PE_PGRS60. Interestingly, some of the PE/PPE genes that have been linked to the pathogenicity of $H_{37}Rv$ also exhibited clear amino acid sequence differences with $H_{37}Ra$ in this study (Kohli et al. 2012).

For example, wag22 protein in $H_{37}Rv$ (Rv1759c) has been shown to be expressed during infection (Espitia et al. 1999). In a previous study, inactivation of wag22 homologue in *Mycobacterium marinum* was linked to defective replication in macrophages and subsequent diminished survival in granulomas (Ramakrishnan et al. 2000). In $H_{37}Rv$, this protein exhibited fibronectin-binding properties and antigenicity at C-terminal end (Espitia et al. 1999); while in $H_{37}Ra$, a truncation at C-terminal end (MRA_1772) resulted in aliphatic index change, globular domain expansion, and loss of phosphorylation sites. The PGRS domain has been shown to protect the PE_PGRS protein from ubiquitin-proteasome-dependent (UPD) proteolysis (Koh et al. 2009). The significance of the PGRS domain in the PE_PGRS protein stability has been previously shown by using PE_PGRS fused to green fluorescent protein (Brennan and Delogu 2002). PE_PGRS22 and wag22 have exhibited C-terminal truncation, which might have resulted in altered in vivo stability in $H_{37}Ra$ (Kohli et al. 2012). In the same study, a single-nucleotide deletion in PE_PGRS59 (Rv3595c) homologue in $H_{37}Ra$ has resulted in the formation of two PE_PGRS proteins, namely, MRA_3634 and MRA_3635 in $H_{37}Ra$. MRA_3635 and Rv3595c are both stable and share the same globular domain position.

Furthermore, both MRA_3635 and MRA_3634 are more hydrophobic and contain additional phosphorylation sites compared to Rv3595c.

Rv3508 (PE_PGRS54) has been shown to play a role in nitric oxide stress response and hypoxia during infection (Yu et al. 2011). Our in silico analyses have revealed substitutions and indels in the amino acid sequence of Rv3508 homologue in $H_{37}Ra$. Furthermore, loss of phosphorylation sites in $H_{37}Ra$ might render this protein unresponsive to such stress responses leading to an attenuated phenotype.

PPE18 has been shown to be expressed on the surface and therefore interacts with TLR2 receptor on macrophages. PPE alters the host innate immune balance by repressing IL-12 production and promoting IL-10 production to help bacteria to persist in the host (Nair et al. 2009). Differential interaction of PPE18 with PE13 and PE31 has been shown to modulate host-pathogen interactions (Mukhopadhyay and Balaji 2011). In $H_{37}Ra$, PE13 and PPE18 expression was suppressed in macrophages (Li et al. 2010).

In silico analysis has shown differences in amino acid sequence of PPE18 protein between $H_{37}Rv$ and $H_{37}Ra$ (Kohli et al. 2012). Specific indels led to alterations in serine/threonine phosphorylation sites in $H_{37}Rv$ and $H_{37}Ra$, which might further cause altered protein-protein interactions. GlobPlot analysis has revealed extension of N-terminal globular domain of $H_{37}Rv$ PE13 homologue in $H_{37}Ra$, which might led to altered function. Similarly, N-terminal extension in MRA_1801 (Rv1787 homologue) caused formation of new serine/threonine phosphorylation sites, leading to alterations in signaling and host–pathogen interactions. Strains of *M. avium* lacking MAV_2928 (homologue of Rv1787) have been linked to compromised virulence by preventing vacuole acidification and phagosome-lysosome fusion (Jha et al. 2010). PPE31 (Rv1807) has been shown to be important for in vivo growth during infection in mice (Sassetti and Rubin 2003). Interestingly, in silico analysis showed this protein to be more stable in $H_{37}Ra$ than $H_{37}Rv$ with obvious physiological implications.

These in silico analyses have revealed differences in many functionally unknown PE/PPE proteins, i.e., PE24, PPE5, PPE9, PPE13, PE_PGRS22, PE_PGRS25, PE_PGRS36, PE_PGRS40, and PE_PGRS47. PPE5 homologue MRA_0313 has N-terminal extension leading to an acquisition of an additional globular domain which might impart an additional role to the protein. Similarly in PE_PGRS40 (Rv2371) homologue MRA_2394, N-terminal extension made the protein globular less hydrophobic in $H_{37}Ra$. Likewise in PPE54 (Rv3343c) homologue MRA_3384, deletion of amino acid stretch in the central region resulted in a globular domain loss, probably causing a loss of function in $H_{37}Ra$. Indels and substitutions in Rv2098c homologue MRA_2113 also caused change in globular domains of the protein in $H_{37}Ra$. In Rv1091 homologue MRA_1102, acquisition of a globular domain resulting from deletions in C-terminal end was observed.

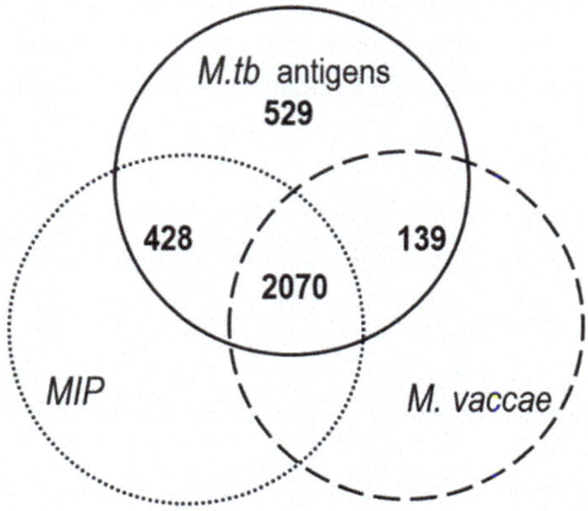

Fig. 10.2 Comparative antigenicity index analysis between *M.tb*, *MIP*, and *M. vaccae* (Singh et al. 2014)

10.3.3 Outcome of Comparative Antigenicity Index Analysis

3166 proteins of *M.tb* ($H_{37}Rv$) exhibited antigenicity index >0.4 and were listed as putative antigens. 152 out of ~ 170 PE/PPE multigene family members displayed antigenicity index >0.4. Further, some of them lied in the high-end range of antigenicity index (> 1) and therefore emphasized their importance in terms of antigenicity in *M.tb* ($H_{37}Rv$). For a meaningful interpretation, another rapidly growing soil mycobacterium *Mycobacterium vaccae* was also included. *Mycobacterium vaccae* has also been explored as a vaccine candidate against TB in some clinical trials (Yang et al. 2010; Yang et al. 2011). 2498 of the putative *M.tb* antigens exhibited amino acid sequence similarity with *MIP* proteins (Fig. 10.2) than *M. vaccae* which had 2209 putative antigen homologues.

10.3.4 Outcome of Gene Duplication in PE/PPE Family

152 PE/PPE antigens with >0.4 antigenicity index (29 PE proteins, 62 PE_PGRS, and 61 PPE proteins) were compared with *MIP* homologues. Out of the 62 PE_PGRS proteins, 52 showed amino acid sequence similarity with MIP_06644 (Singh et al. 2014). MIP_06644 exhibited antigenicity index value of 0.4428 (antigenic) and instability index value of 56.07 (unstable).

After MIP_06644, MIP_03868 exhibited sequence similarity with these 52 proteins. Seven *MIP* PE proteins and 21 *MIP* PPE proteins exhibited amino acid sequence similarity with 26 PE proteins and 58 PPE proteins of *M.tb* ($H_{37}Rv$), respectively. These results have highlighted the prevalence of duplication in PE_PGRS gene family. Along with duplication, increased antigenicity might be crucial for *M.tb* in modulating the host immune response for its persistence.

10.3.5 Comparative Genomic Profiling and Its Prospect in Infectious Diseases Like TB

Comparison of mycobacterial strains or species at genomic, transcriptional, and proteomic level is highly helpful to understand the evolution of differences in various traits like virulence phenotype, growth rate characteristics, host specificity, etc. Pan genomic analysis further broadens the approach by comparing many species together at the same time. At the genus level, pan mycobacterial genomic analysis has revealed that slow growers have lost growth enhancer genes, i.e., *LivFGMH* operon, *shaACDEFG*, and MspA porin (Wee et al. 2017). Such analyses can further address phenotypic differences among *M.tb* strains, between *M.tb* complex, between *M.tb* and NTM (nontuberculous mycobacteria) (Behr 2014).

In vivo or in vitro comparative transcriptional and proteomic analyses are very essential to validate the in silico findings of genomic analysis. For example, *M. bovis* and *M.tb* exhibit more than 99% nucleotide sequence similarity and yet have distinct host preferences. Recently one study with an in-depth transcriptional and translational profiling of *M. bovis* and *M.tb* in bovine macrophages has revealed differential gene expression between these two species (Malone et al. 2018). Similar studies are also needed to understand the precise functional role of PE/PPE multigene family in mycobacterial evolution.

10.4 Conclusion

With the help of comparative genomics studies, it is evident that *Mycobacterium tuberculosis* (*M.tb*) has evolved by gene loss (Ahmed et al. 2008), as well as gene acquisition responsible for pathogenicity and virulence through horizontal gene transfer (Veyrier et al. 2011). Despite reductive evolution, certain gene families in *M.tb* have expanded in size, possibly through various gene duplication events, i.e., *mce*, *pe/ppe*, and *mmpL* gene family. Gene duplication events have impacted nearly 50% of *M.tb* proteome evolution as many multigene families (Tekaia et al. 1999). Our in silico analysis (Kohli et al. 2012) has revealed that well-studied multigene families in the context of *M.tb* virulence, i.e., *mce*, *mmpL*, sigma, and TCSS, might not be adequate to describe the phenotypic difference between $H_{37}Rv$ and $H_{37}Ra$. This demands more investigation of relatively unexplored PE/PPE multigene family as this family exhibited many nucleotide sequence variations upon a comparison between $H_{37}Rv$ and $H_{37}Ra$. *Mycobacterium indicus pranii* (*MIP*) is positioned much above *Mycobacterium tuberculosis* in an evolutionary timeline and is a nonpathogenic soil mycobacterium (Saini et al. 2009); it has shared ~75% of antigens (both putative and validated) of *M.tb* in the in silico analysis (Singh et al. 2014). This study further revealed that *MIP* shared not only a larger number of *M.tb* antigens but also contains larger repertoire of highly antigenic PE/PPE proteins compared to another vaccine candidate *M. vaccae*. Therefore compared to *M. vaccae*, *MIP* might be better in exhibiting antigenic variation and wider antigenic exposure to employ protective immunity against *M.tb*. This comparative antigenic analysis clearly implies gene co-option as another key player in mycobacterial evolution besides gene loss and gene acquisition (horizontal gene transfer). Supporting this notion, an interesting aspect of gene duplication event was observed in highly antigenic PE_PGRS gene family. Our analyses (Singh et al. 2014) revealed PE_PGRS gene family expansion through gene duplication, with improved stability and antigenicity, as a valuable catalyst in the host immune response modulation by *M.tb*. In conclusion, these in silico analyses have attempted to show that PE/PPE multigene family is crucial for virulence, antigenicity, and host-pathogen interactions in tuberculosis. Furthermore, PE/PPE multigene family might also be important for an antituberculosis vaccine efficacy and prominent gene duplication events have highlighted gene co-option as a key player in *M.tb* evolution. This is the first study which has shown gene co-option as another key player in mycobacterial evolution besides gene loss and gene acquisition. Therefore, a systemic functional analysis of proteins from this multigene family will be highly helpful in devising effective preventive (vaccines) and drug therapies against tuberculosis.

References

Abdallah AM, Verboom T, Weerdenburg EM, Gey van Pittius NC, Mahasha PW, Jiménez C, Parra M, Cadieux N, Brennan MJ, Appelmelk BJ, Bitter W (2009) PPE and PE_PGRS proteins of Mycobacterium marinum are transported via the type VII secretion system ESX-5. Mol Microbiol 73(3):329–340. https://doi.org/10.1111/j.1365-2958.2009.06783.x

Agarwal N, Woolwine SC, Tyagi S, Bishai WR (2007) Characterization of the Mycobacterium tuberculosis sigma factor SigM by assessment of virulence and identification of SigM-dependent genes. Infect Immun 75(1):452–461. https://doi.org/10.1128/IAI.01395-06

Ahmad F, Mani J, Kumar P, Haridas S, Upadhyay P, Bhaskar S (2011) Activation of anti-tumor immune response and reduction of regulatory T cells with Mycobacterium indicus pranii (MIP) therapy in tumor bearing mice. PLoS One 6(9):e25424. https://doi.org/10.1371/journal.pone.0025424

Ahmed N, Saini V, Raghuvanshi S, Khurana JP, Tyagi AK, Tyagi AK, Hasnain SE (2007) Molecular analysis of a leprosy immunotherapeutic bacillus provides insights into Mycobacterium evolution. PLoS One 2(10):e968. https://doi.org/10.1371/journal.pone.0000968

Ahmed N, Dobrindt U, Hacker J, Hasnain SE (2008) Genomic fluidity and pathogenic bacteria: applications in diagnostics, epidemiology and intervention. Nat Rev Microbiol 6(5):387–394. https://doi.org/10.1038/nrmicro1889

Akhter Y, Ehebauer MT, Mukhopadhyay S, Hasnain SE (2012) The PE/PPE multigene family codes for virulence factors and is a possible source of mycobacterial antigenic variation: perhaps more? Biochimie 94(1):110–116. https://doi.org/10.1016/j.biochi.2011.09.026

Arruda S, Bomfim G, Knights R, Huima-Byron T, Riley LW (1993) Cloning of an M. tuberculosis DNA fragment associated with entry and survival inside cells. Science 261(5127):1454–1457

Bashyam MD, Hasnain SE (2004) The extracytoplasmic function sigma factors: role in bacterial pathogenesis. Infect Genet Evol 4(4):301–308. https://doi.org/10.1016/j.meegid.2004.04.003

Behr MA (2014) Comparative genomics of mycobacteria: some answers, yet more new questions. Cold Spring Harb Perspect Med 5(2):a021204. https://doi.org/10.1101/cshperspect.a021204

Brennan MJ, Delogu G (2002) The PE multigene family: a 'molecular mantra' for mycobacteria. Trends Microbiol 10(5):246–249

Bretl DJ, Demetriadou C, Zahrt TC (2011) Adaptation to environmental stimuli within the host: two-component signal transduction systems of Mycobacterium tuberculosis. Microbiol Mol Biol Rev 75(4):566–582. https://doi.org/10.1128/MMBR.05004-11

Brosch R, Gordon SV, Marmiesse M, Brodin P, Buchrieser C, Eiglmeier K, Garnier T, Gutierrez C, Hewinson G, Kremer K, Parsons LM, Pym AS, Samper S, van Soolingen D, Cole ST (2002) A new evolutionary scenario for the Mycobacterium tuberculosis complex. Proc Natl Acad Sci U S A 99(6):3684–3689. https://doi.org/10.1073/pnas.052548299

Casali N, Riley LW (2007) A phylogenomic analysis of the Actinomycetales mce operons. BMC Genomics 8:60. https://doi.org/10.1186/1471-2164-8-60

Chitale S, Ehrt S, Kawamura I, Fujimura T, Shimono N, Anand N, Lu S, Cohen-Gould L, Riley LW (2001) Recombinant Mycobacterium tuberculosis protein associated with mammalian cell entry. Cell Microbiol 3(4):247–254

Cole ST, Brosch R, Parkhill J, Garnier T, Churcher C, Harris D, Gordon SV, Eiglmeier K, Gas S, Barry CE 3rd, Tekaia F, Badcock K, Basham D, Brown D, Chillingworth T, Connor R, Davies R, Devlin K, Feltwell T, Gentles S, Hamlin N, Holroyd S, Hornsby T, Jagels K, Krogh A, McLean J, Moule S, Murphy L, Oliver K, Osborne J, Quail MA, Rajandream MA, Rogers J, Rutter S, Seeger K, Skelton J, Squares R, Squares S, Sulston JE, Taylor K, Whitehead S, Barrell BG (1998) Deciphering the biology of Mycobacterium tuberculosis from the complete genome sequence. Nature 393(6685):537–544. https://doi.org/10.1038/31159

Converse SE, Mougous JD, Leavell MD, Leary JA, Bertozzi CR, Cox JS (2003) MmpL8 is required for sulfolipid-1 biosynthesis and Mycobacterium tuberculosis virulence. Proc Natl Acad Sci U S A 100(10):6121–6126. https://doi.org/10.1073/pnas.1030024100

Cox JS, Chen B, McNeil M, Jacobs WR Jr (1999) Complex lipid determines tissue-specific replication of Mycobacterium tuberculosis in mice. Nature 402(6757):79–83. https://doi.org/10.1038/47042

Domenech P, Reed MB, Barry CE 3rd (2005) Contribution of the Mycobacterium tuberculosis MmpL protein family to virulence and drug resistance. Infect Immun 73(6):3492–3501. https://doi.org/10.1128/IAI.73.6.3492-3501.2005

Doytchinova IA, Flower DR (2007) VaxiJen: a server for prediction of protective antigens, tumour antigens and subunit vaccines. BMC Bioinform 8:4. https://doi.org/10.1186/1471-2105-8-4

Espitia C, Laclette JP, Mondragon-Palomino M, Amador A, Campuzano J, Martens A, Singh M, Cicero R, Zhang Y, Moreno C (1999) The PE-PGRS glycine-rich proteins of Mycobacterium tuberculosis: a new family of fibronectin-binding proteins? Microbiology 145 (Pt 12):3487–3495. https://doi.org/10.1099/00221287-145-12-3487

Gioffre A, Infante E, Aguilar D, Santangelo MP, Klepp L, Amadio A, Meikle V, Etchechoury I, Romano MI, Cataldi A, Hernandez RP, Bigi F (2005) Mutation in mce operons attenuates Mycobacterium tuberculosis virulence. Microbes Infect 7(3):325–334. https://doi.org/10.1016/j.micinf.2004.11.007

Gupta A, Geetha N, Mani J, Upadhyay P, Katoch VM, Natrajan M, Gupta UD, Bhaskar S (2009) Immunogenicity and protective efficacy of "Mycobacterium w" against Mycobacterium tuberculosis in mice immunized with live versus heat-killed M. w by the aerosol or parenteral route. Infect Immun 77(1):223–231. https://doi.org/10.1128/IAI.00526-08

Gupta A, Ahmad FJ, Ahmad F, Gupta UD, Natarajan M, Katoch VM, Bhaskar S (2012) Protective efficacy of Mycobacterium indicus pranii against tuberculosis and underlying local lung immune responses in Guinea pig model. Vaccine 30(43):6198–6209. https://doi.org/10.1016/j.vaccine.2012.07.061

Guruprasad K, Reddy BV, Pandit MW (1990) Correlation between stability of a protein and its dipeptide composition: a novel approach for predicting in vivo stability of a protein from its primary sequence. Protein Eng 4(2):155–161

Haile Y, Caugant DA, Bjune G, Wiker HG (2002) Mycobacterium tuberculosis mammalian cell entry operon (mce) homologs in Mycobacterium other than tuberculosis (MOTT). FEMS Immunol Med Microbiol 33(2):125–132

Harboe M, Christensen A, Ahmad S, Ulvund G, Harkness RE, Mustafa AS, Wiker HG (2002) Cross-reaction between mammalian cell entry (Mce) proteins of Mycobacterium tuberculosis. Scand J Immunol 56(6):580–587

Ikai A (1980) Thermostability and aliphatic index of globular proteins. J Biochem 88(6):1895–1898
Jha SS, Danelishvili L, Wagner D, Maser J, Li YJ, Moric I, Vogt S, Yamazaki Y, Lai B, Bermudez LE (2010) Virulence-related Mycobacterium avium subsp hominissuis MAV_2928 gene is associated with vacuole remodeling in macrophages. BMC Microbiol 10:100. https://doi.org/10.1186/1471-2180-10-100
Katoch K, Singh P, Adhikari T, Benara SK, Singh HB, Chauhan DS, Sharma VD, Lavania M, Sachan AS, Katoch VM (2008) Potential of mw as a prophylactic vaccine against pulmonary tuberculosis. Vaccine 26(9):1228–1234. https://doi.org/10.1016/j.vaccine.2007.12.025
Koh KW, Lehming N, Seah GT (2009) Degradation-resistant protein domains limit host cell processing and immune detection of mycobacteria. Mol Immunol 46(7):1312–1318. https://doi.org/10.1016/j.molimm.2008.11.008
Kohli S, Singh Y, Sharma K, Mittal A, Ehtesham NZ, Hasnain SE (2012) Comparative genomic and proteomic analyses of PE/PPE multigene family of Mycobacterium tuberculosis H(3)(7)Rv and H(3)(7)Ra reveal novel and interesting differences with implications in virulence. Nucleic Acids Res 40(15):7113–7122. https://doi.org/10.1093/nar/gks465
Krell T, Lacal J, Busch A, Silva-Jimenez H, Guazzaroni ME, Ramos JL (2010) Bacterial sensor kinases: diversity in the recognition of environmental signals. Annu Rev Microbiol 64:539–559. https://doi.org/10.1146/annurev.micro.112408.134054
Kyte J, Doolittle RF (1982) A simple method for displaying the hydropathic character of a protein. J Mol Biol 157(1):105–132
Lamichhane G, Tyagi S, Bishai WR (2005) Designer arrays for defined mutant analysis to detect genes essential for survival of Mycobacterium tuberculosis in mouse lungs. Infect Immun 73 (4):2533–2540. https://doi.org/10.1128/IAI.73.4.2533-2540.2005
Lee JS, Krause R, Schreiber J, Mollenkopf HJ, Kowall J, Stein R, Jeon BY, Kwak JY, Song MK, Patron JP, Jorg S, Roh K, Cho SN, Kaufmann SH (2008) Mutation in the transcriptional regulator PhoP contributes to avirulence of Mycobacterium tuberculosis H37Ra strain. Cell Host Microbe 3(2):97–103. https://doi.org/10.1016/j.chom.2008.01.002
Letek M, Gonzalez P, Macarthur I, Rodriguez H, Freeman TC, Valero-Rello A, Blanco M, Buckley T, Cherevach I, Fahey R, Hapeshi A, Holdstock J, Leadon D, Navas J, Ocampo A, Quail MA, Sanders M, Scortti MM, Prescott JF, Fogarty U, Meijer WG, Parkhill J, Bentley SD, Vazquez-Boland JA (2010) The genome of a pathogenic rhodococcus: cooptive virulence underpinned by key gene acquisitions. PLoS Genet 6(9):e1001145. https://doi.org/10.1371/journal.pgen.1001145
Li AH, Waddell SJ, Hinds J, Malloff CA, Bains M, Hancock RE, Lam WL, Butcher PD, Stokes RW (2010) Contrasting transcriptional responses of a virulent and an attenuated strain of Mycobacterium tuberculosis infecting macrophages. PLoS One 5(6):e11066. https://doi.org/10.1371/journal.pone.0011066
Linding R, Jensen LJ, Diella F, Bork P, Gibson TJ, Russell RB (2003) Protein disorder prediction: implications for structural proteomics. Structure 11(11):1453–1459
Malone KM, Rue-Albrecht K, Magee DA, Conlon K, Schubert OT, Nalpas NC, Browne JA, Smyth A, Gormley E, Aebersold R, MacHugh DE, Gordon SV (2018) Comparative ‘omics analyses differentiate Mycobacterium tuberculosis and Mycobacterium bovis and reveal distinct macrophage responses to infection with the human and bovine tubercle bacilli. Microb Genom 4. https://doi.org/10.1099/mgen.0.000163
Miller ML, Soufi B, Jers C, Blom N, Macek B, Mijakovic I (2009) NetPhosBac – a predictor for Ser/Thr phosphorylation sites in bacterial proteins. Proteomics 9(1):116–125. https://doi.org/10.1002/pmic.200800285
Mukhopadhyay S, Balaji KN (2011) The PE and PPE proteins of Mycobacterium tuberculosis. Tuberculosis 91(5):441–447. https://doi.org/10.1016/j.tube.2011.04.004
Mulder N, Rabiu H, Jamieson G, Vuppu V (2009) Comparative analysis of microbial genomes to study unique and expanded gene families in Mycobacterium tuberculosis. Infect Genet Evol 9 (3):314–321. https://doi.org/10.1016/j.meegid.2007.12.006

Nair S, Ramaswamy PA, Ghosh S, Joshi DC, Pathak N, Siddiqui I, Sharma P, Hasnain SE, Mande SC, Mukhopadhyay S (2009) The PPE18 of Mycobacterium tuberculosis interacts with TLR2 and activates IL-10 induction in macrophage. J Immunol 183(10):6269–6281. https://doi.org/10.4049/jimmunol.0901367

Nath I (1998) A vaccine for leprosy. Nat Med 4(5):548–550

Rakshit S, Ponnusamy M, Papanna S, Saha B, Ahmed A, Nandi D (2012) Immunotherapeutic efficacy of Mycobacterium indicus pranii in eliciting anti-tumor T cell responses: critical roles of IFNgamma. Int J Cancer 130(4):865–875. https://doi.org/10.1002/ijc.26099

Ramakrishnan L, Federspiel NA, Falkow S (2000) Granuloma-specific expression of Mycobacterium virulence proteins from the glycine-rich PE-PGRS family. Science 288(5470):1436–1439

Rodrigue S, Provvedi R, Jacques PE, Gaudreau L, Manganelli R (2006) The sigma factors of Mycobacterium tuberculosis. FEMS Microbiol Rev 30(6):926–941. https://doi.org/10.1111/j.1574-6976.2006.00040.x

Saini V, Raghuvanshi S, Talwar GP, Ahmed N, Khurana JP, Hasnain SE, Tyagi AK, Tyagi AK (2009) Polyphasic taxonomic analysis establishes Mycobacterium indicus pranii as a distinct species. PLoS One 4(7):e6263. https://doi.org/10.1371/journal.pone.0006263

Saini V, Raghuvanshi S, Khurana JP, Ahmed N, Hasnain SE, Tyagi AK, Tyagi AK (2012) Massive gene acquisitions in Mycobacterium indicus pranii provide a perspective on mycobacterial evolution. Nucleic Acids Res 40(21):10832–10850. https://doi.org/10.1093/nar/gks793

Sandhu P, Akhter Y (2015) The internal gene duplication and interrupted coding sequences in the MmpL genes of Mycobacterium tuberculosis: towards understanding the multidrug transport in an evolutionary perspective. Int J Med Microbiol 305(3):413–423. https://doi.org/10.1016/j.ijmm.2015.03.005

Sassetti CM, Rubin EJ (2003) Genetic requirements for mycobacterial survival during infection. Proc Natl Acad Sci U S A 100(22):12989–12994. https://doi.org/10.1073/pnas.2134250100

Singh IG, Mukherjee R, Talwar GP, Kaufmann SH (1992) In vitro characterization of T cells from Mycobacterium w-vaccinated mice. Infect Immun 60(1):257–263

Singh Y, Kohli S, Sowpati DT, Rahman SA, Tyagi AK, Hasnain SE (2014) Gene cooption in mycobacteria and search for virulence attributes: comparative proteomic analyses of Mycobacterium tuberculosis, Mycobacterium indicus pranii and other mycobacteria. Int J Med Microbiol 304(5–6):742–748. https://doi.org/10.1016/j.ijmm.2014.05.006

Smith I (2003) Mycobacterium tuberculosis pathogenesis and molecular determinants of virulence. Clin Microbiol Rev 16(3):463–496

Steenken W Jr, Gardner LU (1946) History of H37 strain of tubercle bacillus. Am Rev Tuberc 54:62–66

Strong M, Sawaya MR, Wang S, Phillips M, Cascio D, Eisenberg D (2006) Toward the structural genomics of complexes: crystal structure of a PE/PPE protein complex from Mycobacterium tuberculosis. Proc Natl Acad Sci U S A 103(21):8060–8065. https://doi.org/10.1073/pnas.0602606103

Talwar GP (1999) An immunotherapeutic vaccine for multibacillary leprosy. Int Rev Immunol 18 (3):229–249

Tekaia F, Gordon SV, Garnier T, Brosch R, Barrell BG, Cole ST (1999) Analysis of the proteome of Mycobacterium tuberculosis in silico. Tuber Lung Dis 79(6):329–342. https://doi.org/10.1054/tuld.1999.0220

True JR, Carroll SB (2002) Gene co-option in physiological and morphological evolution. Annu Rev Cell Dev Biol 18:53–80. https://doi.org/10.1146/annurev.cellbio.18.020402.140619

Veyrier FJ, Dufort A, Behr MA (2011) The rise and fall of the Mycobacterium tuberculosis genome. Trends Microbiol 19(4):156–161. https://doi.org/10.1016/j.tim.2010.12.008

Wee WY, Dutta A, Choo SW (2017) Comparative genome analyses of mycobacteria give better insights into their evolution. PLoS One 12(3):e0172831. https://doi.org/10.1371/journal.pone.0172831

Yadava A, Suresh NR, Zaheer SA, Talwar GP, Mukherjee R (1991) T-cell responses to fractionated antigens of Mycobacterium w, a candidate anti-leprosy vaccine, in leprosy patients. Scand J Immunol 34(1):23–31

Yang XY, Chen QF, Cui XH, Yu Y, Li YP (2010) Mycobacterium vaccae vaccine to prevent tuberculosis in high risk people: a meta-analysis. J Infect 60(5):320–330. https://doi.org/10.1016/j.jinf.2010.02.005

Yang XY, Chen QF, Li YP, Wu SM (2011) Mycobacterium vaccae as adjuvant therapy to anti-tuberculosis chemotherapy in never-treated tuberculosis patients: a meta-analysis. PLoS One 6 (9):e23826. https://doi.org/10.1371/journal.pone.0023826

Yu G, Fu X, Jin K, Zhang L, Wu W, Cui Z, Hu Z, Li Y (2011) Integrative analysis of transcriptome and genome indicates two potential genomic islands are associated with pathogenesis of Mycobacterium tuberculosis. Gene 489(1):21–29. https://doi.org/10.1016/j.gene.2011.08.019

Zhang F, Xie JP (2011) Mammalian cell entry gene family of Mycobacterium tuberculosis. Mol Cell Biochem 352(1–2):1–10. https://doi.org/10.1007/s11010-011-0733-5

Zheng H, Lu L, Wang B, Pu S, Zhang X, Zhu G, Shi W, Zhang L, Wang H, Wang S, Zhao G, Zhang Y (2008) Genetic basis of virulence attenuation revealed by comparative genomic analysis of Mycobacterium tuberculosis strain H37Ra versus H37Rv. PLoS One 3(6):e2375. https://doi.org/10.1371/journal.pone.0002375

Importance of Cell Wall-Associated Poly-α-L-Glutamine in the Biology of Pathogenic Mycobacteria

11

Rajni Garg, Rajesh Mani, Manish Gupta, Deeksha Tripathi, Harish Chandra, Rakesh Bhatnagar, and Nirupama Banerjee

Abstract

Mycobacterium tuberculosis (Mtb), the formidable scourge known to mankind since ancient times, has remained untamed despite vigorous scientific research in the field. In the last several decades, significant advances have been made to study this pathogen; however, a lot more remains in the realm of unknown. The complex and unique cell wall of the bacterium is a major factor contributing to the unrestrained success of the pathogen in infecting millions around the world. Since the discovery of this bacterium, numerous studies have attempted to unravel the complexities of mycobacterial cell envelop to characterize individual constituents and their importance in pathobiology of Mtb. Major components of the cell envelop of mycobacteria such as lipid-linked polysaccharides-

R. Garg
Jawaharlal Nehru Centre for Advanced Scientific Research, Bangalore, Karnataka, India

Department of Microbiology and Cell Biology, Indian Institute of Science, Bangalore, Karnataka, India

R. Mani · M. Gupta · N. Banerjee (✉)
Molecular Biology and Genetic Engineering Laboratory, School of Biotechnology, Jawaharlal Nehru University, New Delhi, Delhi, India

D. Tripathi
Department of Microbiology, School of Life Sciences, Central University of Rajasthan, Ajmer, Rajasthan, India

H. Chandra
Department of Microbiology, School of Biomedical & Pharmaceutical Sciences, Babasaheb Bhimrao Ambedkar University, Lucknow, India

R. Bhatnagar
Molecular Biology and Genetic Engineering Laboratory, School of Biotechnology, Jawaharlal Nehru University, New Delhi, Delhi, India

Banaras Hindu University, Varanasi, Uttar Pradesh, India

S. E. Hasnain et al. (eds.), Mycobacterium tuberculosis: *Molecular Infection Biology, Pathogenesis, Diagnostics and New Interventions*,
https://doi.org/10.1007/978-981-32-9413-4_11

lipoarabinomannan (LAM), dimycolyl trehalose (cord factor), sulfolipids, and mycolyl-arabinogalactan-peptidoglycan (mAGP) complex have been investigated extensively. However, a lesser known molecule, poly-α-L-glutamine/glutamate (PLG), that constitutes ~10% of dry weight of cell wall has not attracted as much attention. As early as 1990, Hirschfield et al. isolated PLG as insoluble material and showed its association with the Mtb cell wall. In the last few years, our group has been working to identify enzymes that may play a role in the synthesis/assembly and localization of this polymer in the cell wall of mycobacteria. Our recently published work has shown that PLG by itself is weakly immunogenic in mice, but when combined with protein antigens, it can stimulate different arms of the T helper-mediated responses, demonstrating its potential to act as an adjuvant (Mani et al. 2018).

Keywords
Poly-α- L-glutamine/glutamate (PLG) · Mycobacterial virulence · Glutamine synthetase · Adjuvant

Abbreviations

GS	glutamine synthetase
LAM	lipoarabinomannan
mAGP	mycolyl-arabinogalactan-peptidoglycan
MDR	multidrug resistant
Mtb	*Mycobacterium tuberculosis*
PGS	polyglutamate synthase
PLG	poly-α- L-glutamine/glutamate
TB	tuberculosis
XDR	extensively drug resistant

11.1 Introduction

Tuberculosis (TB) continues to be one of the major causes of morbidity and mortality across the globe (Pai et al. 2016). One of the key reasons for unrelenting success of *Mycobacterium tuberculosis* (Mtb), the causative agent of TB, is its distinct cell wall. The latter is an elaborate structure distinct from other eubacteria (Brennan 2003). It is highly impervious in nature, acting as a protective barrier against hydrophilic drugs (Jankute et al. 2015). The emergence of multidrug-resistant (MDR) and extensively drug-resistant (XDR) bacteria has necessitated the need for studying different mycobacterial cell wall components to develop effective new-generation antitubercular therapeutics that can target them individually. Poly-α-L-glutamine/glutamate (PLG) is one such cell wall component found only in pathogenic mycobacteria. In this chapter, we have attempted to review the earlier research related to PLG and present the work done in our laboratory in the context of biology of mycobacteria. Here, we discuss some of the yet unexplored aspects of the glutamine polymer as revealed by our recent research.

11.2 Molecular Structure of Mycobacterial Cell Wall

Mycobacteria possess a cell wall that is markedly distinct from other eubacteria. Mycobacterial cell wall has a complex architecture with an inner and an outer layer (Brennan 2003). The inner layer forming the core of the cell wall contains unique mycolyl-arabinogalactan-peptidoglycan (mAGP) complex that surrounds the cytoplasm (Brennan 2003), with the peptidoglycan facing the plasma membrane, while mycolic acids are present at the distal end. In the outer lamella, proteins are present along with glycolipids and some long- and short-chain fatty acids (Brennan 2003; Alderwick et al. 2015).

11.2.1 The Inner Lamella

The peptidoglycan (PG) in the inner lamella of the mycobacterial cell wall is unique in several ways. It contains N-acetyl glucosamine (NAG) linked to N-modified muramic acid in a β (1–4) configuration (Alderwick et al. 2015). Some modifications in muramic acid are acetyl in nature as in other eubacteria, while others are glycolyl in nature (Mahapatra et al. 2000, 2005; Alderwick et al. 2015). Secondly, the mycobacterial PG is an extensively cross-linked (70–75%) structure. Tetrapeptide side chains consisting of L-alanyl-D-*iso*-glutaminyl-*meso*-diaminopimelyl-D-alanyl-D-alanine are cross-linked with peptides of neighboring glycan chains by a 3–3 bond in addition to the regular 3–4 bond (Wietzerbin-Falszpan et al. 1970; Alderwick et al. 2015). Thirdly, D-glutamate and diaminopimelic acid (DAP) are amidated and lastly some of the D-glutamate or L-alanine residues are modified with glycine (Kotani et al. 1970; Draper 1971; Draper et al. 1987). The arabinogalactan (AG) is a heteropolysaccharide which is covalently attached to 10–12% of muramic acid residues of the PG via a phosphodiester linkage (Amar and Vilkas 1973).

11.2.2 The Outer Lamella

Outer lamella of the wall has soluble proteins and lipids, which primarily function as the signaling molecules of mycobacteria (Brennan 2003). The phthiocerol-containing lipids such as phthiocerol dimycocerosate, glycolipids like lipoarabinomannan (LAM), lipomannan, dimycolyl trehalose (cord factor), sulfolipids, and the phosphatidylinositol mannosides are the typical constituents of the mycobacterial cell wall (Brennan 2003).

In the last few years, substantial work on Mtb cell wall has strengthened our understanding of biogenesis and structure of mAGP complex (Meroueh et al. 2006; Berg et al. 2007). Based on experimental evidences, it was reported that poly-α-L-glutamine (PLG) peptides are non-covalently associated with the peptidoglycan in virulent mycobacteria (Hirschfield et al. 1990). In this chapter, we have described our research elucidating the biochemical and immunological properties of this unique constituent of the mycobacterial cell wall.

11.3 Poly-α-L-Glutamine (PLG)

Though extensive work has been done on the AGP complex and other cell wall components, the poly-α-L-glutamine (PLG) polymer has received little attention despite its discovery several decades ago (Wietzerbin-Falszpan et al. 1973; Juana Wietzerbin et al. 1975; Phiet et al. 1976). In the last few years, our group has focused on decoding the biological significance of PLG.

In 1976, Phiet et al. analyzed cell walls of various strains of mycobacteria, e.g., *M. tuberculosis* (Mtb) W115, five strains of *M. bovis* BCG Pasteur, Glaxo, Phipps, Tice, and Montreal for their PLG content. This study revealed that the amount of PLG in the cell wall positively correlated with degree of virulence of the strain and, secondly, PLG was absent in the avirulent *M. bovis* BCG Montreal strain (Phiet et al. 1976). In an attempt to unveil molecules responsible for immunoreactivity of Mtb cell wall, Hirschfield et al. isolated the insoluble polymer PLG along with a 23 kDa immunogenic protein by trifluoromethanesulfonic acid treatment of the insoluble mAGP complex (Hirschfield et al. 1990). The unique physical properties of the polymer, particularly its insolubility in detergents, organic acids and bases, chaotropes, and many common solvents, resulted in its co-purification with the cell walls. Isolation of the polymer by nondegradative means suggested its non-covalent association with the cell wall. PLG constituted 10% of the mycobacterial cell wall weight (Hirschfield et al. 1990). It was also suggested that PLG might serve as a nitrogen and carbon reserve for the bacterium to survive in adverse conditions (Hirschfield et al. 1990).

11.4 Biogenesis of PLG

11.4.1 Role of Glutamine Synthase

Till date there is no direct evidence showing the specific role of GS (glutamine synthetase) in PLG synthesis; however various reports have suggested a close link between the two. Several studies have shown that amount of PLG in the cell wall is dependent on the expression and activity of glutamine synthetase (GS encoded by *glnA1* gene), secreted in copious amount by virulent mycobacteria (Harth et al. 1994;

Harth and Horwitz 1999; Tullius and Horwitz 2003; Chandra et al. 2010). Harth et al. in 1994 successfully purified the GS enzyme encoded by *glnA1* gene from Mtb to near homogeneity and demonstrated its catalytic activity in the synthesis of glutamine. GS is a 680 kDa dodecameric protein. Since this enzyme is extracellular in nature, it was proposed that it provides glutamine for the assembly of PLG layer (Harth et al. 1994).

Harth and Horwitz investigated the effect of GS inhibitors on the amount of PLG and growth of pathogenic and nonpathogenic mycobacteria. Addition of inhibitor MSO (L-methionine-S-sulfoximine) rapidly inhibited GS activity of Mtb under in vitro conditions (Harth and Horwitz 1999). Moreover, when added to the cultures of pathogenic mycobacteria, it inhibited extracellular GS in a concentration-dependent manner, but had nearly no effect on cellular GS, probably because of its inability to cross the mycobacterial cell wall. Treatment of Mtb with MSO reduced PLG content also in the cell wall, indicating the involvement of GS in PLG synthesis (Harth and Horwitz 1999). The inhibitor selectively reduced growth of all pathogenic mycobacteria, which release GS extracellularly, but did not affect nonpathogenic mycobacteria or other bacteria, which do not secrete GS.

Our group for the first time generated an *M. bovis glnA1* mutant and established direct link between GS activity and PLG formation (Chandra et al. 2010). The mutant strain showed marked reduction in the PLG content, concomitant with decrease in the cell wall strength. The cell wall of the mutant became highly sensitive to a variety of physical and chemical stresses, namely, sonication, SDS, and lysozyme, respectively. Disruption of the *glnA1* gene resulted in attenuation of the mutant in the THP-1 cells and BALB/c mice (Chandra et al. 2010). The mutant showed reduced biofilm formation and increased sensitivity to antitubercular drugs such as rifampicin and D-cycloserine. Further, expression of the mycobacterial *glnA1* in *M. smegmatis* led to synthesis and export of GS in the extracellular medium and concomitant formation of PLG in the cell wall (Chandra et al. 2010). The above results showed involvement of *glnA1* in the synthesis of PLG in virulent mycobacteria.

11.4.2 Role of Rv0574c (Polyglutamate Synthase-like Enzyme)

Formation of PLG in the cell wall as a corollary to glutamine synthetase activity provided potential targets for therapeutic intervention against mycobacterial infection. However, high level of similarity between the catalytic domains of the mycobacterial GS and the human homologue called for identification of alternate enzymes involved in PLG biosynthesis. We approached the problem by mining mycobacterial genome for homologues of genes shown to encode enzymes catalyzing synthesis/localization of polyglutamate polymer in the cell wall of other bacteria. We identified Rv0574c locus annotated as polyglutamate synthase (PGS)-like enzyme showing 95% identity with other capsule synthesis proteins, e.g., CapA from *Streptomyces violaceusniger* Tu 4113, and 25.8% identity with CapA, the polyglutamate synthase of *B. anthracis* (Garg et al. 2015). Based on the above search, we hypothesized that the encoded protein might play a role in the formation of PLG in the cell wall of mycobacteria. The expression pattern of Rv0574c (PGS) through different growth phases correlated with the amount of PLG in Mtb cell wall. The expression levels of the protein peaked in late log phase with concomitant accumulation of PLG which continued to increase till the stationary phase (Garg et al. 2015). The Rv0574c gene is reported to be a member of the dormancy regulon, a 48-gene cluster controlled by the transcriptional regulator DevR (Dev R-Dev S two-component system) in mycobacteria (Park et al. 2003; Sousa et al. 2007; Garg et al. 2015). We observed that signals such as hypoxia, nitric oxide, and carbon dioxide reported to activate DevR-mediated dormancy regulon, upregulated transcription of Rv0574c gene also (Garg et al. 2015). In the light of the above observations, it is proposed that as the cells progress into stationary phase, the above host signals trigger the dormancy regulon (Park et al. 2003; Sousa et al. 2007) and result in PGS-dependent increase in PLG contents in the cell wall to help the bacterium adapt/resist the host-generated stressful environment.

To find the precise role of Rv0574c in mycobacterial biology, a gene deletion mutant was created. The mutant strain contained less PLG in the cell wall than the wild-type strain, though the reduction was not as high as the *glnA1* phenotype (Garg et al. 2015). Whether decrease in the amount of extractable PLG was due to lack of synthesis or localization is a subject of future studies. Like the *glnA1* mutant, the Rv0574c mutant also exhibited increased sensitivity to lysozyme and SDS, which could be due to lower PLG content in the cell wall. The mutant was defective in biofilm formation compared to the parent Mtb strain, which produced biofilm rich in PLG, implying its involvement in maintaining cellular integrity and architecture of the cell wall (Garg et al. 2015).

The importance of Rv0574c in the survival of the bacterium was further highlighted by the loss in virulence of the mutant strain, which was attenuated for growth in the macrophages. Replication of the ΔRv0574c strain was highly reduced (1 $\log_{10}$) in the lungs of BALB/c mice after 8 weeks of infection (Garg et al. 2015). The mutant showed minimal histopathological changes in the lung parenchyma of the infected mice, indicating substantial reduction in pathogenicity (Garg et al. 2015). PLG content in the cell wall of the mutant was also reduced, though not as severely as the *glnA1* phenotype. A small but statistically significant increase in PLG content was seen when Mtb PGS was overexpressed in *M. smegmatis*, suggesting an accessory function of this protein in PLG synthesis/transport. Thus our results suggest that PGS is important for protecting the bacterium against the deleterious effects of host defense mechanisms like NO, CO, hypoxia, etc. and also makes the cell wall resistant to chemical and mechanical stresses with the help of PLG, facilitating its entry into the dormant stage. Further studies are underway to decipher the catalytic properties of this protein in the PLG synthesis.

11.5 Regulation of PLG Synthesis

Nitrogen metabolism plays a critical role in bacterial physiology. In the cell, synthesis of nitrogenous biomolecules depends mainly on two nitrogen donors, glutamate and glutamine (Mehta et al. 2004; Newsholme et al. 2003). In mycobacteria, GS enzyme is responsible for assimilation of inorganic nitrogen to glutamine (Umbarger 1978). Since glutamine synthesis is a high energy consuming process, the *glnA1* gene encoding GS is under stringent control, both at transcriptional and posttranslational levels. It has been reported earlier that the *glnA*1 gene in Mtb has two promoter sequences P1 and P2 in the upstream regulatory region (Harth and Horwitz 1997). The transcript length varied when the bacteria were grown in low and high nitrogen condition, suggesting that transcription initiates from two different sites depending on the availability of nitrogen (Harth and Horwitz 1997). Furthermore, when the nitrogen concentration is low, the GlnR protein acts as a positive regulator by binding to the P1 promoter and activates transcription (Tiffert et al. 2008). While at higher nitrogen concentrations, the GS enzyme is adenylylated by GlnE protein, rendering it inactive to avoid depletion of the cellular glutamate levels by conversion to glutamine (Harper et al. 2008). In the light of these observations, we were motivated to examine the involvement of individual promoters of the *glnA1* gene in regulating PLG synthesis at different nitrogen concentrations in mycobacteria.

For our studies, we used *M. smegmatis* host which produces a nonsecretory and yet functional GS enzyme. Secondly, it does not contain PLG in the cell wall under normal growth conditions (Harth et al. 1994; Harth and Horwitz 1997). For this purpose, we generated several recombinant strains MSFP, MSP1, and MSP2 (*M. smegmatis* expressing Mtb*glnA1* with dual promoter sequence and with individual P1 or P2 promoters, respectively) (Tripathi et al. 2013). The recombinant strains produced extracellular GS enzyme depending on the concentration of nitrogen in the growth medium. Effect on PLG formation was also examined in response to nitrogen availability by isolating and estimating PLG during growth. It was revealed that at low nitrogen concentrations, the P1 promoter was primarily responsible for the synthesis of major amount of the extracellular glutamine synthetase, while the P2 promoter had no apparent role in the production of the enzyme. At high nitrogen condition also, the amounts of both intra- and extracellular GS enzyme produced by the MSP2 strain were lower than the MSP1 strain. Further, the synthesis of PLG directly correlated with the magnitude of extracellular glutamine synthetase activity, implying that it is the P1 promoter that mainly drives the synthesis of both GS and PLG under low nitrogen condition only. Our results with separate individual promoters corroborate earlier studies (Mehta et al. 2004) that high nitrogen concentration acts as a negative regulator of glutamine synthetase in mycobacteria. Secondly, it appears that for P2 promoter, nitrogen may not be the real trigger for activation in the context of glutamine synthetase or PLG synthesis. It would be interesting to examine if there are other signals or factors involved in regulation of the P2 promoter in mycobacteria. The synthesis of PLG during low nitrogen availability and its absence in the cell wall in nitrogen sufficiency indicated that

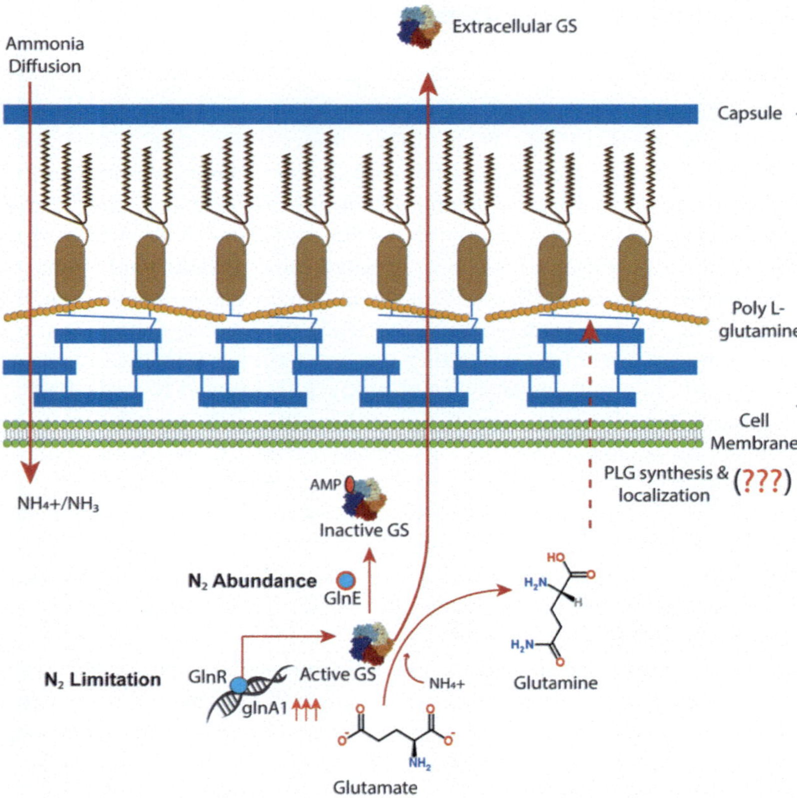

Fig. 11.1 **Regulation of glutamine synthetase (*glnA1*) gene in response to nitrogen availability.** In nitrogen abundance condition, GlnE inhibits activity of GS, leading to reduced glutamine synthesis and thus PLG. In nitrogen limitation condition, GlnR positively regulates GS and leads to formation of glutamine and thus PLG. Discontinuous arrows and ??? indicate steps in PLG synthesis and localization that are still under investigation. (Adapted and modified from Tripathi et al. 2013)

PLG might be a nitrogen storage mechanism in virulent mycobacteria. Figure 11.1 shows pictorial representation of transcriptional and posttranslational regulation of *glnA1* and hence PLG, in response to nitrogen availability.

11.6 Immunostimulatory Properties of PLG

The components of Mtb cell wall are known to interact with a variety of host factors upon infection and hence play a crucial role in the modulation of host immune responses (Karakousis et al. 2004). The strong immune-stimulatory properties of the cell wall components present in complete Freund's adjuvant (CFA), an emulsion of mannide monooleate and heat-killed Mtb, are attributed to muramyl dipeptide (MDP) and trehalose 6,6′-dimycolate (TDM) (Kasmar et al. 2014). Unfortunately, due to multiple toxic effects of CFA, e.g., inflammation and pain at the site of injection, abscess, intraperitoneal granulomas, abdominal adhesions, and lymphoid hyperplasia in the mice, it cannot be used for humans, restricting its use to laboratory studies only (Petrovsky and Aguilar 2004). The adjuvants play a crucial role in enhancing/modulating the immunogenicity of a vaccine candidate (Reed et al. 2013). The "adjuvant-like" properties of synthetic glutamine-rich self-assembling peptides and their derivatives were shown to increase IgG1, IgG2, IgG3, and IgM antibody levels equivalent to that of CFA (Rudra et al. 2010). Taking a cue from these studies, we investigated the adjuvant potential of PLG, using a variety of different model antigens.

Our study involved testing of PLG in combination with various prototypical antigens like the Mtb-secreted antigenic target protein 6 (ESAT-6) and antigen 85B (Ag85B), protective antigen (PA) of *Bacillus anthracis* and BP26 of *Brucella abortus* in the mouse model. The results revealed that no specific antibodies were produced against PLG after repeated immunizations with the peptides alone, indicating its weak immunogenicity in mice. In contrast, adjuvantation of the antigens with PLG triggered a strong antigen-specific immune response compared to antigen alone. A significant increase in the serum IgG levels and its isotypes (IgG1, IgG2a, and IgG2b), especially IgG2a, against all the above antigens occurred by adjuvantation with PLG, suggesting induction of a Th1-biased immune response. Parallel to the humoral response, a robust production of Th1-specific IFN-γ, TNF-α, and IL-2 and Th2-specific IL-6 and IL-10 cytokines was also observed in the cultured splenocytes from PLG adjuvanted mice as a memory response to recall antigen. Higher IgG production and isotype switching to IgG2a in the PLG adjuvanted mice group is consistent with enhanced production of Th1-stimulatory IFN-γ and TNF-α cytokines. A simultaneous induction of antigen-specific IgG and its isotype and cellular memory in PLG adjuvanted mice groups indicated stimulation of both Th1 and Th2 effector arms of immunity.

Recently, Th17 cells have emerged as a major immune mediator against TB due to secretion of high levels of IL-17A. The latter results in increased chemokine production, T- and B-cell localization, and migration of Mtb-infected alveolar macrophages in the lung granulomas, enabling containment of Mtb in the lungs (Van Dis et al. 2018). Earlier, a vaccine formulation containing H1 (a fusion protein of Ag85B and ESAT-6) and H28 (a fusion protein of Rv2660, Ag85B and TB10.4) antigens with CAF01 adjuvant was shown to elicit long-lasting Th17 responses in mice against Mtb infection (Lindenstrøm et al. 2012). Notably, adjuvantation of ESAT-6 with PLG in mice elicited significantly high antigen-specific IL-17 production compared to antigen alone or dimethyldioctadecylammonium/monophosphoryl lipid-A (DDA/MPL, a common Th1 inducer) adjuvanted group, suggesting a strong Th17 response. Thus the ability of PLG to induce a broad, multifaceted immune response makes it a potential candidate adjuvant molecule for exploration, particularly for chronic infectious diseases caused by intracellular pathogens like Mtb and human immunodeficiency virus (HIV), where Th1-type immune response is necessary for effective clearance and protection (Burton and Moore 1998; Simon et al. 2011; Ottenhoff and Kaufmann 2012).

The efficacy of ESAT-6 subunit vaccine candidate was tested after Mtb challenge in mice. Addition of PLG in the ESAT-6 vaccine formulation increased the secretion of Th1-specific cytokines significantly, showing development of immune memory, marked with concomitant reduction of bacterial burden in the lung and spleen of the C57BL/6J mice. Further, the protective efficacy of the formulation was also reflected on the survival of mice. The median survival time of the mice immunized with antigen alone was 66 days compared to 205 days of PLG adjuvanted mice, which was close to 224 days observed with BCG vaccination (Mani et al. 2018). In the above study ESAT-6, a weak immunogen was chosen as the vaccine candidate to allow PLG to manifest its immunomodulatory activity optimally, without being masked by a strong antigen. It would be interesting to evaluate long-term effect on survival/protection with stronger immunogens like Ag85B or CFP10 or a combination of multiple proteins vis-à-vis the BCG vaccine.

The molecular mechanism underlying the action of PLG as a vaccine component remains unclear. Some of the bacterial products are immune-stimulatory by virtue of their similarity with microbial structures, referred to as pathogen-associated molecular patterns (PAMPs), and recognized as foreign element by the host (Reed et al. 2009; O'Hagan et al. 2017). Indeed, it is known that microbial derivatives mediate their action through pattern recognition receptors (PRRs) such as Toll-like receptors or TLRs on the surface of antigen-presenting cells (APCs) mainly macrophages and dendritic cells (Dowling and Mansell 2016). The interaction of PAMPs in a subunit vaccine with the PRRs on the surface of APCs activates the downstream signaling pathways, releasing proinflammatory cytokines along with the expression of MHC class I and II molecules to elicit the adaptive immune response (Liechtenstein et al. 2012). It is likely that PLG may act as a PRR ligand on different APC subsets and stimulate the Th1, Th2, and Th17 pathways of the immune response. Alternatively, PLG may transport antigen proteins by entrapping or by non-covalent association, facilitating their entry into the APCs, thereby augmenting the antigen-specific immune responses. Such a mechanism of adjuvantation has been reported earlier for synthetic poly(lactic-co-glycolic) acid (PLGA) particles (Manish et al. 2013), virus-like particles (Gao et al. 2018), and immune-stimulating complexes also (Sanders et al. 2005).

Our investigations exploring the biological properties of PLG are still in the preliminary stages. The physicochemical properties of PLG, an insoluble, homopolymer of L-glutamine, seem to be largely responsible for the immunomodulatory action of this mycobacterial cell wall component (Phiet et al. 1976; Hirschfield et al. 1990; Mani et al. 2018). In view of its consistent immune-stimulatory profile across protein antigens of diverse bacterial origin, our research has laid the foundation for developing PLG as an efficient Th1-specific adjuvant. However, an essential prerequisite for the development of any biological product is the assessment of its toxicological properties. We found the PLG preparation nontoxic to THP1 macrophages and in C57BL/6J mice, yet comprehensive safety evaluation of the product is imperative in diverse animal models. Secondly, studying the mechanistic orchestration of host immune response and pharmacokinetics of PLG in different hosts will pave the way to exploit its full potential for developing a more precise pathogen–/

vaccine-specific immunomodulator molecule. This study demonstrated that the PLG peptides of Mtb cell wall act like an archetypical adjuvant by stimulating both the humoral and cell-mediated immune response of the host.

In the past couple of decades, extensive research has unraveled the biochemical nature of mycobacterial cell wall, advancing our knowledge regarding individual components of the envelope. However, for some reason poly-α-L-glutamine did not attract as much attention as the others. It was our endeavor to take the existing knowledge about poly-α-L-glutamine forward and explore its significance in the biology of mycobacteria. Our efforts have suggested some interesting aspects of this cell wall constituent preparing the ground for future studies.

References

Alderwick LJ, Harrison J, Lloyd GS, Birch HL (2015) The mycobacterial Cell Wall—peptidoglycan and arabinogalactan. Cold Spring Harb Perspect Med 5:a021113. https://doi.org/10.1101/cshperspect.a021113

Amar C, Vilkas E (1973) Isolation of arabinose phosphate from the walls of Mycobacterium tuberculosis H37Ra. C R Acad Sci Hebd Seances Acad Sci D 277:1949–1951

Berg S, Kaur D, Jackson M, Brennan PJ (2007) The glycosyltransferases of Mycobacterium tuberculosis – roles in the synthesis of arabinogalactan, lipoarabinomannan, and other glycoconjugates. Glycobiology 17:35R–56R. https://doi.org/10.1093/glycob/cwm010

Brennan P (2003) Structure, function, and biogenesis of the cell wall of Mycobacterium tuberculosis. Tuberculosis 83:91–97. https://doi.org/10.1016/S1472-9792(02)00089-6

Burton DR, Moore JP (1998) Why do we not have an HIV vaccine and how can we make one? Nat Med 4:495–498. https://doi.org/10.1038/nm0598supp-495

Chandra H, Basir SF, Gupta M, Banerjee N (2010) Glutamine synthetase encoded by glnA-1 is necessary for cell wall resistance and pathogenicity of Mycobacterium bovis. Microbiology 156:3669–3677. https://doi.org/10.1099/mic.0.043828-0

Dowling JK, Mansell A (2016) Toll-like receptors: the swiss army knife of immunity and vaccine development. 5:e85–e10. https://doi.org/10.1038/cti.2016.22

Draper P (1971) The walls of Mycobacterium lepraemurium: chemistry and ultrastructure. J Gen Microbiol 69:313–324

Draper P, Kandler O, Darbre A (1987) Peptidoglycan and arabinogalactan of Mycobacterium leprae. J Gen Microbiol 133:1187–1194

Gao Y, Wijewardhana C, Mann JFS (2018) Virus-like particle, liposome, and polymeric particle-based vaccines against HIV-1. Front Immunol 9:1–18. https://doi.org/10.3389/fimmu.2018.00345

Garg R, Tripathi D, Kant S et al (2015) The conserved hypothetical protein Rv0574c is required for cell wall integrity, stress tolerance, and virulence of Mycobacterium tuberculosis. Infect Immun 83:120–129. https://doi.org/10.1128/IAI.02274-14

Harper C, Hayward D, Wiid I, van Helden P (2008) Regulation of nitrogen metabolism in *Mycobacterium tuberculosis*: a comparison with mechanisms in *Corynebacterium glutamicum* and *Streptomyces coelicolor*. IUBMB Life 60:643–650. https://doi.org/10.1002/iub.100

Harth G, Horwitz MA (1997) Expression and efficient export of enzymatically active Mycobacterium tuberculosis glutamine synthetase in Mycobacterium smegmatis and evidence that the information for export is contained within the protein. J Biol Chem 272:22728–22735. https://doi.org/10.1074/jbc.272.36.22728

Harth G, Horwitz MA (1999) An inhibitor of exported Mycobacterium tuberculosis glutamine synthetase selectively blocks the growth of pathogenic mycobacteria in axenic culture and in human monocytes: extracellular proteins as potential novel drug targets. J Exp Med 189:1425–1436

Harth G, Clemens DL, Horwitz MA (1994) Glutamine synthetase of Mycobacterium tuberculosis: extracellular release and characterization of its enzymatic activity. Proc Natl Acad Sci U S A 91:9342–9346

Hirschfield GR, McNeil M, Brennan PJ (1990) Peptidoglycan-associated polypeptides of Mycobacterium tuberculosis. J Bacteriol 172:1005–1013

Jankute M, Cox JAG, Harrison J, Besra GS (2015) Assembly of the mycobacterial Cell Wall. Annu Rev Microbiol 69:405–423. https://doi.org/10.1146/annurev-micro-091014-104121

Karakousis PC, Bishai WR, Dorman SE (2004) *Mycobacterium tuberculosis* cell envelope lipids and the host immune response. Cell Microbiol 6:105–116. https://doi.org/10.1046/j.1462-5822.2003.00351.x

Kasmar A, Layre E, Moody DB (2014) Lipid adjuvants and antigens embedded in the mycobacterial cell envelope. In: The art & science of tuberculosis vaccine development, 2nd edn. Oxford University Press, Malaysia, pp 123–149. http://tbvaccines.usm.my/finlay

Kotani S, Yanagida I, Kato K, Matsuda T (1970) Studies on peptides, glycopeptides and antigenic polysaccharide-glycopeptide complexes isolated from an L-11 enzyme lysate of the cell walls of Mycobacterium tuberculosis strain H37Rv. Biken J 13:249–275

Liechtenstein T, Dufait I, Lanna A et al (2012) Modulating co-stimulation during antigen presentation to enhance cancer immunotherapy. Immunol Endocr Metab Agents Med Chem 12:224–235. https://doi.org/10.2174/187152212802001875

Lindenstrøm T, Woodworth J, Dietrich J et al (2012) Vaccine-induced Th17 cells are maintained long-term postvaccination as a distinct and phenotypically stable memory subset. Infect Immun 80:3533–3544. https://doi.org/10.1128/IAI.00550-12

Mahapatra S, Crick DC, Brennan PJ (2000) Comparison of the UDP-N-acetylmuramate:L-alanine ligase enzymes from Mycobacterium tuberculosis and Mycobacterium leprae. J Bacteriol 182:6827–6830

Mahapatra S, Scherman H, Brennan PJ, Crick DC (2005) N Glycolylation of the nucleotide precursors of peptidoglycan biosynthesis of Mycobacterium spp. is altered by drug treatment. J Bacteriol 187:2341–2347. https://doi.org/10.1128/JB.187.7.2341-2347.2005

Mani R, Gupta M, Malik A et al (2018) Adjuvant potential of poly-α-L-glutamine from the cell wall of *Mycobacterium tuberculosis*. Infect Immun 86. https://doi.org/10.1128/IAI.00537-18

Manish M, Rahi A, Kaur M et al (2013) A single-dose PLGA encapsulated protective antigen domain 4 nanoformulation protects mice against Bacillus anthracis spore challenge. PLoS One 8:e61885. https://doi.org/10.1371/journal.pone.0061885

Mehta R, Pearson JT, Mahajan S et al (2004) Adenylylation and catalytic properties of Mycobacterium tuberculosis glutamine synthetase expressed in Escherichia coli versus mycobacteria. J Biol Chem 279:22477–22482. https://doi.org/10.1074/jbc.M401652200

Meroueh SO, Bencze KZ, Hesek D et al (2006) Three-dimensional structure of the bacterial cell wall peptidoglycan. Proc Natl Acad Sci 103:4404–4409. https://doi.org/10.1073/pnas.0510182103

Newsholme P, Procopio J, Lima MMR et al (2003) Glutamine and glutamate--their central role in cell metabolism and function. Cell Biochem Funct 21:1–9. https://doi.org/10.1002/cbf.1003

O'Hagan DT, Friedland LR, Hanon E, Didierlaurent AM (2017) Towards an evidence based approach for the development of adjuvanted vaccines. Curr Opin Immunol 47:93–102. https://doi.org/10.1016/j.coi.2017.07.010

Ottenhoff THM, Kaufmann SHE (2012) Vaccines against tuberculosis: where are we and where do we need to go? PLoS Pathog 8(5):e1002607. https://doi.org/10.1371/journal.ppat.1002607

Pai M, Behr MA, Dowdy D et al (2016) Tuberculosis. Nat Rev Dis Prim 2:16076. https://doi.org/10.1038/nrdp.2016.76

Park H-D, Guinn KM, Harrell MI et al (2003) Rv3133c/dosR is a transcription factor that mediates the hypoxic response of Mycobacterium tuberculosis. Mol Microbiol 48:833–843

Petrovsky N, Aguilar JC (2004) Vaccine adjuvants: current state and future trends. Immunol Cell Biol 82:488–496. https://doi.org/10.1111/j.1440-1711.2004.01272.x

Phiet PH, Wietzerbin J, Zissman E et al (1976) Analysis of the cell wall of five strains of Mycobacterium tuberculosis BCG and of an attenuated human strain, W 115. Infect Immun 13:677–681

Reed SG, Bertholet S, Coler RN, Friede M (2009) New horizons in adjuvants for vaccine development. Trends Immunol 30:23–32. https://doi.org/10.1016/j.it.2008.09.006

Reed SG, Orr MT, Fox CB (2013) Key roles of adjuvants in modern vaccines. Nat Med 19:1597–1608. https://doi.org/10.1038/nm.3409

Rudra JS, Tian YF, Jung JP, Collier JH (2010) A self-assembling peptide acting as an immune adjuvant. Proc Natl Acad Sci 107:622–627. https://doi.org/10.1073/pnas.0912124107

Sanders MT, Brown LE, Deliyannis G, Pearse MJ (2005) ISCOM^{TM}-based vaccines: the second decade. Immunol Cell Biol 83:119–128. https://doi.org/10.1111/j.1440-1711.2005.01319.x

Simon R, Tennant SM, Galen JE, Levine MM (2011) Mouse models to assess the efficacy of non-typhoidal Salmonella vaccines: revisiting the role of host innate susceptibility and routes of challenge. Vaccine 29:5094–5106. https://doi.org/10.1016/j.vaccine.2011.05.022

Sousa EHS, Tuckerman JR, Gonzalez G, Gilles-Gonzalez MA (2007) DosT and DevS are oxygen-switched kinases in Mycobacterium tuberculosis. Protein Sci 16:1708–1719. https://doi.org/10.1110/ps.072897707

Tiffert Y, Supra P, Wurm R et al (2008) The Streptomyces coelicolor GlnR regulon: identification of new GlnR targets and evidence for a central role of GlnR in nitrogen metabolism in actinomycetes. Mol Microbiol 67:861–880. https://doi.org/10.1111/j.1365-2958.2007.06092.x

Tripathi D, Chandra H, Bhatnagar R (2013) Poly-L-glutamate/glutamine synthesis in the cell wall of Mycobacterium bovis is regulated in response to nitrogen availability. BMC Microbiol 13:226. https://doi.org/10.1186/1471-2180-13-226

Tullius MV, Horwitz MA (2003) Glutamine synthetase GlnA1 is essential for growth of Mycobacterium tuberculosis in human THP-1 macrophages and guinea pigs, vol 71, pp 3927–3936. https://doi.org/10.1128/IAI.71.7.3927

Umbarger HE (1978) Amino acid biosynthesis and its regulation. Annu Rev Biochem 47:533–606. https://doi.org/10.1146/annurev.bi.47.070178.002533

Van Dis E, Sogi KM, Rae CS et al (2018) STING-activating adjuvants elicit a Th17 immune response and protect against Mycobacterium tuberculosis infection. Cell Rep 23:1435–1447. https://doi.org/10.1016/j.celrep.2018.04.003

Wietzerbin J, Lederer F, Petit JF (1975) Structural study of poly-L-glutamic acid of Mycobacterium tuberculosis var. hominis Brevannes. Biochem Biophys Res Commun 62:246–252

Wietzerbin-Falszpan J, Das BC, Azuma I et al (1970) Isolation and mass spectrometric identification of the peptide subunits of mycobacterial cell walls. Biochem Biophys Res Commun 40:57–63

Wietzerbin-Falszpan J, Das BC, Gros C et al (1973) The amino acids of the cell wall of Mycobacterium tuberculosis var. bovis, strain BCG. Presence of a poly(L-glutamic acid). Eur J Biochem 32:525–532

Cellular Stress Responses and Immunological Regulations During *Mycobacterium tuberculosis* Infection

12

Nooruddin Khan, Gillipsie Minhas, K. Kala jyothi, and Jyoti Sharma

Abstract

Tuberculosis (TB) is one of the most devastating infectious diseases caused by *Mycobacterium tuberculosis* (MTB). A high percentage of mortality and morbidity associated with TB has been reported globally with the highest number of cases reported in Asia. MTB is known to enter a state of dormancy and phenotypic drug resistance, when it is exposed to multiple stress conditions in the host microenvironment, and thereby it survives asymptomatically in latent phase in the host for decades and even for a lifetime. This raises the need for improved vaccine, drugs, and therapeutics, which could be achieved by a better understanding of the host-microbe interactions as well as immune responses during the infection. Recent studies have highlighted the importance of host cellular stress response pathways, such as unfolded protein response (UPR), oxidative stress response, integrated stress response (ISR), and autophagy during various infections. However, the role of these host stress response pathways during an MTB infection in the modulation of the immune response against the microbe is poorly understood. Therefore, through this chapter, we will highlight the cellular stress response pathways and various molecular mechanisms through which MTB influences the host innate as well as the adaptive immune response during infection, which might aid toward better design and development of therapeutics and vaccine candidates against TB.

Keywords

Mycobacterium tuberculosis · Tuberculosis · Immune response · Stress response · ER stress

N. Khan (✉) · G. Minhas · K. Kala jyothi · J. Sharma
Department of Biotechnology and Bioinformatics, School of Life Sciences, University of Hyderabad, Hyderabad, Telangana, India
e-mail: noor@uohyd.ac.in

S. E. Hasnain et al. (eds.), Mycobacterium tuberculosis: *Molecular Infection Biology Pathogenesis, Diagnostics and New Interventions*,
https://doi.org/10.1007/978-981-32-9413-4_12

Abbreviations

APC	antigen-presenting cells
ATF	activating transcription factor
BCG	bacille Calmette-Guerin
BiP	binding immunoglobulin protein
CHOP	CCAAT-enhancer-binding protein homologous protein
CLR	C-type lectin receptor
CR	complement receptors
DALYs	disability-adjusted life years
DC	dendritic cells
eIF2	eukaryotic initiation factor 2
eNOS	endothelial nitric oxide synthase
ER	endoplasmic reticulum
ESAT6	early secretory antigenic target 6 kDa
IFN	interferon
IL	interleukin
iNOS	induced nitric oxide synthase
IRE1	inositol-requiring enzyme 1
ISR	integrated stress response
JNK	c-Jun N-terminal kinases
LC3	microtubule-associated protein 1A/1B-light chain 3
LPS	lipopolysaccharide
MAPK	mitogen-activated protein kinases
MDR	multidrug resistance
MHC	major histocompatibility complex
MTB	*Mycobacterium tuberculosis*
mTOR	mammalian target of rapamycin or mechanistic target of rapamycin
NK cells	natural killer cells
NLR	NOD-like receptors
NO	nitric oxide
NOD	nucleotide oligomerization domain
NOS	nitric oxide synthase
nNOS	neuronal nitric oxide synthase
PAMPs	pathogen-associated molecular patterns
PERK	protein kinase RNA-like ER kinase
PI3P	phosphatidylinositol 3-phosphate (PI3P)
PKR	protein kinase R
PRR	pattern recognition receptors
RIPE	rifampicin, isoniazid, pyrazinamide, and ethambutol
RLR	RIG-I-like receptors
ROS	reactive oxygen species
SPA	surfactant protein A
STING	stimulator of IFN genes

TB	tuberculosis
TBK1	TANK-binding kinase 1
TLRs	Toll-like receptors
TNF	tumor necrosis factor
UPR	unfolded protein response
WHO	World Health Organization
XBP1	X-box-binding protein 1

12.1 Introduction

Tuberculosis (TB) poses a major health threat in developing nations despite the available therapies. It is one of the top three infectious and fatal diseases after AIDS and malaria. Over one-third of the population is infected with the pathogen worldwide, and it is estimated that over a 100 million people have died of the disease in the past 100 years, while about 35 million are predicted to die between the years 2000 and 2020 (Zumla et al. 2015). With 9.6 million new cases of TB per year being reported worldwide, it becomes even more indispensable to contain the pathogen. In addition, the disease burden of TB way surpasses mortality and contributes to a high level of disability-adjusted life years (DALYs) lost due to illness and outdoes malaria in this aspect (WHO 2016). According to WHO report 2016, 60% of recorded TB cases have been contributed majorly by six developing countries: India, Indonesia, China, Nigeria, Pakistan, and South Africa (WHO 2016).

TB is caused by bacillus *Mycobacterium tuberculosis* (MTB) a notorious pathogen that has become difficult to eradicate with BCG which is the only vaccine available even after a century of its discovery (Luca and Mihaescu 2013). It is acknowledged that only 5% of individuals infected with MTB develop active disease, while the remaining 95% contain the pathogen in an asymptomatic inactive form (dormant/latent form), which is resistant to the current anti-mycobacterial drugs used for treating active disease. Streptomycin was the first effective drug used against active TB. Currently, a combination of rifampicin, isoniazid, pyrazinamide, and ethambutol (RIPE therapy) for 2 months followed by rifampicin and isoniazid for another 4 months is the treatment of choice for uncomplicated active infection (Joshi 2011). However, emergence of multidrug-resistant (MDR) strains has made it difficult to target the pathogen with available drug therapies. Many second-generation recombinant BCG vaccines are in different stages of clinical trials for their immunogenicity to exterminate both forms of the disease (Kaufmann et al. 2010).

TB pathogenesis, in general, is a delicate interplay between the mechanisms employed by the host to evade the pathogen and the strategies employed by the pathogen to survive within the unfavorable host environment. During an MTB infection, the host tries to maintain homeostasis by eliciting various forms of cellular fate mechanisms like necroptosis, apoptosis, and autophagy, among which apoptosis and autophagy have been documented as one of the crucial innate defense mechanisms (Behar et al. 2010, 2011; Bradfute et al. 2013; Du et al. 2013). The pathogen also employs various rescue strategies like induction of anti-inflammatory cytokines and subsequent decrease in TNF-α-mediated apoptosis, upregulation of specific microRNAs that attenuate autophagy to escape the host immune system. A better understanding of the host and pathogen defense mechanisms has therefore become even more important to target the pathogen and to develop new interventions to circumvent the disease.

12.2 *Mycobacterium tuberculosis* Pathogen and Infection

TB is an ancient disease that has been linked with the evolution of human race with reports dating as old as 5000 years. MTB is one of the oldest described bacterial pathogens and is thought to have evolved about 3 million years ago (Gutierrez et al. 2005). With time, various breakthroughs in the field of microbiology have been made, such as the development of BCG vaccine (Calmette and Boquet 1921) and the discovery of drugs like streptomycin, isoniazid, rifampicin, etc. (Van Scoy and Wilkowske 1999). All these developments have led to a boost in treatment and TB research. However, the calamity of TB has reemerged with the advent of different multidrug-resistant strains of MTB.

The pathogen was first described by Robert Koch in 1882, as a nonmotile, non-sporulating, slow-growing pathogenic species. It is facultative, intracellular bacteria that multiply in phagocytic cells like macrophages. The progress of infection depends on the initial interaction between the pathogen and the host. MTB infection commences with inhalation of the pathogen, which then travels to alveoli in the lungs. The bacteria first interact with the resident alveolar macrophages and this step is an absolute prerequisite for MTB to establish an infection in a susceptible host lung. Though the primary role of macrophages is protective, intake of the pathogen by these defense cells is vital for the establishment of infection. The interaction prompts a primary immune response through pro-inflammatory molecules and recruitment of leukocytes. This early encounter in the lungs itself decides the fate of the pathogen, whether the bacteria will remain contained in the macrophages as a latent form or will lead to an active disease. Infection further activates different mechanisms such as fusion of phagolysosomes, respiratory burst, and production of reactive oxygen species (ROS) (Giacomini et al. 2001). The response that is initiated upon binding of the bacteria with the macrophage depends on the type of phagocytic receptor (Ernst 1998). The most common receptors on the macrophages are termed as pattern recognition receptors (PRR), which recognize different bacterial components, termed as pathogen-associated molecular patterns (PAMPs); the complement receptors (C receptors), which are important for bacterial opsonization; mannose receptors, which assist macrophages in phagocytosis of unopsonized bacteria; surfactant protein A (SPA), which modulates the activity of other receptors to enhance macrophage binding and MTB uptake; and scavenger receptors, which act at later stages of infection to further stimulate macrophages and inflammatory responses. The binding of PAMPs to the specific receptors initiates phagocytic pathways, for example, binding to FcR produces reactive oxygen species (ROS) and phagolysosomal fusion, whereas, in the case of CR3 receptor, the binding inhibits the respiratory burst (Armstrong and Hart 1975; Kang et al. 2005; Mortaz et al. 2015; Pahari et al. 2017).

Once internalized, MTB stays in the phagosome and blocks its fusion with the lysosome. The bacteria act by eliminating phosphatidylinositol 3-phosphate (PI3P) from the infected cells, which is used in trafficking of lysosome to phagosome and thus impedes the fusion (Vergne et al. 2005). MTB infection also inhibits the production of phosphatidylinositol 3-phosphate (PI3P), which is essential for induction of autophagy (Vergne and Deretic 2010). The cell wall components of MTB, especially particular lipids, also have an impact as these interact with the cell membrane of phagosomes and affect their maturation and hence its elimination (Axelrod et al. 2008).

12.3 Host Defense Mechanisms: Innate and Adaptive Immune Responses

Each cell has its own defense mechanisms that limit the cellular damage and ultimately eliminate the pathogen. Most of these responses against a pathogen invasion is to maintain the cellular homeostasis by decreasing the cellular functions such as protein translation or inducing cellular processes like apoptosis and autophagy. A successful infection is dependent on the initial interaction between the host and the pathogen. In case of MTB, initial exposure to the pathogen occurs through airborne droplets, which are inhaled and then transported to lungs, where it is taken up by the resident macrophages through phagocytosis for elimination. The initiation of the innate arm of the immune system is through the PRRs, which recognize conserved PAMPs unique to each pathogen and are essential for their survival and pathogenesis. The PRRs are expressed in almost all immune and nonimmune cells and majorly on antigen-presenting cells like macrophages and dendritic cells. They are able to recognize both extracellular and intracellular pathogens as they are expressed both on the cell surface as well as within the endosomes and are involved in activating pro-inflammatory signaling pathways and stimulating phagocytic responses. Based on the function and location, the PRRs have been classified into four families, Toll-like receptors (TLRs), nucleotide-binding oligomerization domain-like receptors (NODs), C-type lectin receptors (CLRs), and RIG-1-like receptors (RLRs) (Takeuchi and Akira 2010).

TLRs are the imperative players with respect to immunity against infectious agents as they play a critical role in initiating an innate and adaptive immune response against vast majority of pathogens. They were first identified in *Drosophila* and later found to be associated with mammals as a response to bacterial infections (Leulier and Lemaitre 2008). TLRs are known to act individually or in sets to recognize molecular signatures of pathogens such as the cell wall components in response to infection. Host response against MTB has been postulated to involve TLR signaling as the pathogen is laden with well-characterized TLR ligands that are potent stimuli to activate the Th1-mediated immune response (Jones et al. 2001; Thoma-Uszynski et al. 2001; Abel et al. 2002; Feng et al. 2003; Basu et al. 2012). Much work done by researchers has shown that majorly TLR-2, TLR-4, and TLR-9 play phenomenal roles in the induction of immunity against MTB infection (Bafica et al. 2005; Bhatt and Salgame 2007; Mukherjee et al. 2016). TLR-2 participates in recognizing the cell wall components of MTB and promoting the macrophages to secrete TNF-α, IL-12, as well as nitric oxide. Mice deficient in TLR-2, when infected with MTB, were not able to elicit such responses (Reiling et al. 2002; Drennan et al. 2004). The GC-rich genome of MTB is a very good ligand for TLR-9, which interacts with CpG motifs that may regulate the progression of granuloma formation (Ito et al. 2007). The role of TLRs in eliciting immunity against TB is understood from the work of Velez et al., where the authors showed that polymorphisms in TLR-2 and TLR-9 are associated with increased susceptibility to TB in different populations (Velez et al. 2010). The role of TLR-4 however is not well understood. Single missense mutation in TLR-4 rendered mice infected with MTB is inefficient in controlling bacterial loads and initiate pro-inflammatory responses, while TLR-4-deficient mice had no such effect (Jo et al. 2007).

Recently, NOD-like receptors (NLRs) have also shown to play a role in immunity against MTB. NLRs contain leucine-rich repeats and, unlike TLRs, recognize intracellular PAMPs in the cytosol. NOD-2 has been shown to play a crucial role in TLR activation through the production of inflammatory cytokines, TNF-α and IL-1β, in human macrophages during TB infection (Brooks et al. 2011). Moreover, NOD-2-deficient mice showed poor innate and adaptive immunity and increased bacterial loads upon MTB infection (Divangahi et al. 2008).

The early response subsequent to phagocytosis involves activation of innate immune and Th1 pathways, which leads to granuloma formation (Salgame 2005). The formation of the granuloma is a protective host response against MTB that can inhibit the spread of bacteria, thus keeping the infection in latent phase. The granuloma consists of infected cells and immune cells (differentiated myeloid cells surrounded by lymphocytes) along with the macrophages and MTB, which is attacked by $CD4^+$ cells and B cells that restrict the bacteria in the capsule-like structure (Peters and Ernst 2003).

Alveolar macrophages facilitate both innate and adaptive immune responses to combat MTB infection. Also, phagocytosis of MTB causes secretion of cytokines that are pro-inflammatory in nature, such as IFN-γ, TNF-α, IL1-β, IL-6, and IL-12 (Brightbill et al. 1999). Th1 cytokine such as IFN-γ is secreted majorly by the CD4+ T cells, which plays a central role as host defense in anti-mycobacterial immunity, such as induction of NOS2 expression and upregulation of MHC class II molecules, which in turn play important roles in antigen processing and presentation (Harding and Boom 2010). IFN-γ thus contributes to MTB control by connecting innate and adaptive immune responses. It has also been shown that IFN-γ enhances the autophagic control of MTB, while the Th2 cytokines like IL-4 have the opposite effects (Harris et al. 2007). In addition, the pro-inflammatory cytokine TNF-α acts synergistically along with IFN-γ during antimicrobial responses against MTB by playing an important role in activation of phagosome maturation and hence the development of granuloma to restrict MTB (Garcia et al. 1997). Further, the role of TNF-α in anti-mycobacterial immunity is evident in case of patients treated with anti-TNF-α antigens in a variety of inflammatory/autoimmune diseases including TB (Harris and Keane 2010). TNF-α can induce apoptosis accompanied by the production of reactive oxygen intermediates/reactive nitrogen intermediates to destroy the engulfed bacteria (Rojas et al. 1999; Ciaramella et al. 2002). Actively involved immune cells such as neutrophils and natural killer (NK) cells, secrete cytokines, such as IFN-γ, which further stimulates secretion of IL-12 by the macrophages (Schierloh et al. 2007). These studies emphasize the role of cytokines in providing protection against various pathogens including MTB. Figure 12.1 shows the innate immune response elicited upon MTB infection.

Development of adaptive immunity needs close interaction between T cells and APCs. Upon macrophage activation via antigen presentation through MHC class II complex, the APCs take up the infected apoptotic macrophages and extracellular MTB to stimulate $CD8^+$ T cells by cross-presenting MTB antigens via MHC class I complex. Thus, the $CD4^+$ and $CD8^+$ T cells activate each other to eliminate MTB. Studies have also highlighted the role of $CD8^+$ cells in infection through cytokine production, such as IFN-γ and TNF-α, which activate the bactericidal mechanisms. Another T-cell subpopulation that has also been associated with an early response toward TB in granuloma formation is the γδT cells (Griffin et al. 1991; Collins and Kaufmann 2001). Figure 12.2 highlights the adaptive immune response produced during MTB infection.

Additionally, costimulatory molecules such as B7.1 (CD80) and B7.2 (CD86) expressed on macrophages and dendritic cells and cytokines produced by these cells stimulate the T cells and thus facilitate the elimination of MTB. MTB, therefore, circumvents the action of these inhibitory components by attenuating the expression of costimulatory molecules or by dampening the production of pro-inflammatory cytokines.

Another primary host defense response is the release of nitric oxide (NO), a gaseous free radical molecule produced by a two-step enzymatic reaction catalyzed by nitric oxide synthases (NOSs). Several free radical studies have revealed that the production of reactive oxygen and nitrogen species is an innate host defense mechanism against various types of pathogens including bacteria, viruses, parasites, and fungi (Nathan and Shiloh 2000; Bermudez 1993). Depending upon the types of cells they are expressed in, NOSs are of three types: endothelial cells (eNOS), neuronal cells (nNOS), and as an inducible form (iNOS). While the first two kinds are constitutively expressed, iNOS is expressed in various types of immune cells and produces excessive NO during infection and inflammation (Moncada et al. 1991; Nathan and Xie 1994; Zaki et al. 2005) or upon induction by the inflammatory cytokines TNF-α, IFN-γ, and IL-1β (Lee et al. 1993). NO and related reactive nitrogen intermediates have been shown to exhibit anti-mycobacterial activities in murine models of TB infection. However, the effect of NO on mycobacterial

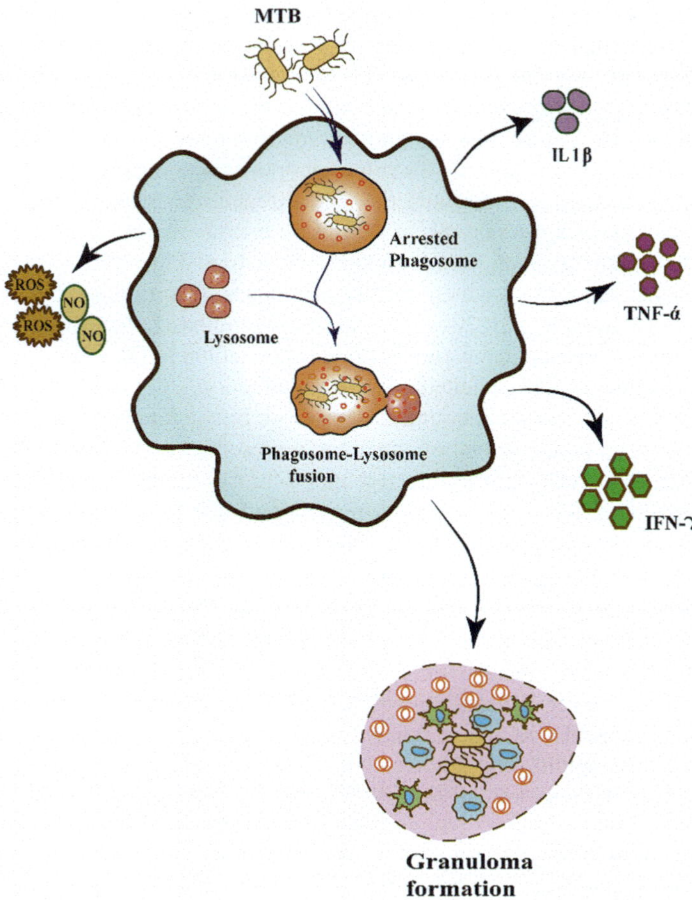

Fig. 12.1 Host innate immune defense mechanisms against *Mycobacterium tuberculosis*
The interaction of MTB pathogen with macrophages in lungs increases ROS and RNS production and leads to phagolysosomal fusion. MTB phagocytosis also releases cytokines, IL1β, TNFα, and IFNγ, which activate different immune responses and granuloma formation to restrict the infection

survival in human macrophages has remained controversial (Rich et al. 1997). Study has shown NO production in human monocyte-derived macrophages upon infection with *Mycobacterium avium*; however, no NO production was demonstrated upon MTB infection in human monocytes (Denis 1991; Weinberg et al. 1995). Study by Rich et al. showed that the infected alveolar macrophages are capable of NO production that can directly or indirectly act as an MTB inhibitor and further impede intracellular MTB growth in alveolar macrophages (Rich et al. 1997). All these studies signify a contributory role of NO in controlling mycobacterial infections in murine macrophages; further investigations are required to ascertain the role of NO in human macrophages.

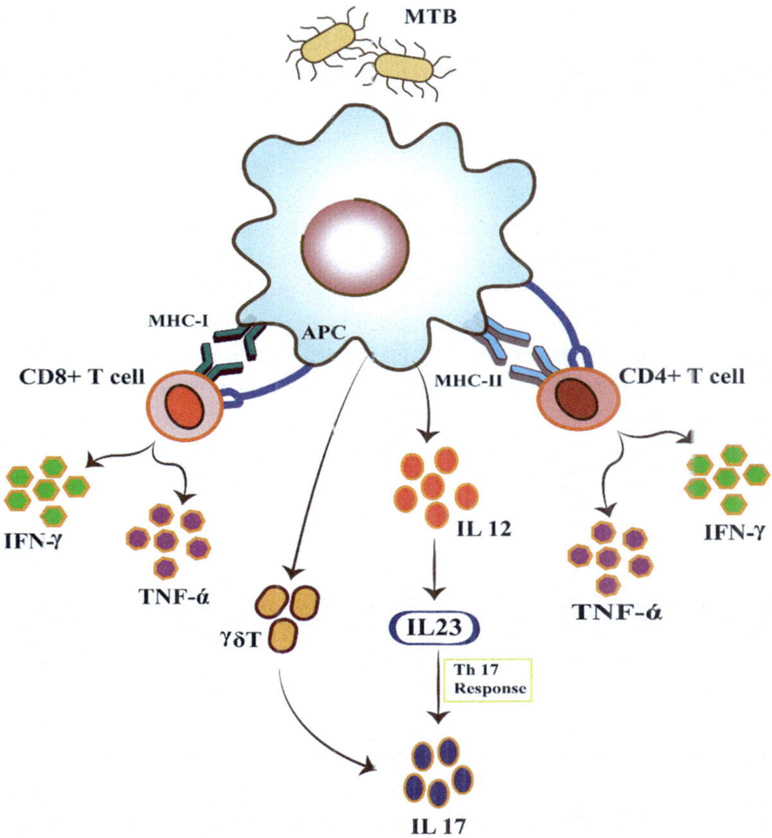

Fig. 12.2 Adaptive immune response against *Mycobacterium tuberculosis*
The host cells upon sensing bacteria trigger series of signaling cascades, which leads to the activation of innate immune response and release of various innate effector molecules including cytokines. These cytokines together with other innate molecules further dictate and shape the effector commitment of the T-cell immune responses

12.4 Host Defense Mechanisms: Stress Response Pathways

Under normal physiology, the body has its own survival mechanisms to maintain the homeostasis. However, various intracellular or extracellular stress conditions, such as endoplasmic reticulum (ER), hypoxia or nutrient starvation, can activate a cascade of stress response signaling events, in the form of unfolded protein response (UPR), integrated stress response (ISR), oxidative stress response, or autophagy (Fulda et al. 2010; Rubartelli et al. 2013). Eukaryotic cells have various sensors that sense stress stimuli, such as nutrient deprivation, endoplasmic reticulum (ER) stress, double-stranded RNA, and heme deprivation. The integration of these stress response pathways with immune regulation has been explored under various inflammatory and metabolic diseases (Ganguly et al. 2016; Battu et al. 2017), which could be investigated further for viral and bacterial infections. Viral proteins are well known to induce stress response due to their dependence for survival on host machinery for protein synthesis. On a similar note, even bacteria and/or their components can induce cellular stress in the host that can evoke different reactions (Celli and Tsolis 2015).

ER not only plays a crucial role in folding and modification of new proteins but also senses different cellular stress (Zhang and Kaufman 2008). Calcium imbalance, hypoxia, or reactive oxidative stress (ROS) in ER are a few instances that can also activate the stress response cascade to return to homeostasis. The cascade initiates through three means: by decreasing the proteins that enter the ER, by elevating the chaperone levels to regulate the protein folding, and by removing misfolded proteins by degradation (Kaufman 2002; Kaufman et al. 2002). These pathways also initiate an immune response. Eventually, if the homeostasis is not retained through these, the ultimate response is apoptosis. Bacterial infection in the body is also known to activate the stress sensors in the ER (Pillich et al. 2016). The stress response mainly increases the phosphorylation of eIF2α that halts the host protein translation and activates the pro-apoptotic response. The ER stress thus regulates many secretory and cellular proteins which ultimately can cause accumulation of misfolded proteins, activating the unfolded protein response (UPR) signaling (Berridge 2002).

MTB infection increases the level of markers indicating toward ER stress. There are primarily three stress sensors in ER: inositol-requiring enzyme 1 (IRE1α), protein kinase RNA-like ER kinase (PERK), and activating transcription factor 6 (ATF6). These three proteins act through a chaperone protein, Ig heavy-chain-binding protein (BiP). Under normal condition, BiP binds to the stress sensors on the luminal side in ER and keeps them in an inactive state. As the unfolded proteins accumulate in ER, BiP is released from the stress sensor proteins and binds to the unfolded proteins. The released sensors act to induce UPR cascade. IRE1 in the unbound form autophosphorylates and splices an intron from mRNA of X-box-binding protein 1 (XBP1); the spliced XBP1 acts as a transcription factor for UPR target genes. Prolonged exposure to ER stress then causes UPR to activate apoptosis signaling through caspases and C/EBP homologous protein (CHOP) (Zhang and Kaufman 2008). On a similar note, PERK activates the kinase domain and phosphorylates the serine residue of eIF2α that inhibits the protein translation and decreases protein load on ER.

Lim et al. investigated the ER stress mechanism in MTB-induced infection in mouse macrophages. The authors found the association with ER stress-induced apoptosis, which confers through caspase 12, present in ER cytoplasm. In response, it was observed at 24 h of MTB infection that the pathogen resists the cessation of host protein translation by decreasing the phosphorylation of eIF2α, which helps in the survival (Lim et al. 2011). The study also found elevated levels of caspase 12 and caspase 9, which is its precursor postinfection. All three stress sensor proteins were found to be activated after the infection (Lim et al. 2011).

There are bacterial components that can also induce UPR signaling, such as lipopolysaccharides (LPS), that is present in the outer membrane of gram-negative bacteria. Similarly, toxins that are released can also activate UPR. The apoptosis of infected macrophages releases antigen of mycobacterial origin, early secretory antigenic target 6 (ESAT6), which can induce ER stress. Choi et al. investigated the ER stress response in the ESAT6-stimulated A549 human epithelial cells. It was noted that the presence of ESAT6 elevated the levels of intracellular calcium, which can cause ROS and ultimately ER stress-induced apoptosis (Choi et al. 2010). ESAT6 is the virulence factor for MTB that is required for pathogenicity. This makes ESAT6 a promising candidate for vaccine development (Wang et al. 2009). ESAT6 treatment also increases phosphorylation of eIF2α and leads to elevated levels of ATF4 and GADD34 and induces apoptosis through the ASK1/JNK pathway. It also acts through cleavage of caspase 12 and 4 into their active forms. Therefore, all these investigations convey that the ER stress triggers apoptosis in infected macrophages through the PERK pathway which involves phosphorylation of eIF2α and activation of CHOP.

Another stress response sensor, dsRNA, activated serine/threonine Protein Kinase R (PKR) is well known to regulate cytokines during viral infections. In case of viral infections, PKR has a well-established role in the phosphorylation of eIF2α and thus halting translation. The large levels of IFN-γ, which are induced upon viral infection, trigger the PKR activity, thus increasing its capacity to recognize viral origin dsRNA. Upon binding to dsRNA, PKR autophosphorylates to an active protein kinase, which phosphorylates eIF2α (De Haro et al. 1995). The role of this kinase has also been investigated in MTB infection. Cheung et al in their study examined the interaction of primary human blood monocytes with BCG. BCG was shown to stimulate expression of cytokines such as TNF-α, IL-6, and IL-10 in monocytes. The authors used PKR inhibitors to demonstrate the involvement of PKR phosphorylation and activation in the stimulation of these cytokines by BCG. PKR activation is known to trigger translocation of NFκβ to the nucleus, through MAPK activation, which ultimately controls the transcription of cytokine genes (Cheung et al. 2005). BCG treatment did not show any effect on the mRNA levels of PKR but had an immediate consequence of its kinase activity. Therefore, this study associates PKR activity with MTB infection, apart from its already known role in viral infections (D'Acquisto and Ghosh 2001; Williams 2001). Further investigations are needed to elicit the mechanism behind the PKR-based anti-mycobacterial activity.

Deprivation of energy, nutrients, or activation of stress conditions such as environmental or oxidative can also initiate autophagy (Ravindran et al. 2014). It is basically a conserved cellular process that prolongs cell survival during stress conditions by initiating the degradation of unnecessary cellular components like nonfunctional macromolecules or damaged organelles into their basic units and redirecting these molecules to be used by the cell. Apart from being a strategy of replenishing essential components to the cell, autophagy has also been established as an innate defense mechanism against various intracellular pathogens (like MTB), viruses, fungi, bacteria, and protozoa (Deretic 2011; Nicola et al. 2012; Beale et al.

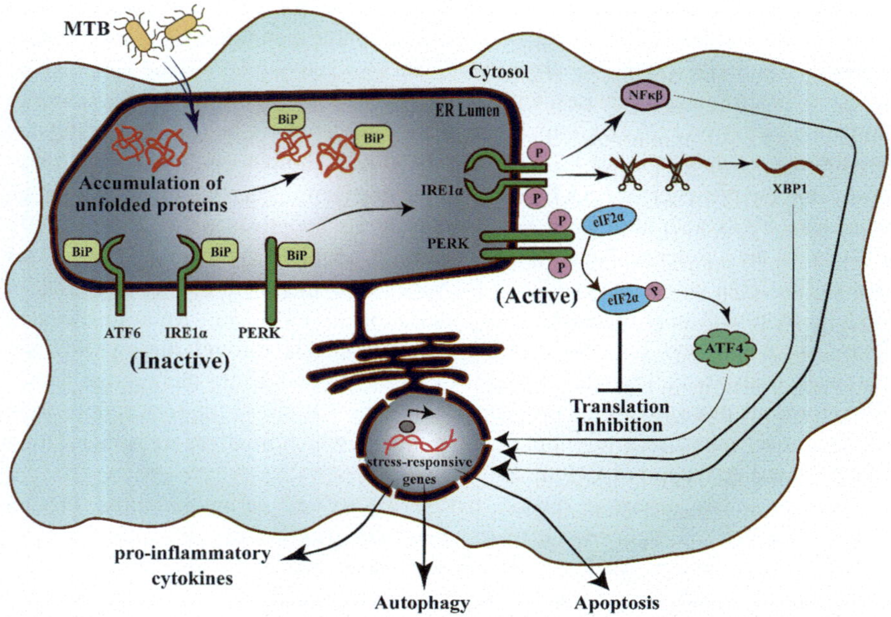

Fig. 12.3 *Mycobacterium tuberculosis* infection evokes host stress response pathways
MTB infection induces ER stress and accumulation of unfolded proteins. ER stress sensors IRE1α, PERK, and ATF6 get activated. Upon activation IRE1α dimerizes and splices XBP1 to a transcription factor for molecular chaperones. It also activates NFκB through JNK pathway, which works as a transcription factor for pro-inflammatory genes. Meanwhile, PERK upon activation dimerizes and phosphorylates eIF2α and inhibits protein translation. Phosphorylated eIF2α induces apoptosis signaling through CHOP. Apoptosis of MTB-infected macrophages releases ESAT6 that further leads to apoptosis by elevating the levels of intracellular calcium in ER, which increases ROS and ER stress

2014), by directing cytoplasmic components to lysosomal degradation (Yoshimori and Noda 2008). Autophagy targets intracellular pathogens resulting in the engulfment of cytoplasmic components by a double-membraned structure called the autophagosomes, which could initiate from membrane sources including the endosome/Golgi system, ER-mitochondrial contact sites (Hailey et al. 2010; Hamasaki et al. 2013), plasma membrane (Ravikumar et al. 2010), ER-Golgi intermediate compartment (Ge et al. 2014), phospholipid precursors (Dupont et al. 2014), or nuclear membrane in certain restricted conditions. It has also been implicated that autophagy can modulate MHC class I-mediated antigen cross-presentation (Oliveira and van Hall 2015). Different stress response pathways evoked during MTB infection are shown in Fig. 12.3.

Several regulators or inducers of autophagy have been reported, such as mammalian target of rapamycin (mTOR), presence of microbes (Tattoli et al. 2012), the TAB2/3-TAK1-IKK signaling axis (Criollo et al. 2011), or events downstream of PRR and immune cytokine activation (Lee et al. 2007). MTB DNA also can act as a signal to induce autophagy through the activation of TBK1 and stimulator of IFN genes (STING)-dependent pathways which flag the pathogen with ubiquitin (Liang et al. 2017). Induction of autophagy can be beneficial in the destruction of MTB as the degradation of organelles releases antibacterial components that destroy the pathogen (Alonso et al. 2007) and enhance peptide presentation (Jagannath et al. 2009). Gutierrez et al. first described in 2004 that a few of the bacilli are destroyed through autophagy and this process is accelerated by factors such as IFN-γ by facilitating phagosome-lysosome fusion and thus the maturation of MTB-containing phagosomes to autophagolysosomes. Gutierrez et al. demonstrated in a murine macrophage cell line stimulated with BCG that the LC3, an autophagy marker, associated with MTB-containing phagosomes decreased the survival of MTB in phagosomes (Gutierrez et al. 2004). Jagannath et al. examined the role of autophagy on the processing of Ag85b, an immunodominant MTB antigen. The authors found that rapamycin- and starvation-induced autophagy enhanced the presentation of the antigen by APCs, macrophages, and DCs. The enhanced antigen presentation through autophagy induction in MTB-infected macrophages and DCs in turn deliberated in vivo protection in MTB-infected mice (Jagannath et al. 2009). Therefore, autophagy can also be targeted to overcome the inhibition of phagolysosome formation by MTB.

Autophagy is also been shown to be induced by anti-TB drugs that are being used to treat active tuberculosis. Kim et al. showed that treatment of MTB-infected macrophages with anti-mycobacterial antibiotics like isoniazid or pyrazinamide robustly activated autophagy in the host cells and that this activation was necessary for their effective drug action through the instigation of oxidative stress (Kim et al. 2012). These drugs subsidize to the formation of cellular and mitochondrial ROS and thus facilitate the activation of autophagy through the recruitment of LC3 to phagosomes (Huang et al. 2009). Treatment of MTB-infected bone marrow-derived macrophages (BMDMs) with isoniazid increased the formation of LC3 punctate which confirmed the autophagosome formation. The induction of autophagy further orchestrated the production of inflammatory cytokines in the macrophages. Thus, antibiotic-induced autophagy progressively targets the bacteria to more acidic compartments and eventually for lysosomal degradation.

Albeit the detailed understanding about the host stress response pathways with respect to MTB infection is at initial stages and requires more in-depth investigations. Current studies highlight a critical role of host stress response pathways in the regulation of innate as well as adaptive immune responses during MTB infection.

12.5 Summary

TB is a prevalent bacterial disease that impacts majorly the developing countries with its long duration and high cost of treatment. The host response to MTB is still not completely understood, which hampers development of better vaccines and therapeutics. Moreover, there are drawbacks in the dissemination of host response toward MTB, as it is also dependent on genetic factors of the host as well as the pathogen. Even the reason behind latency and infection are not well understood. Therefore, it needs a better understanding of the factors involved from both ends, the host and the pathogen, to delineate the pathogenesis of MTB and to design better vaccines and therapeutics. Consequently, new perceptions about ER stress, apoptosis, and autophagy in MTB infection from recent studies can be further explored as targets for the development of a new and better TB vaccine.

Acknowledgments This work has been supported by the Department of Biotechnology, Government of India [BT/PR8624/MED/29/798/2013], the Nano Mission Council, Department of Science and Technology [DST No: SB/YS/LS-163/2013], and the University Grants Commission, Government of India [MRP-MAJOR-BIOT-2013-40689].

Conflict of Interest The authors declare that the research was conducted in the absence of any commercial or financial relationships that could be construed as a potential conflict of interest.

References

Abel B, Thieblemont N, Quesniaux VJF et al (2002) Toll-like receptor 4 expression is required to control chronic Mycobacterium tuberculosis infection in mice. J Immunol 169:3155–3162. https://doi.org/10.4049/JIMMUNOL.169.6.3155

Alonso S, Pethe K, Russell DG, Purdy GE (2007) Lysosomal killing of Mycobacterium mediated by ubiquitin-derived peptides is enhanced by autophagy. Proc Natl Acad Sci USA 104:6031–6036. https://doi.org/10.1073/pnas.0700036104

Armstrong JA, Hart PD (1975) Phagosome-lysosome interactions in cultured macrophages infected with virulent tubercle bacilli. Reversal of the usual nonfusion pattern and observations on bacterial survival. J Exp Med 142:1–16. https://doi.org/10.1084/jem.142.1.1

Axelrod S, Oschkinat H, Enders J et al (2008) Delay of phagosome maturation by a mycobacterial lipid is reversed by nitric oxide. Cell Microbiol 10:1530–1545. https://doi.org/10.1111/j.1462-5822.2008.01147.x

Bafica A, Scanga CA, Feng CG et al (2005) TLR9 regulates Th1 responses and cooperates with TLR2 in mediating optimal resistance to *Mycobacterium tuberculosis*. J Exp Med 202:1715–1724. https://doi.org/10.1084/jem.20051782

Basu J, Shin D-M, Jo E-K (2012) Mycobacterial signaling through toll-like receptors. Front Cell Infect Microbiol 2:145. https://doi.org/10.3389/fcimb.2012.00145

Battu S, Minhas G, Mishra A, Khan N (2017) Amino acid sensing via general control nonderepressible-2 kinase and immunological programming. Front Immunol 8:1719

Beale R, Wise H, Stuart A et al (2014) A LC3-interacting motif in the influenza A virus M2 protein is required to subvert autophagy and maintain virion stability. Cell Host Microbe 15:239–247. https://doi.org/10.1016/j.chom.2014.01.006

Behar SM, Divangahi M, Remold HG (2010) Evasion of innate immunity by Mycobacterium tuberculosis: Is death an exit strategy? Nat Rev Microbiol 8:668–674

Behar SM, Martin CJ, Booty MG et al (2011) Apoptosis is an innate defense function of macrophages against Mycobacterium tuberculosis. Mucosal Immunol 4:279–287

Bermudez LE (1993) Differential mechanisms of intracellular killing of Mycobacterium avium and Listeria monocytogenes by activated human and murine macrophages. The role of nitric oxide. Clin Exp Immunol 91:277–281

Berridge MJ (2002) The endoplasmic reticulum: a multifunctional signaling organelle. Cell Calcium 32:235–249

Bhatt K, Salgame P (2007) Host innate immune response to Mycobacterium tuberculosis. J Clin Immunol 27:347–362

Bradfute SB, Castillo EF, Arko-Mensah J et al (2013) Autophagy as an immune effector against tuberculosis. Curr Opin Microbiol 16:355–365

Brightbill HD, Libraty DH, Krutzik SR et al (1999) Host defense mechanisms triggered by microbial lipoproteins through toll-like receptors. Science 285:732–736. https://doi.org/10.1126/science.285.5428.732

Brooks MN, Rajaram MVS, Azad AK et al (2011) NOD2 controls the nature of the inflammatory response and subsequent fate of Mycobacterium tuberculosis and M. bovis BCG in human macrophages. Cell Microbiol 13:402–418. https://doi.org/10.1111/j.1462-5822.2010.01544.x

Calmette A, Boquet ANL (1921) Contribution à l'ètude du bacille tuberculeux bilié. Ann Inst Pasteur 9:561–570

Celli J, Tsolis RM (2015) Bacteria, the endoplasmic reticulum and the unfolded protein response: Friends or foes? Nat Rev Microbiol 13:71–82

Cheung BKW, Lee DCW, Li JCB et al (2005) A role for double-stranded RNA-activated protein kinase PKR in Mycobacterium-induced cytokine expression. J Immunol 175:7218–7225. https://doi.org/10.4049/jimmunol.175.11.7218

Choi HH, Shin DM, Kang G et al (2010) Endoplasmic reticulum stress response is involved in Mycobacterium tuberculosis protein ESAT-6-mediated apoptosis. FEBS Lett 584:2445–2454. https://doi.org/10.1016/j.febslet.2010.04.050

Ciaramella A, Cavone A, Santucci MB et al (2002) Proinflammatory cytokines in the course of Mycobacterium tuberculosis-induced apoptosis in monocytes/macrophages. J Infect Dis 186:1277–1282. https://doi.org/10.1086/344645

Collins HL, Kaufmann SHE (2001) The many faces of host responses to tuberculosis. Immunology 103:1–9

Criollo A, Niso-Santano M, Malik SA et al (2011) Inhibition of autophagy by TAB2 and TAB3. EMBO J 30:4908–4920. https://doi.org/10.1038/emboj.2011.413

D'Acquisto F, Ghosh S (2001) PACT and PKR: turning on NF-kappa B in the absence of virus. Sci STKE 2001:re1. https://doi.org/10.1126/stke.2001.89.re1

De Haro C, Méndez R, Santoyo J (1996) The eIF-2-alpha kinases and the control of protein synthesis. FASEB J 10:1378–1387. https://doi.org/10.1096/FASEBJ.10.12.8903508

Denis M (1991) Tumor necrosis factor and granulocyte macrophage-colony stimulating factor stimulate human macrophages to restrict growth of virulent Mycobacterium avium and to kill avirulent M. avium: killing effector mechanism depends on the generation of reactive nitrogen intermediates. J Leukoc Biol 49:380–387. https://doi.org/10.1002/jlb.49.4.380

Deretic V (2011) Autophagy in immunity and cell-autonomous defense against intracellular microbes. Immunol Rev 240:92–104

Divangahi M, Mostowy S, Coulombe F et al (2008) NOD2-deficient mice have impaired resistance to Mycobacterium tuberculosis infection through defective innate and adaptive immunity. J Immunol 181:7157–7165. https://doi.org/10.4049/jimmunol.181.10.7157

Drennan MB, Nicolle D, Quesniaux VJF et al (2004) Toll-like receptor 2-deficient mice succumb to Mycobacterium tuberculosis infection. Am J Pathol 164:49–57. https://doi.org/10.1016/S0002-9440(10)63095-7

Du Q, Xie J, Kim HJ, Ma X (2013) Type I interferon: the mediator of bacterial infection-induced necroptosis. Cell Mol Immunol 10:4–6. https://doi.org/10.1038/cmi.2012.52

Dupont N, Chauhan S, Arko-Mensah J et al (2014) Neutral lipid stores and lipase PNPLA5 contribute to autophagosome biogenesis. Curr Biol 24:609–620. https://doi.org/10.1016/j.cub.2014.02.008

Ernst JD (1998) Macrophage receptors for Mycobacterium tuberculosis. Infect Immun 66:1277–1281

Feng C, Scanga C, Collazo-Custodio CM, Cheever A et al (2003) Mice lacking myeloid differentiation factor 88 display profound defects in host resistance and immune responses to Mycobacterium avium infection not exhibited by Toll-like receptor 2 (TLR2)- and TLR4-deficient animals. J Immunol 171:4758–4764

Fulda S, Gorman AM, Hori O, Samali A (2010, 2010) Cellular stress responses: cell survival and cell death. Int J Cell Biol. https://doi.org/10.1155/2010/214074

Ganguly K, Giddaluru J, August A, Khan N (2016) Post-transcriptional regulation of immunological responses through riboclustering. Front Immunol 7:161. https://doi.org/10.3389/fimmu.2016.00161

Garcia I, Miyazaki Y, Marchal G et al (1997) High sensitivity of transgenic mice expressing soluble TNFR1 fusion protein to mycobacterial infections: synergistic action of TNF and IFN-γ in the differentiation of protective granulomas. Eur J Immunol 27:3182–3190. https://doi.org/10.1002/eji.1830271215

Ge L, Zhang M, Schekman R (2014) Phosphatidylinositol 3-kinase and COPII generate LC3 lipidation vesicles from the ER-Golgi intermediate compartment. Elife 3:1–13. https://doi.org/10.7554/eLife.04135

Giacomini E, Iona E, Ferroni L et al (2001) Infection of human macrophages and dendritic cells with Mycobacterium tuberculosis induces a differential cytokine gene expression that modulates T cell response. J Immunol 166:7033–7041. https://doi.org/10.4049/jimmunol.166.12.7033

Griffin JP, Harshan KV, Born WK, Orme IM (1991) Kinetics of accumulation of receptor-bearing T lymphocytes in mice infected with live mycobacteria. Infect Immun 59:4263–4265

Gutierrez MG, Master SS, Singh SB et al (2004) Autophagy is a defense mechanism inhibiting BCG and Mycobacterium tuberculosis survival in infected macrophages. Cell 119:753–766. https://doi.org/10.1016/j.cell.2004.11.038

Gutierrez MC, Brisse S, Brosch R et al (2005) Ancient origin and gene mosaicism of the progenitor of Mycobacterium tuberculosis. PLoS Pathog 1:e5. https://doi.org/10.1371/journal.ppat.0010005

Hailey DW, Rambold AS, Satpute-Krishnan P et al (2010) Mitochondria supply membranes for autophagosome biogenesis during starvation. Cell 141:656–667. https://doi.org/10.1016/j.cell.2010.04.009

Hamasaki M, Furuta N, Matsuda A et al (2013) Autophagosomes form at ER-mitochondria contact sites. Nature 495:389–393. https://doi.org/10.1038/nature11910

Harding CV, Boom WH (2010) Regulation of antigen presentation by Mycobacterium tuberculosis: a role for Toll-like receptors. Nat Rev Microbiol 8:296–307

Harris J, Keane J (2010) How tumour necrosis factor blockers interfere with tuberculosis immunity. Clin Exp Immunol 161:1–9

Harris J, De Haro SA, Master SS et al (2007) T Helper 2 cytokines inhibit autophagic control of intracellular Mycobacterium tuberculosis. Immunity 27:505–517. https://doi.org/10.1016/j.immuni.2007.07.022

Huang J, Canadien V, Lam GY et al (2009) Activation of antibacterial autophagy by NADPH oxidases. Proc Natl Acad Sci USA 106:6226–6231. https://doi.org/10.1073/pnas.0811045106

Ito T, Schaller M, Hogaboam CM et al (2007) TLR9 activation is a key event for the maintenance of a mycobacterial antigen-elicited pulmonary granulomatous response. Eur J Immunol 37:2847–2855. https://doi.org/10.1002/eji.200737603

Jagannath C, Lindsey DR, Dhandayuthapani S et al (2009) Autophagy enhances the efficacy of BCG vaccine by increasing peptide presentation in mouse dendritic cells. Nat Med 15:267–276. https://doi.org/10.1038/nm.1928

Jo EK, Yang CS, Choi CH, Harding CV (2007) Intracellular signalling cascades regulating innate immune responses to Mycobacteria: branching out from Toll-like receptors. Cell Microbiol 9:1087–1098

Jones BW, Means TK, Heldwein KA et al (2001) Different Toll-like receptor agonists induce distinct macrophage responses. J Leukoc Biol 69:1036–1044

Joshi JM (2011) Tuberculosis chemotherapy in the 21st century: back to the basics. Lung India 28:193–200. https://doi.org/10.4103/0970-2113.83977

Kang PB, Azad AK, Torrelles JB et al (2005) The human macrophage mannose receptor directs Mycobacterium tuberculosis lipoarabinomannan-mediated phagosome biogenesis. J Exp Med 202:987–999. https://doi.org/10.1084/jem.20051239

Kaufman RJ (2002) Orchestrating the unfolded protein response in health and disease. J Clin Invest 110:1389–1398

Kaufman RJ, Scheuner D, Schröder M et al (2002) The unfolded protein response in nutrient sensing and differentiation. Nat Rev Mol Cell Biol 3:411–421

Kaufmann SH, Hussey G, Lambert PH (2010) New vaccines for tuberculosis. Lancet 375:2110–2119

Kim JJ, Lee HM, Shin DM et al (2012) Host cell autophagy activated by antibiotics is required for their effective antimycobacterial drug action Cell Host Microbe 11:457–468. https://doi.org/10.1016/j.chom.2012.03.008

Lee SC, Dickson DW, Liu W, Brosnan CF (1993) Induction of nitric oxide synthase activity in human astrocytes by interleukin-1 beta and interferon-gamma. J Neuroimmunol 46:19–24. https://doi.org/10.1016/0165-5728(93)90229-R

Lee HK, Lund JM, Ramanathan B et al (2007) Autophagy-dependent viral recognition by plasmacytoid dendritic cells. Science 315:1398–1401. https://doi.org/10.1126/science.1136880

Leulier F, Lemaitre B (2008) Toll-like receptors–taking an evolutionary approach. Nat Rev Genet 9:165–178. https://doi.org/10.1038/nrg2303

Liang M, Habib Z, Sakamoto K et al (2017) Mycobacteria and autophagy: many questions and few answers. Curr Issues Mol Biol 21:63

Lim YJ, Choi JA, Choi HH et al (2011) Endoplasmic reticulum stress pathway-mediated apoptosis in macrophages contributes to the survival of Mycobacterium tuberculosis. PLoS One 6:e28531. https://doi.org/10.1371/journal.pone.0028531

Luca S, Mihaescu T (2013) History of BCG vaccine. Mædica 8:53–58

Moncada S, Palmer RMJ, Higgs EA (1991) Nitric oxide: physiology, pathophysiology, and pharmacology. Pharmacol Rev 43:109–142. https://doi.org/10.1017/CBO9781107415324.004

Mortaz E, Adcock IM, Tabarsi P et al (2015) Interaction of pattern recognition receptors with Mycobacterium tuberculosis. J Clin Immunol 35. https://doi.org/10.1007/s10875-014-0103-7

Mukherjee S, Karmakar S, Babu SPS (2016) TLR2 and TLR4 mediated host immune responses in major infectious diseases: a review. Brazilian J Infect Dis 20:193–204. https://doi.org/10.1016/j.bjid.2015.10.011

Nathan C, Shiloh MU (2000) Reactive oxygen and nitrogen intermediates in the relationship between mammalian hosts and microbial pathogens. Proc Natl Acad Sci USA 97:8841–8848. https://doi.org/10.1073/pnas.97.16.8841

Nathan C, Xie QW (1994) Regulation of biosynthesis of nitric oxide. J Biol Chem 269:13725–13728

Nicola AM, Albuquerque P, Martinez LR et al (2012) Macrophage autophagy in immunity to Cryptococcus neoformans and Candida albicans. Infect Immun 80:3065–3076. https://doi.org/10.1128/IAI.00358_12

Oliveira CC, van Hall T (2015) Alternative antigen processing for MHC class I: multiple roads lead to Rome. Front Immunol 6:298

Pahari S, Kaur G, Aqdas M et al (2017) Bolstering immunity through pattern recognition receptors: a unique approach to control tuberculosis. Front Immunol 8:906

Peters W, Ernst JD (2003) Mechanisms of cell recruitment in the immune response to Mycobacterium tuberculosis. Microbes Infect 5:151–158. https://doi.org/10.1016/S1286-4579(02)00082-5

Pillich H, Loose M, Zimmer K-P, Chakraborty T (2016) Diverse roles of endoplasmic reticulum stress sensors in bacterial infection. Mol Cell Pediatr 3:9. https://doi.org/10.1186/s40348-016-0037-7

Ravikumar B, Moreau K, Jahreiss L et al (2010) Plasma membrane contributes to the formation of pre-autophagosomal structures. Nat Cell Biol 12:747–757. https://doi.org/10.1038/ncb2078

Ravindran R, Khan N, Nakaya HI et al (2014) Vaccine activation of the nutrient sensor GCN2 in dendritic cells enhances antigen presentation. Science 343:313–317. https://doi.org/10.1126/science.1246829

Reiling N, Holscher C, Fehrenbach A et al (2002) Cutting edge: Toll-Like Receptor (TLR)2- and TLR4-mediated pathogen recognition in resistance to airborne infection with Mycobacterium tuberculosis. J Immunol 169:3480–3484. https://doi.org/10.4049/jimmunol.169.7.3480

Rich EA, Torres M, Sada E et al (1997) Mycobacterium tuberculosis (MTB)-stimulated production of nitric oxide by human alveolar macrophages and relationship of nitric oxide production to growth inhibition of MTB. Tuber Lung Dis 78:247–255. doi: S0962-8479(97)90005-8 [pii]

Rojas M, Olivier M, Gros P et al (1999) TNF-alpha and IL-10 modulate the induction of apoptosis by virulent Mycobacterium tuberculosis in murine macrophages. J Immunol 162:6122–6131

Rubartelli A, Lotze MT, Latz E, Manfredi A (2013) Mechanisms of sterile inflammation. Front Immunol 4:398. https://doi.org/10.3389/fimmu.2013.00398

Salgame P (2005) Host innate and Th1 responses and the bacterial factors that control Mycobacterium tuberculosis infection. Curr Opin Immunol 17:374–380

Schierloh P, Yokobori N, Alemán M et al (2007) Mycobacterium tuberculosis-induced gamma interferon production by natural killer cells requires cross talk with antigen-presenting cells involving Toll-like receptors 2 and 4 and the mannose receptor in tuberculous pleurisy. Infect Immun 75:5325–5337. https://doi.org/10.1128/IAI.00381-07

Takeuchi O, Akira S (2010) Pattern recognition receptors and inflammation. Cell 140:805–820

Tattoli I, Sorbara MT, Vuckovic D et al (2012) Amino acid starvation induced by invasive bacterial pathogens triggers an innate host defense program. Cell Host Microbe 11:563–575. https://doi.org/10.1016/j.chom.2012.04.012

Thoma-Uszynski S, Stenger S, Takeuchi O et al (2001) Induction of direct antimicrobial activity through mammalian toll-like receptors. Science 291(80):1544–1547. https://doi.org/10.1126/science.291.5508.1544

Van Scoy R, Wilkowske C (1999) Antimycobacterial therapy. Mayo Clin Proc 74:1038–1048

Velez DR, Wejse C, Stryjewski ME et al (2010) Variants in toll-like receptors 2 and 9 influence susceptibility to pulmonary tuberculosis in Caucasians, African-Americans, and West Africans. Hum Genet 127:65–73. https://doi.org/10.1007/s00439-009-0741-7

Vergne I, Deretic V (2010) The role of PI3P phosphatases in the regulation of autophagy. FEBS Lett 584:1313–1318. https://doi.org/10.1016/j.febslet.2010.02.054

Vergne I, Chua J, Lee H-H et al (2005) Mechanism of phagolysosome biogenesis block by viable Mycobacterium tuberculosis. Proc Natl Acad Sci USA 102:4033–4038. https://doi.org/10.1073/pnas.0409716102

Wang QM, Kang L, Wang XH (2009) Improved cellular immune response elicited by a ubiquitin-fused ESAT-6 DNA vaccine against Mycobacterium tuberculosis. Microbiol Immunol 53:384–390. https://doi.org/10.1111/j.1348-0421.2009.00138.x

Weinberg J, Misukonis M, Shami P et al (1995) Human mononuclear phagocyte inducible nitric oxide synthase (iNOS): analysis of iNOS mRNA, iNOS protein, biopterin, and nitric oxide production by blood monocytes and peritoneal macrophages. Blood 86:1184–1195

WHO (2016) WHO Global tuberculosis report 2016

Williams BRG (2001) Signal integration via PKR. Sci Signal 2001: re2-re2. https://doi.org/10.1126/stke.2001.89.re2

Yoshimori T, Noda T (2008) Toward unraveling membrane biogenesis in mammalian autophagy. Curr Opin Cell Biol 20:401–407

Zaki MH, Akuta T, Akaike T (2005) Nitric oxide-induced nitrative stress involved in microbial pathogenesis. J Pharmacol Sci 98:117–129. https://doi.org/10.1254/jphs.CRJ05004X

Zhang K, Kaufman RJ (2008) From endoplasmic-reticulum stress to the inflammatory response. Nature 454:455–462. https://doi.org/10.1038/nature07203

Zumla A, Petersen E, Nyirenda T, Chakaya J (2015) Tackling the Tuberculosis epidemic in sub-Saharan Africa – unique opportunities arising from the second European Developing Countries Clinical Trials Partnership (EDCTP) programme 2015-2024. Int J Infect Dis 32:46–49. https://doi.org/10.1016/j.ijid.2014.12.039

Heat Shock Proteins in the Pathogenesis of *Mycobacterium tuberculosis* 13

Prajna Tripathi and Janendra K. Batra

Abstract

Mycobacterium tuberculosis (*M. tb*), the causative agent of tuberculosis, is responsible for immense global suffering taking nearly 1.5 million lives annually (WHO 2016). About one-third of the world's population is estimated to be infected with this obligate pathogen and yet remains asymptomatic (Raviglione and Sulis 2016). *M. tb* is highly adapted for survival in the extremely hostile intracellular environment in host macrophages. The ability of *M. tb* to overcome various host-induced proteotoxic stress conditions relies significantly on its highly efficient chaperone network. Heat shock proteins (Hsps) form a special class of the chaperone network and exhibit an orchestrated repertoire of stress-sensing mechanisms. Hsps, found ubiquitously in most prokaryotes as well as eukaryotes, are very well conserved across the species. In this chapter, we discuss about the recent advances in understanding the myriad number of molecular pathways that Hsps regulate, directly or otherwise in *M. tb*, which highlight the association between Hsps and virulence determination in *Mycobacterium*.

Keywords

Host-pathogen interaction · Caseinolytic proteases · Virulence · Stress · *Mycobacterium* · Hsp

P. Tripathi
National Institute of Immunology, New Delhi, Delhi, India

JH Institute of Molecular Medicine, Jamia Hamdard, New Delhi, Delhi, India

J. K. Batra (✉)
National Institute of Immunology, New Delhi, Delhi, India

Department of Biochemistry, School of Chemical and Life Sciences, Jamia Hamdard, New Delhi, Delhi, India
e-mail: jkbatra@jamiahamdard.ac.in

S. E. Hasnain et al. (eds.), Mycobacterium tuberculosis: *Molecular Infection Biology, Pathogenesis, Diagnostics and New Interventions*,
https://doi.org/10.1007/978-981-32-9413-4_13

Abbreviations

ATP	Adenosine triphosphate
Clp	Caseinolytic protease
IFNγ	Interferon gamma
kDa	Kilodalton
MHC	Major histocompatibility complex
M. tb	*Mycobacterium tuberculosis*
TB	Tuberculosis

13.1 Introduction

Tuberculosis (TB), an infection caused by *Mycobacterium tuberculosis* (*M. tb*), is responsible for immense global suffering taking nearly 1.5 million lives annually (WHO 2016). About one third of the world's population is estimated to be infected with this obligate pathogen and yet remains asymptomatic (Raviglione and Sulis 2016). *M. tb* is highly adapted for survival in the extremely hostile intracellular environment in host macrophages. *M. tb* regulates its gene expression in order to endure a variety of stresses produced by macrophages during infection. It can either persist in a quiescent state or replicate actively within the host without being lethal to the infected host. The outgrowth of bacilli is kept in check by the host defense mechanism which is well suited to handle the invading pathogens. However, when the host defense system becomes compromised, years after remaining latent, these bacilli may start resuscitating, enter into an active phase, and may cause severe disease. A striking example of this is the case of immune deficiencies taking place due to general health deterioration, undernourishment, or HIV infection. Another reason for active tubercular infection is the emergence of drug-resistant bacilli. The increase in occurrence of multidrug-resistant (MDR) and extremely drug-resistant (XDR) strains of *M. tb* in the past decades has made the TB treatment increasingly challenging (WHO 2016). Every year about 0.5 million cases of MDR strains are reported (WHO 2016). The current treatment regimen for TB was developed about 40 years ago with the introduction of four drugs, isoniazid, rifampicin, pyrazinamide, and ethambutol, which formed the first line of defense. However, the complexity of this regimen as well as the propensity of *M. tb* to develop drug resistance has caused an urgent need for the development of novel therapeutics targeting other pathways of the pathogen.

13.2 Mycobacterial Infection, Adaptation, and Survival Within the Host

M. tb bacilli enter the host system primarily through respiratory route in the form of aerosol, and once the bacilli reach alveoli, they are internalized via phagocytosis by the alveolar macrophages. Previous studies have discovered that *M. tb* can also use a number of receptors to access the macrophages (Ernst 1998; Aderem and Underhill 1999; Kang et al. 2005). The phagosomes, containing tubercle bacilli, make contact with early endosomes of the host endocytic machinery but do not succeed in fusing to lysosomes (Pieters and Gatfield 2002; Hestvik et al. 2005). The failure to fuse with the lysosomes results in limited acidification of phagosomes, due to decreased integration of the vacuolar proton ATPases (Sturgill-Koszycki et al. 1994). A seminal study by Armstrong and Hart (1975) suggested that the ability of *M. tb* to inhibit phagolysosomal fusion could be one of the many mechanisms that this pathogen uses to sustain within the host. Some of the later studies reiterated that *M. tb* infection leads to the arrest of phagosome maturation which is an important strategy of these bacilli to survive within macrophages that are otherwise potent microbicidal cells (Sturgill-Koszycki et al. 1994; MacMicking et al. 2003).

M. tb has been shown to survive as well as occasionally replicate within the macrophages even after not being able to stall the phagosome-lysosome fusion (Pethe et al. 2004; MacGurn and Cox 2007), and it can also endure highly acidified environments (Gomes et al. 1999). Host macrophages attempt to contain these bacilli within the phagolysosomes in the form of caseous granulomas. A granuloma can be pathologically defined as an organized collection of differentiated macrophages with a characteristic morphology. In case of mycobacterial infection, a granuloma is an organized structure composed of a large number of multinucleated giant cells, epithelioid cells with tight interdigitations linking adjacent cells, and foamy macrophages surrounded on the outside by lymphocytes, neutrophils, fibroblasts, dendritic cells (DCs), and other matrix components (Peters and Ernst 2003; Grosset 2003). A necrotic zone develops in the center of the granuloma due to the aggregation of acellular debris and is called "caseum" because of its milky appearance. The rupture of this caseous center, a liquefied mass of bacilli, is crucial for the transmission of pulmonary tuberculosis (Flynn and Chan 2001). In some cases these granulomas may become calcified and fibrotic, mostly containing dead bacilli (Flynn and Chan 2001).

Though the formation of granuloma appears to be a host strategy to wall off or eliminate mycobacterial infection, some bacilli still become successful in surviving within the granuloma in a quiescent state. These lesions are widely assumed to be the site of persistence of dormant or metabolically less active tubercle bacilli, where they remain for years together. Somewhere in this deadlock between the host and the pathogen, a shift may take place in favor of the pathogen, allowing its reactivation. Intact *M. tb* bacilli have been recovered long after the infection from phagolysosomes (Jordao et al. 2008) indicating that the bacteria must have adapted to live in hostile environments. Persisting bacilli experience nutritionally deprived, highly anoxic, and acidic environment in the macrophage. Infection with *M. tb* also leads to stimulation of macrophages with IFNγ (Schaible et al. 1998; Via et al. 1998) which in turn activates the production of reactive oxygen intermediates (ROI) and reactive nitrogen intermediates (RNI).

Once the bacilli enter host macrophages, they encounter a plethora of stresses, due to environmental fluctuation and host immune response, which include pH stress, nutrient-depleted or carbon-starved environment, heat stress, and generation of reactive oxygen species (ROS) by the phagosomal NADPH oxidases (Beste 2007; Axelrod 2008; Ehrt and Schnappinger 2009; Farhana et al. 2010; Butler 2010). Adaptation of *M. tb* to such extreme conditions is facilitated by a number of mechanisms which include the manipulation of host cell response as well as tolerance to the stress subjected by host defense system (Lowrie 1983; Chan et al. 1992; Wayne and Sohaskey 2001; Firmani and Riley 2002; Farhana et al. 2010). To survive the lethal effects of the abovementioned antimicrobial molecules released by activated macrophages, *M. tb* enters a stationary phase or a non-replicating dormant phase (Wayne and Sohaskey 2001). The bacilli also alter their gene expression pattern to cope with stressful conditions. In particular, the induction of a set of evolutionarily highly conserved genes encoding heat shock proteins (Hsps) plays a crucial role in conferring stress tolerance to microbial pathogens (Lindquist and Craig 1988).

13.3 Heat Shock Proteins (Hsps): A Brief Introduction

Hsps, found ubiquitously in most prokaryotes as well as eukaryotes, are very well conserved across the species. Hsps were first discovered as chromosomal puffs in *Drosophila melanogaster* larvae when they were subjected to temperature stress (Tissiéres et al. 1974). Hsps are significantly expressed in most organisms even under normal conditions and take part in various fundamental cellular activities along with facilitating protein quality control (Dobson and Karplus 1999). Classically, Hsps are defined as those proteins that bind to the exposed hydrophobic patches on non-natively folded/aggregated proteins in a non-covalent manner (Csermely et al. 1998). This underscores the role of many Hsps as molecular chaperones, which are mostly ATP-dependent proteins that assist in quality control, folding, and refolding of cellular proteins (Fig. 13.1). The conventional classification of Hsps groups them into the following five different classes on the basis of their molecular weights (Hartl and Hayer-Hartl 2002).

(i) **Small heat shock proteins (sHSPs):** They have low molecular weights but form large oligomeric structures, usually composed of 24 subunits, and function in an ATP-independent manner (Van Montfort et al. 2001; Haslbeck et al. 2005). sHSPs are the least conserved class of Hsps and generally contain an alpha-crystallin domain as their characteristic feature. sHSPs are involved in protecting the partially unfolded or folded proteins from aggregation (Haslbeck et al. 2005).

(ii) **HSP60 family or GroE chaperone:** The 60-kDa GroEL proteins are highly conserved and form a barrel-shaped assembly composed of two homoheptameric rings (Bukau and Horwich 1998). The double-ringed GroEL chaperone interacts with its heptameric co-chaperone GroES, to constitute the ATP-dependent GroEL-ES chaperone machinery (Hartl 1996; Sigler et al. 1998). This is the major chaperone machinery that assists protein folding in bacterial cells (Houry et al. 1999).

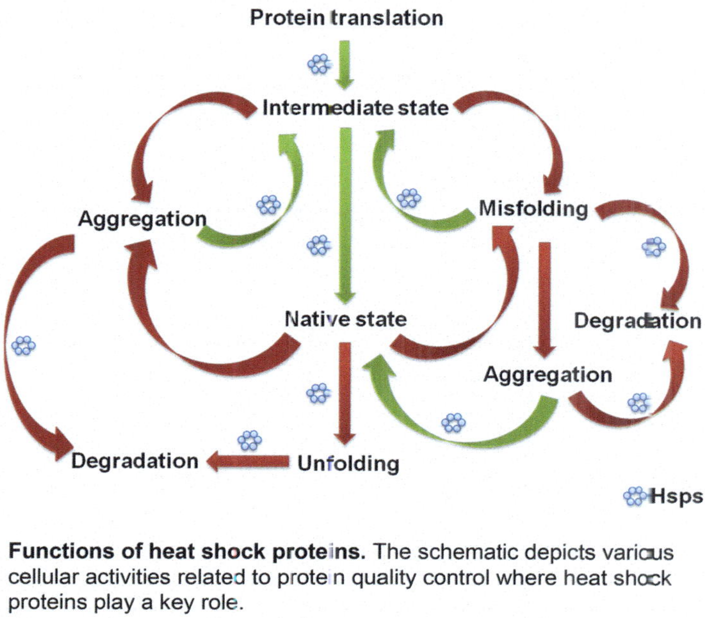

Fig. 13.1 Functions of heat shock proteins. The schematic depicts various cellular activities related to protein quality control where heat shock proteins play a key role

(iii) **HSP70 family or DnaK chaperone:** Members of the HSP70 family of chaperones are conserved across all three surviving domains of life (Powers and Balch 2013). DnaK plays a crucial role in the protein homeostasis network by maintaining the native conformation of newly synthesized polypeptides as well as existing cellular proteins, at the expense of ATP (Bukau and Horwich 1998). DnaK is known to associate with its co-chaperone Hsp40 (also known as DnaJ in *E. coli*) and a nucleotide exchange factor GrpE, to form a complex called KJE and perform chaperonic activity (Bukau and Horwich 1998; Wickner et al. 1999).

(iv) **HSP90 family:** These are a conserved group of ATP-dependent cytosolic chaperones. Hsp90 chaperone interacts with other chaperones and facilitates the late-stage maturation and folding of proteins (Li and Buchner 2013).

(v) **HSP100 family or caseinolytic proteases (Clps):** The Clp proteins are an integral part of the ATPases associated with diverse cellular activities (AAA +) superfamily of proteins and, hence, contain the characteristic ATPase domain(s) (Schirmer et al. 1996). The Clp proteins, ClpA, ClpX, and ClpC, can act as stand-alone chaperones that prevent aggregation of substrates or they can also function as the protein unfolding component of the chambered Clp protease machinery by associating with the peptidase component ClpP, which does not contain any ATPase domains. Clp proteins can prevent protein aggregation and can also unfold the non-native proteins or polypeptides for their subsequent degradation (Gottesman and Hendrickson 2000). At the structural level, Clp proteins form homohexameric ring-shaped structure with a central cavity. They are broadly classified into two classes on the basis of number of nucleotide-binding domains present in them. Each monomer of the class I Clp proteins which includes ClpA, ClpB, and ClpC possesses two nucleotide-binding domains (NBDs), whereas that of class II Clp proteins that include ClpX and ClpY possesses only a single NBD (Schirmer et al. 1996; Gottesman and Hendrickson 2000). ClpB is a unique member of the Clp family as it does not take part in protein degradation; instead it mediates the resolubilization of protein aggregates in association with the KJE chaperone system (Weibezahn et al. 2003).

13.4 Hsps: A Suit of Armor

In order to survive within the host, a microbe must outcompete the environmental fluctuations and perturbations, including the exposure to various cellular stresses. It is well established that different environmental conditions can have profoundly different impacts on the microbial pathogens, like *Shigella*, *Yersinia*, *Agrobacterium*, fungi, viruses, etc., and could result in either elevated or repressed growth and/or virulence of these organisms (O'Reilly and Zak 1992; Banta et al. 1998; Konkel and Tilly 2000; Klein and Tebbets 2007; Hatta et al. 2007). This indicates that maintaining homeostatic balance is important for precise coordination of microbial virulence and pathogenesis. It is therefore contingent upon the pathogen integrating the environmental cues and initiating a fitting response to any stress-related hostility. The prime adaptive response to elevation in stress levels includes the induction of Hsps along with alternative sigma factors and other regulatory proteins (Henderson et al. 2006).

The potential correlation between Hsps and bacterial pathogenesis or virulence has been shown by many studies. In many bacteria, Hsps have been found to be directly involved in pathogenesis. A 66-kDa Hsp has been shown to be involved in the binding of *Salmonella typhimurium* cells to mucosal cells (Ensgraber and Loos 1992). One of the most striking examples of the involvement of Hsps in bacterial pathogenesis is the infection of macrophages with *Listeria monocytogenes*. Ideally, in an infected macrophage, the fusion of phagosome with an endosomal compartment would lead to the digestion of the phagocytosed *Listeria* species. However, *Listeria* averts this phenomenon by inducing a member of the Hsp100 family called ClpP which mediates its release into the cytosol and hence multiplication (Gaillot et al. 2000). The deletion of ClpP from the *Listeria* genome shows strong reduction in bacterial virulence in mouse models of infection. The stress-induced expression of ClpP has been linked to the rise in production of listeriolysin, a toxin responsible for the phagosomal escape of *Listeria* (Gaillot et al. 2000). Additionally, another member of the Hsp100 family, ClpC, is also shown to facilitate early phagosomal escape of the bacteria (Nair et al. 2000) as well as its intracellular survival and virulence in mice (Rouquette et al. 1996, 1998). ClpC in *L. monocytogenes* also modulates various virulence factors, like InlA, InlB, and ActA, which are required for adhesion and invasion into the host cells, while loss of ClpE leads to reduced virulence (Nair et al. 1999). In *Leishmania donovani*, ClpB is required for full amastigote development and also during the initial stages of mammalian infection (Krobitsch and Clos 1999). Deletion of ClpB in *Leptospira interrogans* hampers the general stress response and reduces virulence of the bacteria (Lourdault et al. 2011).

Porphyromonas gingivalis, the causative agent of periodontal disease, fails to exhibit any phenotype with respect to bacterial adherence or invasion of cultured human epithelial cells upon deletion of *hsp90* from its genome (Sweier et al. 2003). Similarly, the deletion of Hsp70 or DnaK from many bacterial pathogens, such as *Brucella suis*, *Campylobacter jejuni*, and *Salmonella enterica* serovar *typhimurium*, results in adverse effects on the intracellular growth and colonization of these bacteria in mouse models of infection. Hsp70 has also been shown to be crucial for the survival and virulence of malarial parasite, *Plasmodium falciparum* (Külzer et al. 2012). The ability to evade internalization by phagocytes forms an important aspect of pathogenesis of the human pathogen *Staphylococcus*. This phenomenon of immune evasion has been reported to be regulated by the heat shock cognate protein, Hsc70 (Hirschhausen et al. 2010). Cpn60, a protein of the HSP60 family, has been reported to be crucial for epithelial cell invasion by *Neisseria gonorrhoeae* (Tauschek et al. 1997; Du et al. 2005).

13.5 Contribution of Hsps Toward Pathogenesis and Virulence of *Mycobacterium*

The ability of *M. tb* to overcome various host-induced proteotoxic stress conditions relies significantly on its highly efficient chaperone network. Hsps form a special class of the chaperone network and exhibit an orchestrated repertoire of stress sensing mechanisms. Recent advances in understanding the myriad number of molecular pathways that Hsps regulate, directly or otherwise in *M. tb*, highlight the association between Hsps and virulence determination in *Mycobacterium*. Some specific examples of the same will be discussed in this section.

The genome of *M. tb* encodes several heat shock proteins which mainly assist in protein folding, refolding, degradation, and stress regulation. Some of the major chaperone systems in *Mycobacteria* are discussed below.

(i) KJE complex which includes the Hsp70 homolog called DnaK, Hsp40 homologs called DnaJ1 and DnaJ2, and a nucleotide exchange factor called GrpE. DnaK is reported to perform chaperonic activity as well as reactivation of aggregated proteins in conjunction with ClpB (Lupoli et al. 2016). While both the paralogs of *M. tb* DnaJ contain properties characteristic of a J-domain protein and assist in chaperoning (Stewart et al. 2004; Lupoli et al. 2016), only one of these is essential at a time for proper functioning of the KJE machinery and survival of bacilli (Lupoli et al. 2016).

(ii) GroEL-GroES complex which includes the Hsp65 homolog called Cpn60.2 or GroEL2 and its co-chaperone Cpn10 or GroES. Another paralog of GroEL2, named GroEL1 or Cpn60.1, is known to exist in *M. tb* and is reported to be a homolog of Hsp60. Both the GroEL proteins have ATPase activity. However, unlike GroEL2, GroEL1 is nonessential for cell survival (Hu et al. 2008). Though several studies speculate its function as a chaperonin (Sielaff et al. 2011) or in stress adaptation (Sharma et al. 2016; Bhat et al. 2017), the precise biological role of GroEL1 still remains elusive.

(iii) HSP100 or the caseinolytic protease (Clp) machinery includes ClpX, ClpB, and the two paralogs of ClpP, ClpP1, and P2 and of ClpC, ClpC1, and C2. All these Clp proteins, except for ClpB, collaborate to form the Clp proteolytic machinery (Raju et al. 2012; Schmitz and Sauer 2014), wherein homohexamers of ClpC1 or ClpX act as the unfoldase subunit and the hetero-hexameric ClpP1P2 complex acts as the peptidase unit (Akopian et al. 2012). The Clp proteolytic machinery is responsible for the turnover of proteins whose accumulation is deleterious for the viability of mycobacterial cells. ClpC1 and ClpX are also known to perform chaperonic activity on their own (Kar et al. 2008). ClpB acts as a disaggregase and assists in bringing aggregated proteins back to their native folded state. ClpB can perform stand-alone chaperonic activity as well as it can associate with the KJE complex to enhance its activity (Lupoli et al. 2016).

(iv) Small heat shock proteins (sHSPs), which are ATP-independent chaperones, form large oligomers and resolve partially folded proteins (Haslbeck et al. 2005). The most common member of this family is α-crystallin or Acr protein which, due to its enhanced expression, is also considered as one of the markers of non-replicating persistent phase of tubercle bacilli (Sherman et al. 2001; Voskuil et al. 2003). *M. tuberculosis* genome contains two paralogs of the *acr* gene, *acr1* (previously referred to as *acr* or *hspX*) and an *acr2* gene, whereas *M. leprae* genome has a single homolog called *acr3*, and *M. avium* genome has one *acr2* and three functional *acr3* genes.

Studies on the general stress response in *M. tuberculosis* and *M. leprae* indicate a global upregulation of the operons which encode the Hsp70 chaperone protein, DnaK; its co-chaperone Hsp40 protein, DnaJ1; and Hsp65 proteins, GroEL2 and GroES (Stewart et al. 2001; Raman et al. 2001; Williams et al. 2007). The expression of *acr* gene encoding Hsp20 protein has been found to increase in the TB bacilli by as much as ~18–55 folds within 24 h in the mouse models of infection (Wilkinson et al. 2005). In addition, Acr is reported to be important for the pathogenesis of *M. tb* (Stewart et al. 2005). GroEL2, one of the major chaperones responsible for folding linear amino acid chains into three-dimensional structures in *M. tb*, also acts as a major adhesin for the binding of *M. tb* to monocytes (Hickey et al. 2009). Initial investigations on the functionality of *M. tb* GroEL2 protein led to the discovery of its pivotal role in controlling the pathogenesis of adjuvant arthritis – a rat model of rheumatoid arthritis (van Eden et al. 1988). Since this seminal study, there have been an enormous number of studies reiterating the potent immunomodulatory actions of *M. tb* GroEL2. The host immune subversion phenomenon in *M. tb* is a major contributor to its success as a human pathogen. In addition to its regular niche in macrophages, these bacilli have also been known to interfere with the response of DCs, which act as primary antigen-presenting cells in host. The cleaved form of GroEL2, which predominates in mycobacterium, was recently shown to associate with this phenomenon. In contrast to full-length GroEL2, the cleaved form of GroEL2 was found to be poorly immunostimulatory and it dampened the process of DC maturation (Georgieva et al. 2018). Additionally, it also contributes in blocking the mitochondria-dependent apoptosis of infected macrophages which is eventually implicated in the *M. tb* evasion of host innate immune response (Joseph et al. 2017). The other paralog, GroEL1 or Cpn60.1, is implicated in the formation of biofilms by mycobacterial bacilli and in caseous necrosis (Ojha et al. 2005). Biofilms have long been associated with the robustness of *M. tb* and its pathogenesis. The unique characteristics of mycobacterial biofilms impart resistance from environmental aggressions as well as antibiotics (Ojha et al. 2008; Esteban et al. 2008). Cpn10 or GroES, the co-chaperone of GroELs, is shown to be responsible for the bone resorption activity of *M. tb* (Meghji et al. 1997).

Hsp70 or DnaK is another major chaperone protein which assists in protein quality control, and transposon mutagenesis studies have predicted Hsp70 to be essential for cell viability and native protein folding in mycobacterial species (Griffin et al. 2011; Fay and Glickman 2014). Interestingly, apart from its normal intracellular localization, DnaK has also been found to localize on the cell surface of virulent *M. tb* bacilli which is indicative of its potential role in virulence (Hickey et al. 2009). The chaperonic functions of DnaK are facilitated by either of the two homologs of Hsp40 found in *M. tb* called DnaJ1 and DnaJ2 (Lupoli et al. 2016). The Hsp40 homologs have also been shown to be interacting partners of MBP64 protein (Chen et al. 2011) which has long been implicated in virulence and pathogenesis of *M. tb* (Harboe et al. 1996; Pym et al. 2002; Majlessi et al. 2005). Like GroEL2, Hsp40 also exhibits immunomodulatory effects on rheumatoid arthritis patients (Tukaj et al. 2010), again indicating its potential in signaling within the host cells. Some of the abovementioned Hsps are also immunodominant antigens found in the virulent tubercle bacilli. Newer developments in the field of tuberculosis show that the production of CC and CXC chemokines is responsible for the formation of granuloma (Méndez-Samperio et al. 2008). Hsp70 protein is reported to act as a major inducer of these chemokines and hence aid driving the granuloma formation (Wang et al. 2001).

Hsp22.5, another heat shock protein that gets activated under stress, has been shown to be important for the survival of TB bacilli in murine models in the later stages of chronic tuberculosis. Deletion of Hsp22.5 from the *M. tb* genome resulted in altered transcription of some genes vital for ATP synthesis, protein synthesis, dormancy, and lipid metabolism (Abomoelak et al. 2010).

ClpB was recently shown to function as an active disaggregase that resolves protein aggregates in *M. tb* (Lupoli et al. 2016). Further, ClpB was shown to be involved in sequestering and dividing the irreversibly oxidized proteins of stressed mycobacteria within and between the cells (Vaubourgeix et al. 2015). Moreover, the deletion of ClpB coding gene led to defects in growth and persistence of *M. tb* in mice (Vaubourgeix et al. 2015).

A functional Clp protease has been shown to be essential for survival of mycobacteria in vitro as well as during infection in mice (Raju et al. 2012). Interestingly, ClpX, which is the substrate recognition subunit of Clp protease assembly, has also been shown to interact with FtsZ and interfere with FtsZ assembly (Dziedzic et al. 2010). FtsZ is a highly conserved homolog of the eukaryotic protein tubulin and plays essential role in mycobacterial cell division (Romberg and Levin 2003; Dziadek et al. 2003). FtsZ is known to localize at the mid-cell division site in the form a Z-like ring which leads to initiation of the division process. The Z-ring assembly is modulated in vivo and under antibiotic stress as a consequence of interaction of ClpX with FtsZ (Chauhan et al. 2006; Dziedzic et al. 2010). In addition, the GTP-dependent polymerization activity of FtsZ is also hindered (Dziedzic et al. 2010). Thus, it becomes evident that ClpX also acts as an important regulator of cell division and helps the tubercle bacilli cope with intracellular stress. The Clp protease gene regulator, *clgR*, is reported to be strongly induced under multiple stress conditions, such as hypoxia, redox stress, heat shock, SDS stress, etc. (Stewart et al. 2001; Sherrid et al. 2010; McGillivray et al. 2014, 2015), which in turn leads to a condition-dependent differential activation of downstream *clp* genes that encode ClpB, ClpP1, P2, and ClpC1 (Mehra and Kaushal 2009; Barik et al. 2010; Personne et al. 2013). These reports further demonstrate that ClgR-regulated transcriptional activation of *clp* genes is essential for *M. tb* to replicate within macrophages and has important implications on the pathogenesis of mycobacteria. As much as the heat shock proteins are shown to be critical for the viability and virulence of *M. tb*, elevated expression of these proteins proves to be deleterious for the organism in chronic phase of infection. Stewart et al. (2001) demonstrated that a strain of *M. tb* which constitutively overexpressed Hsp70 because of deletion of the transcriptional repressor HspR, though fully virulent in initial phase of infection, showed significantly compromised persistence in subsequent stages of infection. In the same study, it was reported that infection of mice with *ΔhspR* BCG strain led to an enhanced induction of $CD8^+$ IFN-γ-secreting T cells in the spleen indicating the immunogenic potential of Hsps.

13.6 Therapeutic Potential of Hsps for TB Control

In order to thwart tuberculous infection timely, an effective diagnosis of the same is of vital importance. At present, detection of tubercle bacilli in the body is based mainly on two types of tests: Mantoux tuberculin skin test (TST) and interferon gamma release assays (IGRAs) or the TB blood test. A positive patient is further confirmed by sputum smear microscopy, which is usually followed by culturing the bacteria. However, neither of these tests is able to discriminate between an active TB disease and a latent TB infection (LTBI). The entire process of TB diagnosis is quite consuming in terms of time and resources, and hence there is a need to develop newer and faster diagnostic approaches and also to identify efficacious biomarkers of TB. Since Hsps have already been elucidated to be involved in the progression of TB infection, it is imperative to discuss their potential as biomarkers of the disease.

Hsps have recently started gaining attention as putative biomarkers of tuberculous infections. Profiling of the cerebrospinal fluid of tuberculous meningitis (TBM) patients has been carried out in various studies, where *M. tb* Hsp65, Hsp16, and Hsp71 stand out as promising biomarkers (Kashyap et al. 2005; Mudaliar et al. 2006; Tang et al. 2008). A similar profiling was done with the serum and sputum samples of TB-positive patients, and it was found that fluctuations in the immune response of host led to alterations in the profile of *M. tb* Hsps (Shekhawat et al. 2014). A dramatic increase in the levels of Hsp16, Hsp65, and Hs71 was seen in patient samples. A follow-up study, where monocytes were infected with the same human clinical samples, was also found to corroborate the abovementioned findings. The expression of Hsp16, Hsp65, and Hs71 was recorded to be as high as threefold in only 48 h post infection (Shekhawat et al. 2014). These observations are also indicative of the probable role of Hsps in host-pathogen interaction and ultimately in altering the host immune response. Further, Hsp16 is reported to be highly specific to latency-like conditions (Rajpal et al. 2011). Households with high risk of exposure to TB also form an important part of the target group for interventions. Usually household exposure to confirmed TB patients results in latent TB infection (LTBI) which eludes normal diagnostic approach (Lienhardt et al. 2010). The levels of Hsp16 have been shown to differentially increase in people with high risk of TB exposure versus the low-risk exposure group or in the low-risk exposure group versus the noninfected, healthy group of people (Shekhawat et al. 2016). The Hsp profile in general was also found to be differential among the high-risk, low-risk, and healthy subjects. Another study has shown sHsp18 as a major immunodominant antigen resulting from *M. leprae* infection (Lini et al. 2008). Thus, the potential of Hsps as biomarkers of the disease has been highlighted by both bedside and bench-side investigations. They may also prove to be helpful in designing better diagnostics for LTBI.

Apart from sharing the previously described diverse cellular functions like protein turnover, stress tolerance, regulation of survival, etc., Hsps are also known to perform a number of moonlighting functions owing to their secretion in the extracellular milieu (Henderson et al. 2010). The extracellular secretion of Hsps is a favorable feature as it enables the bacilli to influence the host immune response. However, the exact mechanism of secretion is not yet clear. Released mycobacterial Hsps bind to host cellular receptors and initiate an innate immune response, through Toll-like receptors mostly, and result in secretion of pro-inflammatory molecules (Lehner et al. 2004); hence they are also termed as "chaperokines" due to their dual activity as chaperones as well as cytokines (Asea et al. 2000). Hsps, along with the cargo of proteins that they carry, get processed and presented on host MHCs. Hsps apparently act as a link between the innate and acquired immune responses and thus have the ability to function as vaccine adjuvants as well (Segal et al. 2006). Recombinant *M. tb* Hsp70, for example, shows significant adjuvant effect in combination with recombinant outer membrane protein 31 (Dhakal et al. 2013). Further, even significantly small quantities of Hsp70 bound peptides, approximately 120 pM, is sufficient to generate a cytotoxic T-cell response in vivo, while as high as 2000-folds of free peptide was unable to produce similar effect (Minton 2004; Javid et al. 2004). Mycobacterial Hsp60 and Hsp70 have been used as effective carrier molecules for conjugated vaccines. Cross-linking mycobacterial Hsp60 or Hsp70 with a synthetic peptide $(NANP)_{40}$, an epitope from a major surface protein of *P. falciparum*, and administering them without adjuvant to different strains of mice as well as squirrel monkeys resulted in heightened production of anti-$(NANP)_{40}$ antibody (Lussow et al. 1991; Barrios et al. 1992; Perraut et al. 1993). It has also been shown that mycobacterial Hsp70 dramatically enhances the immunogenicity of HIV-1 gag p24 antigen (Suzue and Young 1996). Another promising approach for control of TB is the use of DNA vaccines. Such a plasmid DNA vaccine against TB was constructed by fusing mycobacterial Hsp65 and human IL-2 genes (Okada et al. 2007; Changhong et al. 2009). Administration of the Hsp-based DNA vaccine to mice resulted in increased Th-1-type cellular response, and interferon-γ and IL-2 were produced in greater amounts with a higher titer value of antigen-specific anti-Hsp65 IgG2a. It is already established that the Th-1-type response plays crucial role in controlling TB (Kim et al. 2000). The vaccine was also able to provide significant protection against TB infection in challenge experiments in mice. Another study also reports that *M. tb* Hsp65 DNA vaccine, designed initially to control infection, can also have pronounced therapeutic effects (Lowrie et al. 1999). Hsp16.3, a protein well documented to play key role in TB persistence, has also been reported to have potential to be developed as new TB vaccine component (Shi et al. 2009). Immunization of mice with Hsp16.3 or its T-cell epitope was shown to produce antibodies that were significantly stronger and much more specific as compared to the classical TB vaccine, BCG (Shi et al. 2009). Colonization in the lungs and spleen of immunized mice was reported to have dramatically reduced indicating the potential of Hsp16.3 as a prophylactic vaccine candidate (Shi et al. 2009).

Taking into account the regulatory role of Hsps in physiological as well as pathological conditions, they are beginning to emerge as potential candidates for drug targeting and could be a breakthrough in the future. Many eukaryotic heat shock proteins have already been targeted to treat cancer and numerous inhibitory molecules of the same are under clinical trials (Taldone et al. 2009, 2014a, b). Targeting Hsps is an active area of research currently in bacterial drug discovery as well; however, targeting them against TB is still in its initial stages. Nevertheless, development of inhibitors of GroEL/ES, Hsp90, GrpE, and DnaK in other bacteria (Cellitti et al. 2009; Chapman et al. 2009; Minagawa et al. 2011; Abdeen et al. 2016) gives us an indication that Hsps could also be potentially targeted in mycobacteria. Mycobacterial Hsp70 or DnaK has been targeted with two of the terphenyls isolated from the fungus *Hypoxylon rickii*, Rickenyl A and D, and they were reported to function as inhibitors of Hsp70 (Mohammadi-Ostad-Kalayeh et al. 2017). They compete with ATP for binding to DnaK in the range of 29 μM (Rickenyl D) and 49 μM (Rickenyl A). Structural differences in the bacterial and human Hsp70 protein were shown to evidently affect the affinity of inhibitor molecules making them more specific and selective for bacterial Hsp.

Clp proteins which belong to HSP100 family have been a major focus among the Hsps for TB drug discovery efforts in the past decade. ClpP has been demonstrated to be an unusual drug target, as both inhibition and hyperactivation of this gene can lead to cell death. In *Caulobacter crescentus*, cyclic peptides have been found to be able to bind to ClpXP-substrate complex, thereby preventing degradation of the substrate and acting as bactericidal agents (Cheng et al. 2007). Compounds of β-lactone family have also been reported to inhibit ClpP activity and attenuate production of other virulence factors in *S. aureus* and *L. monocytogenes* (Gaillot et al. 2000; Zeiler et al. 2011). ADEPs (acyldepsipeptides), on the other hand, are known to activate the protease function of ClpP in *M. tuberculosis* (Brotz-Oesterhelt et al. 2005). ADEPs also act against various other Gram-positive bacteria and their multidrug-resistant isolates (Socha et al. 2010). ADEPs bind to ClpP on the same site as the ATPase partner and lead to conformational changes in the oligomer which eventually results in unregulated proteolysis and cell death (Lee et al. 2010; Li et al. 2010). Bortezomib, a peptidyl boronate antineoplastic drug, has been shown to target ClpP1P2 of *M. tb* (Moreira et al. 2015). It also inhibits the *M. tb* 20S proteasome system (Lin et al. 2008). Clp ATPase ClpC1 has been reported to be a target for two natural inhibitor molecules, cyclomarin A (Schmitt et al. 2011) and rufomycin (Gao et al. 2015), which act on the ClpC1-mediated ATP-dependent proteolysis pathway. Cyclomarin A is a natural product that kills both replicating as well as non-replicating mycobacteria. Another naturally occurring cyclic peptide known as lassomycin has been demonstrated to have a similar spectrum of activity against ClpC of *M. tb* resulting in stimulation of its activity without having any effect on the proteolytic activity of ClpP. The end result of such activation is accumulation of toxic proteins leading to death of the pathogen (Gavrish et al. 2014). Ecumicin, another anti-mycobacterial natural peptide, also targets ClpC1 in a similar manner as lassomycin (Choules et al., 2015). The peptides targeting ClpC1 display nanomolar to micromolar potency against *M. tb*, with limited cytotoxicity to host cells. The activities of the abovementioned molecules have been shown to be highly specific to *M. tb* as they were found to be inactive against other Gram-positive bacteria (Gavrish et al. 2014; Gao et al. 2015; Moreira et al. 2017). Moreover, several small organic molecules displaying specific antimicrobial activity have also been identified against *E. coli* ClpB (Martin et al. 2013).

13.7 Conclusion

Continuous surge in the drug resistance phenomenon in *M. tb* and problems due to the current lengthy and complex treatment regimens have led to an urgent need to develop new anti-tubercular agents. The crucial role of mycobacterial heat shock proteins in the survival and maintenance of virulence of this pathogen makes them attractive potential targets for newer therapeutic interventions. A great deal of work has been done in the structure function analysis of heat shock proteins Studies have also implicated them in the pathogenesis of tuberculosis. Small organic molecules inhibiting the in vitro activity of some of the mycobacterial Hsps have been designed and characterized. Future in vivo studies will determine their real potential as therapeutics against the TB bacterium. Inhibition of Hsps as a combination drug therapy also offers a promising prospect. Further evaluation and understanding of inhibitory molecules as well as their interaction with Hsps will hold immense potential in designing novel and improved anti-tubercular drugs and vaccines.

References

Abdeen S, Salim N, Mammadova N et al (2016) GroEL/ES inhibitors as potential antibiotics. Bioorg Med Chem Lett 26:3127–3134

Abomoelak B, Marcus SA, Ward SK et al (2010) Characterization of a novel heat shock protein (Hsp22.5) involved in the pathogenesis of *Mycobacterium tuberculosis*. J Bacteriol 193:3497–3505

Aderem A, Underhill DM (1999) Mechanisms of phagocytosis in macrophages. Annu Rev Immunol 17:593–623

Akopian T, Kandror O, Raju RM et al (2012) The active ClpP protease from M. tuberculosis is a complex composed of a heptameric ClpP1 and a ClpP2 ring. EMBO J 31:1529–1541

Armstrong JA, Hart PA (1975) Phagosome-lysosome interactions in cultured macrophages infected with virulent tubercle bacilli. Reversal of the usual nonfusion pattern and observations on bacterial survival. J Exp Med 142:1–16

Asea A, Kraeft SK, Kurt-Jones EA et al (2000) HSP70 stimulates cytokine production through a CD14-dependant pathway, demonstrating its dual role as a chaperone and cytokine. Nat Med 6:435–442

Axelrod S, Oschkinat H, Enders J et al (2008) Delay of phagosome maturation by a mycobacterial lipid is reversed by nitric oxide. Cell Microbiol 10:1530–1545

Banta LM, Bohne J, Lovejoy SD et al (1998) Stability of the *Agrobacterium tumefaciens* VirB10 protein is modulated by growth temperature and periplasmic osmoadaption. J Bacteriol 180:6597–6606

Barik S, Sureka K, Mukherjee P et al (2010) RseA, the SigE specific anti-sigma factor of *Mycobacterium tuberculosis*, is inactivated by phosphorylation-dependent ClpC1P2 proteolysis. Mol Microbiol 75:592–606

Barrios C, Lussow AR, Embden JV (1992) Mycobacterial heat-shock proteins as carrier molecules. II: The use of the 70-kDa mycobacterial heat-shock protein as carrier for conjugated vaccines can circumvent the need for adjuvants and Bacillus Calmette-Guérin priming. Eur J Immunol 22:1365–1372

Beste DJ, Hooper T, Stewart G et al (2007) GSMN-TB: a web-based genome-scale network model of *Mycobacterium tuberculosis* metabolism. Genome Biol 8:R89

Bhat AH, Pathak D, Rao A (2017) *The alr-groEL1* operon in *Mycobacterium tuberculosis*: an interplay of multiple regulatory elements. Sci Rep 7:43772

Brotz-Oesterhelt H, Beyer D, Kroll HP et al (2005) Dysregulation of bacterial proteolytic machinery by a new class of antibiotics. Nat Med 11:1082–1087

Bukau B, Horwich AL (1998) The Hsp70 and Hsp60 chaperone machines. Cell 92:351–366

Butler RE, Cihlarova V, Stewart GR (2010) Effective generation of reactive oxygen species in the mycobacterial phagosome requires K+ efflux from the bacterium. Cell Microbiol 12:1186–1193

Cellitti J, Zhang Z, Wang S et al (2009) Small molecule DnaK modulators targeting the beta-domain. Chem Biol Drug Des 74:349–357

Chan J, Xing Y, Magliozzo RS et al (1992) Killing of virulent *Mycobacterium tuberculosis* by reactive nitrogen intermediates produced by activated murine macrophages. J Exp Med 175:1111–1122

Changhong S, Hai Z, Limei W et al (2009) Therapeutic efficacy of a tuberculosis DNA vaccine encoding heat shock protein 65 of *Mycobacterium tuberculosis* and the human interleukin 2 fusion gene. Tuberculosis 89:54–61

Chapman E, Farr GW, Furtak K et al (2009) A small molecule inhibitor selective for a variant ATP-binding site of the chaperonin GroEL. Bioorg Med Chem Lett 19:811–813

Chauhan A, Madiraju MVVS, Fol M et al (2006) *Mycobacterium tuberculosis* cells growing in macrophages are filamentous and deficient in FtsZ rings. J Bacteriol 188:1856–1865

Chen L, He X, Zhao X et al (2011) Interaction of Mycobacterium tuberculosis MPB64 protein with heat shock protein 40. Afr J Microbiol Res 5(4):394–398

Cheng L, Naumann TA, Horswill AR et al (2007) Discovery of antibacterial cyclic peptides that inhibit the ClpXP protease. Protein Sci 16:1535–1542

Choules M, Yu Y, Cho SH et al (2015) A rufomycin analogue is an anti-tuberculosis drug lead targeting CLPC1 with no cross resistance to ecumicin. Planta Med 81:871

Csermely P, Schnaider T, So C et al (1998) The 90-kDa molecular chaperone family: structure, function, and clinical applications. A comprehensive review. Pharmacol Ther 79:129–168

Dhakal J, Brah GS, Agrawal RK et al (2013) Over-expression of gene encoding heat shock protein 70 from *Mycobacterium tuberculosis* and its evaluation as vaccine adjuvant. Ind J Med Microbiol 31:123–129

Dobson CM, Karplus M (1999) The fundamentals of protein folding: bringing together theory and experiment. Curr Opin Struct Biol 9:92–101

Du Y, Lenz J, Arvidson CG (2005) Global gene expression and the role of sigma factors in *Neisseria gonorrhoeae* in interactions with epithelial cells. Infect Immun 73:4834–4845

Dziadek J, Rutherford SA, Madiraju MV et al (2003) Conditional expression of *Mycobacterium smegmatis* ftsZ, an essential cell division gene. Microbiology 149:1593–1603

Dziedzic R, Kiran M, Plocinski P et al (2010) *Mycobacterium tuberculosis* ClpX interacts with FtsZ and interferes with FtsZ assembly. PLoS One 5:e11058

Ehrt S, Schnappinger D (2009) Mycobacterial survival strategies in the phagosome: defence against host stresses. Cell Microbiol 11:1170–1178

Ensgraber M, Loos M (1992) A 66-kilodalton heat shock protein of *Salmonella typhimurium* is responsible for binding of the bacterium to intestinal mucus. Infect Immun 60:3072–3078

Ernst JD (1998) Macrophage receptors for *Mycobacterium tuberculosis*. Infect Immun 66:1277–1281

Esteban J, Martín-de-Hijas NS, Kinnari TJ et al (2008) Biofilm development by potentially pathogenic non-pigmented rapidly growing mycobacteria. BMC Microbiol 8:184

Farhana A, Guidry L, Srivastava A et al (2010) Reductive stress in microbes: implications for understanding *Mycobacterium tuberculosis* disease and persistence. Adv Microbial Physiol 57:43–117

Fay A, Glickman MS (2014) An essential nonredundant role for mycobacterial DnaK in native protein folding. PLoS Genet 10:e1004516

Firmani MA, Riley LW (2002) Reactive nitrogen intermediates have a bacteriostatic effect on *Mycobacterium tuberculosis* in vitro. J Clin Microbiol 40:3162–3166

Flynn JL, Chan J (2001) Immunology of tuberculosis. Ann Rev Immunol 19:93–129

Gaillot O, Pellegrini E, Bregenholt S et al (2000) The ClpP serine protease is essential for the intracellular parasitism and virulence of *Listeria monocytogenes*. Mol Microbiol 35:1286–1294

Gao W, Kim JY, Anderson JR et al (2015) The cyclic peptide ecumicin targeting ClpC1 is active against *Mycobacterium tuberculosis* in vivo Antimicrob Agents Chemother 59 880–889

Gavrish E, Sit CS, Cao S et al (2014) Lassomycin, a ribosomally synthesized cyclic peptide, kills *Mycobacterium tuberculosis* by targeting the ATP-dependent protease ClpC1P1P2. Chem Biol 21:509–518

Georgieva M, Sia JK, Bizzell E et al (2018) *Mycobacterium tuberculosis* GroEL2 modulates dendritic cell responses. Infect Immun 86:e00387–e00317

Gomes MS, Paul S, Moreira AL et al (1999) Survival of Mycobacterium avium and Mycobacterium tuberculosis in acidified vacuoles of murine macrophages. Infect Immun 67(7):3199–3206

Gottesman ME, Hendrickson WA (2000) Protein folding and unfolding by Escherichia coli chaperones and chaperonins. Curr Opin Microbiol 3:197–202

Griffin JE, Gawronski JD, De Jesus MA et al (2011) High-resolution phenotypic profiling defines genes essential for mycobacterial growth and cholesterol catabolism. PLoS Pathog 7:e1002251

Grosset J (2003) *Mycobacterium tuberculosis* in the extracellular compartment: an underestimated adversary. Antimicrob Agents Chemother 47:833–836

Harboe M, Oettinger T, Wiker HG et al (1996) Evidence for occurrence of the ESAT-6 protein in *Mycobacterium tuberculosis* and virulent *Mycobacterium bovis* and for its absence in *Mycobacterium bovis* BCG. Infect Immun 64:16–22

Hartl FU (1996) Molecular chaperones in cellular protein folding. Nature 381:571–579

Hartl FU, Hayer-Hartl M (2002) Molecular chaperones in the cytosol: from nascent chain to folded protein. Science 295:1852–1858

Haslbeck M, Franzmann T, Weinfurtner D et al (2005) Some like it hot: the structure and function of small heat-shock proteins. Nat Struct Mol Biol 12:842–846

Hatta M, Hatta Y, Kim JH et al (2007) Growth of H5N1 influenza A viruses in the upper respiratory tracts of mice. PLoS Pathog 3:e133

Henderson B, Allan E, Coates AR (2006) Stress wars: the direct role of host and bacterial molecular chaperones in bacterial infection. Infect Immun 74(7):3693–3706

Henderson B, Lund PA, Coates AR (2010) Multiple moonlighting functions of mycobacterial molecular chaperones. Tuberculosis 90:119–124

Hestvik ALK, Hmama Z, Av-Gay Y (2005) Mycobacterial manipulation of the host cell. FEMS Microbiol Rev 29:1041–1050

Hickey TB, Thorson LM, Speert DP et al (2009) *Mycobacterium tuberculosis* Cpn60.2 and DnaK are located on the bacterial surface, where Cpn60.2 facilitates efficient bacterial association with macrophages. Infect Immun 77:3389e401

Hirschhausen N, Schlesier T, Schmidt MA et al (2010) A novel staphylococcal internalization mechanism involves the major autolysin Atl and heat shock cognate protein Hsc70 as host cell receptor. Cell Microbiol 12:1746–1764

Houry WA, Frishman D, Eckerskorn C et al (1999) Identification of in vivo substrates of the chaperonin GroEL. Nature 402:147–154

Hu Y, Henderson B, Lund PA et al (2008) A *Mycobacterium tuberculosis* mutant lacking the groEL homologue cpn60. 1 is viable but fails to induce an inflammatory response in animal models of infection. Infect Immun 76:1535–1546

Javid B, MacAry PA, Oehlmann W et al (2004) Peptides complexed with the protein HSP70 generate efficient human cytolytic T-lymphocyte responses. Biochem Soc Trans 32:622–625

Jordao L, Bleck CK, Mayorga L et al (2008) On the killing of mycobacteria by macrophages. Cell Microbiol 10:529–548

Joseph S, Yuen A, Singh V et al (2017) *Mycobacterium tuberculosis* Cpn60.2 (GroEL2) blocks macrophage apoptosis via interaction with mitochondrial mortalin. Biol Open 6:481–488

Kang PB, Azad AK, Torrelles JB et al (2005) The human macrophage mannose receptor directs Mycobacterium tuberculosis lipoarabinomannan-mediated phagosome biogenesis. J Exp Med 202:987–999

Kar NP, Sikriwal D, Rath P et al (2008) *Mycobacterium tuberculosis* ClpC1. FEBS J 275:6149–6158

Kashyap RS, Dobos KM, Belisle JT et al (2005) Demonstration of components of antigen 85 complex in cerebrospinal fluid of tuberculous meningitis patients. Clin Diagnostic Lab Immunol 12:752–758

Kim JJ, Yang JS, Montaner L et al (2000) Coimmunization with IFN-gamma or IL-2, but not IL-13 or IL-4 cDNA can enhance Th1-type DNA vaccine-induced immune responses in vivo. J Interferon Cytokine Res 20:311–319

Klein BS, Tebbets B (2007) Dimorphism and virulence in fungi. Curr Opin Microbiol 10:314–319

Konkel ME, Tilly K (2000) Temperature-regulated expression of bacterial virulence genes. Microbes Infect 2:157–166

Krobitsch S, Clos J (1999) A novel role for 100 kD heat shock proteins in the parasite *Leishmania donovani*. Cell Stress Chaperones 4:191–198

Külzer S, Charnaud S, Dagan T et al (2012) *Plasmodium falciparum*-encoded exported hsp70/hsp40 chaperone/co-chaperone complexes within the host erythrocyte. Cell Microbiol 14:1784–1795

Lee BG, Park EY, Lee KE et al (2010) Structures of ClpP in complex with acyldepsipeptide antibiotics reveal its activation mechanism. Nat Struct Mol Biol 17:471–478

Lehner T, Wang Y, Whittall T et al (2004) Functional domains of HSP70 stimulate generation of cytokines and chemokines, maturation of dendritic cells and adjuvanticity. Biochem Soc Trans 32:629–632

Li J, Buchner J (2013) Structure, function and regulation of the hsp90 machinery. Biomed J 36:106–117

Li DHS, Chung YS, Gloyd M et al (2010) Acyldepsipeptide antibiotics induce the formation of a structured axial channel in ClpP: a model for the ClpX/ClpA-bound state of ClpP. Chem Biol 17:959–969

Lienhardt C, Fielding K, Hane AA et al (2010) Evaluation of the prognostic value of IFN-γ release assay and tuberculin skin test in household contacts of infectious tuberculosis cases in Senegal. PloS one 5:e10508

Lin G, Tsu C, Dick L et al (2008) Distinct specificities of *Mycobacterium tuberculosis* and mammalian proteasomes for N-acetyl tripeptide substrates. J Biol Chem 283:34423–34431

Lindquist S, Craig EA (1988) The heat-shock proteins. Annu Rev Genet 22:631–677

Lourdault K, Cerqueira GM, Wunder EA et al (2011) Inactivation of clpB in the pathogen *Leptospira interrogans* reduces virulence and resistance to stress conditions. Infect Immun 79:3711–3717

Lowrie DB (1983) How macrophages kill tubercle bacilli. J Med Microbiol 16:1–12

Lowrie DB, Tascon RE, Bonato VL et al (1999) Therapy of tuberculosis in mice by DNA vaccination. Nature 400:269–271

Lupoli TJ, Fay A, Adura C, Glickman MS, Nathan CF (2016) Reconstitution of a proteostasis network highlights essential cofactor interactions with chaperone DnaK. Proc Natl Acad Sci 113(49):E7947–E7956

Lussow AR, Barrios C, van Embden J et al (1991) Mycobacterial heat-shock proteins as carrier molecules. Eur J Immunol 21:2297–2302

MacGurn JA, Cox JS (2007) A genetic screen for *Mycobacterium tuberculosis* mutants defective for phagosome maturation arrest identifies components of the ESX-1 secretion system. Infect Immun 75:2668–2678

MacMicking JD, Taylor GA, McKinney JD (2003) Immune control of tuberculosis by IFN-γ-inducible LRG-47. Science 302:654–659

Majlessi L, Brodin P, Brosch R et al (2005) Influence of Esat-6 secretion system 1 (RD1) of *Mycobacterium tuberculosis* on the Interaction between mycobacteria and the host immune system. J Immunol 174:3570–3579

Martin I, Underhaug J, Celaya G et al (2013) Screening and evaluation of small organic molecules as ClpB inhibitors and potential antimicrobials. J Med Chem 56:7177–7189

McGillivray A, Golden NA, Gautam US et al (2014) The *Mycobacterium tuberculosis* Rv2745c plays an important role in responding to redox stress. PLoS One 9:e93604

McGillivray A, Golden NA, Kaushal D (2015) The *Mycobacterium tuberculosis* Clp gene regulator is required for *in Vitro* reactivation from hypoxia-induced dormancy. J Biol Chem 290:2351–2367

Meghji S, White PA, Nair SP et al (1997) *Mycobacterium tuberculosis* chaperonin 10 stimulates bone resorption: a potential contributory factor in Pott's disease. J Exp Med 186:1241–1246

Mehra S, Kaushal D (2009) Functional genomics reveals extended roles of the *Mycobacterium tuberculosis* stress response factor σ H. J Bacteriol 191:3965–3980

Méndez-Samperio P (2008) Expression and regulation of chemokines in mycobacterial infection. J Infect 57:374–384

Minagawa S, Kondoh Y, Sueoka K et al (2011) Cyclic lipopeptide antibiotics bind to the N-terminal domain of the prokaryotic Hsp90 to inhibit the chaperone activity. Biochem J 435:237–246

Minton K (2004) Antigen presentation: Shocking stimulation. Nat Rev Immunol 4:162

Mohammadi-Ostad-Kalayeh S, Hrupins V, Helmsen S et al (2017) Development of a microarray-based assay for efficient testing of new HSP70/DnaK inhibitors. Bioorg Med Chem 25:6345–6352

Moreira W, Ngan GJ, Low JL et al (2015) Target mechanism-based whole-cell screening identifies bortezomib as an inhibitor of caseinolytic protease in mycobacteria. mBio 6:e00253–e00215

Moreira W, Santhanakrishnan S, Ngan GJY et al (2017) Towards selective Mycobacterial clpp1p2 inhibitors with reduced activity against the human proteasome. Antimicrob Agents Chemother 61:e02307–e02316

Mudaliar AV, Kashyap RS, Purohit HJ et al (2006) Detection of 65 kD heat shock protein in cerebrospinal fluid of tuberculous meningitis patients. BMC Neurol 6:34

Nair S, Frehel C, Nguyen L et al (1999) ClpE, a novel member of the HSP100 family, is involved in cell division and virulence of *Listeria monocytogenes*. Mol Microbiol 31:185–196

Nair S, Derré I, Msadek T et al (2000) CtsR controls class III heat shock gene expression in the human pathogen *Listeria monocytogenes*. Mol Microbiol 35:800–811

Ojha A, Anand M, Bhatt A et al (2005) GroEL1: a dedicated chaperone involved in mycolic acid biosynthesis during biofilm formation in mycobacteria. Cell 123:861–873

Ojha AK, Baughn AD, Sambandan D et al (2008) Growth of *Mycobacterium tuberculosis* biofilms containing free mycolic acids and harbouring drug-tolerant bacteria. Mol Microbiol 69:164–174

Okada M, Kita Y, Nakajima T et al (2007) Evaluation of a novel vaccine (HVJ–liposome/HSP65 DNA+ IL-12 DNA) against tuberculosis using the cynomolgus monkey model of TB. Vaccine 25:2990–2993

O'Reilly T, Zak O (1992) Enhancement of the effectiveness of antimicrobial therapy by muramyl peptide immunomodulators. Clin Infect Dis 14(5):1100–1109

Perraut R, Lussow AR, Gavoille S et al (1993) Successful primate immunization with peptides conjugated to purified protein derivative or mycobacterial heat shock proteins in the absence of adjuvants. Clin Exp Immunol 93:382–386

Personne Y, Brown AC, Schuessler DL et al (2013) *Mycobacterium tuberculosis* ClpP proteases are co-transcribed but exhibit different substrate specificities. PLoS One 8:e60228

Peters W, Ernst JD (2003) Mechanisms of cell recruitment in the immune response to *Mycobacterium tuberculosis*. Microb Infect 5(2):151–158

Pethe K, Swenson DL, Alonso S et al (2004) Isolation of *Mycobacterium tuberculosis* mutants defective in the arrest of phagosome maturation. Proc Natl Acad Sci USA 101:13642–13647

Pieters J, Gatfield J (2002) Hijacking the host: survival of pathogenic mycobacteria inside macrophages. Trends Microbiol 10:142–146

Powers ET, Balch WE (2013) Diversity in the origins of proteostasis networks — a driver for protein function in evolution. Nat Rev Mol Cell Biol 14(4):237–248

Pym AS, Brodin P, Brosch R et al (2002) Loss of RD1 contributed to the attenuation of the live tuberculosis vaccines *Mycobacterium bovis* BCG and *Mycobacterium microti*. Mol Microbiol 46:709–717

Rajpal SK, Snehal SW, Milind SP et al (2011) *Mycobacterium tuberculosis* heat shock protein 16 as a potential marker for latent TB: a preliminary findings. J Clin Cell Immunol 2:2

Raju RM, Unnikrishnan M, Rubin DHF et al (2012) *Mycobacterium tuberculosis* ClpP1 and ClpP2 function together in protein degradation and are required for viability *in vitro* and during infection. PLoS Pathog 8:e1002511

Raman S, Song T, Puyang X et al (2001) The alternative sigma factor SigH regulates major components of oxidative and heat stress responses in *Mycobacterium tuberculosis*. J Bacteriol 183:6119–6125

Raviglione M, Sulis G (2016) Tuberculosis 2015: burden, challenges and strategy for control and elimination. Infect Dis Rep 8:6570

Romberg L, Levin PA (2003) Assembly dynamics of the bacterial cell division protein FtsZ: poised at the edge of stability. Annu Rev Microbiol 57:125–154

Rouquette C, Ripio MT, Pellegrini E et al (1996) Identification of a ClpC ATPase required for stress tolerance and in vivo survival of *Listeria monocytogenes*. Mol Microbiol 21:977–987

Rouquette C, De Chastellier C, Nair S et al (1998) The ClpC ATPase of *Listeria monocytogenes* is a general stress protein required for virulence and promoting early bacterial escape from the phagosome of macrophages. Mol Microbiol 27:1235–1245

Schaible UE, Sturgill-Koszycki S, Schlesinger PH et al (1998) Cytokine activation leads to acidification and increases maturation of Mycobacterium avium-containing phagosomes in murine macrophages. J Immunol 160:1290–1296

Schirmer EC, Glover JR, Singer MA et al (1996) HSP100/Clp proteins: a common mechanism explains diverse functions. Trends Biochem Sci 21:289–296

Schmitt EK, Riwanto M, Sambandamurthy V et al (2011) The natural product cyclomarin kills Mycobacterium tuberculosis by targeting the ClpC1 subunit of the caseinolytic protease. Angew Chem Int Ed 50:5889–5891

Schmitz KR, Sauer RT (2014) Substrate delivery by the AAA+ ClpX and ClpC1 unfoldases activates the mycobacterial ClpP1P2 peptidase. Mol Microbiol 93:617–628

Segal BH, Wang XY, Dennis CG et al (2006) Heat shock proteins as vaccine adjuvants in infections and cancer. Drug Discov Today 11:534–540

Sharma A, Rustad T, Mahajan G et al (2016) Towards understanding the biological function of the unusual chaperonin Cpn60. 1 (GroEL1) of *Mycobacterium tuberculosis*. Tuberculosis 97:137–146

Shekhawat SD, Jain RK, Gaherwar HM et al (2014) Heat shock proteins: possible biomarkers in pulmonary and extrapulmonary tuberculosis. Hum Immunol 75:151–158

Shekhawat SD, Purohit HJ, Taori GM et al (2016) Evaluation of heat shock proteins for discriminating between latent tuberculosis infection and active tuberculosis: a preliminary report. J Infect Public Health 9:143–152

Sherman DR, Voskuil M, Schnappinger D et al (2001) Regulation of the *Mycobacterium tuberculosis* hypoxic response gene encoding alpha-crystallin. Proc Natl Acad Sci USA 98:7534–7539

Sherrid AM, Rustad TR, Cangelosi GA et al (2010) Characterization of a Clp protease gene regulator and the reaeration response in *Mycobacterium tuberculosis*. PLoS One 5:e11622

Shi C, Zhang H, Zhang T et al (2009) New alternative vaccine component against *Mycobacterium tuberculosis* heat shock protein 16.3 or its T-cell epitope. Scand J Immunol 70 465–474

Sielaff B, Lee KS, Tsai FTF (2011) Structural and functional conservation of *Mycobacterium tuberculosis* GroEL Paralogs suggests that GroEL1 is a Chaperonin. J Mol Biol 405:831–839

Sigler PB, Xu Z, Rye HS et al (1998) Structure and function in GroEL-mediated protein folding. Annu Rev Biochem 67:581–608

Socha AM, Tan NY, LaPlante KL et al (2010) Diversity-oriented synthesis of cyclic acyldepsipeptides leads to the discovery of a potent antibacterial agent. Bioorg Medic Chem 18:7193–7202

Stewart GR, Snewin VA, Walzl G et al (2001) Overexpression of heat-shock proteins reduces survival of *Mycobacterium tuberculosis* in the chronic phase of infection. Nat Med 7:732–737

Stewart GR, Robertson BD, Young DB (2004) Analysis of the function of mycobacterial DnaJ proteins by overexpression and microarray profiling. Tuberculosis 84:180–187

Stewart GR, Newton SM, Wilkinson KA et al (2005) The stress-responsive chaperone alpha-crystallin 2 is required for pathogenesis of *Mycobacterium tuberculosis*. Mol Microbiol 55:1127–1137

Sturgill-Koszycki S, Schlesinger PH, Chakraborty P et al (1994) Lack of acidification in Mycobacterium phagosomes produced by exclusion of the vesicular proton-ATPase. Science 263:678–681

Suzue K, Young RA (1996) Adjuvant-free hsp70 fusion protein system elicits humoral and cellular immune responses to HIV-1 p24. J Immunol 156:873–879

Sweier DG, Combs A, Shelburne CE et al (2003) Construction and characterization of a *Porphyromonas gingivalis* htpG disruption mutant. FEMS Microbiol Lett 225:101–106

Taldone T, Sun W, Chiosis G (2009) Discovery and development of heat shock protein 90 inhibitors. Bioorg Med Chem 17:2225–2235

Taldone T, Ochiana SO, Patel PD et al (2014a) Selective targeting of the stress chaperome as a therapeutic strategy. Trends Pharmacol Sci 35:592–603

Taldone T, Patel HJ, Bolaender A et al (2014b) Protein chaperones: a composition of matter review (2008–2013). Expert Opin Ther Pat 24:501–518

Tang D, Kang R, Cao L et al (2008) A pilot study to detect high mobility group box 1 and heat shock protein 72 in cerebrospinal fluid of pediatric patients with meningitis. Crit Care Med 36:291–295

Tauschek M, Hamilton CW, Hall LA et al (1997) Transcriptional analysis of the groESL operon of *Neisseria gonorrhoeae*. Gene 189:107–112

Tissiéres A, Mitchell HK, Tracy UM (1974) Protein synthesis in salivary glands of Drosophila melanogaster: relation to chromosome puffs. J Mol Biol 84(3):389–398

Tukaj S, Kotlarz A, Jozwik A et al (2010) Hsp40 proteins modulate humoral and cellular immune response in rheumatoid arthritis patients. Cell Stress Chaperones 15:555–566

van Eden W, Thole JE, van der Zee R et al (1988) Cloning of the mycobacterial epitope recognized by T lymphocytes in adjuvant arthritis. Nature 331:171–173

Van Montfort R, Slingsby C, Vierlingt E (2001) Structure and function of the small heat shock protein/α-crystallin family of molecular chaperones. Adv Prot Chem 59:105–156

Vaubourgeix J, Lin G, Dhar N et al (2015) Stressed mycobacteria use the chaperone ClpB to sequester irreversibly oxidized proteins asymmetrically within and between cells. Cell Host Microbe 17:178–190

Via LE, Fratti RA, McFalone M et al (1998) Effects of cytokines on mycobacterial phagosome maturation. J Cell Sci 111(7):897–905

Voskuil MI, Schnappinger D, Visconti KC et al (2003) Inhibition of respiration by nitric oxide induces a *Mycobacterium tuberculosis* dormancy program. J Exp Med 198:705–713

Wang Y, Kelly CG, Karttunen JT et al (2001) CD40 is a cellular receptor mediating mycobacterial heat shock protein 70 stimulation of CC-chemokines. Immunity 15:971–983

Wayne LG, Sohaskey CD (2001) Nonreplicating persistence of *Mycobacterium tuberculosis*. Annu Rev Microbiol 55:139–163

Weibezahn J, Schlieker C, Bukau B et al (2003) Characterization of a trap mutant of the AAA+ chaperone ClpB. J Biol Chem 278:32608–32617

WHO (2016). Global tuberculosis report 2016. Available from: http://www.who.int/gho/tb/epidemic/cases_deaths/en/

Wickner S, Maurizi MR, Gottesman S (1999) Posttranslational quality control: folding, refolding, and degrading proteins. Science 286:1888–1893

Wilkinson KA, Stewart GR, Newton SM et al (2005) Infection biology of a novel alpha-crystallin of *Mycobacterium tuberculosis*: Acr2. J Immunol 174:4237–4243

Williams DL, Pittman TL, Deshotel M et al (2007) Molecular basis of the defective heat stress response in *Mycobacterium leprae*. J Bacteriol 189:8818–8827

Zeiler E, Braun N, Bottcher T et al (2011) Vibralactone as a tool to study the activity and structure of the ClpP1P2 complex from *Listeria monocytogenes*. Angew Chem Int Ed Engl 50:11001–11004

Endoplasmic Reticulum Stress: Importance in Pathogenesis of *Mycobacterium tuberculosis*

14

Tarina Sharma, Sonam Grover, Nasreen Z. Ehtesham, and Seyed E. Hasnain

Abstract

Mycobacterium tuberculosis (M.tb), the cause of deadly disease tuberculosis, is an opportunistic pathogen that primarily infects host alveolar macrophages. *M.tb* has developed several mechanisms to persist in infected host cells and to disseminate the disease. Its pathogenesis mainly depends on its competence to modulate the host machinery for its own benefit. One of the cellular responses triggered by *M.tb* is endoplasmic reticulum stress in infected macrophages, which eventually disturbs the physiological functioning of the ER. Uncontrolled ER stress activates IRE1, PERK, and ATF6 pathway to induce apoptosis of infected cells. Although apoptosis is known to control and clear the infection in primary stages of infection, in case of *M.tb*-infected ER stressed macrophages, apoptosis is able to disseminate the pathogen during its advanced stages. *M.tb*-infected lung granulomas are the preferential site of accumulation of apoptotic macrophages,

Tarina Sharma and Sonam Grover have equally contributed to this chapter.

T. Sharma
Molecular Infection and Functional Biology Lab, Kusuma School of Biological Sciences, Indian Institute of Technology-Delhi, New Delhi, Delhi, India

S. Grover
JH Institute of Molecular Medicine, Jamia Hamdard, New Delhi, Delhi, India

N. Z. Ehtesham
Inflammation Biology and Cell Signaling Laboratory, ICMR-National Institute of Pathology, Safdarjung Hospital Campus, New Delhi, Delhi, India
e-mail: nasreenzehtesham.nip@gov.in

S. E. Hasnain (✉)
JH Institute of Molecular Medicine, Jamia Hamdard, New Delhi, Delhi, India

Dr Reddy's Institute of Life Sciences, University of Hyderabad Campus, Hyderabad, India
e-mail: vc@jamiahamdard.ac.in

S. E. Hasnain et al. (eds.), Mycobacterium tuberculosis: *Molecular Infection Biology, Pathogenesis, Diagnostics and New Interventions*,
https://doi.org/10.1007/978-981-32-9413-4_14

thereby increasing the risk of disease dissemination. The present chapter will describe the mechanism for ER stress response generated by known *M.tb* virulence factors such as ESAT-6, HBHA, 38-kDa antigen, and PE_PGRS5. Future insights to describe *M.tb* infection in respect of eliciting ER stress response-mediated apoptosis and host interacting partners has the potential in identifying novel targets for vaccination and drugs to combat the disease.

Keywords
Mycobacterium tuberculosis · Endoplasmic reticulum stress · Unfolded protein response · Apoptosis · Oxidative stress · Calcium homeostasis

Abbreviations

ATF4	Activating transcription factor 4
ATF6	Activating transcription factor 6
bZiP	Basic leucine zipper
CHOP	C/EBP homologous protein
CREB	cAMP response element binding
ER	Endoplasmic reticulum
ERAD	ER-associated degradation
ERSE	ERS response elements
GRP78	78-kDA glucose regulated protein
HBHA	Heparin-binding hemagglutinin antigen
Hsp70	Heat shock protein family
IP3R	Inositol 1,4,5-triphate receptor
IRE1	Inositol requiring enzyme 1a
JNK	Jun-N-terminal kinase
M.tb	*Mycobacterium tuberculosis*
MAPK	Mitogen-activated protein kinases
MCPIP	MCP-1- induced protein
MPTP	Mitochondrial permeability transition pore
NO	Nitric oxide
PERK	Protein kinase-like ER kinase
ROS	Reactive oxygen species
SERCA	Sarco-endoplasmic reticulum calcium ATPase
TB	Tuberculosis
TLR	Toll-like receptor
TRAF2	TNF-receptor associated factor 2
UPR	Unfolded protein response
UPRE	UPR element
XBP1	X-box binding protein 1

14.1 Introduction

Tuberculosis (TB), caused by the opportunistic pathogen *Mycobacterium tuberculosis (M.tb)*, remains the most common deadly disease. According to the 2017 WHO TB report, 6.3 million new cases were reported with 1.3 million deaths in HIV-negative people with an additional 3,74,000 deaths among HIV-positive cases worldwide. Occurrence of multi-drug-resistant and extensive drug-resistant cases has further worsened the scenario, which has become further complicated due to HIV co-infections (WHO 2017).

M.tb, a diplomatic pathogen, is easily adapted in human host and employs multiple strategies to survive and multiply inside macrophages. The pathogenesis of TB depends on the interplay between host defense mechanisms and survival strategies employed by the bacteria. Host mechanisms employed to clear the infection includes phagolysosome formation, apoptosis of host cells, generation of immune responses, endoplasmic reticulum (ER) stress, etc. On one hand, as the immune system tries to detect and clear the infection, the bacterium on the other hand resists the immune response through several mechanisms that are the primary cause of death during persistent infections. Once infected, *M.tb* secretes several proteins in the macrophages for its survival in highly acidic phagolysosomal compartments (Brodin et al. 2010). *M.tb* secretes several proteins, which are largely in disordered state and cause the host cells to evoke an unfolded protein response (UPR) (Ahmad et al. 2018; Grover et al. 2018). ER stress is one such response generated by host cells to cope up with unwanted proteins produced by the pathogen. On the contrary, *M.tb* employs this mechanism for its own survival as well. Thus, back and forth interplay between these mechanisms such as ER stress-dependent apoptosis defines the outcome of disease.

Endoplasmic reticulum stress is usually a mechanism provoked in response to unfolded or misfolded proteins in ER. During some infectious diseases, the ER stress response is generated by the interaction of pathogen proteins with ER stress sensor proteins. As a consequence of the ER stress response, UPR is initiated and favors the host to overcome the deleterious effects of stress accumulation. ER stress-mediated response is a general phenomenon occurring in host cells; however in some cases pathogens may employ this as a strategy to initiate signaling cascades leading to disease pathogenesis. During different stages of infection, the ER stress response leads to different outcomes either as a pro-host or pro-pathogen.

M.tb also employs various strategies to combat host responses generated upon infection. An important factor in *M.tb* pathogenesis is induction of ER stress by *M.tb* virulence factors. Some of the key *M.tb* virulence factors involved in ER stress are ESAT-6 (Rv3875) (Choi et al. 2010), 38-kDa antigen (Rv0934) (Lim et al. 2015), heparin-binding hemagglutinin antigen (HBHA) (Rv0475) (Choi et al. 2013), and PE_PGRS5 (Rv0297) (Grover et al. 2018). Prolonged or uncontrolled ER stress leads to apoptosis of the cell and clearance of stress responses. Previous reports have shown that necrosis may favor the spread of mycobacteria, while apoptosis facilitates to control the infection. Contradictory to this, more recent reports have suggested a role for apoptosis in mycobacterial pathogenesis (Srinivasan et al. 2014).

Apoptosis, thought to play a significant role in defense mechanism employed by macrophages to clear the infection, can also be hired by bacteria for its own survival and dissemination. A role of ER stress-mediated apoptosis by live *M.tb* $H_{37}Rv$ in bacterium survival has been demonstrated, where infection of macrophages with *M. tb* initiates the ER stress and subsequent apoptosis of infected cells to disseminate the bacterium (Lim et al. 2011).

14.2 Endoplasmic Reticulum Stress-Mediated Pathways in Mycobacterial-Infected Macrophages

ER, an organelle containing chaperones and enzymes, is the site for protein folding and transportation of proteins. Properly folded proteins from ER get modified in the Golgi apparatus and are subsequently targeted to their destined sites in the cell (Smith et al. 2011). Immature protein aggregates and misfolded proteins in ER are targeted to ER-associated degradation (ERAD) machinery for clearance or by autophagic degradation (Araki and Nagata 2012). Protein aggregation is a phenomenon in which misfolded or unfolded proteins form clumps and aggregates due to their exposed hydrophobic amino acids. Protein aggregation in the ER lumen creates a state of ER stress in cells, which contributes to pathological conditions in several diseases such as diabetes mellitus, Alzheimer's disease, Parkinson's disease, atherosclerosis, and a few infectious diseases (Bhandary et al. 2013).

Mycobacterial infection also triggers ER stress (Lim et al. 2011), which leads to disturbance in protein folding in infected cells. It results in the activation of UPR to maintain regular functioning of the cells. Apart from performing protein folding and transportation after protein synthesis, ER also serves as major reservoir of intracellular Ca^{2+} ions, cholesterol steroids, and lipid synthesis. Depletion in calcium levels from ER due to mycobacterial infection leads to misfolded and unfolded protein accumulation. This accumulation of unfolded proteins results in ER stress and subsequent apoptosis of infected cells. Induction of apoptosis in ER stress-dependent manner by mycobacterial infection may either help in clearing the infection in early stages or may allow the bacterium to disseminate through apoptotic bodies during later stages of infection. An enhanced level of rough ER and smooth ER in macrophages was observed when infected with the virulent strain of *M.tb* $H_{37}Rv$ and attenuated strain of *M.tb* $H_{37}Ra$, respectively (Saquib et al. 2015).

Four underlying mechanisms by host cells in response to ER stress are (a) inhibition of protein synthesis to reduce protein aggregation; (b) production of ER chaperones to increase protein folding; (c) activation of ERAD genes to enhance ERAD machinery functioning; and (d) apoptosis of ER stressed cells (Yoshida 2007).

A mycobacterial-infected cell under uncontrolled ER stress activates signaling pathways that subsequently direct the cells to undergo apoptosis. Failure to perform this, host cells continue to harbor mycobacteria, thus facilitating its survival. The three ER stress sensor proteins localized in ER are (a) protein kinase-like ER kinase (PERK), a double-stranded RNA-dependent protein kinase; (b) inositol-requiring enzyme 1a (IRE1a); and (c) activating transcription factor 6 (ATF6). Another important factor in ER stress response is regulation of Ca^{2+} homeostasis. Disruption in calcium levels affect the protein folding in ER stressed cells and induce apoptosis (Cui et al. 2016).

14.2.1 IRE1a-Initiated ER Stress Response

IRE1a is a type I transmembrane sensor involved in activation of UPR for maintaining physiological conditioning of cells. Generally, mammalian IRE1a is involved in survivability of cells, but it is also known to degrade anti-apoptotic miRNAs to induce apoptosis (Sano and Reed 2013). Its luminal domain forms homo-dimers upon sensing ER stress, and its cytosolic domain gets auto-phosphorylated stimulating the kinase activity. Due to this kinase activity, the basic leucine zipper (bZiP) containing transcription factor, cAMP response element binding (CREB)/activating transcription factor (ATF) gets activated (Hassler et al. 2012).

X-box binding protein 1 (XBP1), a transcriptional activator interacts with UPR element UPRE and ERS response elements (ERSE-I and ERSE-II) of promoter of target genes. It thus induces the expression of other related proteins to reduce ER stress. During *M.tb* infection, induction of ER stress leads to overstimulation of Toll-like receptors (TLRs) in macrophages in an XBP1-dependent manner thereby controlling mycobacterial infection (Lim et al. 2015; Martinon et al. 2010). RNase domain of activated IRE1a interacts with TNF-receptor-associated factor 2 (TRAF2) and forms the IRE1-TRAF2-ASK1 complex, which further stimulates the mitogen-activated protein kinases (p38 MAPK) and Jun-N-terminal kinase (JNK) (Ron and Hubbard 2008). *M. kansasii*-infected macrophages have undergone apoptosis through an ER stress-dependent IRE1a/ASK1/JNK cascade (Lim et al. 2015).

14.2.2 The PERK Pathway

PERK, a serine/threonine protein kinase, is a type I transmembrane protein. Its destined function is to attenuate the protein synthesis machinery to reduce protein aggregation in ER stressed cells. PERK/eIF2a/CHOP pathway is activated by macrophages in response to *M.tb* infection and 38-kDa antigen (Lim et al. 2015). Upon sensing ER stress, PERK releases BiP (chaperone molecule), which afterwards gets oligomerized and auto-phosphorylated. PERK when activated phosphorylates eukaryotic initiation factor (eIF2a) at Ser51 residue. This phosphorylation inhibits the initiation complex of translational machinery thus haltering the new polypeptide synthesis. It reduces the global protein accumulation and the burden on ERAD process for protein folding thereby maintaining physiological conditioning of the cell (Bertolotti et al. 2000).

Activating transcription factor 4 (ATF4) is a transcription factor that belongs to the CREB protein family. Selective induction of ATF4 mRNA by eIF2a leads to its overexpression, which in turn induces the transcription of UPR transcription factors and ER stress response folding proteins. These factors then translocate to the nucleus for initiating ER stress response. ATF4 also initiates the transfer of activating transcription factor 6 (ATF6) from ER to Golgi apparatus.

C/EBP homologous protein (CHOP) is another component of ER stress-mediated apoptosis. During *M.tb* infection, CHOP levels considerably increase in the granulomatous regions (Seimon et al. 2010). Infection studies have shown that *M.tb* exploits the reduced levels of CHOP due to a decreased eIF2a phosphorylation for its own survival. It indicates that *M.tb* modulates host cell ER stress response for its own survival (Lim et al. 2011). In TB granulomas, ER stress is evident in specific areas of accumulated apoptotic macrophages (Seimon et al. 2010). Initially, overall protein synthesis was halted by PERK using eIF2a to promote cell survival. Later, due to an uncontrolled ER stress caused by *M.tb* infection, downstream CHOP is activated by PERK to induce apoptosis.

14.2.3 ATF6 Induces ER Stress Response

ATF6, a 90 kDa ER transmembrane protein, is translocated from ER to Golgi apparatus during conditions of ER stress in the cell. Its N-terminal region consists of the CREB/bZip transcription factor. ATF6a and ATF6b are two subtypes of ATF6 in mammalian cells. It gets activated in response to ER stress and cleaved to a 50 kDa protein fragment by proteolytic cleavage at its N-terminal site (Shen et al. 2002). Cleaved fragments get translocated to the nucleus for promoting transcription of ER stress response-related genes such as cAMP response element (CRE), ERSE-I, ERSE- II, and UPR element (UPRE). Binding of ATF6 to XBP1 induces UPR-related proteins expression. These proteins altogether relieve the stress from cells by reducing misfolded protein accumulation (Wang et al. 2000).

14.2.4 Chaperones in ER Stress Response

Unfolded protein response is activated to protect cells against adverse effects of ER stress. ER provides environment of proper protein folding by the presence of many enzymes and chaperones. Appropriate protein folding is an important phenomenon in maintaining physiological functioning of cells as only correctly folded proteins can be modified and translocated to the destined location (Smith et al. 2011). ER chaperones play a major role in protein synthesis regulation as well as in cell death induction (Tabas and Ron 2011). One of the molecular chaperones from the heat shock protein family (Hsp70) also known as GRP78/BiP is employed in rescuing cells from stressed conditions. BiP functions by participating in folding and assembly of proteins as well as targeting the misfolded/unfolded proteins towards degradation by 26S ubiquitination pathway. Bip is also a resident of mitochondria and nucleus apart from ER and upregulated in tumor cells (Ortiz and Cardemil 2001; Zhang et al. 2010). Bip is found to be overexpressed under various stress conditions such as depleted calcium/oxygen levels and glycopenia. Upregulated expression of Bip during *M.tb* and other mycobacterial species has also been reported in several studies (Choi et al. 2013; Lim et al. 2011).

Under resting conditions, inactive Bip is generally found to be non-covalently associated with ER stress receptors such as IRE1, PERK, and ATF6. On sensing stress conditions, Bip dissociates from the stress receptors and UPR is triggered. Dissociation of Bip from IRE1, PERK, and ATF6 leads to signaling cascades required for neutralizing the deleterious effects of ER stress and brings the ER functions back to normal (Harding et al. 1999). Few of the mycobacterial proteins such as ESAT-6, HBHA, and 38-kDa antigen are known to upregulate expression of ER chaperones, thus inducing ER stress responses. Increased expression of the pro-inflammatory cytokine, MCP-1-induced protein (MCPIP) in response to Mtb 38-kDa antigen, leads to ROS generation and subsequent misfolding of proteins. These events trigger Bip to be released from ER stress sensors and initiate protein folding. Other ER chaperones including CHOP and activated phosphorylated eIF2 has also been shown to be expressed by cells in response *M.tb* 38-kDa Ag (Lim et al. 2015).

14.2.5 Oxidative Stress in ER Stress

ER provides a suitable oxidative environment for the protein folding process. A change in oxidative environment and intracellular calcium levels induce the production of reactive oxygen species (ROS) in cells. ROS production in stressed cells is linked to protein folding competence of ER. ROS and nitric oxide (NO) generation are not only the consequences of ER stress but also stand as an integral process during ER stress. ROS may be generated as a by-product in electron transfer reactions or because of reduced glutathione (GSH) depletion during protein misfolding (Santos et al. 2009)

Mitochondrial ROS generation also tunes the downstream effects during ER stress. Mitochondrial ROS accumulation due to depleted GSH can enhance ER stress and also leads to cell mortality during persistent ER stress (Yoon et al. 2011). The electron transport chain and mitochondrial phosphorylation plays an important role in ROS generation. Mitochondrial respiration interference in cells reduces ROS accumulation (Bravo et al. 2011). Cells devoid of cytochrome C gene have mitigated response in context to UPR induction by hypoxia (Liu et al. 2008).

During ER stress induction, ER and mitochondria appear to function together to cope up this situation. Firstly, ER and mitochondria lie in close vicinity to each other that suggests a physical interaction between the two organelles (Raturi and Simmen 2013). Secondly, Ca^{2+} levels affect the mitochondrial membrane potential and ROS production (Bravo et al. 2011). Thirdly, there is expression of Lon protease and NIX (regulating mitochondrial membrane potential and ER Ca^{2+}). In response to these signaling molecules, the mitochondrial permeability transition pore (MPTP) gets opened (Chen et al. 2010).

Enhanced oxidative stress in cells leads to an inflow of Ca^{2+} from the extracellular environment and also an outflow from the ER. Ca^{2+} ions henceforth move to the mitochondria and nuclei stimulating the mitochondrial process and further ROS generation (Murphy 2009). An increase in ROS levels in the mitochondria in return proliferates the cascade of Ca^{2+} release from the ER (Moserova and Kralova 2012) and subsequent deterioration in ER functioning. Thus, the interplay between ER and mitochondrial-dependent oxidative stress decides the overall fate of stressed cells. *M.tb* live infection has been shown to induced ROS elevation during ER stress response forcing the infected cells to undergo apoptosis. This ER stress-mediated apoptosis has been linked with mycobacterial survival (Lim et al. 2011). ER stress-mediated apoptosis of *M. kansasii* SM-1-infected cells (Lim et al. 2013) and *M.tb* ESAT-6-stimulated cells (Choi et al. 2010) majorly depends on significant ROS production as a consequence of ER stress.

14.2.6 Calcium Homeostasis in ER Stress

ER, a major reservoir of Ca^{2+} inside cells, maintains calcium homeostasis for proper folding during stressed conditions. An alteration in their levels leads to apoptosis of the stressed cells due to improper folding or uncontrolled stress. Release of Ca^{2+} from ER in response to *M.tb* infection has been shown to reduce the protein-folding capacity of ER in stressed cells. This reduction in Ca^{2+} levels is linked to CHOP activation by *M.tb* infection. Another consequence of Ca^{2+} release is its entry in mitochondria and subsequent activation of oxidative stress in the form of ROS and NO production (Sah et al. 1999). Altogether these effects allow the stressed cells to undergo apoptosis (Chen et al. 2013; Lim et al. 2015). *M.tb* infection leads to an increase in Ca^{2+} results in ROS-mediated apoptosis. Activation of CHOP by *M.tb* infection initiates the disintegration of mitochondrial membrane leading to release Ca^{2+}. The release of cytochrome C from mitochondria then causes the apoptosis of cells. Calcium-regulated proteins will also get affected by this increase in Ca^{2+} levels. Calpain is a proteinase, which gets activated by Ca^{2+}, and induces apoptosis by activating caspase12 expression. ESAT-6 of *M.tb* has been shown to induce calpain activation, which subsequently leads to apoptosis in caspase12-dependent manner (Choi et al. 2010). Calreticulin, a glycoprotein chaperone of ER, is Ca^{2+} binding protein that preforms folding of unfolded proteins. The overexpression of calreticulin sensitizes the cells to undergo apoptosis (Ferri and Kroemer 2001).

14.3 Mycobacterial Proteins Involved in ER Stress

Mycobacterial pathogenesis is linked to ER stress in host cells (Cui et al. 2016). Macrophages infected with live *M.tb* $H_{37}Rv$ have increased levels of rough ER (Saquib et al. 2015). The ER stress-mediated pathway in *M.tb*-infected macrophages has proved to be a survival advantage to the pathogen (Lim et al. 2011). Although apoptosis has been thought to be a protective mechanism against the pathogen, it may not prove beneficial in later stages of *M.tb* infection. At the sites of TB granulomas, where an accumulation of apoptotic macrophages has been observed, an increased level of ER stress is evident (Seimon et al. 2010). During *M.tb* infection, this ER stress-mediated apoptosis may favor the mycobacteria to disseminate.

Few of the mycobacterial proteins listed in Table 14.1 have been identified to be involved in ER stress response and ER stress-mediated apoptosis (Fig. 14.1). The mode of action for generating ER stress responses by *M.tb* proteins are illustrated in Fig. 14.2.

Table 14.1 Mycobacterial proteins involved in ER stress-mediated cellular responses

Sl. No.	Mycobacterial protein	ER stress response	References
1	ESAT-6	eIF2a/ATF4/CHOP upregulation	Choi et al. (2010)
		Enhanced Ca^{2+} and ROS levels	
		Apoptosis	
2	HBHA	CHOP upregulation	Choi et al. (2013)
		Enhanced Ca^{2+} and ROS levels	
		Apoptosis	
3	38-kDa antigen	PERK/eIF2a/CHOP upregulation	Lim et al. (2013)
		Enhanced Ca^{2+} and ROS levels	
		Apoptosis	
		TLR-2/4 dependent	
4	PE_PGRS5	GRP78/GRP94 and CHOP/ATF4 upregulation	Grover et al. (2018)
		Enhanced Ca^{2+}, ROS, and NO levels	
		Apoptosis	
		TLR-4 dependent	

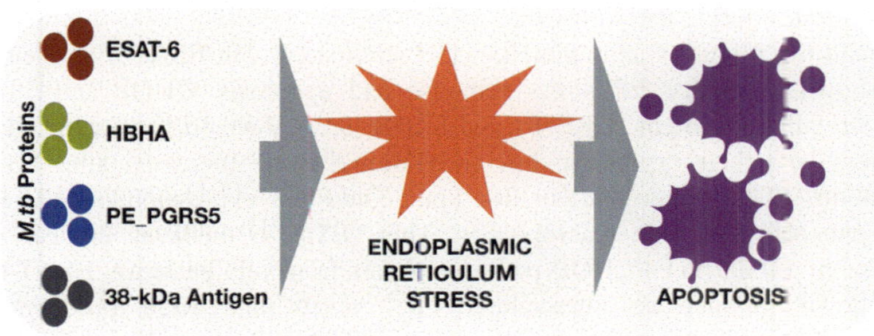

Fig. 14.1 Endoplasmic stress-mediated apoptosis by proteins of *Mycobacterium tuberculosis*

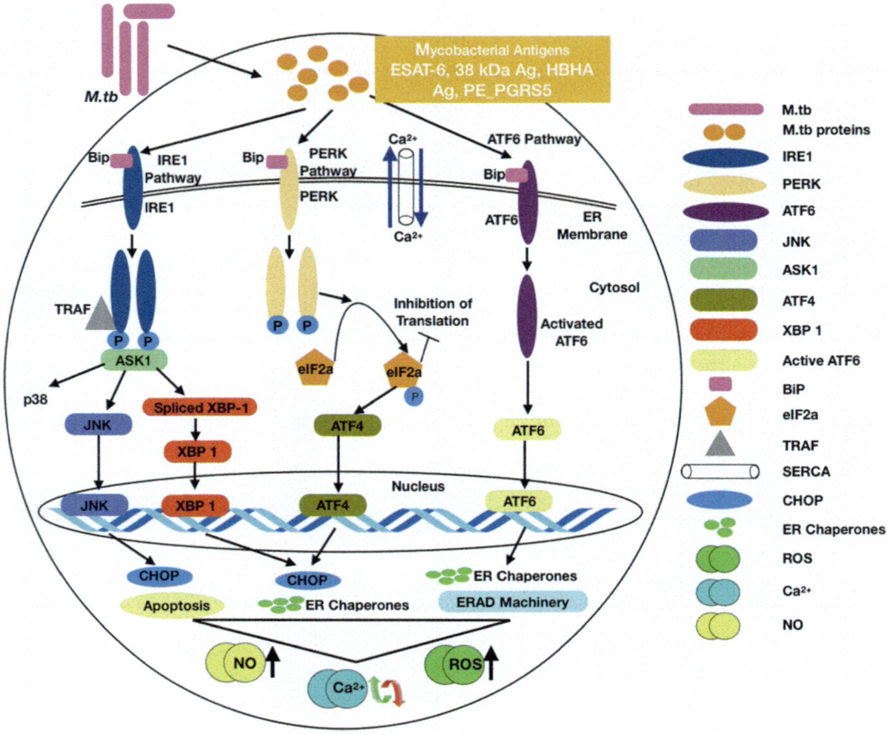

Fig. 14.2 *Mycobacterium tuberculosis* evokes endoplasmic reticulum stress-mediated response and apoptosis by ESAT-6, HBHA antigen, 38-kDa antigen, and PE_PGRS5 proteins

14.3.1 ESAT-6

ESAT-6 is an important virulence factor of *M.tb* involved in the early phases of TB pathogenesis. ESAT-6, an important candidate for vaccine development, is known to induce Th-1 immune responses (Wang et al. 2009). *M.tb* secretes ESAT-6 into culture filtrates during early growth.

Human macrophages and epithelial cells stimulated with ESAT-6 antigen have been shown to induce ER stress response and apoptosis. CHOP and 78-kDA glucose-regulated protein (GRP78) and CHOP are known to be associated with ER stress, and their expression has been known to be increased after ESAT-6 stimulation. ESAT-6 stimulation also upregulates the expression of ATF4 and eIF2a phosphorylation in macrophages. Thus, ESAT-6 mediates the ER stress response by eIF2a/ATF4/CHOP pathway. Apart from this pathway, ESAT-6 has also shown to increase intracellular Ca^{2+} levels and ROS production in macrophages. ER stress has been known to induce caspase12 activation for promoting apoptosis in mammalian cells. Macrophages undergo apoptosis in a caspase12-, caspase4-, and caspase3-dependent manner by mycobacterial ESAT-6 antigen. Also, ASK1-JNK pathway is involved in the ER stress response leading to apoptosis related to ER stress response. ESAT-6 antigen, secreted from *M.tb*, induces ER stress-induced apoptosis through the CHOP/ASK1/JNK-mediated pathway (Choi et al. 2010).

14.3.2 HBHA Antigen

HBHA antigen is one of virulence factor of *M.tb* involved in pathogenesis and infectivity. It is a 28 kDa, methylated surface protein (Parra et al. 2004). ROS and NO production by ER stressed cells are associated with apoptosis and other downstream stress responses (Gotoh and Mori 2006).

HBHA antigen has been reported to increase cytosolic Ca^{2+} levels with an enhanced ROS levels and promote ER stress-mediated apoptosis by inducing CHOP-dependent caspases (Sohn et al. 2011). HBHA also leads to CHOP-dependent expression of IL-6 and MCP-1 via Nf-kB activation. All these have shown the role of ER stress in HBHA antigen-mediated apoptosis (Choi et al. 2013).

14.3.3 38-kDa Antigen

One of the most important *M.tb* antigen is the 38-kDa phosphate transport protein (PstS-1) antigen, secreted by *M.tb* cultures (Harboe and Wiker 1992). The 38-kDa antigen is shown to elevate the expression of the pro-inflammatory cytokine MCP-1-induced protein (MCPIP). MCPIP subsequently increases ROS generation enhancing unfolded protein accumulation. Bip is then released from IRE1, PERK, and ATF6 and binds misfolded proteins to reduce ER stress. The 38-kDa antigen-induced CHOP production is reduced in case of JNK pathway inhibition. The p38 MAPK inhibition leads to activation of JNK and enhanced expression of CHOP by 38-kDa antigen stimulation. Suggested evidence points towards the role of 38-kDa antigen in PERK/eIF2a/CHOP pathway activation during ER stress response (Lim et al. 2015). The 38-kDa antigen provokes the ROS production and ER stress response by ERO1a, which normalizes Ca^{2+} levels by the Sarco-endoplasmic Reticulum Calcium ATPase (SERCA) inositol 1,4,5-triphate receptor (IP3R) (Ferri and Kroemer 2001).

14.3.4 PE_PGRS5

Ten percent *M.tb* genome codes for PE/PPE/PE_PGRS proteins exclusively present in pathogenic strains of mycobacteria (Akhter et al. 2012). Dissimilarities in

PE/PPE/PE_PGRS proteins from virulent $H_{37}Rv$ and avirulent $H_{37}Ra$ strains of *M.tb* are responsible for virulence competence of the strain (Kohli et al. 2012). One of the proteins of this family is PE_PGRS5 encoded by the Rv0297 gene, which has been shown by transcriptomic studies to be expressed in granulomas in later stages of infection (Kruh et al. 2010). The PGRS domain is highly disordered and includes putative ER retention signals, which assist in its localization to the ER. Rv0297PGRS leads to the activation of UPR and ER stress markers GRP78/GRP94 and CHOP/ATF4, and subsequent disruption in intracellular Ca^{2+} levels (Grover et al. 2018). Alteration of calcium levels leads to the production of ROS (Hasnain et al. 1999), and ROS generation is related to the UPR pathway. ROS production and oxidative stress are important components of the ER stress response. Rv0297PGRS has also been shown to increase ROS and NO levels in macrophages. The final ER stress response induced by PE_PGRS5 forces the cells to undergo apoptosis via caspase8 activation. The role of TLR-4 in ER stress-mediated apoptosis by Rv0297PGRS has been evident (Grover et al. 2018).

14.4 Conclusion

ER is a site for folding and transportation of synthesized proteins. It also serves as major site for storage of intracellular Ca^{2+}. Infectious diseases may induce ER stress response and aid in apoptosis of cells to clear the infection. In case of mycobacterial infections, ER stress response is mediated through several virulence factors of *M.tb*. Several of the *M.tb* proteins after interacting with ER stress sensor proteins of host cells initiate the cascade of events leading to the unfolded protein response. UPR serves to control or reduce the ER stress-related damage to maintain physiological functioning of host cells. Upon prolonged or uncontrolled ER stress, the response generated transforms into apoptosis, which subsequently helps the host to clear the infection. In early stages of mycobacterial infection, apoptosis may help in clearing the infection. But in later stages of infection, ER stress-mediated apoptosis may favor the *Mycobacterium* to disseminate the disease. In *M.tb*-infected fibrocaseous lung granulomas, apoptotic macrophages accumulate at sites of ER stress. Apoptosis in infected foamy macrophages leads to caseum accumulation and progression of disease pathogenesis. *M.tb*, a smart pathogen, modulates the host responses for its own benefit and employs the host machinery to transmit the disease. Thus, ER stress-mediated apoptosis may not always prove beneficial to the host and may rather aid in pathogen survival and dissemination of disease. There is a need to investigate *M.tb* proteins involved in the molecular mechanisms employed for mediating ER stress response or apoptosis to target the pathogen. The host partners involved in these survival strategies would also need to be elucidated in depth to provide insights for the development of host directed interventions against the disease.

References

Ahmad J, Farhana A, Pancsa R, Arora S, Srinivasan A, Tyagi A, Babu M, Ehtesham N, Hasnain S (2018) Contrasting function of structured N-terminal and unstructured C-terminal segments of *Mycobacterium tuberculosis* PPE37 protein. MBio 9(1):e01712-01717. https://doi.org/10.1128/mBio.01712-17

Akhter Y, Ehebauer M, Mukhopadhyay S, Hasnain S (2012) The PE/PPE multigene family codes for virulence factors and is a possible source of mycobacterial antigenic variation: perhaps more? Biochimie 94(1):110–116. https://doi.org/10.1016/j.biochi.2011.09.026

Araki K, Nagata K (2012) Protein folding and quality control in the ER. Cold Spring Harb Perspect Biol 4(8):a015438. https://doi.org/10.1101/cshperspect.a015438

Bertolotti A, Zhang Y, Hendershot LM, Harding HP, Ron D (2000) Dynamic interaction of BiP and ER stress transducers in the unfolded-protein response. Nat Cell Biol 2:326. https://doi.org/10.1038/35014014

Bhandary B, Marahatta A, Kim HR, Chae HJ (2013) An involvement of oxidative stress in endoplasmic reticulum stress and its associated diseases. Int J Mol Sci 14(1):434–456. https://doi.org/10.3390/ijms14010434

Bravo R, Vicencio JM, Parra V, Troncoso R, Munoz JP, Bui M, Quiroga C, Rodriguez AE, Verdejo HE, Ferreira J, Iglewski M, Chiong M, Simmen T, Zorzano A, Hill JA, Rothermel BA, Szabadkai G, Lavandero S (2011) Increased ER-mitochondrial coupling promotes mitochondrial respiration and bioenergetics during early phases of ER stress. J Cell Sci 124 (Pt 13):2143–2152. https://doi.org/10.1242/jcs.080762

Brodin P, Poquet Y, Levillain F, Peguillet I, Larrouy-Maumus G, Gilleron M, Ewann F, Christophe T, Fenistein D, Jang J, Jang MS, Park SJ, Rauzier J, Carralot JP, Shrimpton R, Genovesio A, Gonzalo-Asensio JA, Puzo G, Martin C, Brosch R, Stewart GR, Gicquel B, Neyrolles O (2010) High content phenotypic cell-based visual screen identifies *Mycobacterium tuberculosis* acyltrehalose-containing glycolipids involved in phagosome remodeling. PLoS Pathog 6(9):e1001100. https://doi.org/10.1371/journal.ppat.1001100

Chen Y, Lewis W, Diwan A, Cheng EH, Matkovich SJ, Dorn GW 2nd (2010) Dual autonomous mitochondrial cell death pathways are activated by Nix/BNip3L and induce cardiomyopathy. Proc Natl Acad Sci U S A 107(20):9035–9042. https://doi.org/10.1073/pnas.0914013107

Chen T, Zhao Q, Li W, Xie J (2013) *Mycobacterium tuberculosis* PE_PGRS17 promotes the death of host cell and cytokines secretion via Erk kinase accompanying with enhanced survival of recombinant Mycobacterium smegmatis. J Interf Cytokine Res 33(8):452–458. https://doi.org/10.1089/jir.2012.0083

Choi HH, Shin DM, Kang G, Kim KH, Park JB, Hur GM, Lee HM, Lim YJ, Park JK, Jo EK, Song CH (2010) Endoplasmic reticulum stress response is involved in *Mycobacterium tuberculosis* protein ESAT-6-mediated apoptosis. FEBS Lett 584(11):2445–2454. https://doi.org/10.1016/j.febslet.2010.04.050

Choi JA, Lim YJ, Cho SN, Lee JH, Jeong JA, Kim EJ, Park JB, Kim SH, Park HS, Kim HJ, Song H (2013) Mycobacterial HBHA induces endoplasmic reticulum stress-mediated apoptosis through the generation of reactive oxygen species and cytosolic Ca2+ in murine macrophage RAW 264.7 cells. Cell Death Dis 4. https://doi.org/10.1038/cddis.2013.489

Cui Y, Zhao D, Barrow PA, Zhou X (2016) The endoplasmic reticulum stress response: a link with tuberculosis? Tuberculosis 97:52–56. https://doi.org/10.1016/j.tube.2015.12.009

Ferri KF, Kroemer G (2001) Organelle-specific initiation of cell death pathways. Nat Cell Biol 3 (11):E255–E263. https://doi.org/10.1038/ncb1101-e255

Gotoh T, Mori M (2006) Nitric oxide and endoplasmic reticulum stress. Arterioscler Thromb Vasc Biol 26(7):1439–1446. https://doi.org/10.1161/01.ATV.0000223900.67024.15

Grover S, Sharma T, Singh Y, Kohli S, P M, Singh A, Semmler T, Wieler LH, Tedin K, Ehtesham NZ, Hasnain SE (2018) The PGRS domain of *Mycobacterium tuberculosis* PE_PGRS protein Rv0297 is involved in endoplasmic reticulum stress-mediated apoptosis through toll-like receptor 4. MBio 9(3). https://doi.org/10.1128/mBio.01017-18

Harboe M, Wiker HG (1992) The 38-kDa protein of *Mycobacterium tuberculosis*: a review. J Infect Dis 166(4):874–884

Harding HP, Zhang Y, Ron D (1999) Protein translation and folding are coupled by an endoplasmic-reticulum-resident kinase. Nature 397:271. https://doi.org/10.1038/16729

Hasnain SE, Taneja TK, Sah NK, Mohan M, Pathak N, Sahdev S, Athar M, Totey SM, Begum R (1999) Invitro culturedSpodoptera frugiperda insect cells: model for oxidative stress-induced apoptosis. J Biosci 24(1):13–19. https://doi.org/10.1007/bf02941101

Hassler J, Cao Stewart S, Kaufman Randal J (2012) IRE1, a double-edged sword in pre-miRNA slicing and cell death. Dev Cell 23(5):921–923. https://doi.org/10.1016/j.devcel.2012.10.025

Kohli S, Singh Y, Sharma K, Mittal A, Ehtesham NZ, Hasnain SE (2012) Comparative genomic and proteomic analyses of PE/PPE multigene family of *Mycobacterium tuberculosis* H(3)(7)Rv and H(3)(7)Ra reveal novel and interesting differences with implications in virulence. Nucleic Acids Res 40(15):7113–7122. https://doi.org/10.1093/nar/gks465

Kruh NA, Troudt J, Izzo A, Prenni J, Dobos KM (2010) Portrait of a pathogen: the *Mycobacterium tuberculosis* proteome in vivo. PLoS One 5(11). https://doi.org/10.1371/journal.pone.0013938

Lim Y-J, Choi J-A, Choi H-H, Cho S-N, Kim H-J, Jo E-K, Park J-K, Song C-H (2011) Endoplasmic reticulum stress pathway-mediated apoptosis in macrophages contributes to the survival of *Mycobacterium tuberculosis*. PLoS One 6(12):e28531. https://doi.org/10.1371/journal.pone.0028531

Lim Y-J, Choi H-H, Choi J-A, Jeong JA, Cho S-N, Lee J-H, Park JB, Kim H-J, Song C-H (2013) Mycobacterium kansasii-induced death of murine macrophages involves endoplasmic reticulum stress responses mediated by reactive oxygen species generation or calpain activation. Apoptosis 18(2):150–159. https://doi.org/10.1007/s10495-012-0792-4

Lim Y-J, Choi J-A, Lee J-H, Choi CH, Kim H-J, Song C-H (2015) *Mycobacterium tuberculosis* 38-kDa antigen induces endoplasmic reticulum stress-mediated apoptosis via toll-like receptor 2/4. Apoptosis 20(3):358–370. https://doi.org/10.1007/s10495-014-1080-2

Liu L, Wise DR, Diehl JA, Simon MC (2008) Hypoxic reactive oxygen species regulate the integrated stress response and cell survival. J Biol Chem 283(45):31153–31162. https://doi.org/10.1074/jbc.M805056200

Martinon F, Chen X, Lee A-H, Glimcher LH (2010) TLR activation of the transcription factor XBP1 regulates innate immune responses in macrophages. Nat Immunol 11:411. https://doi.org/10.1038/ni.1857. https://www.nature.com/articles/ni.1857#supplementary-information

Moserova I, Kralova J (2012) Role of ER stress response in photodynamic therapy: ROS generated in different subcellular compartments trigger diverse cell death pathways. PLoS One 7(3): e32972. https://doi.org/10.1371/journal.pone.0032972

Murphy MP (2009) How mitochondria produce reactive oxygen species. Biochem J 417:1–13. https://doi.org/10.1042/BJ20081386

Ortiz C, Cardemil L (2001) Heat-shock responses in two leguminous plants: a comparative study. J Exp Bot 52(361):1711–1719. https://doi.org/10.1093/jexbot/52.361.1711

Parra M, Pickett T, Delogu G, Dheenadhayalan V, Debrie AS, Locht C, Brennan MJ (2004) The mycobacterial heparin-binding hemagglutinin is a protective antigen in the mouse aerosol challenge model of tuberculosis. Infect Immun 72(12):6799–6805. https://doi.org/10.1128/IAI.72.12.6799-6805.2004

Raturi A, Simmen T (2013) Where the endoplasmic reticulum and the mitochondrion tie the knot: the mitochondria-associated membrane (MAM). Biochim Biophys Acta 1833(1):213–224. https://doi.org/10.1016/j.bbamcr.2012.04.013

Ron D, Hubbard SR (2008) How IRE1 reacts to ER stress. Cell 132(1):24–26. https://doi.org/10.1016/j.cell.2007.12.017

Sah NK, Taneja TK, Pathak N, Begum R, Athar M, Hasnain SE (1999) The baculovirus antiapoptotic p35 gene also functions via an oxidant-dependent pathway. Proc Natl Acad Sci U S A 96(9):4838–4843

Sano R, Reed JC (2013) ER stress-induced cell death mechanisms. BBA-Mol Cell Res 1833 (12):3460–3470. https://doi.org/10.1016/j.bbamcr.2013.06.028

Santos CXC, Tanaka LY, Wosniak J, Laurindo FRM (2009) Mechanisms and implications of reactive oxygen species generation during the unfolded protein response: roles of endoplasmic reticulum Oxidoreductases, mitochondrial Electron transport, and NADPH oxidase. Antioxid Redox Signal 11(10):2409–2427. https://doi.org/10.1089/ars.2009.2625

Saquib NM, Jamwal S, Midha MK, Verma HN, Manivel V (2015) Quantitative proteomics and lipidomics analysis of endoplasmic reticulum of macrophage infected with *Mycobacterium tuberculosis*. Int J Proteomics 2015:270438. https://doi.org/10.1155/2015/270438

Seimon TA, Kim MJ, Blumenthal A, Koo J, Ehrt S, Wainwright H, Bekker LG, Kaplan G, Nathan C, Tabas I, Russell DG (2010) Induction of ER stress in macrophages of tuberculosis granulomas. PLoS One 5(9). https://doi.org/10.1371/journal.pone.0012772

Shen J, Chen X, Hendershot L, Prywes R (2002) ER stress regulation of ATF6 localization by dissociation of BiP/GRP78 binding and unmasking of Golgi localization signals. Dev Cell 3 (1):99–111. https://doi.org/10.1016/S1534-5807(02)00203-4

Smith MH, Ploegh HL, Weissman JS (2011) Road to ruin: targeting proteins for degradation in the endoplasmic reticulum. Science 334(6059):1086

Sohn H, Kim JS, Shin SJ, Kim K, Won CJ, Kim WS, Min KN, Choi HG, Lee JC, Park JK, Kim HJ (2011) Targeting of *Mycobacterium tuberculosis* heparin-binding hemagglutinin to mitochondria in macrophages. PLoS Pathog 7(12). https://doi.org/10.1371/journal.ppat.1002435

Srinivasan L, Ahlbrand S, Briken V (2014) Interaction of *Mycobacterium tuberculosis* with host cell death pathways. Cold Spring Harb Perspect Med 4:a022459

Tabas I, Ron D (2011) Integrating the mechanisms of apoptosis induced by endoplasmic reticulum stress. Nat Cell Biol 13(3):184–190. https://doi.org/10.1038/ncb0311-184

Wang Y, Shen J, Arenzana N, Tirasophon W, Kaufman RJ, Prywes R (2000) Activation of ATF6 and an ATF6 DNA binding site by the endoplasmic reticulum stress response. J Biol Chem 275 (35):27013–27020. https://doi.org/10.1074/jbc.M003322200

Wang X, Barnes PF, Dobos-Elder KM, Townsend JC, Chung YT, Shams H, Weis SE, Samten B (2009) ESAT-6 inhibits production of IFN-gamma by *Mycobacterium tuberculosis*-responsive human T cells. J Immunol 182(6):3668–3677. https://doi.org/10.4049/jimmunol.0803579

WHO (2017) Global tuberculosis report 2017. World Health Organization

Yoon H, Kim DS, Lee GH, Kim KW, Kim HR, Chae HJ (2011) Apoptosis induced by manganese on neuronal SK-N-MC cell line: endoplasmic reticulum (ER) stress and mitochondria dysfunction. Environ Health Toxicol 26:e2011017. https://doi.org/10.5620/eht.2011.26.e2011017

Yoshida H (2007) ER stress and diseases. FEBS J 274(3):630–658. https://doi.org/10.1111/j.1742-4658.2007.05639.x

Zhang Y, Liu R, Ni M, Gill P, Lee AS (2010) Cell surface relocalization of the endoplasmic reticulum chaperone and unfolded protein response regulator GRP78/BiP. J Biol Chem 285 (20):15065–15075. https://doi.org/10.1074/jbc.M109.087445

Toxin-Antitoxin (TA) Systems in Stress Survival and Pathogenesis

15

Ashutosh Kumar, Anwar Alam, Pranami Bharadwaj, Sharmistha Tapadar, Mamta Rani, and Seyed E. Hasnain

Abstract

Mycobacterium tuberculosis (*M.tb*), by virtue of its ability to evolve, has developed mechanisms that enable it to modulate its growth through regulation of replication, transcription, translation, generation of heterogeneous population of persister cells, etc. for survival in different stressful environment during its infection cycle. Toxin-antitoxin (TA) systems are ubiquitous in prokaryotic genomes that enable them to survive in various unfavourable conditions. A toxin protein may inhibit the growth, whereas an antitoxin may neutralize the effect of toxin in different ways. TA systems are involved in stress adaptation, antimicrobial tolerance or resistance, modification in the physiological state of organisms, biofilms formation, growth regulation for survival, plasmid maintenance, anti-phage activities, virulence, and programmed cell death.

Ashutosh Kumar and Anwar Alam have equally contributed to this chapter.

A. Kumar · P. Bharadwaj · S. Tapadar
Department of Microbiology, Tripura University (A Central University), Agartala, Tripura, India

A. Alam
JH Institute of Molecular Medicine, Jamia Hamdard, New Delhi, Delhi, India

Inflammation Biology and Cell Signaling Laboratory, ICMR-National Institute of Pathology, Safdarjung Hospital Campus, New Delhi, Delhi, India

M. Rani
Plant Microbe Interactions Laboratory, National Institute of Plant Genome Research, New Delhi, Delhi, India

S. E. Hasnain (✉)
JH Institute of Molecular Medicine, Jamia Hamdard, New Delhi, Delhi, India

Dr Reddy's Institute of Life Sciences, University of Hyderabad Campus, Hyderabad, India
e-mail: vc@jamiahamdard.ac.in

S. E. Hasnain et al. (eds.), Mycobacterium tuberculosis: *Molecular Infection Biology, Pathogenesis, Diagnostics and New Interventions*,
https://doi.org/10.1007/978-981-32-9413-4_15

Environmental microorganisms express a wider repertoire of TA systems as compared to intracellular human pathogens due to a higher probability to encounter different environmental stresses within their ecosystem. However, the presence of high level of TA systems in *M.tb* is due to the fact that *M.tb* has to endure several types of stresses including acidic, hypoxic, oxidative, and immune surveillance within the host for its survival. TA systems are also present in pathogenic bacteria infecting plants. Based on the mechanism of action, different types of TA systems are classified within the microorganisms. Recently, genes related to type II TA systems have been proposed to be useful in genotyping of tuberculosis caused by different strains of *M.tb*.

Keywords
Toxin-antitoxin systems · *Mycobacterium tuberculosis* · Stress survival · Growth regulation · Drug tolerance

Abbreviations

ATP	Adenosine triphosphate
DATIN	Dormancy-associated translation inhibitor
DNA	Deoxyribonucleic acid
DR	Direct repeat
E. coli	*Escherichia coli*
IS	Insertion sequence
M.tb	*Mycobacterium tuberculosis*
mRNA	Messenger RNA
MTBC	*M.tb* complex
PCD	Programmed cell death
PSK	Post-segregational killing
SNPs	Single-nucleotide polymorphisms
sRNA	Small regulatory RNAs
TA	Toxin-antitoxin
TAC	Toxin-antitoxin-chaperone
Vap	Virulence-associated protein
VNTR	Variable number tandem repeats

15.1 Introduction

Toxin-antitoxin (TA) systems in bacteria were first recognized as plasmid-borne loci which help in plasmid maintenance through elimination of daughter cells lacking TA encoding plasmid (Guglielmini and Melderen 2011). A set of linked genes, together encoding a protein ‘poison’ and a corresponding ‘antidote’, forms the TA system (Gerdes 2000). The TA systems present on plasmids make sure that only the daughter cells inheriting the plasmid survive after cell division. In daughter cells devoid of plasmid, unstable antitoxin is degraded while the stable toxic protein kills the new cell, and this phenomenon is known as 'post-segregational killing' (PSK). TA systems are present in multiple copies in prokaryotes (Yamaguchi et al. 2011). Various microbial genome analyses have comprehensively highlighted the diversity in the distribution of TA systems.

Previous studies have shown that the genomes of *Nitrosomonas europaea*, *Sinorhizobium meliloti*, and *Mycobacterium bovis* contain more than 50 presumptive TA systems, whereas *Rickettsia prowazekii, Bacillus subtilis, Campylobacter jejuni*, etc. contain no or very few TA systems (Pandey and Gerdes 2005; Sevin and Barloy-Hubler 2007). However, there is little consensus to prove correlation between the number of TA systems and the growth rate of the members within a phylum. Additionally, diversity in the distribution of TA systems among different isolates of the same species is also observed.

Detailed study of phylogenetic patterns of TA loci in several prokaryotic genomes suggests presence of multiple TA loci in free-living prokaryotes and few or no TA loci in obligate intracellular prokaryotes (Pandey and Gerdes 2005). TA loci are beneficial to organisms that confront stressful environment. TA systems, also referred to as junk, are considered to be constituents of plasmids and have been retained within the cells due to their addictive nature (Kroll et al. 2010) Some toxins act as general repressors of gene expression, while others are more specific in regulating gene expression (Engelberg-Kulka et al. 2006; Pimentel et al. 2005). Some TA systems that act as bacteriostatic toxins play a key role in growth regulation and may restrict growth rather than kill the host cell (Diago-Navarro et al. 2010). 'Persisters' are slow-growing population of cells, which can survive stress and later grow into actively dividing cells when the environment is favourable (Kussell et al. 2005). It has been demonstrated that an imbalance between the level of toxin and its antitoxin due to overexpression or mutations in either of them results in high persistence (Fridman et al. 2014; Korch and Hill 2006). Interruption of transcription and translation machinery of host cells due to bacteriophage may also trigger activation of TA system that in turn limits phage replication, termed as antiphage mechanism (Hazan and Engelberg-Kulka 2004).

M.tb faces different stresses in its pathogenesis and possesses several proteins such as two component systems, sigma factors, TA systems, acid response, halophilic proteins, etc. for its survival (Kumar et al. 2018). As compared to other mycobacteria, *M.tb* shows presence of a significant number of TA systems in its genome, which during the state of persistence are induced by active toxins, which may largely contribute towards its pathogenesis (Ramage et al. 2009). Transcriptomic analyses of antibiotic-induced *M.tb* persisters showed that about 10 TA systems were significantly upregulated, pointing to the importance of TA system in *M.tb* persistence (Keren et al. 2011). Interestingly, *M.tb* possesses abundant number of TA loci, while *M. leprae* has none, possibly due to the fact that *M. leprae* has evolved from *M.tb* through reductive evolution (Ramage et al. 2009; Cole et al. 2001). The presence of TA system provides an evolutionary edge to *M.tb* in terms of aiding its survival in both extra- and intracellular conditions as compared to *M. leprae* which can survive only as obligate intracellular pathogen. Similarly, obligate intracellular microorganisms such as *Rickettsia* and *Buchnera spp*. have either very few or no TA loci, although several exceptions are also in existence (Pandey and Gerdes 2005; Leplae et al. 2011). Analyses of bacterial gene sequences by Shao et al. (Shao et al. 2011) conclusively point to the presence of TA systems in symbiotic bacteria and overrules previous studies by Pandey et al. that reported absence of TA systems in symbiotic bacteria (Pandey and Gerdes 2005; Shao et al. 2011).

15.2 Classes of Toxin-Antitoxin Systems

The TA loci are classified into different groups such as vapBC, parDE, relBE, ccd, phd/doc, mazEF, and higBA due of differences in mechanism of action (Gerdes et al. 2005). In *M.tb*, majority of TA systems belong to the class of VapBC (virulence-associated protein) (Gerdes and Maisonneuve 2012). VapC induces dormancy by suppressing translation and induction of vapB transcription which later leads to revival of cells (Winther and Gerdes 2009). It has been shown recently by deletion and overexpression studies that some members of the VapBC TA systems of *M.tb* are involved in bacteriostasis, morphological changes, growth arrest, and mycobacterial pathogenesis (Agarwal et al. 2018). *M.tb* VapBC30 system has been shown to be involved in growth regulation through ribonuclease activity (Deep et al. 2018). *M. tb* toxin VapC30 inhibits the growth of *Escherichia coli (E. coli)* when expressed without its cognate antitoxin VapB30. There is no effect of *M.tb* VapC30 on *E. coli* when co-expressed with *M.tb* VapB30. *M.tb* VapC30 degrades RNA molecules that are magnesium and manganese ion dependent (Lee et al. 2015).

M.tb MazF toxin members, when expressed in *E. coli* or *Mycobacterium smegmatis,* affect their growth (Gupta 2009; Zhu et al. 2006). The overexpression of MazF3, MazF6, and MazF9 of *M.tb* in *Mycobacterium bovis BCG* induces bacteriostasis (Tiwari et al. 2015). *M.tb* MazF toxins are also involved in the drug tolerance, virulence, and stress adaptation (Tiwari et al. 2015).

The RelBE system is among the most characterized TA systems that bind with the A site of the ribosome and affect protein synthesis by cleaving the mRNAs preferably between the second and third nucleotides of the termination codon (Pedersen 2003). In contrast, RelE binds to initial coding region and cleaves the first 100 codons of mRNA and inhibits growth (Hurley et al. 2011). Variety of stress conditions such as nitrosative stress, oxidative stress, and antibiotic stress affect the transcript profiles of RelE toxins of *M.tb*. The overexpression of toxin RelE of *M.tb* affected growth of the *E. coli* and *M.tb*. The three RelE toxins of *M.tb* are involved in individual antibiotic specific tolerance (Singh et al. 2010).

The YefM antitoxin is highly unstable as it is prone to degradation by Lon, an ATP-dependent serine protease. YefM is co-expressed with the YoeB toxin and the resultant complex so formed consists of dimer of YefM and single molecule of YoeB (Kamada and Hanaoka 2005). Rv3357–Rv3358 of *M.tb* codes for YefM/YoeB system.

The HigBA family, HigB toxin (Rv1955) and HigA antitoxin (Rv1956), are part of the operon comprising of Rv1954A and Rv1957. Rv1957 is found as a SecB-like chaperone required for antitoxin stabilization. HigB inhibits protein synthesis by cleaving mRNAs that are being translated in *E. coli* (Smollett et al. 2009; Fivian-Hughes and Davis 2010; Christensen-Dalsgaard et al. 2010; Bordes et al. 2011).

The tripartite toxin-antitoxin-chaperone system (TAC) complex is induced during heat shock, hypoxia, nutrient starvation, and persistence. Within the TAC complex, the chaperone directly binds to HigA antitoxin and prevents it aggregation or degradation, thereby aiding in HigA folding and successive interaction with HigB toxin (Bordes et al. 2011).

M.tbH_{37}Rv also possesses two ParDE systems, and ParDE2 operon has been investigated recently and was found that toxin MParE2 interacts with GyrB subunit and inhibits bacterial growth by inhibiting DNA gyrase, thereby blocking DNA replication (Gupta et al. 2016).

15.3 Types of Toxin-Antitoxin Systems

TA system is characterized by neutralization of toxin by the antitoxin (Fig. 15.1). In case of a type I TA system, translation of messenger RNA encoding the toxin is inhibited by binding of a non-coding RNA antitoxin to the mRNA. The protein toxin in case of type II TA system is inhibited post-translationally through binding of another protein antitoxin. In case of type III TA systems, a small RNA binds directly to the toxin protein. Type IV-VI TA systems are also reported but are relatively less common. Toxin-antitoxin genes, transferred predominantly through horizontal gene transfer, are mostly associated with pathogenic bacteria and with plasmids conferring antibiotic resistance or virulence (Mine et al. 2009; Van Melderen and Saavedra De Bast 2009).

15.3.1 Type I TA System

In this type of TA system, antitoxins consist of small regulatory RNAs (sRNA) comprising of 50–200 nucleotides. A non-coding RNA antitoxin complementarily binds to the toxin-encoding mRNA resulting in mRNA degradation or inhibition of toxin translation (Brielle et al. 2016; Brantl and Jahn 2015). Type I toxin and antitoxins are transcribed from their own promoter, while in other types of TA systems, they are part of operon with other genes. The translation of the type I toxin mRNA is inhibited by base pairing with the antitoxin sRNA that prevents interaction with ribosome resulting in inhibition of translation of toxin mRNA (Brantl 2012; Fozo et al. 2008). Most of the type I toxins are small hydrophobic proteins that create pores in the inner membrane leading to breakdown of membrane potential to stop ATP synthesis, thereby blocking energy-demanding activities such as protein synthesis (Wen et al. 2014; Lee and Lee 2016). *sym*R/*sym*E module of *E. coli* is considered as an example of type I TA system (Kawano et al. 2007).

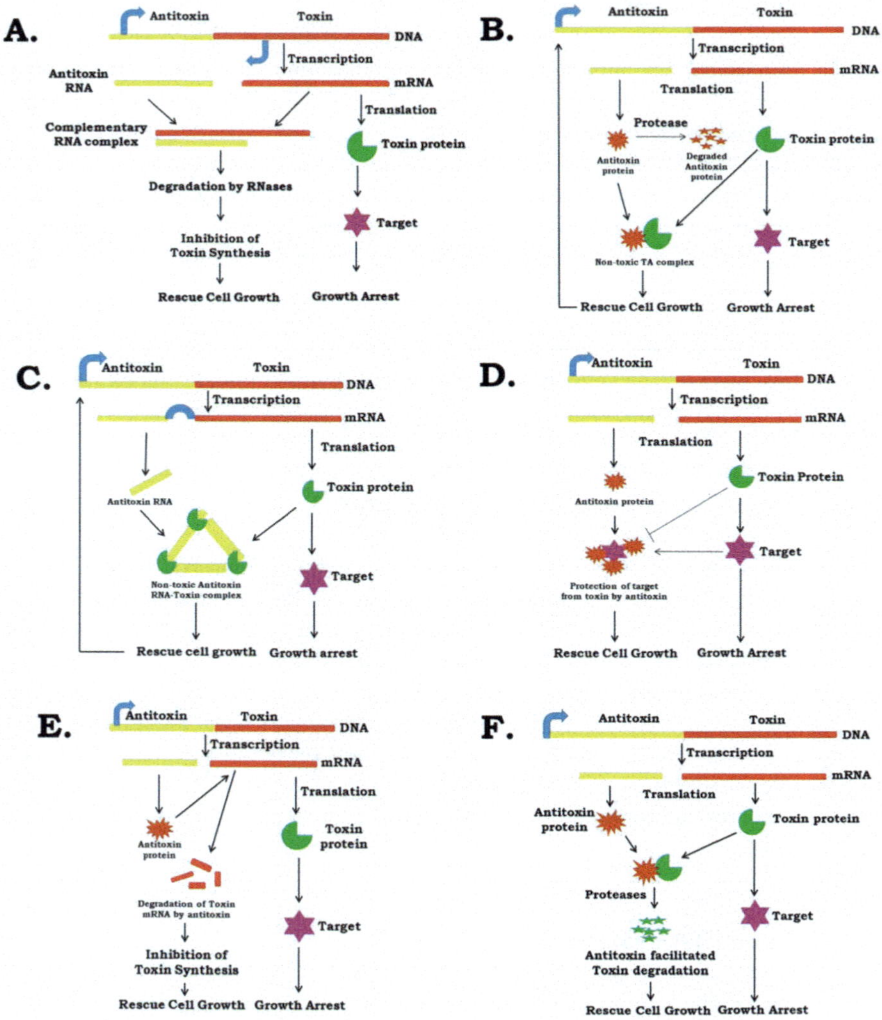

Fig. 15.1 Types of toxin-antitoxin systems: Toxin proteins inhibit the growth of cells and there are different mechanisms through which antitoxins neutralize the effect of toxins: (**a**) Suppression in translation of toxins by complementary RNA complex formation between toxin and antitoxin mRNA. (**b**) Non-functional toxin-antitoxin protein complex neutralizes the effect of toxins on growth. (**c**) Non-functional complex formation due to interaction of antitoxin mRNA with toxin proteins. (**d**) Interaction of antitoxin with target proteins stops toxin activities. (**e**) Antitoxin protein degrades mRNA of toxin resulting in suppression of toxin protein synthesis. (**f**) Antitoxins facilitate degradation of toxins by proteinases resulting in rescue of cell growth

15.3.2 Type II TA System

Type II TA systems are most extensively characterized in both prokaryotes and archaea. In type II TA system, the functional activity of toxin proteins is inhibited due to interaction between stable toxin and labile antitoxin. Toxin and antitoxin proteins are expressed simultaneously as both are organized into operons. The activity of different toxin protein regulates various mechanisms within the cell. For example, CcdB protein of *E. coli* (strain K12) inhibit the function of DNA gyrase by inactivating DNA topoisomerase II, whereas MazF cleaves cellular mRNAs for the inhibition of protein synthesis (Bernard and Couturier 1992; Zhang et al. 2003). The TA complex acts as a repressor for TA operon system as it binds to the palindromic sequence of the promoter region. Due to antitoxin degradation, the concentration of TA complex reduces thereby leading to production of more toxins and antitoxin.

15.3.3 Type III TA System

Type III TA system is characterized by direct interaction between a toxic protein and RNA antitoxin. The RNA gene involved totally neutralizes the effects of the toxic protein. ToxI-ToxN TA system from a plant pathogen named *Erwinia carotovora* subspecies *atrosepticum* (*Pectobacterium atrosepticum*) is a perfect example of type III TA system. The function of toxin protein (ToxN) is directly suppressed by interaction with antitoxin RNA, forming aToxN-RNA (an RNA antitoxin) complex (Fineran et al. 2009; Blower et al. 2012).

15.3.4 Type IV TA System

In type IV TA system, there is no direct interaction between toxin and antitoxin proteins. Antitoxin protein interacts with the target of toxin protein, thereby suppressing the activity of toxin on its target. The functional aspect of type IV TA system is exemplified in case of toxin CbeA of *E. coli* (1303) that prevents polymerization of the cytoskeletal proteins (MreB and FtsZ) and inhibits cell division. The antitoxin CbeA (YeeU) protein inhibits by binding directly with the target, namely, MreB and FtsZ toxin, instead of forming a toxin-antitoxin complex (Masuda et al. 2012).

15.3.5 Type V TA System

In the type V TA system, antitoxin protein retards synthesis of toxin protein by degrading the mRNA transcribed to code toxin protein. This is exemplified by GhoT of *E. coli* (strain K12), a protein that induces persistence and ghost cell formation with damaged membrane. The antitoxin GhoS is an endoribonuclease and precisely cleaves the mRNA encoding for membrane-lytic peptide toxin GhoT (Wang et al. 2013).

15.3.6 Type VI TA System

In the type VI TA system, antitoxin protein facilitates the proteolytic degradation of toxin protein. There is no degradation of toxin protein by proteases in the absence of antitoxin, and if toxin-antitoxin proteins are together, toxin degradation occurs. In SocAB TA system in *Caulobacter crescentus*, SocB protein inhibits DNA elongation by intercalating with DnaN, whereas antitoxin SocA facilitates degradation of SocB in the presence of protease ClpXP (Aakre et al. 2013).

15.4 Biological Roles of Toxin-Antitoxin Systems

TA systems regulate bacterial survival in different types of unfavourable conditions. TA systems are involved in several biological functions such as growth regulation, physiological changes of the cells, programmed cell death, etc. (Fig. 15.2). A more detailed importance of the TA systems in various physiological conditions is described below:

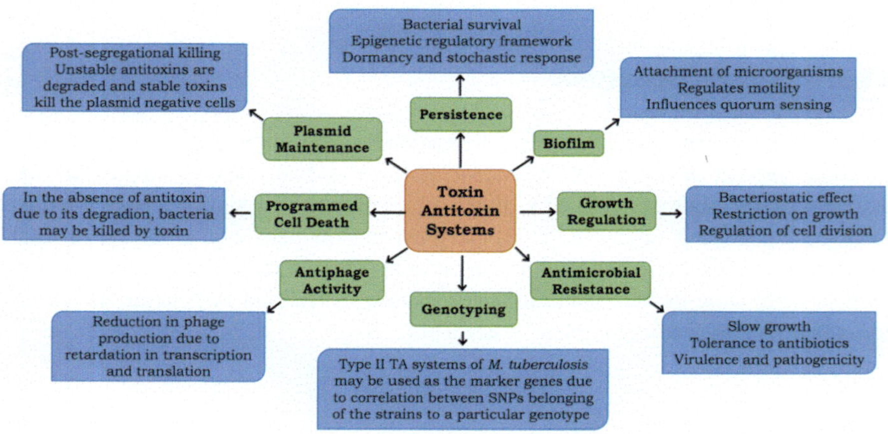

Fig. 15.2 Roles and applications of toxin-antitoxin systems: Toxin-antitoxin systems play important roles such as plasmid maintenance, drug resistance/tolerance, biofilm formation, growth regulation, antiphage activity, genotyping, programmed cell death, etc.

15.4.1 Stress Survival

TA systems act on regulatory machinery that control mechanisms critical for survival of bacteria (Engelberg-Kulka et al. 2006; Yamaguchi et al. 2011). For example, TA systems are important for adaptation of *M.tb* to unfavourable environmental conditions inside the host and maybe required for triggering a non-replicating state (Lewis 2007). Mycobacteria often adopt a non-replicative persistent, inactive state, to avoid unfavourable stress conditions like hypoxia, oxidative stress, nutritional limitations, acidic pH, etc., within the host macrophages (Wu et al. 2012). There are 88 putative TA systems present in *M.tb*H_{37}Rv that are conserved in *M.tb* complex (MTBC) but are few in other non-pathogenic mycobacteria, indicating a potential contribution to the pathogenic lifestyle of *M.tb* (Ramage et al. 2009). In response to starvation, several toxins within the cells are upregulated leading to inhibition of translation and selective degradation of mRNA. Contrary to starvation which leads to generalized upregulation of TA loci, low pH exposure leads to downregulation of few TA genes (Gupta et al. 2017).

15.4.2 Persistence

In majority of bacteria, there are set of genes that are involved in growth inhibition, and their overexpression may result in cell death, similar to programmed cell death of eukaryotes (Wen et al. 2014). It has been reported that TA systems are important for persister phenotype in *E. coli* (Tsilibaris et al. 2007). TA systems present in *M.tb* may regulate cell division during infection (Warner and Mizrahi 2006). After infection, *M.tb* initially grows and then acquires latency, a state of non-replicating cells in unfavourable conditions that can survive long periods with the potential to reactivate itself later whenever environment is favourable (Stewart et al. 2003, North and Jung 2004). In the case of persistence, a subpopulation of non-replicating bacteria becomes tolerant to antibiotics (Gomez and McKinney 2004). The presence of TA system in *M.tb*, as in other bacteria, imparts antibiotic tolerance and is one of the main reasons why long term of antibiotic therapy is required to cure tuberculosis (Keren et al. 2004; Ramage et al. 2009). It has been reported that mycobacterial MazF ribonucleases are involved in the drug tolerance, adaptation in oxidative stress, and nutrient depletion and virulence (Tiwari et al. 2015). *M.tb* RelE toxins are involved in formation of persisters specific to individual antibiotic and involved in drug-specific tolerance (Singh et al. 2010).

15.4.3 Biofilms

Planktonic bacteria can aggregate and attach themselves on biotic or abiotic surfaces to form biofilms. Formation of biofilms by pathogens is considered to be one of the main survival strategies to counter the host defence (Rybtke et al. 2011; Kumar et al. 2017a). TA systems have been shown to be involved in biofilm formation, with exceptions. It has been shown that type II TA system of *E. coli* involving MqsR protein is induced during biofilm formation and deletion of this gene resulted in the absence of biofilm formation (Kasari et al. 2010). The role of m*qsRA* TA system in biofilm formation is associated with motility and as autoinducer-2 quorum sensing system (Gonzalez Barrios et al. 2006). The yefM-yoeB and relBE TA systems of *Streptococcus pneumonia* are involved in the biofilm formation. It has been reported that mutant strains lacking yefM-yoeB or both yefM-yoeB and relBE show reduction in biofilm formation (Chan et al. 2018). It has also been shown that deletion of mutants of *mazF* and *relE* gene homologue did not affect biofilm formation in *Streptococcus mutans* (Lemos et al. 2005).

15.4.4 Antiphage Activity

The competitive edge of pathogen against the host is driven by the enormous diversity and multiplicity of TA systems. Bacteria have evolved several defence mechanisms to protect themselves and survive against the onslaught of phage infection. Bacteria exhibit a wide array of mechanisms to resist bacteriophages that include inhibition of adsorption, exclusion of superinfection, cleavage of nucleic acids of the phages through restriction-modification or CRISPR-Cas systems, and abortive infection (Stern and Sorek 2012; Seed 2015). The abortive infection systems trigger premature death of phage-infected bacteria and restrict the phage to replicate or spread, thereby protecting the uninfected bacterial population within the niche. Thus, there is a link between abortive infection and TA systems. Several TA systems have anti-phage activity including Hok-Sok, LsoAB and MazEF, ToxIN, and AbiEG (Short et al. 2018).

15.4.5 Bacterial Virulence and Pathogenecity

Presence of TA modules in the genome is directly related to the virulence of bacteria (Georgiades and Raoult 2011), for example, type I toxins are involved in lysis of host cell. The PepA1 toxin of *S. aureus* is a pore-forming peptide that causes bacterial cell death. When there is oxidative stress inside host cell, the PepA1 toxin is typically released from its SprA antitoxin. This is an example of altruistic behaviour, as the peptide drives erythrocyte lysis, resulting in release of slowly dividing cells that escape the immune system (Sayed et al. 2012). Deletion of VapBC homologues in *Haemophilus influenzae* results in remarkable decrease of virulence in animal models for otitis media, tissue, etc. (Ren et al. 2012).

15.4.6 Growth Regulation

M.tb encounters different types of unfavourable conditions during infection. To survive in stressful conditions, bacteria regulate its growth. There are several proteins that are involved in growth regulation other than TA systems such as DATIN, IciA, MSMEG_1878 (*M.tb*Rv3241c orthologue in *M. smegmatis*), etc. *M. tb* DATIN and MSMEG_1878 inhibit protein synthesis by interacting with ribosome (Kumar et al. 2012; Li et al. 2018). *M.tb* IciA inhibits bacterial growth by inhibiting the opening of two strands of DNA during replication (Kumar et al. 2009; Kumar et al. 2017b). Toxin-antitoxin systems are activated during starvation stress. The RNase toxins, instead of being bactericidal, are usually bacteriostatic in nature. There is quick arrest of growth in response to starvation or other environmental stresses which help in their survival, and quicker resumption of growth occurs when the situation improves (Gerdes 2000; Gerdes et al. 2005).

15.4.7 Programmed Cell Death

Programmed cell death (PCD or apoptosis) is a physiological process and occurs mainly in multicellular, eukaryotic organisms during the process of embryonic development or tissue turnover. Dysregulation of PCD results in diseases like tumour formation, autoimmune diseases, or lysosomal disorders (Hayes 2003). Bacteria, being unicellular, do undergo PCD. However in natural environment, bacteria exist as biofilms that represent multicellular colonies and display coordination as in multicellular organisms. Such immobilized bacteria maintain discrete and ordered spatial structures within the biofilm niche. There are several genes in bacteria that are homologues to eukaryotic genes involved in PCD (Koonin and Aravind 2002), and TA modules of *E. coli* are either PCD genes or mediators of reversible growth arrest, which alternatively might allow the cells to enter a dormant or a semi-dormant state. It has been shown that PCD in bacteria might allow surviving cells to scavenge nutrients from dead ones and may prevent spread of bacteriophages (Engelberg-Kulka and Glaser 1999).

15.5 Applications of Toxin-Antitoxin Systems

There is a growing need for new antimicrobial agents due to decrease in the effectivity of drugs being used currently arising out of increase in multi-drug resistance. The toxins of TA systems usually target various biological processes such as replication, transcription, translation, macromolecular synthesis, cell wall synthesis, phage infection, and cytoskeletal polymerization. The fact that some of these toxins also overlap with the targets of the antibiotics (Wen et al. 2014) provides an option to explore TA systems as drug target in bacteria. There are several antibiotics that act indirectly against the TA systems. The detailed study of the interaction between the toxin and the antitoxin may help in the formulation of new drugs. Mapping the precise location of different TA systems in bacterial chromosome or plasmid will uncover fundamental insight into their possible applications as drug targets.

Previously, IS-elements, DR-elements, variable number tandem repeats (VNTR), and single-nucleotide polymorphisms (SNPs) were used as genetic markers in housekeeping genes or other genes and were used for genotyping. The type II TA systems are involved in virulence, persistence, and survival of *M.tb* inside host macrophages (Zaychikova et al. 2015). Analyses of 173 sequenced genomes of *M. tb* for the genes of type II TA systems show genetic diversity (SNPs) that correlates with the specific genotype of *M.tb* strains (Zaychikova et al. 2015). This correlation between a genotype of particular strain and SNPs in different genes of type II TA system paved way to consider TA systems as a new biomarker for genotyping of tuberculosis caused by different strains of *M.tb*.

15.6 Conclusion

Genome sequencing and its annotation have suggested presence of significant number of TA systems in microorganisms. The classification of TA system is based on similarity of primary sequences and on the specificity of interaction between pair of toxin and antitoxin molecules. Type II system is structurally most flexible in terms of location of antitoxin gene either upstream or downstream of the toxin gene or regulation of repression activities that is encoded by different genes. The various types of TA system may shuffle among themselves in terms of function and can exhibit 'mix and match' phenomenon. This is exemplified by the structural similarity between GinI solitary toxins with type II HicA toxin. In some cases, toxins can also evolve from the antitoxin under selective pressure. As in the case of VapD sequences, which mediate defence against phages, TA system points to a common origin with CRISPR-cas system. There is consensus that the evolution of TA system in bacteria enabled these unicellular organisms into robust molecular machines that could withstand the onslaught of environment and various stress. The importance of TA system is also underlined by the fact that while many pathogenic and obligatory intracellular pathogens opted for reductive evolution in genome size and shed off the extra burden of many genes, yet they retained the genes associated with TA system. The distribution of TA system varies even among strains of the same species of bacteria. TA systems are highly mobile in nature and moves between genomes through horizontal gene transfer. The presence of TA genes within the transposons makes them refractory to gene efflux and stabilizes the TA system within the cells. The additive property of TA genes within the transposons facilitates increased stability and exclusion of foreign DNA. Although the activity of TA system may be subdued by host cell machinery, it may rescue its activity similar to the restriction modification system. TA systems are preferentially associated with genomic islands or plasmids which serve as a mechanism to maintain their structural integrity during stress and survival through post-segregational killing. *M.tb* possesses a huge number of TA systems that are usually located within distinct genomic islands and trigger decrease in metabolic activity. In case the TA genes are integrated within the core genome, these accumulate mutation that may lead to loss of addictive property or deletion of TA system. In case of *E. coli*, type I *hok-sok* system are inactivated by IS sequences, gene rearrangements, and point mutations. TA system can evolve through integration with the host regulatory machinery and may replace the antitoxin molecules with signal transduction molecules.

Toxins have many cellular functions including inhibition of protein synthesis, DNA replication, and synthesis of cell wall in response to unfavourable conditions. Some toxins in TA system act as ribonucleases, while some other toxins act as gyrase inhibitors and kinases. The TA genes express in different stressed conditions such as nutrient deficiency, antibiotic treatment, bacteriophage infection, host immune responses, oxidative stress, and high temperature. They are involved in persistence, slow cell growth, cell cycle arrest, or cell death. Proteins involved in TA systems may act as important targets for drug development that may help in reduction of treatment duration of tuberculosis and other infectious diseases. In spite of the diversity in structure of TA system, the function of the TA system is tightly regulated by other cellular networks such that the prolific activity of TA system is activated only as a response to cellular physiology and the toxins are unleashed for minimal activity. The activity of TA locus that regulate activation of signaling pathways involved in persister cell formation in biofilms are modules acting as effectors of persister cell formation. These are usually deeply integrated into cellular signaling pathways that tightly control their activation and use their characteristic auto-regulatory features to tune the induction, duration and intensity of the phenotypic switch into dormancy.

It is speculated that our current understanding of the toxin-antitoxin system is still redundant and a deeper understanding of the mechanistic significance of TA system will enable us to unravel the mysteries of how these unicellular organisms could dominate the environment. It is envisaged that unlocking the mechanism of TA system could allow us to shape better strategies for overcoming the harmful effects of TA system in clinical pathogenesis and in dealing with microbial drug tolerance.

Funding This project has been funded by "UGC-BSR Research Start-Up-Grant project No. F. 30-487/2019(BSR) sanctioned to Ashutosh Kumar"

References

Aakre CD, Phung TN, Huang D, Laub MT (2013) A bacterial toxin inhibits DNA replication elongation through a direct interaction with the β sliding clamp. Mol Cell 52(5):617–628. https://doi.org/10.1016/j.molcel.2013.10.014

Agarwal S, Tiwari P, Deep A, Kidwai S, Gupta S, Thakur KG, Singh R (2018) System-wide analysis unravels the differential regulation and in vivo essentiality of virulence-associated proteins B and C toxin-antitoxin Systems of *Mycobacterium tuberculosis*. J Infect Dis 217 (11):1809–1820. https://doi.org/10.1093/infdis/jiy109

Bernard P, Couturier M (1992) Cell killing by the F plasmid CcdB protein involves poisoning of DNA-topoisomerase II complexes. J Mol Biol 226(3):735–745. https://doi.org/10.1016/0022-2836(92)90629-X

Blower TR, Short FL, Rao F, Mizuguchi K, PeiXY FPC, Luisi BF, Salmond GPC (2012) Identification and classification of bacterial type III toxin–antitoxin systems encoded in chromosomal and plasmid genomes. Nucleic Acids Res 40:6158–6173

Bordes P, Cirinesi AM, Ummels R, SalaA SS, Bitter W, Genevaux P (2011) SecB like chaperone controls a toxin–antitoxin stress-responsive system in *Mycobacterium tuberculosis*. Proc Natl Acad Sci U S A 108:8438–8443

Brantl S (2012) Bacterial type I toxin-antitoxin systems. RNA Biol 9:1488–1490. https://doi.org/10.4161/rna.23045

Brantl S, Jahn N (2015) sRNAs in bacterial type I and type III toxin–antitoxin systems. FEMS Microbiol Rev 39:413–427

Brielle R, Pinel-Marie ML, Felden B (2016) Linking bacterial type I toxins with their actions. Curr Opin Microbiol 30:114–121

Chan WT, Domenech M, Moreno-Córdoba I, Navarro-Martínez V, Nieto C, Moscoso M, García E, Espinosa M (2018) The Streptococcus pneumoniae yefM-yoeB and relBE toxin-antitoxin operons participate in oxidative stress and biofilm formation. Toxins 10(9):E378. https://doi.org/10.3390/toxins10090378

Christensen-Dalsgaard M, Jorgensen MG, Gerdes K (2010) Three new RelE-homologous mRNA interferases of *Escherichia coli* differentially induced by environmental stresses. Mol Microbiol 75:333–348

Cole ST, Eiglmeier K, Parkhill J, James KD, Thomson NR (2001) Massive gene decay in the leprosy bacillus. Nature 409:1007–1011

Deep A, Tiwari P, Agarwal S, Kaundal S, Kidwai S, Singh R, Thakur KG (2018) Structural, functional and biological insights into the role of *Mycobacterium tuberculosis* VapBC11 toxin–antitoxin system: targeting a tRNase to tackle mycobacterial adaptation. Nucleic Acids Res 46 (21):11639–11655. https://doi.org/10.1093/nar/gky924

Diago-Navarro E, Hernandez-Arriaga AM, López-Villarejo J, Muñoz-Gómez AJ, Kamphuis MB, Boelens R, Lemonnier M, Díaz-Orejas R (2010) parD toxin-antitoxin system of plasmid R1 – basic contributions, biotechnological applications and relationships with closely-related toxin-antitoxin systems. FEBS J 277(15):3097–3117. https://doi.org/10.1111/j.1742-4658.2010.07722.x

Engelberg-Kulka H, Glaser G (1999) Addiction modules and programmed cell death and antideath in bacterial cultures. Annu Rev Microbiol 53:43–70

Engelberg-Kulka H, Amitai S, Kolodkin-Gal I, Hazan R (2006) Bacterial programmed cell death and multicellular behavior in bacteria. PLoS Genet 2(10):e135. https://doi.org/10.1371/journal.pgen.0020135

Fineran PC, Blower TR, Foulds IJ, Humphreys DP, Lilley KS, Salmond GP (2009) The phage abortive infection system, ToxIN, functions as a protein-RNA toxin-antitoxin pair. Proc Natl Acad Sci U S A 106:894–899

Fivian-Hughes AS, Davis EO (2010) Analyzing the regulatory role of the HigA antitoxin within *Mycobacterium tuberculosis*. J Bacteriol 192:4348–4356

Fozo EM, Hemm MR, Storz G (2008) Small toxic proteins and the antisense RNAs that repress them. Microbiol Mol Biol Rev 72:579–589. https://doi.org/10.1128/MMBR.00025-08

Fridman O, Goldberg A, Ronin I, Shoresh N, Balaban NQ (2014) Optimization of lag time underlies antibiotic tolerance in evolved bacterial populations. Nature 513(7518):418–421. https://doi.org/10.1038/nature13469

Georgiades K, Raoult D (2011) Genomes of the most dangerous epidemic bacteria have a virulence repertoire characterized by fewer genes but more toxin–antitoxin modules. PLoS One 6:e17962

Gerdes K (2000) Toxin-antitoxin modules may regulate synthesis of macromolecules during nutritional stress. J Bacteriol 182(3):561–572. https://doi.org/10.1128/JB.182.3.561-572.2000

Gerdes K, Maisonneuve E (2012) Bacterial persistence and toxin-antitoxin loci. Annu Rev Microbiol 66:103–123

Gerdes K, Christensen SK, Lobner-Olesen A (2005) Prokaryotic toxin-antitoxin stress response loci. Nat Rev Microbiol 3:371–382

Gomez JE, McKinney JD (2004) M. tuberculosis persistence, latency, and drug tolerance. Tuberculosis (Edinb) 84:29–44

Gonzalez Barrios AF, Zuo R, Hashimoto Y, Yang L, Bentley WE, Wood TK (2006) Autoinducer 2 controls biofilm formation in *Escherichia coli* through a novel motility quorum-sensing regulator (MqsR, B3022). J Bacteriol 188(1):305–316. https://doi.org/10.1128/JB.188.1.305-316.2006

Guglielmini J, Melderen LV (2011) Bacterial toxin-antitoxin systems- translation inhibitors everywhere. Mob Genet Elem 1(4):283–290. https://doi.org/10.4161/mge.18477

Gupta A (2009) Killing activity and rescue function of genome-wide toxin-antitoxin loci of *Mycobacterium tuberculosis*. FEMS Microbiol Lett 290:45–53

Gupta M, Nayyar N, Chawla M, Sitaraman R, Bhatnagar R, Banerjee N (2016) The chromosomal *parDE2* toxin–antitoxin system of *Mycobacterium tuberculosis* H37Rv: genetic and functional characterization. Front Microbiol 7:886. https://doi.org/10.3389/fmicb.2016.00886

Gupta A, Venkataraman B, Vasudevan M, Gopinath-Bankar K (2017) Co-expression network analysis of toxin-antitoxin loci in *Mycobacterium tuberculosis* reveals key modulators of cellular stress. Sci Rep 7(1):5868. https://doi.org/10.1038/s41598-017-06003-7

Hayes F (2003) Toxins-antitoxins: plasmid maintenance, programmed cell death and cell cycle arrest. Science 301(5639):1496–1499

Hazan R, Engelberg-Kulka H (2004) *Escherichia coli* mazEF-mediated cell death as a defense mechanism that inhibits the spread of phage P1. Mol Gen Genomics 272(2):227–234. https://doi.org/10.1007/s00438-004-1048-y

Hurley JM, Cruz JW, Ouyang M, Woychik NA (2011) Bacterial toxin RelE mediates frequent codon-independent mRNA cleavage from the 5′ end of coding regions in vivo. J Biol Chem 286 (17):14770–14778. https://doi.org/10.1074/jbc.M110.108969

Kamada K, Hanaoka F (2005) Conformational change in the catalytic site of the ribonuclease YoeB toxin by YefM antitoxin. Mol Cell 19:497–509

Kasari V, Kurg K, Margus T, Tenson T, Kaldalu N (2010) The *Escherichia coli* mqsR and ygiT genes encode a new toxin-antitoxin pair. J Bacteriol 192(11):2908–2919. https://doi.org/10.1128/JB.01266-09

Kawano M, Aravind L, Storz G (2007) An antisense RNA controls synthesis of an SOS-induced toxin evolved from an antitoxin. Mol Microbiol 64:738–754. https://doi.org/10.1111/j.1365-2958.2007.05688.x

Keren I, Shah D, Spoering A, Kaldalu N, Lewis K (2004) Specialized persister cells and the mechanism of multidrug tolerance in *Escherichia coli*. J Bacteriol 186:8172–8180

Keren I, Minami S, Rubin E, Lewis K (2011) Characterization and transcriptome analysis of *Mycobacterium tuberculosis* persisters. MBio 2(3):e00100–e00111. https://doi.org/10.1128/mBio.00100-11

Koonin EV, Aravind L (2002) Origin and evolution of eukaryotic apoptosis: the bacterial connection. Cell Death Differ 9(4):394–404

Korch SB, Hill TM (2006) Ectopic overexpression of wild-type and mutant hipA genes in *Escherichia coli*: effects on macromolecular synthesis and persister formation. J Bacteriol 188 (11):3826–3836. https://doi.org/10.1128/JB 01740-05

Kroll J, Klinter S, Schneider C, Voß I, Steinbüchel A (2010) Plasmid addiction systems: perspectives and applications in biotechnology. Microb Biotechnol 3(6):634–657. https://doi.org/10.1111/j.1751-7915.2010.00170.x

Kumar S, Farhana A, Hasnain SE (2009) In-vitro helix opening of M. tuberculosis oriC by DnaA occurs at precise location and is inhibited by IciA like protein. PLoS One 4(1):e4139. https://doi.org/10.1371/journal.pone.0004139

Kumar A, Majid M, Kunisch R, Rani PS, Qureshi IA, Lewin A (2012) *Mycobacterium tuberculosis* DosR regulon gene Rv0079 encodes a putative, 'Dormancy Associated Translation Inhibitor (DATIN)'. PLoS One 7(6):e38709. https://doi.org/10.1371/journal.pone.0038709

Kumar A, Alam A, Rani M, Ehtesham NZ, Hasnain SE (2017a) Biofilms: survival and defense strategy for pathogens. Int J Med Microbiol 307(8):481–489. https://doi.org/10.1016/j.ijmm.2017.09.016

Kumar A, Rani M, Ehtesham NZ, Hasnain SE (2017b) Commentary: modification of host responses by mycobacteria. Front Immunol 8:466. https://doi.org/10.3389/fimmu.2017.00466. eCollection 2017

Kumar A, Alam A, Tripathi D, Rani M, Khatoon H, Pandey S, Ehtesham NZ, Hasnain SE (2018) Protein adaptations in extremophiles: an insight into extremophilic connection of mycobacterial proteome. Semin Cell Dev Biol 84:147–157. https://doi.org/10.1016/j.semcdb.2018.01.003

Kussell E, Kishony R, Balaban NQ, Leibler S (2005) Bacterial persistence: a model of survival in changing environments. Genetics 169(4):1807–1814. https://doi.org/10.1534/genetics.104.035352

Lee KY, Lee BJ (2016) Structure, biology, and therapeutic application of toxin–antitoxin systems in pathogenic bacteria. Toxins 8(10):305. https://doi.org/10.3390/toxins8100305

Lee IG, Lee SJ, Chae S, Lee KY, Kim JH, Lee BJ (2015) Structural and functional studies of the *Mycobacterium tuberculosis* VapBC30 toxin-antitoxin system: implications for the design of novel antimicrobial peptides. Nucleic Acids Res 43(15):7624–7637. https://doi.org/10.1093/nar/gkv689

Lemos JA, Brown TA Jr, Abranches J, Burne RA (2005) Characteristics of Streptococcus mutans strains lacking the MazEF and RelBE toxin-antitoxin modules. FEMS Microbiol Lett 253 (2):251–257. https://doi.org/10.1016/j.femsle.2005.09.045

Leplae R, Geeraerts D, Hallez R, Guglielmini J, Dreze P, Van ML (2011) Diversity of bacterial type II toxin-antitoxin systems: a comprehensive search and functional analysis of novel families. Nucleic Acids Res 39:5513–5525

Lewis K (2007) Persister cells, dormancy and infectious disease. Nat Rev Microbiol 5:48–56. https://doi.org/10.1038/nrmicro1557

Li Y, Sharma MR, Koripella RK, Yang Y, Kaushal PS, Lin Q, Wade JT, Gray TA, Derbyshire KM, Agrawal RK, Ojha AK (2018) Zinc depletion induces ribosome hibernation in mycobacteria. Proc Natl Acad Sci 115(32):8191–8196. https://doi.org/10.1073/pnas.1804555115

Masuda H, Tan Q, Awano N, Wu KP, Inouye M (2012) YeeU enhances the bundling of cytoskeletal polymers of MreB and FtsZ, antagonizing the CbtA (YeeV) toxicity in Escherichia coli. Mol Microbiol 84:979–989

Mine N, Guglielmini J, Wilbaux M, Van Melderen L (2009) The decay of the chromosomally encoded ccdO157 toxin-antitoxin system in the Escherichia coli species. Genetics 181 (4):1557–1566. https://doi.org/10.1534/genetics.108.095190

North RJ, Jung YJ (2004) Immunity to tuberculosis. Annu Rev Immunol 22:599–623

Pandey DP, Gerdes K (2005) Toxin-antitoxin loci are highly abundant in free-living but lost from host-associated prokaryotes. Nucleic Acids Res 33(3):966–976. https://doi.org/10.1093/nar/gki201

Pedersen K (2003) The bacterial toxin RelE displays codon-specific cleavage of mRNAs in the ribosomal A site. Cell 112:131–140

Pimentel B, Madine MA, de la Cueva-Méndez G (2005) Kid cleaves specific mRNAs at UUACU sites to rescue the copy number of plasmid R1. EMBO J 24(19):3459–3469. https://doi.org/10.1038/sj.emboj.7600815

Ramage HR, Connolly LE, Cox JS (2009) Comprehensive functional analysis of *Mycobacterium tuberculosis* toxin-antitoxin systems: implications for pathogenesis, stress responses, and evolution. PLoS Genet 5:e1000767

Ren D, Walker AN, Daines D (2012) Toxin–antitoxin loci vapBC-1 and vapXD contribute to survival and virulence in nontypeable Haemophilus influenza. BMC Microbiol 12:263
Rybtke MT, Jensen PO, Hoiby N, Givskov M, Tolker-Nielsen T, Bjarnsholt T (2011) The implication of *Pseudomonas aeruginosa* biofilms in infections. Inflamm Allergy Drug Targets 10:141–157
Sayed N, Nonin-Lecomte S, Réty S, Felden B (2012) Functional and structural insights of a *Staphylococcus aureus* apoptotic-like membrane peptide from a toxin–antitoxin module. Biol Chem 287:43454–43463
Seed KD (2015) Battling phages: how bacteria defend against viral attack. PLoS Pathog 11(6): e1004847
Sevin EW, Barloy-Hubler F (2007) RASTA-Bacteria: a web-based tool for identifying toxin-antitoxin loci in prokaryotes. Genome Biol 8:R155
Shao Y, Harrison EM, Bi D, Tai C, He X (2011) TADB: a web-based resource for type 2 toxin antitoxin loci in bacteria and archaea. Nucleic Acids Res 39:D606–D611
Short FL, Akusobi C, Broadhurst WR, Salmond GPC (2018) The bacterial type III toxin-antitoxin system, ToxIN, is a dynamic protein-RNA complex with stability-dependent antiviral abortive infection activity. Sci Rep 8:1013. https://doi.org/10.1038/s41598-017-18696-x
Singh R, Barry CE, Boshoff HI (2010) The three RelE homologs of *Mycobacterium tuberculosis* have individual, drug-specific effects on bacterial antibiotic tolerance. J Bacteriol 192:1279–1291
Smollett KL, Fivian-Hughes AS, Smith JE, Chang A, Rao T, Davis EO (2009) Experimental determination of translational start sites resolves uncertainties in genomic open reading frame predictions – application to *Mycobacterium tuberculosis*. Microbiology 155:186–197
Stern A, Sorek R (2012) The phage-host arms-race: shaping the evolution of microbes. BioEssays 33:43–51
Stewart GR, Robertson BD, Young DB (2003) Tuberculosis: a problem with persistence. Nat Rev Microbiol 1:97–105
Tiwari P, Arora G, Singh M, Kidwai S, Narayan OP, Singh R (2015) MazF ribonucleases promote *Mycobacterium tuberculosis* drug tolerance and virulence in Guinea pigs. Nat Commun 6:6059. https://doi.org/10.1038/ncomms7059
Tsilibaris V, Maenhaut-Michel G, Mine N, Melderen LV (2007) What is the benefit to Escherichia coli of having multiple toxin-antitoxin Systems in its Genome? J Bacteriol 189(17):6101–6108. https://doi.org/10.1128/JB.00527-07
Van Melderen L, Saavedra De Bast M (2009) Bacterial toxin-antitoxin systems: more than selfish entities? PLoS Genet 5(3):e1000437. https://doi.org/10.1371/journal.pgen.1000437
Wang X, Lord DM, Hong SH, Peti W, Benedik MJ, Page R, Wood TK (2013) Type II toxin/antitoxin MqsR/MqsA controls type V toxin/antitoxin GhoT/GhoS. Environ Microbiol 15 (6):1734–1744. https://doi.org/10.1111/1462-2920.12063
Warner DF, Mizrahi V (2006) Tuberculosis chemotherapy: the influence of bacillary stress and damage response pathways on drug efficacy. Clin Microbiol Rev 19:558–570
Wen Y, Behiels E, Devreese B (2014) Toxin–antitoxin systems: their role in persistence, biofilm formation, and pathogenicity. Pathog Dis 70:240–249. https://doi.org/10.1111/2049-632X.12145
Winther KS, Gerdes K (2009) Ectopic production of VapCs from Enterobacteria inhibits translation and trans-activates YoeB mRNA interferase. Mol Microbiol 72:918–930
Wu Y, Vulić M, Keren I, Lewis K (2012) Role of oxidative stress in persister tolerance. Antimicrob Agents Chemother 56:4922–4926. https://doi.org/10.1128/AAC.00921-12
Yamaguchi Y, Park JH, Inouye M (2011) Toxin-antitoxin systems in bacteria and archaea. Annu Rev Genet 45:61–79. https://doi.org/10.1146/annurev-genet-110410-132412
Zaychikova MV, Zakharevich NV, Sagaidak MO, Bogolubova NA, Smirnova TG, Andreevskaya SN, Larionova EE, Alekseeva MG, Chernousova LN, Danilenko VN (2015) *Mycobacterium tuberculosis* type II toxin antitoxin systems: genetic polymorphisms and functional properties

and the possibility of their use for genotyping. PLoS One 10(12):e0143682. https://doi.org/10.1371/journal.pone.0143682

Zhang Y, Zhang J, Hoeflich KP, Ikura M, Qing G, Inouye M (2003) MazF cleaves cellular mRNAs specifically at ACA to block protein synthesis in *Escherichia coli*. Mol Cell 12(4):913–923. https://doi.org/10.1016/S1097-2765(03)00402-7

Zhu L, Zhang Y, Teh JS, Zhang J, Connell N, Rubin H, Inouye M (2006) Characterization of mRNA interferases from *Mycobacterium tuberculosis*. J Biol Chem 281:18638–18643

Nucleotide Excision Repair Pathway in Mycobacteria

16

Manoj Thakur and K. Muniyappa

Abstract

Nucleotide excision repair (henceforth abbreviated as NER) plays a pivotal role in all organisms to protect their genetic material against radiations, toxic chemicals, and normal by-products of cellular metabolism. In humans, defects in excision repair causes inherited diseases, and NER-related human diseases are associated with cancer and aging. Much of our current understanding of NER has emerged from experimental evidence in model systems including *Escherichia coli*, yeast, and mammalian cells. Considering the importance of NER in the maintenance of genome integrity, it is surprising that only a few studies have investigated NER in mycobacteria. Here we provide a brief overview of the mechanism of the DNA repair in mycobacteria. A detailed understanding of structure-function relationship of DNA repair proteins in tubercle bacillus could facilitate the identification and development of novel therapeutic targets for tuberculosis therapy.

Keywords

DNA damage · DNA helicase · *Mycobacterium tuberculosis* UvrA · UvrB · UvrC · UvrD · Nucleotide excision repair · SOS response

Abbreviations

ATP	adenosine triphosphate
BCG	Bacillus Calmette–Guerin
BER	base excision repair;
CFU	colony-forming units

M. Thakur · K. Muniyappa (✉)
Department of Biochemistry, Indian Institute of Science, Bangalore, India
e-mail: kmbc@iisc.ac.in

S. E. Hasnain et al. (eds.), Mycobacterium tuberculosis: *Molecular Infection Biology Pathogenesis, Diagnostics and New Interventions*,
https://doi.org/10.1007/978-981-32-9413-4_16

DSBs	double-strand breaks
dsDNA	double-stranded DNA
EcRecA	*Escherichia coli* RecA
EcUvrA	*Escherichia coli* UvrA
EcUvrB	*Escherichia coli* UvrB
HR	homologous recombination
iNOS	inducible nitric oxide synthase
MDR	multidrug resistant
MMR	mismatch repair
MtRecA	*Mycobacterium tuberculosis* RecA
MtUvrB	*Mycobacterium tuberculosis* UvrB
NER	nucleotide excision repair
NHEJ	non-homologous end joining
RNI	reactive nitrogen intermediates
ROI	reactive oxygen intermediates
ssDNA	single-stranded DNA
TCR	transcription-coupled repair
UV-light	ultraviolet-light
XDR	extensively drug resistant

16.1 Overview of DNA Damage Response Mechanisms

Numerous studies have established that the genomic DNA undergoes damage inflicted by radiation, chemicals, environmental agents, and internal metabolic (endogenous) processes (Hoeijmakers 2001; Lindahl and Wood 1999). These produce nucleotide modifications such as bulky adducts, single- and double-strand breaks, inter- or intra-strand cross-links, DNA-protein cross-links, apurinic or apyrimidinic base-free sites, misincorporation of nucleotide or ribonucleotide by DNA polymerases, and DNA replication slippage (Vaisman et al. 2013; Dexheimer 2012; Lindahl and Wood 1999; Van Houten and Sancar 1987). If these lesions remain unrepaired, they lead to genomic instability and various diseases like cancer, neurological disorders, immunodeficiency, premature aging, and Fanconi anaemia in humans (Iyama and Wilson 2013). Therefore, maintenance of genome stability under normal and stressed conditions becomes an important task to the cellular repair machinery.

To circumvent the harmful or deleterious effects of myriad types of DNA damage, cells have evolved specialized types of DNA repair systems. These can be classified based on the protein components and mechanism(s) involved as well as different types of DNA lesions being processed. The different DNA damage repair pathways include direct reversal of damaged base, base excision repair (BER), mismatch repair (MMR), nucleotide excision repair (NER), and single- and double-strand breaks (DSBs) repair (Fig. 16.1). The repair of both single- and

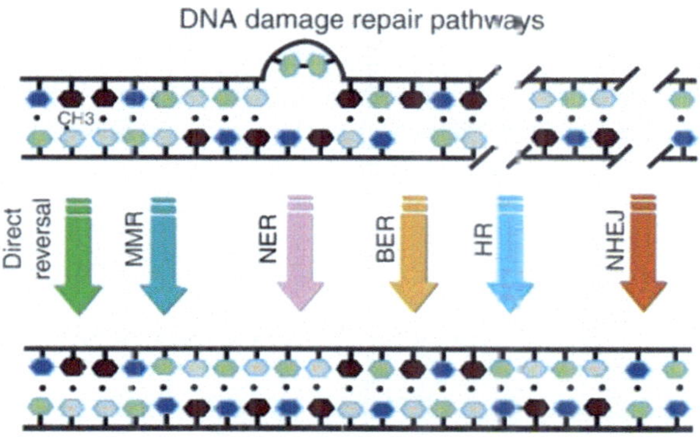

Fig. 16.1 Overview of DNA damage and repair mechanisms. The figure illustrates common types of DNA lesions and the specific mechanism to repair each type of lesion (this Figure is cited from Hakem (2008). (Reproduced with the permission of EMBO J)

double-strand breaks involves homologous recombination (HR) and non-homologous DNA end-joining (NHEJ) pathways (Kuzminov 2001; Cannan and Pederson 2016; Friedberg 2016; Yi and He 2013; Hakem 2008). The repair of damaged bases involves three mechanistically distinct processes: (1) photolyase catalysed reversal of UV light-induced photolesions; (2) O^6-alkylguanine-DNA alkyltransferases (AGTs) catalysed reversal of various O-alkylated DNA damage; and (3) the AlkB family dioxygenases mediated reversal of N-alkylated base adducts (Yi and He 2013). The photolyase enzyme reactivates DNA by direct repair of T<>T lesions, and BER is used to repair single-stranded breaks and to correct the damaged bases resulting from spontaneous DNA deamination, hydroxylation of bases, or exposure of DNA to alkylating agents. On the other hand, MMR recognizes and repairs erroneous insertions, deletions, and misincorporation of bases arising from DNA replication, and base deamination. This system recognizes the presence of hemimethylated dGATC sequences that provides molecular basis to distinguish between the newly synthesized daughter strand and parental DNA strand (Radman and Wagner 1986). This hemi-methylated dGATC site 5′ or 3′ to the mismatch then cleaved by the concerted action of MutS, MutL, and MutH in an ATP-dependent manner (Acharya et al. 2003). ExoI or ExoX (3′ → 5′ exonuclease), or ExoVII or RecJ (5′ → 3′ exonuclease) excise the cleaved strand up to and slightly past the mismatch. The resulting single-stranded gap is repaired by DNA synthesis using intact complimentary strand as a template and ligated by DNA polymerase III or DNA ligase (Li 2008).

The double-stranded breaks are repaired via either HR or NHEJ, but what determines the pathway preference remains elusive (Lieber 2010). In eubacteria, HR comprises of four stages: (1) when DSBs occur, a process called DNA end resection is activated to generate 3′ single-stranded overhangs; (2) RecA polymerizes onto 3′-single-stranded DNA (ssDNA) to form a nucleoprotein filament; (3) RecA nucleoprotein filament initiates search for homology, pairing, and strand exchange; and (4) finally, the extension of heteroduplex DNA by RuvAB followed by resolution of the Holliday junction by RuvC produces crossover and non-crossover products (Cox 2007; Kowalczykowski et al. 1994). The bacterial cells also depend on NHEJ during pre-replicative stage when only single copy of the genome is available. Genetic and biochemical evidences have also shown that mycobacteria possess NHEJ pathway that depends on DNA ligase (LigD) and Ku protein (Matthews and Simmons 2014; Shuman and Glickman 2007). Although the biochemical mechanisms of all these pathways have been extensively studied in considerable depth in bacteria, how these repair events are modulated within the nucleoid is not fully understood.

16.2 DNA Damage and SOS Response

The induction of the SOS response in bacteria is a global response to DNA damage in which cell division is arrested and DNA repair is activated by the expression of DNA damage repair genes (Adikesavan et al. 2011; Schlacher and Goodman 2007; Radman 1975). In *Escherichia coli*, the SOS response involves two key proteins: LexA (a repressor) and RecA (an inducer) as shown in Fig. 16.2 (Radman 1975). Under normal growth conditions, LexA dimer binds to a 20-base pair consensus palindromic DNA sequence (also known as the SOS box) and represses the transcription of a regulon (comprising of more than 50 genes, including *lexA* and *recA*)

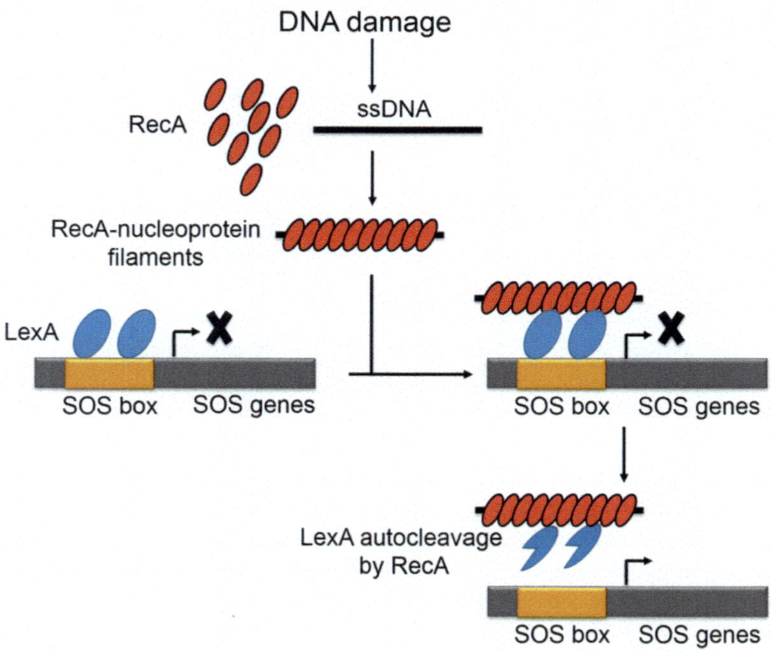

Fig. 16.2 DNA damage and SOS response. Initially the LexA repressor binds to SOS box, present upstream of SOS genes. Under DNA-damaging conditions, the single-stranded DNA (ssDNA) that originates from double-strand breaks (DSBs) repair pathway binds RecA proteins leading to the formation of RecA-nucleoprotein filaments. These filaments induce the autocatalytic cleavage of LexA allowing derepression of SOS genes. (Figure adapted from Butala et al. 2009)

(Adikesavan et al. 2011). Upon DNA damage, RecA forms a nucleoprotein filament on ssDNA (Cox 2007; Kowalczykowski et al. 1994). The RecA nucleoprotein filament facilitates the self-cleavage of LexA, leading to the derepression of SOS regulon genes (Kowalczykowski et al. 1994). The *uvr* genes are among the first to be induced, followed by *lexA* and *recA* genes, while genes encoding the low fidelity, error-prone repair DNA polymerases Pol II (*polB*), Pol IV (*dinB*), and Pol V (*umuC* and *umuD*) are induced under extensive and persistent DNA-damaging conditions (Butala et al. 2009; Smith and Sharma 1987; Smith et al. 1987). These DNA polymerases allow DNA replication over the DNA lesions (translesion DNA synthesis) that block the primary replicative DNA polymerase Pol III and leads to mutations in the newly replicated and repaired regions. The *uvrA*, *uvrB*, and *ydjQ* (*cho*) genes are SOS-induced genes, whereas *uvrC* gene is not regulated by SOS response (Janion 2001, 2008; Simmons et al. 2008; Bohr 1991; Calsou and Salles 1985; Youngs et al. 1974). Although the mechanism of UvrABC mediated repair of UV-induced photoproducts have been studied extensively under in vitro conditions, how this system is organized in vivo is unclear. Some estimates indicate that at least 17 protein components are recruited to the inner membrane of *E. coli* after UV irradiation. The colocalization of UV-induced 6–4 photoproducts and NER proteins to the inner membrane following UV-irradiation suggests that a part of the repair process may associate with the inner membrane (Lin et al. 1997).

The *Mycobacterium tuberculosis recA* gene encodes an 85 kDa precursor, which is processed to generate a 38 kDa functionally active RecA and 47 kDa intein (Davis et al. 1991; Kumar et al. 1996). The X-ray structure of *M. tuberculosis* RecA (MtRecA) is very similar to that of *E. coli* RecA (EcRecA) (Datta et al. 2000). The transcriptional regulation of *M. tuberculosis recA* differs from that of *E. coli recA: M. tuberculosis recA* is regulated by two promoters (Movahedzadeh et al. 1997b). One of the promoters is regulated by LexA repressor, while the second promoter is DNA damage-inducible, but independently of both LexA and RecA (Davis et al. 2002b). The *M. tuberculosis* LexA-binding site (SOS box) is similar to that of *Bacillus subtilis* (Cheo box) (Movahedzadeh et al. 1997a). Further investigations revealed the existence of several LexA-regulated genes in *M. tuberculosis* (Davis et al. 2002a). The apparent lack of correlation between the expression of DNA damage-inducible genes and the LexA repressor in *M. tuberculosis* supports an alternative mechanism of LexA-independent DNA damage induction (Brooks et al. 2001). Despite this, *M. tuberculosis* genome possesses the canonical core of the SOS system, except *polB* and *umuD* (Cole 1998; Mizrahi and Andersen 1998). The global gene expression following DNA damage in both wild-type and *recA*$^-$ strain of *M. tuberculosis* revealed that most of the genes induced by DNA damage do not have any known roles in DNA repair. However, another set of the inducible genes with a known or predicted function in DNA repair was found to be independent of *recA* (Rand et al. 2003). This study also revealed yet another group of genes, *lexA* itself, *ruvA*, *ruvB*, and *dnaE2* whose expression was strictly dependent on *recA* for induction and group of genes whose expression appear to be regulated by both *recA*-dependent and *recA*-independent mechanisms, e.g. *recA* itself, *radA*, *ruvC*, and *ssb*.

16.3 Discovery of Nucleotide Excision Repair Genes

Briefly, the survival of *E. coli* K-12 cells following UV radiation led to the discovery of NER genes (Boyce and Howard-Flanders 1964; Setlow and Carrier 1964; Howard-Flanders et al. 1962a). In early studies, thymine dimers were found to be the main photoproducts after UV irradiation (Setlow and Setlow 1962, 1963; Setlow et al. 1963). The identification of repair-deficient mutants was based on contra-selection of irradiated T1 bacteriophage particles (Howard-Flanders and Theriot 1962). These studies identified three loci in *E. coli* K-12 that control the excision of pyrimidine dimers and these genes were designated as *uvrA*, *uvrB*, and *uvrC* (Setlow and Carrier 1964; Howard-Flanders et al. 1962a, b, 1966).

16.4 Overview of Nucleotide Excision Repair Pathway

Studies over the past four decades have identified distinct DNA repair pathways and molecular mechanisms underlying the myriad types of biochemical reactions of various repair systems. Among various forms of genomic DNA damage, "bulky" DNA lesions that cause helical distortion in the double helix can lead to replication fork arrest, and mutations (Reardon and Sancar 2005). The bulky chemical adducts and interstrand cross-links are the outcomes of UV irradiation and various other mutagens (Kisker et al. 2013; Truglio et al. 2006; Jeggo et al. 2015). In all three domains of life, various proteins that belong to distinct DNA repair pathways recognize these DNA alterations and then either repair them by direct reversal (photoreactivation system) or target them for removal by a series of DNA repair systems (Reardon and Sancar 2005; Kisker et al. 2013; Truglio et al. 2006; Jeggo et al. 2015). One of the most studied and best understood DNA repair machinery in eubacteria and archaebacteria is the NER pathway (Jeggo et al. 2015; Kisker et al. 2013; Truglio et al. 2006; Reardon and Sancar 2005; Ogrunc et al. 1998; Grogan 2000). In these two domains, the NER pathway is dedicated to the removal of a variety of chemically and structurally diverse DNA lesions, including UV-induced photoproducts (cyclobutane dimers, 6–4 photoproducts, and thymine glycol), bulky adducts, apurinic or apyrimidinic (AP) sites, and cross-links, albeit with variable efficiencies (Goosen 2010; Truglio et al. 2006; Reardon and Sancar 2005; Alekseyev et al. 2004; Selby and Sancar 1991; Snowden et al. 1990; Lin and Sancar 1989; Van Houten et al. 1988; Van Houten and Sancar 1987; Thomas et al. 1986; Sancar et al. 1985; Seeberg et al. 1983). In addition to the structural distortions in the DNA, UvrABC can also repair cross-links between protein and DNA as well as the damages caused by reactive oxygen and nitrogen species (Salem et al. 2009; Imoto et al. 2008; Minko et al. 2002, 2005; Nakano et al. 2005).

In all three domains of life, NER comprises of discrete steps: DNA scanning, damage detection, damage verification, incision or removal of damage, DNA synthesis, and ligation as depicted in Fig. 16.3 (Hu et al. 2017; Kisker et al. 2013;

Fig. 16.3 (continued) is required for the removal of incised lesion containing small oligonucleotide. Repair is then completed by the action of DNA polymerase I and DNA ligase which synthesizes the new strand using intact complementary strand as a template and ligates the nick, respectively. (Figure adapted from Kisker et al. 2013)

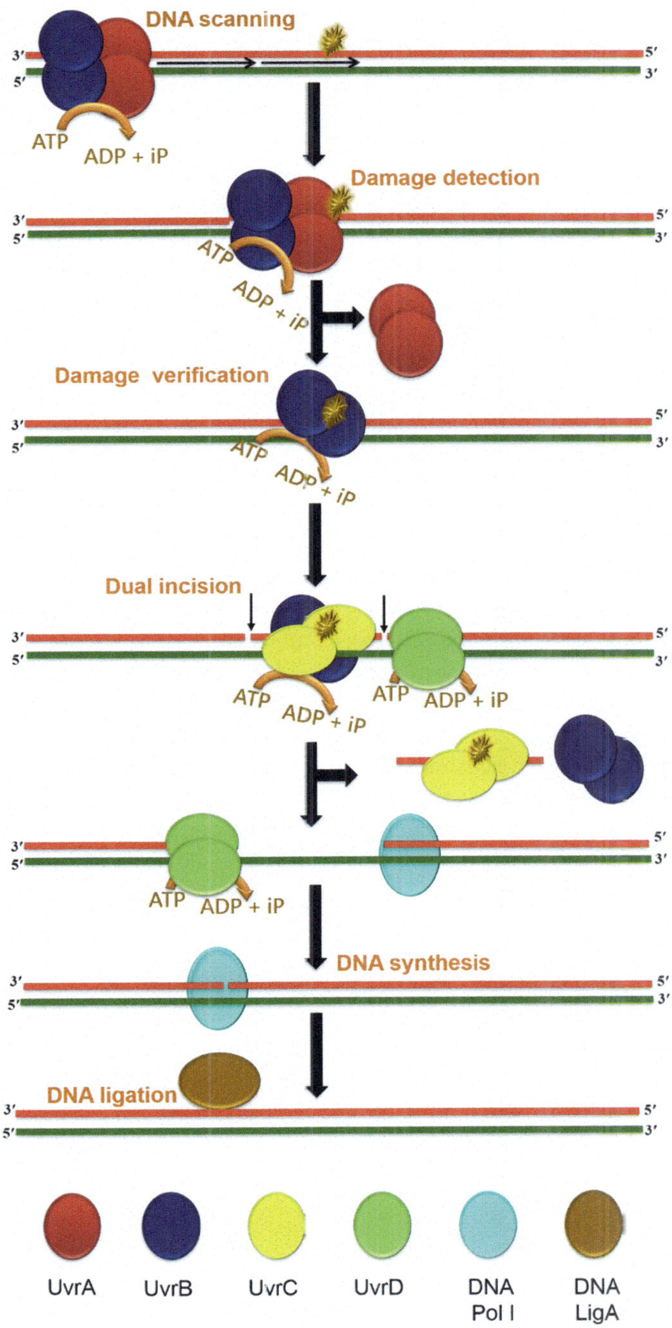

Fig. 16.3 Schematic representation of the prokaryotic NER pathways. Genome is scanned by the heterotetrameric $UvrA_2$–$UvrB_2$ complex in search for damaged nucleotides in an ATP-dependent manner. After the damage detection by UvrA component, pathway proceeds with damage verification by UvrB followed by 3′ and 5′ incisions catalysed through UvrC. The helicase activity of UvrD

Truglio et al. 2006; Ogrunc et al. 1998;). In prokaryotes, NER can be initiated by two major pathways often referred to as global genome repair (GGR) and transcription-coupled repair (TCR). GGR can take place at any region of the genome and is generally initiated by the direct binding of UvrAB complex to the DNA under DNA-damaging conditions or during the SOS response (Simmons et al. 2008; Janion 2001, 2008; Crowley et al. 2006; Hanna et al. 2001). The transcription coupled NER is initiated by stalled RNA polymerase on the template strand with the help of TCR specific factor, Mfd (Pani and Nudler 2017; Van Houten and Kad 2014). The TCR not only leads to repair of lesions on the DNA but also accomplishes efficient and correct transfer of information from genes to RNA. Although the first step of initiating NER is different in GGR and TCR, both the processes require the same molecular protein complexes to complete the excision repair (Van Houten and Kad 2014). Specifically, UvrAB complex bound to ATP scans the genome for lesions. Once a lesion is detected by UvrA, the lesion containing strand is transferred to the UvrB for damage verification and UvrA is released from the process (Pakotiprapha et al. 2012; Jaciuk et al. 2011; Kad et al. 2010; Malta et al. 2007; Truglio et al. 2006). The UvrB subunit binds specifically to the lesion containing DNA strand by inserting its beta-hairpin domain between the two strands of the DNA for verification of the lesion. UvrC binds to this tight UvrB-DNA preincision complex and mediates the incisions on both 3′ and 5′ sides of the lesion on the single-stranded DNA (Pu et al. 1989). Many bacterial species such as *E. coli*, *Mycobacterium tuberculosis, Salmonella typhimurium, Salmonella enteritidis, Klebsiella pneumoniae, and Clostridium acetobutylicum* harbour another protein called YdjQ (renamed as Cho) which incises only at the 3′ side of the lesion. It was found that Cho cleaves the poor DNA substrates of UvrC very efficiently and thus further increases the substrate range of NER (Moolenaar et al. 2002). Subsequently, UvrD (DNA helicase II) removes the excised lesion containing the oligonucleotide by its helicase action and DNA polymerase I fills the gap using intact complimentary strand as a template (Brosh 2014; Lee and Yang 2006; Runyon et al. 1990; Caron et al. 1985). Finally, DNA ligase seals the nick; thus genome integrity is restored (Paul-Konietzko et al. 2015; Orren et al. 1992; Van Houten et al. 1988; Husain et al. 1985).

Much of the available knowledge regarding NER has come through investigations in model organisms such as *E. coli*, yeast, and cultured mammalian cells. A detailed discussion on the mechanistic aspects of NER is beyond the scope of this review; however, the reader is referred to excellent reviews that deal with the mechanistic aspects of NER (Hu et al. 2017; Pani and Nudler 2017; Kisker et al. 2013; Van Houten et al. 2005). While the NER pathway has been studied extensively over the years, the comparison between *E. coli* and mycobacteria is important for identifying possible genetic and mechanistic differences that might contribute to the observed differences. This review focuses on the function of NER in mycobacteria and on the role of this pathway in the cellular response to UV-induced DNA damage.

16.5 DNA Repair in *Mycobacterium tuberculosis*

The macrophages mount an effective host defence by recognizing, engulfing, and killing *M. tuberculosis* (Levitte et al. 2016). The recruitment of macrophages is through Toll-like receptor (TLR)-mediated signalling that is triggered by pathogen-activated molecular patterns (PAMPs) present on bacterial surface (Fig. 16.4). The infected macrophages migrate across the lung epithelium and then transport the

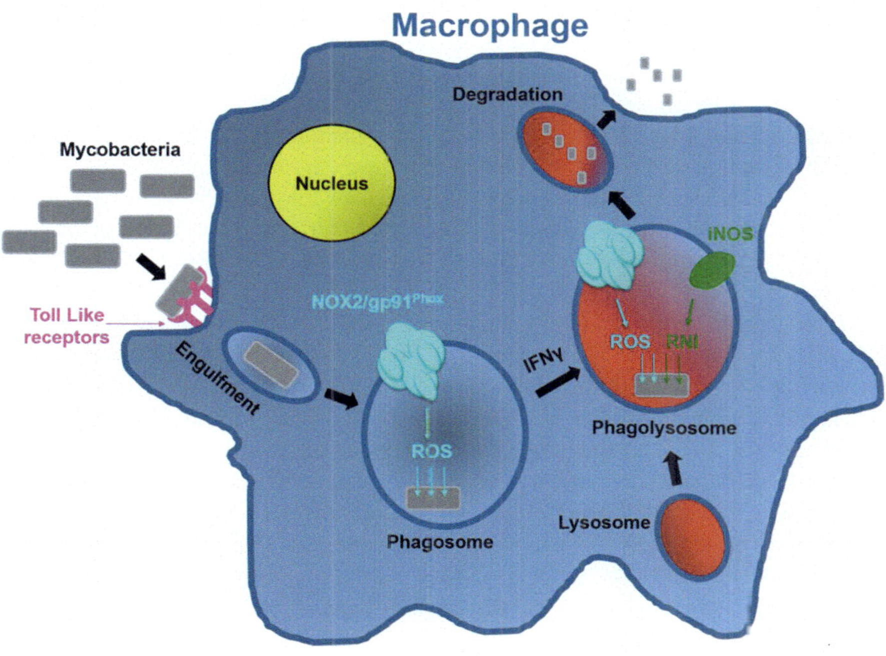

Fig. 16.4 Simplified diagram of the phagocytosis and degradation of a mycobacterial cell. Phagocytosis is triggered by the binding of Toll-like receptors to the PAMPs on the cell surface of mycobacteria. After the particle is taken up by the phagocyte, mycobacterial DNA is exposed to the deleterious effect of ROI generated by the action of phagocyte oxidase ($NOX2/gp91^{Phox}$). Upon immunological activation with IFNγ, the phagosome matures and fuses with lysosomes leading to the production of RNI from iNOS (inducible nitric oxide synthase 2) and ROI from NOX2. These species react with a wide range of molecules, including nucleic acids, proteins, lipids, and carbohydrates. If the damaged DNA of *M. tuberculosis* left unrepaired, the mycobacterial cells inside the macrophages will undergo apoptosis inside phagolysosomes, and subsequently the cell debris will be secreted out by the phagocyte. (Figure adapted from Ehrt and Schnappinger 2009)

bacteria to other tissues (Cambier et al. 2014a). Subsequently, new macrophage recruitment takes place and the original infected macrophages form granuloma, an aggregation of differentiated macrophages, and other immune cells. In these granulomas the infection expands by spreading bacteria to newly arrived macrophages (Cambier et al. 2014b). The experimental evidences suggest a link between the association of Toll-like receptor signalling, reactive oxygen intermediates (ROI), and reactive nitrogen intermediates (RNI) (Marcato et al. 2008; Ryan et al. 2004). Such signalling cascades lead to the activation of macrophages as an innate immune response to kill *M. tuberculosis*. Studies suggest that both ROI and RNI can permeabilize the cell wall/membrane of the pathogen (Fig. 16.4) (Schlosser-Silverman et al. 2000; Fang 1997). Most common damages that occur in DNA are the base modifications, generation of abasic sites, and strand breaks in response to RNI and ROI (Wink et al. 1991). Specifically, ROS produces single pyrimidine and purine base lesions, intrastrand cross-links, purine 5′,8-cyclonucleosides, DNA-protein adducts, and interstrand cross-links (Cadet and Wagner 2013). Therefore, inability to rectify these damages in the DNA inside the macrophages can compromise its survival and ability to cause pathogenesis. Several studies suggest that *M. tuberculosis* is well equipped to counteract the harmful effects of ROS, RNI, and low pH (<3.5) generated by the host immune system (Ehrt and Schnappinger 2009). But how the genome integrity of *M. tuberculosis* is maintained under these conditions continues to be an important area of research. To this end, genome-based studies in *M. tuberculosis* showed the involvement of various DNA repair proteins and nucleoid-associated proteins (NAPs) to circumvent such DNA damaging conditions (Philipp et al. 1998a, b; Bergh and Cole 1998; Brosch et al. 1998).

M. tuberculosis harbours all the vital genes of NER, base excision repair (BER), HR, NHEJ, SOS repair, and mutagenesis, as deduced from bioinformatics and genome-based studies, but the function of these proteins and enzymes is not fully understood. Moreover, studies on DNA repair components is important considering that *M. tuberculosis* lacks functional mismatch repair pathway (MMR) (Cole 1998). In contrast to *Helicobacter pylori*, in which a correlation exists between the absence of MMR and high level of genetic diversity, *M. tuberculosis* genomes are very stable (Kang and Blaser 2006). This observation suggests that other DNA repair mechanisms could be very efficient in this pathogen. Recently, NucS-dependent DNA repair system that mimics the MutS/MutL-based MMR has been identified in *Mycobacterium smegmatis* (Castaneda-Garcia, et al. 2017). Thus, characterization of *M. tuberculosis* DNA repair pathways in vivo and in vitro may contribute to the understanding of the generation of clonal populations of *M. tuberculosis* (Sreevatsan et al. 1997).

The adaptive nature of *M. tuberculosis* is significant in the clinical context because of its ability to supress central metabolism, halt DNA replication, and enter into a stage of dormancy rendering itself extremely resistant to host defence and drug treatments (Gengenbacher and Kaufmann 2012). Thus, the physiological changes that take place during the shift from dormancy to reactivation of the tubercle bacillus should be investigated in detail (Gopinath et al. 2015; Ehrt et al. 2015;

Gorna et al. 2010). To this end, Affymetrix GeneChip System Microarray (based on a specific oligonucleotide array format) revealed that almost half of the DNA repair genes are continuously expressed during log phase of growth, indicating that the bacteria continuously counteract the constant exposure to DNA-damaging conditions (Fu and Fu-Liu 2007). Additionally, transposon site hybridization (TraSH) technique identified the genes, Rv2554c, *dut*, *ligA*, *polA*, and *adnB* and those encoding the NER-related proteins UvrD1, UvrD2, and UvrC, which are essential for optimal growth by *M. tuberculosis* (Sassetti et al. 2003). The *uvrD2* and *ligA* genes were also found to be essential for the survival of *M. smegmatis* (Sinha et al. 2008; Korycka-Machala et al. 2006). In another study, every nonessential gene of *M. tuberculosis* was disrupted and their effect on the growth rate and virulence was determined in mice models. This study also revealed a set of DNA repair proteins required for survival in mice, apart from a variety of proteins from different pathways. These include three BER proteins (Ung, Nfo/End, and XthA), MazG, and RecN (Sassetti and Rubin 2003). The proteome profiling of *M. tuberculosis* also indicated the role of DNA repair genes during dormancy and reactivation (Gopinath et al. 2015). Altogether these studies suggest that most of the *M. tuberculosis* DNA repair genes are nonessential for survival in the broth, disruption of these genes lead to attenuation of virulence in mice. A possible link between survival and spatial-temporal regulation of DNA repair proteins has been established by the isolation of *M. tuberculosis* W-Beijing strains. These strains belong to a drug resistance family which could be divided into several branches based on unique alterations in three putative antimutator genes: *mutT4, mutT2*, and *ogt* (Dos Vultos et al. 2008; Nouvel et al. 2007; Ebrahimi-Rad et al. 2003).

M. tuberculosis infection is transmitted by aerosol droplets containing bacteria, coughed out by individuals with active disease, into the lower lungs of a new host. Various studies show that *M. tuberculosis* viability in the aerosols decline to 55%, 10%, and almost zero after 5, 30, and 60 min, respectively (Lever et al. 2000; Loudon et al. 1969). Studies based on artificially generated droplets carrying BCG pointed out that UV radiation and desiccation sterilize mycobacterial contaminated air (Peccia and Hernandez 2004; Ko et al. 2002; Riley et al. 1976). Nevertheless, from these studies it is evident that the viability of mycobacteria drastically reduces in the aerosols. It is known that intracellular dehydration (desiccation) leads to DSBs in DNA, whereas UV exposure leads to the generation of cyclobutane pyrimidine dimers and pyrimidine (6–4) pyrimidine photoproducts, in which two adjacent pyrimidines are covalently linked to each other. In mycobacteria, the HR and NHEJ pathways are involved in the repair of DSBs. Interestingly, *M. smegmatis* mutant strains lacking RecA and Ku and/or LigD showed that neither of these proteins are essential for its growth in vitro (Korycka-Machala et al. 2006). Despite this, both processes are required for survival under desiccation in the broth culture (Pitcher et al. 2007). Similar studies were also performed by generating mutants that influence NER in *M. smegmatis* and *M. tuberculosis*. Mutations in *polA*, *uvrB*, and *uvrD1* of *M. smegmatis* and *uvrB* in *M. tuberculosis* lead to increased susceptibility to UV radiation in vitro (Guthlein et al. 2009; Darwin and Nathan 2005; Gordhan et al. 1996). The deletion of both *uvrB* and *uvrD1* in *M. smegmatis* showed additive effect on survival in response to UV radiation indicating that one of the proteins is involved in another DNA repair pathway. The microarray analysis of *M. tuberculosis* genes induced under UV treatment in broth cultures indicated that the levels of expression of HR and NER gene products increase significantly. These proteins include UvrA, UvrB, UvrD2, PolA, RecA, RuvC, RuvA, LexA, RecX, and AdnA (Boshoff et al. 2003, 2004). These studies also highlighted a role for several other proteins like Ogt, Ada/AlkA, DinF, Lhr, DinX, and DnaE2 in DNA repair. Besides this, neither the relationship between HR and NER proteins nor the effect of double mutations in NER and HR proteins is completely understood in mycobacteria.

16.6 DNA Repair Pathways Are Active During Macrophage Infection

The genome of *M. tuberculosis* is exposed to DNA-damaging conditions inside the granulomas (Adams et al. 1997; Moller et al. 2001). Mice with a non-functional allele of gp91phox subunit of phagocyte oxidase cytochrome b and iNOS KO mice lacking an inducible nitric oxide synthase (iNOS) gene fail to produce RNS. Both these strains of mice experience exaggerated growth of mycobacterial infection (Adams et al. 1997; MacMicking et al. 1997; Chan et al. 1995). The microarray analysis revealed that the expression levels of genes involved in NER and BER are upregulated in the activated macrophages. Apart from these, *M. tuberculosis ada/alkA* transcript levels increase significantly in human or murine macrophages as well as in response to H_2O_2 in broth culture (Fontan et al. 2008; Schnappinger et al. 2003). Moreover, deletion of *ada/alkA* gene in *M. tuberculosis* leads to increased susceptibility to RNIs, but survival was found to be unaffected (Durbach et al. 2003). Similarly, deletion of *uvrB* gene in *M. smegmatis and M. tuberculosis* also lead to increased susceptibility to RNIs in broth cultures (Guthlein et al. 2009; Darwin and Nathan 2005).

The characterization of human granuloma is an active area of research (Tsai et al. 2006). Human granulomas are generally comprised of a central core with $CD68^+$ epithelioid cells, surrounded by macrophages, T lymphocytes (predominantly $CD4^+$), and multi-nucleate giant Langerhans cells. Another characteristic feature of human granulomas, which is different from murine granulomas, is the presence of central acellular eosinophilic region formed due to necrosis and apoptosis of cells. Because this central region does not possess any vascularization, the environment surrounding these cells becomes hypoxic (Aly et al. 2006). Proteome profiling, microarray analysis, and molecular genetic studies have shown that *M. tuberculosis* survives within granuloma by varying the expression levels of DNA repair genes, the expression of some of these genes were either upregulated or downregulated significantly under hypoxic conditions (Gorna et al. 2010; Kim et al. 2008; Saunders and Britton 2007; Bacon et al. 2004; Voskuil et al. 2004; Muttucumaru et al. 2004; Smeulders et al. 1999). In vitro studies show that *M. smegmatis* strains deleted for *ung* and *uvrB* genes display increased sensitivity to hypoxia (Kurthkoti et al. 2008; Kurthkoti and Varshney 2012). Studies with artificial granulomas of mice based on hollow-fibre model revealed that *uvrA* and *recF* were upregulated during dormancy stages (Karakousis et al. 2004). Mice exposed to aerosolized virulent strain of *M. tuberculosis* develop persistent infection and become chronic after 45–60 days post-infection. Transcriptional profiling during chronic and reactivation phases of murine tuberculosis using in vivo microarray analysis (IVMA) shows that even after 28 days of aerosol infection, CFU counts were not changed, although the bacilli were metabolically active with a 50% active transcriptome. A total of 137 genes were significantly regulated in mid-chronic tuberculosis (45 and 60 days) compared to an early stage (14 days) where the expression of genes involved in lipid and carbohydrate metabolism was significantly enriched. Additional genes, including the virulence regulator *virS* and the DNA repair protein *ung*, were upregulated during the reactivation stage, and *hupB* and *recO* were downregulated, indicating their possible roles in mycobacterial resurgence (Talaat et al. 2007). Altogether, these findings along with genome-wide expression analyses from human lung samples indicate that, while residing in granulomatous tissue, *M. tuberculosis* undergoes several changes in the expression of DNA repair genes.

The reactivation of infection begins with the disintegration of granulomas because of the failure of macrophages or T cells and with the decline in the host's immunity. Although little is known about reactivation of infection, various studies based on the effect of gene and/or protein expression levels or their activity on survival implicate a role for mycobacterial DNA repair genes. In vivo studies with immunocompromised BALB/C mice suggested upregulation in the expression levels of *M. tuberculosis* genes *ogt*, *dinG*, *ung*, and *recA* during reactivation. Of note, even though the gene expression profiles of all isolates were found to be similar, the profiles of upregulated DNA repair genes were different in granuloma, pericavity, and distant lung tissue (Talaat et al. 2007). The observed differences could be due to dissimilar physiology of cells; thus additional studies are needed to understand the complexity and reactivation of infection mounted by this respiratory pathogen.

16.7 *Mycobacterium tuberculosis* and Chemotherapy

The continuous exposure to the antibiotics leads to the emergence of drug resistance strains in *M. tuberculosis* (Karunakaran and Davies 2000; Martinez and Baquero 2000). Drug resistance strains can be divided into two subgroups: MDR (multidrug-resistant) and extensively drug-resistant (XDR) strains. These notations are based on the fact that MDR strains are resistant to two front-line drugs, isoniazid and rifampicin, whereas XDR strains are resistant to isoniazid and rifampicin including fluoroquinolones and second-line drugs (capreomycin, kanamycin, or amikacin). In general, antibiotic resistance is thought to arise from horizontal gene transfer carried by plasmids or transposons and random mutations in the chromosomal genes (Davies 1997). The genome-wide studies have indicated that the occurrence of drug-resistant strains is due to the generation of spontaneous mutations or gene inactivation (Cubillos-Ruiz et al. 2008; Filliol et al. 2006; Tang et al. 2005; Arnold et al. 2005). For instance, resistance towards anti-tubercular agents has been linked to SNPs in drug target genes, including *rpoB*, *katG*, and *uvrB* (Eldholm et al. 2014; Ramaswamy et al. 2003; Zaczek et al. 2009). Like all other organisms, DNA repair genes directly influence the mutation frequency in mycobacteria. For example, deletion of *ung*, *uvrB*, *uvrD*, *fpg*, and *udgB* in *M. smegmatis* and *ada/alkA* together with *ogt* in *M. tuberculosis* leads to increase in mutation rates (Dos Vultos et al. 2009). In view of these results, it was speculated that the use of antibiotics in combination with new drugs targeting the DNA repair genes might increase the effectiveness of chemotherapy. For example, it was found that *recA*-deficient mutant of BCG had increased susceptibility to metronidazole in vitro and an *ogt* mutant of BCG had increased susceptibility to isoniazid (Wiid et al. 2002; Sander et al. 2001). However, with the use of each drug, it would be important to ascertain that targeting DNA repair proteins do not result in an increased mutation rate that could lead to the emergence of drug-resistant strains (Gorna et al. 2010).

16.8 Why Study NER in *Mycobacterium tuberculosis*?

Although much is known about NER in *E. coli*, gaps remain in our understanding of this pathway in other bacteria. Several studies have provided experimental evidences that the damage search/recognition complex is comprised of two each of UvrA and UvrB subunits, but exactly how these subunits distinguish damaged from undamaged DNA remains unknown. Among many unanswered questions, the oligomeric status of UvrB and the nature of the preincision complex are unclear. Do two UvrB subunits bind at the site of a lesion? If UvrB exists as a dimer in solution, then what is the mechanism underlying the formation of $UvrA_2B_2$ complex as seen in *Geobacillus stearothermophilus* crystal structure? In this structure, the two UvrB subunits are located ~145 Å apart from each other and positioned ~80 Å away from the lesion (Pakotiprapha et al. 2012). This raises another question: which UvrB subunit will remain bound to the lesion during the successive steps of NER? Further, how UvrA dimer specifically transfers damaged strand to UvrB? How damage is detected among structurally diverse lesions and substrates by $UvrA_2B_2$? How UvrA is dislodged from the repair complex and how does UvrB recruit UvrC? What is the stoichiometry of UvrB and UvrC in the complex? What is the functional relevance of UvrC homodimers and homotetramers? There exist significant differences in the NER pathway between *E. coli* and other bacteria, but very less is known about NER, especially in bacterial pathogens such as *M. tuberculosis*. Furthermore, many species like *M. smegmatis*, *Streptomyces* species, *D. radiodurans*, and *Pseudomonas putida* possess a second copy of UvrA, but its actual role in vivo is unclear (Timmins et al. 2009; Tark et al. 2008).

The deletion of *Yersinia pseudotuberculosis* and *Borrelia burgdorferi uvrA* genes or disabling UvrA activity leads to their decreased virulence (Darwin and Nathan 2005; Garbom et al. 2004; Graham and Clark-Curtiss 1999). Additionally, the *uvrA* gene transcript levels are upregulated in *M. tuberculosis* grown in human macrophages (Graham and Clark-Curtiss 1999). Trans-complementation of *M. tuberculosis uvrB* demonstrated a crucial role in mounting resistance against nitrosative, oxidative, and UV exposure-induced DNA damage (Darwin and Nathan 2005; Graham and Clark-Curtiss 1999). *M. tuberculosis uvrB* has been shown to be important for attenuation of virulence in iNOS wild-type and iNOS knockout mice (Darwin and Nathan 2005). Notably, clinical isolates of XDR strains of *M. tuberculosis* harbour a unique *uvrB*^{A582V} mutation, an important (catalytic) active site residue (Eldholm et al. 2014). An inhibitor, [2-(5-amino-1,3,4-thiadiazol-2-ylbenzo[f]chromen-3-one) (ATBC)], against UvrABC complex works at a micromolar concentration, but the molecular basis of its activity and the specific target within the complex has not been deciphered (Mazloum et al. 2011). Thus, efforts are underway to identify *M. tuberculosis* key determinants of infection and vulnerable targets from the NER system whose structures and biochemical functions could be exploited for the development of new antitubercular agents (Ferraris et al. 2018).

Genetic analysis revealed that *M. tuberculosis* UvrA is important for survival against DNA-damage, including nitrosative and oxidative DNA damage, implicating this gene in the persistence of the bacillus in the host. Comparative analysis of the crystal structures of *M. tuberculosis* UvrA in its ligand free form with *B. stearothermophilus* UvrA•UvrB complex, *B. stearothermophilus* UvrA•ADP, and *Thermotoga maritima* UvrA•DNA showed the conformational plasticity displayed by UvrB binding domain of UvrA (Rossi et al. 2011). Altogether, combined genetic, biochemical, bioinformatics, and structural investigations of *M. tuberculosis* UvrA have provided significant insights into the mechanistic details of UvrA from *M. tuberculosis*. In-depth biochemical characterization revealed that the *M. tuberculosis* UvrB subunit behaves as a DNA-stimulated ATPase also possessing an ATP-dependent DNA helicase activity (Fig. 16.5) (Thakur et al. 2016). Recent structural and biochemical studies have demonstrated direct interaction between *M. tuberculosis* UvrA and UvrB proteins in the absence of any ligands and further suggested that this protein–protein interaction may be exploited as a target for blocking NER in *M. tuberculosis* (Ferraris et al. 2018; Lahiri et al. 2018; Singh 2017). Although biochemical and structural studies are limited in the case of *M. tuberculosis* UvrC, it is identified as an essential gene for survival in mouse model of infection in transposon mutagenesis screen, and mutations in this gene result in increased sensitivity against UV and hydrogen peroxide (Sassetti et al. 2003; Prammananan et al. 2012).

M. tuberculosis possesses two UvrD homologues: UvrD1 and UvrD2 (Cole 1998). *M. tuberculosis* UvrD1 is a DNA helicase with 3′–5′ polarity and unwinds nicked DNA substrates resembling NER intermediates but displays maximum DNA unwinding activity on replication forks (Curti et al. 2007; Singh et al. 2010). The strains in which *uvrD1* was inactivated either alone or in conjunction with *uvrA* showed increased susceptibility to UV irradiation, mitomycin C, RNI, and ROI. Moreover, *uvrD1* and *uvrA uvrD1* mutants showed attenuation following infection of mice either by aerosol or intravenous route (Houghton et al. 2012). *M. tuberculosis* UvrD1 and UvrA also suppress DNA strand exchange catalysed

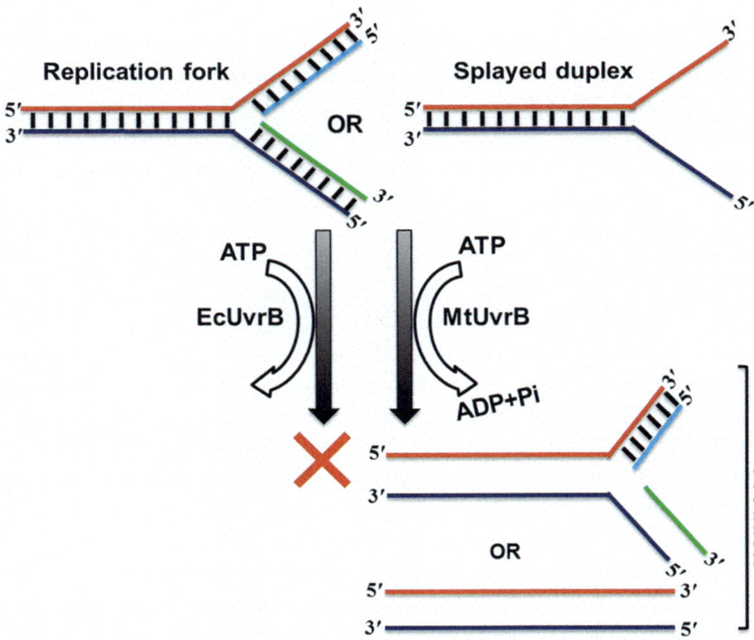

Fig. 16.5 Model depicting differences in the biochemical properties of MtUvrB and EcUvrB subunit. Each line (red and blue colour) represents a DNA strand with polarity as indicated. Experimental evidences support the notion that MtUvrB can unwind DNA replication intermediates like splayed duplex or replication fork DNA structures in an ATP-dependent manner. In contrast, EcUvrB neither hydrolyses ATP nor unwinds these structures of its own. EcUvrB needs EcUvrA to activate its ATPase as well as helicase activity. (Figure adapted with permission from Thakur et al. 2016) [Copyright 2016 American Chemical Society]

by *M. tuberculosis*, *M. smegmatis*, and *E. coli* RecA suggesting molecular cross talks with homologous pathway involved in chromosome stability (Singh et al. 2010). Inactivation of *uvrB* and *uvrD1* genes increased marker integration frequencies, further suggesting a role for these proteins in other DNA transaction processes beside NER (Guthlein et al. 2009). Additionally, mycobacterial Ku interacts with UvrD1 in a genome-wide yeast two-hybrid screen and stimulates UvrD1 to catalyse its ATP-dependent unwinding of 3′-tailed DNA. This interaction was further supported by the formation of a stable ternary complex of UvrD1, Ku, and DNA in the absence of ATP (Sinha et al. 2007). Altogether, these findings revealed structural and functional cross talk between NHEJ and NER. UvrD2 is the product of an essential gene which is comprised of N-terminal ATPase domain (Pfam00580) typical of superfamily I helicases and C-terminal HRDC domain (Pfam00570) typical of superfamily II helicases. These two domains are connected by a CxxC-(14)-CxxC tetracysteine module, which is present in helicases from *Actinomycetales* (Sinha et al. 2007). The C-terminal HRDC domain is not required for in vitro DNA helicase and ATPase activity, whereas tetracysteine module (but not the tetracysteines) is required for DNA unwinding. Surprisingly, the helicase activity of a truncated UvrD2 lacking the tetracysteine and HRDC domain can be restored by the addition of Ku protein (Sinha et al. 2007). These findings suggest that the Ku-dependent unwinding activity together with unusual domain structure of UvrD2 may have a biological function, distinct from that of *M. smegmatis* UvrD1. In *M. tuberculosis*, complementation by N-terminal domain of UvrD2 was able to compensate for the loss of *uvrD2* deletion suggesting that the C-terminal domains are not essential. Moreover, an ATPase dead form of UvrD2 protein was unable to translocate along DNA and cannot compensate for lack of the wild-type protein. These studies indicate that the role of UvrD2 is not to unwind DNA, but rather in DNA translocation and protein displacement (Williams et al. 2011). Since mycobacterial genome is rich in GC content, it is predicted that many genes involved in persistence and pathogenesis have G-quadruplex forming motifs in their promoter region, which might regulate their expression (Perrone et al. 2017). More recently, MtUvrD1 and MtUvrD2 were subjected to extensive in vitro biochemical studies, which revealed that these proteins have the potential to resolve G-quadruplex structures, implicating their important roles in maintenance of genome integrity and gene expression via G-quadruplex DNA resolution (Saha et al. 2019). The *polA* mutants of *M. smegmatis* displayed hypersensitivity to DNA damage induced by UV irradiation and hydrogen peroxide. Using transposon site hybridization (TraSH), it was found that *polA* gene is required for the optimal growth of *M. tuberculosis* (Gordhan et al. 1996; Sassetti et al. 2003). Mycobacteria possess *ligA* (encodes a NAD(+)-dependent enzyme), which is essential for its viability (Korycka-Machala et al. 2007; Srivastava et al. 2005). In addition, *M. tuberculosis* contains three additional ATP-dependent ligases, LigB, LigC, and LigD which are shown to be nonessential for cell growth under in vitro conditions (Gong et al. 2004). Interestingly LigD interacts with Prim-PolC in the presence of single nucleotide gap containing dsDNA substrate. Deletion of *ligD* gene was found to be sensitive to oxidative damage, suggesting the role of LigD in the excision repair other than NHEJ.

16.9 Concluding Remarks

M. tuberculosis, an exquisitely adapted human pathogen, encounters hostile environments during transmission, infection, and clinical latency. *M. tuberculosis* encodes a complex array of enzymes of several lipid pathways, DNA metabolic pathways, cell surface proteins, and regulators, which are essential for genome stability and cell survival in infected host cells. The attenuation of virulence of *M. tuberculosis uvrB* , *uvrA*, and *uvrD1* mutant strains in mice suggests a possible role for NER in pathogenesis. Although the genes encoding proteins involved in the NER pathway are conserved between E. coli and M. tuberculosis, the observation that NER components in tubercle bacillus exhibit significantly different mechanistic properties from the E. coli UvrABC proteins is worthy of further investigation. Additionally, it would be interesting to investigate the role of NER in the generation of MDR and XDR strains of *M. tuberculosis* and phenotypic heterogeneity among these strains. These studies might help to guide the development of more efficacious treatments targeting the NER components of *M. tuberculosis*. In conclusion, we envision that further research on the role(s) of NER components in tubercle bacillus may provide insights into the processes critical for genome maintenance and new opportunities for therapeutic intervention.

Acknowledgements K.M. acknowledges funding from J. C. Bose National Fellowship (DST, New Delhi), Bhatnagar Fellowship (CSIR, New Delhi) and a grant (BT/CoE/34/SP15232/2015) under the Center of Excellence from the Department of Biotechnology, New Delhi. M.T. was supported by a Senior Research Fellowship [09/079(2548)/2012-EMR-I] from the Council of Scientific and Industrial Research, New Delhi. We apologize to our colleagues whose work could not be cited due to space constraints.

References

Acharya S, Foster PL, Brooks P et al (2003) The coordinated functions of the *E. coli* MutS and MutL proteins in mismatch repair. Mol Cell 12:233–246

Adams LB, Dinauer MC, Morgenstern DE et al (1997) Comparison of the roles of reactive oxygen and nitrogen intermediates in the host response to *Mycobacterium tuberculosis* using transgenic mice. Tuber Lung Dis 78:237–246

Adikesavan AK, Katsonis P, Marciano DC et al (2011) Separation of recombination and SOS response in *Escherichia coli* RecA suggests LexA interaction sites. PLoS Genet 7:e1002244

Alekseyev YO, Hamm ML, Essigmann JM (2004) Aflatoxin B1 formamidopyrimidine adducts are preferentially repaired by the nucleotide excision repair pathway in vivo. Carcinogenesis 25:1045–1051

Aly S, Wagner K, Keller C et al (2006) Oxygen status of lung granulomas in *Mycobacterium tuberculosis*-infected mice. J Pathol 210:298–305

Arnold C, Westland L, Mowat G et al (2005) Single-nucleotide polymorphism-based differentiation and drug resistance detection in *Mycobacterium tuberculosis* from isolates or directly from sputum. Clin Microbiol Infect 11:122–130

Bacon J, James BW, Wernisch L et al (2004) The influence of reduced oxygen availability on pathogenicity and gene expression in *Mycobacterium tuberculosis*. Tuberculosis 84:205–217

Bergh S, Cole ST (1998) Using MycDB on the world-wide web. Methods Mol Biol 101:109–118

Bohr VA (1991) Gene specific damage and repair after treatment of cells with UV and chemotherapeutical agents. Adv Exp Med Biol 283:225–233

Boshoff HI, Reed MB, Barry CE et al (2003) DnaE2 polymerase contributes to in vivo survival and the emergence of drug resistance in *Mycobacterium tuberculosis*. Cell 113a:183–193

Boshoff HI, Myers TG, Copp BR et al (2004) The transcriptional responses of *Mycobacterium tuberculosis* to inhibitors of metabolism: novel insights into drug mechanisms of action. J Biol Chem 279:40174–40184

Boyce RP, Howard-Flanders P (1964) Release of ultraviolet light-induced thymine dimers from DNA in *E. coli* K-12. Proc Natl Acad Sci U S A 51:293–300

Brooks PC, Movahedzadeh F, Davis EO (2001) Identification of some DNA damage-inducible genes of *Mycobacterium tuberculosis*: apparent lack of correlation with LexA binding. J Bacteriol 183:4459–4467

Brosch R, Gordon SV, Billault A et al (1998) Use of a *Mycobacterium tuberculosis* H37Rv bacterial artificial chromosome library for genome mapping, sequencing, and comparative genomics. Infect Immun 66:2221–2229

Brosh RM Jr (2014) UvrD helicase: the little engine that could. Cell Cycle 13:1213–1215

Butala M, Zgur-Bertok D, Busby SJ (2009) The bacterial LexA transcriptional repressor. Cell Mol Life Sci 66:82–93

Cadet J, Wagner JR (2013) DNA base damage by reactive oxygen species, oxidizing agents, and UV radiation. Cold Spring Harb Perspect Biol 5:pii: a012559. https://doi.org/10.1101/cshperspect.a012559

Calsou P, Salles B (1985) Regulation of the SOS response analyzed by RecA protein amplification. J Bacteriol 162:1162–1165

Cambier CJ, Takaki KK, Larson RP et al (2014a) Mycobacteria manipulate macrophage recruitment through coordinated use of membrane lipids. Nature 505:218–222

Cambier CJ, Falkow S, Ramakrishnan L (2014b) Host evasion and exploitation schemes of *Mycobacterium tuberculosis*. Cell 159:1497–1509

Cannan WJ, Pederson DS (2016) Mechanisms and consequences of double-Strand DNA break formation in chromatin. J Cell Physiol 231:3–14

Caron PR, Kushner SR, Grossman L (1985) Involvement of helicase II (uvrD gene product) and DNA polymerase I in excision mediated by the UvrABC protein complex. Proc Natl Acad Sci U S A 82:4925–4929

Castaneda-Garcia A, Prieto AI, Rodriguez-Beltran J et al (2017) A non-canonical mismatch repair pathway in prokaryotes. Nat Commun 8:14246

Chan J, Tanaka K, Carroll D et al (1995) Effects of nitric oxide synthase inhibitors on murine infection with *Mycobacterium tuberculosis*. Infect Immun 63:736–740

Cole ST (1998) Comparative mycobacterial genomics. Curr Opin Microbiol 1:567–571

Cox MM (2007) Regulation of bacterial RecA protein function. Crit Rev Biochem Mol Biol 42:41–63

Crowley DJ, Boubriak I, Berquist BR et al (2006) The *uvrA, uvrB* and *uvrC* genes are required for repair of ultraviolet light induced DNA photoproducts in *Halobacterium sp.* NRC-1. Saline Syst 2:11. https://doi.org/10.1186/1746-1448-2-11

Cubillos-Ruiz A, Morales J, Zambrano MM (2008) Analysis of the genetic variation in *Mycobacterium tuberculosis* strains by multiple genome alignments. BMC Res Notes 1:110. https://doi.org/10.1186/1756-0500-1-110. PMID: 18992142

Curti E, Smerdon SJ, Davis EO (2007) Characterization of the helicase activity and substrate specificity of *Mycobacterium tuberculosis* UvrD. J Bacteriol 189:1542–1555

Darwin KH, Nathan CF (2005) Role for nucleotide excision repair in virulence of *Mycobacterium tuberculosis*. Infect Immun 73:4581–4587

Datta S, Prabu MM, Vaze MB et al (2000) Crystal structures of *Mycobacterium tuberculosis* RecA and its complex with ADP-AlF(4): implications for decreased ATPase activity and molecular aggregation. Nucleic Acids Res 28:4964–4973

Davies JE (1997) Origins, acquisition and dissemination of antibiotic resistance determinants. Ciba Found Symp 207:15–27

Davis EO, Sedgwick SG, Colston MJ (1991) Novel structure of the recA locus of Mycobacterium tuberculosis implies processing of the gene product. J Bacteriol 173:5653–5662

Davis EO, Dullaghan EM, Rand L (2002a) Definition of the mycobacterial SOS box and use to identify LexA-regulated genes in *Mycobacterium tuberculosis*. J Bacteriol 184:3287–3295

Davis EO, Springer B, Gopaul KK et al (2002b) DNA damage induction of recA in *Mycobacterium tuberculosis* independently of RecA and LexA. Mol Microbiol 46:791–800

Dexheimer TS (2012) DNA repair pathways and mechanisms. DNA repair of cancer stem cells. In: Mathews et al (eds) Springer Science+Business Media, Dordrecht, pp 19–32. https://doi.org/10.1007/978-94-007-4590-2_2

Dos Vultos T, Mestre O, Rauzier J et al (2008) Evolution and diversity of clonal bacteria: the paradigm of *Mycobacterium tuberculosis*. PLoS One 3:e1538

Dos Vultos T, Mestre O, Tonjum T et al (2009) DNA repair in Mycobacterium tuberculosis revisited. FEMS Microbiol Rev 33:471–487

Durbach SI, Springer B, Machowski EE et al (2003) DNA alkylation damage as a sensor of nitrosative stress in *Mycobacterium tuberculosis*. Infect Immun 71:997–1000

Ebrahimi-Rad M, Bifani P, Martin C et al (2003) Mutations in putative mutator genes of *Mycobacterium tuberculosis* strains of the W-Beijing family. Emerg Infect Dis 9:838–845

Ehrt S, Schnappinger D (2009) Mycobacterial survival strategies in the phagosome: defence against host stresses. Cell Microbiol 11:1170–1178

Ehrt S, Rhee K, Schnappinger D (2015) Mycobacterial genes essential for the pathogen's survival in the host. Immunol Rev 264:319–326

Eldholm V, Norheim G, von der Lippe B et al (2014) Evolution of extensively drug-resistant *Mycobacterium tuberculosis* from a susceptible ancestor in a single patient. Genome Biol 15:490. https://doi.org/10.1186/s13059-014-0490-3

Fang FC (1997) Perspectives series: host/pathogen interactions. Mechanisms of nitric oxide-related antimicrobial activity. J Clin Invest 99:2818–2825

Ferraris DM, Miggiano R, Rossi F, Rizzi M (2018) *Mycobacterium tuberculosis* molecular determinants of infection, survival strategies, and vulnerable targets. Pathogens 7:pii: E17. https://doi.org/10.3390/pathogens7010017

Filliol I, Motiwala AS, Cavatore M et al (2006) Global phylogeny of *Mycobacterium tuberculosis* based on single nucleotide polymorphism (SNP) analysis: insights into tuberculosis evolution, phylogenetic accuracy of other DNA fingerprinting systems, and recommendations for a minimal standard SNP set. J Bacteriol 188:759–772

Fontan P, Aris V, Ghanny S et al (2008) Global transcriptional profile of *Mycobacterium tuberculosis* during THP-1 human macrophage infection. Infect Immun 76:717–725

Friedberg EC (2016) A history of the DNA repair and mutagenesis field: the discovery of base excision repair. DNA Repair 37:A35–A39

Fu LM, Fu-Liu CS (2007) The gene expression data of *Mycobacterium tuberculosis* based on Affymetrix gene chips provide insight into regulatory and hypothetical genes. BMC Microbiol 7:37. PMID: 17501996

Garbom S, Forsberg A, Wolf-Watz H et al (2004) Identification of novel virulence-associated genes via genome analysis of hypothetical genes. Infect Immun 72:1333–1340

Gengenbacher M, Kaufmann SH (2012) *Mycobacterium tuberculosis*: success through dormancy. FEMS Microbiol Rev 36:514–532

Gong C, Martins A, Bongiorno P et al (2004) Biochemical and genetic analysis of the four DNA ligases of mycobacteria. J Biol Chem 279:20594–20606

Goosen N (2010) Scanning the DNA for damage by the nucleotide excision repair machinery. DNA Repair 9:593–596

Gopinath V, Raghunandanan S, Gomez RL et al (2015) Profiling the proteome of *Mycobacterium tuberculosis* during dormancy and reactivation. Mol Cell Proteomics 14:2160–2176

Gordhan BG, Andersen SJ, De Meyer AR et al (1996) Construction by homologous recombination and phenotypic characterization of a DNA polymerase domain *polA* mutant of *Mycobacterium smegmatis*. Gene 178:125–130

Gorna AE, Bowater RP, Dziadek J (2010) DNA repair systems and the pathogenesis of *Mycobacterium tuberculosis:* varying activities at different stages of infection. Clin Sci 119:187–202

Graham JE, Clark-Curtiss JE (1999) Identification of *Mycobacterium tuberculosis* RNAs synthesized in response to phagocytosis by human macrophages by selective capture of transcribed sequences (SCOTS). Proc Natl Acad Sci U S A 96:11554–11559

Grogan DW (2000) The question of DNA repair in hyperthermophilic archaea. Trends Microbiol 8:180–185

Guthlein C, Wanner RM, Sander P et al (2009) Characterization of the mycobacterial NER system reveals novel functions of the UvrD1 helicase. J Bacteriol 191:555–562

Hakem R (2008) DNA-damage repair; the good, the bad, and the ugly. EMBO J 27:589–605

Hanna MN, Ferguson RJ, Li YH et al (2001) *uvrA is* an acid-inducible gene involved in the adaptive response to low pH in *Streptococcus* mutants. J Bacteriol 183:5964–5973

Hoeijmakers JH (2001) Genome maintenance mechanisms for preventing cancer. Nature 411:366–374

Houghton J, Townsend C, Williams AR et al (2012) Important role for *Mycobacterium tuberculosis* UvrD1 in pathogenesis and persistence apart from its function in nucleotide excision repair. J Bacteriol 194:2916–2923

Howard-Flanders P, Theriot LA (1962) A method for selecting radiation-sensitive mutants of *Escherichia coli*. Genetics 47:1219–1224

Howard-Flanders P, Boyce RP, Theriot L (1962a) Mechanism of sensitization to ultra-violet light of T1 bacteriophage by the incorporation of 5-bromodeoxyuridine or by pre-irradiation of the host cell. Nature 195:51–54

Howard-Flanders P, Boyce RP, Simson E et al (1962b) Genetic locus in *E. coli* K12 that controls the reactivation of UV-photoproducts associated with thymine in DNA. Proc Natl Acad Sci U S A 48:2109–2115

Howard-Flanders P, Boyce RP, Theriot L (1966) Three loci in *Escherichia coli* K-12 that control the excision of pyrimidine dimers and certain other mutagen products from DNA. Genetics 53:1119–1136

Hu J, Selby CP, Adar S et al (2017) Molecular mechanisms and genomic maps of DNA excision repair in *Escherichia coli* and humans. J Biol Chem 292:15588–15597

Husain I, Chaney SG, Sancar A (1985) Repair of cis-platinum-DNA adducts by ABC excinuclease in vivo and in vitro. J Bacteriol 163:817–823

Imoto S, Bransfield LA, Croteau DL et al (2008) DNA tandem lesion repair by strand displacement synthesis and nucleotide excision repair. Biochemistry 47:4306–4316

Iyama T, Wilson DM III (2013) DNA repair mechanisms in dividing and non-dividing cells. DNA Repair (Amst) 12:620–636

Jaciuk M, Nowak E, Skowronek K et al (2011) Structure of UvrA nucleotide excision repair protein in complex with modified DNA. Nat Struct Mol Biol 18:191–197

Janion C (2001) Some aspects of the SOS response system-a critical survey. Acta Biochim Pol 48:599–610

Janion C (2008) Inducible SOS response system of DNA repair and mutagenesis in *Escherichia coli*. Int J Biol Sci 4:338–344

Jeggo PA, Pearl LH, Carr AM (2015) DNA repair, genome stability and cancer: a historical perspective. Nat Rev Cancer 16:35–42

Kad NM, Wang H, Kennedy GG, Warshaw DM et al (2010) Collaborative dynamic DNA scanning by nucleotide excision repair proteins investigated by single- molecule imaging of quantum-dot-labeled proteins. Mol Cell 37:702–713

Kang J, Blaser MJ (2006) Bacterial populations as perfect gases: genomic integrity and diversification tensions in *Helicobacter pylori*. Nat Rev Microbiol 4:826–836

Karakousis PC, Yoshimatsu T, Lamichhane G et al (2004) Dormancy phenotype displayed by extracellular *Mycobacterium tuberculosis* within artificial granulomas in mice. J Exp Med 200:647–657

Karunakaran P, Davies J (2000) Genetic antagonism and hypermutability in *Mycobacterium smegmatis*. J Bacteriol 182:3331–3335

Kim SY, Lee BS, Shin SJ et al (2008) Differentially expressed genes in *Mycobacterium tuberculosis* H37Rv under mild acidic and hypoxic conditions. J Med Microbiol 57:1473–1480

Kisker C, Kuper J, Van Houten B (2013) Prokaryotic nucleotide excision repair. Cold Spring Harb Perspect Biol 5:a012591

Ko G, First MW, Burge HA (2002) The characterization of upper-room ultraviolet germicidal irradiation in inactivating airborne microorganisms. Environ Health Perspect 110:95–101

Korycka-Machala M, Brzostek A, Rozalska S et al (2006) Distinct DNA repair pathways involving RecA and nonhomologous end joining in *Mycobacterium smegmatis*. FEMS Microbiol Lett 258:83–91

Korycka-Machala M, Rychta E, Brzostek A et al (2007) Evaluation of NAD(+) -dependent DNA ligase of mycobacteria as a potential target for antibiotics. Antimicrob Agents Chemother 51:2888–2897

Kowalczykowski SC, Dixon DA, Eggleston AK et al (1994) Biochemistry of homologous recombination in *Escherichia coli*. Microbiol Rev 58:401–465
Kumar RA, Vaze MB, Chandra NR et al (1996) Functional characterization of the precursor and spliced forms of RecA protein of *Mycobacterium tuberculosis*. Biochemistry 35:1793–1802
Kurthkoti K, Varshney U (2012) Distinct mechanisms of DNA repair in mycobacteria and their implications in attenuation of the pathogen growth. Mech Ageing Dev 133:138–146
Kurthkoti K, Kumar P, Jain R, Varshney U (2008) Important role of the nucleotide excision repair pathway in Mycobacterium smegmatis in conferring protection against commonly encountered DNA-damaging agents. Microbiology 154(9):2776–2785
Kuzminov A (2001) Single-strand interruptions in replicating chromosomes cause double-strand breaks. Proc Natl Acad Sci U S A 98:8241–8246
Lahiri S, Rizzi M, Rossi F et al (2018) *Mycobacterium tuberculosis* UvrB forms dimers in solution and interacts with UvrA in the absence of ligands. Proteins 86:98–109
Lee JY, Yang W (2006) UvrD helicase unwinds DNA one base pair at a time by a two-part power stroke. Cell 127:1349–1360
Lever MS, Williams A, Bennett AM (2000) Survival of mycobacterial species in aerosols generated from artificial saliva. Lett Appl Microbiol 31:238–241
Levitte S, Adams KN, Berg RD et al (2016) Mycobacterial acid tolerance enables phagolysosomal survival and establishment of tuberculous infection in vivo. Cell Host Microbe 20:250–258
Li GM (2008) Mechanisms and functions of DNA mismatch repair. Cell Res 18:85–98
Lieber MR (2010) The mechanism of double-strand DNA break repair by the nonhomologous DNA end-joining pathway. Annu Rev Biochem 79:181–211
Lin JJ, Sancar A (1989) A new mechanism for repairing oxidative damage to DNA: (A)BC excinuclease removes AP sites and thymine glycols from DNA. Biochemistry 28:7979–7984
Lin CG, Kovalsky O, Grossman L (1997) DNA damage-dependent recruitment of nucleotide excision repair and transcription proteins to *Escherichia coli* inner membranes. Nucleic Acids Res 25:3151–3158
Lindahl T, Wood RD (1999) Quality control by DNA repair. Science 286:1897–1905
Loudon RG, Bumgarner LR, Lacy J et al (1969) Aerial transmission of mycobacteria. Am Rev Respir Dis 100:165–171
MacMicking JD, North RJ, LaCourse R et al (1997) Identification of nitric oxide synthase as a protective locus against tuberculosis. Proc Natl Acad Sci U S A 94:5243–5248
Malta E, Moolenaar GF, Goosen N (2007) Dynamics of the UvrABC nucleotide excision repair proteins analyzed by fluorescence resonance energy transfer. Biochemistry 46:9080–9088
Marcato LG, Ferlini AP, Bonfim RC et al (2008) The role of toll-like receptors 2 and 4 on reactive oxygen species and nitric oxide production by macrophage cells stimulated with root canal pathogens. Oral Microbiol Immunol 23:353–359
Martinez JL, Baquero F (2000) Mutation frequencies and antibiotic resistance. Antimicrob Agents Chemother 44:1771–1777
Matthews LA, Simmons LA (2014) Bacterial nonhomologous end joining requires teamwork. J Bacteriol 196:3363–3365
Mazloum N, Stegman MA, Croteau DL et al (2011) Identification of a chemical that inhibits the mycobacterial UvrABC complex in nucleotide excision repair. Biochemistry 50:1329–1335
Minko IG, Zou Y, Lloyd RS (2002) Incision of DNA-protein crosslinks by UvrABC nuclease suggests a potential repair pathway involving nucleotide excision repair. Proc Natl Acad Sci U S A 99:1905–1909
Minko IG, Kurtz AJ, Croteau DL et al (2005) Initiation of repair of DNA-polypeptide cross-links by the UvrABC nuclease. Biochemistry 44:3000–3009
Mizrahi V, Andersen SJ (1998) DNA repair in *Mycobacterium tuberculosis*. What have we learnt from the genome sequence? Mol Microbiol 29:1331–1339
Moller P, Loft S, Lundby C et al (2001) Acute hypoxia and hypoxic exercise induce DNA strand breaks and oxidative DNA damage in humans. FASEB J 15:1181–1186

Moolenaar GF, van Rossum-Fikkert S, van Kesteren M et al (2002) Cho, a second endonuclease involved in *Escherichia coli* nucleotide excision repair. Proc Natl Acad Sci U S A 99:1467–1472

Movahedzadeh F, Colston MJ, Davis EO (1997a) Characterization of *Mycobacterium tuberculosis* LexA: recognition of a Cheo (Bacillus-type SOS) box. Microbiology 143:929–936

Movahedzadeh F, Colston MJ, Davis EO (1997b) Determination of DNA sequences required for regulated *Mycobacterium tuberculosis* RecA expression in response to DNA-damaging agents suggests that two modes of regulation exist. J Bacteriol 179:3509–3518

Muttucumaru DG, Roberts G, Hinds J et al (2004) Gene expression profile of *Mycobacterium tuberculosis* in a non-replicating state. Tuberculosis 84:239–246

Nakano T, Katafuchi A, Terato H, Suzuki T et al (2005) Activity of nucleotide excision repair enzymes for oxanine cross-link lesions. Nucl Acids Symp Ser (Oxf) 49:293–294

Nouvel LX, Dos Vultos T, Kassa-Kelembho E et al (2007) A non-sense mutation in the putative anti-mutator gene *ada/alkA* of *Mycobacterium tuberculosis* and *M. bovis* isolates suggests convergent evolution. BMC Microbiol 7:39. PMID: 17506895

Ogrunc M, Becker DF, Ragsdale SW et al (1998) Nucleotide excision repair in the third kingdom. J Bacteriol 180:5796–5798

Orren DK, Selby CP, Hearst JE et al (1992) Post-incision steps of nucleotide excision repair in *Escherichia coli*. Disassembly of the UvrBC-DNA complex by helicase II and DNA polymerase I. J Biol Chem 267:780–788

Pakotiprapha D, Samuels M, Shen K et al (2012) Structure and mechanism of the UvrA-UvrB DNA damage sensor. Nat Struct Mol Biol 19:291–298

Pani B, Nudler E (2017) Mechanistic insights into transcription coupled DNA repair. DNA Repair (Amst) 56:42–50

Paul-Konietzko K, Thomale J, Arakawa H et al (2015) DNA ligases I and III support nucleotide excision repair in DT40 cells with similar efficiency. Photochem Photobiol 91:1173–1180

Peccia J, Hernandez M (2004) UV-induced inactivation rates for airborne *Mycobacterium bovis BCG*. J Occup Environ Hyg 1:430–435

Perrone R, Lavezzo E, Riello E et al (2017) Mapping and characterization of G-quadruplexes in *Mycobacterium tuberculosis* gene promoter regions. Sci Rep 7:5743. https://doi.org/10.1038/s41598-017-05867-z

Philipp WJ, Gordon S, Telenti A et al (1998a) Pulsed field gel electrophoresis for mycobacteria. Methods Mol Biol 101:51–63

Philipp WJ, Schwartz DC, Telenti A et al (1998b) Mycobacterial genome structure. Electrophoresis 19:573–576

Pitcher RS, Green AJ, Brzostek A et al (2007) NHEJ protects mycobacteria in stationary phase against the harmful effects of desiccation. DNA Repair 6:1271–1276

Prammananan T, Phunpruch S, Jaitrong S et al (2012) *Mycobacterium tuberculosis uvrC* essentiality in response to UV-induced cell damage. Southeast Asian J Trop Med Public Health 43:370–375

Pu WT, Kahn R, Munn MM et al (1989) UvrABC incision of N-methylmitomycin A-DNA monoadducts and cross-links. J Biol Chem 264:20697–20704

Radman M (1975) SOS repair hypothesis: phenomenology of an inducible DNA repair which is accompanied by mutagenesis. Basic Life Sci 5A:355–367

Radman M, Wagner R (1986) Mismatch repair in *Escherichia coli*. Annu Rev Genet 20:523–538

Ramaswamy SV, Reich R, Dou SJ et al (2003) Single nucleotide polymorphisms in genes associated with isoniazid resistance in *Mycobacterium tuberculosis*. Antimicrob Agents Chemother 47:1241–1250

Rand L, Hinds J, Springer B et al (2003) The majority of inducible DNA repair genes in *Mycobacterium tuberculosis* are induced independently of RecA. Mol Microbiol 50:1031–1042

Reardon JT, Sancar A (2005) Nucleotide excision repair. Prog Nucleic Acid Res Mol Biol 79:183–235

Riley RL, Knight M, Middlebrook G (1976) Ultraviolet susceptibility of BCG and virulent tubercle bacilli. Am Rev Respir Dis 113:413–418

Rossi F, Khanduja JS, Bortoluzzi A et al (2011) The biological and structural characterization of Mycobacterium tuberculosis UvrA provides novel insights into its mechanism of action. Nucleic Acids Res 39:7316–7328

Runyon GT, Bear DG, Lohman TM (1990) *Escherichia coli* helicase II (UvrD) protein initiates DNA unwinding at nicks and blunt ends. Proc Natl Acad Sci U S A 87:6383–6387

Ryan KA, Smith MF Jr, Sanders MK et al (2004) Reactive oxygen and nitrogen species differentially regulate toll-like receptor 4-mediated activation of NF-kappa B and interleukin-8-expression. Infect Immun 72:2123–2130

Saha T, Shukla K, Thakur RS et al (2019) *Mycobacterium tuberculosis* UvrD1 and UvrD2 helicases unwind G-quadruplex DNA. FEBS J. https://doi.org/10.1111/febs.14798. PMID: 30821905

Salem AM, Nakano T, Takuwa M et al (2009) Genetic analysis of repair and damage tolerance mechanisms for DNA-protein cross-links in *Escherichia coli*. J Bacteriol 191:5657–5668

Sancar A, Franklin KA, Sancar G et al (1985) Repair of psoralen and acetylaminofluorene DNA adducts by ABC excinuclease. J Mol Biol 184:725–734

Sander P, Papavinasasundaram KG, Dick T et al (2001) *Mycobacterium bovis BCG recA* deletion mutant shows increased susceptibility to DNA-damaging agents but wild-type survival in a mouse infection model. Infect Immun 69:3562–3568

Sassetti CM, Rubin EJ (2003) Genetic requirements for mycobacterial survival during infection. Proc Natl Acad Sci U S A 100:12989–12994

Sassetti CM, Boyd DH, Rubin EJ (2003) Genes required for mycobacterial growth defined by high density mutagenesis. Mol Microbiol 48:77–84

Saunders BM, Britton WJ (2007) Life and death in the granuloma: immunopathology of tuberculosis. Immunol Cell Biol 85:103–111

Schlacher K, Goodman MF (2007) Lessons from 50 years of SOS DNA-damage-induced mutagenesis. Nat Rev Mol Cell Biol 8:587–594

Schlosser-Silverman E, Elgrably-Weiss M, Rosenshine I et al (2000) Characterization of *Escherichia coli* DNA lesions generated within J774 macrophages. J Bacteriol 182:5225–5230

Schnappinger D, Ehrt S, Voskuil MI et al (2003) Transcriptional adaptation of Mycobacterium tuberculosis within macrophages: insights into the phagosomal environment. J Exp Med 198:693–704

Seeberg E, Steinum AL, Nordenskjold M et al (1983) Strand-break formation in DNA modified by benzo[alpha]pyrene diolepoxide. Quantitative cleavage by Escherichia coli uvrABC endonuclease. Mutat Res 112:139–145

Selby CP, Sancar A (1991) Noncovalent drug-DNA binding interactions that inhibit and stimulate (A)BC excinuclease. Biochemistry 30:3841–3849

Setlow RB, Carrier WL (1964) The disappearance of thymine dimers from DNA: an error-correcting mechanism. Proc Natl Acad Sci U S A 51:226–231

Setlow RB, Setlow JK (1962) Evidence that ultraviolet-induced thymine dimers in DNA cause biological damage. Proc Natl Acad Sci U S A 48:1250–1257

Setlow JK, Setlow RB (1963) Nature of photoreactivable ultra-violet lesion in deoxyribonucleic acid. Nature 197:560–562

Setlow RB, Swenson PA, Carrier WL (1963) Thymine dimers and inhibition of DNA synthesis by ultraviolet irradiation of cells. Science 142:1464–1466

Shuman S, Glickman MS (2007) Bacterial DNA repair by non-homologous end joining. Nat Rev Microbiol 5:852–861

Simmons LA, Foti JJ, Cohen SE et al (2008) The SOS regulatory network. EcoSal Plus 3:1. https://doi.org/10.1128/ecosalplus.5.4.3. PMID:26443738

Singh A (2017) Guardians of the mycobacterial genome: a review on DNA repair systems in *Mycobacterium tuberculosis*. Microbiology 163:1740–1758

Singh P, Patil KN, Khanduja JS et al (2010) *Mycobacterium tuberculosis* UvrD1 and UvrA proteins suppress DNA strand exchange promoted by cognate and noncognate RecA proteins. Biochemistry 49:4872–4883

Sinha KM, Stephanou NC, Gao F et al (2007) Mycobacterial UvrD1 is a Ku-dependent DNA helicase that plays a role in multiple DNA repair events, including double-strand break repair. J Biol Chem 282:15114–15125

Sinha KM, Stephanou NC, Unciuleac MC et al (2008) Domain requirements for DNA unwinding by mycobacterial UvrD2, an essential DNA helicase. Biochemistry 47:9355–9364

Smeulders MJ, Keer J, Speight RA et al (1999) Adaptation of Mycobacterium smegmatis to stationary phase. J Bacteriol 181:270–283

Smith KC, Sharma RC (1987) A model for the recA-dependent repair of excision gaps in UV-irradiated *Escherichia coli*. Mutat Res 183:1–9

Smith KC, Wang TV, Sharma RC (1987) recA-dependent DNA repair in UV-irradiated *Escherichia coli*. J Photochem Photobiol B 1:1–11

Snowden A, Kow YW, Van Houten B (1990) Damage repertoire of the *Escherichia coli* UvrABC nuclease complex includes abasic sites, base-damage analogues, and lesions containing adjacent 5′ or 3′ nicks. Biochemistry 29:7251–7259

Sreevatsan S, Pan X, Stockbauer KE et al (1997) Restricted structural gene polymorphism in the *Mycobacterium tuberculosis* complex indicates evolutionarily recent global dissemination. Proc Natl Acad Sci U S A 94:9869–9874

Srivastava SK, Tripathi RP, Ramachandran R (2005) NAD+-dependent DNA ligase (Rv3014c) from *Mycobacterium tuberculosis*. Crystal structure of the adenylation domain and identification of novel inhibitors. J Biol Chem 280:30273–30281

Talaat AM, Ward SK, Wu CW et al (2007) Mycobacterial bacilli are metabolically active during chronic tuberculosis in murine lungs: insights from genome-wide transcriptional profiling. J Bacteriol 189:4265–4274

Tang X, Morris SL, Langone JJ et al (2005) Microarray and allele specific PCR detection of point mutations in *Mycobacterium tuberculosis* genes associated with drug resistance. J Microbiol Methods 63:318–330

Tark M, Tover A, Koorits L et al (2008) Dual role of NER in mutagenesis in *Pseudomonas putida*. DNA Repair 7:20–30

Thakur M, Kumar MB, Muniyappa K (2016) *Mycobacterium tuberculosis* UvrB is a robust DNA-stimulated ATPase that also possesses structure-specific ATP-dependent DNA helicase activity. Biochemistry 55:5865–5883

Thomas DC, Kunkel TA, Casna NJ et al (1986) Activities and incision patterns of ABC excinuclease on modified DNA containing single-base mismatches and extrahelical bases. J Biol Chem 261:14496–14505

Timmins J, Gordon E, Caria S et al (2009) Structural and mutational analyses of *Deinococcus radiodurans* UvrA2 provide insight into DNA binding and damage recognition by UvrAs. Structure 17:547–558

Truglio JJ, Croteau DL, Van Houten B et al (2006) Prokaryotic nucleotide excision repair: the UvrABC system. Chem Rev 106:233–252

Tsai MC, Chakravarty S, Zhu G et al (2006) Characterization of the tuberculous granuloma in murine and human lungs: cellular composition and relative tissue oxygen tension. Cell Microbiol 8:218–232

Vaisman A, McDonald JP, Huston D et al (2013) Removal of misincorporated ribonucleotides from prokaryotic genomes: an unexpected role for nucleotide excision repair. PLoS Genet 9: e1003878

Van Houten B, Kad N (2014) Investigation of bacterial nucleotide excision repair using single-molecule techniques. DNA Repair 20:41–48

Van Houten B, Sancar A (1987) Repair of N-methyl-N′-nitro-N-nitrosoguanidine-induced DNA damage by ABC excinuclease. J Bacteriol 169:540–545

Van Houten B, Gamper H, Hearst JE et al (1988) Analysis of sequential steps of nucleotide excision repair in *Escherichia coli* using synthetic substrates containing single psoralen adducts. J Biol Chem 263:16553–16560

Van Houten B, Croteau DL, DellaVecchia MJ et al (2005) 'Close-fitting sleeves': DNA damage recognition by the UvrABC nuclease system. Mutat Res 577:92–117

Voskuil MI, Visconti KC, Schoolnik GK (2004) *Mycobacterium tuberculosis* gene expression during adaptation to stationary phase and low-oxygen dormancy. Tuberculosis 84:218–227

Wiid I, Grundlingh R, Bourn W et al (2002) O(6)-alkylguanine-DNA alkyltransferase DNA repair in mycobacteria: pathogenic and non-pathogenic species differ. Tuberculosis 82:45–53

Williams A, Guthlein C, Beresford N et al (2011) UvrD2 is essential in *Mycobacterium tuberculosis,* but its helicase activity is not required. J Bacteriol 193:4487–4494

Wink DA, Kasprzak KS, Maragos CM et al (1991) DNA deaminating ability and genotoxicity of nitric oxide and its progenitors. Science 254:1001–1003

Yi C, He C (2013) DNA repair by reversal of DNA damage. Cold Spring Harb Perspect Biol 5(1): a012575. https://doi.org/10.1101/cshperspect.a012575

Youngs DA, Van der Schueren E, Smith KC (1974) Separate branches of the *uvr* gene-dependent excision repair process in ultraviolet-irradiated *Escherichia coli* K-12 cells; their dependence upon growth medium and the *polA, recA, recB,* and *exrA* genes. J Bacteriol 117:717–725

Zaczek A, Brzostek A, Augustynowicz-Kopec E et al (2009) Genetic evaluation of relationship between mutations in *rpoB* and resistance of *Mycobacterium tuberculosis* to rifampin. BMC Microbiol 9:10. https://doi.org/10.1186/1471-2180-9-10

Mesenchymal Stem Cells: A Hidden Arsenal for Mtb Persistence, Resuscitation, and Reactivation

17

Jaishree Garhyan, Bikul Das, and Rakesh Bhatnagar

Abstract

Mycobacterium tuberculosis (Mtb), the causative agent of tuberculosis, is a global public health threat and remains the leading cause of death among infectious diseases. The pathogen has developed strategies to survive inside and outside diverse host cells under wide range of unfavorable conditions, leading to persistence, and also undergo reactivation in the event of favorable conditions. Several sources of endogenous reactivation are proposed, including specific host cells that could harbor viable and dormant Mtb in the protective intracellular compartments. However, it is not clear how these host cells could survive the hypoxic/oxidative stress microenvironment prevalent in the post-drug treated granulomas. Recently Das et al. identified human CD271+ bone marrow mesenchymal stem cells (CD271+ BM-MSCs) as niche host cells that can maintain viable, non-replicating dormant Mtb. Most importantly, from patients who had successfully completed anti-tubercular therapy, viable Mtb could be isolated from

J. Garhyan
Department of Microbiology, University of Massachusetts, Amherst, MA, USA

Molecular Biology and Genetic Engineering Laboratory, School of Biotechnology, Jawaharlal Nehru University, New Delhi, Delhi, India

B. Das
Department of Stem Cell and Infectious Diseases, Kavikrishna Laboratory, Guwahati Biotech Park, Indian Institute of Technology, Guwahati, India

Department of Stem Cells and Infectious Diseases, Thoreau Laboratory for Global Health, M2D2, University of Massachusetts, Lowell, MA, USA

R. Bhatnagar (✉)
Molecular Biology and Genetic Engineering Laboratory, School of Biotechnology, Jawaharlal Nehru University, New Delhi, Delhi, India
e-mail: rakeshbhatnagar@jnu.ac.in

S. E. Hasnain et al. (eds.), Mycobacterium tuberculosis: *Molecular Infection Biology, Pathogenesis, Diagnostics and New Interventions*,
https://doi.org/10.1007/978-981-32-9413-4_17

CD271+ BM-MSCs, and not in other types of cells including macrophages. The follow-up study showed that Mtb could escape anti-tubercular drug therapy by hiding intracellular to CD271+ BM-MSCs serving as source for endogenous reactivation of TB. With these studies at the forefront, a novel model of granuloma reactivation could be proposed where CD271+ BM-MSCs harboring viable Mtb migrate to inflammatory granulomas for disease reactivation. Moreover, CD271+BM-MSCs also are able to resuscitate the viable but non-culturable (VBNC) Mtb inside the granuloma to reactivate the disease. Studies show that pulmonary tuberculosis (PTB) patients with culture negative sputum contain an occult population of VBNC Mtb in their sputum. CD271+ host-Mtb interaction studies suggest that these VBNC may be resuscitate back to culturable state by CD271+MSCs. Study of mechanism of Mtb survival and resuscitation inside MSCs is challenging and ongoing. Hypoxia may play an important role in the survival and persistence of Mtb inside the MSCs in TB patients post therapy as well in mice drug-induced dormancy model indicated by upregulation of HIF-1alpha. Latent TB has been challenging and difficult to study because of the dynamic capability of Mtb to adapt easily for over 200 years in vivo and in vitro. Despite all the ongoing efforts, so far, targeting dormant TB seems difficult. Although there are drugs which target cell wall synthesis, transcription, translation, and ATPase activity in actively dividing stage, their efficacy against dormant bacilli is unclear. Isoniazid treatment for 9 months is recommended for latent or dormant TB; however, non-adherence to such long regimen is a problem; moreover in light of current studies the efficacy of isoniazid against dormant bacilli is questionable. Studies on Mtb residing inside MSCs may shed light into new therapeutics to target non-replicating bacteria as well as VBNC Mtb and prevent endogenous reactivation of the disease.

Keywords

Mesenchymal stem cells · Mtb persistence · Dynamic granuloma model · Resuscitation

17.1 Introduction

Mycobacterium tuberculosis (Mtb) causes the deadly disease tuberculosis (TB) in millions of humans, killing nearly 2 million yearly, despite a vigorous host immune response and months of effective drug treatment, which transform active granulomas into fibrotic and hypoxic sterile granulomas. Mtb has the spectacular ability of long-term persistence and reactivation despite vigorous host immunity and prolonged anti-tubercular therapy (Bloom and McKinney 1999). Studies indicate that after successful drug treatment, Mtb persists in a non-replicating dormant state in the sterile granulomas and/or extra-granuloma sites. Epidemiological studies indicate that these endogenous dormant Mtb may cause disease recurrence. Although TB recurrence may also occur due to exogenous re-infection with a new Mtb strain, the reactivation of endogenous dormant Mtb is a significant source of recurrent TB. Most clinical cases are not due to new infections but through reactivation of dormant *Mtb* residing in LTBI hosts (Gengenbacher and Kaufmann 2012). Hence, understanding the endogenous reactivation is imperative to develop novel management strategies against recurrent TB. However, the source of endogenous Mtb reactivation is not clearly known. Several sources of endogenous reactivation are proposed, including specific host cells that could harbor viable and dormant Mtb in the protective intracellular compartments. However, it is not clear how these host cells could survive the hypoxic/oxidative stress microenvironment prevalent in the post-drug-treated granulomas. Recently Das et al. identified human CD271+ bone marrow mesenchymal stem cells as niche cells that can maintain viable, non-replicating dormant Mtb (Das et al. 2013). Most importantly, in patients who had successfully completed anti-tubercular therapy, viable Mtb could be isolated from CD271+ BM-MSCs, and not in other types of cells including macrophages, which was presumed to be the primary host cells for dormant Mtb. Moreover, CD271+ BM-MSCs protects viable Mtb from anti-TB drug therapy. Mtb escapes anti-tubercular drug therapy by hiding intracellular to CD271+ BM-MSCs serving as a depot for endogenous reactivation. Therefore, a novel model of granuloma reactivation could be proposed where CD271+ BM-MSCs harboring viable Mtb migrate to inflammatory granulomas for disease reactivation (Fig. 17.1). Moreover, ongoing studies show that CD271+BM-MSCs also are able to resuscitate the viable but non-culturable Mtb inside the granuloma to reactivate the disease.

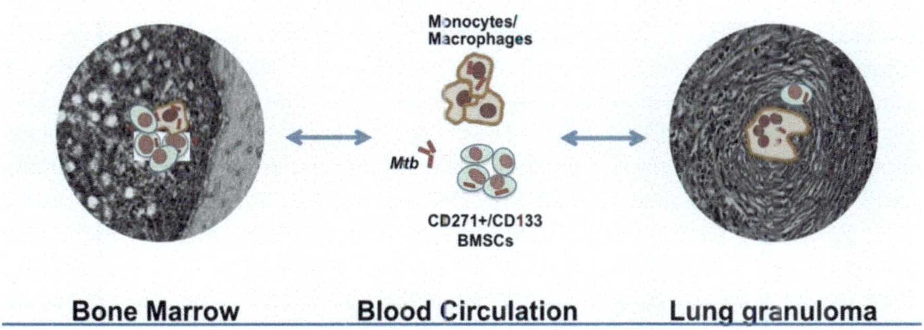

Fig. 17.1 Hypothetical model of CD271+ BM-MSC role in tubercular dormancy and reactivation Schematic representation of a novel model of granuloma reactivation where CD271+ BM-MSCs harboring viable Mtb migrate to inflammatory granulomas for disease reactivation. We propose that CD271+ BM-MSCs travel from bone marrow through the circulatory system to the lungs as they are CD271+ BM-MSCs like most stem cells are programmed to migrate to the inflamed organs

17.2 Latency, Reactivation, and Post-treatment Relapse of Tuberculosis

Clinical Relevance of Stem Cell Niche Strikingly, about one-third of the world's population has latent TB, where people are infected by TB bacteria but not (yet) ill with active disease and cannot transmit the disease. Approximately 5–10% of the latently infected patients are estimated to develop TB at some point in their lifetime, while 10–15% of patients who have undergone complete anti-TB drug treatment show relapse. However, it is puzzling that relapse occurs in patients where lung granulomas are found sterile (Sia and Wieland 2011; Lin et al. 2014). Some autopsies detect viable bacteria, but these subjects may have had "active" disease (Altaf Bachh et al. 2010; Lin et al. 2014; Shleeva et al. 2002). Additionally, autopsies in the Cornell murine model tend to support the human autopsy findings and the speculation that persistent bacteria could somehow appear in relatively sterile granuloma to reactivate the disease. Human studies found evidence of *M. tuberculosis* inside cells like adipocytes, macrophages, epithelial cells, endothelial cells, etc. (Mayito et al. 2019; Neyrolles et al. 2006; Antony et al. 1983; Mariotti et al. 2013). But so far, none of these studies found the viable Mtb; only molecular footprints in the form of DNA could be detected (Sia and Wieland 2011; Lin et al. 2014; Shi et al. 2005; Shleeva et al. 2002). However, these findings do indicate that there are extra granuloma sites for Mtb, yet the question remained to be answered conclusively: Where do they reside? Do they need niche cells or not?

Extra-Granuloma Sites of Dormant Mtb There are numerous studies on extra-granuloma sites that serve as homing sites for Mtb. Studies describe wide arrays of sites like the lung epithelial cells (outside granuloma), extrapulmonary sites like adipocytes, endothelial cells, macrophages, etc. which could be probable extra granulomatous location of Mtb in dormant stage (Mayito et al. 2019). One study concluded that the bacilli can persist intracellularly in the lung tissue without causing clinical lesions (Grosset 2003). *M. tuberculosis* DNA was found not only in macrophages but also in other non-phagocytic cells like type II pneumocytes, endothelial cells, and fibroblasts (Mariotti et al. 2013). Neyrolles et al. found adipocytes in adipose tissue act as host for Mtb DNA (Neyrolles et al. 2006). Similarly, other studies also found Mtb DNA in extra-granuloma sites (Shi et al. 2005; Reece and Kaufmann 2012; Shleeva et al. 2002). In light of all the above autopsy and other reports, more questions may be raised: If granulomas are sterile, where do the persistent Mtb reside? Where is their protective niche? How many of them are required to activate the disease? Moreover, if robust why did not Mtb reside in the primary granuloma? And, why does the reactivation take place in the apical part of the lung?

Persistent Non-replicating Phenotype At present neither in vitro nor in vivo models of dormancy provide adequate argument for correlation between numeric quantity and survival of persistent non-replicating Mtb and reactivation. The well-known murine Cornell model of dormancy tends to support that perhaps persistent bacteria are rare and not plenty. Real-time PCR analysis detected only 1000 bacteria per ml tissue, but eventually, this small number of bacteria seems enough for reactivation of the disease in the animal (Mukamolova et al. 2010; Russell 2007; de Wit et al. 1995). So, it may be assumed that only a few bacteria are required to reactivate the disease.

In human, due to a more vigorous immune response to primary infection, the number of persistent TB bacilli is probably smaller in number, and most of the time they may not be culturable in the laboratory. So far, the attempt to culture persistent Mtb from patients remains unsuccessful. There are reports of culturing Mtb from latent TB individuals, but whether these subjects had the quiescent phase of latent TB or the active phase of latent TB could not be ascertained. None of these studies confirm the phenotype of latent TB.

In vitro models do not take into account the host elicited immune response responsible for persistence. The immune system shapes the phenotype and might maintain an equilibrium between host defense and the number of bacilli metamorphosing into non-replicative type (persisting) with prospects of reactivation at a later time point. However, most animal models do not confer dormancy, and in most models a severe acute infection develops and the animal eventually succumbs to death. Here we address the two known animal dormancy models which are instrumental for reactivation studies:

Cornell Model of Dormancy The most clinically relevant in vivo model is the Cornell model. In this model, anti-TB drug treatment leads to elimination of almost all the bacteria in the infected animal indicated by zero CFU from lung or other tissues. However, on immunosuppression, lung granuloma is observed, "sleeper cells" get reactivated, and the disease reappears (Scanga et al. 1999). In clinical settings, TB patients go through couple of diagnostic tests and are prescribed several anti-TB drugs for nearly a year that kills most of the bacilli. However, similar to the Cornell model, after a few months, or year, in about 10% cases, the disease reappears, but the bacilli remain sensitive to drugs, indicating that the relapse was not because of drug resistance but because of reactivation of the persistent bacilli that was not cleared by chemotherapy. These relapse cases undergo another round of anti-TB drug treatment; however many of them do not survive. The Cornell model closely mimics the relapse clinical cases confirming dormant or persistent infection reappearing from a non-pulmonary site to pulmonary site.

Forsyth Model of Dormancy In this model a mutant of *M. tuberculosis* strain (auxotrophic for streptomycin), 18b strain, is cultured with regular addition of streptomycin for 4 weeks. On streptomycin withdrawal for a week, Western blot panels show synthesis of increased amount of alpha crystalline by the bacilli, indicating conversion to a non-replicative form. In other words, the *M. tuberculosis* strain 18b gets into dormant phase on streptomycin withdrawal. Therefore, this mouse model could represent the putative dormant infection model, where bacteria is presumed to stay in a non-replicating dormant state (Das et al. 2013).

The two above animal models support niche-to-niche interaction hypothesis. There appears to be niche-to-niche interaction from bone marrow to lung granulomatous tissue for persistent bacilli to reactivate the disease. It appears that not all persistent bacteria are culturable or viable. Thus, the granulomas in autopsies may have bacteria detected under microscope, but not all of those are culturable.

Post-drug treatment, where granulomas are relatively sterile, Mtb may employ a "sleeper cell" strategy, where a few viable bacteria hide among extrapulmonary healthy tissues for years, instead of hiding in the granuloma to be eventually killed by the immune cells. Residing within a host cell that sustains Mtb viability is crucial for Mtb survival, dormancy, and reactivation Macrophages which are known to be the primary host for Mtb may not be suitable for Mtb survival and persistence and dissemination. Only 1 out of 10 bacteria retain viability in macrophages. Thus, macrophages may not be an ideal host for the long term persistent of Mtb. What would be the ideal host then, the adipocytes, endothelial cells, or other host cells?

17.3 Significance of the Non-granuloma Model of Tubercular Dormancy

In animal models, Mtb acquires a dynamic equilibrium where some bacteria continue to proliferate and some acquire a persistent phenotype or a non-replicating state. These non-replicating *M. tuberculosis* (NRM) are supposed to reside inside the foamy macrophages and then periodically replicate to initiate a fresh cycle of replication and infection of fresh macrophages to keep a granuloma alive. This model of persistence is called dynamic granuloma model. Dynamic granuloma model considers high persistent Mtb which are eliminated by anti-TB drug treatment. Dynamic granuloma model is also shown in Guinea pigs which form persistent granuloma lesion containing high number of persistent bacteria surviving in the necrotic core of the granuloma (Lenaerts et al. 2007, 2015). Moreover, it is assumed that a similar situation may also occur in human, where active secondary granuloma may contain both the replicating and non-replicating bacteria. Clinical studies show that subjects with latent TB benefit from long-term treatment, therefore supporting the dynamic granuloma model of latency in human.

However, extrapolating model of dynamic granuloma to human can be debatable. The first argument against it is that animals succumb to primary infection, whereas TB patients rarely die of primary infection due to strong immune system. Second, clinical studies support that chemotherapy could eliminate persistent TB, but that does not confirm that a number of persistent Mtb are high in those patients. Moreover, no diagnostic test confirms that those subjects who were benefitted from chemotherapy belonged to pure latent state (or the quiescent state of the latent TB), instead of the early stage of reactivation (or active state of latent TB). It is expected that subjects with the active latency would benefit from drug treatment, because it is expected to have replicating bacteria in the granuloma. On the other hand, subject with quiescent state of latency may not get any benefit, because the existing TB drugs are not known to target the persistent dormant bacteria. Overall, the basic question remains unanswered: Are persistent TB plenty or rare? Do they replicate during the quiescent latent stage or not?

Reactivation of Non-replicating Dormant Phenotype It is known that LTBI (latent tuberculosis infection) patients show negative sputum culture and smear, positive IGRA (Interferon-Gamma Release Assays) or TST (tuberculin skin test), and no clinical symptoms whatsoever. Reactivation of LTBI can happen during the patient lifetime and can turn into active TB indicated by clinical symptom along with positive sputum, IGRA or TST, and chest x-ray. The fundamental mechanism involved in reactivation of latent/dormant tuberculosis is not very clear. This could be largely attributed to the difficulty in developing and manipulating animal models of latent tuberculosis. Although few studies have provided some insights on the role of T lymphocytes, gamma interferon (IFN-γ), tumor necrosis factor alpha (TNF-alpha), interleukin-12, reactive nitrogen intermediate (RNI), etc. on reactivation (Islam et al. 2004), further research is impending.

17.4 Bone Marrow-Mesenchymal Stem cell Niche of Tubercle Bacilli

The correlation between Mtb and stem cells was discovered by Das et al. (Das et al. 2013). A population of Mesenchymal stem cells, CD271+ MSCs, harbor tubercle bacilli in bone marrow in animals as wells as human patients. CD271 is a type of low affinity nerve growth factor receptor (LNGFR), which serves as a co-receptor for neurotrophins, a family of nerve growth factors. LNGFR is also a member of the TNF receptor superfamily. Interestingly, a subset of naïve MSCs express CD271 and are localized in the hypoxic niche of human BM. Das et al. developed an in vitro culture system to maintain the naïve CD271+ BM-MSCs (Pal and Das 2017), where knockdown of CD271 was found to reduce glutathione level (unpublished data) suggesting that the receptor may be involved in regulating the cellular redox level. Whether this putative redox regulating activity of CD271 allows Mtb to hide intracellularly to this type of MSC is not yet clear. Subsequently, Garhyan et al. found out that non-replicating Mtb could be recovered post-antibiotic treatment (Garhyan et al. 2015; Beamer et al. 2014). These results showed that long-lasting quiescent nature of the CD271+ BM-MSC could provide long-term shelter to Mtb in spite of anti-TB drugs treatment. MSCs have properties which provide long-term protective niche for long-term survival of the Mtb bacilli. The following reasons support the correlation between stem cells and Mtb: (a) Stem cells are present in human lung granuloma and they are capable of self-renewal. (b) MSCs have drug efflux pumps to efflux out the antibiotics in order to protect the bacilli. (c) The fact that MSCs are quiescent and have low reactive oxygen species also allow them to be a perfect niche for Mtb. It is worth noting that undifferentiated MSCs harbor Mtb, while differentiated ones don't harbor the bacilli.

Although CD271+ BM-MSC population provides a protective niche to dormant Mtb (5), there are two fundamental questions which remain unanswered. First, what is the direct role of CD271 in making MSCs suitable and long-term stable hosts for Mtb? Second, what is the significance of CD271 in Mtb reactivation in the lung? High expression of the drug efflux pump ABCG2 in CD271+ BM-MSCS indicates that CD271+ marker is linked with offering antibiotic protection to the pathogen (Das et al. 2013). Expression of CD271 followed by CD271-responsive genes is responsible for genes associated with DNA repair (in p53-dependent manner) and drug response in stem cells (Redmer et al. 2017; Li et al. 2015). In a separate study, Fan et al. demonstrated that lethal oxidative DNA damage occurs in nongrowing Mtb cells treated with different bactericidal antibiotics (Fan et al. 2018). With these two studies, it is speculated that since CD271 is needed for a proper p53-dependent response to DNA-damaging drugs, it may offer protection to intracellular nongrowing Mtb from lethal oxidation-induced DNA damage caused by bactericidal antibiotics and thus may offer enhance protection from antibiotics. To address the second question of significance of CD271 in Mtb reactivation, we should keep in mind that forced differentiation of Mtb-infected BM-MSCs leads to loss of Mtb viability and loss of CD271+marker (Das et al. 2013) indicating that BM-MSCs' undifferentiated state along with CD271+ marker may be the requirement for Mtb to remain dormant. Thus, we can refute the possibility of differentiation of BM-MSCs and reaching various sites to reactivate the disease. This further leads to the speculation that BM-MSCs have to travel in their undifferentiated state with CD271+marker (stem state) to the site to reactivate the disease (Fig. 17.1). Moreover, since CD271 also has a role in migration to various tissue (Li et al. 2015), it may provide an edge to the CD271+ BM-MSCs to disseminate Mtb to various organs and reactivate in the most favorable environment. Additionally, human MSCs are known to express several Toll-like receptors, NOD-2 and Rig-I (Khan et al. 2017), which can be significant in Mtb protection. Studies show that human mesenchymal stem cells internalize Mtb through scavenger receptors MARCO (macrophage receptor with collagenous structure) and SR-B (a.k.a. CD36). Unlike macrophages, mannose receptors on MSCs are not responsible for Mtb uptake (Khan et al. 2017). MSCs do not allow Mtb replication due to its intrinsic nitric oxide.

In our evolutionary history, mammalian ancestors might have had a strong BM-to-lung route of regeneration: Mtb transmitted to its replicating niche from its protective niche through its depot BM-MSCs, and thus, only a few Mtb bacteria (~ 60 Mtb infected cells in mice) (Garhyan et al. 2015) is successful in establishing infection in mice. Indeed, such atavistic reactivation of ancient primitive defense mechanism is anticipated (Chen et al. 2000; Das 2000). Hence, further studies may shed more light about the remnant of BM-lung regenerative mechanism, which of course will greatly benefit the infectious disease and stem cell regenerative research alike. It is important to note that recently hematopoietic stem cells have also been found to be depot for dormant Mtb (Gengenbacher and Kaufmann 2012), further supporting potential evolutionary aspect of the stem cell basis of TB latency and reactivation.

Resuscitation of Non-replicating Dormant Phenotype by BM-MSCs We found that PTB patients (from northeast region of India) treated with anti-TB drugs have viable but non-culturable (VBNC) Mtb in sputum that show culturability only in the presence of EPSN (resuscitation factor containing supernatant of mid log phase H37Rv culture). Importantly, VBNC Mtb resuscitation can also be attained and enhanced by culturing in CD271+BM-MSCs (unpublished work).

Pulmonary tuberculosis (PTB) worldwide is diagnosed by identifying the presence of acid-fast bacillus (AFB) under microscopy. Mtb culture from expectorated sputum or tissue is the confirmatory gold standard test for active TB (Sia and Wieland 2011). However, there is a huge pool of sputum smear-negative PTB patients who show chest x-ray abnormal/positive while having a negative sputum smear/culture. Sputum-negative tuberculosis poses a challenge in diagnosis, treatment, and eradication of tuberculosis (Granville et al. 1953; Wayne and Sohaskey 2001; Hobby et al. 1954; Beck and Yegian 1952; Falk et al. 1954; Hurford and Valentine 1957; Salkin and Wayne 1956; Dutt and Stead 1994; Mc 1959). Numerous reports have shown the presence of non-culturable tubercle bacilli in human sputum (Mukamolova et al. 1998; Garton et al. 2008; Gao and Fan 1986; Dhillon et al. 2014) which are undetectable through culture and/or smear. CDC (Center for Disease Control and Prevention) reports that only 50–80% of PTB cases are sputum smear positive (Dooley et al. 1990). Other studies indicate that in India 28.7% of TB cases are new smear-negative cases (Vengattaraman 2010; Altaf Bachh et al. 2010; Ministry of Health and Family Welfare 2007). "With the sharp rise of PTB in countries which are worst affected by the HIV epidemics, the number of patients with suspected PTB who are sputum smear negative and chest x-ray/radiography (CXR) positive has increased" (WHO, Geneva. TB Control as an Integral part of primary health care. 1988, 11). The non-culturable Mtb in human tissue are non-replicating, non-susceptible to chemotherapy with added ability to resuscitate back to culturable cells, remaining potential source of reactivation of the disease (Mc 1959; Russel et al. 1955).

Reactivation of dead and sterile granuloma is a major issue in pulmonary tuberculosis management (Lin et al. 2014). The relevance of VBNC Mtb and the mechanism of VBNC cell reactivation in the latent TB remain unknown to most extent. Understanding the VBNC reactivation mechanism may provide new insight about latent TB, and developing new therapy. Most of the latent TB granuloma are sterile as indicated by autopsy studies in primates (Hobby et al. 1954; Lin et al. 2014; Shi et al. 2005). In cases where Mtb is recovered from latent granulomas, they are considered to be in VBNC state (Gengenbacher and Kaufmann 2012; Reece and Kaufmann 2012). Numerous studies show that addition of spent media from exponentially growing Mtb resuscitates the non-culturable bacteria culturable (Shleeva et al. 2002), suggesting that the sterile granuloma contains viable but non-culturable (VBNC) Mtb, also known as persisters. VBNC state is a general adaptive strategy of many bacteria that evolved specific resuscitation-promoting factor genes, rpf including Mtb (Gengenbacher and Kaufmann 2012) (Reece and Kaufmann 2012). Strikingly, RPF-dependent VBNC Mtb were recovered from sputum of patients with active TB and also from autopsies indicating the potential clinical significance of VBNC Mtb (Mukamolova et al. 2010). Although extraneous addition of RPF may resuscitate VBNC cells in vitro, the unavailability/inaccessibility of RPF in latent granuloma could be argument against the hypothesis that the latent granuloma are physically separated by fibrous wall or caseum and devoid of actively replicating bacteria (Gengenbacher and Kaufmann 2012; Russell 2007). However, as Mtb is heterogeneously persistent in different dynamic stages, the slowly-replicating ones (Gong et al. 2015; Lenaerts et al. 2015; Cadena et al. 2017) may be able to produce RPF for the replication of VBNC bacilli in hypoxic/necrotic granuloma. Reactivation occurs even in immunocompetent individual with LTBI where lungs are paucibacillary Mtb state (Dutta and Karakousis 2014) through unknown mechanism (Nunes-Alves et al. 2014; Stewart et al. 2003). To add to this complexity, VBNC Mtb is highly immunogenic and, therefore, may pose additional challenge for the survival of resuscitated VBNC Mtb in immunocompetent subjects. Hence, it could be presumed that a yet unknown mechanism may resuscitate the VBNC Mtb in latent granuloma by evading immune system.

VBNC can be studied in in vitro by Wayne model where Mtb cultured under hypoxia (to mimic tubercle bacilli in vivo) progressively switches to two culturable states of non-replicating persistence, NRPI and NRP II (Wayne and Hayes 1996; Wayne and Sohaskey 2001). On longer incubation, NRP II progresses further to true dormant bacteria requiring resuscitation to return back to culturable phenotype (Wayne and Hayes 1996; Wayne 1977), closest to VBNC state in tissues. This viable, non-replicating dormant bacteria can be converted to replicative form using enriched media. For example, early stationary phase culture supernatant of Mtb has resuscitation activity which could resuscitate dormant bacilli in old batch cultures. Zhang et al. showed that dormant Mtb could be resuscitated in solid plates, using culture supernatant of TB bacilli (Sun and Zhang 1999). VBNC in vitro model may add on tools to solve the puzzle of reactivation.

17.5 Stem Cells and Tuberculosis Therapy

All past, present, and ongoing studies indicate that stem cells can either reactivate TB or protect from the disease. One study shows that autologous infusion of stem cells acts as an adjunct therapy for MDR and XdR Mtb patients (Gengenbacher and Kaufmann 2012). When administered with second-line drugs, stem cell therapy improves the clinical condition of the patients. Moreover, hematopoietic stem cells (HSCs) have recently been shown to play an important role in protective immunity against Mtb (Kaufmann et al. 2018). Divangahi et al. interestingly found that Bacillus Calmette-Guerin (BCG) exposure to bone marrow HSCs can epigenetically modify the macrophage progeny from the BCG-educated HSCs (Kaufmann et al. 2018). More importantly, these epigenetically modified macrophages are far better in protection against Mtb compared to the unmodified ones. This study also suggests that stem cells of BM specifically hematopoietic in origin can be used for vaccine development. Overall, bone marrow-residing stem cells, both mesenchymal and hematopoietic, provide promising novel strategy for treatment and/or prevention of pulmonary tuberculosis.

Summary Mtb has co-evolved with human host and therefore has several strategies to survive against vigorous immune reaction. Mtb-host coexistence for decades can be either symbiotic where each is supporting the survival of the other or a perpetual race against each other where one has to survive over the other. Bone marrow stem cell niche is emerging as an important niche for Mtb dormancy and reactivation. Recently Das et al. identified human CD271+ bone marrow mesenchymal stem cells (CD271+ BM-MSCs) as niche host cells that can maintain viable, non-replicating dormant Mtb. Moreover, CD271+BM-MSCs also are able to resuscitate the viable but non-culturable (VBNC) Mtb inside the granuloma to reactivate the disease. Study of mechanism of Mtb survival and resuscitation inside CD271+ BM-MSCs is ongoing. In CD271+ BM-MSC, hypoxia-responsive genes (HIF-1alpha signaling pathway) along with DNA damage-responsive genes (CD271 signaled p53-dependent signaling pathway) may play an important role in the Mtb survival and persistence. The self-renewing and metabolically quiescent plus low ROS state of CD271+ stem cells may provide a favorable environment for the pathogen to remain in the CD271+ BM-MSC niche. In addition to MSCs, other stem cell types in BM niche, including HSCs, may also provide host support for Mtb dormancy and reactivation.

References

Altaf Bachh A, Gupta R, Haq I, Varudkar HG (2010) Diagnosing sputum/smear-negative pulmonary tuberculosis: Does fibre-optic bronchoscopy play a significant role? Lung India 27 (2):58–62. https://doi.org/10.4103/0970-2113.63607

Antony VB, Sahn SA, Harada RN, Repine JE (1983) Lung repair and granuloma formation. Tubercle bacilli stimulated neutrophils release chemotactic factors for monocytes. Chest 83 (5 Suppl):95S–96S

Beamer G, Major S, Das B, Campos-Neto A (2014) Bone marrow mesenchymal stem cells provide an antibiotic-protective niche for persistent viable Mycobacterium tuberculosis that survive antibiotic treatment. Am J Pathol 184(12):3170–3175. https://doi.org/10.1016/j.ajpath.2014.08.024

Beck F, Yegian D (1952) A study of the tubercle bacillus in resected pulmonary lesions. Am Rev Tuberc 66(1):44–51

Bloom BR, McKinney JD (1999) The death and resurrection of tuberculosis. Nat Med 5 (8):872–874. https://doi.org/10.1038/11309

Cadena AM, Fortune SM, Flynn JL (2017) Heterogeneity in tuberculosis. Nat Rev Immunol 17 (11):691–702. https://doi.org/10.1038/nri.2017.69

Chen Y, Zhang Y, Jiang TX, Barlow AJ, St Amand TR, Hu Y, Heaney S, Francis-West P, Chuong CM, Maas R (2000) Conservation of early odontogenic signaling pathways in Aves. Proc Natl Acad Sci USA 97(18):10044–10049. https //doi.org/10.1073/pnas.160245097

Das B (2000) The science behind squalene (The human antioxidant). ICBR

Das B, Kashino SS, Pulu I, Kalita D, Swami V, Yeger H, Felsher DW, Campos-Neto A (2013) CD271(+) bone marrow mesenchymal stem cells may provide a niche for dormant Mycobacterium tuberculosis. Sci Transl Med 5(170):170ra113. https://doi.org/10.1126/scitranslmed.3004912

de Wit D, Wootton M, Dhillon J, Mitchison DA (1995) The bacterial DNA content of mouse organs in the Cornell model of dormant tuberculosis. Tuber Lung Dis 76(6):555–562

Dhillon J, Fourie PB, Mitchison DA (2014) Persister populations of Mycobacterium tuberculosis in sputum that grow in liquid but not on solid culture media. J Antimicrob Chemother 69 (2):437–440. https://doi.org/10.1093/jac/dkt357

Dooley SW Jr, Castro KG, Hutton MD, Mullan RJ, Polder JA, Snider DE Jr (1990) Guidelines for preventing the transmission of tuberculosis in health-care settings, with special focus on HIV-related issues. MMRW Morb Mortal Wkly Rep 39(RR-17):1–29

Dutt AK, Stead WW (1994) Smear-negative pulmonary tuberculosis. Semin Respir Infect 9 (2):113–119

Dutta NK, Karakousis PC (2014) Latent tuberculosis infection: myths, models, and molecular mechanisms. Microbiol Mol Biol Rev 78(3):343–371. https://doi.org/10.1128/MMBR.00010-14

Falk A, Tucker WB, Kaufman JE (1954) A clinical and bacteriologic correlation of resected pulmonary tuberculous lesions. Am Rev Tuberc 70(4):689–700

Fan XY, Tang BK, Xu YY, Han AX, Shi KX, Wu YK, Ye Y, Wei ML, Niu C, Wong KW, Zhao GP, Lyu LD (2018) Oxidation of dCTP contributes to antibiotic lethality in stationary-phase mycobacteria. Proc Natl Acad Sci USA 115(9):2210–2215. https://doi.org/10.1073/pnas.1719627115

Gao DZ, Fan RL (1986) Smear-positive and culture-negative results in detecting acid-fast bacilli in the sputum and their clinical significance. Zhonghua Jie He He Hu Xi Xi Ji Bing Za Zhi 9 (6):374–376

Garhyan J, Bhuyan S, Pulu I, Kalita D, Das B, Bhatnagar R (2015) Preclinical and clinical evidence of Mycobacterium tuberculosis persistence in the hypoxic niche of bone marrow mesenchymal stem cells after therapy. Am J Pathol 185(7):1924–1934. https://doi.org/10.1016/j.ajpath.2015.03.028

Garton NJ, Waddell SJ, Sherratt AL, Lee SM, Smith RJ, Senner C, Hinds J, Rajakumar K, Adegbola RA, Besra GS, Butcher PD, Barer MR (2008) Cytological and transcript analyses reveal fat and lazy persister-like bacilli in tuberculous sputum. PLoS Med 5(4):e75. https://doi.org/10.1371/journal.pmed.0050075

Gengenbacher M, Kaufmann SH (2012) Mycobacterium tuberculosis: success through dormancy. FEMS Microbiol Rev 36(3):514–532. https://doi.org/10.1111/j.1574-6976.2012.00331.x

Gong C, Linderman JJ, Kirschner D (2015) A population model capturing dynamics of tuberculosis granulomas predicts host infection outcomes. Math Biosci Eng 12(3):625–642. https://doi.org/10.3934/mbe.2015.12.625

Granville GE, Jenkins DE, Cooley DA, Debakey ME, Whitcomb FC, Halpert B (1953) Cultural properties of M. tuberculosis in resected pulmonary lesions of patients treated with streptomycin and para-aminosalicylic acid. Am Rev Tuberc 68(5):727–733

Grosset J (2003) Mycobacterium tuberculosis in the extracellular compartment: an underestimated adversary. Antimicrob Agents Chemother 47(3):833–836

Hobby GL, Auerbach O, Lenert TF, Small MJ, Comer JV (1954) The late emergence of M. tuberculosis in liquid cultures of pulmonary lesions resected from humans. Am Rev Tuberc 70(2):191–218

Hurford JV, Valentine WH (1957) Viable tubercle bacilli in closed lesions. Tubercle 38(3):194–198

Islam N, Kanost AR, Teixeira L, Johnson J, Hejal R, Aung H, Wilkinson RJ, Hirsch CS, Toossi Z (2004) Role of cellular activation and tumor necrosis factor-alpha in the early expression of Mycobacterium tuberculosis 85B mRNA in human alveolar macrophages. J Infect Dis 190 (2):341–351. https://doi.org/10.1086/421522

Kaufmann E, Sanz J, Dunn JL, Khan N, Mendonca LE, Pacis A, Tzelepis F, Pernet E, Dumaine A, Grenier JC, Mailhot-Leonard F, Ahmed E, Belle J, Besla R, Mazer B, King IL, Nijnik A, Robbins CS, Barreiro LB, Divangahi M (2018) BCG Educates Hematopoietic Stem Cells to

Generate Protective Innate Immunity against Tuberculosis. Cell 172 (1–2):176–190 e119. https://doi.org/10.1016/j.cell.2017.12.031

Khan A, Mann L, Papanna R, Lyu MA, Singh CR, Olson S, Eissa NT, Cirillo J, Das G, Hunter RL, Jagannath C (2017) Mesenchymal stem cells internalize Mycobacterium tuberculosis through scavenger receptors and restrict bacterial growth through autophagy. Sci Rep 7(1):15010. https://doi.org/10.1038/s41598-017-15290-z

Lenaerts AJ, Hoff D, Aly S, Ehlers S, Andries K, Cantarero L, Orme IM, Basaraba RJ (2007) Location of persisting mycobacteria in a Guinea pig model of tuberculosis revealed by r207910. Antimicrob Agents Chemother 51(9):3338–3345. https://doi.org/10.1128/AAC.00276-07

Lenaerts A, Barry CE 3rd, Dartois V (2015) Heterogeneity in tuberculosis pathology, microenvironments and therapeutic responses. Immunol Rev 264(1):288–307. https://doi.org/10.1111/imr.12252

Li S, Yue D, Chen X, Wang L, Li J, Ping Y, Gao Q, Wang D, Zhang T, Li F, Yang L, Huang L, Zhang Y (2015) Epigenetic regulation of CD271, a potential cancer stem cell marker associated with chemoresistance and metastatic capacity. Oncol Rep 33(1):425–432. https://doi.org/10.3892/or.2014.3569

Lin PL, Ford CB, Coleman MT, Myers AJ, Gawande R, Ioerger T, Sacchettini J, Fortune SM, Flynn JL (2014) Sterilization of granulomas is common in active and latent tuberculosis despite within-host variability in bacterial killing. Nat Med 20(1):75–79. https://doi.org/10.1038/nm.3412

Mariotti S, Sargentini V, Pardini M, Giannoni F, De Spirito M, Gagliardi MC, Greco E, Teloni R, Fraziano M, Nisini R (2013) Mycobacterium tuberculosis may escape helper T cell recognition by infecting human fibroblasts. Hum Immunol 74(6):722–729. https://doi.org/10.1016/j.humimm.2013.02.005

Mayito J, Andia I, Belay M, Jolliffe DA, Kateete DP, Reece ST, Martineau AR (2019) Anatomic and cellular niches for mycobacterium tuberculosis in latent tuberculosis infection. J Infect Dis 219(5):685–694. https://doi.org/10.1093/infdis/jiy579

Mc DW (1959) In apparent infection: relation of latent and dormant infections to microbial persistence. Public Health Rep 74(6):485–499

Ministry of Health and Family Welfare DGoHS, Central TB Division (ed) (2007) TB India: RNTCP status report. Chapter 7, Performance of RNTCP

Mukamolova GV, Kaprelyants AS, Young DI, Young M, Kell DB (1998) A bacterial cytokine. Proc Natl Acad Sci USA 95(15):8916–8921

Mukamolova GV, Turapov O, Malkin J, Woltmann G, Barer MR (2010) Resuscitation-promoting factors reveal an occult population of tubercle Bacilli in Sputum. Am J Respir Crit Care Med 181(2):174–180. https://doi.org/10.1164/rccm.200905-0661OC

Neyrolles O, Hernandez-Pando R, Pietri-Rouxel F, Fornes P, Tailleux L, Barrios Payan JA, Pivert E, Bordat Y, Aguilar D, Prevost MC, Petit C, Gicquel B (2006) Is adipose tissue a place for Mycobacterium tuberculosis persistence? PLoS One 1:e43. https://doi.org/10.1371/journal.pone.0000043

Nunes-Alves C, Booty MG, Carpenter SM, Jayaraman P, Rothchild AC, Behar SM (2014) In search of a new paradigm for protective immunity to TB. Nat Rev Microbiol 12(4):289–299. https://doi.org/10.1038/nrmicro3230

Pal B, Das B (2017) in vitro culture of naive human bone marrow mesenchymal stem cells: a Stemness based approach. Front Cell Dev Biol 5:69. https://doi.org/10.3389/fcell.2017.00069

Redmer T, Walz I, Klinger B, Khouja S, Welte Y, Schafer R, Regenbrecht C (2017) The role of the cancer stem cell marker CD271 in DNA damage response and drug resistance of melanoma cells. Oncogenesis 6(1):e291. https://doi.org/10.1038/oncsis.2016.88

Reece ST, Kaufmann SH (2012) Floating between the poles of pathology and protection: can we pin down the granuloma in tuberculosis? Curr Opin Microbiol 15(1):63–70. https://doi.org/10.1016/j.mib.2011.10.006 S1369-5274(11)00176-7 [pii]

Russel WF, Dressler SH, Middlebrook G, Denst J (1955) Implications of the phenomenon of open cavity healing for the chemotherapy of pulmonary tuberculosis. Am Rev Tuberc 71(3, Part 1):441–446

Russell DG (2007) Who puts the tubercle in tuberculosis? Nat Rev Microbiol 5(1):39–47. https://doi.org/10.1038/nrmicro1538
Salkin D, Wayne LG (1956) The bacteriology of resected tuberculous pulmonary lesions. I. The effect of interval between reversal of infectiousness and subsequent surgery. Am Rev Tuberc 74 (3):376–387
Scanga CA, Mohan VP, Joseph H, Yu K, Chan J, Flynn JL (1999) Reactivation of latent tuberculosis: variations on the Cornell murine model. Infect Immun 67(9):4531–4538
Shi L, Sohaskey CD, Kana BD, Dawes S, North RJ, Mizrahi V, Gennaro ML (2005) Changes in energy metabolism of Mycobacterium tuberculosis in mouse lung and under in vitro conditions affecting aerobic respiration. Proc Natl Acad Sci USA 102(43):15629–15634. https://doi.org/10.1073/pnas.0507850102
Shleeva MO, Bagramyan K, Telkov MV, Mukamolova GV, Young M, Kell DB, Kaprelyants AS (2002) Formation and resuscitation of "non-culturable" cells of Rhodococcus rhodochrous and Mycobacterium tuberculosis in prolonged stationary phase. Microbiology 148 (Pt 5):1581–1591. https://doi.org/10.1099/00221287-148-5-1581
Sia IG, Wieland ML (2011) Current concepts in the management of tuberculosis. Mayo Clin Proc 86 (4):348–361. https://doi.org/10.4065/mcp.2010.0820 S0025-6196(11)60013-X [pii]
Stewart GR, Robertson BD, Young DB (2003) Tuberculosis: a problem with persistence. Nat Rev Microbiol 1(2):97–105. https://doi.org/10.1038/nrmicro749
Sun Z, Zhang Y (1999) Spent culture supernatant of Mycobacterium tuberculosis H37Ra improves viability of aged cultures of this strain and allows small inocula to initiate growth. J Bacteriol 181(24):7626–7628
Vengattaraman S (2010) Diagnosing sputum/smear negative pulmonary tuberculosis. Lung India 27(4):263. https://doi.org/10.4103/0970-2113.71979
Wayne LG (1977) Synchronized replication of Mycobacterium tuberculosis. Infect Immun 17 (3):528–530
Wayne LG, Hayes LG (1996) An in vitro model for sequential study of shiftdown of Mycobacterium tuberculosis through two stages of nonreplicating persistence. Infect Immun 64 (6):2062–2069
Wayne LG, Sohaskey CD (2001) Nonreplicating persistence of mycobacterium tuberculosis. Annu Rev Microbiol 55:139–163. https://doi.org/10.1146/annurev.micro.55.1.139

Biofilms: A Phenotypic Mechanism of Bacteria Conferring Tolerance Against Stress and Antibiotics

18

Anwar Alam, Ashutosh Kumar, Prajna Tripathi, Nasreen Z. Ehtesham, and Seyed E. Hasnain

Abstract

Biofilm, as a heterogenous congregation of microbial cells enclosed within a pellicle, has largely gained attention due to their historical importance in environment as sludges, flocs, slimes, etc. Biofilms in medical research have been an active area of research in periodontics, in wounds, and in surgical implants. With the availability of whole genome sequences, it is now evident that the mechanisms that control biofilm formation have largely remained conserved during the course of evolution, pointing to the fact that biofilm formation is an

Anwar Alam, Ashutosh Kumar and Prajna Tripathi have contributed equally to this chapter.

A. Alam
JH Institute of Molecular Medicine, Jamia Hamdard, New Delhi, Delhi, India

Inflammation Biology and Cell Signaling Laboratory, ICMR-National Institute of Pathology, Safdarjung Hospital Campus, New Delhi, Delhi, India

A. Kumar
Department of Microbiology, Tripura University (A Central University), Agartala, Tripura, India

P. Tripathi
National Institute of Immunology, New Delhi, Delhi, India

JH Institute of Molecular Medicine, Jamia Hamdard, New Delhi, Delhi, India

N. Z. Ehtesham
Inflammation Biology and Cell Signaling Laboratory, ICMR-National Institute of Pathology, Safdarjung Hospital Campus, New Delhi, Delhi, India
e-mail: nasreenzehtesham.nip@gov.in

S. E. Hasnain (✉)
JH Institute of Molecular Medicine, Jamia Hamdard, New Delhi, Delhi, India

Dr Reddy's Institute of Life Sciences, University of Hyderabad Campus, Hyderabad, India
e-mail: vc@jamiahamdard.ac.in

S. E. Hasnain et al. (eds.), Mycobacterium tuberculosis: *Molecular Infection Biology, Pathogenesis, Diagnostics and New Interventions*,
https://doi.org/10.1007/978-981-32-9413-4_18

integral part in the lifecycle of any unicellular organism. The ability to easily inter-switch between planktonic to a sessile life cycle is an important armor for these unicellular organisms to overcome stress. The matrix not only acts a physical barrier that protects the bacteria but also provides an ecological niche for close interaction and communication among these unicellular entities. This coordinated community-like behavior synchronizes metabolic upregulation or downregulation, both in time and space, and allows these microorganisms to achieve physiological proficiency in terms of ability to tolerate stress that might not be possible as a single cell. Research on biofilms from the perspective to explore the mechanisms of drug tolerance is now considered an apt model as compared to the use of planktonic microorganisms. The increasing use of medical implants further necessitates the need to accelerate research in anti-biofilm strategies for these medical devices. This chapter presents an overview of the mechanisms of biofilm formation and the various interventions for prevention of biofilms.

Keywords
Biofilm · Persister · Drug tolerance · Mycobacterium tuberculosis

Abbreviations

ABC	ATP-binding cassette transporters
$CaCl_2$	Calcium chloride
cAMP	Cyclic adenosine monophosphate
DNA	Deoxyribonucleic acid
FDA	Food and Drug Administration
GMP	Guanosine monophosphate
kHz	Kilohertz
$MgSO_4$	Magnesium sulfate
NaCl	Sodium chloride
RNA	Ribonucleic acid

18.1 Introduction

Microbial biofilm refers to assemblage of monospecies or poly-species microbial cells that is irreversibly attached with surface or with each other and is embedded within an extracellular matrix of polymeric substances. Cells within the microbial biofilms exhibit an altered phenotype in terms of growth rate or gene transcription and are therefore distinct as compared to the colonies of bacteria growing on agar plate macromolecular polymer matrix.

Antony van Leeuwenhoek is credited with the first account of biofilm when he reported (1683–1708) the presence of aggregated microbes in the scurf of his teeth and tongue. Bacterial adhesion to surfaces is a common phenomenon that has routinely been observed in the environment. The term film has been used in marine microbiology to distinguish sessile microorganisms from the planktonic forms. The "bottle effect" observed for marine microorganisms pointed that bacterial growth and activity increased significantly due to attachment of these organisms on the surface. Initial investigation about the composition of biofilm using "ruthenium red" dye followed by fixation with osmium tetroxide showed that biofilms consist of polysaccharides. The medical relevance of biofilm emanated from the studies carried out on sputum samples of patients suffering from lung infection due to cystic fibrosis (CF) caused by *Pseudomonas aeruginosa*. The first image of medical biofilms, published in 1977, was obtained from lung biopsies and showed aggregated bacteria surrounded by "glycocalyx." The term glycocalyx was soon replaced with "biofilms" to emphasize the in vivo sessile growth of colonized microorganisms that were considered to cause antimicrobial resistance as compared to planktonic microorganisms (Costerton et al. 1987). The focus of research that was initially directed toward testing medicaments on planktonic forms of microorganisms soon diverged as it was observed that organisms in the biofilms showed altered physiology and gene expression. Nearly 80% of the hospital-borne infections is attributed to biofilm-forming microorganism and hence studies on biofilm-forming microorganism have taken the front seat of all major research on drugs (Kumar et al. 2017). The major thrust in biofilm research has been due to the advent of instrumentation such as scanning electron microscopy (SEM), confocal laser scanning microscopy, Robbins device biofilm sampler, etc.; novel techniques such as that have complemented standard microbiological culture techniques for biofilm characterization.

Bacteria ubiquitously form biofilm and the mechanism of adhesion is highly conserved across different species of bacteria during the course of evolution. Biofilm formation is an inherent and distinct phenomena among bacteria surviving in the environment, where adhesion and protection is of higher importance and hence a sessile life cycle is favored. The evolution of nonpathogenic environmental bacteria to pathogenic strains has led to shedding of the biofilm machinery and enhancement of motility machinery required for pathogenicity-invasion of host cells and escape from immune cells. This is evident from the observation that subculturing of planktonic laboratory strains of bacterium leads to shunting off of the genetic machinery involved in expression of cellular components required for adhesion, thereby rendering these cells incapable to form biofilms. In the presence of selective pressure such as surfactants and antibiotics, genetic machinery essential for formation of protective surface glycocalyx is triggered. Cells with surface-bound glycocalyx continue to survive while planktonic cells lacking protective surface structures fail to survive. The interaction of bacteria with the environment, within and outside the community, and the outcome of the response toward drugs are predominantly influenced by the phenotype as well. In drug discovery strategies, biofilm model could provide more relevant data that may address core questions as compared to the use of planktonic laboratory strains.

18.2 Biofilm Formation by Bacteria

Bacterial species are known to inhabit a variety of ecosystems and persist even under extremely adverse conditions. The survival mechanisms of some pathogenic microbes can be clearly credited to their ability to form biofilms. In addition to acting as a physical barrier against shear forces and antimicrobial agents generated by the host system, biofilms also provide a protective niche for the survival of bacteria (Jefferson 2004). Bacteria use their ability to form biofilms as a defense strategy against host-mediated assaults like perturbations in pH, generation of reactive oxygen intermediates, nutrient deprivation, attack by phagocytic cells, etc. Biofilms are considered as the product of evolution against adverse conditions. The fact that certain stress regulators and chaperones significantly contribute to the phenomenon of biofilm formation indicates that a major motivation behind biofilm formation is defense. Stress response mediators such as σ^B, RpoS, GroEL, DnaK, and DnaJ have been implicated in microbial biofilm formation. In addition to stress, there are other factors as well that trigger the formation of such a sessile microbial community. Pathogenic and commensal bacteria also use biofilm development as a means to colonize and persist in environments that seem propitious within the host and thus flourish well with time. Bacterial species such as *Staphylococcus*, *Streptococcus*, *Pseudomonas*, *Vibrio cholerae*, etc., display elaborate biofilm development under conditions that offer readily utilizable carbon sources in abundance suggesting that this may also act as a trigger factor. Biofilms formed at high shear locations, such as blood or saliva stream, are highly viscous and display remarkable tensile strength, thus conferring tremendous survival advantage to the constituent microbial cells. Moreover, microbes also reap other benefits from this interactive community such as division of metabolic labor. Close proximity of microorganisms within the biofilm also fosters horizontal exchange of genetic information as well as imparts reproductive advantage to these organisms and may eventually act as a driving force for biofilm formation. Taken together, the community approach of growth endows microorganisms with many benefits and there may be different forces that motivate them to enter into this state.

18.3 Process of Biofilm Formation

The formation of this close-knit community of bacteria called biofilm is a complex process involving multiple steps (Fig. 18.1). The series of events that occur during this process, which eventually leads to adaptation under diverse environmental and nutritional conditions, include (i) initial adherence to surface (ii) microcolony formation (aggregation and encapsulation), (iii) formation of a three-dimensional matrix and maturation, and (iv) detachment and dispersal. Initiation of each of the above stages may be contingent upon regulation of the expression of different sets of genes.

Biofilm development is initiated by a reversible attachment of motile bacterial cells to a biotic or abiotic surface. This initial adherence is mediated by physical forces like hydrophobic interactions, van der Waals forces, etc., and thus is a noncommittal stage where microbial cells can detach themselves from the surface. Various environmental and genetic factors contribute to the process of adhesion. Bacteria cells may choose to come off if the surface doesn't provide requisite anchorage or nutrients to them or if other variables such as bacterial orientation, local temperature, pressure, etc., do not seem favorable. However, once the cell

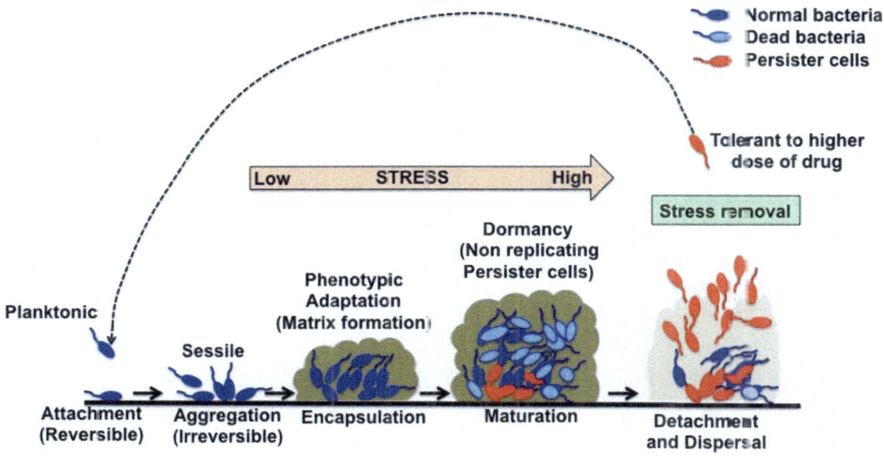

Fig. 18.1 Process of biofilm formation. Planktonic forms of bacteria reversibly attach to the substratum and release quorum sensing molecules that allow them to irreversibly aggregate. The sessile bacteria endogenously secrete matrix around the cells, which provide a niche for growth and cell division. In presence of stress, such as drugs, few bacteria cells undergo dormancy and become non-replicating persister cells. Within the mature biofilm the cells communicate and show community-like behavior. As soon as stress is withdrawn, the dormant bacteria become active and are released as planktonic forms to start the next cycle of biofilm formation

surface structures like pili, fimbriae, flagella, etc., come into play, the bacterial cells become immobilized and proceed to a state of irreversible attachment. Evidence shows that microbial adhesion extensively depends on the surface properties of substratum. Additionally, it is reported that the bacterial appendages promote electrostatic interactions and chemical reactions in order to consolidate the bond between surface and bacterial cells, e.g., surface pili type IV, have been implicated in adherence and movement of *P. aeruginosa* colonies through viscous cell surfaces (Klausen et al. 2003). Attachment of bacteria to biotic surfaces is largely governed by the interaction of host surface proteins and extracellular carbohydrate moieties with that of the invading bacterial cell. *E. coli* is reported to adhere to host cells using its pili type I adhesions like FimH that bind to mannosylated moieties on the surface of host (Wright et al. 2007). Similarly, *Enterococcus* species have been shown to adhere to host cells and aggregate further with the help of adhesion molecules like Sag (Mohamed and Huang 2007).

Irreversible adhesion is followed by aggregation of bacterial cells leading to the formation of a microcolony. This results in multiplication of bacteria and colonization of the site. The formation of such a community demands for the development of methods to communicate and exchange information between the constituent microbial cells. Crosstalk between bacterial cells induces the production of exopolysaccharides, signaling molecules, and peptides that enhance structural integrity of the biofilm and coordinate bacterial growth via quorum sensing. Quorum sensing can be described as a mechanism where a variety of autoinducers stimulate genetic expression of regulators of cell density.

The stimulation of cell signaling molecules and selective regulation of certain genes responsible for extracellular polysaccharide (EPS) secretion triggers the formation of a three-dimensional extracellular matrix. EPSs have been found to play a pivotal role in determining the phenotype of a bacterial biofilm. Many bacteria use cyclic-di-GMP as the second messenger for production of various EPS components of the matrix. Cellulose has been shown to be an important component of the extracellular matrix formed by *E. coli* and *Salmonella typhimurium*. The complex three-dimensional architecture of a biofilm has channels that aid in the uptake of nutrients and release of waste material. The *Staphylococcus* species produce polysaccharide intercellular adhesin (PIA) and poly-N-acetyl glucosamine (PNAG) polymers as the major components of their biofilm matrices, whereas, in the case of *P. aeruginosa*, polysaccharides such as Psl, Pel, and alginate are the main constituents of extracellular matrix (Orgad et al. 2011). Thus, great diversity exists in the phenotype and nature of different bacterial biofilms. Similarly, dramatic differences in the matrix architecture may result from slightest of environmental perturbations. Other constituents of the EPSs may include proteins, amyloids, DNA, lipids, etc.

The concluding stage of biofilm development is characterized by detachment of cells leading to dispersal of bacteria. Bacteria switch from sessile to motile lifestyle either by passive sloughing due to fluid shear or by specific dispersal due to nutrient limitations, environmental cues, and other signals. These signals stimulate the expression of EPS-degrading enzymes and motility structures while downregulating the genes mediating bacterial adherence (McDougald et al. 2012). Detached cells may become quickly adapted to planktonic life or may disseminate further to colonize newer sites.

18.4 Factors Involved in Regulation of Biofilm Formation

The initiation of the complex process of biofilm development relies largely on the dynamics of bacterial attachment to a substratum. Bacterial cell surface properties such as hydrophobicity, presence of motility appendages, and surface molecules play an important role in initial attachment to a surface. Certain features of a surface, such as roughness, may favor biofilm formation. Rough surfaces present higher surface area and arrest shear forces. The rate and extent of bacterial adherence may also be affected by the physicochemical properties of a surface. Interaction with hydrophobic nonpolar surfaces results in stronger attachment as they appear to enable bacteria to defy repulsive forces.

Fluctuations in the nutrient pool or ionic concentrations in the biotic or abiotic habitat of bacteria may alter the properties of biofilm. In some strains of *Lactobacillus*, nutrient limitation (especially low carbon to nitrogen ratio) may trigger biofilm formation. Contrary to this, in *P. aeruginosa* nutrient deprivation induces detachment of the biofilm. Similarly, availability of peptides and amino acids in host tissues and body fluids has been reported to be crucial for several bacteria to form biofilm.

Biofilm formation in *Listeria monocytogenes* proceeds only in the presence of carbohydrates like mannose and trehalose and appropriate levels of phosphate in the local environment. The presence of salt such as NaCl, $MgSO_4$, and $CaCl_2$ in appropriate concentrations has also been reported to be crucial for biofilm formation (Marsden et al. 2017).

Environmental factors also play a significant role throughout the process. The concentration of second messengers such as cyclic-di-GMP and cAMP, crucial for surface attachment of bacteria, is controlled by the levels of carbon and oxygen in the local environment (McDonough and Rodriguez 2012). Perturbations in the environmental conditions result in dynamic changes in biofilm structure; e.g., under oxygen-rich conditions *P. aeruginosa* displays a mushroom-shaped morphology with water channels between macro-colonies consisting of rodlike cells, whereas, under anoxic conditions, it has a three-dimensional mesh-like morphology with channels between macro-colonies of elongated filamentous cells. These morphological changes possibly alter the diffusion of nutrients and aid the bacteria to adapt metabolically. Insufficient oxygen may abrogate the process of biofilm formation in *E. coli*. Temperature is reported to be an important factor for the regulation of EPS production in *Clostridium perfringens* and hence it dramatically affects the biomass, density, and thickness of the biofilm. Similarly, *L. monocytogenes* loses its ability to form biofilms at higher temperature. In addition to temperature, variations in pH have also been shown to influence biofilm formation in the clinical isolates of *Burkholderia pseudomallei*. While alkaline environment is required by *Vibrio cholerae* cells to multiply and form biofilm, in the case of *Lactobacillus rhamnosus*, biofilm formation is modulated by the presence of bile and mucins, low pH, and high osmolarity (Lebeer et al. 2007). Overall, such changes in the morphology of bacterial biofilms may be seen as a fitting response to the environmental cues and a practice for better adaptation to the surrounding environment.

The intrinsic biofilm formation ability is also contingent upon genetic factors. Studies on the regulatory mechanism driving bacterial biofilm formation have identified several genes such as the flagella regulator *sinR* in *Bacillus subtilis* and transcription activator *prfA* in *L. monocytogenes* as the major orchestrators of the process (Newman et al. 2013). Becker et al. (2001) reported that *Staphylococcus aureus* residing in biofilms showed enhanced expression of enzymes such as phosphoglycerate mutase and alcohol dehydrogenase that are involved in glycolysis and fermentation. In addition, the transcript levels of two other genes *sarA* and *sigB* were found to be distinctly upregulated in *S. aureus* biofilms. Bacteria deficient in *sigB* were found to have lost the ability to form biofilm. Gene expression studies in *E. coli* biofilms revealed enhanced expression of *ompC* and *slp*, the genes encoding outer membrane protein and lipoprotein, respectively, which have recently been associated with the initial steps of *E. coli* biofilm formation. The microarray analysis of biofilms also exhibited differential expression of genes under hypoxic and nutrition-deprived conditions. RpoS, the σ^S subunit of RNA polymerase, is considered to play a pivotal role in biofilm formation. Majority of the differentially regulated genes identified till date are involved in bacterial motility, chemotaxis, attachment, membrane bioenergetics, glycolysis, and phage-related functions.

18.5 Interaction of Microbial Community Within Biofilms

Within the biofilm, bacterial cells can coexist with members of same species of bacteria or with other species of microorganisms. The metabolic potential of each species of bacteria govern the relative composition of bacteria within the biofilms. The substrate concentration can influence the population of the bacteria; the fast-growing population of bacteria can dominate over the slow-growing bacteria. The interaction between cells may be competitive in terms of nutrient, space, metabolites, etc.; however in certain conditions cells may act cooperatively to facilitate survival of bacteria from same or other species. The distribution of bacterial population is also dependent on oxygen tension; the core of the biofilm is dominated by cells that are anaerobic while the outer layer of biofilms support aerobic bacteria. This spatial distribution of cells within biofilms is evident in the case of biofilms inhabitated by aerobic *Burkholderia cepacia* and facultative aerobic *Klebsiella oxytoca*. At an initial microcolony formation stage when the biofilm matrix is minimum, *B. cepacia* dominates the population due to its higher growth rate. As the biofilm matrix accumulates it hinders diffusion of oxygen and creates a gradient of oxygen tension. Under low oxygen tension, *K. oxytoca* gains advantage and grows at a faster rate as compared to *B. cepacia*. Bacteria produce toxins and metabolites that confer them with competitive edge over inter−/intraspecies of bacteria and allow them to gain foothold within the matrix. *P. aeruginosa* species can produce growth inhibitory toxins that allow it to invade the biofilms caused by *B. cepacia* and eventually displace the competitor cells. When both species of bacteria produce inhibitory toxins, then the net effect on competitive survival of cells depends on timing of release of toxins and the relative toxicity response of bacteria from each species. In a mixed biofilm, toxins produced by *Pseudoalteromonas tunicata* suppress the growth of sensitive species like *Cytophaga fucicola* but not of insensitive species such as *Roseobacter gallaeciensis*. The protective efficacy of biofilms is enhanced due to the synergistic presence of multiple species of bacteria within biofilms (Tolker-Nielsen 2015). A study showed that marine bacteria, when co-cultured with other species of bacteria, achieved nearly twofold higher levels of survival within biofilms as compared to monospecies (Burmølle et al. 2006). *Staphylococcus aureus* and *Candida albicans* are often found to coexist in biofilms and exhibit higher virulence and tolerance to antimicrobial drugs such as vancomycin. *C. albicans* is usually absent on teeth of health individuals and does not cause symptomatic disease. However, when *C. albicans* partners with dental bacteria such as *Streptococcus gordonii* or *S. oralis*, it enhances bacterial aggregation, colonization, and biofilm formation. These cross-kingdom interactions between fungi and bacteria can be either cooperative or competitive and mediate synergistic or antagonistic associations that are further modulated by host immune responses and local environmental factors.

18.6 Antibiotic Resistance of Bacterial Biofilms

Majority of the pathogens and environmental organisms are found to be present in biofilm and as single entity in nature. Majority of the microbial and chronic infections are associated with biofilms that increase chances of nosocomial infections. Bacterial biofilm resists phagocytosis and immune response on pathogens. It is known that the aquatic biofilms are reservoirs of resistant bacteria that present very close to each other increase chances of genetic exchanges such as antibiotic resistance genes (Balcázar et al. 2015). There are bacterial biofilms present on the root and other parts of plants that may protect from plant pathogens (Bogino et al. 2013). It has been reported that during maturation of biofilms there are increases in the expression of genes related to energy metabolism and transporter genes whereas decrease in the expression of genes related to cell structure, central intermediary metabolism, energy metabolism, nucleotide biosynthesis, and metabolism that indicates that biofilm cells are metabolically active but retard its cell division. There are abnormal conditions present inside biofilms where depletion of oxygen concentration, slow growth, increased doubling times due to nutrient limitation, gradients in pH, change in the metabolic activity of microorganisms, presence of compounds that may inactivate antibacterial compounds, and slow diffusion of antibiotics make microorganisms resistant to antibiotics and drugs. The antibiotic resistance in biofilms may be divided into innate biofilm factors such as oxidative stress, microenvironments within biofilms, biofilm matrix as a diffusion barrier, etc., or induced resistance factors such as changes in the protein expression at transcriptional level, stress protein expression, increased horizontal gene transmission resulting in drug resistance, etc. Reduced antibiotic susceptibility due to biofilm formation at the implanted devices causes persistent infections. It has been reported that *P. aeruginosa* isolates from cystic fibrosis (CF) patients are sensitive to antibiotics in planktonic form but difficult to eradicate due to the presence in biofilms (Hengzhuang et al. 2012). In the case of tuberculosis, *M. tuberculosis* face unfavorable conditions inside hosts that are similar to biofilms that increase the chances of drug tolerance by pathogen.

18.7 Tolerance to Antibiotics and Efflux Pumps

More than 80% of nosocomial infections are related to biofilms that may be present at the medical devices. In the biofilms, bacteria encounter several types of stresses; because of that there are upregulation of toxin-antitoxin systems, different stress proteins, proteins involved in growth regulation, and downregulation of metabolic proteins that causes tolerance to antimicrobial drugs. There are bacteria of different physiological states present in the biofilm due to differential gene expression resulting in differences in the susceptibility of antimicrobial compounds. There are different mechanisms that may be activated during drug tolerance such as modification in drug target site, inactivation of drugs, upregulation of efflux pumps, etc. It is known that expression of efflux pumps found in both bacteria and eukaryotic organisms changes in response to different physiological and environmental conditions (Sun et al. 2014), and these pumps may efflux antibiotics, heavy metals, pollutants, chemicals involved in quorum sensing, compounds produced by plants, bacterial metabolites, etc. (Blanco et al. 2016). In the biofilm, there is activation of efflux pumps and stress response toxin-antitoxin systems that increase tolerance to drugs, so these may be effective targets for anti-biofilm strategies. It has been reported that there are different efflux pumps in *P. aeruginosa* that are involved in resistance to different antibiotics such as ciprofloxacin, gentamicin, and tobramycin. MacABCsm, an ABC-type tripartite efflux pump, is involved in drug resistance and biofilm formation in *Stenotrophomonas maltophilia*. The multidrug resistance efflux pumps of *Salmonella enterica* serovar Typhimurium are involved in biofilm formation and drug resistance. There is increase in the cases of drug resistance in the tuberculosis infection. *M. tuberculosis* survives inside macrophages that induces efflux pumps resulting in increases in drug tolerance and treatment duration (Adams et al. 2011). Inhibition of these efflux pumps may be effective strategies to decrease treatment duration.

18.8 Impact of Biofilm Formation and Phenotypic Behavior

The genes that regulate biofilm formation have largely remained conserved along the course of evolution, pointing to the fact that biofilm formation is an essential component in the life cycle of the unicellular organism. Biofilm formation offers an ecological niche for the unicellular microorganisms to come in contact with each other and offers the unicellular organisms with a great chance to adopt a lifestyle where the functions of many cells cooperatively act to drive across a process that might be impossible to carry as a single cell. The close proximity of unicellular microorganisms within the matrix allows cells to transfer genes through conjugation. It has been established that the conjugation rates in biofilms are as high as 1000-fold as compared to planktonic forms (Stalder and Top 2016). Biofilms are thus considered "hot spots" for horizontal gene transfer and allow bacteria to evolve and adapt to changing environment and stress such as antibiotics, pollutants, and heavy metals. Within the matrix of biofilms, cells attain dormancy and undergo several phenotypic changes including a subpopulation of non-replicating cells called persisters. Persister cells are endowed with higher drug tolerance and can survive nearly 1000-fold higher minimum inhibitory concentration of drugs. The vast diversity of mechanism for biofilm formation among different species of bacteria has posed a serious challenge in terms of devising strategies to combat biofilm. A recent study has implicated the role of chaperonic proteins in biofilm formation. Chaperons aid in protein folding and shape the three-dimensional structure of many proteins simultaneously. Peptidyl-prolyl isomerases (PPIase) are a class of cyclophilins that are ubiquitously expressed in prokaryotes and eukaryotes and are involved in interconversion of cis and trans isomers of peptide bonds with proline residue. PPIase possess chaperonic activity and regulate several key functions in bacteria including iron regulation, signaling, and immune defense. In *Mycobacterium tuberculosis*, PPIase is present in two forms – PpiA and PpiB. *M. tuberculosis* PpiA share structural and phylogenetic similarity with cyclophilins present in eukaryotes. *M. tuberculosis* PpiB is essential for the survival of the pathogen and is involved in biofilm formation (Pandey et al. 2016, 2017). *M. smegmatis* overexpressing *M. tuberculosis* PpiB showed enhanced biofilm formation and exhibits drug tolerance to anti-TB drugs. As a result, higher dosage of anti-TB drugs are required for treatment of patients suffering from tuberculosis (TB), thereby causing toxicity in patients. The involvement of *M. tuberculosis* PpiB in biofilm formation as well as in several key pathways of the bacterial metabolism makes it a "choke point" enzyme for inhibitors. It has been shown that cyclophilin inhibitors such as cyclosporine A and several FDA-approved drugs including acarbose can bind with PpiB. Cyclosporine A, acarbose, and gallium nanoparticle can inhibit the activity of *M. tuberculosis* PpiB, thereby suppressing formation of biofilm. The reduction in biofilm formation results in higher penetrance of drugs through biofilm matrix. Tackling biofilm-related infection is a cumbersome task due to the fact that several heterogenous species of bacteria exist within biofilms. The use of inhibitors can selectively eliminate a group of bacteria from the heterogenous population; however a subpopulation of bacteria that are resilient to the activity of these inhibitors can transfer gene to other bacteria. As a result, several inhibitors that have been tested against biofilm-forming microorganism either failed or their efficacy gradually reduced with time. The bottleneck in devising strategy for targeting biofilms has been the lack of a target that is essential across a wide spectrum of microorganism. PpiB can prove to be a novel target for biofilm-forming organisms due to its presence in all biofilm-forming bacteria. Besides, PpiB homologues in several biofilm-forming bacteria possess similar amino residues in the domain that binds with cyclosporine A, acarbose, and gallium nanoparticle (Kumar et al. 2019). Targeting PpiB could prove to be the masterstroke against a wide genre of biofilm-forming bacteria involved in multiple pathogenic symptoms.

18.9 Pathology Associated with Biofilm-Forming Bacteria

Biofilms are a major reason behind the recalcitrant nature of pathogenic microbes toward antimicrobial agents and host immune responses, as biofilms act as a physical barrier against drugs and provide a protective niche for mycobacterial survival. Conventional biofilms are extremely resistant to host inflicted insults and can endure antimicrobial agents as much as 10–1000 times of that required to eradicate planktonic microbes. Biofilms have also been established as a crucial factor in the pathogenesis of a large number of bacteria. The nature of biofilms also makes it formidable for the host immune cells to attack or opsonize bacteria harbored within them (Piatek et al. 2013). *S. aureus* evades opsonization by phagocytic cells in blood by producing coagulase that activates prothrombin and promotes production of insoluble fibrin which eventually leads to strengthening of the biofilm (Cheng et al. 2010). In cystic fibrosis patients, alginate and rhamnolipid production is enhanced by *P. aeruginosa* as a strategy to escape macrophages and stimulate necrosis of polymorphonuclear leukocytes, respectively (Van Gennip et al. 2009).

The oral cavity provides an excellent niche for survival of nearly 800 species of microorganisms that form a complex-multispecies biofilm or plaque. Anaerobic bacteria such as *Porphyromonas gingivalis* colonize the biofilm at a later stage, erode the dental surface with acidic toxins, and cause periodontitis and peri-implantitis. *P. gingivalis* differentially express nearly 20% of its genome during the various stages of biofilm formation and virulence. The versatility of *P. gingivalis* in attaching to oral soft tissue, dental implant, and other oral bacteria is due to the presence of fimbriae on the cell surface. Loss of Fim A fimbriae results in inhibition of adherence to epithelial cells and dental fibroblasts. Recent studies showed that arginine plays a key role in fimbriae-mediated biofilm formation. The activity of streptococcal Arc A, which catalyzes hydrolysis of arginine to citrulline, suppresses *fim A* expression in *P. gingivalis* and prevents biofilm formation. Unlike other bacteria that utilize siderophores for sequestration of iron, *P. gingivalis* possess proteases such as gingipains for acquiring iron from host heme. Several species of dental microbiota exhibit mutualistic coexistence with *P. gingivalis*. Isobutyric acid produced by *P. gingivalis* aid in growth of *Treponema denticola*, while succinate produced by *T. denticola* enhances growth of *P. gingivalis*. *P. gingivalis* infections lead to chronic inflammatory disease of oral cavity characterized by loss of teeth and increased risk of cardiovascular diseases.

M. tuberculosis is endowed with persistence and drug tolerance, typical of biofilm-associated infections caused by pathogenic bacteria. It has been reported that *M. tuberculosis* develops biofilms within host cells that allow it to evade the immune surveillance. *M. tuberculosis* biofilms are associated with several other non-tubercular bacteria and opportunistic pathogens. *M. tuberculosis* biofilms are mainly composed of methoxy-free mycolic acids and its biosynthesis is modulated by fatty acid synthase complex II-GroEL-mediated pathway. Disruption of polyketide synthases, *pks16* and *pks1/15*, is associated with reduced biofilm formation in *M. tuberculosis* and suggests the role of surface lipids in initial aggregation of mycobacteria. As such the high lipid content of the bacterial cell wall limits the diffusion of drugs; biofilm-encapsulated mycobacterial cells show enhanced drug refractory physiology. *M. tuberculosis* exhibit enhanced tolerance to up to 1000 times higher minimum inhibitory concentration of drugs due to the presence of persister cells within the biofilms. *S. epidermidis* biofilm provides resistance to phagocytosis by interfering with the activation of complement protein C3b (Kristian et al. 2008). Biofilm infections have also been implicated in various other diseases such as endocarditis, chronic otitis media, obstructive pulmonary diseases, and chronic (diabetes) wound infections. Biofilms also dwell in patients with internal medical devices for treatment (Table 18.1). Similarly, lung infections by *P. aeruginosa* can be significantly attenuated by targeting the quorum sensing pathway (O'Loughlin et al. 2013). Staphylococcal biofilm formation is demonstrated to be suppressed in vivo by using RNAIII-inhibiting peptide.

Table 18.1 List of clinically relevant biofilm-associated diseases

Organ/ body part affected	Organism	Disease/symptom	Drugs/ interventions
Eye (contact lens)	*Staphylococcus epidermidis, Staphylococcus aureus, Staphylococcus saprophyticus, Serratia marcescens, Fusarium solani*	Endophthalmitis, keratitis, scleral buckle infection, lacrimal infection, periorbital infections	Tobramycin, polymyxin, bacitracin, levofloxacin
Wound (burns)	*Bacillus, Clostridium, Staphylococcus epidermidis, Staphylococcus aureus*	PASH syndrome, delayed healing in burns	Infliximab, clindamycin, metronidazole, Amikacin
Dental	*Porphyromonas gingivalis, Lactobacillus casei, Streptococcus sobrinus*	Dental plaque, dental caries, pulpitis, apical periodontitis, gingivitis	Sodium hypochlorite, acetaminophen, corticosteroid
Prosthetic heart valve	*Enterococci, d*iphtheroids, *candida* spp.	Endocarditis	Ampicillin, gentamicin, penicillin G
Breast implant	*Staphylococcus epidermidis, Corynebacterium, Mycobacterium fortuitum*	Chronic infection of breast	Chloramex, Fucidin, Terramycin-coated implant
Urinary catheter	*Neisseria gonorrhoeae, Staphylococcus aureus, Enterococcus faecalis*	Urinary tract infections	Fluoroquinolone, cephalosporin, aminoglycoside
Orthopedic metal implants	*Pseudomonas, streptococci, Proteus mirabilis*	Chronic osteomyelitis, peri-prosthetic joint infections	Titanium nitride coating, gentamicin-coated polyurethane sleeves
Lung	*Burkholderia cepacia, Pseudomonas aeruginosa, Aspergillus fumigatus*	Cystic fibrosis, ventilator-associated pneumonia	Ivacaftor, bronchodilators, ureicopenicillin, carbapenem
Ear	*Haemophilus influenza*	Influenza, chronic otitis media	Ofloxacin ear drop, amoxicillin
Vagina	*Corynebacterium* spp., *Micrococcus* spp., *Candida albicans, Lactobacillus plantarum, group B streptococci*	Intrauterine device-associated infections, Gardnerella vaginitis	Metronidazole, gentamycin

18.10 Anti-biofilm Strategies

Due to the adverse impact of biofilms on human health, various methods such as mechanical disruption, disinfectants, chlorination, and ultraviolet have been used traditionally. Recent advances in mechanism of biofilms have provided new insights to control biofilm-based infections. Table 18.2 shows the list of various anti-biofilm strategies that are either used alone or in combination. Extracts from plants such as *Rhodiola crenulata, Malus pumila, Dolichos lablab*, etc., show potent anti-biofilm activity. Chemical derivatives of active ingredients of plant extracts such as icariin and resveratrol can be synthesized and exhibit significant anti-biofilm activity. Green tea possesses high level of anti-oxidants and can suppress biofilm caused by *Streptococcus mutans*. Honey possesses antimicrobial property due to the presence of antimicrobial peptide, defensins, and is effective in inhibiting *Enterococcus* spp. biofilm formation. Similarly, several essential oils such as cumin oil, cinnamon oil, and oregano oil are effective against biofilm caused by *Klebsiella pneumonia* and mixed-species biofilms of enteropathogenic *E. coli* and *S. aureus*, respectively (Ong et al. 2018). The emergence of antibiotic resistance has highlighted the use of bacteriophages in controlling bacterial infections. Phages are highly specific toward bacterial targets, persist as long as the host bacterium is present, and do not show side effects on normal microbiota. Bacteriophages can potentially release enzymes such as polysaccharide depolymerase that liquefy the biofilm matrix and allow the phage to penetrate the biofilm. Genetically engineered phages such as *E. coli*-specific phage (T7), cloned with dispersin (dspB) gene, have shown higher efficacy in biofilm degradation. The major technological challenges in using phages for anti-biofilm strategy are i) reducing the release of toxins by phage, ii) overcoming bacterial resistance to phages, and iii) neutralizing the virulence genes of phages from incorporating into the host bacterial genome. The quorum sensing (QS) molecules such as acetyl homoserine lactone (AHL), autoinducer 2 (AI-2), etc., aid in aggregation of bacteria. Studies have shown that inhibitors that quench the activity of quorum sensing molecules can disrupt the aggregation of cells, considered as an essential step in biofilm life cycle. Naturally occurring quorum sensing inhibitors such as furanone, ajoene, naringin, curcumin, epigallocatechin gallate, etc., have shown anti-biofilm properties against several gram-negative and gram-positive bacterial species. Quorum quenching enzymes such as AHL lactonases, AHL acylases, and oxidoreductases are effective in disruption of biofilm without affecting the microorganism. Therefore treatment with antibiotics in combination with QS molecules or quorum quenching enzymes can be effective in targeting bacteria within biofilms. QS inhibitors such as furanone and penicillic acid increased the efficacy of antibiotic tobramycin against *P. aeruginosa* biofilms. Blockage of cell to cell communication by using peptides such as RNA III-inhibiting peptide (RIP) that inhibits phosphorylation of TRAP (target of RNA III) proteins can prevent initial stages of biofilm formation. The formation of biofilms on medical devices and implants poses a unique challenge for engineering devices that are inert

Table 18.2 List of anti-biofilm strategies

Strategy	Mode of action	Efficacy against
Plant extracts	Act as antioxidants	*Streptococcus mutans*, *Lactobacillus plantarum*, *S. epidermidis*
Green tea,		
Cuminum cyminum		
Cinnamon oil		
Oregano oil		
Thyme oil		
Bacteriophage	Infects and lyse bacteria cells	*P. aeruginosa*, *P. fluorescens*
Engineered with dispersion B gene		
Phage PhilBB-PF 7A		
Phage in combination with antibiotics		
Quorum sensing inhibitors	Blocks intercellular communication carried out through quorum sensing molecules	*P. aeruginosa*, *B. cepacia*, *S. aureus*, *Y. enterocolitica*
AHL cyclase		
AHL lactonases		
AIP inhibitors		
Oxidoreductases		
Surface modification	Prevents attachment and aggregation of cells on surface	*S. epidermidis*, biofilms on heart devices
Coating with furanone		
Zirconium oxide		
Nanocrystalline silicon carbide		
Polyethylene glycol		
Nanoparticle	Generation of reactive oxygen species, destabilizing cell membrane and enzymes	*Candida* spp., *E. coli*, *Enterococcus* spp., *Salmonella* spp.
Silver		
Gallium nitride		
Titanium oxide		
Silica nanoparticle		
Enzymes	Breakdown of components of biofilm matrix	*Streptococcus pyogenes*, *Acinetobacter baumannii*, *Haemophilus influenza*
DNase 1		
Lysostaphin		
Lyase		
Photodynamic	Activation of photosensitizing drugs due to light	*C. albicans*, *Enterococcus faecalis*
Therapy		
Methylene blue		
Toluidine blue		

(continued)

Table 18.2 (continued)

Strategy	Mode of action	Efficacy against
Bacteriocin	Formation of pores on cell membrane of bacteria	*Listeria monocytogenes*, *S. aureus*
Nisin A		
Lacticin Q		
Ultrasonic treatment	Mechanical disruption	*E. coli*, *P. aeruginosa*
>20 kHz frequency of sound waves		
Anti-biofilm agents	Disruption of initial stages of biofilm matrix	Wound colonizing *P. aeruginosa*, peridontal colonizing *Aggregatibacter actinomycetemcomitans*, *Salmonella* spp., *Porphyromonas gingivalis*
Molsidomine		
Diethylamine nonoate diethylammonium		
Povidone iodine		
Xylitol		
Dispersin B		

to biofilm formation (Koo et al. 2017). Since surface attachment is the first step of biofilm formation, coating the surface with inert substances or chemical modification of surface is an innovative strategy for preventing biofilm formation on medical devices. Surface adsorption of furanone and quaternary ammonium salts prevents *Staphylococcus epidermidis* biofilm formation on catheters. Nanoparticles such as silver possess antimicrobial properties. Surface medication with silver coating prevents biofilm formation and studies have shown that silver nanoparticles can reduce 97% of biofilm formation caused by *Candida* spp. Similarly, titanium dioxide nanomaterials are effective against fungal biofilms and are widely used in dental implants (Roy et al. 2018). Gallium shares similar charge and shape with ferric ion and binds with iron siderophores. Iron acts as cofactor of key enzymes involved in bacterial metabolism and disruption of iron metabolism can significantly inhibit growth of bacteria. Gallium competes with ferric ion and inactivates Fe siderophores, thereby leading to suppression of growth and metabolism of bacteria. Recent studies show that gallium nanoparticles not only inhibit biofilm formation but also improve the efficacy of anti-tuberculosis drugs against mycobacterial biofilms (Kumar et al. 2019). Several constituents of biofilms such as DNA, proteins, and polysaccharides can be targeted by using enzymes. A mixture of deoxyribonucleotides, glycosidases, and proteases has shown significant anti-biofilm potential. Treatment with DNase1 (5 μg/ml) either alone or in combination with antibiotics (rifampicin, ampicillin, azithromycin, etc.) reduced biofilm formation of *S. aureus*, *S. pyogenes*, *Acinetobacter baumannii*, *Haemophilus influenza*, etc. Lysostaphin (LS) in combination with nafcillin and enzymes such as amylase, lyase, and lactonase are effective against methicillin-resistant *S. aureus* (MRSA).

Photodynamic therapy is the process of activating drugs, called photosensitizer, using light of particular wavelength. During the exposure of light, photosensitive drugs produce reactive oxygen that destroys adjacent cells. Oral biofilms that consist of multispecies microorganism are treated with helium laser light in the presence of toluidine blue.

18.11 Conclusion

Combating microbial biofilm infections and getting rid of it completely is a daunting task for clinicians. Tackling biofilms is difficult as there are several proteins that are involved in biofilm formation and the involvement of these proteins also varies across biofilm-forming organism. Prolonged treatment with combinations of antibiotics that possess different killing mechanisms can be used to eliminate the infection. However, the efficacy of this method may still be questionable. The mutation rate of bacterial genome is high and offers it a unique advantage to adapt to stress. As a result of mutation the protein targets change their conformation or the active site amino acids get altered, thereby reducing the interaction of inhibitors to target genes or proteins. Apart from this genotypic adaptation, bacteria can temporarily adapt phenotypically by altering the secretion of extracellular molecules that in turn form an impregnable coating. The extracellular coating or matrix can deposit over biotic or abiotic surfaces. In other cases, removal of medical devices and abscesses is necessary for a successful outcome. Treatment only with antibiotics is not a viable option in the case of biofilm-associated infection, partly because the mutation in bacterial genes renders the protein targets inaccessible to drugs and the gradient of drugs across the matrix decreases toward the core of the biofilm. A successful medicament against biofilm would involve a combination of several anti-biofilm strategies that target different steps of biofilm formation. In the light of the fact that dissemination of persister cells from the mature biofilm is preceded by dissolution or loosening of the matrix, the auto-dissolution of matrix is governed by mechanisms that are inherently controlled by the persister cells. The signal for stress removal is communicated from the cells on the outer surface of the matrix to the persister cells within the core of the matrix. It may inevitably lead to change in gene expression of persister cells that in turn result in generation of "factors" that eventually dissolve the matrix. RNA sequencing of persister cells may shed light about "wake up" genes that regulate the expression of these "factors" that can potentially dissolve the matrix. The usefulness of these "factors" as a futuristic tool against biofilm-associated infections is an open question. Other future strategies for biofilm control may include targeting the essential regulatory pathways such as nucleotide signaling and damaging the amyloid structures which play a key role in biofilm formation and adaptation (Romero et al. 2010). Alternatively, chaperonic PpiB proteins, considered as a "choke point protein" involved in biofilm formation across many biofilm-forming pathogens and which regulate other proteins in the cell, could pave the way for a successful target against biofilm-borne infections.

References

Adams KN, Takaki K, Connolly LE, Wiedenhoft H, Winglee K, Humbert O, Edelstein PH, Cosma CL, Ramakrishnan L (2011) Drug tolerance in replicating mycobacteria mediated by a macrophage-induced efflux mechanism. Cell 145(1):39–53

Balcázar JL, Subirats J, Borrego CM (2015) The role of biofilms as environmental reservoirs of antibiotic resistance. Front Microbiol 6:1216

Becker P, Hufnagle W, Peters G, Herrmann M (2001) Detection of differential gene expression in biofilm-forming versus planktonic populations of *Staphylococcus aureus* using micro-representational-difference analysis. Appl Environ Microbiol 67(7):2958–2965

Blanco P, Hernando-Amado S, Reales-Calderon JA, Corona F, Lira F, Alcalde-Rico M, Bernardini A, Sanchez MB, Martinez JL (2016) Bacterial multidrug efflux pumps: much more than antibiotic resistance determinants. Microorganisms 4(1):pii. E14

Bogino PC, Oliva Mde L, Sorroche FG, Giordano W (2013) The role of bacterial biofilms and surface components in plant-bacterial associations. Int J Mol Sci 214(8):15838–15859

Burmølle M, Webb JS, Rao D, Hansen LH, Sørensen SJ, Kjelleberg S (2006) Enhanced biofilm formation and increased resistance to antimicrobial agents and bacterial invasion are caused by synergistic interactions in multispecies biofilms. Appl Environ Microbiol 72:3916–3923

Cheng AG, McAdow M, Kim HK, Bae T, Missiakas DM, Schneewind O (2010) Contribution of coagulases towards *Staphylococcus aureus* disease and protective immunity. PLoS Pathog 6(8): e1001036

Costerton JW, Cheng KJ, Geesey GG, Ladd TI, Nickel JC, Dasgupta M, Marrie TJ (1987) Bacterial biofilms in nature and disease. Annu Rev Microbiol 41:435–464

Hengzhuang W, Wu H, Ciofu O, Song Z, Høiby N (2012) In vivo pharmacokinetics/pharmacodynamics of colistin and imipenem in *Pseudomonas aeruginosa* biofilm infection. Antimicrob Agents Chemother 56(5):2683–2690

Jefferson KK (2004) What drives bacteria to produce a biofilm? FEMS Microbiol Lett 236 (2):163–173

Klausen M, Heydorn A, Ragas P, Lambertsen L, Aaes-Jorgensen A, Molin S, Tolker-Nielsen T (2003) Biofilm formation by *Pseudomonas aeruginosa* wild type, flagella and type IV pili mutants. Mol Microbiol 48(6):1511–1524

Koo H, Allan RN, Howlin RP, Stoodley P, Hall-Stoodley L (2017) Targeting microbial biofilms: current and prospective therapeutic strategies. Nat Rev Microbiol 15(12):740–755

Kristian SA, Birkenstock TA, Sauder U, Mack D, Götz F, Landmann R (2008) Biofilm formation induces C3a release and protects *Staphylococcus epidermidis* from IgG and complement deposition and from neutrophil-dependent killing. J Infect Dis 197(7):1028–1035

Kumar A, Alam A, Rani M, Ehtesham NZ, Hasnain SE (2017) Biofilms: survival and defense strategy for pathogens. Int J Med Microbiol 307(8):481–489

Kumar A, Alam A, Grover S, Pandey S, Tripathi D, Kumari M, Rani M, Singh A, Akhter Y, Ehtesham NZ, Hasnain SE (2019) Peptidyl-prolyl isomerase-B is involved in *Mycobacterium tuberculosis* biofilm formation and a generic target for drug repurposing-based intervention. NPJ Biofilms Microbiomes 5:3

Lebeer S, De Keersmaecker SC, Verhoeven TL, Fadda AA, Marchal K, Vanderleyden J (2007) Functional analysis of luxS in the probiotic strain *Lactobacillus rhamnosus* GG reveals a central metabolic role important for growth and biofilm formation. J Bacteriol 189(3):860–871

Marsden AE, Grudzinski K, Ondrey JM, DeLoney-Marino CR, Visick KL (2017) Impact of salt and nutrient content on biofilm formation by Vibrio fischeri. PLoS One 12(1):e0169521

McDonough KA, Rodriguez A (2012) The myriad roles of cyclic AMP in microbial pathogens: from signal to sword. Nat Rev Microbiol 10(1):27–38

McDougald D, Rice SA, Barraud N, Steinberg PD, Kjelleberg S (2012) Should we stay or should we go: mechanisms and ecological consequences for biofilm dispersal. Nat Rev Microbiol 10 (1):39–50

Mohamed JA, Huang DB (2007) Biofilm formation by enterococci. J Med Microbiol 56 (12):1581–1588
Newman JA, Rodrigues C, Lewis RJ (2013) Molecular basis of the activity of SinR, the master regulator of biofilm formation in Bacillus subtilis. J Biol Chem 288(15):10766–10778
O'Loughlin CT, Miller LC, Siryaporn A, Drescher K, Semmelhack MF, Bassler BL (2013) A quorum-sensing inhibitor blocks *Pseudomonas aeruginosa* virulence and biofilm formation. Proc Natl Acad Sci 110(44):17981–17986
Ong KS, Mawang CI, Daniel-Jambun D, Lim YY, Lee SM (2018) Current anti-biofilm strategies and potential of antioxidants in biofilm control. Expert Rev Anti-Infect Ther 16(11):855–864
Orgad O, Oren Y, Walker SL, Herzberg M (2011) The role of alginate in *Pseudomonas aeruginosa* EPS adherence, viscoelastic properties and cell attachment. Biofouling 27(7):787–798
Pandey S, Sharma A, Tripathi D, Kumar A, Khubaib M, Bhuwan M, Chaudhuri TK, Hasnain SE, Ehtesham NZ (2016) *Mycobacterium tuberculosis* peptidyl-prolyl isomerases also exhibit chaperone like activity in-vitro and in-vivo. PLoS One 11(3):e0150288
Pandey S, Tripathi D, Khubaib M, Kumar A, Sheikh JA, Sumanlatha G, Ehtesham NZ, Hasnain SE (2017) *Mycobacterium tuberculosis* peptidyl-prolyl isomerases are immunogenic, alter cytokine profile and aid in intracellular survival. Front Cell Infect Microbiol 7:38
Piatek R, Zalewska-Piatek B, Dzierzbicka K, Makowiec S, Pilipczuk J, Szemiako K, Cyranka-Czaja A, Wojciechowski M (2013) Pilicides inhibit the FGL chaperone/usher assisted biogenesis of the Dr fimbrial polyadhesin from uropathogenic *Escherichia coli*. BMC Microbiol 13:131
Romero D, Aguilar C, Losick R, Kolter R (2010) Amyloid fibers provide structural integrity to Bacillus subtilis biofilms. Proc Natl Acad Sci 107(5):2230–2234
Roy R, Tiwari M, Donelli G, Tiwari V (2018) Strategies for combating bacterial biofilms: a focus on anti-biofilm agents and their mechanisms of action. Virulence 9(1):522–554
Stalder T, Top E (2016) Plasmid transfer in biofilms: a perspective on limitations and oppurtunities. NPJ Biofilms Microbiomes 2:16022
Sun J, Deng Z, Yan A (2014) Bacterial multidrug efflux pumps: mechanisms, physiology and pharmacological exploitations. Biochem Biophys Res Commun 453(2):254–267
Tolker-Nielsen T (2015) Biofilm development. Microbiol Spectr 3(2):MB-0001-2014
Van Gennip M, Christensen LD, Alhede M, Phipps R, Jensen PØ, Christophersen L, Pamp SJ, Moser C, Mikkelsen PJ, Koh AY, Tolker-Nielsen T, Pier GB, Høiby N, Givskov M, Bjarnsholt T (2009) Inactivation of the rhlA gene in *Pseudomonas aeruginosa* prevents rhamnolipid production, disabling the protection against polymorphonuclear leukocytes. APMIS 117 (7):537–546
Wright KJ, Seed PC, Hultgren SJ (2007) Development of intracellular bacterial communities of uropathogenic *Escherichia coli* depends on type 1 pili. Cell Microbiol 9(9):2230–2241

Part III

Vaccines, Diagnostics and New Interventions

Best Practices in Mycobacterial Research Laboratories

19

Noman Siddiqi

Abstract

Research on *Mycobacterium tuberculosis* (MTB), the bacterium causing tuberculosis (TB), has seen a significant increase in the last decade. Advances in molecular biology techniques now provide scientists with a plethora of approaches to investigate bacterial pathogenesis in the laboratory and in animal models. Here we outline some good laboratory practices for working with MTB in a research setting. The risk of bacterial transmission via aerosols or through accidental exposure requires that we implement rigorous practices to mitigate this risk. Most laboratory research on MTB is performed in specialized laboratories at biological safety level 3. We discuss primary and secondary barriers to MTB exposure (infection), controls that reduce the risk of exposure to the bacterium during experimental manipulation, provide guidelines for the prudent use of biosafety cabinets and personal protective equipment, and present examples of standard operating protocols, including setup of biosafety cabinets, waste management practices, equipment use, management of spills, occupational health programs, and facility decontamination.

Keywords

Mycobacterium tuberculosis · BSL3 · Respirators · Tuberculosis · Risk assessment · PPE · Laboratory design · Engineering control · Occupational health · Bacterial inactivation · Spill cleanup

N. Siddiqi (✉)
Harvard TH Chan School of Public Health, Boston, MA, USA
e-mail: nsiddiqi@hsph.harvard.edu

S. E. Hasnain et al. (eds.), Mycobacterium tuberculosis: *Molecular Infection Biology, Pathogenesis, Diagnostics and New Interventions*,
https://doi.org/10.1007/978-981-32-9413-4_19

Abbreviations

A/BSL3	Animal biological safety level 3
BCG	Bacillus Calmette-Guerin
BI	Biological indicators
BSC	Biosafety cabinet
BSL1	Biological safety level 1
BSL2	Biological safety level 2
BSL3	Biological safety level 3
DPM	Differential pressure monitors
HEPA	High-efficiency particulate air
IGRA	Interferon -gamma release assay
MTB	*Mycobacterium tuberculosis*
OH	Occupational health
PAPR	Powered air-purifying respirators
PCR	Polymerase chain reaction
PPE	Personal protective equipment
RT-PCR	Reverse transcription polymerase chain reaction
SOP	Standard operating procedure
TB	Tuberculosis
TST	Tuberculin skin test
VHP	Vapor Hydrogen Peroxide

19.1 Introduction

In the past two decades, mycobacterial research has seen immense growth (Bloom et al. 2017). With the development of tools to target mycobacterium at the genetic level (Hatfull and Jacob Jr 2014), laboratory-based research has expanded and flourished, globally. As molecular biological technologies have proliferated, so has our understanding of the risks associated with performing research on this pathogen. Accordingly, biological safety considerations have evolved over the years, to ensure science is conducted in a safe environment (Fleming 2006; Wooley and Byers 2017). This has led to a transformation in research practices, biosafety program management, and risk assessment (World Health Organization 2012). While researchers have innovated in the laboratory, architects and engineers have worked on improving laboratory systems and building design. Likewise, equipment manufacturers have addressed safety concerns by improving design and safety features. However, although most institutions and governments have instituted policies to regulate the operation of high containment laboratories, there is still a need to define global operating standards for such laboratories and work practices.

A discussion on the risks associated with manipulating *Mycobacterium tuberculosis* (MTB) and risk mitigation strategies in research laboratories is presented below.

19.2 Risk Assessment

The following factors are generally considered when performing risk assessment (Fleming 2006):

1. Characteristics of the agent/material (Fig. 19.1).
2. Activities that may lead to exposure (Fig. 19.2).
3. Potential of the exposure to cause infection or injury.
4. Consequences of the ensuing infection or injury.

Transmission:	Airborne, parental, via mucous membranes
Infective Dose:	1-10 bacteria, in nuclei droplets (5μm-20μm in size)
Host:	Humans, non-human primates, cattle, rodents
Reservoir:	Humans, cattle
Virulence:	Multiple strains with varying virulence
Viability:	Survives for weeks in dry state on surfaces
Geographical Spread:	Endemic in many parts of the world
Treatment:	Treatable, although multi-drug resistant strains pose a challenge.
Sensitive to UV light, disinfectants like bleach, phenolic acids, vapor hydrogen peroxide chlorine dioxide and quaternary ammonium chloride.	

Fig. 19.1 Characteristics of *Mycobacterium tuberculosis*

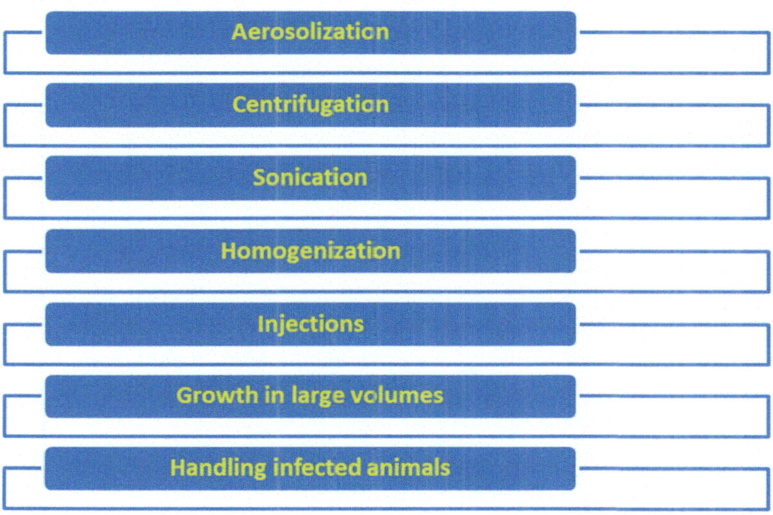

Fig. 19.2 Activities leading to exposure

Mycobacterium tuberculosis (MTB) is a nonmotile bacterium, typically transmitted via aerosols (Cole et al. 2005; Pfyffer 2007). Figure 19.1 highlights some salient characteristics of the bacterium (Public Health Agency of Canada 2010; Hirai 1991; Kramer et al. 2006; Riley and Nardell 1989). Figure 19.2 describes activities routinely performed in a research setting, involving manipulation of MTB. The site and level of exposure contributes to the establishment of a potential infection. Whether this infection leads to disease is dependent on the health status of the individual, on whether prophylactics (e.g., vaccines) were administered, and whether treatment was received following infection. Host factors such as age, immune status, immunodeficiency, etc., also contribute to the clinical disease outcome (Bloom et al. 2017).

Based on their risk assessment, pathogens are classified into distinct risk groups and their research falls under specific laboratory designations (World Health Organization 2004; Centers for Disease Control and Prevention 2009) (Fig. 19.3).

Members of the *Mycobacterium tuberculo*sis complex, including *M. bovis*, *M. africanum*, *M. microti*, *M. canetti*, *M. caprae*, and *M. pinnipedii*, are considered risk group 3 pathogens and are usually assigned a biological safety level 3 (BSL3) laboratory designation, for research purposes. The *Biosafety in Microbiological and Biomedical Laboratories* (BMBL) definition for biological safety level 3 (BSL3), "...is applicable to clinical, diagnostic, teaching, research, or production facilities where work is performed with indigenous or exotic agents that may cause serious or potentially lethal disease through inhalation route exposure". Laboratory personnel must receive specific training in handling pathogenic and potentially lethal agents, and must be supervised by scientists competent in handling infectious agents and associated procedures. All procedures involving the manipulation of infectious materials must be conducted within Biosafety Cabinets (BSC), other physical containment devices, or by personnel wearing appropriate personal protective equipment. A BSL-3 laboratory has special engineering and design features." (Centers for Disease Control and Prevention 2009).

19.3 Risk Control

Risk mitigation strategies can be divided into two groups – primary and secondary barriers of control.

Primary barriers provide protection to the researcher and include equipment like biosafety cabinets (BSC) and personal protective equipment (PPE).

Secondary barriers encompass engineering and administrative controls, which provide protection to the facility users and the surrounding environment. Engineering controls cover laboratory design, equipment containment, and placement, while administrative controls include the training requirements, biosafety program, occupational health requirements, standard operating procedures (SOP), and protocol risk assessment.

	NIH GUIDELINES	WHO BIOSAFETY MANUAL
1	Agents are not associated with disease in healthy adult humans	(No or low individual and community risk) A microorganism that is unlikely to cause human or animal disease
2	Agents that are associated with human disease which is rarely serious and for which preventive or therapeutic interventions are often available	(Moderate individual risk; low community risk) A pathogen that can cause human or animal disease but is unlikely to be a serious hazard to laboratory workers, the community, livestock or the environment. Laboratory exposures may cause serious infection, but effective treatment and preventive measures are available and risk of spread is limited
3	Agents that are associated with serious or lethal human disease for which preventive or therapeutic interventions may be available (high individual risk but low community risk)	(High individual risk ; low community risk) A pathogen that usually causes serious human or animal disease but does not ordinarily spread from one infected individual to another. Effective treatment and preventive measures are available
4	Agents that are likely to cause serious or lethal human disease for which preventive or therapeutic interventions are not usually available (high individual risk and high community risk)	(High individual and community risk) A pathogen that usually causes serious human or animal disease and that can be readily transmitted from one individual to another, directly or indirectly. Effective treatment and preventive measures are not usually available

Fig. 19.3 Risk group definition

19.4 Biosafety Cabinet

A laminar flow biological safety cabinet (BSC) provides three kinds of protection: a) protection to personnel, b) protection to the product, and c) protection to the environment. A basic understanding of BSC operation is essential for researchers working in a BSL3 facility. A common mistake includes covering the air intake grills

Class of BSC	Typical Biosafety Level laboratory	Application
Class I	1,2,3	low to moderate risk biological agents. Protection to user or product
Class II	1,2,3	low to moderate risk biological agents. Protection to user and product.
Class III	4	high risk biological agents. Protection to user and product.

Fig. 19.4 Different types of biosafety cabinets. (Source: Baker Company)

in the front and back of the BSC; this blocks the air flow, compromising the aseptic environment. Other operational errors include placing the BSC under supply or exhaust air ducts, next to entry doors, overcrowding the cabinet with equipment/samples, and simultaneous use of the cabinet by multiple researchers.

BSCs are grouped into three classes based on their functionality (Fig. 19.4).

BSC class II type A2 cabinets are the most commonly used types in the research laboratory. When working with large amounts of volatile chemicals, it is preferable to use type B cabinets. Class III cabinets offer the highest levels of protection and containment.

19.5 Respirators

Although different classes of respirators are available, not all provide protection suitable for an MTB research laboratory. The two most favored respirators for MTB research are N-95 and powered air-purifying respirators (PAPR).

N-95 respirators are one of the more frequently used respirators. Researchers get fit tested for the make and model of the respirator. Ideally, the fit test is repeated annually, as the fit of the mask is molded to the contours of the face (NIOSH 2013). Facial hair – even a stubble – can interfere with the efficiency of the respirator.

PAPRs use air displacement from the wearer's mask to provide protection. This involves a battery-operated motor blowing filtered air into the mask, which could be a partial- or full-face mask (Roberts 2014).

Individual respirators provide varying degrees of protection from aerosolized contaminants. Usually a PAPR with a full-face mask offers better protection than

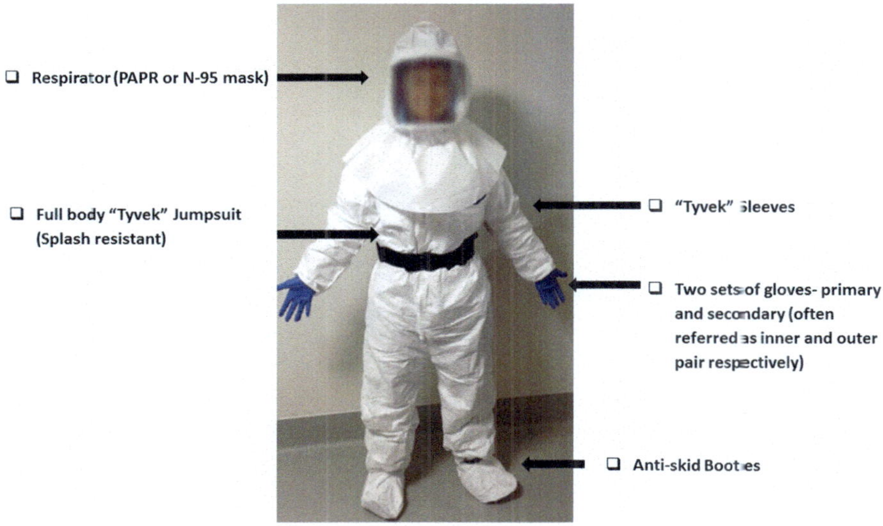

Fig. 19.5 Personal protective equipment (PPE)

an N-95 respirator. A PAPR with a full-face mask is recommended for experiments involving a high risk of aerosolization of the pathogen, such as flow cytometry, infection, or homogenization. The PAPR airflow is battery-powered, so it is critical to confirm the battery is adequately charged before use.

19.6 Other PPEs

To prevent contamination of clothes it is best to wear a Tyvek jumpsuit (over street clothes) when working in the BSL3, while other individual protective coverings consisting of disposable gowns, hoodies, booties, and long sleeves can also be used. Some laboratories require users to change into scrubs and then wear a Tyvek suit. It is also important to follow the "two pairs of gloves" rule when handling the pathogen. Extra Tyvek sleeves and booties (shoe covers; see Fig. 19.5) could be worn to provide extra protection to arms and feet, as these are more likely to get contaminated during work. Figure 19.6 outlines the steps for wearing and removing PPE while entering or exiting the BSL3 laboratory.

19.7 Engineering Control – Design of the Laboratory

The BSL3 laboratory builds on features present in BSL1 and BSL2 laboratories. Specifically, it has a specialized ventilation system allowing for directional airflow, airtight seals and finishes, access controls, waste management, and many other

a

A: Wearing PPE

In the ante room of the BSL3:

- Wear gloves (1st pair). (Put on the hair cap and wear N-95 respirator)
- Wear Tyvek Jumpsuit
- Pull cuffs of the gloves over the suit sleeve and seal together with tape (prevents skin exposure)
- If wearing PAPR.--Test PAPR. Wear PAPR and mask.
- Put on extra Tyvek sleeves.
- Wear booties
- Wear 2nd pair of gloves.
- Ready to step inside the BSL3 lab. Check Air flow and Differential Pressure Monitor (DPM) at entry door.

b

Fig. 19.6 (A&B): Typical donning and doffing procedure for PPE

features. Figure 19.7 illustrates some recommendations for laboratory design, drawn from the NIH handbook (Memarzadeh and Wheeland 2016).

Directional airflow is set up to allow clean air to enter the facility and to exit through exhaust ducts with high-efficiency particulate air (HEPA) filters. This prevents any contaminated air from escaping the laboratory into adjoining clean areas. There is no recirculation of air in these laboratories. It is best to have a

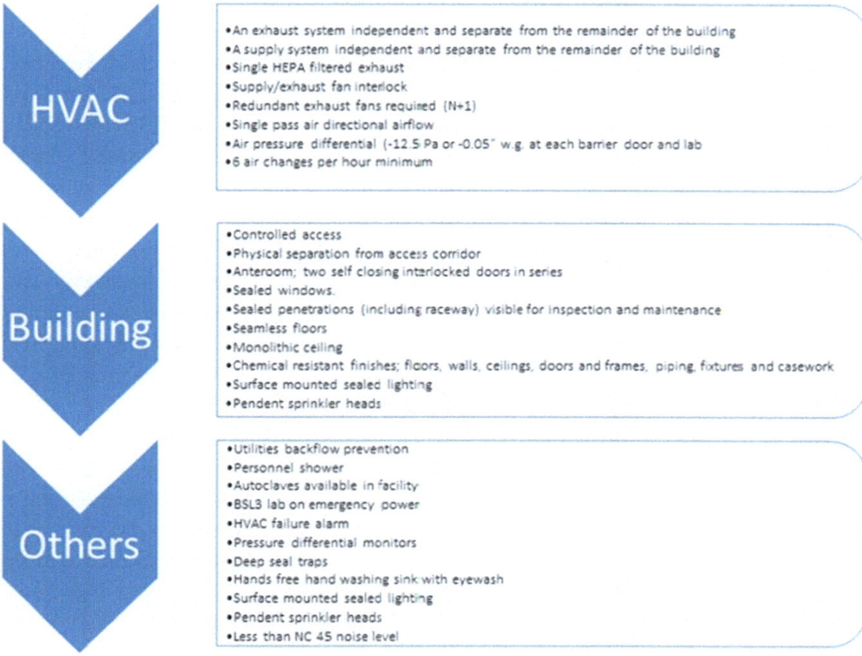

Fig. 19.7 Laboratory design recommendations

dedicated air supply and exhaust system serving the BSL3. These systems should be hardwired so that the supply fan automatically shuts down if the exhaust fan fails, preventing the space from "going positive," i.e., allowing air to flow out of the facility. Differential pressure monitors (DPM) should be installed for each door and wired to an alarm (visual and audible) to indicate a failure in the system. Researchers need to be trained to monitor the DPMs and exit the laboratory if the alarms turn on. Installing redundant HEPA filter units on the exhaust system will allow for easy maintenance of the filters. The air supply handler and exhaust fan systems supplying the BSL3 are critical for the safe operation of the laboratory and should have a backup power source.

BSCs and autoclaves are essential to the functioning of the BSL3 laboratory. Ideally, a pass-through autoclave, with one door opening inside the laboratory and another outside, should be installed. This allows for loading of contaminated waste inside the laboratory. Once the sterilization cycle is complete, the autoclave door outside the laboratory can then be used to remove waste. It is best practice to confirm the sterilization performance of the autoclave periodically (Association for the Advancement of Medical Instrumentation 2017), e.g., by running challenge loads with biological indicators (BI) every fortnight.

19.8 Waste Management

It is important to set up procedures for handling waste generated in the laboratory. Waste streams are defined based on methods of disinfection and/or sterilization. Broadly, waste can be classified into two streams: either created inside the biological safety cabinet or outside it. Waste generated within the BSC, being potentially more contaminated and with a higher risk of personnel exposure, needs to be disinfected and sterilized, while waste/trash generated outside the BSC carries less risk and may be treated less stringently. One should add multiple steps of disinfection and a final sterilization cycle for waste generated inside the BSC. Our laboratory typically uses chemical disinfection procedures to treat waste inside the BSC before it is bought out and autoclaved for sterilization. Waste generated outside the cabinet is packed and autoclaved without undergoing a chemical disinfection process first.

The cabinet waste stream can include solid, liquid, or sharps waste, each of which needs to be processed differently. One practice is to pack solid waste in a single biohazard bag, liquids in plastic containers or bottles, and sharps in approved cardboard or plastic containers. Consider adding disinfectant at a defined concentration to the liquid waste, to allow for initial disinfection. Sharps can be dipped in disinfectant before being discarded in approved sharp containers. Solids can be wiped with disinfectant and packed in a bag.

All waste generated in the laboratory should be sterilized before being removed from the laboratory. It is recommended to have a double door autoclave to facilitate this process. The double door allows the user to define a "dirty" and "clean" side in the laboratory. Each sterilization run is typically monitored by chemical indicators and built-in sensors in the autoclave which measure pressure, temperature, and exposure time (Association for the Advancement of Medical Instrumentation 2017).

19.9 Occupational Health, Disease Surveillance, and Response to Exposure

An occupational health (OH) program can cover (1) prejob offer and annual health screening, (2) annual fit test for N-95 respirators, (3) identification of high-risk individuals (e.g., immune suppressed or pregnant), and (4) annual updates of an exposure control plan.

Annual health screens help in disease surveillance and can potentially identify unreported or unidentified exposure. This can help in re-evaluating risk associated with specific procedures and suggest PPE or engineering controls to mitigate the risk of exposure.

Occupational health providers can also provide information to researchers on host factors which influence infection and clinical features of tuberculosis (TB). Some of these include age, vaccination (BCG) status, immune/immunodeficiency status, coexisting diseases, and drug treatments.

If potential exposure is suspected, contacting the occupational health provider immediately and following their recommendations is critical. Many laboratories keep pathogen information cards which can be presented to health providers by the researcher after potential exposure in the laboratory. These cards list information about the pathogen, disease, and potential testing and treatment options. Diagnostic testing may include the tuberculin skin test (TST) or the interferon gamma release assay (IGRA) test for determining a basal value of MTB antigen-specific immune response (Mazurek et al. 2007; Kimura et al. 1999). This is then used for evaluating infection progression following subsequent testing and for recommending treatment options.

19.10 Administrative Controls – Training Requirement

An important way to reduce incidents and exposure in the laboratory is to train users in best laboratory practices. A combination of classroom teaching and hands-on training is recommended. As an example, we require all our researchers to undertake 15–20 h of hands-on training and 1 h of didactic training. Beyond this, our researchers also get a chance to shadow other scientists working in the BSL3. Annual training refreshers reinforce various biosafety practices and correct use of equipment in the BSL3.

The hands-on training module is designed to simulate a typical experiment, including BSC setup, culturing bacteria (in liquid and solid media), waste management, dealing with spills, a potential exposure situation, decontamination procedures, and cleaning of the BSC post-experiment. These are offered as individual sessions with expert trainers. After three to four rounds (at least 3 h each) of such sessions, the trainee is required to demonstrate their understanding and ability to work independently in the BSL3, as judged by their peers, who have experienced working in the BSL3, to confirm their readiness to work unsupervised in the laboratory. Investigators are required to undertake similar training for any new procedure that they need to perform in the laboratory.

BSL3 users participate in reviewing protocols and procedures before they are implemented in the laboratory. This ensures that all users are familiar with the protocols and associated risks. The group vetting of procedures also allows us to maintain stringent safety standards across the gamut of experimental procedures.

19.10.1 Access

Once a trainee has met all training requirements they are allowed access to the laboratory. Users are encouraged to work in pairs when working off-hours and during holidays. All entry/exit doors into BSL3 should have card access control. Researchers with access are trained to not share their access cards and to bar visitors or untrained people from the laboratory.

Casual visitors should not be allowed into the biocontainment facility. When scientifically necessary or approved, untrained individuals may enter the facility only when accompanied by an approved user. This is especially important for tradesmen and engineers who need to enter the facility to perform repairs.

19.11 Standard Practices and Precautions

The majority of work practices for BSL3 builds on practices and precautions for BSL2 laboratories (Emmert 2013). Additional practices are listed below:

- Stored items should be packed correctly and clearly labeled with investigator's name, date, and content information.
- When bringing hands out of the BSC:
  - Wipe gloves with disinfectant.
  - Discard the second pair (outer pair) of gloves.
  - Remove hands from cabinet without touching anything.
  - Wear a new pair of gloves before carrying out further procedure.
- All vessels containing pathogens should rest on the floor of the BSC while transferring cultures from one container to other to avoid any spillages or splashes as a result of these containers slipping out of the hands.
- To bring items out of the BSC during an experiment/procedure, a space should be identified as "clean." A surface would be considered clean only after it is wiped down with disinfectant.
- Before an item is removed from the cabinet it should be wiped (surface decontamination) and placed on the clean side.
- Any manipulation of a pathogen in the BSC is considered to render the BSC "dirty," necessitating the need to clean a side before removing items from the BSC.
- Ideally, all pathogen manipulation should occur only in the BSC. If this is not possible, a risk assessment should be performed and steps to minimize risk implemented. An example of minimizing aerosol risk during centrifugation is to use aerosol caps on rotor buckets.

19.12 Use of the Biosafety Cabinet

The biosafety cabinet is part of the primary barrier and offers protection to users and the environment when correctly set up and used. Generally, BSCs in the BSL3 should be tested and certified every 6 months.

- Ensure that start-up and purge cycles of BSC are properly functioning and BSC was disinfected by previous user.
- The BSC floor can be lined with an absorbent sheet (to contain liquid spills) and is removed at the end of each experiment as part of BSC cleanup.
- Containers/bags to capture different kinds of waste generated during the experiment should be placed in the cabinet. Disinfecting wipes and a container of disinfectant should be placed inside (Fig. 19.8). The BSC should be set up to allow one side to be used for bringing items out (the "clean" side). Manipulations

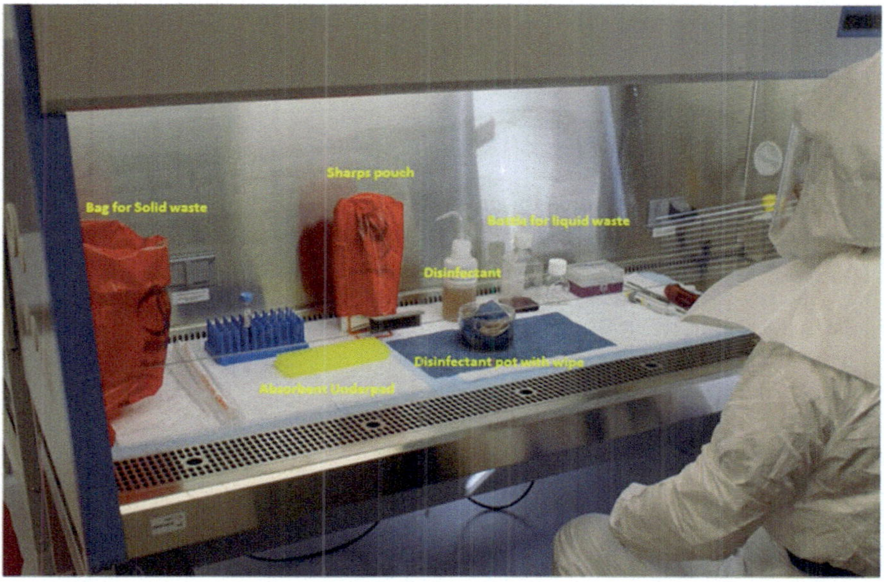

Fig. 19.8 Biosafety cabinet setup for experiment

should occur on one side of the BSC, being careful to maintain a "clean to dirty/contaminated" direction of work flow.

- Pathogen-containing items must always be in double containment (boxes/bags) when taken outside of a BSC.
- Items must be cleaned/disinfected when moving from the dirty to the clean side of the BSC.
- Objects must be cleaned/disinfected before being moved out of the BSC.
- A secondary container should be installed on the clean side to move liquid cultures out of the BSC.

19.13 Exposure

Failure to wear the right PPE or breakdown of PPE may lead to exposure to the pathogen. The risk is elevated when working with sharps and animals. Needlesticks, cuts, animal bites, and scratches are potential exposure routes. In some instances, PAPR failure or the N 95 mask coming off may expose the investigator. Rips and nicks in primary and secondary gloves also pose a high risk of exposure. Equipment failure, e.g., loss of power to the biosafety cabinet, spills in a centrifuge or incubators, and sprays/splashes from a flow cytometer or plate reader could also potentially lead to exposure.

The laboratory should have a plan in place to manage exposures to the investigator(s) or the environment. These plans should consider occupational health

(OH) intervention for the exposed employee(s), containment of the material within the lab, and steps to mitigate the risk associated with cleanup.

Following potential exposure, an investigator(s) should take the following steps:

- Exit the facility as per BSL3 SOPs.
- To get help, or if unable to exit, use the phone or panic alarm/distress button.
- Inform the principal investigator, laboratory director, or a colleague.
- Call the occupational health physician and follow their instructions.
- *If directed by the physician or if unable to reach a healthcare provider, report to the nearest hospital emergency room.*
- Create a summary (if possible) of the incident, including information about the pathogen, possible route of exposure, any drug resistance strain, OH physician's name, and phone number.

19.14 Inactivation of Bacteria for Sample Removal

Occasionally, researchers may need to remove samples from the BSL3 laboratory to a BSL2 laboratory to conduct follow-on experiments; e.g., bacterial samples might need to be processed for nucleic acid/protein/metabolite analysis or cell lines may be removed for downstream experiments involving PCR, RT-PCR, proteomics, cytometry, or mass spectrometry.

Before removal, samples need to be processed to clear them of any live bacteria. We can use physical or chemical means to sterilize samples. The choice of method is determined largely by the downstream experiments to be performed on these samples. Physical methods such as heat inactivation and filtration are widely used. Chemical methods include the use of formalin, glutaraldehyde, paraformaldehyde, phenol, methanol-chloroform, etc. (Schwebach et al. 2001; Doig et al. 2002). Figure 19.9 describes some methods we use for transferring samples from BSL3 to BSL2. It is important that users validate each method that they intend to use for bacterial inactivation. Validation would involve following the exact steps in a planned experiment, exposing the sample to the inactivation reagent, washing the sample, and plating it to check for bacterial growth. A strict standard applies here: there should be no bacterial growth on the plates after 6 weeks of incubation. Positive and negative controls should be included in this testing.

19.15 Decontamination

Various methods are used for decontaminating work surfaces in the laboratory. An adequate concentration of disinfectant and sufficient exposure time are important factors that determine the efficacy of decontamination. Certain surfaces may be

SAMPLE	METHOD	COMMENT	APPLICATION
Tissue culture	Incubate with 10% Formalin at 4°C O/N	1 hour at RT is acceptable too	Histology
Protein extraction	Bead beating with 1X SDS PAGE buffer and heating at 100°C for 20 minutes		Western blots
Supernatant (bacterial or TC)	2X filtration through a 0.2um filter	Do the second round of filtration on a clean side	Soluble protein studies, cytokine profiles
Bacteria from broth or plates	Heat inactivate at 100°C for 1 hours	Small volumes up to 50ul can be heat inactivated at 80°C for 2 hours in PCR plates.	DNA prep
Bacteria from broth or plates or homogenized tissue	Trizol reagent, with bead beating, incubate at RT for 15 minutes	Can be used for TC cells as well	RNA prep
Tissue culture	2%Paraformaldehyde (final concentration) incubate at RT for 1hr	For bacterial pellets and animal tissue samples, the incubation is done at 4°C O/N and or use 4% PFA final concentration	Fixes cell- for FACS, fluorescence imaging
Bacteria from Broth or plate	Incubate for 24 hours in Chloroform/methanol 2:1 at 4C		DNA prep, Lipid extraction

Fig. 19.9 Commonly employed inactivation methods

harder to decontaminate than others (e.g., porous versus nonporous material) may. Surface decontamination can be performed using disinfectants (wipe down). Newer methods of decontamination, such as the use of vapor hydrogen peroxide (VHP) (Hall et al. 2007) and chlorine dioxide (Lowe et al. 2012), are very effective for decontaminating large areas, such as rooms and suites. However, these methods require specialized equipment and trained professionals, unlike surface wipe down methods, which can be done by laboratory staff. One also needs to verify there are no leaks from the laboratory before using VHP or chlorine dioxide. The choice of one method over another is based on risk assessment and biosafety considerations.

Equipment in need of repair (or being disposed of) can be removed from the laboratory after gaseous decontamination. Smaller equipment can be decontaminated inside a BSC, while larger pieces need to be isolated, either in a room or covered by a flexible sheet (plastic tent), for decontamination.

19.16 Spills and Cleanup

Strict precautions should be taken when working, transporting, or removing samples containing MTB. Spills can occur when researchers fail to follow safe practices as outlined in the above sections. Spills can happen inside a biosafety cabinet when a tube is dropped while pipetting samples. Spills can also occur outside of the BSC, in a storage area, or within equipment (for example, tubes breaking in the centrifuge, a culture flask breaking in the incubator shaker, or a leak in a primary container containing MTB). Spills in the BSC are contained and so are easily dealt with, unlike spills in equipment or general laboratory areas.

TIP: After starting the run, wait for the centrifuge to get to the desired speed before walking away. Usually, any breaking or imbalance occurs within the first 30 seconds of operation. If you hear any breaking and/or alarm sounds- shut the centrifuge and exit the facility. Follow the steps described below. If you open the centrifuge after a completed run and discover a spill, hold your breath (when wearing N-95 mask), shut the centrifuge lid and exit the facility. Follow the steps listed below.

1. In the ante room put on a PAPR with full face hood and go back in to investigate the spill. If the centrifuge is located in the common equipment room, make sure everyone exits the facility prior to opening the centrifuge.
2. Remove the buckets to a BSC. Any broken parts should be removed to the BSC as well. Use a plastic Tupperware box to transport the buckets and other pieces. Use forceps to pick up broken pieces. Switch off the machine. Leave the centrifuge lid open.
3. Exit the facility. Discard your suit and PAPR hood. Post signage to prevent people from accidentally entering the room. Wait for sufficient air changes to happen before re-entry.
4. Wear PPE, including PAPR with full face hood.
5. Make a waste barrel and put it next to centrifuge and use it for discarding all spill cleanup material. Take two pots full of disinfectant, gloves, blue wipes and brown towels for the cleanup.
6. Open the lid, wipe the drum with disinfected soaked wipes or paper towels. Discard these into pot with disinfectant. Change gloves.
7. Wipe the rotor and inside of the lid with more disinfectant wipes. Discard in the disinfectant pot. Change gloves.
8. Do a final wipe down with 70% ethanol. Let it air dry before use.
9. In the BSC, soak the buckets and lids in disinfectant in a Tupperware container for adequate time. Remove the buckets and wipe down with disinfectant wipes.
10. Do a final rinse with 70% ethanol. Let the buckets and tops air dry on the clean side of the hood. Once dry remove to the outside. Autoclave the buckets and rotors if possible.
11. File an accident report.

Fig. 19.10 Protocol for cleaning a spill in centrifuge

Spills in the BSC can be cleaned with a disinfectant-soaked wipes or paper towels. It is important to have an absorbent underpad covering the floor of the BSC. These underpads can be replaced if a spill occurs. Always disinfect a spill area first before continuing with the experiment.

Spills in equipment and general work areas need more intensive cleanup. Consider isolating the area and evacuating the laboratory for an appropriate period of time. The time will depend on the number of air changes needed in the space for removal of any aerosolized particulates (Morbidity and Mortality Weekly Report (MMWR) 2005). Discard any disposable PPE worn at the time of the spill and decontaminate the rest before reuse. Wear adequate PPE and use respiratory protection which offers maximum protection (e.g., PAPR with a full-face hood), when coming in to clean the spill. It is advisable to work in pairs when cleaning a spill. An example of spill cleanup procedure in a centrifuge is shown in Fig. 19.10.

19.17 Summary

A BSL3 laboratory requires both engineering and administrative controls. A thorough risk assessment should drive the setup of standard operating procedures (SOPs) and experimental practices. The description of facility and practices outlined here are based on risk assessment of *Mycobacterium tuberculosis* and its manipulation in a research laboratory setting. Clinical laboratories may share some practices mentioned here but are typically not held to the same standards.

The emphasis on biological safety in research ensures that pathogens are handled with utmost caution. The best practices outlined here ensure the prevention of laboratory-acquired infections (Kao et al. 1997) or accidental release of the pathogen to the environment. While some practices can be modified based on facility design or equipment, others form the bedrock of MTB research. An example of the former case would be the choice of PPE, which could be N95 or a PAPR. An example of the latter case would be the inactivation of pathogen-containing samples before removal from the BSL3 laboratory.

These practices should not be viewed as static but as requiring constant re-evaluation and updating. Research procedures are evolving continuously and it follows that we should likewise continually improve and adapt best practices for handling mycobacterium in the laboratory.

References

Association for the Advancement of Medical Instrumentation (2017) Good hospital practice: steam sterilization and sterility assurance. ANSI/AAMI ST79. AAMI, Arlington

Bloom BR, Atun R, Cohen T, et al (2017) Tuberculosis. In: Holmes KK, Bertozzi S, Bloom BR, et al (eds) Major infectious diseases, 3rd edn. The International Bank for Reconstruction and Development/The World Bank, Washington, DC; Chapter 11

Centers for Disease Control and Prevention (2009) Biosafety in microbiological and biomedical laboratories. Section IV—laboratory biosafety level criteria, 5th edn. U.S. Department of Health and Human Services, Washington, DC, pp 30–59. http://www.cdc.gov/biosafety/publications/bmbl5/

Cole ST et al (2005) Tuberculosis and the tubercle bacillus. ASM Press, Washington, DC

Doig C, Seagar AL, Watt B, Forbes KJ (2002) The efficacy of the heat killing of *Mycobacterium tuberculosis*. J Clin Pathol 55(10):778–779

Emmert E (2013) Biosafety guidelines for handling microorganisms in the teaching laboratory: development and rationale. J Microbiol Biol Educ 14:78–83

Fleming DO (2006) Risk assessment of biological hazards. In: Fleming DO, Hunt DL (eds) Biological safety: principles and practices, 4th edn. ASM Press, Washington, DC, pp 81–91

Hall L, Otter JA, Chewins J, Wengenack NL (2007) Use of hydrogen peroxide vapor for deactivation of Mycobacterium tuberculosis in a biological safety cabinet and a room. J Clin Microbiol 45(3):810–815

Hatfull GF, Jacob WR Jr (2014) Molecular genetics of mycobacteria. ASM Press, Washington, DC

Hirai Y (1991) Survival of bacteria under dry conditions; from a viewpoint of nosocomial infection. J Hosp Infect 19(3):191–200

Kao AS, Ashford DA, McNeil MM, Warren NG, Good RC (1997) Descriptive profile of tuberculin skin testing programs and laboratory-acquired tuberculosis infections in public health laboratories. J Clin Microbiol 35(7):1847–1851

Kimura M, Converse PJ, Astemborski J, Rothel JS, Vlahov D, Comstock GW, Graham NM, Chaisson RE, Bishai W (1999) Comparison between a whole blood interferon-gamma release assay and tuberculin skin testing for the detection of tuberculosis infection among patients at risk for tuberculosis exposure. J Infect Dis 179(5):1297–1300

Kramer A, Schwebke I, Kampf G (2006) How long do nosocomial pathogens persist on inanimate surfaces? A systematic review. BMC Infect Dis 6:130

Lowe JJ, Gibbs SG, Iwen PC, Smith PW (2012) A case study on decontamination of a biosafety level-3 laboratory and associated ductwork within an operational building using gaseous chlorine dioxide. J Occup Environ Hyg 9(12):D196–D205

Mazurek GH, Weis SE, Moonan PK, Daley CL, Bernardo J, Lardizabal AA, Reves RR, Toney SR, Daniels LJ, LoBue PA (2007) Prospective comparison of the tuberculin skin test and 2 whole-blood interferon-gamma release assays in persons with suspected tuberculosis. Clin Infect Dis 45(7):837–845

Memarzadeh F, Wheeland DG (2016) Biocontainment facility design. In: NIH design requirements: manual for biomedical laboratories and animal research facilities. Revised 2018, NIH, pp155–170

Morbidity and Mortality Weekly Report (MMWR) (2005) Guidelines for preventing the transmission of Mycobacterium tuberculosis in health-care settings. Vol. 54/No. RR-17

NIOSH (2013) Respirator awareness: your health may sepend on it. NIOSH Publication, Pittsburgh No. 2013-138

Pfyffer GE (2007) *Mycobacterium*: general characteristics, laboratory detection, and staining procedures. In: Murray PR (ed) Manual of clinical microbiology, 9th edn. ASM Press, Washington, DC, pp 543–572

Public Health Agency of Canada (2010) Mycobacterium tuberculosis complex pathogen safety data sheet, http://www.phac-aspc.gc.ca/lab-bio/res/psds-ftss/tuber-eng.php

Riley RL, Nardell EA (1989) Clearing the air. The theory and application of ultraviolet air disinfection. Am Rev Respir Dis 139(5):1286–1294

Roberts V (2014) To PAPR or not to PAPR? Can J Respir Ther 50(3):87–90

Schwebach JR, Jacobs WR Jr, Casadevall A (2001) Sterilization of Mycobacterium tuberculosis Erdman samples by antimicrobial fixation in a biosafety level 3 laboratory. J Clin Microbiol 39 (2):769–771

Wooley D, Byers K (eds) (2017) Biological safety: principles and practices, 5th edn. ASM press, Washington, DC

World Health Organization (2004) Laboratory biosafety manual, 3rd ed. Geneva

World Health Organization (2012) Tuberculosis laboratory biosafety manual. Geneva. Available from: https://www.ncbi.nlm.nih.gov/books/NBK179135/

Clinical Aspects and Principles of Management of Tuberculosis

20

Ashfaq Hasan, Sai Haranath Praveen, Chandrakant Tarke, and Fahad Abdullah

Abstract

Tuberculosis over the ages, has killed more people than any other infection has. Notwithstanding the advances in modern science, clinical diagnosis sometimes remains elusive, owing principally to the frequent paucibacillary occurrence of the disease and the slow doubling time of the organism; empiric treatment is often fraught with risks in the era of increasing drug resistance.

This chapter attempts to provide an overview of the disease, beginning with the pathogenesis and its protean clinical presentations. It also discusses the recent evolution of molecular methods that have lately provided an impetus to early diagnosis with a clear opportunity to unmask drug resistance before initiating "blind", potentially ineffective, and sometimes harmful treatment with standard therapy.

The chapter also provides insight into tuberculosis in special situations, and discusses briefly the treatments in uncomplicated cases as well as in special situations, and in instances of drug resistance. Preventive methods including current and upcoming vaccines are mentioned.

Finally, a short discussion of the sequelae of tuberculosis—which have the potential to be confused with active disease—is presented.

Keywords

Pulmonary · Infection · Tuberculosis

A. Hasan (✉) · F. Abdullah
Department of Respiratory Medicine, Deccan College of Medical Sciences, Hyderabad, India
e-mail: principal@deccancollegeofmedicalsciences.com

S. H. Praveen
Department of Pulmonary and Critical Care Medicine, Apollo Hospitals, Hyderabad, India

C. Tarke
Department of Pulmonary and Sleep Medicine, Apollo Hospitals, Hyderabad, India

S. E. Hasnain et al. (eds.), Mycobacterium tuberculosis: *Molecular Infection Biology, Pathogenesis, Diagnostics and New Interventions*,
https://doi.org/10.1007/978-981-32-9413-4_20

Abbreviations

ADA	Adenosine deaminase
ADRs	Adverse drug reactions (ADRs)
AFB	Acid-fast bacilli
AIDS	Acquired immunodeficiency syndrome
ARDS	Acute respiratory distress syndrome
ART	Antiretroviral therapy
ATT	Anti-tubercular therapy
BAL	Broncho-alveolar lavage
BCG	Bacillus Calmette-Guerin
DOTS	Directly observed treatment
EMB	Ethambutol
GABA	Gamma-aminobutyric acid
HIV	Human immunodeficiency virus
INH	Isonicotinylhydrazide
IRIS	Immune reconstitution syndrome
LAM	Lipoarabinomannan
LPA	Line probe assays
LTBI	Latent TB infection
MDR-TB	Multiple drug-resistant TB
MGIT	Mycobacteria growth indicator tube
MICs	Minimum inhibitory concentrations
NAAT	Nucleic acid amplification technology
PAS	Para-amino salicylic acid
PPD	Purified protein derivative
SAT	Self-administered therapy modified
TB	Tuberculosis
WHO	World Health Organization
XDR	Extensive drug resistant

20.1 Introduction

TB can involve all organs except the nails, hair and teeth [Ekaterina Kulchavenya. Ther Adv Infect Dis. 2014 Apr;2(2): 61–70]. Pulmonary TB is the most common form of TB. The symptomatology of TB has to do with its pathogenesis, the specific organ system involved and the duration of the disease.

Primary TB, which occurs in the *Mycobacterium*-naive individual, usually presents with a frequently occult primary focus in the peripheral lung parenchyma (Stead's focus) accompanied by a pleural effusion; or by a primary focus—frequently inapparent—in the lower part of the upper lobe or the upper part of the lower lobe (Ghon's focus) accompanied by lymphadenopathy. Other manifestations of primary TB such as erythema nodosum or phlyctenular conjunctivitis may be present. In young children with immature cellular immunity, or in immunocompromised individuals, the process may evolve to progressive primary disease.

Symptoms and signs relate to the compressive effect of enlarged lymph nodes on adjacent structures (collapse or hyperinflation of a lobe or segment), or rupture of caseating lymph nodes into a bronchus, with bronchiectasis later supervening in such a lobe.

Post-primary disease, the typical adult form of TB, usually occurs by reactivation of a dormant focus–or less frequently, by reinfection; the better aerated lung apices are preferentially involved. The classical triad of the constitutional symptoms of TB (fever, night sweats, weight loss)—most usually of several weeks' duration—is present in a high proportion of patients (Heemskerk et al. 2015).

The extent of disease at the time of first diagnosis may be variable, from patchy consolidation to a large cavity. This not only reflects the protean pulmonary forms of the process but also has to do with host immunity, the virulence of the strain and, most of all, with the time elapsed between disease onset and diagnosis.

Cavitation is an important milestone in the timeline of the post-primary pulmonary tuberculosis. With ingress of O2-rich air into a fresh cavity (caused by the coughing out of a caseous collection), the aerobic *Mycobacterium* is presented with an opportunity to multiply manifold in an aerobic environment: the infectiousness of the subject (and the possibility of a positive diagnosis on sputum testing) commensurately increases (Golub et al. 2006). Cavitation also increases the propensity to haemoptysis (which may sometimes be massive) and that of cavitary colonization with a fungal ball (mycetoma) later on in healed disease. One third of all individuals affected with post-primary TB die within a few months ("galloping consumption") (Reported tuberculosis in the United States, 2013 2014). The remainder either go on to develop a chronic pattern with a more gradual progressive decline or undergo spontaneous remission of the process and restoration to health. With chronicity, fibrosis is common.

Disseminated TB, which often casts "miliary" shadows in the lungs, occurs due to haematogenous dissemination following either primary or post-primary TB. The name derives from the resemblance of the widespread granulomatous lesions to millet seeds (seen in pathological specimens as 1–2 mm yellow granules). Symptoms may be nonspecific and depend on the site and degree of organ involvement. Hepatosplenomegaly, lymphadenopathy and pleural effusions may be present. Fundus examination, often neglected, will frequently show choroid tubercles, which are virtually diagnostic of the condition in a likely setting.

Diagnosis often hinges on microbiology from broncho-alveolar lavage (BAL) or on biopsy of involved organs (Lewinsohn et al. 2017) though it may occasionally be made on sputum (sponaneously expectcrated or induced).

Disseminated TB Disseminated TB should always raise suspicion for an accompanying immune deficiency such as HIV disease. Sometimes an occult disseminated form of TB ("cryptic miliary TB") presents as a "fever of unknown origin", usually in the elderly or immunocompromised patients. In difficult cases (as with the classic miliary form), bone marrow or liver biopsy with Mycobacterial cultures may be diagnostic.

Unlike the post-primary form of disease, untreated miliary TB is predictably fatal.

Extra-Pulmonary Tuberculosis

Extra-pulmonary tuberculosis has been increasingly seen since the advent of the HIV epidemic. TB can involve any organ of the body (other than nails, hair and teeth). Organ systems typically involved include pleura (pleural effusions), lymph nodes (generally cervical or mediastinal), airway (endobronchial tuberculosis), skeletal (joint or vertebral involvement, sometimes with a psoas abscess), gastrointestinal, genitourinary, meningeal and pericardial involvement. Symptoms pertain to the system involved, but diagnosis may be difficult unless a high degree of suspicion is maintained. Tissue or fluid samples from the involved organs with staining, Mycobacterial cultures or molecular probe assays are often diagnostic although radiological patterns in specific cases may be very suggestive of the process (e.g. vertebral involvement with a paraspinal abscess) (Lewinsohn et al. 2017).

Diagnosis

The diagnosis of tuberculosis, especially in the developed nations, rests on the keystone of suspicion. Physical examination is generally unrewarding. Auscultation of the lung typically underestimates the lung problem relative to that seen on imaging. With a pleural effusion, breath entry is typically decreased, with a flat note over the effusion. Cavernous bronchial breathing may be detected over a cavity, and tubular bronchial breathing over an area of consolidation or dense fibrosis.

The radiologic features in the various forms of pulmonary TB have already been discussed above under the relevant heads.

Microbiological isolation of the *Mycobacterium tuberculosis* on culture has been the only definitive test for TB thus far, though it could be argued that molecular probes for specific mycobacterial genes and other technologies (discussed below and also elsewhere in this book) offer viable and more attractive options for diagnosis. Presumptive diagnosis is typically achieved by microscopy (usually of sputum, but also of fluid obtained by BAL and other means) (Pai et al. 2016).

Mycobacteria, not being stainable by "conventional" stains, require a modified acid stain (the Ziehl-Neelsen method) in order to be visualized; LED microscopy though relatively expensive, by making the bacilli easy to see under low power, enables more fields to be scanned in a shorter period of time, and thus increases the sensitivity of microscopy several-fold (Bhalla et al. 2013).

Conventional culture methods are slow (4–6 weeks, less by radiometric assays), and frequently there is a compelling need to start appropriate anti-tubercular therapy as early as possible. Rapid culture techniques are now in universal use. They use liquid rather than solid media. The Mycobacteria growth indicator tube (MGIT) method takes days rather than weeks and facilitates drug sensitivity testing.

Since relapses (see below) occur in approximately 5% of patients after 6 months of standard therapy, it is a moot point whether establishment of minimum inhibitory concentrations (MICs) in the presence of a "critical concentration' of drug is relevant in selected cases (Colangeli et al. 2018).

Automated nucleic acid amplification technology (NAAT) has the potential of amplifying even a fragment of mycobacterial DNA rapidly (in 2 h) and has revolutionized TB diagnosis. In its currently available form, the GeneXpert test is intended specifically to diagnose *Mycobacterium tuberculosis* (and thereby differentiate it from other Mycobacteria such as environmental Mycobacteria), and also presumptively diagnose rifampicin resistance.

Though not invariable, rifampicin resistance also implies INH resistance. Together, rifampicin and INH resistance fulfil the diagnosis of multiple drug-resistant TB (MDR-TB). Thus a positive GeneXpert test makes a presumptive diagnosis of MDR-TB (Steingart et al. 2009). The test is more sensitive than direct microscopy and extremely specific (Steingart et al. 2014). The test is not infallible, however, and any result that is discordant with the clinical picture should either be repeated or further information sought using another modality. False-positive NAAT results can occur due to laboratory contamination. Importantly, NAAT can detect nucleic acid from dead non-viable bacilli, and so a positive test after completion of the course of treatment has no relevance in that situation (Boyles et al. 2014) nor is it of relevance in monitoring response to treatment (Friedrich et al. 2013). It is important to realize that NAAT is not a substitute for AFB smear and culture testing; culture is essential for species identification and for drug susceptibility testing (Cheng et al. 2005). A more sensitive version of Xpert has been developed.

The line probe assays (LPA) can permit rapid identification of specific gene markers associated with rifampicin resistance alone or in combination with isoniazid, and provide clinically relevant information about the level of INH resistance (low level associated with the INH-A gene; versus high level associated with the kat-G gene) (WHO treatment guidance for drug resistant tuberculosis 2016).

Definitive diagnosis has traditionally hinged on the culture characteristics of MTB: species discrimination can be achieved on the basis of colony morphology as well as on the basis of biochemical tests performed on the culture. Conventional culture (Lowenstein-Jensen medium and its variants) is capable of detecting as few as ten bacteria per millilitre with the sensitivity and specificity of sputum culture approximating 80 and 98%, respectively (Morgan et al. 1983). Following initial culture, species identification is done by any of the following techniques: biochemical methods (testing, nucleic acid hybridization probes, high-pressure liquid chromatography or mass spectrophotometry). Drug sensitivity testing is subsequently carried out, usually at a reference laboratory (Shinnick and Good 1995) to first-line anti-tubercular agents, but sometimes to second-line agents as well when resistance to components of first-line anti-tubercular agents has been documented. In individuals living with HIV who have CD4 counts <100 cells/mm^3, or those who are immunocompromised, Mycobacterial cultures of blood and urine should complement conventional testing. Point-of-care urinary detection of lipoarabinomannan, a component of the Mycobacterial cell wall (urine LAM test) may be useful for HIV-infected individuals, and its use may possibly confer a mortality benefit (Reported tuberculosis in the United States, 2013 2014). The future of point-of-care devices is next-generation sequencing, and, perhaps, innovations like detection of volatile organic compounds in exhaled breath (Garcia-Basteiro et al. 2018)

Serological tests (tests for antibodies against TB) have poor sensitivity and specificity; false-positive tests are ubiquitous, and serologic testing for the diagnosis of TB is discouraged by the WHO.

A pleural fluid adenosine deaminase (ADA) >35 U/L has been reported to be 93% sensitive and 90% specific for the diagnosis of pleural TB in lymphocytic exudates (Garcia-Zamalloa and Taboada-Gomez 2012).

20.2 TB in Special Circumstances

20.2.1 TB in Childhood

TB infection can rapidly progress to disease in infants and young children (progressive primary TB). Mortality is high in early childhood because of the severe forms of TB (such as miliary TB and meningeal TB) that occur more frequently in young children. Progressive primary TB is a serious consequence of primary tuberculosis. Consolidation or bronchopneumonia is more common than cavitary disease in children. Haematogenous dissemination leading to miliary TB is also more common in infants and young children. Tubercular pleural effusion, however, is less common in children below 5 years of age.

Central nervous TB is the most serious form of childhood TB (Prevention 2014). The presentation is usually with nonspecific symptoms such as fever, headache, drowsiness and irritability, but in advanced cases, vomiting, neck rigidity, seizures, focal neurologic deficit and coma may supervene.

Since sputum samples for diagnosis are difficult to obtain in young children (children tend to swallow sputum rather than expectorate), gastric washings can rather be sent for microbiological examination. A PPD test that is greater than 10 mm in BCG-unvaccinated children and > 15 mm in BCG-vaccinated children is considered positive. (Targeted tuberculin testing and treatment of latent tuberculosis infection. This official statement of the American Thoracic Society was adopted by the ATS Board of Directors, July 1999. This is a Joint Statement of the American Thoracic Society (ATS) and the Centers for Disease Control and Prevention (CDC). This statement was endorsed by the Council of the Infectious Diseases Society of America. (IDSA), September 1999, and the sections of this statement 2000.)

The principles of treatment of tuberculosis in children are similar to that of adults. Paediatric TB patients should be monitored carefully for ethambutol-induced ocular toxicity, owing to the fact that children are less likely to report visual impairment than adults.

In a prospective study of children on multidrug-resistant tuberculosis, children had lower serum concentrations in spite of higher dosing of moxifloxacin, a fact that was ascribed to increased drug elimination in children. Quinolones have also been known to rarely cause arthropathy and tendon rupture. More recent studies have shown that most quinolones are quite well tolerated but very specific and narrow indications are present for quinolone use in children (Patel and Goldman 2016).

20.2.2 TB in Pregnancy

Tuberculosis in pregnancy has important implications for the mother and child (Loto and Awowole 2012).

The diagnosis of tuberculosis in pregnancy can be challenging, since many of the symptoms of early TB may be confounded by pregnancy; for instance, the normal weight gain in pregnancy may temporarily mask the associated weight loss. Chest radiograph is not an absolute contraindication and when necessary may be carried out with "abdominal shielding" of the fetus.

The complications of TB in pregnancy include early spontaneous abortion, intrauterine growth retardation, preterm delivery, low birth weight and increased neonatal mortality. Congenital TB is rare, but can be associated with high perinatal mortality.

First-line oral drugs, i.e. isoniazid, rifampicin, ethambutol and pyrazinamide, can all be used during pregnancy safely. Streptomycin and other aminoglycosides have been proven to be potentially teratogenic in pregnancy. They are also capable of causing eighth nerve paralysis, with deficits ranging from mild hearing loss to bilateral deafness (Loto and Awowole 2012). Concerns have been raised about the use of fluoroquinolones in pregnancy even though the weight of the evidence seems to point toward safety (Acar et al. 2019)

20.2.3 TB in HIV

TB is considered an AIDS-defining disease [Centers for Disease Control and Prevention. Appendix A: AIDS-Defining Conditions. MMWR. 2008;57(RR10):9]. The risk of developing TB is estimated to be between 16 and 27 times greater in people living with HIV than among those without HIV infection. HIV and TB have a salutary impact upon each other and expedite each other's progression. The risk of progressing from latent to active TB is estimated to be between 12 and 20 times greater in people living with HIV than among those without HIV infection (Global TB report 2017).

Against the backdrop of HIV infection, the rate of progression of TB from the acquisition of infection to full-blown disease may take weeks, rather than years—as is the case in HIV-negative individuals (Mayer and Dukes Hamilton 2010). It is important to realize that TB can occur at all "stages" of HIV disease. The clinical presentation of HIV patients who have relatively high CD4 counts is no different to that which occurs in individuals without HIV. Such individuals manifest with the classical manifestations of post-primary tuberculosis such as upper lobe infiltrates. On the other hand, lower lobe involvement, pleural effusions and intrathoracic lymphadenopathy are common in individuals with low CD4 counts (typically <200/μL). Indeed the chest X-ray may sometimes only show vague interstitial infiltrates or even be normal (Mayer and Dukes Hamilton 2010). Taken together, the fact that in advanced HIV disease (a) the sputum is less likely to be positive for AFB (b) the PPD is usually non-reactive (c) the X-ray abnormalities are likely to represent other opportunistic infections or even non-infective conditions associated with HIV disease and (d) histopathology of involved tissues is characterized by lack of typical granulomata; these considerations make the diagnosis extremely challenging in this setting. In addition, there is increased frequency of extra-pulmonary TB among HIV-positive individuals.

Frequently, a positive NAAT (GeneXpert) test is the justification for starting treatment. However it must be cautioned that even a negative NAAT test cannot convincingly rule out disease, and, given that delays in treatment may be fatal, the decision to initiate treatment is sometimes taken on clinical grounds.

20.3 Treatment of TB

The treatment of tuberculosis differs from the treatment of most other infections in certain respects. Multiple drugs are required in a regimen since spontaneous mutations can lead to resistance to the action of one or more drugs. Also, because the bacterium is relatively slow growing, it takes longer to achieve bacteriologic sterility. Multidrug therapy, the norm for TB, has been very effective, with a modest rate of relapse (Colangeli et al. 2018).

The standard four-drug regime ("short course") comprises a 2-month intensive phase with the four front-line drugs (rifampicin, isoniazid, ethambutol and pyrazinamide), followed by a continuation phase with the drugs (minus pyrazinamide and possibly ethambutol) for a further 4 months.

The four front-line drugs have been chosen based on their efficacy in rapidly reducing the number of organisms initially (during the initial phase), thus reducing infectivity; and on their sterilizing potential (the ability to eradicate viable bacteria completely and so prevent relapses); and their relative lack of side effects (see Table 20.1). These drugs are combined into regimens.

The six classes of second-line agents (fluoroquinolones, the aminoglycosides, capreomycin, ethionamide and prothionamide, para-amino salicylic acid (PAS), cycloserine and terizidone) have a rather lower efficacy and a higher propensity for side effect and drug intolerance. Several other antibiotics of uncertain efficacy have also been categorized as anti-tubercular agents and have a role in treating multidrug-resistant TB (MDR-TB) (see Table 20.2).

Two new agents, bedaquiline (a diaryl-quinolone) and delamanid (a nitroimidazole), are now approved for the treatment of MDR-TB.

The WHO recommends using a daily regimen and fixed-dose combinations (Guidelines for treatment of drug-susceptible tuberculosis and patient care 2017).

Table 20.1 First-line agents for standard first-line anti-tubercular therapy

Drug	Adult daily dose	Comment
Isoniazid	5 mg/kg	Give pyridoxine 10 mg daily to prevent peripheral neuropathy; if established peripheral neuropathy, give pyridoxine 50–75 mg daily
		Protect tablets from light
Rifampicin	10 mg/kg	Preferably given 30 min before food
Pyrazinamide	25 mg/kg	Avoid with active hepatic disease (see further discussion on drug-induced hepatitis, below); adjust dose for renal function
Ethambutol	15 mg/kg	Adjust dose for renal function
		Ocular examination to be done before and periodically during treatment

Administration of the drugs on a daily basis both during the intensive and during the continuation phase is preferable although daily administration during the initial phase and intermittent (thrice weekly) administration during the continuation phase is also acceptable (Guidelines for treatment of drug-susceptible tuberculosis and patient care 2017). Intermittent administration of drugs (especially in the initial phase) is not offered in the presence of HIV disease.

If at the end of the intensive phase the bacteria are sensitive to the three most important agents (isoniazid, rifampicin and pyrazinamide), ethambutol may be discontinued (Nahid et al. 2016). However, drug testing for three of the four front-line drugs for each patient at the end of the intensive phase is impractical, and limited testing (GeneXpert) is usually performed initially in those patients whose sputum remains positive. There is evidence to show that extending the continuation phase by 7 months (making up a total duration of treatment of 9 months) helps in preventing relapse in those patients who have cavitation on initial chest X-rays; and also in those who show positive cultures at the end of the initial phase of treatment (Benator et al. 2002); as well as in those whose medication given during the initial phase did not include pyrazinamide.

Anti-tubercular drugs require to be supplemented with carefully monitored steroid therapy in two circumstances: tubercular meningitis (a short course of dexamethasone or prednisolone is typically given, tapered over 6 to 8 weeks) and tuberculous pericarditis (Guidelines for treatment of drug-susceptible tuberculosis and patient care 2017).

Since many patients cannot be relied upon to regularly take their drugs, and irregular intake of drugs has the potential to result in drug resistance (see *Treatment Failure and Relapse;* and *Drug Resistant TB,* below), it is imperative to ensure adherence to the regimen. Directly observed therapy, short-course (DOTS) is the direct observation by another person, of the patient taking the medication (Chaudhuri 2017). The use of DOTS has improved adherence in subgroups of patients such as TB with HIV or if DOTS was at the site of DOTS centre or provider location.

Table 20.2 First and second-line anti-tubercular agents and their adverse drug reactions (ADRs)

Drugs	Common ADRs	Uncommon ADRs	Rare ADRs
Isoniazid*	Asymptomatic elevation of serum hepatic enzymes	Hepatitis, cutaneous hypersensitivity, peripheral neuropathy	Fever, giddiness, convulsions, optic neuritis, mental symptoms, haemolytic anaemia, aplastic anaemia, lupoid reactions, arthralgia, gynaecomastia
Rifampicin*	Pruritus	Hepatitis, cutaneous hypersensitivity, gastrointestinal reactions, thrombocytopenia, febrile reaction, "flu syndrome"	Shortness of breath, shock, haemolytic anaemia, acute renal failure, thrombotic thrombocytopenic purpura
Pyrazinamide*	Anorexia, nausea, flushing, photosensitization	Hepatitis, vomiting, arthralgia, cutaneous reactions	Sideroblastic anaemia, gout
Ethambutol*		Retrobulbar neuritis, cutaneous reactions, arthralgia	Peripheral neuropathy
Streptomycin	Cutaneous hypersensitivity, giddiness, numbness, tinnitus	Vertigo, ataxia, deafness	Clinical renal failure, aplastic anaemia
Thioacetazone	Gastrointestinal reactions, cutaneous hypersensitivity, vertigo, conjunctivitis	Hepatitis, erythema multiforme, exfoliative dermatitis, haemolytic anaemia	Agranulocytosis
Amikacin/ kanamycin/ capreomycin	Hearing damage, vestibular disturbance, deranged renal function tests	Clinical renal failure	
Ofloxacin/ ciprofloxacin	Gastrointestinal reactions, insomnia	Anxiety, dizziness, headache, paresthesia, tremor	Convulsion, cutaneous hypersensitivity
Ethionamide/ prothionamide	Gastrointestinal reactions	Hepatitis, peripheral neuropathy	Convulsion, depression, alopecia, hypothyroidism, impotence, gynaecomastia
Cycloserine	Dizziness, headache, depression, memory loss	Psychosis, convulsion	Sideroblastic anaemia
Para-aminosalicylic acid	Gastrointestinal reactions	Hepatitis, drug fever	Hypothyroidism, haematological reactions, generalized hypersensitivity, malabsorption

*First line anti-tubercular agents

Self-administered therapy (SAT) modified with some form of monitoring incorporated into the process shows some promise. Although DCTS could be administered by a family member, the preferred person is a trained lay provider or healthcare worker (Guidelines for treatment of drug-susceptible tuberculosis and patient care 2017).

The treatment of TB in HIV disease merits special consideration. Antiretroviral therapy (ART) should be started irrespective of the clinical stage of HIV disease and CD4 count. Anti-tubercular therapy (ATT) should precede the commencement of ART. ART should then be commenced as soon as it is evident that ATT is being tolerated (typically between 2 weeks and 2 months).

The *immune reconstitution inflammatory syndrome (IRIS)* is sometimes seen with the returning immune competence that follows ART; an increased immune response to tubercle bacilli or antigens occurs. IRIS usually occurs between 1 and 3 months of the commencement of ART and is associated with a decrease in viral load and an increase in CD4+ T cell count. There is a worsening in the clinical picture weeks or months into treatment, with an increase in the size of lymph nodes, development of pleural effusions and features of noncommunicating hydrocephalus in a patient with TB meningitis.

Although most patients with IRIS have mild to moderate symptoms, IRIS, especially when related to TB meningitis, can be life-threatening. An "unmasking IRIS" may occur in patients with unsuspected TB (Raviglione 2017).

Treatment of IRIS includes addition of steroids (which probably work by decreasing the levels of pro-inflammatory cytokines) and the continuation of ATT and ART.

With both ATT and ART on board, drug-drug interactions especially involving rifampicin (by induction of the cytochrome P450 system) may occur. Rifamycins induce the metabolism of non-nucleoside reverse transcriptase inhibitors and protease inhibitors. Rifampicin is a more potent inducer of the cytochrome P450 system than rifapentine, which in turn is more potent than rifabutin. Despite potential drug interactions, rifamycins should nevertheless be included in TB regimens, with dosage adjustment if necessary. Rifabutin is the preferred rifamycin in HIV-positive individuals.

20.4 Alternative Treatment Approaches

Host-directed therapy to modulate the immune response to TB is an active area of research (Kaufmann et al. 2014). The use of an anti-diabetic drug—metformin, for example—is being investigated. Also, studies of autologous mesenchymal stromal cell infusions is being investigated in MDR and XDR-TB (Skrahin et al. 2014). An approach using high-altitude sanatoria was also explored (Murray 2014) in a throwback to the pre-anti-tubercular chemotherapy era. Study using atypical Mycobacterial injection in tuberculous did not show any major outcome differences from placebo although steroids did decrease the subsequent development of constrictive pericarditis (Mayosi et al. 2014).

Given the role that vitamin D plays at the level of the antigen-presenting cell, the role of vitamin D deficiency as a possible player in TB predisposition has recently come into focus; indeed a recent meta-analysis has revealed that vitamin D supplementation in fact did not affect time to sputum culture conversion overall, though it did accelerate sputum culture conversion in patients with multidrug-resistant pulmonary tuberculosis (Jolliffe et al. 2019).

20.5 Critical Care of Patients with TB

Patients with TB may be admitted to an ICU for a variety of reasons, including respiratory failure, multiorgan failure, decreased consciousness associated with tubercular meningitis, pneumothorax, cardiogenic shock (due to massive pericardial effusion) and liver or kidney failure due to a drug reaction induced by ATT. Confluent tuberculous bronchopneumonia and TB ARDS are less common causes of respiratory failure (Hagan and Nathani 2013).

Tuberculosis treatment is especially challenging in critically ill patients due to potential for poor gastric absorption and the high rates of organ dysfunction and drug toxicity in this subset of patients. Deciding the optimal regimen and dosing becomes challenging in the presence of concomitant hepatic and renal failure. In a recent study done on TB patients requiring ICU admission, mortality was 72.5% (Tatar et al. 2018).

20.6 Treatment Failure and Relapse

Treatment failure is suspected when the sputum remains positive for AFB at the end of the first 3 months of treatment. It occurs due to factors that may involve incorrect regimens or dosing or indeed due to malabsorption or poor-quality drugs (Lambregts-van Weezenbeek and Veen 1995). Patients are regarded as treatment failures when a microbiologic indicator like sputum smear/culture for TB remains positive after 3–4 months of starting treatment or comes back positive after a period of remaining negative (fall and rise phenomenon) (Prasad et al. 2018). Patients may respond for a short time after starting treatment or do not respond to treatment from the inception. In any case the implication of treatment failure is drug-resistant TB (see below), and strenuous efforts must be made to look for microbiological resistance to first—and, in the proper context, to second-line agents—by the available tools. Relapse, on the other hand, is the recurrence of TB after completion of treatment that has been successful, with documented culture negativity. Relapse may be due either to a reinfection or reactivation of the dormant bacillus. Drug-resistant TB is to be considered in these settings. The implications of relapse are not as grave, the isolates generally being drug sensitive.

In either case, the exact regimen chosen should be based upon the drug sensitivity pattern. Specific recommendations upon the construction of such a regimen are available at WHO treatment guidelines for drug-resistant tuberculosis 2016 update (http://www.who.int).

20.7 Drug-Resistant TB

Drug-resistant TB is a major global problem and is a threat to the success of the End TB strategy. MDR-TB or multidrug-resistant TB (MDR-TB) is defined by resistance of *M. tuberculosis* against to at least rifampicin and isoniazid. Since these two drugs are the most effective drugs available against TB, resistance to both considerably shortens the odds of complete cure.

XDR-TB or extensively drug-resistant TB is defined as resistance to rifampicin and INH plus resistance to at least one fluoroquinolone and one second-line injectable agent (amikacin, kanamycin or capreomycin).

Some of the reasons that lead to drug resistance have been mentioned in the section above. However the most important of these are failure to comply with the treatment regimen on the part of the patient and the addition of a single drug ("monotherapy") to a failing regimen; or a situation of unrecognized primary (that which is present at the start of therapy) or secondary (that which develops during treatment) drug resistance.

INH mono-resistance of itself probably does not impact the outcome of therapy provided that ethambutol and possibly pyrazinamide are used for the entire duration of therapy. Consideration should be given in this case to extending the duration of treatment to 9 months from the standard 6 (Sadanand 2011). For reasons already discussed, rifampicin-resistant individuals should be managed as MDR-TB patients. In such patients it is not unusual to encounter resistance to additional drugs (e.g. ethambutol).

It has been shown that clinical cure is possible in patients with MDR and even XDR-TB (WHO drug resistant TB 2016). However, owing to a variety of reasons not the least of which is drug intolerance, only about half of MDR-TB and a third of XDR-TB patients ever complete therapy. The WHO estimates there are over 600,000 MDR-TB or rifampicin-resistant cases annually (Global TB report, WHO update 2017; wwww.who.int). The treatment of MDR-TB and XDR-TB is complex and requires specialized knowledge and close monitoring to ensure rational and safe therapy (Lange et al. 2018).

The choice of drugs for treatment of MDR-TB is frequently difficult. Formulation of drugs into regimens is facilitated by the organization of drugs into groups:

Group A = levofloxacin/moxifloxacin, bedaquiline, linezolid.
Group B = clofazimine, cycloserine/terizidone.
Group C = ethambutol, delamanid, pyrazinamide, imipenem-cilastatin, meropenem, amikacin (streptomycin), ethionamide/prothionamide, p-aminosalicylic acid.

The WHO makes specific recommendations for short and long MDR-TB regimens. In the WHO's 2018 update (WHO treatment guidelines for multidrug-

and rifampicin-resistant tuberculosis 2018), for those MDR-TB patients "who have not been previously treated for more than one month with second-line medicines used in the shorter MDR regimen or in whom resistance to fluoroquinolones and second-line injectable agents has been excluded," the WHO permits a shorter MDR-TB regimen of 9–12 months.

For MDR/RR (RR= rifampicin resistant) TB patients on longer regimens, the WHO recommends that "all three Group-A agents and at least one Group-B agent (should) be included to ensure that treatment starts with at least 4 TB agents likely to be effective, and that at least three agents (are) included for the rest of treatment after bedaquiline is stopped. If only one or two Group A agents are used, both Group B agents are to be included. If the regimen cannot be composed with agents from Groups A and B alone, Group C agents are added to complete it. Kanamycin and capreomycin are not to be included in the treatment of MDR/RR-TB patients on longer regimens" (WHO treatment guidelines for multidrug- and rifampicin-resistant tuberculosis 2018). A detailed discussion about the treatment of MDR-TB is available at the foregoing resource.

Novel anti-tubercular agents (e.g. bedaquiline and delamanid) have been discussed elsewhere in this text.

20.8 The Sequelae of Tuberculosis

The remodelling of the lung that follows TB can significantly contribute to morbidity and mortality. Haemoptysis is frequently one of the remote sequelae and can on occasion be life-threatening. An unresolved cavity is frequently the cause, and bleeding from an unsupported blood vessel in the cavity wall or in a post tubercular bronchiectatic segment can be a source of troublesome bleed. Cavities colonized by fungi (aspergillomas), broncholiths eroding into the bronchial wall and rarely expansion and subsequent rupture of an unsupported blood vessel in the cavity wall (Rasmussen's aneurysm) can cause massive haemoptysis (van den Heuvel and van Rensburg 2006). A scar carcinoma sometimes develops in an area of fibrosis and may sometimes be difficult to distinguish from a dense fibrotic infiltrate (Gao et al. 2015). Post-tubercular bronchiectasis is predisposed to when caseation necrosis and inflammation with retention of secretions leads to destruction of bronchial walls with ensuing permanent dilatation. Sometimes compression of bronchial lumen by enlarged lymph node can produce bronchiectasis by a different mechanism. Post-primary TB involves the upper lobes of the lung, and so the gravitationally advantaged upper lobes do not usually undergo a process of chronic and recurrent infection unlike the lower lobes that are involved by bronchiectasis of other aetiologies.

Mycetoma (fungal ball) is a mass of fungal hyphal material that grows within a cavity. *Aspergillus fumigatus* is the most common aetiological agent, but other fungi such as *Mucor* or *Fusarium* can also cause a fungal ball to develop (Rippon 1988). Such a mycetoma is usually asymptomatic but can sometimes lead to haemoptysis.

No treatment required in asymptomatic cases, but surgical treatment needs to be considered in patients with massive or recurrent haemoptysis.

20.9 Side Effects of Treatment

A list of adverse effects associated with anti-tubercular agents appears in Table 20.2. Side effects are possible with any of the anti-tubercular agents, and on rare occasions, they may be serious and even fatal. The risk of liver injury with anti-TB therapy is always a concern. Among the four first-line agents INH, rifampicin and pyrazinamide are all capable of causing hepatotoxicity. Isoniazid is capable of causing asymptomatic mild elevation in transaminases in 15–20% as well as fatal severe acute hepatitis in 0.05–1%. Jaundice with liver injury can occur in 0.5–1%. Peripheral neuropathy due to pyridoxine deficiency can occur with INH therapy in up to 6.5% of elderly individuals (Ruan et al. 2018); pyridoxine supplementation is required for all patients taking INH (John 2019). Factors increasing the chances of INH hepatotoxicity include age, female gender, alcohol use, prior hepatitis, concurrent use of cytochrome-inducing agents and slow acetylator status (Sharma et al. 2016). Acute INH toxicity can present as seizures due to decreased GABA levels. INH is metabolized to several metabolites including the hepatotoxic metabolite hydrazine, toxic-free radicals and mono-methylhydrazine. Degradation by acetylation occurs via N-acetyltransferase-2 gene (NAT2); a deficiency of the NAT enzyme can lead to accumulation of hepatotoxic metabolite. Ten percent of patients with anti-tubercular drug-induced hepatotoxicity show progression to acute liver failure; the rest are self-limited (Badrinath and John 2018).

If the serum bilirubin rises to ≥3 mg/dL or serum transaminases exceed more than five times the upper limit of normal (or exceed three times the upper limit of normal in a symptomatic patient), all drugs should be stopped (Nahid et al. 2016). The decision on what regimen to use on restarting should be individualized. The regimen, even if retailed, might then still include one or more potentially hepatotoxic drugs, and these should be restarted one at a time, only when liver function tests return to their baseline (e.g. the transaminases fall to less than twice normal). In general, a cholestatic type of derangement would prompt addition of INH first; with a non-cholestatic picture, rifampicin could first be added (Sterling). Needless to say, careful monitoring with frequent testing of serum liver function is required. Most physicians would omit pyrazinamide when restarting ATT except in the mildest of cases or with other compelling reasons (Sterling).

Patients receiving ethambutol (EMB) should be questioned regarding their vision at each monthly visit. Monthly visual acuity and colour vision discrimination testing is recommended in those patients in whom EMB doses of greater than 15–20 mg/kg are being used, those who have been taking ethambutol for longer than 2 months and those with renal insufficiency in whom the levels of drug may be expected to rise (Blumberg et al. 2003).

For patients who require a regimen with no hepatotoxic agents, potential agents include ethambutol, levofloxacin or moxifloxacin, an injectable agent and other second-line oral drugs. The optimal choice of agents and duration of treatment (at least 18–24 months) is uncertain. (See "Antituberculous drugs: An overview", section on "Second-line agents".)

20.10 The Prevention of TB

In the end, efforts to finally eradicate TB will need to hinge upon preventative aspects as much as upon its treatment. These strategies would include currently established practices, as well as novel strategies. The former include the following.

20.10.1 Vaccines

Tuberculosis is now the primary cause of death from infectious diseases exceeding HIV, AIDS and malaria. The WHO estimates 10.4 million new cases and 1.7 million deaths every year (WHO Global TB report 2017). Of these, 0.4 million are HIV patients who die mostly in LMIC (low- and middle-income countries) (WHO Global TB report 2017; www.who.int). Within the next 16 years, the goal is to reduce TB incidence by 90% but the progress has been painfully slow. The most effective way to reduce the burden of TB in the community, to date, has been to identify patients affected with TB as early as possible and to treat them effectively with anti-tubercular therapy.

The role of vaccines in this regard, though potentially vital, has been disappointing. Of all vaccine tested to date, the bacillus Calmette-Guerin (BCG)—a live attenuated vaccine that was made in 1919 after an incredible 11 years and 231 sub-cultures (Hasan 2014)—is still the most effective vaccine we have today. Yet its effect has been inconsistent, ranging from 80% to zero protection; several factors including geographical differences appear to play a role (British adolescents had better responses than South Indian participants perhaps due to background exposure to MTB or atypical mycobacteria). The efficacy of BCG has been most manifested in its protection of infants and young children from the dangerous forms of TB (such as disseminated "miliary" TB and TB meningitis). Also, its effects appear to wane with time (Hasan 2014).

The BCG is not a single vaccine. Several strains—all derived from the original 1921 BGC strain—are in use, which also partly explains its variable efficacy.

In 1998, the Sanger Centre in Cambridge, Britain and Paris's Pasteur Institute in France cracked the genetic code for the old H37Rv strain of tubercle bacillus, and then, the genome of highly virulent CDC1551 (the "Oshkosh" strain) was mapped at the TIGR (Fine 1989). These developments have led to the possibility of exciting new candidate vaccines, several of which are in fact in the pipeline.

20.10.2 Treatment of Latent TB Infection (LTBI)

It has been estimated that one-third of the world has been infected with TB bacteria. A relatively small number of immunocompetent individuals affected with LTBI will ever progress to active disease. Reactivation of TB occurs when the immune system is compromised by external factors such as steroids or other immunosuppressants or by immune deficiency syndromes, which enables the latent (dormant) bacteria to "awaken" causing active disease. There is unfortunately no test that can currently predict which of these individuals will progress to active disease immunocompromised individuals have a high possibility of doing so.

Given that perspective, treatment for latent TB infection is most relevant for children under 5 years, the elderly and individuals with HIV infection. To eliminate TB from the planet, a multipronged strategy incorporating early and effective treatment, vaccination and elimination of dormant reservoir of latent TB will be essential.

LTBI treatment with INH ("chemoprophylaxis") is still the mainstay for treating latent TB infection but given that treatment extends to 6 months, and that there is one death from hepatotoxicity for every 25,000–40,000 cases with such treatment, alternate strategies including rifamycin-based regimens are being explored (Rangaka et al. 2015).

References

Acar S, Keskin-Arslan E, Erol-Coskun H, Kaya-Temiz T, Kaplan YC (2019) Pregnancy outcomes following quinolone and fluoroquinolone exposure during pregnancy: a systematic review and meta-analysis. Reprod Toxicol 85:65–74. https://doi.org/10.1016/j.reprotox.2019.02.002

Badrinath B, John S. (2018) Isoniazid toxicity. StatPearls Publishing

Benator D, Bhattacharya M, Bozeman L, Burman W, Cantazaro A, Chaisson R, Gordin F, Horsburgh CR, Horton J, Khan A, Lahart C, Metchock B, Pachucki C, Stanton L, Vernon A, Villarino ME, Wang YC, Weiner M, Weis S (2002) Rifapentine and isoniazid once a week versus rifampicin and isoniazid twice a week for treatment of drug-susceptible pulmonary tuberculosis in HIV-negative patients: a randomised clinical trial. Lancet 360(9332):528–534

Bhalla M, Sidiq Z, Sharma P, Singhal R, Myneedu V, Sarin R (2013) Performance of light-emitting diode fluorescence microscope for diagnosis of tuberculosis. Int J Mycobacteriol 2(3):174–178. https://doi.org/10.1016/j.ijmyco.2013.05.001

Blumberg HM, Burman WJ, Chaisson RE, Daley CL, Etkind SC, Friedman LN, Fujiwara P, Grzemska M, Hopewell PC, Iseman MD, Jasmer RM, Koppaka V, Menzies RI, O'Brien RJ, Reves RR, Reichman LB, Simone PM, Starke JR, Vernon AA (2003) American Thoracic Society/Centers for Disease Control and Prevention/Infectious Diseases Society of America: treatment of tuberculosis. Am J Respir Crit Care Med 167(4):603–662. https://doi.org/10.1164/rccm.167.4.603

Boyles TH, Hughes J, Cox V, Burton R, Meintjes G, Mendelson M (2014) False-positive Xpert (R) MTB/RIF assays in previously treated patients: need for caution in interpreting results. Int J Tuberc Lung Dis 18(7):876–878. https://doi.org/10.5588/ijtld.13.0853

Chaudhuri A (2017) Recent changes in technical and operational guidelines for tuberculosis control programme in India – 2016: a paradigm shift in tuberculosis control. J Assoc Chest Phys 5 (1):1–9. https://doi.org/10.4103/2320-8775.196644

Cheng VC, Yew WW, Yuen KY (2005) Molecular diagnostics in tuberculosis. Eur J Clin Microbiol Infect Dis 24(11):711–720. https://doi.org/10.1007/s10096-005-0039-1

Colangeli R, Jedrey H, Kim S, Connell R, Ma S, Chippada Venkata UD, Chakravorty S, Gupta A, Sizemore EE, Diem L, Sherman DR, Okwera A, Dietze R, Boom WH, Johnson JL, Mac Kenzie WR, Alland D (2018) Bacterial factors that predict relapse after tuberculosis therapy. N Engl J Med 379(9):823–833. https://doi.org/10.1056/NEJMoa1715849

Fine PE (1989) The BCG story: lessons from the past and implications for the future. Rev Infect Dis 11(Suppl 2):S353–S359

Five-year follow-up of a controlled trial of five 6-month regimens of chemotherapy for pulmonary tuberculosis. Hong Kong Chest Service/British Medical Research Council (1987) Am Rev Respir Dis 136(6):1339–1342. https://doi.org/10.1164/ajrccm/136.6.1339

Friedrich SO, Rachow A, Saathoff E, Singh K, Mangu CD, Dawson R, Phillips PP, Venter A, Bateson A, Boehme CC, Heinrich N, Hunt RD, Boeree MJ, Zumla A, McHugh TD, Gillespie SH, Diacon AH, Hoelscher M (2013) Assessment of the sensitivity and specificity of Xpert MTB/RIF assay as an early sputum biomarker of response to tuberculosis treatment. Lancet Respir Med 1(6):462–470. https://doi.org/10.1016/s2213-2600(13)70119-x

Gao F, Ge X, Li M, Zheng X, Xiao L, Zhang G, Hua Y (2015) CT features of lung scar cancer. J Thorac Dis 7(3):273–280. https://doi.org/10.3978/j.issn.2072-1439.2015.02.07

Garcia-Basteiro AL, DiNardo A, Saavedra B, Silva DR, Palmero D, Gegia M, Migliori GB, Duarte R, Mambuque E, Centis R, Cuevas LE, Izco S, Theron G (2018) Point of care diagnostics for tuberculosis. Pulmonology 24(2):73–85. https://doi.org/10.1016/j.rppnen.2017.12.002

Garcia-Zamalloa A, Taboada-Gomez J (2012) Diagnostic accuracy of adenosine deaminase and lymphocyte proportion in pleural fluid for tuberculous pleurisy in different prevalence scenarios. PLoS One 7(6):e38729. https://doi.org/10.1371/journal.pone.0038729

Global TB report 2017. vol 15/01/2019

Golub JE, Bur S, Cronin WA, Gange S, Baruch N, Comstock GW, Chaisson RE (2006) Delayed tuberculosis diagnosis and tuberculosis transmission. Int J Tuberc Lung Dis 10(1):24–30

Guidelines for treatment of drug-susceptible tuberculosis and patient care (2017) World Health Organization, Geneva

Hagan G, Nathani N (2013) Clinical review: tuberculosis on the intensive care unit. Crit Care 17 (5):240. https://doi.org/10.1186/cc12760

Hasan A. (2014) BCG—The waning of a vaccine. J Med Allied Sci

Heemskerk D, Caws M, Marais B, Farrar J (2015) Tuberculosis in adults and children. Springer, London

John MBS (2019) Isoniazid toxicity. StatPearls Publishing

Jolliffe DA, Ganmaa D, Wejse C, Raqib R, Haq MA, Salahuddin N, Daley PK, Ralph AP, Ziegler TR, Martineau AR (2019) Adjunctive vitamin D in tuberculosis treatment: meta-analysis of individual participant data. Eur Respir J:1802003. https://doi.org/10.1183/13993003.02003-2018

Kaufmann SH, Lange C, Rao M, Balaji KN, Lotze M, Schito M, Zumla AI, Maeurer M (2014) Progress in tuberculosis vaccine development and host-directed therapies – a state of the art review. Lancet Respir Med 2(4):301–320. https://doi.org/10.1016/s2213-2600(14)70033-5

Lambregts-van Weezenbeek CS, Veen J (1995) Control of drug-resistant tuberculosis. Tuber Lung Dis 76(5):455–459

Lange C, Chesov D, Heyckendorf J, Leung CC, Udwadia Z, Dheda K (2018) Drug-resistant tuberculosis: an update on disease burden, diagnosis and treatment. Respirology 23 (7):656–673. https://doi.org/10.1111/resp.13304

Lewinsohn DM, Leonard MK, LoBue PA, Cohn DL, Daley CL, Desmond E, Keane J, Lewinsohn DA, Loeffler AM, Mazurek GH, O'Brien RJ, Pai M, Richeldi L, Salfinger M, Shinnick TM, Sterling TR, Warshauer DM, Woods GL (2017) Official American Thoracic Society/Infectious Diseases Society of America/Centers for Disease Control and Prevention clinical practice guidelines: diagnosis of tuberculosis in adults and children. Clin Infect Dis 64(2):111–115. https://doi.org/10.1093/cid/ciw778

Loto OM, Awowole I (2012) Tuberculosis in pregnancy: a review. J Pregnancy 2012:379271. https://doi.org/10.1155/2012/379271
Mayer KH, Dukes Hamilton C (2010) Synergistic pandemics: confronting the global HIV and tuberculosis epidemics. Clin Infect Dis 50(Suppl 3):S67–S70. https://doi.org/10.1086/651475
Mayosi BM, Ntsekhe M, Bosch J, Pandie S, Jung H, Gumedze F, Pogue J, Thabane L, Smieja M, Francis V, Joldersma L, Thomas KM, Thomas B, Awotedu AA, Magula NP, Naidoo DP, Damasceno A, Chitsa Banda A, Brown B, Manga P, Kirenga B, Mondo C, Mntla P, Tsitsi JM, Peters F, Essop MR, Russell JB, Hakim J, Matenga J, Barasa AF, Sani MU, Olunuga T, Ogah O, Ansa V, Aje A, Danbauchi S, Ojji D, Yusuf S (2014) Prednisolone and mycobacterium indicus pranii in tuberculous pericarditis. N Engl J Med 371(12):1121–1130. https://doi.org/10.1056/NEJMoa1407380
Morgan MA, Horstmeier CD, DeYoung DR, Roberts GD (1983) Comparison of a radiometric method (BACTEC) and conventional culture media for recovery of mycobacteria from smear-negative specimens. J Clin Microbiol 18(2):384–388
Murray JF (2014) Tuberculosis and high altitude. Worth a try in extensively drug-resistant tuberculosis? Am J Respir Crit Care Med 189(4):390–393. https://doi.org/10.1164/rccm.201311-2043OE
Nahid P, Dorman SE, Alipanah N, Barry PM, Brozek JL, Cattamanchi A, Chaisson LH, Chaisson RE, Daley CL, Grzemska M, Higashi JM, Ho CS, Hopewell PC, Keshavjee SA, Lienhardt C, Menzies R, Merrifield C, Narita M, O'Brien R, Peloquin CA, Raftery A, Saukkonen J, Schaaf HS, Sotgiu G, Starke JR, Migliori GB, Vernon A (2016) Official American Thoracic Society/Centers for Disease Control and Prevention/Infectious Diseases Society of America clinical practice guidelines: treatment of drug-susceptible tuberculosis. Clin Infect Dis 63(7): e147–e195. https://doi.org/10.1093/cid/ciw376
Pai M, Nicol MP, Boehme CC (2016) Tuberculosis diagnostics: state of the art and future directions. Microbiol Spectr 4(5). https://doi.org/10.1128/microbiolspec.TBTB2-0019-2016
Patel K, Goldman JL (2016) Safety concerns surrounding quinolone use in children. J Clin Pharmacol 56(9):1060–1075. https://doi.org/10.1002/jcph.715
Prasad R, Gupta N, Banka A (2018) Multidrug-resistant tuberculosis/rifampicin-resistant tuberculosis: principles of management. Lung India 35(1):78–81. https://doi.org/10.4103/lungindia.lungindia_98_17
Prevention CfDCa (2014) Reported tuberculosis in the United States, 2013 (trans: Services UDoHaH). Atlanta, GA
Rangaka MX, Cavalcante SC, Marais BJ, Thim S, Martinson NA, Swaminathan S, Chaisson RE (2015) Controlling the seedbeds of tuberculosis: diagnosis and treatment of tuberculosis infection. Lancet 386(10010):2344–2353. https://doi.org/10.1016/S0140-6736(15)00323-2
Raviglione MC (2017) Tuberculosis. In: Kasper DLaF, A.S (ed) Harrison's infectious diseases, 9th edn. McGraw-Hill Education, New York
Reported tuberculosis in the United States, 2013. (2014)
Rippon JW (1988) Medical mycology: the pathogenic fungi and the pathogenic actinomycetes, 3rd edn. Saunders, Philadelphia
Ruan LY, Fan JT, Hong W, Zhao H, Li MH, Jiang L, Fu YH, Xing YX, Chen C, Wang JS (2018) Isoniazid-induced hepatotoxicity and neurotoxicity in rats investigated by (1)H NMR based metabolomics approach. Toxicol Lett 295:256–269. https://doi.org/10.1016/j.toxlet.2018.05.032
Sadanand S (2011) Harrison's infectious diseases. Yale J Biol Med 84(3):327–323
Sharma S, Mohan A, Sharma A (2016) Miliary tuberculosis: a new look at an old foe. J Clin Tuberc Other Mycobact Dis 3. https://doi.org/10.1016/j.jctube.2016.03.003
Shinnick TM, Good RC (1995) Diagnostic mycobacteriology laboratory practices. Clin Infect Dis 21(2):291–299
Skrahin A, Ahmed RK, Ferrara G, Rane L, Poiret T, Isaikina Y, Skrahina A, Zumla A, Maeurer MJ (2014) Autologous mesenchymal stromal cell infusion as adjunct treatment in patients with multidrug and extensively drug-resistant tuberculosis: an open-label phase 1 safety trial. Lancet Respir Med 2(2):108–122. https://doi.org/10.1016/s2213-2600(13)70234-0

Steingart KR, Schiller I, Horne DJ, Pai M, Boehme CC, Dendukuri N (2014) Xpert(R) MTB/RIF assay for pulmonary tuberculosis and rifampicin resistance in adults. Cochrane Database Syst Rev (1):Cd009593. https://doi.org/10.1002/14651858.CD009593.pub3

Sterling TR. Treatment of drug-susceptible pulmonary tuberculosis in HIV-uninfected adults. https://www.uptodate.com/contents/treatment-of-drug-susceptible-pulmonary-tuberculosis-in-hiv-uninfected-adults. Accessed 21 Feb 2019

Targeted tuberculin testing and treatment of latent tuberculosis infection. This official statement of the American Thoracic Society was adopted by the ATS Board of Directors, July 1999. This is a Joint Statement of the American Thoracic Society (ATS) and the Centers for Disease Control and Prevention (CDC). This statement was endorsed by the Council of the Infectious Diseases Society of America. (IDSA), September 1999, and the sections of this statement (2000) Am J Respir Crit Care Med 161(4 Pt 2):S221–S247. https://doi.org/10.1164/ajrccm.161.supplement_3.ats600

Tatar D, Senol G, Kirakli C, Edipoglu O, Cimen P (2018) Contributing factors to mortality rates of pulmonary tuberculosis in intensive care units. J Chin Med Assoc 81(7):605–610. https://doi.org/10.1016/j.jcma.2018.02.003

van den Heuvel MM, van Rensburg JJ (2006) Images in clinical medicine. Rasmussen's aneurysm. N Engl J Med 355(16):e17. https://doi.org/10.1056/NEJMicm050783

WHO Global TB report 2017. www.who.int.

WHO treatment guidance for drug resistant tuberculosis (2016)

WHO treatment guidelines for multidrug- and rifampicin-resistant tuberculosis (2018) https://www.who.int/tb/publications/2018/WHO.2018.MDR-TB.Rx.Guidelines.prefinal.text.pdf

Ziganshina LE, Squire SB (2008) Fluoroquinolones for treating tuberculosis. Cochrane Database Syst Rev (1):Cd004795. https://doi.org/10.1002/14651858.CD004795.pub3

Tuberculosis Vaccine: Past Experiences and Future Prospects

21

Gurpreet Kaur, Deepjyoti K. Das, Sanpreet Singh, Junaid Khan, Mohammad Sajid, Hilal Bashir, Mohammad Aqdas, Shikha Negi, Uthaman Gowthaman, and Javed N. Agrewala

Abstract

Vaccines are the best prophylactic measure that have eradicated several diseases like polio, measles, and small pox. Currently, efficient vaccines are unavailable for many dreaded diseases, including tuberculosis (TB). TB is caused by an intracellular pathogen *Mycobacterium tuberculosis* (*Mtb*) and is responsible for about two million deaths and nine million new cases annually. The WHO has announced TB as a global emergency in the wake of emerging multidrug-resistant, extremely drug-resistant, and totally drug-resistant strains of *Mtb*. An alarming increase in the number of TB cases around the world and its co-occurrence with HIV has further complicated the problem. Additionally, BCG has failed in reducing the global TB burden, despite its widespread usage. Interestingly, BCG protects children from TB, indicating that it has sufficient antigenic repertoire to protect against *Mtb*. In contrast, failure to protect adults, suggests BCG inability to generate long-lasting immunological memory. Further, protection rendered by BCG against pulmonary TB in adults is highly inconsistent, varying from 0% to 85% (Andersen and Doherty, Nat Rev Microbiol 3:656–62, 2005). Furthermore, its efficacy is least in TB-endemic countries. Studies conducted in Malawi, India, and other endemic countries have concluded that BCG induces inadequate protection. The probable reasons suggested are the

G. Kaur · J. N. Agrewala (✉)
Immunology Laboratory, Institute of Microbial Technology, Chandigarh, India

Centre for Biomedical Engineering, Indian Institute of Technology-Ropar, Rupnagar, India
e-mail: jagrewala@iitrpr.ac.in

D. K. Das · S. Singh · J. Khan · M. Sajid · H. Bashir · M. Aqdas · S. Negi
Immunology Laboratory, Institute of Microbial Technology, Chandigarh, India

U. Gowthaman
Department of Laboratory Medicine, Immunobiology and Orthopedics,
Yale University School of Medicine, New Haven, CT, USA

S. E. Hasnain et al. (eds.), Mycobacterium tuberculosis: *Molecular Infection Biology, Pathogenesis, Diagnostics and New Interventions*,
https://doi.org/10.1007/978-981-32-9413-4_21

interference by nontuberculous mycobacteria (NTM) in antigen processing and presentation of antigen-presenting cells (APCs), latent TB infection (LTBI), and high prevalence of helminth infestation. There are currently 12 vaccine candidates in clinical trials, including recombinant BCG vaccines, attenuated *Mtb* strains, recombinant viral-vectored platforms, protein/adjuvant combinations, and mycobacterial extracts (WHO, Global tuberculosis report, 2017). However, the whole-cell vaccine candidates may encounter same problem in TB-endemic population, as mentioned above for BCG. Therefore, radically novel strategies of vaccination against *Mtb* are urgently needed. Peptide vaccines can be a possible alternative to overcome the failures of BCG because it comprises of epitopes that can directly bind to MHC molecules, circumventing the interference of NTM and latent *Mtb* in antigen processing. Furthermore, epitopes from multiple stages like LTBI, active, chronic, and drug-resistant *Mtb* can be selected. In essence, peptide-based prophylactic and therapeutic vaccines represent a promising future strategy for efficient management of TB.

Keywords
Mycobacterium tuberculosis · Immune response · Vaccines · Peptide vaccines

Abbreviations

APCs	Antigen-presenting cells
DCs	Dendritic cells
DIM	Phthiocerol dimycocerosates
Hly	Listeriolysin
LTBI	Latent TB infection
Mtb	*Mycobacterium tuberculosis*
NTM	Nontuberculous mycobacteria
PBMCs	Peripheral blood mononuclear cells
pfo	Perfringolysin
SNP	Single nucleotide polymorphisms
TB	Tuberculosis
Tregs	T regulatory cells
WHO	World Health Organization

21.1 Vaccine Development

Vaccination is considered as one of the cornerstones in the history of modern medicine (Plotkin and Plotkin 2011). In case of various medical interventions, this is the most widely implemented field that has gained considerable attention worldwide. In the USA alone, vaccines have protected 100 million people from different diseases (Van Panhuis et al. 2013). Further, about 2.5 million deaths annually are prevented globally (WHO 2013). Vaccines are quite important for protecting present and future generations. Well-planned vaccination schedules have helped to save hundreds of millions of people from death, apart from savings in public health expenditures. An effective vaccine should be able to provide lifelong immunity against a particular pathogen with no adverse side effects. Vaccines prime our immune system by exposing an individual to a nonlethal, milder version of the pathogen's antigens in order to develop memory T cells, without causing disease. Ideally, a vaccine is made of pathogen-derived components that are critical for induction of protective immunity. Consequently, vaccines are composed of either an attenuated or killed version of the bacteria, inactivated bacterial toxins, subunits of the pathogen, or peptides that mimic naturally occurring proteins from pathogens (Table 21.1).

21.2 BCG: The Only Available Vaccine Against TB

BCG is the only successful vaccine developed so far against TB infection. Despite its widespread use, BCG delivers only minimal protection and has failed to eradicate or reduce TB. BCG is derived from *Mycobacterium bovis*, a bovine mycobacterium passaged 230 times in the Pasteur laboratory in Lyon by Albert Calmette and Guérin between 1908 and its first human use in 1921. First administration of BCG vaccine in humans was done by Benjamin Weill-Halle assisted by Raymond Turpin at the Charité Hospital, Paris. On July 18, 1921, a healthy infant was immunized with BCG whose mother died immediately after his birth because of TB. BCG was fed orally in a spoon, followed by ingestion of milk. The child remained healthy and showed no signs of TB. Furthermore, 114,000 infants were vaccinated with BCG from 1924 to 1928 and no complications were observed (Luca and Mihaescu 2013). Similar trials were conducted in Europe, Africa, South America, and Asia. Earlier, BCG was given orally but now WHO recommends intradermal route of vaccination, as it allows more consistent dosing. Other reason can be that the efficacy of BCG reduces one- to two-fold on exposure to gastric enzymes and pH (Tacquet et al. 1962). Moreover, it has been shown that intradermal vaccination reduces the risk of TB by 50%. Interestingly, BCG not only protects TB; rather it has also shown good results against leprosy (Orege et al. 1993). Intriguingly, BCG has exhibited quite convincing results as an immunotherapeutic agent against bladder cancer. It has since then become the most widely used vaccine in human history, with over 3 billion doses administered. Currently, more than 100 million children are annually administered BCG within hours or days of birth. The protective efficacy of BCG is highest (85%) in TB non-endemic population, whereas it is lowest (0%) in TB-endemic regions. Further, it protects the childhood but not the adult manifestation of TB. Nonetheless, due to the overall variability in its efficacy, BCG remains the most controversial vaccine.

Table 21.1 Outline of the development of human vaccines (Plotkin 2014)

Live attenuated	Killed whole organisms	Purified proteins or polysaccharides	Genetically engineered
Eighteenth century			
Smallpox (1798)			
Nineteenth century			
Rabies (1885)	Typhoid (1896)		
	Cholera (1896)		
	Plague (1897)		
Early twentieth century, first half			
Tuberculosis (bacillus Calmette-Guérin) (1927)	Pertussis (1926)	Diphtheria toxoid (1923)	
Yellow fever (1935)	Influenza (1936)	Tetanus toxoid (1926)	
	Rickettsia (1938)		
Twentieth century, second half			
Polio (oral) (1963)	Polio (injected) (1955)	Anthrax secreted proteins (1970)	Hepatitis B surface antigen recombinant (1986)
Measles (1963)	Rabies (cell culture) (1980)	Meningococcus polysaccharide (1974)	Lyme OspA (1998)
Mumps (1967)	Japanese encephalitis (mouse brain) (1992)		Cholera (recombinant toxin B) (1993)
Rubella (1969)	Tick-borne encephalitis (1981)	Pneumococcus polysaccharide (1977)	
Adenovirus (1980)	Hepatitis A (1996)		
Typhoid (SalmonellaTY21a) (1989)	Cholera (WC-rBS) (1991)	*Haemophilus influenzae* type B polysaccharide (1985)	
Varicella (1995)	Meningococcal conjugate (group C) (1999)	*H. influenzae* type b conjugate (1987)	
Rotavirus reassortants (1999)	Japanese encephalitis (2009) (Vero cell)	Typhoid (Vi) polysaccharide (1994)	
Cholera (attenuated) (1994)	Cholera (WC only) (2009)	Acellular pertussis (1996)	
Cold-adapted influenza (1999)	Vaxchora for the prevention of cholera (2016)	Hepatitis B (plasma derived) (1981)	

(continued)

Table 21.1 (continued)

Live attenuated	Killed whole organisms	Purified proteins or polysaccharides	Genetically engineered
Twenty-first century			
Rotavirus (attenuated and new reassortants) (2006)		Pneumococcal conjugates (heptavalent) (2000)	Human papillomavirus recombinant (quadrivalent) (2006)
Zoster (2006)		Meningococcal conjugates (quadrivalent) (2005)	Human papillomavirus recombinant (bivalent) (2009)
		Pneumococcal conjugates (13-valent) (2010)	Meningococcal group B proteins (2013)
		VSV-EBOV vaccine against Ebola (2014)	Dengvaxia live attenuated tetravalent chimeric vaccine for Dengue (2015)
		Malaria vaccine, *P. falciparum* protein fused with hepatitis B surface antigen (2015)	Shingrix (recombinant zoster vaccine) to prevent shingles (2017)
		Injectable influenza vaccine, Fluad (2015)	Heplisav-B, the new hepatitis B vaccine (2017)

21.3 Modulation of Immunity by BCG

Macrophages, dendritic cells, and neutrophils are the major cells of immune system and first line of defense to combat *Mtb*. Upon vaccination, epidermal macrophages interact with BCG through the pattern recognition receptors (PRRs), such as toll-like receptor 2/4 (TLR2/4), macrophage-inducible $Ca2^{+}$-dependent lectin (MINCLE), complement receptor 3 (CR3), and mannose receptor (MR) present on their surface (de Koning et al. 2012; Bugarcic et al. 2008). In contrast, alveolar macrophages interact with *Mtb*.

BCG provokes antigen-presenting cells (APCs) to secrete cytokines chemokines and recruit various cells of the immune system. The APCs process and present BCG antigens to T cells. Activated CD4 T cells then release cytokines such as IFN-γ, IL-2, and TNF-α that participate in protection against *Mtb* infection. IFN-γ-secreting CD4 T cells express $CD45RA^{-}$ $CCR7^{+/-}$ $CD62L^{-}$ phenotypic markers, indicating central and effector memory cells in BCG-vaccinated donors (Li et al. 2011). Apart from CD4 T cells, NK and γδ T cells produce IFN-γ on BCG vaccination (Ritz et al. 2012).

21.4 Activation of Innate Immunity by BCG

BCG activates dendritic cells (DCs) and macrophages mainly through TLR-2 and TLR-4 by signaling cascade involving MyD88 adaptor molecule, which recruits IRAK, TRAF6, and MAPK, eventually activating NF-κB. This leads to their maturation and activation. This results in the release of proinflammatory cytokines IL-12, TNF-α, and IL-1β (Takeda and Akira 2004). Various chemokines, matrix metalloproteinases (MMPs), and adhesion molecules ICAM-1, VCAM-1, p-selectins, L-selectins (CD62L), and E-selectins are upregulated (Kupper and Fuhlbrigge 2004).

21.5 Activation of Adaptive Immunity by BCG

Th1 immunity is crucial for protection against *Mtb*. BCG vaccination in newborn is reported to elicit Th1 cells that show effector memory phenotype ($CD45RA^{-}CCR7^{-}CD27^{+}$) and release mainly IFN-γ, while central memory cells secrete chiefly IL-2 (Soares et al. 2008). Further, recent studies have illustrated the potent role of Th17 cells in protection against *Mtb* (Khader and Cooper 2008). Furthermore, CD8 T cells play an important role in protection against TB. BCG taken up by APCs is presented by DCs or macrophages in context with MHC class I molecules to CD8 T cells. MHC-I-deficient ($\beta_2m^{-/-}$) mice immunized with BCG survived upon exposure to *Mtb*, indicating the importance of CD8 T cell immunity (Flynn et al. 1992). Further, protective CD8 T cell response elicited by BCG in newborn children was attributed to either cytotoxic degranulation or IFN-γ cytokine secretion (Murray et al. 2006). Although BCG-induced CD4 T cells are more crucial than CD8 T cells, both coordinate with each other to protect against *Mtb*. Th1 response generated against *Mtb* can be suppressed by Th2 cells. Thus, a predominance of Th1 cells is essential for protection against *Mtb* infection (Rook et al. 2004).

21.6 Reasons for Failure of BCG Vaccine

The Chingleput trial in India has established 0% protective efficiency of BCG (Tuberculosis Research Centre) (Datta et al. 1999). Results have shown that the efficacy of BCG lasts up to 15 years in the United Kingdom (Sterne et al. 1998), 30–40 years in Norway (Nguipdop-Djomo et al. 2016), and even higher up to 50–60 years in Alaska (Aronson et al. 2004). Exposure to nontuberculous mycobacteria (NTMs) (Chiang and Riley 2005), genetically varied human population (Brosch et al. 2007), different BCG strains (Osborn 1983), helminth infestation, and exposure to nontuberculous mycobacteria (NTM) are among the main causes of variable BCG efficacy. Despite its wide usage, BCG has not adequately reduced the global TB burden. Due to its inconsistent performance across the globe, BCG remains a debatable vaccine.

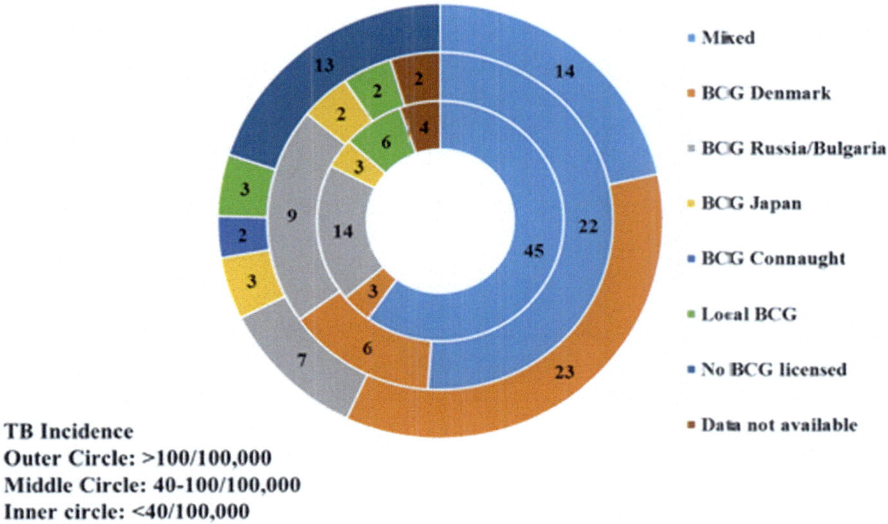

Fig. 21.1 Different BCG strains used globally and incidence of TB in the region (Ritz and Curtis 2009). Each circle represents group of countries. The outer circle signifies countries with a TB incidence of >100/100,000 individuals; middle circle represents 40–100/100,000 and the inner circle represents <40/100,000. The numbers in the doughnut graph symbolize the number of countries within that particular category. Mixed means countries that use 2 or > 2 strains of BCG. "Local BCG" implies to those countries, which employ indigenously produced BCG

Several reasons have been attributed for the variable efficacy of BCG vaccine worldwide:

(i) *Variation in BCG strains:* The inconsistency of BCG vaccine might be attributed to the variability in the seed strains produced/cultured by various countries (Brosch et al. 2007). The original attenuated BCG was kept in running cultures for several years until seed lots were preserved and stored. These lyophilized strains were then distributed to various countries (Dockrell and Smith 2017). Now with improved sequencing and omics technologies, genetic and proteomic variations among these strains have been proved (Abdallah et al. 2015). These variations may be due to single nucleotide polymorphisms (SNP), duplication, and deletions at genetic levels. Furthermore, some countries have used more than one strain in their immunization program, which has added complications in the variability of results (Ritz and Curtis 2009). A majority of countries (105 of 188 countries) have used more than one strain of BCG in the same country (Fig. 21.1). Genomic comparison and analysis of 13 different BCG strains used for vaccination have shown a greater degree of variation in the T cell epitopes due to higher deletion rates among them (Zhang et al. 2013).

(ii) *Interference of nontuberculous mycobacteria (NTM):* BCG protects neonates against pulmonary TB but it fails to protect adults (Dark 1978). This is due to inability of BCG to generate long-lasting memory T cells. This is evident from a

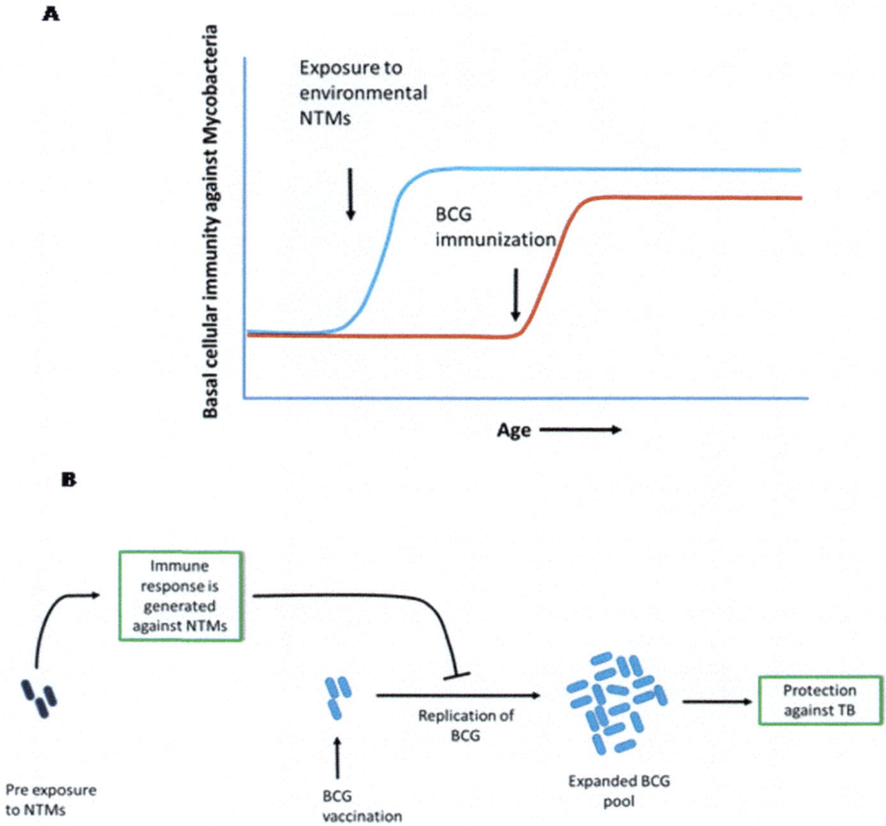

Fig. 21.2 (**a**) Masking hypothesis: exposure to NTMs at an early age increased the individual's basal immunity against the mycobacteria (blue line). The same individual when vaccinated with BCG at a later age (red line), BCG-induced immunity could not be further enhanced due to masking of immune response by prior exposure to NTMs. (**b**) Blocking hypothesis: Pre-exposure to NTMs at an earlier age increased the individual's immunity against mycobacteria. The same individual when vaccinated with BCG, the immune response generated against NTM obstructed BCG replication and therefore prevented the optimum exposure of BCG antigens to the immune system. Therefore, the person is not protected from *Mtb* infection, as noted in the case of TB-endemic population. In contrast, in the absence of NTMs, BCG generates sufficient immunity to protect against *Mtb*, as observed in TB non-endemic regions

study where mice that have been given NTMs by oral gavage following BCG vaccination showed decreased long-lasting protection (Flaherty et al. 2006). The possible reason suggested was that in TB-endemic regions like India, BCG efficacy was lower because of exposure to NTM (Palmer and Long 1966). Nontuberculous mycobacteria hinders the proliferation of BCG. Consequently, the immune system is not optimally exposed to BCG antigens and therefore fails to elicit adequate immunity. Majorly, two hypothesis has been attributed to this cause (Fig. 21.2):

Masking hypothesis. According to this hypothesis, exposure of human to NTMs generates a baseline immunity. Thus, the baseline immunity already generated by NTMs masks the BCG-induced response (Fig. 21.2a). A clinical study encompassing school children from two entirely different populations, viz., UK (a country with lowest TB incidence and with highest BCG success) and Malawi (a region in which BCG vaccination has shown no effect against pulmonary TB), was reported. Subjects from UK had a low baseline cellular immunity to mycobacteria, which was increased upon BCG immunization. On the contrary, subjects from Malawi had a higher level of basal cellular immunity against mycobacteria, due to pre-exposure of NTMs in this region, and thus the immunity was not enhanced by BCG (Black et al. 2002).

Blocking hypothesis. In this case, the preexisting immune response generated due to NTMs prevented the replication of BCG (Fig. 21.2b). Mice sensitized with environmental NTMs isolated from soil and sputum samples from Malawi hampered the growth of BCG without any effect on TB incidence (Fine et al. 2001).

Studies have also suggested that BCG itself hampers antigen processing and presentation of its own antigens, consequently generating an inadequate immune response (Singh et al. 2006). BCG evades lysosomal fusion in macrophages (Ullrich et al. 2000). This results in reduction of antigen processing and downregulating the expression of MHC-II molecules. These macrophages have lower capability to activate Th1 immunity, which is a prerequisite for protection against *Mtb*. Further, NTMs can enhance T regulatory cell (Tregs) expansion and recruitment, thus retarding BCG efficacy (Ho et al. 2010).

(iii) *Helminth infestation:* Another important cause for BCG failure in the TB-endemic regions is the prevalence of helminth infestation. Helminth enhances the Th2-type immunity and higher IgE profile and increases the level of Treg cells (Elias et al. 2007). Clinical study has proved that the infection with *Ascaris lumbricoides* can hamper the generation of a proper Th1 immunity (Van Soelen et al. 2012). Similarly, *Strongyloides stercoralis* infection skews the immunity toward Th2 type (Carvalho and Da Fonseca Porto 2004). These worms interfere with the performance of BCG. Infection with *Schistosoma mansoni* results in an elevated level of IL-4/IFN-γ ratio (Zwingenberger et al. 1991). A lower IL-4/IFN-γ ratio in the bronchoalveolar lavage and pleural fluid has been correlated with a better immunity against TB (Ashenafi et al. 2014). However, it has shown that helminth infestations do not reduce the efficacy of BCG by increasing the Th2 cells but by enhancing TGF-β production (Elias et al. 2008).

(iv) *Route of administration:* The route of administration of BCG is disputed. Originally, BCG was administered orally and a study showed that oral route involves lesser pathology and similar level of protection (Dockrell and Smith 2017). While dose responsiveness of BCG vaccination is not prominent in subcutaneous route of immunization, intratracheal route offers better dose responsiveness and better reduction in *Mtb* CFU count in lungs (Aguilo et al. 2014). Further, advantage in intratracheal and intranasal BCG vaccination includes elicitation of T effector memory (TEM) and T resident memory (TRM) cells in the lung tissue. Contrary to this, subcutaneous and intradermal routes of BCG immunization fail to offer memory T cell response (Verreck et al. 2017).

(v) *Batch to batch variation of BCG strain:* The culturing of BCG involves various steps and thus differences have been observed in the growth pattern among batches. The batches respond differently toward sensitivity against isoniazid, which correlates with the immunity they provide in mice. Further, randomized clinical trials in neonates have supported dissimilarity in the immune response (Biering-Sørensen et al. 2015).

(vi) *BCG triggers the generation of Tregs and Th2 cells:* Another dark side of BCG vaccine is that it can directly dampen anti-TB response by generating Tregs and Th2 cells (Lagranderie and Guyonvarc'h 2014). The dose of BCG vaccine influences the type of T cell response generated. A lower dose of BCG generates Th1 cells, while a higher dose of BCG leads to mixed generation of Th1 and Th2 cells (Power et al. 1998).

(vii) *Nonspecific activity of BCG:* BCG shows promising results in bladder cancer, melanoma cancer, asthma, leprosy therapies, and probably against *Buruli ulcer* (Fuge et al. 2015; El-Zein et al. 2017). Consequently, this casts a doubt on BCG reliability as a vaccine. Rather it works as an immunomodulator to bolster host immunity against array of diseases (Fig. 21.3). This can be due to the fact that the generation of nonspecific immune response by BCG may be due to the activation of mononuclear phagocytic system via PAMPS like TLR-2, TLR-4, and TLR-8 signaling, consequently initiating a cascade of inflammatory responses and thereby recruiting inflammatory cells to the site of infection. Further, it provides feedback signals for maturation of DCs, macrophages, and neutrophils.

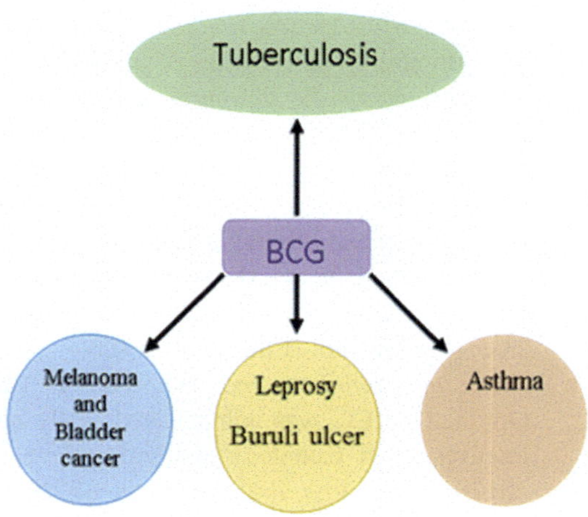

Fig. 21.3 *Diverse roles of BCG.* BCG not only protects against TB but also from cancers, Buruli ulcer, asthma, and leprosy. This activity of BCG may be due to its immunomodulatory activity, since it can trigger cascade of signaling events through TLR-2, TLR-4, and TLR-8

21.7 Prospects for New TB Vaccine

For complete eradication of TB, an effective vaccine is urgently needed. Despite the fact that TB can be cured, 10 million people developed TB disease in 2017: 5.8 million men, 3.2 million women, and 1.0 million children. Globally, 3.5% of new TB cases and 18% of previously treated cases had MDR/RR-TB (WHO 2018). Therefore, it is a daunting task to reduce TB burden. Currently, it is important to develop drugs or vaccines to control the emergence and spread of drug-resistant TB. The treatment for drug-resistant TB is not only costly but has devastating toxic effects. The new generation of vaccines being developed should target TB and its drug-resistant forms. An effective TB vaccine may be prophylactic and/or therapeutic. An ideal therapeutic vaccine should shorten the dose and duration of treatment and improve the potency of drugs. A prophylactic vaccine would be safer in pre- or postexposure to *Mtb* infection and effective in averting primary and latent infection and reactivation of latent forms. This can be achieved by two ways, either replacing BCG with an alternative live mycobacterial vaccine or boosting the efficacy of BCG with the subunit vaccines. The second approach may be to combine both approaches by priming with a novel live mycobacterial vaccine and boosting with a subunit vaccine. During last few years, there has been a tremendous pressure on researchers to develop a safe and effective vaccine against TB. Some of the vaccine candidates have entered various stages of clinical trials.

21.8 Recombinant BCG Vaccines (rBCG)

Importantly, an ideal immune phenotype of protection should be taken into consideration while designing an effective vaccine. A successful vaccine should generate appropriate and adequate immunity, enduring memory T cells and protection from the pathogen on subsequent encounter. In mouse model, currently accepted memory phenotypes are $CD44^{hi}CD62L^{lo}CCR7^{lo}$ for effector memory T cells, $CD44^{hi}CD62L^{hi}CCR7^{hi}$ for central memory T cells, and $CD69^{hi}CD103^{hi}$ $CD62L^{lo}$ for resident memory T cells (Henao-Tamayo et al. 2010). IFN-γ-secreting CD4 T cells are one of the major targets of TB vaccine research. IFN-γ-producing CD8 T cells too play a crucial role in protection against *Mtb* (Flynn et al. 1993). CD8 T cells play a significant role in controlling latent TB infection (LTBI) (van Pinxteren et al. 2000). Potent rBCG vaccines should induce a balanced number of *Mtb*-specific CD4 and CD8 T cells. Studies have shown that desired memory T cells should have polyfunctional characteristics, i.e., ability to secrete IFN-γ, tumor necrosis factor (TNF)-α, and IL-2 to provide protection against *Mtb* infection (Beuneaud et al. 2005). Combination of biomarkers for human memory T cells with effector phenotype includes $CD45RA^{hi}CD45RO^{-}CCR7^{-}$ and for central memory T cells $CD45RA^{hi/lo}CD45RO^{-}$ $CCR7^{+}$. Long-lasting memory T cells expressing CD127 (IL-7 receptor) are associated with protection against *Mtb* infection (Seder and Ahmed 2003). Beyond doubt, BCG has imparted significant protection against

Mtb in the Western population and childhood manifestation of the disease globally. BCG vaccine is cost-effective, stable, and easy to manufacture. It has an ability to express foreign antigens in their functional form. These attributes have driven the development of various forms of rBCG. We discuss here some potential rBCG vaccine candidates and their ability to protect against TB.

21.9 Different Strategies Used in the Development of rBCG

Over the past several years, various strategies have been strived in the advancement of rBCG (Table 21.2). One of the major strategies adopted is the overexpression of immunodominant antigens of *Mtb* such as antigen 85 (Ag85) complex and α-crystalline protein (HspX). Other strategy is the reintroduction of antigens such as CFP-10, ESAT-6, INV and MPT64, HBHA, PE/PPE family, ApA, MDP1, MPT64, PepA, Rv0111, and Rv2520c that might have been deleted during

Table 21.2 Description of strains and antigens/modifications used in the rBCG development

Vaccine	Strain	Description	Model	References
rBCG30	BCG tice	Overexpression of Ag85B	Human	Hoft et al. (2008)
rBCG-Ag85A	BCG Tokyo	Overexpression of Ag85A	Monkey	Sugawara et al. (2009)
rBCG-85C	BCG Danish	Overexpression of Ag85C	Guinea pigs	Jain et al. (2008)
rBCG:X	BCG Danish	Expresses HspX protein	Mice	Shi et al. (2010)
rBCG:PhspX-85B	BCG Pasteur	Expresses Ag85B under control of promoter HspX	Mice	Kong et al. (2011)
rBCG-AE	BCG China	Expresses fusion protein Ag85A-ESAT-6	Mice	Deng et al. (2014)
rBCG-Aeras 403	BCG Danish	Endosome escape activity with overexpression of Ag85A, Ag85B, TB10.4	Monkey	Rahman et al. (2012)
rBCGΔureC: hlyC	BCG Parague	Endosome escape activity	Mice	Desel et al. (2011)
BCG:RDI	BCG Pasteur	Reintroduction of RD1 gene regions	Mice	Pym et al. (2003)
rBCG-E6	BCG Danish	Reintroduction of ESAT-6	Guinea pigs	Dey et al. (2009)
rBCG:Ag85B-ESAT-6-Rv2608	BCG Danish	Co-expresses Ag85B, ESAT-6, and Rv2608	Mice	Lu et al. (2012)
rBCG-AMM	BCG Danish	Expresses Ag85B-MPT64-*Mtb*8.4	Mice	Qie et al. (2009)
Pro-apoptotic BCG	BCG tice	Deletion *nlaA* gene that inhibits apoptosis	Mice	Chattergoon et al. (2000)

attenuation procedure of BCG but are crucial in imparting protective immunity (Jain et al. 2008). Ag85A (Rv3804c), Ag85B (Rv1886c), and Ag85C (Rv0129c) are important antigens from Ag85 complex used to develop rBCG vaccines. These proteins have mycolyl-transferase activity that is necessary for the synthesis of mycobacterial cell wall (Belisle et al. 1997). Horwitz et al. have developed rBCG30 that overexpresses Ag85B. Ag85A is the most abundant secretory protein of *Mtb* and potent inducer of IFN-γ production and Th1 immunity (Horwitz et al. 2000). In guinea pigs, rBCG30 showed increased protection against *Mtb*, as compared to wild-type BCG. Phase I clinical trials with rBCG30 in around 30 healthy individual upshots increased immunogenicity in terms of central and effector memory CD4 T cells and CD8 T cells, as compared to parental BCG (Hoft et al. 2008). Unfortunately, the outcome of rBCG30 clinical trial did not meet the expected immunogenicity and further testing was discontinued. Tullius et al. developed a mutant strain rBCG(mbt)30 that was unable to produce mycobactin and exoquelin molecules required for iron acquisition. In another strategy, rBCG pantothenate auxotroph rBCG(PanCD)30 was designed (Tullius et al. 2008). Both these vaccines showed higher degree of attenuation than BCG and instigated potent protective immunity in guinea pigs. These vaccine candidates may provide better and safer protection. Ag85A strongly induced T cells to secrete IL-2 and IFN-γ (Lozes et al. 1997). rBCG:Ag85A-vaccinated mice and guinea pigs provided better protection than BCG against *Mtb* in pulmonary and spleen infection (Sugawara et al. 2007). Further, this vaccine was investigated in rhesus monkeys (*Macaca mulatta*). rBCG: Ag85A induced greater efficacy than BCG Tokyo strain. rBCG:Ag85C-inoculated guinea pigs imparted better protection, as evidenced by reduced lung infection and granuloma formation (Jain et al. 2008).

Combination of two or more immunodominant antigens of *Mtb* as fusion protein in rBCG was hypothesized to attain better efficacy. Wang et al. developed three vaccines rBCG:Ag85A, rBCG:Ag85B, and rBCG:Ag85AB. The rBCG expressing fusion protein (Ag85A-Ag85B) exhibited greater protection than rBCG:Ag85A and rBCG:Ag85B alone (Wang et al. 2012). HspX, known as α-crystalline, is a 16 kDa heat-shock protein that is potentially used in the development of rBCG. *Mtb* produces HspX protein in high amounts during latent and persistence metabolic conditions. Shi et al. developed a rBCG:X vaccine that overexpresses HspX. rBCG: X provided greater efficacy and long-lasting protection against *Mtb* than parental BCG. rBCG:X induced high level of IFN-γ and reduced mycobacterium load in the lungs and other tissues (Shi et al. 2010). Kong et al. generated rBCG:HspX-Ag85B that overexpressed Ag85B under the control of HspX promoter. This vaccine induced IFN-γ and T cell proliferation (Kong et al. 2011).

BCG favorably resides inside endosomes of macrophages and DCs. This localization supports the presentation of BCG antigens in association with MHC-II to CD4 T cells, resulting in their activation, proliferation, and differentiation. CD8 T cells are activated when antigen is presented in context with MHC-I. Thus to activate *Mtb*-specific CD8 T cells, BCG must gain access from endosome into cytosol of the host cells. In this context, Kaufmann et al. developed rBCGΔUreC:Hly$^+$ (VPM1002) vaccine (Nieuwenhuizen et al. 2017). It secretes listeriolysin (Hly) and contains an inactivated urease C gene to ensure optimal Hly activity in an acidic phagosomal environment (Albert et al. 1998). Listeriolysin of rBCGΔUreC:Hly$^+$ punches holes into phagosomal membrane and enables BCG to escape from endosome into the cytosol of cells. Further, it induces apoptosis and autophagy that facilitates the cross-presentation of antigens to CD8 T cells. Further, robust activation of Th1 cells and Th17 cells was noticed. Furthermore, it successfully expands central memory CD4 T cells and CD8 T cells. In essence, rBCGΔUreC:Hly$^+$ provokes immunity that is crucial for protection against *Mtb*. Moreover, the vaccine is cleared more rapidly from tissues than BCG, thus resulting in reduced persistence in the host. Sun et al. developed a rBCG-pfo that expresses perfringolysin in a pH-independent manner with overexpression of Ag85A, Ag85B, and TB10.4. Here, BCG escapes endosome using perfringolysin. This vaccine showed superior protection over rBCG30 with enhanced protection and safety of the endosome escape construct (Fig. 21.4) (Portnoy et al. 1992). Jacob et al. proposed a pro-apoptotic vaccine by deleting nlaA genes (Rv3238c) that inhibits apoptosis of infected host cells. rBCG vaccine that promotes caspase-dependent apoptosis of host cells leads to the cross-priming of CD8 T cells. Caspase-mediated apoptosis is very well known for the induction of powerful cellular immunity (Chattergoon et al. 2000).

Reintroduction of genes that were deleted during the attenuation procedure of BCG is also one of the important strategies for rBCG development. Virulence region RD1 is absent in all sub-strains of BCG, but present in virulent mycobacteria (Brosch et al. 2000). Cole et al. made an effort to improve the efficacy of the vaccine by reintroduction of some lost genes into BCG. This strategy used the collection of T cell epitopes, which were lost during attenuation of BCG. These gene regions include RD1 locus, which encodes immunodominant proteins CFP-10 (Rv3874), ESAT-6, INV (Rv1474), and MPT64.

Cytokines mediate important roles in regulation of homeostasis of naive and memory T cells and their response to pathogens. Introducing genes encoding mammalian cytokines, or cytokine antagonists such as the latency activating peptide (LAP) that regulates transforming growth factor-β (TGF-β), can increase immunogenicity of BCG (Wilkinson et al. 2000; Slobbe et al. 1999). However, the limitation of this strategy is that enhanced Th1 response may cause clearance of vaccine and therefore will reduce memory response. Further, memory enhancing cytokines like IL-7 and IL-15, play a dominant role in the generation and homeostasis of memory T cells (Singh et al. 2010). Numerous studies have shown the adjuvant nature of IL-7 and IL-15 and their role in the generation and sustenance of memory T cells (Singh et al. 2011; Nanjappa et al. 2008). Therefore, supplementing vaccines with IL-7 and IL-15 may enhance the long-term sustenance of memory T cells. Interestingly, substantial increase in the enduring memory T cells in mouse against *Mtb* infection by supplementing BCG with rIL-7 and rIL-15 was observed. Further, it significantly reduced the mycobacterial load and alleviated the pathology in the lungs, thus providing better protection than BCG (Singh et al. 2010). Similarly, they showed that cytokines like IL-1, IL-6, and TNF-α bolstered BCG potency by generating long-lasting protective memory T cell response (Singh et al. 2011). More recently, rBCG expressing Ag85B-IL-7 fusion protein was developed, which increased the

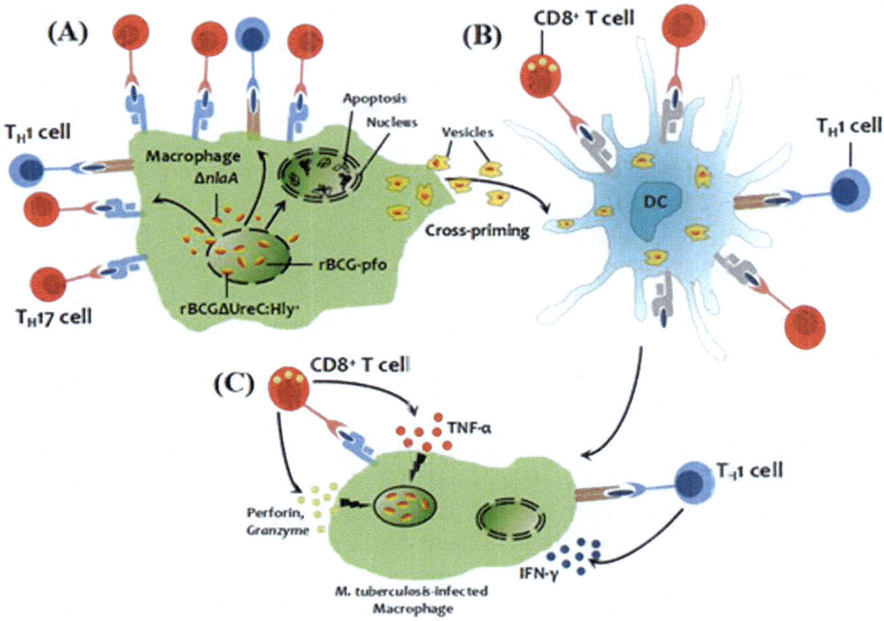

Fig. 21.4 *Mechanisms of the immune responses generated by rBCGΔUreC:Hly⁺*. (**a**) rBCGΔUreC:Hly⁺ (with a deletion of UreC and expression of listeriolysin:Hly⁺) or rBCG-pfo (expressing pH-independent perfringolysin) enable the vaccine to form pores in the early phagosome that facilitate the leakage of rBCG into the cytoplasm of host cells. This results in strong presentation of peptide in association with MHC class I molecules and cross-priming of CD8 T cells, generation of Th1 cells and Th17 cells, and induction of apoptosis. (**b**) Cross-priming and enhanced apoptosis increase the uptake of rBCG antigens in the form of vesicles by DCs and generates CD4 T cell and CD8 T cell immunity. (**c**) CD8 T cells coordinate with CD4 T cells in eliminating *Mtb*. CD8 T cell utilizes its perforin, granzyme, and TNF-α to lyse *Mtb*-infected macrophages, whereas CD4 Th1 cells produce IFN-γ to activate macrophages to kill *Mtb*

frequency of IL-17A⁺ γδ T cells and enhanced Th1 response against *Mtb* (Hatano et al. 2016). Thereby, developing a recombinant BCG expressing memory enhancing cytokines can be an effective strategy to successfully improve the efficacy of BCG in protecting not only childhood but also adult manifestation of TB.

ChAdOx1.PPE15 is a recombinant, replication-deficient chimpanzee adenovirus, expressing the mycobacterial PPE15 protein of *Mtb* (WHO 2017). This candidate is part of the overall UOXF TB vaccine program to work toward a BCG booster vaccination regimen in adults and is being evaluated in guinea pigs for improved protective efficacy.

CysVac2 is a fusion protein vaccine composed of Ag85B and CysD, a major component of the sulfate activation pathway of *Mtb*. It was designed to target both active and chronic phase of infection. Preclinical experiments in mice demonstrated that the fusion protein provides strong protection when combined with the novel adjuvant MPL/DDA and induced the generation of multifunctional CD4 T cells expressing IFN-γ, IL-2, and TNF, as well as double-positive IFN-γ^+TNF$^+$ and IL-2$^+$TNF-α^+ CD4 T cells in the lung (Counoupas et al. 2016). The vaccine was particularly effective at controlling late-stage infection, a major problem in case of BCG vaccination. The vaccine is being tested in guinea pigs to strengthen its case for deployment in clinical trials.

MVA85A is a recombinant vaccine expressing Ag85A in modified vaccinia Ankara virus (McShane et al. 2005). In rhesus macaques, the vaccine delivered by aerosol route showed a high immunogenic effect (White et al. 2013). This approach generated strong antigen-specific responses in the lungs and was well tolerated. Moreover, this route of immunization did not generated antibodies to the viral vehicle. In phase I studies, a well-tolerated and immunogenic effect with only minor adverse reaction was noticed in HIV-infected adults in Senegal (Dieye et al. 2013). This was the first virally vectored vaccine to be tested in humans delivered through aerosol route. Furthermore, in a randomized phase I trial, MVA85 administration via intramuscular and intradermal routes in 24 healthy human volunteers proved safe and well tolerated. Both routes of immunization led to generation of strong and sustained Ag-specific CD4 T cell responses. MV85A was well tolerated and immunogenic in a phase IIb trials conducted on 2797 healthy infants who had previously been vaccinated with BCG. However, the vaccine showed no significant protection against TB infection (Tameris et al. 2014). The results from this study suggested that CD4 T cells induced by MVA85A do not correlate with protection against *Mtb* infection. MVA85A induces polyfunctional CD4 T cells expressing IFN-γ, IL-2, TNF-α, and IL-17 (Satti et al. 2014; Scriba et al. 2010). Detailed future studies will be required to determine whether the protective efficacy of MVA85A can be improved.

Ad35/AERAS-402 is an adenovirus-vectored vaccine formed by non-replicating adenovirus serotype 35 (Ad35), expressing fusion protein composed of three major antigens of *Mtb* Ag85A, Ag85B, and TB10.4 (WHO 2014). It is primarily designed as a booster vaccine. The vaccine was well tolerated and found to be safe in a phase I study (Churchyard et al. 2015) [NCT Identifier 01017536]. Vaccination with AERAS-402 with MV85A booster showed increase in antigen-specific T cell responses and polyclonal CD8 T cells (IFN-γ^+TNF-α^+IL-2$^+$) [NCT 01683773] (Sheehan et al. 2015). The studies have shown the safety and immunogenicity of vaccine in healthy infants, previously vaccinated with BCG (Kagina et al. 2014). Aerosolized delivery of AERAS402 in rhesus macaques induced potent Ag85A/B-specific, cytokine-producing effector CD4 T cell and CD8 T cell response. Unfortunately, no enhanced survival or decrease in bacterial burden was observed (Darrah et al. 2014).

M. indicus pranii (MIP) is a progenitor of *M. avium* complex, sharing cross-reactive antigens with *Mtb* and *M. leprae*. Studies in mouse and guinea pigs have shown its prophylactic and therapeutic potential, resulting in a significant decrease in bacillary burden, reduced pathology, and enhanced survival (Singh et al. 2017; Gupta et al. 2012). Comparative genomic analysis of MIP has attributed its immunomodulatory properties to its high antigenic potential. Recently concluded multicentric clinical trial in category-II TB patients further confirmed its immunotherapeutic role in difficult to treat patients having advanced disease (Sharma et al. 2017). MIP was found to be safe without any adverse effects. Significantly higher number of patients in the MIP group showing sputum culture conversion, as early as 4 weeks after initiation of therapy, suggests elimination of detectable bacilli, which may help in shortening the treatment duration, indicating its role in the prevention of TB.

Although *Mtb* infects one-third of global population, only 5–10% develop active TB and 90–95% remain protected. This fact depicts that individuals infected with *Mtb* develop effective and enduring protective immunity, thereby suggesting that growing bacterium may produce intracellular antigens in the macrophage that may be crucial for protective immunity. Based on this rational, Sharma and Agrewala developed a vaccine by culturing live *Mtb* in the macrophages to an extent so that the bacilli release sufficient antigens. The mycobacterium was then killed by anti-TB drugs and the preparation was made safe by gamma irradiation. The cells were adoptively transferred in the *Mtb*-exposed animals. The results showed optimum immunity and better protection than BCG (Sharma and Agrewala 2004). In future, macrophage-based vaccine may generate same results as noticed in the case of DCs used for successful treatment of cancer (Sabado and Bhardwaj 2010).

21.10 Peptide Vaccines

Peptide vaccines consist of epitopes capable of inducing both T cell- and B cell-mediated immune responses. Peptides used in these vaccines are generally 20–30 amino acid sequences that are synthesized to form an immunogenic peptide molecule representing the specific epitope of an antigen. The structural and proteolytic stability of the synthetic peptides is further improved by linking them to adjuvants that subsequently act as activators of the immune system. Peptides give an advantage of selecting immunodominant CD4 T cell and CD8 T cell epitopes of the pathogen and avoiding moieties that can provoke autoreactivity/autoimmunity. Peptides are synthetic in nature and do not possess infectious material. Therefore, there is no possibility of reverting to pathogenic status, a problem that is usually associated with attenuated or mutated strains of pathogens. Conversely, peptide vaccines are often weakly immunogenic and therefore require adjuvant for their immunogenicity. Manufacturing of peptide vaccines is generally a safe and cost-effective option, when compared to conventional vaccines. There are many peptide vaccines under developmental stage for multiple diseases such as HIV, hepatitis C virus (HCV), malaria, swine fever, influenza, anthrax, human papilloma virus (HPV), diabetes mellitus, Alzheimer's disease, and therapeutic anticancer vaccines. Majority of candidate peptide vaccines are under phase I (270 studies) and phase II (224 studies) stage of development. For instance, phase I clinical trials with GP2, a HER2/neu-derived peptide, demonstrated that GP2-based vaccines are safe and effective in stimulating peptide-specific immunity. The GP2 peptide vaccine is currently being evaluated in a phase II trial against breast cancer (Clive et al. 2012). Only 12 studies have progressed to phase III clinical stage and are being indicated for treatment of multiple types of cancers (Li et al. 2014). However, the commercialization of a successful peptide vaccine is still an unfulfilled dream.

21.11 Lipidated Promiscuous Peptide Vaccine (L91)

As mentioned above, there is an urgent demand for a radically different vaccine for TB-endemic population, where BCG has totally failed (Datta et al. 1999). Thus, a successful vaccine should meet the following attributes: (i) does not require extensive antigen processing, (ii) not affected by the presence of preexisting antibodies/immunity against mycobacteria or BCG, (iii) composed of appropriate CD4 and CD8 T cell epitopes, and (iv) deliver suitable stimulatory signals for elicitation of innate and adaptive immunity. Keeping these points in mind, Agrewala et al. synthesized a chimeric lipidated peptide vaccine comprising of MHC-I and MHC-II promiscuous epitopes selected from latent, early, and chronic stages of *Mtb* conjugated to TLR-2 agonist Pam2Cys. This is a totally synthetic vaccine and therefore can be used in AIDS and immunocompromised subjects. Most of the peptides are poor immunogens. Therefore, the peptides were conjugated to Pam2Cys to make it more immunogenic.

A peptide vaccine can be an effective and successful stratagem in TB-endemic regions, since peptides do not require extensive antigen processing, as they can directly bind to MHC molecules for being presented to T cell for their activation, proliferation, and differentiation. Lipidated peptide vaccine has self-adjuvanting property due to the presence of Pam2Cys and therefore does not require any exogenous adjuvant. Pam2Cys binds to TLR-2 and induces local inflammation and generates long-lasting protective T cell immunity against *Mtb*. The construct can be targeted to DCs through TLR-2, which are copiously expressed on DCs. DCs are the only APCs, capable of activating naïve T cells. Signaling through TLR2 triggers robust release of cytokines like IL-12, which induces protective Th1 immunity. In addition, it stimulates nitric oxide (NO) release by macrophages that limit the intracellular survival of *Mtb* (Thoma-Uszynski et al. 2001). Further, peptides have no fear of obstruction in their processing by NTMs or antigen released by *Mtb* present in latency inside the macrophages. Hence, lipidated promiscuous peptides do not entail extensive processing and overcome the blocking effect of preexisting antibodies in the sera of individuals of TB-endemic area. Furthermore, inclusion of non-cross-reactive epitopes helps to evade the inappropriate immune responses such as autoimmunity through cross-reactivity of pathogenic components with the host proteins (Gowthaman et al. 2012). In human, HLA polymorphism renders majority of peptides ineffective because they bind to only one or two HLA alleles. On the other hand, promiscuous peptides permissively bind to several HLA alleles, therefore overcoming HLA polymorphism. Promiscuous CD4 T cell epitopes were identified from 16 kDa antigen (α-crystalline protein) of *Mtb* that provokes strong Th1 immune response after being tested using PBMCs of individuals with diverse HLA alleles (Agrewala et al. 1999). In addition, CD8 T cell epitopes were selected from TB10.4 antigen of *Mtb*. Vaccination with the construct elicits enduring memory T cell response and provides significantly better protection than BCG against *Mtb*. Furthermore, the chimera works as a prophylactic and therapeutic vaccine and reduces the dose and duration of TB drug regimen (Rai et al. 2016). These findings suggest a potent role of lipidated peptide vaccines in the prevention of disease, particularly in TB-endemic areas.

21.12 TB Vaccine Candidates in Clinical Trials

The status of new TB vaccines in pipeline in various stages of development, as published by WHO in August 2017, is shown in Fig. 21.5. There are 12 vaccines in phase I, II, or III trials. Their main characteristics are summarized below. Ad5 Ag85A is a human recombinant replication-deficient adenovirus serotype 5 vaccine vector expressing Ag85A (fbpA, Rv3804c, secreted antigen 85A, mycolyltransferase 85A), an immunodominant antigen produced by all mycobacterial species. The vaccine was evaluated in a phase I trial in 12 BCG-naive and 12 previously BCG-immunized, healthy Canadian adults, which demonstrated no vaccine-related serious adverse events. The study showed enhanced CD4 T cell and CD8 T cell immunity in previously BCG-vaccinated individuals compared to BCG-naive volunteers, supporting its further clinical development as a booster vaccine after BCG priming (Smaill and Xing 2014). The vaccine also showed safe, immunogenic, and enhanced protection against *Mtb* challenge in murine, bovine, and guinea pig models [NCT00800670]. BCG-primed guinea pigs when intranasally boosted with AdAg85A showed significant enhancement in their survival rate following pulmonary *Mtb* challenge (Xing et al. 2009). A safety and immunogenicity study of this vaccine by aerosol route has begun in BCG-vaccinated healthy volunteers.

MTBVAC is a live attenuated *Mtb* strain with deleted phoP and fadD26 genes. PhoP is a transcription factor that controls 2% of the genome of *Mtb* including production of immunomodulatory cell wall lipids and ESAT-6 secretion (Gonzalo-Asensio et al. 2008). Deletion of fadD26 leads to complete abrogation of synthesis of virulence surface lipids phthiocerol dimycocerosates (DIM) (Camacho et al. 1999). The main advantage of using attenuated live *Mtb* as vaccine is that many genetic regions encoding important immunodominant antigens that are absent in BCG are still present in attenuated *Mtb*. Moreover, chromosomal deletions in virulence genes provide an assurance for safety and genetic stability. The primary target population is newborns (as replacement for BCG) and secondary target includes adolescences (booster with BCG). MTBVAC exhibits safety and bio-distribution profiles similar

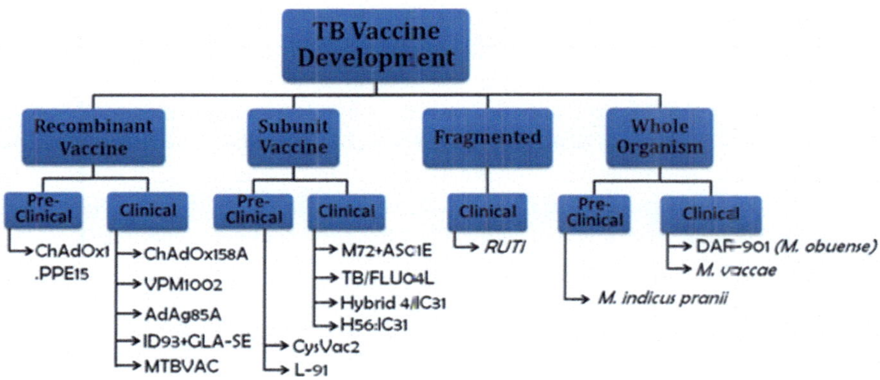

Fig. 21.5 Flowchart showing TB vaccines in pipeline

to BCG and confers superior protection in preclinical studies (Arbues et al. 2013). MTBVAC has been the first and only live attenuated *Mtb* vaccine approved to enter into clinical trials. A first-in-human MTBVAC clinical trial was recently conducted successfully in healthy adults in Lausanne (Switzerland) [NCT02013245]. In September 2015, MTBVAC entered phase Ib trial in infants under the South African Tuberculosis Vaccine Initiative (SATVI) program. Phase IIa trials in both the target populations are expected to start in 2018. Highly attenuated MTBVAC constructed with inactivation of an additional gene generated in exported repeated protein (Erp) could be a potential TB vaccine candidate for use in a high-risk immunosuppressive population (Solans et al. 2014).

ChAdOx185A-MVA85A is a simian adenovirus and MVA85A is a recombinant pox virus vaccine, expressing antigen 85A. These candidate vaccines aim to generate a joint heterologous prime-boost regimen delivered through both systemic and mucosal routes. MVA85A induces primarily CD4T cell responses and recombinant adenoviral vectors induce primarily CD8 T cell responses. Intranasal administration of ChAdOx1.85A induced strong immune responses in the lungs but failed to protect animals against aerosol *Mtb* challenge. In contrast, ChAdOx1.85A followed by MVA85A administered either mucosally or systemically induced strong immune responses and was able to improve the protective efficacy of BCG (Stylianou et al. 2015). Phase I trial of ChAdOx185A by i.m. route in BCG-vaccinated adults in UK, both alone and as part of a prime-boost strategy with MVA85A, has been completed [WHO 2017, NCT01829490].

As per WHO report 2017, there are currently following nine vaccines in phase II or phase III trials (Fig. 21.5). DAR-901 booster is a whole-cell vaccine, prepared from nontuberculous mycobacteria (*M. obuense*). It is developed at the Geisel School of Medicine, Dartmouth University, USA. DAR-901 represents broth-grown, scalable variant of SRL172 (agar-grown). SRL172 is the only TB vaccine which has showed reduction in the culture-confirmed *Mtb* in a phase III randomized clinical trials and was found to be safe, well-tolerated, and immunogenic in humans. A randomized controlled phase III trial in Tanzania showed protection against *Mtb* upon boosting with SRL172 in HIV-infected adults, who had received BCG at birth (Solans et al. 2014). A recent study showed that a DAR-901 booster is more effective at providing protection against *Mtb* than a BCG booster (Lahey et al. 2016).

Hybrid 4/IC31 (AERAS-404) contains a fusion protein of Ag85B and TB10.4, with IC31 as adjuvant. IC-31 is a two-component adjuvant comprising an 11-mer antibacterial peptide (KLK) and a synthetic oligodeoxynucleotide (ODN1a), a TLR-9 agonist (Aichinger et al. 2011). The vaccine candidate has been shown to be immunogenic and protective in a preclinical animal model of TB. It is known to induce persistent polyfunctional CD4 T cell responses in adults. A phase II study of its ability to prevent infection in healthy adults has been completed (NCT02075203). It is also in a phase I/II trial in infants.

H56:IC31 is a protein subunit vaccine comprised of fusion protein of Ag85B, ESAT-6 (two of the *Mtb* antigens secreted in the acute phase of infection), and the nutrient stress-induced antigen Rv2660c, formulated in IC31 adjuvant (WHO 2014).

In mice, H56:IC31 could control late-stage infection and reduced bacterial burden more efficiently than BCG, 24 weeks postinfection. It can work both as preventive and postexposure vaccine. As a prime-boost with BCG, H56:IC31 has delayed and reduced clinical disease in *Mtb*-challenged cynomolgus macaques and prevented reactivation of latent infection (Lin et al. 2012). The vaccine showed high immunogenicity in healthy adults with or without previous disease (Luabeya et al. 2015). H56:IC31 completed a phase I trial at SATVI, South Africa. The vaccine was safe and tolerated at all tested doses. It was immunogenic in adult pulmonary TB patients, who have successfully completed TB treatment [NCT02375698]. A phase II study is underway to evaluate the safety profile of H56:IC31 in HIV-uninfected, remotely BCG-vaccinated adolescents (NCT03265977).

ID93 + GLA-SE consists of four antigens of *Mtb* (Rv2608, Rv3619, and Rv3620, which are associated with virulence, and Rv1813, the latency antigen) in a novel glucopyranosyl lipid adjuvant-stable emulsion (GLA-SE; TLR4 agonist). This fusion protein-adjuvant combination induced antigen-specific polyfunctional CD4 Th1 cells (IFN-γ^+ TNF-α^+ IL-2^+) and reduced the number of *Mtb* in the lungs of mice and guinea pigs that were infected with virulent or multidrug-resistant *Mtb* strains. Further, the vaccine elicited polyfunctional effector CD4 T cell and CD8 T cell response in BCG-vaccinated or *Mtb*-exposed human PBMCs (Bertholet et al. 2010). GLA-SE is known to induce polyclonal Th1 response and signals through both MyD88 and TRIF in vivo (Orr et al. 2013). Synergistic interaction between MyD88 and TRIF is required for Th1 cell polarization with a synthetic TLR4 agonist. A phase IIa trial for this vaccine was completed in South Africa in HIV-negative TB patients that have recently finished treatment for pulmonary TB disease [NCT02465216]. A future phase IIb trial is being planned to investigate the prevention of recurrence of disease in the same population.

M72 + AS01E is a subunit vaccine, consisting of fusion protein of *Mtb* antigens *Mtb*32A (pepA, probable serine protease pepA, Rv0125) and *Mtb*39A (Rv1196, a PPE family protein) in adjuvant AS01E, which enhances cell-mediated immunogenicity. AS01E is an adjuvant in liposome formulation with monophosphoryl lipid A (MPL) and QS21 immunostimulants. Several phase IIa studies were completed and phase IIb studies are planned for this vaccine candidate. Safety and immunogenicity of M72 was tested in different populations, as a booster to BCG in the Gambian infants, in HIV-infected adults in combination with antiretroviral therapy in the Switzerland, and in healthy PPD-positive adults in the Philippines (Idoko et al. 2014; Thacher et al. 2014). A randomized phase II study was conducted for safety, reactogenicity, and immunogenicity of M72 + AS01E in three cohorts: (i) TB-naïve adults, (ii) adults previously treated for TB, and (iii) adults who have completed the intensive phase of TB treatment. In addition, phase II trials on safety and immunogenicity of M72 + AS01 were also conducted on adults with TB and AIDS patients [NCT01755598]. The study ended prematurely because of high incidence of redness/swelling reactions at the site of inoculation (Gillard et al. 2016).

VPM1002 is a recombinant BCG vaccine developed at the Max Planck Institute of Infection Biology (Grode et al. 2013; Nieuwenhuizen et al. 2017). Phase I study confirmed the safety and efficacy of the vaccine candidate. The vaccine was safe and well tolerated in healthy volunteers and led to the induction of multifunctional CD4 T cell and CD8 T cell immune response [NCT01113281; NCT00749034]. Phase II study to evaluate safety and immunogenicity of VPM 1002 in comparison with BCG in newborn infants in South Africa was found to be as safe and well tolerated as BCG [NCT01479972]. VPM1002 has teamed up with Serum Institute of India Pvt. Ltd. (SIIPL). Together they are conducting a phase II trial in South Africa to evaluate efficacy and safety in HIV-exposed and HIV-unexposed infants. In addition, a phase III trial is being planned in India to assess the potential of VPM1002 as a postexposure vaccine in prevention of TB recurrence after successful drug therapy.

RUTI is a polyantigenic liposomal vaccine made of detoxified, fragmented *Mtb* cells (FCMtb). It facilitates a balanced Th1/Th2 response to a wide range of antigens. Standard treatment for latent TB infection (LTBI) requires isoniazid (INH) from 6 to 9 months, which presents important compliance problems. RUTI was developed to generate a polyantigenic immune response and to reduce the treatment duration of latent TB (Cardona and Amat 2006). Immunization with RUTI vaccine further pronounced the decrease in the colony-forming units, as compared to treatment with rifampicin and isoniazid (Cardona et al. 2005). Treatment with RUTI showed no signs of toxicity following chemotherapy in mice and guinea pigs. Phase I studies showed dose-dependent local adverse reactions and elevated the T cell response in healthy volunteers (Vilaplana et al. 2010). In another study, RUTI was shown to have potential for both prophylaxis and immunotherapy for TB infections (Vilaplana et al. 2011). A phase II clinical trial demonstrated reasonable tolerability of RUTI at three different doses with the induction of local abscess in HIV-infected and HIV-noninfected individuals (Nell et al. 2014). RUTI induced stronger activation of IFN-γ-secreting CD4 T cells and CD8 T cells against tuberculin purified protein derivative, ESAT6 and Ag85B (Guirado et al. 2008). The main target for RUTI is MDR-TB and it is being implemented in phase IIa study in patients with MDR-TB.

TB/FLU-04L is a mucosal vectored vaccine based on an attenuated replication-deficient influenza virus vector expressing antigens Ag85A and ESAT-6 and influenza virus strain A/Puetro Rico/8/34 H1N1. It was designed as a prophylactic boost vaccine for infants, adolescents, and adults. It has completed a phase I study [NCT02501421] and a phase IIa trial in people with LTBI is being planned for this vaccine (WHO 2017).

21.13 Vaccines in Phase III Clinical Trials

Some of the vaccines against TB have reached phase III clinical trials. *M. vaccae* is a whole-cell-based heat-killed immunotherapeutic vaccine. It consists of heat-inactivated *M. vaccae*. It is the only therapeutic vaccine recommended for TB immunotherapy by the WHO (2017; Stanford et al. 2004). It is licensed by the China Food and Drug Administration as an immunotherapeutic agent to help shorten the treatment for patients with drug-susceptible TB. A phase III trial to study efficacy and safety in China found vaccine to be well tolerated with no serious adverse effects (WHO 2014). Heat-killed *M. vaccae* when added to chemotherapy of TB patients showed improved sputum conversion and x-ray appearances. Immunostimulating effects of *M. vaccae* were attributed to their cell wall, which elicit more production of Th1 cytokines. Phase III trial to assess its immunogenicity and safety in preventing latent TB is completed. It is the largest TB vaccine clinical trial in the past decade, involving 10,000 people aged 15–65 years. The MIP was safe when used as a therapeutic vaccine in combination with drugs in sputum-spear positive pulmonary TB patients (Sharma et al. 2017).

21.14 Summary and Future Prospects

Vaccination is the most effective and economically viable strategy to improve public health. The use of vaccine has saved millions of lives. Some of the diseases such as smallpox, polio, and measles have almost been eradicated through vaccination. Unfortunately, this could not be achieved in the case of TB. Tremendous efforts have been made by scientific community world-wide in understanding the protective efficacy of vaccines. A reliable vaccine that can eliminate TB globally is yet far from reality. The appearance of drug-resistant strains of *Mtb* is a stern warning to further expand our horizon of understanding molecular mechanism of how 85–95% of *Mtb*-infected individuals remains protected throughout their life and only 5–15% develop active TB. Further, it is quite crucial to deeply evaluate the sequence of events being generated during host-pathogen interaction. More importantly, the variable protective efficacy of BCG and its failure in TB-endemic areas create a more complex situation for the development of universal vaccine. Therefore, it demands additional knowledge of interfering factors such as NTMs, LTBI, and helminths that affect the host immunity against pathogens.

Recently gut microbiota, environmental and nutritional factors have been suggested to contribute substantially in preventing or provoking diseases (Sathyabama et al. 2014; Khan et al. 2016; Feng et al. 2018). Furthermore, in TB-endemic population, prior immunity developed due to exposure of NTMs and LTBI interferes with antigen processing and presentation of BCG. Thereby, the immune system fails to optimally expoilt BCG antigens to elicit long-lasting memory T cells. Consequently, a vaccine that does not require extensive antigen processing and induces enduring memory T cell response would be an ideal vaccine for TB-endemic world. To gain an in-depth understanding of disease pathogenesis, a rationalized vaccine should be multi-epitope, composed of promiscuous CD4 T cell and CD8 T cell epitopes from latent, active, chronic, and drug-resistant forms of *Mtb*. The vaccine should bear an adjuvanting moiety through which it can be targeted to APCs, in particular DCs. Based on these justifications, peptide vaccines may be quite successful in the future against TB for endemic and non-endemic inhabitants.

References

Abdallah AM, Hill-Cawthorne GA, Otto TD, Coll F, Guerra-Assunção JA, Gao G, Naeem R, Ansari H, Malas TB, Adroub SA, Verboom T (2015) Genomic expression catalogue of a global collection of BCG vaccine strains show evidence for highly diverged metabolic and cell-wall adaptations. Sci Rep 5:15443

Agrewala JN, Deacock S, Jurcevic S, Wilkinson R (1999) Peptide recognition by T-cell clones of an HLA-DRB11501/0901 heterozygous donor is promiscuous only between parental alleles. Hum Immunol 55:34–38

Aguilo N, Toledo AM, Lopez-Roman EM, Perez-Herran E, Gormley E, Rullas-Trincado J, Angulo-Barturen I, Martin C (2014) Pulmonary Mycobacterium bovis BCG vaccination confers dose-dependent superior protection compared to that of subcutaneous vaccination. Clin Vaccine Immunol 21:594–597

Aichinger MC, Ginzler M, Weghuber J, Zimmermann L, Riedl K, Schütz G, Nagy E, Von Gabain A, Schweyen R, Henics T (2011) Adjuvating the adjuvant: facilitated delivery of an immunomodulatory oligonucleotide to TLR9 by a cationic antimicrobial peptide in dendritic cells. Vaccine 29:426–436

Albert ML, Sauter B, Bhardwaj N (1998) Dendritic cells acquire antigen from apoptotic cells and induce class I-restricted CTLs. Nature 392:86–89

Andersen P, Doherty TM (2005) The success and failure of BCG – implications for a novel tuberculosis vaccine. Nat Rev Microbiol 3:656–662

Arbues A, Aguilo JI, Gonzalo-Asensio J, Marinova D, Uranga S, Puentes E, Fernandez C, Parra A, Cardona PJ, Vilaplana C, Ausina V (2013) Construction, characterization and preclinical evaluation of MTBVAC, the first live-attenuated M. tuberculosis-based vaccine to enter clinical trials. Vaccine 31:4867–4873

Aronson NE, Santosham M, Comstock GW, Howard RS, Moulton LH, Rhoades ER, Harrison LH (2004) Long-term efficacy of BCG vaccine in American Indians and Alaska Natives: a 60-year follow-up study. JAMA 291:2086–2091

Ashenafi S, Aderaye G, Bekele A, Zewdie M, Aseffa G, Hoang AT, Carow B, Habtamu M, Wijkander M, Rottenberg M, Aseffa A (2014) Progression of clinical tuberculosis is associated with a Th2 immune response signature in combination with elevated levels of SOCS3. Clin Immunol 151:84–99

Belisle JT, Vissa VD, Sievert T, Takayama K, Brennan PJ, Besra GS (1997) Role of the major antigen of Mycobacterium tuberculosis in cell wall biogenesis. Science 276:1420–1422

Bertholet S, Ireton GC, Ordway DJ, Windish HP, Pine SO, Kahn M, Phan T, Orme IM, Vedvick TS, Baldwin SL, Coler RN (2010) A defined tuberculosis vaccine candidate boosts BCG and protects against multidrug-resistant Mycobacterium tuberculosis. Sci Transl Med 2:53ra74

Biering-Sørensen S, Jensen KJ, Aamand SH, Blok B, Andersen A, Monteiro I, Netea MG, Aaby P, Benn CS, Hasløv KR (2015) Variation of growth in the production of the BCG vaccine and the association with the immune response. An observational study within a randomised trial. Vaccine 33:2056–2065

Black GF, Weir RE, Floyd S, Bliss L, Warndorff DK, Crampin AC, Ngwira B, Sichali L, Nazareth B, Blackwell JM, Branson K (2002) BCG-induced increase in interferon-gamma response to mycobacterial antigens and efficacy of BCG vaccination in Malawi and the UK: two randomised controlled studies. Lancet 359:1393–1401

Bouneaud C, Garcia Z, Kourilsky P, Pannetier C (2005) Lineage relationships, homeostasis, and recall capacities of central- and effector-memory CD8 T cells in vivo. J Exp Med 201:579–590

Brosch R, Gordon SV, Pym A, Eiglmeier K, Garnier T, Cole ST (2000) Comparative genomics of the mycobacteria. Int J Med Microbiol 290:143–152

Brosch R, Gordon SV, Garnier T, Eiglmeier K, Frigui W, Valenti P, Dos Santos S, Duthoy S, Lacroix C, Garcia-Pelayo C, Inwald JK (2007) Genome plasticity of BCG and impact on vaccine efficacy. Proc Natl Acad Sci 104:5596–5601

Bugarcic A, Hitchens K, Beckhouse AG, Wells CA, Ashman RB, Blanchard H (2008) Human and mouse macrophage-inducible C-type lectin (Mincle) bind Candida albicans. Glycobiology 18:679–685

Camacho LR, Ensergueix D, Perez E, Gicquel B, Guilhot C (1999) Identification of a virulence gene cluster of Mycobacterium tuberculosis by signature-tagged transposon mutagenesis. Mol Microbiol 34:257–267

Cardona PJ, Amat I (2006) Origin and development of RUTI, a new therapeutic vaccine against Mycobacterium tuberculosis infection. Archivos de Bronconeumología (English Edition) 42:25–32

Cardona PJ, Amat I, Gordillo S, Arcos V, Guirado E, Díaz J, Vilaplana C, Tapia G, Ausina V (2005) Immunotherapy with fragmented Mycobacterium tuberculosis cells increases the effectiveness of chemotherapy against a chronical infection in a murine model of tuberculosis. Vaccine 23:1393–1398

Carvalho EM, Da Fonseca Porto A (2004) Epidemiological and clinical interaction between HTLV-1 and Strongyloides stercoralis. Parasite Immunol 26:487–497

Chattergoon MA, Kim JJ, Yang JS, Robinson TM, Lee DJ, Dentchev T, Wilson DM, Ayyavoo V, Weiner DB (2000) Targeted antigen delivery to antigen-presenting cells including dendritic cells by engineered Fas-mediated apoptosis. Nat Biotechnol 18:974–979

Chiang CY, Riley LW (2005) Exogenous reinfection in tuberculosis. Lancet Infect Dis 5:629–636

Churchyard GJ, Snowden MA, Hokey D, Dheenadhayalan V, McClain JB, Douoguih M, Pau MG, Sadoff J, Landry B (2015) The safety and immunogenicity of an adenovirus type 35-vectored TB vaccine in HIV-infected, BCG-vaccinated adults with CD4(+) T cell counts >350 cells/mm. Vaccine 33:1890–1896

Clive KS, Tyler JA, Clifton GT, Holmes JP, Ponniah S, Peoples GE, Mittendorf EA (2012) The GP2 peptide: a HER2/neu-based breast cancer vaccine. J Surg Oncol 105:452–458

Counoupas C, Pinto R, Nagalingam G, Hill-Cawthorne GA, Feng CG, Britton WJ, Triccas JA (2016) Mycobacterium tuberculosis components expressed during chronic infection of the lung contribute to long-term control of pulmonary tuberculosis in mice. NPJ Vaccines 1:16012

Dark WD (1978) A 'denturist' is born. Dent Manage 18:57–61

Darrah PA, Bolton DL, Lackner AA, Kaushal D, Aye PP, Mehra S, Blanchard JL, Didier PJ, Roy CJ, Rao SS, Hokey DA (2014) Aerosol vaccination with AERAS-402 elicits robust cellular immune responses in the lungs of rhesus macaques but fails to protect against high-dose Mycobacterium tuberculosis challenge. J Immunol 193:1799–1811

Datta M, Vallishayee SR, Diwakara SA (1999) Fifteen year follow up of trial of BCG vaccines in South India for tuberculosis prevention. Indian J Med Res 110:56–69

de Koning HD, Simon A, Zeeuwen PL, Schalkwijk J (2012) Pattern recognition receptors in infectious skin diseases. Microbes Infect 14:881–893

Deng YH, He HY, Zhang BS (2014) Evaluation of protective efficacy conferred by a recombinant Mycobacterium bovis BCG expressing a fusion protein of Ag85A-ESAT-6. J Microbiol Immunol Infect 47:48–56

Desel C, Dorhoi A, Bandermann S, Grode L, Eisele B, Kaufmann SH (2011) Recombinant BCG Δ ure C hly+ induces superior protection over parental BCG by stimulating a balanced combination of type 1 and type 17 cytokine responses. J Infect Dis 204:1573–1584

Dey B, Jain R, Khera A, Rao V, Dhar N, Gupta UD, Katoch VM, Ramanathan VD, Tyagi AK (2009) Boosting with a DNA vaccine expressing ESAT-6 (DNAE6) obliterates the protection imparted by recombinant BCG (rBCGE6) against aerosol Mycobacterium tuberculosis infection in guinea pigs. Vaccine 28:63–70

Dieye TN, NDiaye BP, Dieng AB, Fall M, Britain N, Vermaak S, Camara M, Diop-Ndiaye H, Ngom-Gueye NF, Diaw PA, Toure-Kane C (2013) Two doses of candidate TB vaccine MVA85A in antiretroviral therapy (ART) naive subjects gives comparable immunogenicity to one dose in ART+ subjects. PLoS One 8:e67177

Dockrell HM, Smith SG (2017) What have we learnt about BCG vaccination in the last 20 years? Front Immunol 8:1134

El-Zein M, Conus F, Benedetti A, Menzies D, Parent ME, Rousseau MC (2017) Association between Bacillus Calmette-Guerin vaccination and childhood asthma in the Quebec birth cohort on immunity and health. Am J Epidemiol 186:344–355

Elias D, Britton S, Kassu A, Akuffo H (2007) Chronic helminth infections may negatively influence immunity against tuberculosis and other diseases of public health importance. Expert Rev Anti-Infect Ther 5:475–484

Elias D, Britton S, Aseffa A, Engers H, Akuffo H (2008) Poor immunogenicity of BCG in helminth infected population is associated with increased in vitro TGF-beta production. Vaccine 26:3897–3902

Feng Q, Chen WD, Wang YD (2018) Gut microbiota: an integral moderator in health and disease. Front Microbiol 9:151

Fine PE, Floyd S, Stanford JL, Nkhosa P, Kasunga A, Chaguluka S, Warndorff DK, Jenkins PA, Yates M, Ponnighaus JM (2001) Environmental mycobacteria in northern Malawi: implications for the epidemiology of tuberculosis and leprosy. Epidemiol Infect 126:379–387

Flaherty DK, Vesosky B, Beamer GL, Stromberg P, Turner J (2006) Exposure to Mycobacterium avium can modulate established immunity against Mycobacterium tuberculosis infection generated by Mycobacterium bovis BCG vaccination. J Leukoc Biol 80:1262–1271

Flynn JL, Goldstein MM, Triebold KJ, Koller BB, Bloom BR (1992) Major histocompatibility complex class I-restricted T cells are required for resistance to Mycobacterium tuberculosis infection. Proc Natl Acad Sci 89:12013–12017

Flynn JL, Chan J, Triebold KJ, Dalton DK, Stewart TA, Bloom BR (1993) An essential role for interferon gamma in resistance to Mycobacterium tuberculosis infection. J Exp Med 178:2249–2254

Fuge O, Vasdev N, Allchorne P, Green JS (2015) Immunotherapy for bladder cancer. Res Rep Urol 7:65–79

Gillard P, Yang PC, Danilovits M, Su WJ, Cheng SL, Pehme L, Bollaerts A, Jongert E, Moris P, Ofori-Anyinam O, Demoitié MA (2016) Safety and immunogenicity of the M72/AS01E candidate tuberculosis vaccine in adults with tuberculosis: a phase II randomised study. Tuberculosis 100:118–127

Gonzalo-Asensio J, Mostowy S, Harders-Westerveen J, Huygen K, Hernández-Pando R, Thole J, Behr M, Gicquel B, Martín C (2008) PhoP: a missing piece in the intricate puzzle of Mycobacterium tuberculosis virulence. PLoS One 3:e3496

Gowthaman U, Rai PK, Khan N, Jackson DC, Agrewala JN (2012) Lipidated promiscuous peptides vaccine for tuberculosis-endemic regions. Trends Mol Med 18:607–614

Grode L, Ganoza CA, Brohm C, Weiner J 3rd, Eisele B, Kaufmann SH (2013) Safety and immunogenicity of the recombinant BCG vaccine VPM1002 in a phase 1 open-label randomized clinical trial. Vaccine 31:1340–1348

Guirado E, Gil O, Cáceres N, Singh M, Vilaplana C, Cardona PJ (2008) Induction of a specific strong polyantigenic cellular immune response after short-term chemotherapy controls bacillary reactivation in murine and Guinea pig experimental models of tuberculosis. Clin Vaccine Immunol 15:1229–1237

Gupta A, Ahmad FJ, Ahmad F, Gupta UD, Natarajan M, Katoch VM, Bhaskar S (2012) Protective efficacy of Mycobacterium indicus pranii against tuberculosis and underlying local lung immune responses in guinea pig model. Vaccine 30:6198–6209

Hatano S, Tamura T, Umemura M, Matsuzaki G, Ohara N, Yoshikai Y (2016) Recombinant Mycobacterium bovis bacillus Calmette-Guerin expressing Ag85B-IL-7 fusion protein enhances IL-17A-producing innate gammadelta T cells. Vaccine 34:2490–2495

Henao-Tamayo MI, Ordway DJ, Irwin SM, Shang S, Shanley C, Orme IM (2010) Phenotypic definition of effector and memory T-lymphocyte subsets in mice chronically infected with Mycobacterium tuberculosis. Clin Vaccine Immunol 17:618–625

Ho P, Wei X, Seah GT (2010) Regulatory T cells induced by Mycobacterium chelonae sensitization influence murine responses to bacille Calmette-Guerin. J Leukoc Biol 88:1073–1080

Hoft DF, Blazevic A, Abate G, Hanekom WA, Kaplan G, Soler JH, Weichold F, Geiter L, Sadoff JC, Horwitz MA (2008) A new recombinant bacille Calmette-Guerin vaccine safely induces significantly enhanced tuberculosis-specific immunity in human volunteers. J Infect Dis 198:1491–1501

Horwitz MA, Harth G, Dillon BJ, Masleša-Galić S (2000) Recombinant Bacillus Calmette-Guerin (BCG) vaccines expressing the Mycobacterium tuberculosis 30-kDa major secretory protein induce greater protective immunity against tuberculosis than conventional BCG vaccines in a highly susceptible animal model. Proc Natl Acad Sci 97:13853–13858

Idoko OT, Owolabi OA, Owiafe PK, Moris P, Odutola A, Bollaerts A, Ogundare E, Jongert E, Demoitié MA, Ofori-Anyinam O, Ota MO (2014) Safety and immunogenicity of the M72/AS01 candidate tuberculosis vaccine when given as a booster to BCG in Gambian infants: an open-label randomized controlled trial. Tuberculosis 94:564–578

Jain R, Dey B, Dhar N, Rao V, Singh R, Gupta UD, Katoch VM, Ramanathan VD, Tyagi AK (2008) Enhanced and enduring protection against tuberculosis by recombinant BCG-Ag85C and its association with modulation of cytokine profile in lung. PLoS One 3:e3869

Kagina BM, Tameris MD, Geldenhuys H, Hatherill M, Abel B, Hussey GD, Scriba TJ, Mahomed H, Sadoff JC, Hanekom WA, Mansoor N (2014) The novel tuberculosis vaccine, AERAS-402, is safe in healthy infants previously vaccinated with BCG, and induces dose-dependent CD4 and CD8T cell responses. Vaccine 32:5908–5917

Khader SA, Cooper AM (2008) IL-23 and IL-17 in tuberculosis. Cytokine 41:79–83

Khan N, Vidyarthi A, Nadeem S, Negi S, Nair G, Agrewala JN (2016) Alteration in the gut microbiota provokes susceptibility to tuberculosis. Front Immunol 7:529

Kong CU, Ng LG, Nambiar JK, Spratt JM, Weninger W, Triccas JA (2011) Targeted induction of antigen expression within dendritic cells modulates antigen-specific immunity afforded by recombinant BCG. Vaccine 29:1374–1381

Kupper TS, Fuhlbrigge RC (2004) Immune surveillance in the skin: mechanisms and clinical consequences. Nat Rev Immunol 4:211–222

Lagranderie M, Guyonvarc'h PM (2014) The interplay between bacillus Calmette-Guerin and Treg cells and its role to prevent or cure inflammatory diseases. Expert Rev Clin Immunol 10:741–745

Lahey T, Laddy D, Hill K, Schaeffer J, Hogg A, Keeble J, Dagg B, Ho MM, Arbeit RD, von Reyn CF (2016) Immunogenicity and protective efficacy of the DAR-901 booster vaccine in a murine model of tuberculosis. PLoS One 11:e0168521

Li L, Qiao D, Zhang X, Liu Z, Wu C (2011) The immune responses of central and effector memory BCG-specific CD4+ T cells in BCG-vaccinated PPD+ donors were modulated by Treg cells. Immunobiology 216:477–484

Li W, Joshi M, Singhania S, Ramsey K, Murthy A (2014) Peptide vaccine: progress and challenges. Vaccine 2:515–536

Lin PL, Dietrich J, Tan E, Abalos RM, Burgos J, Bigbee C, Bigbee M, Milk L, Gideon HP, Rodgers M, Cochran C, Guinn KM, Sherman DR, Klein E, Janssen C, Flynn JL, Andersen P (2012) The multistage vaccine H56 boosts the effects of BCG to protect cynomolgus macaques against active tuberculosis and reactivation of latent Mycobacterium tuberculosis infection. J Clin Invest 122:303–314

Lozes E, Huygen K, Content J, Denis O, Montgomery DL, Yawman AM, Vandenbussche P, Van Vooren JP, Drowart A, Ulmer JB, Liu MA (1997) Immunogenicity and efficacy of a tuberculosis DNA vaccine encoding the components of the secreted antigen 85 complex. Vaccine 15:830–833

Lu Y, Xu Y, Yang E, Wang C, Wang H, Shen H (2012) Novel recombinant BCG coexpressing Ag85B, ESAT-6 and Rv2608 elicits significantly enhanced cellular immune and antibody responses in C57BL/6 mice. Scand J Immunol 76:271–277

Luabeya AK, Kagina BM, Tameris MD, Geldenhuys H, Hoff ST, Shi Z, Kromann I, Hatherill M, Mahomed H, Hanekom WA, Andersen P (2015) First-in-human trial of the post-exposure tuberculosis vaccine H56:IC31 in Mycobacterium tuberculosis infected and non-infected healthy adults. Vaccine 33:4130–4140

Luca S, Mihaescu T (2013) History of BCG vaccine. Maedica (Buchar) 8:53–58

McShane H, Pathan AA, Sander CR, Goonetilleke NP, Fletcher HA, Hill AV (2005) Boosting BCG with MVA85A: the first candidate subunit vaccine for tuberculosis in clinical trials. Tuberculosis 85:47–52

Murray RA, Mansoor N, Harbacheuski R, Soler J, Davids V, Soares A, Hawkridge A, Hussey GD, Maecker H, Kaplan G, Hanekom WA (2006) Bacillus Calmette Guerin vaccination of human newborns induces a specific, functional CD8+ T cell response. J Immunol 177:5647–5651

Nanjappa SG, Walent JH, Morre M, Suresh M (2008) Effects of IL-7 on memory CD8 T cell homeostasis are influenced by the timing of therapy in mice. J Clin Invest 118:1027–1039

Nell AS, D'lom E, Bouic P, Sabaté M, Bosser R, Picas J, Amat M, Churchyard G, Cardona PJ (2014) Safety, tolerability, and immunogenicity of the novel antituberculous vaccine RUTI: randomized, placebo-controlled phase II clinical trial in patients with latent tuberculosis infection. PLoS One 9:e89612

Nguipdop-Djomo P, Heldal E, Rodrigues LC, Abubakar I, Mangtani P (2016) Duration of BCG protection against tuberculosis and change in effectiveness with time since vaccination in Norway: a retrospective population-based cohort study. Lancet Infect Dis 16:219–226

Nieuwenhuizen NE, Kulkarni PS, Shaligram U, Cotton MF, Rentsch CA, Eisele B, Grode L, Kaufmann SH (2017) The recombinant Bacille Calmette-Guerin vaccine VPM1002: ready for clinical efficacy testing. Front Immunol 8:1147

Orege PA, Fine PE, Lucas SB, Obura M, Okelo C, Okuku P (1993) Case-control study of BCG vaccination as a risk factor for leprosy and tuberculosis in western Kenya. Int J Lepr Other Mycobact Dis 61:542–549

Orr MT, Duthie MS, Windish HP, Lucas EA, Guderian JA, Hudson TE, Shaverdian N, O'Donnell J, Desbien AL, Reed SG, Coler RN (2013) MyD88 and TRIF synergistic interaction is required for TH1-cell polarization with a synthetic TLR4 agonist adjuvant. Eur J Immunol 43:2398–2408

Osborn TW (1983) Changes in BCG strains. Tubercle 64:1–13

Palmer CE, Long MW (1966) Effects of infection with atypical mycobacteria on BCG vaccination and tuberculosis. Am Rev Respir Dis 94:553–568

Plotkin S (2014) History of vaccination. Proc Natl Acad Sci 111:12283–12287

Plotkin SA, Plotkin SL (2011) The development of vaccines: how the past led to the future. Nat Rev Microbiol 9:889–893

Portnoy DA, Tweten RK, Kehoe M, Bielecki J (1992) Capacity of listeriolysin O, streptolysin O, and perfringolysin O to mediate growth of Bacillus subtilis within mammalian cells. Infect Immun 60:2710–2717

Power CA, Wei G, Bretscher PA (1998) Mycobacterial dose defines the Th1/Th2 nature of the immune response independently of whether immunization is administered by the intravenous, subcutaneous, or intradermal route. Infect Immun 66:5743–5750

Pym AS, Brodin P, Majlessi L, Brosch R, Demangel C, Williams A, Griffiths KE, Marchal G, Leclerc C, Cole S (2003) Recombinant BCG exporting ESAT-6 confers enhanced protection against tuberculosis. Nat Med 9:533–539

Qie YQ, Wang JL, Liu W, Shen H, Chen JZ, Zhu BD, Xu Y, Zhang XL, Wang HH (2009) More vaccine efficacy studies on the recombinant Bacille Calmette-Guerin co-expressing Ag85B, Mpt64190–198 and Mtb8. 4. Scand J Immunol 69:342–350

Rahman S, Magalhaes I, Rahman J, Ahmed RK, Sizemore DR, Scanga CA, Weichold F, Verreck F, Kondova I, Sadoff J, Thorstensson R (2012) Prime-boost vaccination with rBCG/rAd35 enhances CD8+ cytolytic T-cell responses in lesions from Mycobacterium tuberculosis-infected primates. Mol Med 18:647

Rai PK, Chodisetti SB, Nadeem S, Maurya SK, Gowthaman U, Zeng W, Janmeja AK, Jackson DC, Agrewala JN (2016) A novel therapeutic strategy of lipidated promiscuous peptide against Mycobacterium tuberculosis by eliciting Th1 and Th17 immunity of host. Sci Rep 6:23917

Ritz N, Curtis N (2009) Mapping the global use of different BCG vaccine strains. Tuberculosis 89:248–251

Ritz N, Dutta B, Donath S, Casalaz D, Connell TG, Tebruegge M, Robins-Browne R, Hanekom WA, Britton WJ, Curtis N (2012) The influence of Bacille Calmette-Guerin vaccine strain on the immune response against tuberculosis: a randomized trial. Am J Respir Crit Care Med 185:213–222

Rook GA, Hernandez-Pando R, Dheda K, Seah GT (2004) IL-4 in tuberculosis: implications for vaccine design. Trends Immunol 25:483–488

Sabado RL, Bhardwaj N (2010) Directing dendritic cell immunotherapy towards successful cancer treatment. Immunotherapy 2:37–56

Sathyabama S, Khan N, Agrewala JN (2014) Friendly pathogens: prevent or provoke autoimmunity. Crit Rev Microbiol 40:273–280

Satti I, Meyer J, Harris SA, Thomas ZR, Griffiths K, Antrobus RD, Rowland R, Ramon RL, Smith M, Sheehan S, Bettinson H (2014) Safety and immunogenicity of a candidate tuberculosis vaccine MVA85A delivered by aerosol in BCG-vaccinated healthy adults: a phase 1, double-blind, randomised controlled trial. Lancet Infect Dis 14:939–946

Scriba TJ, Tameris M, Mansoor N, Smit E, van der Merwe L, Isaacs F, Keyser A, Moyo S, Brittain N, Lawrie A, Gelderbloem S (2010) Modified vaccinia Ankara-expressing Ag85A, a novel tuberculosis vaccine, is safe in adolescents and children, and induces polyfunctional CD4 + T cells. Eur J Immunol 40:279–290

Seder RA, Ahmed R (2003) Similarities and differences in CD4+ and CD8+ effector and memory T cell generation. Nat Immunol 4:835–842

Sharma N, Agrewala JN (2004) Potent role of vaccines prepared from macrophages infected with live bacteria in protection against Mycobacterium tuberculosis and Salmonella typhimurium infections. J Infect Dis 190:107–114

Sharma SK, Katoch K, Sarin R, Balambal R, Jain NK, Patel N, Murthy KJ, Singla N, Saha PK, Khanna A, Singh (2017) Efficacy and safety of Mycobacterium indicus pranii as an adjunct therapy in category II pulmonary tuberculosis in a randomized trial. Sci Rep 7 3354

Sheehan S, Harris SA, Satti I, Hokey DA, Dheenadhayalan V, Stockdale L, Thomas ZR, Minhinnick A, Wilkie M, Vermaak S, Meyer J (2015) A phase I, open-label trial, evaluating the safety and immunogenicity of candidate tuberculosis vaccines AERAS-402 and MVA85A, administered by prime-boost regime in BCG-vaccinated healthy adults. PLoS One 10:e0141687

Shi C, Chen L, Chen Z, Zhang Y, Zhou Z, Lu J, Fu R, Wang C, Fang Z, Fan X (2010) Enhanced protection against tuberculosis by vaccination with recombinant BCG over-expressing HspX protein. Vaccine 28:5237–5244

Singh CR, Moulton RA, Armitige LY, Bidani A, Snuggs M, Dhandayuthapani S, Hunter RL, Jagannath C (2006) Processing and presentation of a mycobacterial antigen 85B epitope by murine macrophages is dependent on the phagosomal acquisition of vacuolar proton ATPase and in situ activation of cathepsin D. J Immunol 177:3250–3259

Singh V, Gowthaman U, Jain S, Parihar P, Banskar S, Gupta P, Gupta UD, Agrewala JN (2010) Coadministration of interleukins 7 and 15 with bacille Calmette-Guerin mounts enduring T cell memory response against Mycobacterium tuberculosis. J Infect Dis 202:480–489

Singh V, Jain S, Gowthaman U, Parihar P, Gupta P, Gupta UD, Agrewala JN (2011) Co-administration of IL-1+IL-6+TNF-alpha with Mycobacterium tuberculosis infected macrophages vaccine induces better protective T cell memory than BCG. PLoS One 6:e16097

Singh B, Saqib M, Gupta A, Kumar P, Bhaskar S (2017) Autophagy induction by Mycobacterium indicus pranii promotes Mycobacterium tuberculosis clearance from RAW 264.7 macrophages. PLoS One 12:e0189606

Slobbe L, Lockhart E, O'Donnell M, Makintosh C, Buchan G (1999) An in vitro comparison of BCG and cytokine secreting BCG vaccines. Immunology 96:517–523

Smaill F, Xing Z (2014) Human type 5 adenovirus-based tuberculosis vaccine: is the respiratory route of delivery the future? Expert Rev Vaccines 13:927–930

Soares AP, Scriba TJ, Joseph S, Harbacheuski R, Murray RA, Gelderbloem SJ, Hawkridge A, Hussey GD, Maecker H, Kaplan G, Hanekom WA (2008) Bacillus Calmette-Guerin vaccination of human newborns induces T cells with complex cytokine and phenotypic profiles. J Immunol 180:3569–3577

Solans L, Uranga S, Aguilo N, Arnal C, Gomez AB, Monzon M, Badiola JJ, Gicquel B, Martin C (2014) Hyper-attenuated MTBVAC erp mutant protects against tuberculosis in mice. Vaccine 32:5192–5197

Stanford J, Stanford C, Grange J (2004) Immunotherapy with Mycobacterium vaccae in the treatment of tuberculosis. Front Biosci 9:1701–1719

Sterne JA, Rodrigues LC, Guedes IN (1998) Does the efficacy of BCG decline with time since vaccination? Int J Tuberc Lung Dis 2:200–207

Stylianou E, Griffiths KL, Poyntz HC, Harrington-Kandt R, Dicks MD, Stockdale L, Betts G, McShane H (2015) Improvement of BCG protective efficacy with a novel chimpanzee adenovirus and a modified vaccinia Ankara virus both expressing Ag85A. Vaccine 33:6800–6808

Sugawara I, Udagawa T, Taniyama T (2007) Protective efficacy of recombinant (Ag85A) BCG Tokyo with Ag85A peptide boosting against Mycobacterium tuberculosis-infected guinea pigs in comparison with that of DNA vaccine encoding Ag85A. Tuberculosis 87:94–101

Sugawara I, Sun L, Mizuno S, Taniyama T (2009) Protective efficacy of recombinant BCG Tokyo (Ag85A) in rhesus monkeys (Macaca mulatta) infected intratracheally with H37Rv Mycobacterium tuberculosis. Tuberculosis 89:62–67

Tacquet A, Gaudier B, Gernez-Rieux C (1962) Etude expérimentale de la vitalité du BCG au cours de la traversee gastro-intestinale chez des enfants non allergiques vaccinés par voie digestive. Pharmacology 6:373–381

Takeda K, Akira S (2004) TLR signaling pathways. Semin Immunol 16:3–9

Tameris M, Geldenhuys H, Luabeya AK, Smit E, Hughes JE, Vermaak S, Hanekom WA, Hatherill M, Mahomed H, McShane H, Scriba TJ (2014) The candidate TB vaccine, MVA85A, induces highly durable Th1 responses. PLoS One 9:e87340

Thacher EG, Cavassini M, Audran R, Thierry AC, Bollaerts A, Cohen J, Demoitié MA, Ejigu D, Mettens P, Moris P, Ofori-Anyinam O (2014) Safety and immunogenicity of the M72/AS01 candidate tuberculosis vaccine in HIV-infected adults on combination antiretroviral therapy: a phase I/II, randomized trial. AIDS 28:1769–1781

Thoma-Uszynski S, Stenger S, Takeuchi O, Ochoa MT, Engele M, Sieling PA, Barnes PF, Röllinghoff M, Bölcskei PL, Wagner M, Akira S (2001) Induction of direct antimicrobial activity through mammalian toll-like receptors. Science 291:1544–1547

Tullius MV, Harth G, Masleša-Galić S, Dillon BJ, Horwitz MA (2008) A replication-limited recombinant Mycobacterium bovis BCG vaccine against tuberculosis designed for human immunodeficiency virus-positive persons is safer and more efficacious than BCG. Infect Immun 76:5200–5214

Ullrich HJ, Beatty WL, Russell DG (2000) Interaction of Mycobacterium avium-containing phagosomes with the antigen presentation pathway. J Immunol 165:6073–6080

Van Panhuis WG, Grefenstette J, Jung SY, Chok NS, Cross A, Eng H, Lee BY, Zadorozhny V, Brown S, Cummings D, Burke DS (2013) Contagious diseases in the United States from 1888 to the present. N Engl J Med 369:2152

van Pinxteren LA, Cassidy JP, Smedegaard BH, Agger EM, Andersen P (2000) Control of latent Mycobacterium tuberculosis infection is dependent on CD8 T cells. Eur J Immunol 30:3689–3698

Van Soelen N, Mandalakas AM, Kirchner HL, Walzl G, Grewal HM, Jacobsen M, Hesseling AC (2012) Effect of Ascaris Lumbricoides specific IgE on tuberculin skin test responses in children in a high-burden setting: a cross-sectional community-based study. BMC Infect Dis 12:211

Verreck FA, Tchilian EZ, Vervenne RA, Sombroek CC, Kondova I, Eissen OA, Sommandas V, van der Werff NM, Verschoor E, Braskamp G, Bakker J (2017) Variable BCG efficacy in rhesus populations: pulmonary BCG provides protection where standard intra-dermal vaccination fails. Tuberculosis 104:46–57

Vilaplana C, Montané E, Pinto S, Barriocanal AM, Domenech G, Torres F, Cardona PJ, Costa J (2010) Double-blind, randomized, placebo-controlled phase I clinical trial of the therapeutical antituberculous vaccine RUTI®. Vaccine 28:1106–1116

Vilaplana C, Gil O, Cáceres N, Pinto S, Díaz J, Cardona PJ (2011) Prophylactic effect of a therapeutic vaccine against TB based on fragments of Mycobacterium tuberculosis. PLoS One 6:e20404

Wang C, Fu R, Chen Z, Tan K, Chen L, Teng X, Lu J, Shi C, Fan X (2012) Immunogenicity and protective efficacy of a novel recombinant BCG strain overexpressing antigens Ag85A and Ag85B. Clin Dev Immunol 2012:563838

White AD, Sibley L, Dennis MJ, Gooch K, Betts G, Edwards N, Reyes-Sandoval A, Carroll MW, Williams A, Marsh PD, McShane H (2013) Evaluation of the safety and immunogenicity of a candidate tuberculosis vaccine, MVA85A, delivered by aerosol to the lungs of macaques. Clin Vaccine Immunol 20:663–672

Wilkinson KA, Martin TD, Reba SM, Aung H, Redline RW, Boom WH, Toossi Z, Fulton SA (2000) Latency-associated peptide of transforming growth factor beta enhances mycobacteriocidal immunity in the lung during Mycobacterium bovis BCG infection in C57BL/6 mice. Infect Immun 68:6505–6508

World Health Organization (Ed.) (2013) Global tuberculosis report

World Health Organization (Ed.) (2014) Global tuberculosis report

World Health Organization (Ed.) (2017) Global tuberculosis report

World Health Organization (Ed.) (2018) Global tuberculosis report

Xing Z, McFarland CT, Sallenave JM, Izzo A, Wang J, McMurray DN (2009) Intranasal mucosal boosting with an adenovirus-vectored vaccine markedly enhances the protection of BCG-primed Guinea pigs against pulmonary tuberculosis. PLoS One 4:e5856

Zhang W, Zhang Y, Zheng H, Pan Y, Liu H, Du P, Wan L, Liu J, Zhu B, Zhao G, Chen C (2013) Genome sequencing and analysis of BCG vaccine strains. PLoS One 8:e71243

Zwingenberger K, Hohmann A, de Brito MC, Ritter M (1991) Impaired balance of interleukin-4 and interferon-gamma production in infections with Schistosoma mansoni and intestinal nematodes. Scand J Immunol 34:243–251

Immunotherapeutic Potential of *Mycobacterium indicus pranii* Against Tuberculosis

22

Sangeeta Bhaskar and Bindu Singh

Abstract

Mycobacterium indicus pranii (MIP), a nonpathogenic saprophytic mycobacterium, is potential tuberculosis (TB) vaccine candidate as it shares a large number of cross-reactive antigens with *M.tb* and has demonstrated unique ability for immunomodulation. Studies in animal models of TB have demonstrated its significant prophylactic and therapeutic efficacy against TB. The immunological aspects of MIP-mediated protection include induction of Th1 and Th17 type of response, activation of macrophage effector functions, and $CD4^+$ T cell activation along with enhanced $CD8^+$ T cell cytotoxic activity, thereby resulting in inhibition of the intracellular growth and multiplication of *M.tb*. When used as a booster to BCG vaccine, MIP conferred higher protection in animal models of TB. Protective efficacy of killed MIP vaccine has been established in clinical trials. Immunotherapeutic potential of this vaccine is confirmed in a recently conducted multicentric clinical trial in category II TB patients, having advanced disease, and was difficult to treat. Moreover, practical implementation of MIP as TB vaccine has a major advantage as it is a cost-effective vaccination strategy for effective control of TB with no adverse effects.

Keywords

MIP · Tuberculosis · Vaccine

S. Bhaskar (✉) · B. Singh
Product Development Cell-I, National Institute of Immunology, New Delhi, Delhi, India
e-mail: sangeeta@nii.ac.in

S. E. Hasnain et al. (eds.), Mycobacterium tuberculosis: *Molecular Infection Biology, Pathogenesis, Diagnostics and New Interventions*,
https://doi.org/10.1007/978-981-32-9413-4_22

List of Abbreviations

MIP	*Mycobacterium indicus pranii*
MW	*Mycobacterium leprae*
Mw	Mycobacteria
MAC	*Mycobacterium avium* complex
PLC	Phospholipase
BCG	*Mycobacterium bovis*
RIF	Rifampicin
INH	Isoniazid
PYZ	Pyrazinamide
ETB	Ethambutol
ATT	Antitubercular treatment
PTB	Pulmonary TB

22.1 *Mycobacterium indicus pranii*

There are more than 150 recognized species in the genus "*Mycobacterium*." Many of these species are pathogenic that poses serious health threats to humans such as *M. tuberculosis* (causing tuberculosis) and *M. leprae* (leprosy) while others are saprophytic. In the late 1970s, Talwar et al. identified *Mycobacterium w* (Mw) from a group of 16 strains of atypical mycobacteria which gave highest immune response against *M. leprae*. The nomenclature "Mw" initiated as an outcome of the coding given to those atypical strains of mycobacteria investigated for the development of vaccine against leprosy, where mycobacteria coded as "w" was found to be promising. At that time, no overlap was found with the name Mw and names of other known mycobacterial species, but later on *Mycobacterium w* (Mw) was renamed as *Mycobacterium indicus pranii* (MIP) as there was confusion with hyper-virulent *M. tuberculosis*-W strain of Beijing genotype family. This name was given on the basis of location of its isolation from India (***indicus***), discovered by Prof. Pran Talwar (***pra****nii*), and characterization at the National Institute of Immunology (pra***nii***).

In 1980s, immunotherapeutic activity of MIP was evaluated by clinical trials in multi-bacillary leprosy patients along with chemotherapy. Immunotherapy with MIP once every 3 months combined with chemotherapy significantly enhanced bacillary clearance and shortened the recovery time (Zaheer et al. 1993). It was found to be effective in multi-bacillary leprosy patients with high bacteriological index (Talwar et al. 1990; Talwar 1999). Besides therapeutic benefits, it showed immunoprophylactic efficacy in household contacts of leprosy patients (Sharma et al. 2005). Killed MIP vaccine subsequently got approval of the Drug Controller General of India and is now manufactured and marketed by an Indian pharmaceutical company under the name "Immuvac."

Later on molecular phylogenetic analysis and genome-wide comparisons revealed that MIP is a predecessor of *Mycobacterium avium* complex (MAC) which comprises of opportunistic pathogens. Comprehensive analysis of cellular and biochemical features together with chemotaxonomic markers and when comparison with other mycobacterial species establishes specific attributes of MIP (Ahmed et al. 2007; Saini et al. 2012). Analysis of the study suggested that MIP can be placed at a transitory position, in between saprophytes and pathogens. This unique placement has given it the advantage of sharing a large number of antigens with *M.tb* as compared to other nonpathogenic vaccine candidates. MIP has moderate growth rate (time of colony appearance 6–8 days) that is in between slow-growing (3 weeks for *M.tb*) and fast-growing (3 days for *M. smegmatis*) mycobacterial species. Whole-genome sequence analyses provided evidence of substantive and interesting differences between MIP and *M.tb* as well as other members of the *M. tuberculosis* complex. Comparison of "antigenicity index" of MIP with *M.tb* antigens, based on the analyses by VaxiJen, suggested that MIP possess unique ability for immunomodulation (Saini et al. 2012).

MIP has shown immunomodulatory activity in various diseases (Ahmad et al. 2011; Katoch et al. 2008; Rakshit et al. 2012; Talwar 1999). Killed MIP vaccine is approved for human use against leprosy. Further in animal studies, MIP was found to be protective against TB and human clinical trials as immunotherapeutic/immunoprophylactic candidate vaccine are underway (Gupta et al. 2009, 2012a, b; Saini et al. 2012). MIP has broad-spectrum immunomodulatory potential as it shares a large number of B and T cell antigens with other mycobacteria, viz., *M.tb* and *M. leprae* (Saini et al. 2012; Yadava et al. 1991).

22.1.1 Genomic and Proteomic Characteristics

Whole-genome sequencing of MIP could explain the attributes, responsible for its unique lifestyle and also molecular basis of immunomodulation. Its genome size is approximately 5,589,007 base pair and has mosaic architecture. The genome consists of 5270 ORFs. The mean G + C content of MIP genome measures up to 68%. Approximately, 34% genes of MIP are laterally acquired and these acquisitions are found to be crucial to MIP physiology. It has 66 genes from PE/PPE family (50 PPE and 16 PE), out of which five genes are unique to MIP (Saini et al. 2009, 2012). Crucial role of PE/PPE family in host immunomodulation was highlighted by antigenic analyses between *M.tb* and MIP. One hundred fifty-two members of PE/PPE multigene family exhibit antigenicity index > 0.4, out of which 141 proteins have amino acid sequence similarity with MIP proteins.

MIP proteome was elucidated by BLAST analysis. Till now, 80% of MIP proteins have been annotated. Out of these, 7.5% proteins were found to be unique in MIP with no remarkable homology with proteins present in *M.tb* proteomes.

Analysis of protein coding genes revealed that about 82% of *M.tb* proteins have homology with MIP proteins (Rahman et al. 2014; Singh et al. 2014).

22.1.2 Nonpathogenic Characteristics of MIP

Various attributes of MIP were revealed by annotation of genome which explained nonpathogenic nature of MIP. The presence of PE-PPE genes and mce1 operon enables MIP to invade the host cells, but it does not have mce2 and mce3 operons which are required for persistent macrophage infection, whereas pathogenic mycobacteria *M. avium* and *M.tb* have both mce2 and mce3 operons which enable their survival inside the macrophages. MIP also lacks phospholipase (*plc*) ABCD genes, which help mycobacteria to use host fatty acids as a potential source of carbon during persistent infections. The *dev*S/*dev*R two-component system, essential for survival in low oxygen conditions, is also absent in MIP. MIP also lacks gene *trp*D, which is important for tryptophan biosynthesis. Due to the absence of these genes, MIP cannot survive in the host for prolonged periods of time and thus, it is not able to cause latent infection. On the other hand, MIP possesses abundant catalases and superoxide dismutases, which help in mitigating oxidative stress. Complete cyanide and thiocyanate biodegradation machinery is present in MIP which is responsible for degradation of thiocyanate to produce CO_2 and NH_3, which in turn are used by MIP as a source of carbon and nitrogen for limited survival intracellularly. Ten ORFs of MIP genome having significant similarity to hemerythrin genes could be responsible for the survival of MIP for a short period in the anoxic conditions.

These nonpathogenic attributes of MIP were confirmed by in vivo experiments in different animal models which established that infection with MIP is self-limiting and clears off within 4–6 weeks, which probably makes an ideal scenario. A whole bacterial vaccine which can persist for a restricted duration in the host to induce enough memory T cells but also in due course of time removed by the immune response of the host without causing any adverse pathology is an important desired feature of a vaccine for its efficacy as well as from safety point of view.

22.2 Protective Efficacy of MIP Against Tuberculosis

As tuberculosis continues to be a serious health problem, there are intensive efforts to find new vaccines to control this disease. The present vaccine against tuberculosis (TB), *Mycobacterium bovis* (BCG), has poor protection against the pulmonary TB in adults. It is general perception that active TB occurs when immune system is weakened, which otherwise keeps *M.tb* in check when fully competent. This suggests that natural immunity fails to control the pathogen for long time in subset of individuals and is not sufficient. Whole bacterial vaccines are one of the important vaccine strategies. This approach has advantage of multiple antigens and built-in adjuvant activity. Mycobacterial strains sharing antigens with *M. tuberculosis* are being evaluated as alternatives to *M. bovis* for vaccine use. Being whole organisms, these vaccines are able to induce broad-spectrum immune response (both humoral and cellular immune responses) as compared to subunit vaccines. Following this approach, MIP has been studied for its immunomodulatory potential against TB and leprosy.

Earlier studies have reported that MIP gives protection against tuberculosis in both BCG responder and also in non-responder mice strains (Singh et al. 1991). Protective potential of MIP and the underlying immune responses were further studied in the animal models of tuberculosis. Generally, live bacteria are known to give higher protection than killed one as live bacteria persist in the host for some time which results in strong memory response. Route of immunization also has an important role. Aerosol inhalation is a non-invasive delivery method that physically targets the lung for its pharmacological effect and induces both local and systemic immune response. Protective efficacy of live and killed forms of MIP via parenteral route and through aerosol immunization was compared with BCG (Gupta et al. 2009). MIP immunization by both the routes gave higher protection than BCG in the animal models of tuberculosis. Live MIP vaccine by aerosol route gave significantly higher protection as compared to BCG. More importantly, MIP immunization also provided significantly higher long-term protection as compared to BCG against TB.

MIP induces Th1 and Th17 type of response whether given in killed or live form. It activates macrophage effector functions as well as lymphocytes activity. Besides activation of $CD4^+$ T cells and IFN-γ secretion, MIP also induces significant $CD8^+$ T cell cytotoxic activity. Activated macrophages are able to restrict the intracellular growth and multiplication of *M.tb* (Gupta et al. 2009). Protective efficacy and modulation of immune response post-*M.tb* infection was further evaluated in susceptible guinea pig model. Pathogenesis in guinea pig model resembles human tuberculosis. At different postinfection time points, reduction in bacterial load, improvement in lung pathology, and organized granulomatous response were observed in the MIP-vaccinated group as compared to BCG-immunized group (Gupta et al. 2012b). In MIP-immunized group, higher protective Th1 response was observed as compared to BCG group at early time point after *M.tb* infection and moderate Th1 response at late chronic stage of infection. Inside the granuloma of infected lungs higher activation of antigen-presenting cells was observed in the MIP-immunized group.

22.2.1 MIP Immunotherapy as an Adjunct to Chemotherapy for Treatment of TB

The current chemotherapy [the multidrug regimen comprising of rifampicin (RIF), isoniazid (INH), pyrazinamide (PYZ), and ethambutol (ETB)] available for TB usually results in poor compliance as these drugs are to be taken for a minimum of

6–9 months. Although, new cases of TB can be cured by currently available drugs and to improve the compliance, WHO has introduced the "directly observed therapy" but still patients who default on therapy develop severe risk of relapse and acquiring drug resistance. To effectively eliminate the replicating and persistent bacteria from the infected lungs, multipronged strategy is needed.

During the course of tuberculosis infection, the protective Th1 immune response is suppressed and immunosuppressive Th2 response gets elevated which facilitates the survival of *M.tb* and progression of disease. Immunotherapy strategy which could boost the Th1 immune response in infected person might prove to be effective as it can act synergistically to chemotherapy. Efficacy of MIP as an adjunct to standard chemotherapy for TB was evaluated in animal models, when given by aerosol or subcutaneous route. MIP immunotherapy by intranasal route resulted in increased infiltration of antigen-specific lymphocytes and macrophages in the *M.tb*-infected lungs which resulted in killing of bacteria in synergistic way along with chemotherapy. MIP treatment along with standard drugs resulted in further reduction in bacterial loads and less pulmonary pathology as compared to "only drugs" treated group (Gupta et al. 2012a). A balanced inflammatory and suppressive immune response was observed in the late stage of "drug+MIP immunotherapy" treatment, which helped in the control of initial inflammatory reaction and restoration of normal lung tissue. Recent study suggested that autophagy induced by MIP plays an important role in reversal of phagosome maturation block in *M.tb*-infected macrophages, which leads to accelerated clearance of *M.tb* when MIP immunotherapy is given by nasal route (Singh et al. 2017). Combined results of different studies suggested that MIP could be potentially very useful in eliminating the persistent bacteria when given as an adjunct to standard chemotherapy. For induction of early local immune response in the lung, MIP immunotherapy by intranasal route can play a very crucial role.

22.2.2 MIP as a Booster to BCG Vaccine Confers Higher Protection in Animal Models

BCG, the current vaccine against TB, protects against childhood tuberculosis but is poorly effective in adults. Since BCG has given significant protection against pediatric TB in endemic areas, prime-boost strategy is the most practical approach for control of TB. MIP shares a large number of antigens with *M.tb* and BCG and also had been shown to give higher protection as compared to BCG in animal models of TB. With this background, efficacy of MIP was evaluated as a booster to BCG vaccine in animal models of tuberculosis. MIP booster enhanced the BCG-mediated immune response and resulted in higher protection as evident from the reduction in bacterial load in the lungs of infected animals of BCG-MIP group as compared to only BCG-immunized group (Fig. 22.1). Evaluation of pulmonary pathology

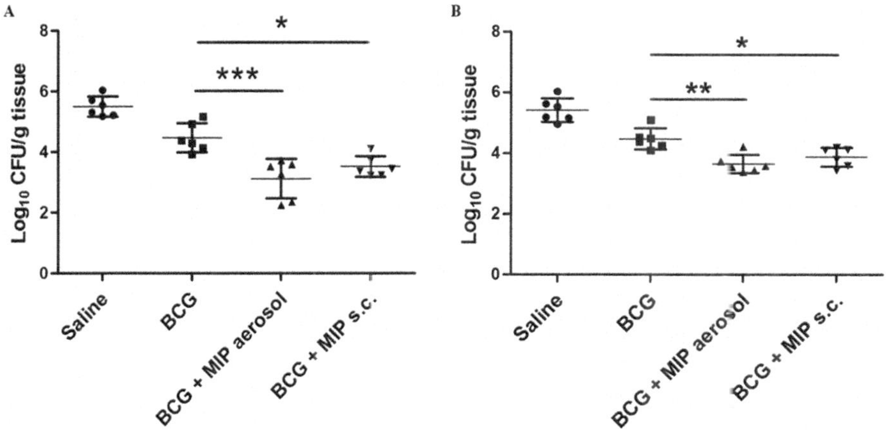

Fig. 22.1 Bacterial loads at 4 and 8 weeks after *M.tb* challenge. Bacterial load in different groups was evaluated in lungs after 4 weeks (**a**) and 8 weeks (**b**) of *M.tb* challenge. Significantly reduced colony-forming units (CFU) per gm of tissue were found in "BCG-MIP" regimen as compared to "only BCG"-vaccinated group. Data represents the mean CFU ± SD of six guinea pigs in each group [$*P \leq 0.05$; $**P \leq 0.01$; $***P \leq 0.001$]

demonstrated a reduced number of macroscopic lesions, cavities, and hemorrhagic spots in BCG-MIP group as compared to unvaccinated and BCG only groups. Aerosol route of MIP booster immunization proves to be more effective in protection than subcutaneous route (Saqib et al. 2016). It was observed that Th1 and Th17 cytokines were upregulated in infected lungs of "BCG-MIP" group in contrast to "only BCG" group. Higher levels of multifunctional T cells with high MFI for TNF-α and IFN-γ were observed in the BCG-MIP group after *M.tb* infection. Findings of the study demonstrated that MIP can be given as a booster to BCG vaccine for efficient protection against TB, which might be a very cost-efficient approach for effective control of TB.

22.3 Clinical Trials

In 1990s, prophylactic efficacy of MIP against leprosy was evaluated in a population of about 30,000 people from 272 villages in Kanpur (India). The population was vaccinated with two doses of MIP at six-month interval. Healthy contacts of leprosy patients who had no evidence of TB disease were given MIP vaccine/placebo. After follow-up for 13 years, retrospective analysis of the data showed that incidence of tuberculosis was reduced significantly in the MIP-vaccinated group as compared to placebo group in that randomized placebo-controlled study (Katoch et al. 2008).

Further higher protection of MIP was observed in the subgroup, vaccinated with BCG in the childhood suggesting that booster dose of MIP could enhance the protective efficacy of BCG. That observation formed the basis of detailed animal studies where MIP was evaluated as booster to BCG vaccine.

Later, in one of the recently conducted clinical trials, MIP was given as an adjunct to chemotherapy in Cat II pulmonary TB patients. This was double-blind, placebo-controlled, multicentric trial where MIP demonstrated significantly higher cure rate in difficult to treat cases, i.e., those with high bacillary load, drug resistance, and/or with bilateral cavities. At 4 week after the start of therapy, sputum culture conversion in a significantly higher number of patients (67.1%) was observed in the MIP group compared to the placebo (57%) group ($p = 0.0002$). Early sputum conversion in the MIP group suggested that it played an important role in clearance of *M.tb*. This has important implications in controlling the spread of *M.tb* as live bacteria in the sputum are major contributors for sustained incidence of tuberculosis. MIP has come a long way. Recently a phase III, double-blind, placebo-controlled clinical study has started to evaluate its efficacy in preventing TB in healthy household contacts of pulmonary TB patients.

22.4 Safety of MIP Immunization

Toxicity studies were done with heat-killed MIP in mice as well as guinea pigs and no untoward reactions were observed in these experimental animals. Killed MIP vaccine was found to be safe and well tolerated in clinical trials. Safety profile of live MIP was also studied and no detectable MIP was observed in any organs after 30 days of immunization, provided by any route. No pathological symptoms were observed in various organs at any time point after immunization. A normal lung parenchyma was retained, the interalveolar septae were thin, and no cellular exudates were observed in the alveolar spaces at any time point post-aerosol immunization (Gupta et al. 2012a, b). Infection with live MIP in mice and guinea pigs is self-limiting, which is crucial for the induction of robust memory response. A desired feature of a vaccine is that it should persist long enough for generation of memory T-cells but still should be gradually removed by the host immune response without developing any adverse pathology. MIP, being a nonpathogenic bacterial vaccine with exceptional properties of immunomodulation, has an advantage for the practical application of this vaccine to masses and could be very cost-effective strategy for effective control of TB (Fig. 22.2).

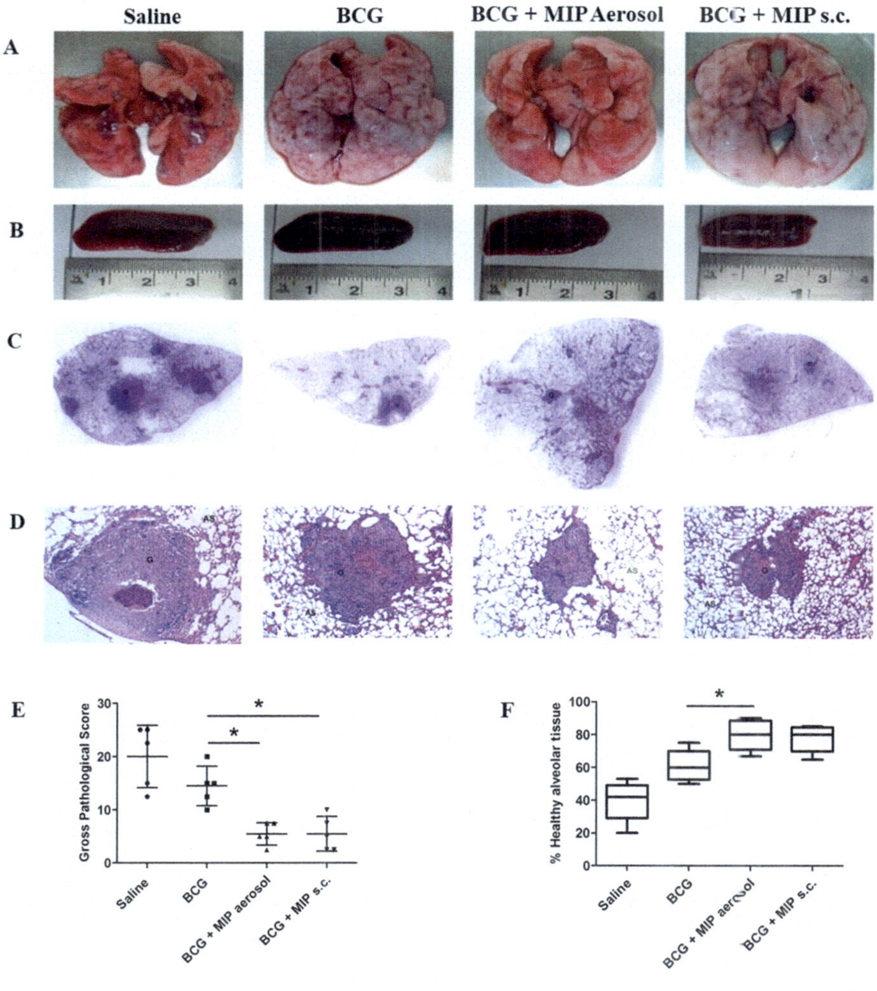

Fig. 22.2 Pulmonary pathology at 4 weeks after infection. Shown are the representative images of lungs (**a**) and spleen (**b**) of guinea pigs from different groups at 4 weeks postinfection. The presence of pathological lesions like nodules, hemorrhagic spots, and cavities in the lungs was macroscopically examined. Spleen size was also measured and compared among the groups. (**c**) Pictures showing H&E-stained whole lung sections. (**d**) H&E-stained sections visualized at 40 X magnification. (**e**) Gross pathological score and (**f**) percent healthy alveolar tissue in infected lungs. Data represents the mean ± SD [$*P \leq 0.05$; $**P \leq 0.01$; $***P \leq 0.001$; comparisons were made with "only BCG"-immunized group]

References

Ahmad F, Mani J, Kumar P, Haridas S, Upadhyay P, Bhaskar S (2011) Activation of anti-tumor immune response and reduction of regulatory T Cells with *Mycobacterium indicus pranii* (MIP) therapy in tumor bearing mice. PLoS ONE 6(9):e25424

Ahmed N, Saini V, Raghuvanshi S, Khurana JP, Tyagi AK (2007) Molecular analysis of a leprosy immunotherapeutic bacillus provides insights into mycobacterium evolution. PLoS ONE 2(10): e968

Gupta A, Geetha N, Mani J, Upadhyay P, Katoch VM, Natrajan M, Gupta UD, Bhaskar S (2009) Immunogenicity and protective efficacy of "*Mycobacterium w*" against *Mycobacterium tuberculosis* in mice immunized with live versus heat-killed M.w by the aerosol or parenteral route. Infect Immun 77(1):223–231

Gupta A, Ahmad FJ, Ahmad F, Gupta UD, Natarajan M, Katoch V, Bhaskar S (2012a) Efficacy of *Mycobacterium indicus pranii* immunotherapy as an adjunct to chemotherapy for tuberculosis and underlying immune responses in the lung. PLoS ONE 7(7):e39215

Gupta A, Ahmad FJ, Ahmad F, Gupta UD, Natarajan M, Katoch VM, Bhaskar S (2012b) Protective efficacy of *Mycobacterium indicus pranii* against tuberculosis and underlying local lung immune responses in guinea pig model. Vaccine 30(43):6198–6209

Katoch K, Singh P, Adhikari T, Benara SK, Singh HB, Chauhan DS, Sharma VD, Lavania M, Sachan AS, Katoch VM (2008) Potential of *Mw* as a prophylactic vaccine against pulmonary tuberculosis. Vaccine 26:1228–1234

Rahman SA, Singh Y, Kohli S, Ahmad J, Ehtesham NZ, Tyagi AK, Hasnain SE (2014) Comparative analyses of nonpathogenic, opportunistic and totally pathogenic mycobacteria reveal genomic and biochemical variabilities and highlight the survival attributes of *Mycobacterium tuberculosis*. mBio 5(6):02020–02014

Rakshit S, Ponnusamy M, Papanna S, Saha B, Ahmed A, Nandi D (2012) Immunotherapeutic efficacy of *Mycobacterium indicuspranii* in eliciting anti-tumor T cell responses: critical roles of IFNγ. Int J Cancer 130(4):865–875

Saini V, Raghuvanshi S, Talwar GP, Ahmed N, Khurana JP, Hasnain SE, Tyagi A, Tyagi AK (2009) Polyphasic taxonomic analysis establishes *Mycobacterium indicus pranii* as a distinct species. PLoS ONE 4(7):e6263

Saini V, Raghuvanshi S, Niyaz A, Khurana JP, Hasnain SE, Tyagi A, Tyagi AK (2012) Massive gene acquisitions in *Mycobacterium indicus pranii* provide a perspective on mycobacterial evolution. Nucleic Acids Res 40:10832–10850

Saqib M, Khatri R, Singh B, Gupta A, Bhaskar S (2016) M*ycobacterium indicus pranii* as a booster vaccine enhances BCG induced immunity and confers higher protection in animal models of tuberculosis. Tuberculosis 101:164e173

Sharma P, Mukherjee R, Talwar GP, Sarathchandra KG, Singh P (2005) Immunoprophylactic effects of the anti-leprosy Mw vaccine in household contacts of leprosy patients: clinical field trials with a follow up of 8–10 years. Lepr Rev 76:127–143

Singh IG, Mukherjee R, Talwar GP (1991) Resistance to intravenous inoculation of Mycobacterium tuberculosis H37Rv in mice of different inbred strains following immunization with a leprosy vaccine based on *Mycobacterium w*. Vaccine 9:10–14

Singh Y, Kohli S, Sowpati DT, Rahman SA, Tyagi AK, Hasnain SE (2014) Gene cooption in mycobacteria and search for virulence attributes: comparative proteomic analyses of *Mycobacterium tuberculosis*, *Mycobacterium indicus pranii* and other mycobacteria. Int J Med Microbiol 304(5–6):742–748

Singh B, Saqib M, Gupta A, Kumar P, Bhaskar S (2017) Autophagy induction by *Mycobacterium indicus pranii* promotes *Mycobacterium tuberculosis* clearance from RAW 264.7 macrophages. PLoS ONE 12(12):e0189606

Talwar GP, Zaheer SA, Mukherjee R, Walia R, Suresh NR (1990) Immunotherapeutic effects of a vaccine based on a saprophytic cultivable mycobacterium, *Mycobacterium w* in multi-bacillary leprosy patients. Vaccine 8:121–129

Talwar GP (1999) An immunotherapeutic vaccine for multi-bacillary leprosy. Int Rev Immunol 18(3):229–249

Yadava A, Suresh NR, Zaheer S, Talwar GP, Mukherjee R (1991) T-Cell responses to fractionated antigens of *Mycobacterium w*, a candidate anti-leprosy vaccine, in leprosy patients. Scand J Immunol 34(1):23–31

Zaheer SA, Mukherjee R, Ramkumar B, Misra RS, Talwar GP (1993) Combined multidrug and *Mycobacterium w* vaccine therapy in patients with multibacillary leprosy. J Infect Dis 167:401–410

TB Diagnostics: Journey from Smear Microscopy to Whole Genome Sequencing 23

Himanshu Vashistha and K. K. Chopra

Abstract

Tuberculosis is not just a basic clinical and general issue of the whole world. It has attracted worldwide attention of the entire scientific majority because of its alarming rate of incidence and mortality. Since history of early examination of tuberculosis and its drug resistance, it has been proved that the diagnosis upgrades that enhances the survival and early case detection related to early treatment initiation. Careful bacteriological assurance and general prosperity approach to manage TB rely upon initial microscopic examination, solid & liquid culture methods, rapid molecular and standard DST assays.

For each assay of detection, there are new procedures and advancements adopted since their inception. Developing diagnostic techniques having access to quick accurate results, near-patient point-of-care diagnostics are now vital to curb and control TB and HIV-TB burden in resource-poor countries This chapter has focused on the journey of those diagnostic techniques which remain the mainstay of TB diagnosis from early diagnostic work of smear microscopy to latest techniques involving whole genome sequencing. Starting late it is basic to comprehend that at present, there is no autonomous test for the quick acknowledgment of tuberculosis in all patients. While some new frameworks are direct, others have complex components and require sophisticated infrastructure and trained manpower. It is therefore imperative to choose from different diagnostic assays individually or to combine them inside a country's national TB control program.

H. Vashistha
Viral Hepatitis Division, National Centre for Disease Control, New Delhi, Delhi, India

K. K. Chopra (✉)
New Delhi Tuberculosis Center, New Delhi, Delhi, India

S. E. Hasnain et al. (eds.), Mycobacterium tuberculosis: *Molecular Infection Biology, Pathogenesis, Diagnostics and New Interventions*,
https://doi.org/10.1007/978-981-32-9413-4_23

Keywords
Tuberculosis · Diagnostics · Culture · Next-generation sequencing

Abbreviations

AFB	Acid-fast bacilli
ATT	Anti-tubercular therapy
BCG	Bacillus Calmette-Guerin
CBNAAT	Cartridge-based nucleic acid amplification test
CDC	Center for disease control
CFP	Culture filtrate protein
CFU	Colony-forming unit
CRI	Colorimetric redox indicator
DB	Database
DNA	Deoxyribonucleic acid
DR	Direct repeat
DST	Drug susceptibility testing
ESAT	Early secreted antigenic target
FDA	Food and Drug Administration
FIND	Foundation for Innovative Newer Diagnostics
FM	Fluorescent microscopy
HIV	Human immunodeficiency virus
IFNg	Interferon gamma
IGRA	Interferon-gamma release assays
INH	Isoniazid
IS6110	Insertion sequence
IVD	In vitro diagnostics
LAM	Lipoarabinomannan
LAMP	Loop-mediated amplification PCR
LED	Light-emitting diode
LJ	Lowenstein Jensen
LM	Liquid medium
LPA	Line probe assay
LTBI	Latent tuberculosis infection
MDR	Multidrug resistance
MIRU-VNTR	Mycobacterium interspersed repeated units-variable number of tandem repeats
MODS	Microscopic observation of drug susceptibility
MTB	*Mycobacterium tuberculosis*
MTBC	Mycobacterium *tuberculosis* complex
NGS	Next-generation sequencing
NRA	Nitrate reductase assay
NTM	Nontuberculous *Mycobacterium*

OADC	Oleic acid, albumin, dextrose, catalase
PANTA	Polymyxin B, azlocillin, nalidixic acid, trimethoprim, amphotericin B
PPD	Purified protein derivative
PPV	Positive predictive value
RFLP	Restriction fragment length polymorphism
RIF	Rifampicin
RNTCP	Revised National Tuberculosis Control Program
rRNA	Ribosomal ribonucleic acid
SLID	Second-line injectable drugs
SPR	Surface plasmon resonance
TAT	Turnaround time
TB	Tuberculosis
TST	Tuberculin skin test
TTD	Time to detection
VOC's	Volatile organic compounds
WGS	Whole genome sequencing
WHO	World Health Organization
XDR	Extensive drug resistance
ZN	Ziehl Neelsen

23.1 Background

Tuberculosis (TB) has been a torment of the mankind, an ancient scourge. Historically, the consummate immensity inflicted by TB is matchless with other diseases on the human race in terms of morbidity and mortality. Consequences of TB have momentous outcome on human existence especially in terms of social and economic part. In history, TB has ruled over in terms of killing millions of people till date; however other diseases like smallpox and plague have also killed many people, but their reign has been relatively short-lived. TB has changed its face as the time has lapsed and evolved itself to dominate over the entire cure. However, even today TB remains as a daunting enemy frightening to exterminate humans.

Despite many advances in the twentieth century, tuberculosis still executes somewhere in the range of 2–3 million individuals worldwide every year and is rejuvenated in many parts of our world (Drobniewski et al. 2003). *Mycobacterium tuberculosis* is the source of infection which is responsible for tierce of the earth's human population (nearly 2 billion people) (Drobniewski et al. 2003). The current pandemic is alarming due to the publishing of multiple drug resistance and increasing comorbidities. In this chapter, we discuss about the most accepted diagnostic techniques innovated in laboratory diagnosis of the opportunistic pathogen and a devastating killer known as "the *Mycobacterium tuberculosis*".

23.2 Epidemiology

TB is among the top ten leading causes of mortality worldwide especially from solitary infectious agent. Approximately, 10.4 million people fell ill with TB in 2016: adults counted for 90%, males 65%, people with HIV (74% in Africa) 10%, and five countries contributing to 56%, India, Indonesia, China, the Philippines, and Pakistan (WHO Global TB report 2017).

In 2016 alone, 6.3 million new instances of TB were accounted for (up from 6.1 million in 2015), proportionate to 61% of the evaluated occurrence of 10.4 million. In 2016, 41% of the 3.4 million new bacteriologically affirmed and recently treated TB cases told internationally were reported to have used rifampicin, up from 31% in 2015 (WHO Global TB Report. 2017). Immediate and exact determination of TB, HIV-related TB, and drug-resistant TB, trailed by arrangement of treatment in accordance with global norms, anticipates well-being among individuals who build up the infection. Expanding access to diagnosis ahead of schedule and precise determination utilizing a WHO-prescribed fast demonstrative is one of the three principle destinations of TB lab fortifying endeavors under the End TB Strategy (WHO Global TB Report 2017).

23.3 Diagnosis

Diagnosis reveals a disease or condition by laboratory-based investigations. To comprehend why presently our world is impairing to end TB, we need to understand tuberculosis (TB) diagnosis. Treating a TB-infected patient involves finding and diagnosing the cause first. In the field of TB, challenge starts at the very first step, i.e., diagnosing the case. TB is identified by detecting *Mycobacterium tuberculosis* in a patient clinical specimen. Prompt confirmation of tuberculosis is still a demanding problem in paucibacillary and extrapulmonary forms.

Traditional TB diagnosis banks on medical history, tuberculin skin test, chest X-rays, smear microscopy, and bacteriological examination. For survival, rapid determination of tuberculosis and its susceptibility toward drugs is a requisite and by screening infectious cases fosters contact tracing, enactment of institutional cross-infection plan of action, and other public health operations.

Therefore, the urgent need for rapid diagnosis and confirmed recognition of active *M. tuberculosis* disease that can be achieved at the point of care, i.e., outside the traditional laboratory setting, is still in quest even after several diagnostic modalities. The fresh dynamism of copious policy makers, funding agencies, and grants promotes the need to foster innovations that deliver better diagnostic tools having higher confidence while using affordable approaches. Following detection, treatment and complete eradication of the TB menace and its associated risks can be taken with confidence to fulfill End TB Strategy goals.

23.3.1 Laboratory Diagnosis of Tuberculosis

Tuberculosis history is equally matured to human history. Hints of it have been found from Neolithic entombment locales somewhere in the range of 7000 years of age and in antiquated Egyptian mummies. Hippocrates distinguished the ailment, calling it "phthisis" or "consumption," and thought that it was lethal. Anyway until the seventeenth century because of progressively exact neurotic and anatomical illustration of the illness being recorded, TB started to be distinguished. *Early detection work*: During 1720, Benjamin Marten (an English clinician) recommended that consumption, purported in light of the fact that it appeared to devour individuals from inside, could be brought about by "wonderfully minute living creatures." It would at long last be left to German doctor Robert Koch to really "see" them by utilization of a recoloring procedure called staining. Prior work on isolation of the tubercle bacilli by the bacteriologist was first introduced in Berlin on 24 March 1882. Utilizing exceptionally arduous techniques, Koch could disengage *Mycobacterium tuberculosis* by utilizing Tindall's blood serum medium (Brock 1999).

TB Laboratory Diagnostic Tool Development: The Objectives
The general objective for enhancing TB diagnostics is to help the advancement of economical understanding of patient-focused applications on basic innovation stages, fitting to various levels of developing and underdeveloped nation well-being frameworks. This should prompt generally speaking enhanced access to proper medicines and/or treatments.

Objectives for TB Diagnostic Test Development

- Ease the process of TB case detection, including smear-negative, extrapulmonary, and pediatric TB, by raising the sensitivity and specificity and expanding improved approachability
- Development of simple, accurate, safe, rapid, and inexpensive assays that are targeted for point-of-care level usage in the healthcare system and reducing the turnaround time (TAT) to few hours
- To authorize more efficacious TB treatment monitoring
- To screen drug resistance rapidly in both first- and second-line anti-TB drugs
- To dependably recognize latent TB infection and decide the danger of movement to dynamic (active form) infection, empowering the reasonable utilization of preventative treatment

23.4 See the Bug

23.4.1 Smear Microscopy

Microscopy stood as a mainstay of tuberculosis control since TB diagnosis initiation, due to its ability to capture sputum-smear-positive (most infectious) cases, is less time consuming and costs less, although found to be low in specificity. Sputum smear microscopy is developed more than a century ago; the presence and/or absence of acid-fast bacilli is determined in a sputum sample using a microscope in this technique.

It is for the most part hard to analyze TB when the bacterial load is under 10,000 for each milliliter of the sputum test, giving wrong negative outcomes for a few patients (Mudur 2017). From this time forward, smear microscopy affectability is expanded by utilizing different fluorochrome dyes, for example, auramine and rhodamine, and fluorescent-recoloring procedures. Not only fluorescence microscopy sensitivity is 10% higher than conventional Ziehl-Neelsen (ZN) microscopy, but also the examination of fluorochrome-stained smears takes less time. Major shortcomings in fluorescence microscopy are its high cost, due to expensive mercury vapor light sources, periodic maintenance, and need of a dark room. The combination of light-emitting diodes (LED) and fluorescent light output for fluorescent microscope (FM) has developed a simple yet robust LED FM microscope, which requires less power along with no dark room requirement. The endorsement of WHO has established LED FM as an alternative to light microscopes (LMs) especially in resource-limited settings (Minion et al. 2011).

Smear microscopy detects most infectious pulmonary tuberculosis cases. The quality of smear microscopy is crucial in TB diagnostics in resource-limited settings as the performance is dependent on its execution (Rieder et al. 2007; CDC Homepage [https://stacks.cdc.gov/view/cdc/31282] 2000).

Low sensitivity (25–75% in respect of culture) and more bacilli requirement for positivity (approx. 5×10^3–10^4 bacilli per ml) are the major limitations. Sensitivity and the positive predictive value (PPV) are affected by scores of aspects (CDC Homepage [https://stacks.cdc.gov/view/cdc/31282] 2000; WHO Roadmap 2010; Stop TB Partnership 2008), namely, prevalence and infectiousness of the disease, the type, consistency and quality of the sample provided, mycobacteria number in the sample and smear preparation quality, and staining and observing process. Mycobacterial species are not identified and also the bacilli viability is not detected in the sample through smear microscopy. In case of HIV-TB-coinfected patients and patients with paucibacillary samples, smear microscopy result is mostly negative and requires diligent screening to identify scanty AFB bacilli (Figs. 23.1 and 23.2).

23.4.1.1 Histopathology

Laboratory diagnosis of extrapulmonary tuberculosis is done through histopathological examination of tissue specimens. The sensitivity mainly depends on ease of sampling and site-specific sample aspiration. Reduced granuloma formation in patients with HIV infection hampers the specificity. Trained experts and resource facilities for histopathology are commonly unavailable in many low-resource country settings.

Advantages Sputum smear microscopy is an easy, inexpensive, and rapid method in quickly detecting infectious cases of pulmonary TB, having sufficient number of AFBs captured easily by sputum microscopy.

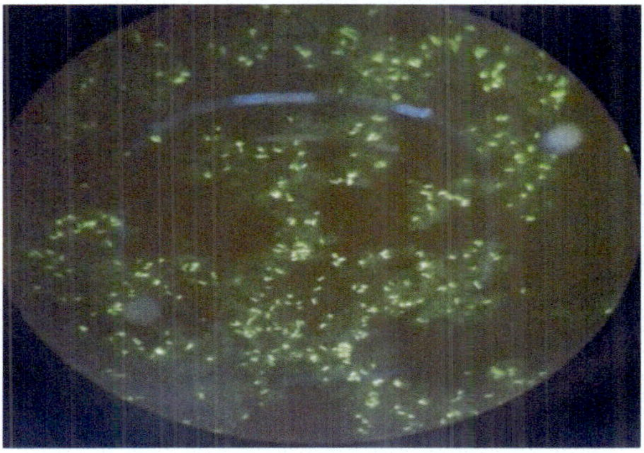

Fig. 23.1 Fluorescent microscopy result from a direct sputum sample in a (3+) AFB-positive TB patient

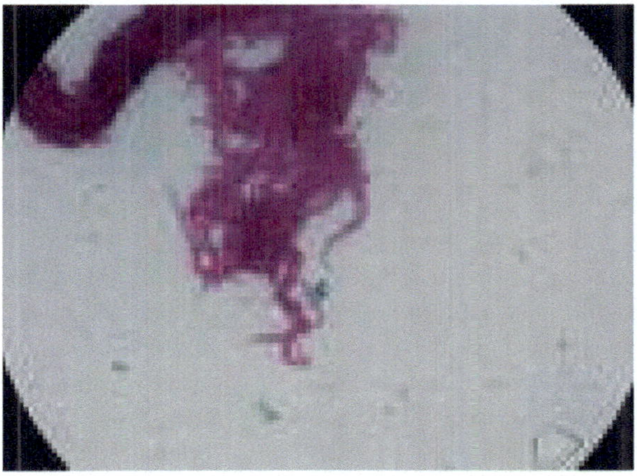

Fig. 23.2 ZN smear microscopy of *M. tuberculosis* culture positive, showing characteristic cord formation

Disadvantages Smear microscopy requires at least 5000 bacilli per milliliter of sample for a positive result in direct microscopy. In patients suffering from extrapulmonary TB, along with HIV coinfection and also those with nontuberculous mycobacteria (NTM) infection, it is documented that sensitivity of smear microscopy is further reduced.

Limitations NTM are not differentiated from MTBC, while dead and viable bacilli are also not distinguished. Moreover, drug-resistant forms are not differentiated from susceptible strains.

23.4.2 Grow the Bug

23.4.2.1 Culture

Sputum mycobacterial culture served as the "gold standard" diagnostic assay in TB diagnosis, and it also facilitates DST for definitive diagnosis of drug-resistant forms

Fig. 23.3 Characteristic colony morphology of *M. tuberculosis*
(**a**) Solid culture on LJ medium. (**b**) Image showing rough, buff, tough, ivory-colored MTB colony morphology. (c) Liquid culture MGIT-positive tube showing flaky growth

of tuberculosis. Because of the higher sensitivity of culture, capturing of only few numbers of bacilli (approximately 10 bacilli/ml in contrast to a minimum of 5000 bacilli/ml of sputum for microscopy) is also possible. Detection of a number of TB cases is increased by 30–50% with the use of culture techniques. Moreover, for species identification and drug susceptibility testing (DST), cultures are more useful (Migliori et al. 2012; De Kantor et al. 1998).

One of the major touchstones for accurate recognition of *Mycobacterium tuberculosis* is culture, but more time consumption and often no recovery in paucibacillary specimens are significant constraints. However, requirement of dedicated laboratory with technical expertise for mycobacterial culture and DST are often not routinely available in high-burden countries with TB, limiting their usefulness. The turnaround time for conventional sputum mycobacterial culture is 6–8 weeks and therefore the definitive diagnosis is delayed.

Solid Media

Egg-based media: Lowenstein-Jensen (LJ) medium is an egg-based enrichment media, having malachite green (inhibitor of non-mycobacterial organism), especially for sputum culture. Glycerol in LJ favors *M. tuberculosis* growth, while LJ lacking glycerol instead consisting sodium pyruvate benefits *M. bovis* growth (Leao et al. 2004) (Fig. 23.3a, b).

Ogawa medium, an LJ lacking asparagine, nonselective, egg-based medium, is refrigerated for long-term usage even after few months, if tube caps are tightly closed to minimize evaporation.

A drawback is that if contamination occurs, it engulfs the whole surface of the slant, resulting in culture loss. For specimens with less bacillary load, it may take three to eight weeks for culture to be positive.

Agar-based media: Most commonly used, Middlebrook 7H10 and 7H11 media are often formulated in the laboratory from powdered agar bases, by adding Middlebrook oleic acid-albumin-dextrose-catalase (OADC) enrichment. Typical cord formation is a characteristic feature of *M. tuberculosis* and can be counted as early as one week after incubation because of the clearness in 7H10 and 7H11 plates. In addition, colonial morphology on agar plates is better visible as compared to egg-containing slants, which is helpful for the identification of mycobacteria. Slant tubes and/or plates are prepared and are less likely to become contaminated. Middlebrook 7H11 has little advantage over 7H10 as it contains 0.1% casein hydrolysate, which helps in recovery of isoniazid-resistant mycobacteria. In addition, 7H11 supports better growth of multidrug-resistant (MDR) strains, lacking 7H10 agar plates (Iseman 2000).

Liquid Media

Liquid media support rapid growth than solid media. TAT for Middlebrook 7H9 liquid medium takes 7–14 days, Middlebrook 7H11 agar takes 18–28 days, and LJ medium takes 21–42 days (Mase et al. 2007).

For the fulfillment of the aim of early growth of *M. tuberculosis*, numerous diagnostic methods and modifications have been developed during the last four decades. In the 1980s, semiautomated and automated liquid culture systems became available, which facilitated rapid growth and detection of *M. tuberculosis* with a TAT of about 10 days (Palomino 2012).

Few essential methods, adopted by RNTCP in India and also endorsed through WHO, are the following:

(i) *BACTEC system*: This assay depends on utilization of substrate palmitic acid converting to radioactive carbon dioxide. In this framework, a colorimetric sensor recognizes the creation of CO_2 broke down in the culture medium. This assay is based on generation of radioactive carbon dioxide. In this system, a colorimetric sensor detects the production of CO_2 dissolved in the culture medium. TAT is 5–10 days.

(ii) *MGIT (mycobacteria growth indicator tube)*: MGIT tube contains modified Middlebrook 7H9 broth, OADC growth supplement, and mixture of antimicrobial compounds known as PANTA (polymyxin B, amphotericin B, nalidixic acid, trimethoprim, azlocillin). Growth is detected, based on O_2 consumption by aerobic microorganisms, through a detection system (nonradioactive in nature) which utilizes fluorochromes for detection and drug screening. Early growth detection (7–12 days) is done and is useful for rapid phenotypic drug susceptibility (Fig. 23.3c).

(iii) *ESP culture system II*: In this method, *M. tuberculosis* growth detection occurs due to estimation of rate of oxygen consumption which is present within the upper headspace of the culture vials.

Advantages Identification of *M. tuberculosis* from culture confirms TB diagnosis, which significantly increases (30–50%) the number of cases. Due to its high sensitivity, it can detect cases even in smear-negative cases. Culture not only provides the necessary isolates but also serves as a base for putting up the conventional DST.

Disadvantages Culture medium are highly prone to contamination and expensive. It requires dedicated facilities and infrastructure for media preparation, specimen processing, and growth of organisms, along with skilled laboratory personnel and specialized biosafety conditions.

Limitations Specimen decontamination is necessary prior to be inoculated in culture to prevent overgrowth by other microorganisms. Even culture is not 100% sensitive, as all decontamination methods are also harmful to mycobacteria. Culture is less sensitive in specimens with paucibacillary load. Although gene amplification methods are nowadays much more sensitive, they pose chances of false positivity and reason being primarily contamination (Rodrigues et al. 2009).

23.5 Identify the Bug

23.5.1 Identification of *Mycobacterium tuberculosis* Complex

Since biochemical tests are time taking in identifying *M. tuberculosis* complex, in 2007, WHO recommended liquid culture assays, DST, and rapid speciation (strip speciation) tests based on the detection of a TB-specific antigen from positive cultures, for confirmation of *M. tuberculosis* complex (Capilia TB; Tauns Laboratories Inc., Shizuoka, Japan (Shen et al. 2009) helping in rapid diagnosis of TB and MDR-TB).

23.5.1.1 Chromatographic Immunoassay

A qualitative identification method for differentiating *Mycobacterium tuberculosis* complex from positive cultures. Certain diagnosis of TB actually occurs when *M. tuberculosis* complex is differentiated from other organisms in a clinical sample which is culture positive in solid or liquid media. As *M. tuberculosis* complex strains (except some sub-strains of *M. bovis* BCG and nontuberculous mycobacteria) predominantly secrete the MPB64 protein, diagnostic discrimination can be done between *M. tuberculosis* complex and nontuberculous mycobacteria. The principle reaction using monoclonal antibodies against MPB64 has been developed and evaluated as the basis of such assays (Abe et al. 1999; Hillemann et al. 2005; Hirano et al. 2004).

Lateral Flow Assays Available BD (Becton, Dickinson and Company) Diagnostic Systems, Sparks, MD, USA; SD Bioline MPT64 assay, SD Bioline assay, Capilia TB-neo, TAUNS Laboratories Co, Numazu, Japan; and BD MGIT TBc identification test (Fig. 23.4a, b).

a

b

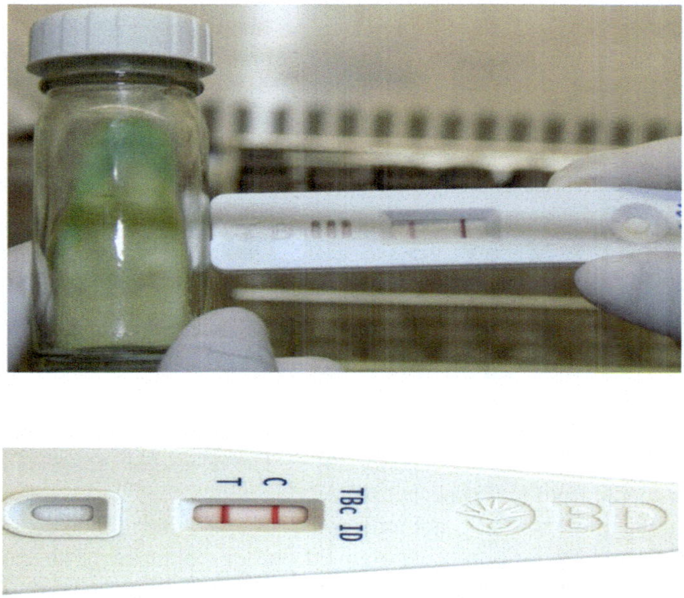

Fig. 23.4 Lateral flow assays showing positive bands for confirmation of MTB complex (**a**) Positive LJ culture with positive *M. tuberculosis complex* result in SD Bioline MPT64 assay (**b**) Positive *M. tuberculosis complex* result in BD TBcID assay

23.5.1.2 Serological Methods

Serological methods use blood serum and can also be easily applied in resource-limited settings, therefore being an attractive choice as a diagnostic tool. In contrast, false positivity remains a major limitation in antibody detection with less defined preparations in high-burden countries like India.

Screening of people having latent TB infection requires the development of new diagnostic tools and standards to curb the scourge (Thillai et al. 2014; Diel et al. 2012). Interferon-gamma release assays (IGRAs) were developed to diagnose LTBI; however tuberculin skin test (TST) proved to be the least expensive test (Thillai et al. 2014). The TST and IGRA work on the principle of capturing the interaction of T cells to TB antigens (Cruz-Knight and Blake-Gumbs 2013).

Tuberculin Skin Testing (TST)

MTB protein, namely, tuberculin, a derivative mix of proteins (PPD), is injected into a person in subcutaneous tissue intradermally. With previous exposure to mycobacterial proteins in the vaccine or having any latent MTB infection, a skin reaction occurs due to type IV delayed hypersensitivity as an immunologic response (Thillai et al. 2014). Skin induration reaction size is measured with readings at 48 and 72 h and a standard cutoff value of 10 mm. However, false-positive responses are documented in persons who are BCG vaccinated previously during infancy and false-negative responses also documented in immunosuppressed individuals. In India, TST is being used in diagnostic algorithm for pediatric age group as a supportive evidence.

IGRAs

Interferon-gamma release assay test was developed, as the percentage of false reporting is high in interpretation of tuberculin skin test. Two available examples of IGRA assays are Quantiferon-TB Gold and T-Spot TB test. The IGRA assay diagnosis detects MTB antigens, including the early secretory antigen target 6 (ESAT-6) and culture filtrate protein 10 (CFP-10), both which stimulate interferon-gamma production in the host.

The IGRAs are more sensitive (81%) and specific (88%) which is 10% more than 70% sensitivity noted in TST (Diel et al. 2012). Detection of cytokine IFN-g is made which is released from T cells, reacting antigens not present in the BCG vaccine (Diel et al. 2012; Goldman and Schafer 2011). The release of IFN-g is measured in the blood sample from the suspected patient.

Immunodiagnostic tests are not universally accepted due to their cross reactivity and poor sensitivity especially in high-TB-burden countries (with exception of tuberculin skin test). Indian government has banned the use and also the sale of these diagnostic tests because of their high false reporting issues.

Transformation of latent infection to active TB form in healthy individuals and its overall understanding overlay foundations in better future diagnostics. Identifying risk factors involved in high and/or low-TB-burden settings will definitely solve the equation of the complex immune response in TB, and its better understanding will lead to the invention of more useful diagnostic tools which can change today's scenario.

23.5.1.3 Drug Resistance Detection Using Phenotypic Assays

Rapid reporting of drug susceptibility testing to first-line drugs and/or second-line drugs requires detection of drug resistance and is the key for the earlier implementation of treatment in a bacteriologically confirmed case. Phenotypic drug susceptibility testing (DST) methods are performed in solid or liquid media as direct or indirect tests.

Direct testing involves a batch of drug-containing and a drug-free media, on which a concentrated specimen is inoculated directly. In contrast, in indirect testing, a pure culture cultivated from the original specimen is inoculated on the drug-containing media.

Phenotypic methods such as indirect testing assays are extensively validated and are honored as the gold standard. Three methods are commonly used: *proportion, absolute concentration, and resistance ratio*. Moreover, DST results among the three methods for first-line anti-TB drugs usually did not vary.

DST methods are based on growth in solid media using a manual system and using both automated and manual systems in case of liquid medium. For predicting the results, direct observation and calculation of growth by colony count method in solid media and in liquid broth media were made through oxygen consumption due to CO_2 production or enzymatic modification of the media.

First-line DST Drug sensitivity testing techniques forecast result to one medication tested as a delegate from a group of medications (e.g., rifampicin results cover rifampicin and rifapentine, prothionamide covers ethionamide, and the other way around). First-line anti-TB drugs incorporate isoniazid (H), rifampicin (R), ethambutol (E), and streptomycin (S), despite the fact that streptomycin is currently not part of most standard medications, and pyrazinamide (Z), which is normally part of the standard treatment yet, isn't viewed as a first-line testing drug because of the trouble in getting dependable outcomes. Rifampicin and isoniazid results are reproducible most accurately than streptomycin, ethambutol, and pyrazinamide results. After confirmation of MDR-TB, rapid susceptibility testing of additional first- and second-line drugs needs to be performed following current WHO recommendations (WHO Policy Statement 2008).

Second-Line DST Assay It is a challenging, multiplex, and high-priced method. Well-promoted and endorsed liquid culture and DST methods and the standardized solid proportion method are validated. Automated liquid systems are recommended for second line drugs, as the current gold standard phenotypic DST method. MDR-TB cases are often treated by more potent second-line group of anti-TB drugs, e.g., the injectables (kanamycin, capreomycin, and amikacin), the fluoroquinolones (ofloxacin, levofloxacin, and moxifloxacin), ethionamide or prothionamide, clofazimine, cycloserine, and linezolid. Aminoglycosides, cyclic peptides, and fluoroquinolones have already proved their reliability and reproducibility, promising a quality assured diagnosis of XDR-TB.

After almost four and a half decades, news regarding novel drugs for MDR/XDR-TB treatment have raised some considerable hope. These drugs have been developed following several trials and they have also succeeded in receiving provisional approval. Two examples are *bedaquiline*, a drug from diarylquinolone family (developed by Janssen Pharmaceutica), and *delamanid*, a dihydro-nitroimidazooxazole derivative; both are also being used in MDR TB treatment in some countries after approval. Regulatory approval for use of bedaquiline and delamanid in MGIT DST method is still in trial phase in high-burden countries.

DST Using Noncommercial Methods These DST methods are less expensive and having lack of standardization as it is vulnerable to errors and demographic modifications in adopting the methodology. Hands-on expertise is mandatory in performing these methods following good laboratory practices (GLP), best microbiological techniques, and with more appropriate quality assurance, backed up with enough training. WHO has selectively reviewed the performance for available noncommercial culture and DST methods that were providing adequate results under controlled laboratory protocols in reference/national laboratories (WHO Policy Statement 2010a, b).

The most common methods incorporate microscopic observation of drug susceptibility (MODS), colorimetric redox indicator (CRI) methods, and the nitrate reductase assay (NRA).

MODS Specimens inoculated to drug-free and drug-containing liquid culture media are compared. Following this, early growth is observed in microcolony method through microscopic examination. Hence this method is also recommended for rapid screening of patients suspected with MDR-TB.

CRI Methods Liquid culture medium having a specific colored indicator in a microtiter plate is used and reduced after in vitro exposure of *M. tuberculosis* strains to anti-TB drugs. Moreover, this method has got several recommendations as an indirect test on *M. tuberculosis* isolates from MDR-TB suspects. Time to detection (TTD) of MDR-TB is not quicker than conventional DST methods using commercial liquid culture or molecular line probe assays but certainly cheaper.

NRA The ability of *M. tuberculosis* to reduce nitrates forms a basis of detection through a colored reaction in this assay. It is therefore recommended as a direct and/or indirect method on solid culture, for screening of MDR-TB suspects. It acknowledges that time to detection of MDR-TB is comparably slower than conventional DST methods using solid culture.

23.6 Multiply the Bug's DNA

23.6.1 Drug Resistance Detection Using Genotypic Assays

The use of molecular biology is imperative in the rapid diagnosis of mycobacteria. Methods based on capturing nucleic acids have taken the place of most of the classical methods as a rapid screening tool. These assays adaptability is advantageous primarily, as directly used on the sample material, or culture, and even can be used as a rapid and unequivocal species differentiation assay from culture material has made them successful screening platforms. Molecular genetic tests offer significant time advancements in the identification of mycobacteria, thereby permitting quick initiation of resistance tests and specific treatment. They are useful tools for the detection and differentiation of mycobacteria from cultures and can have a high specificity and sensitivity. Also, they must be placed in the reinforcement of the conventional diagnostic workup, and therefore, assay results should always correlate by standard methods also.

PCR-Based Assays

Biochemical tests are mostly used for identification of mycobacteria for a long time now (traditional practice), but due to longer TAT, the results are ambiguous; hence, PCR-based assays emerged and were implemented later on. PCR is specific for gene amplification assays, needs careful interpretation, and therefore must be in strong correlation with other parameters such as clinical, histological, cytological, biochemical, imaging, and therapeutic specifications while making final conclusion, especially in drug-resistant TB diagnosis.

Line probe assays (LPA)

LPA have been developed and implemented in most of the countries in their National Tuberculosis Control Program worldwide after endorsement from WHO. LPA, a PCR-based multiplex genotypic DST assay, fulfills the objectives of screening of rifampicin and isoniazid drug resistance and simultaneously identifies *M. tuberculosis* complex in the patient sample directly.

(a) *INNO-LiPA*: INNO-LiPA (RIF & MYCOBACTERIA v2) are two different line probe assay kits used for the drug resistance detection and identification of the genus *Mycobacterium* and 16 distinct mycobacterial species, respectively. Abiding to the manufacturer's interpretation criteria, the following *Mycobacterium* species can be differentiated simultaneously: *M. tuberculosis complex, M. kansasii, M. xenopi, M. gordonae, M. genavense, M. simiae M. marinum, M. ulcerans, M. celatum, MAIS, M. avium, M. intracellulare, M. scrofulaceum, M. malmoense, M. haemophilum, M. chelonae complex, M. fortuitum complex,* and *M. smegmatis.* This kit performs on the basis of nucleotide differences in the 16S–23S rRNA spacer region and can be done on either liquid or solid cultures (Scarparo et al. 2001; Tortoli et al. 2003). The INNO-LiPA test comprises three major steps: DNA extraction, DNA amplification, and hybridization.

(b) *Genotype line probe assay*: The GenoType series is based on the patented *DNA strip technology* and it permits the genus *Mycobacterium* to be genetically differentiated to different species.

The *Hain CM test* kit is designed to identify the most common mycobacteria, and another AS assay differentiates additional *Mycobacterium* species further. Species differentiation among the *M. tuberculosis* complex is done by GenoType MTBC. Only after the culture turns out to be positive, the CM, AS, and MTBC strips can be able to detect and analyze mycobacteria but not from patient material directly.

The *GenoType MTBDRplus* V2 test simultaneously detects *M. tuberculosis* complex and its inherent drug resistance to rifampicin and/or isoniazid by screening most common mutations in the rpoB and katG/inhA (high/low isoniazid resistance) genes, respectively (Nikolayevskyy et al. 2009) (Fig. 23.5a).

Fig. 23.5 In figure (**a**), DNA strip is interpreting MDR results in an MTBDRplus V 2.0 kit
In figure (**b**), DNA strip is showing interpretation of MTBDRsl results of an XDRTB patient

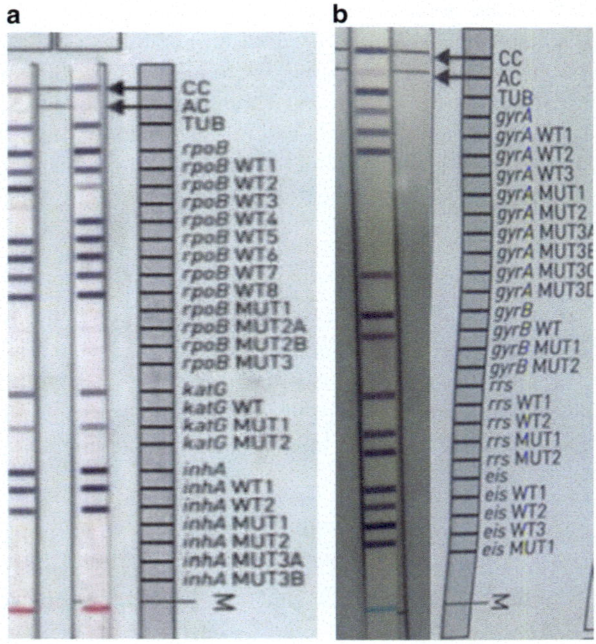

The *GenoType MTBDRsl* V2 detects not only the confirmation of *M. tuberculosis* complex but also its resistance to second-line anti-TB drug groups, namely, fluoroquinolones by screening mutations in the *gyrA* and *gyrB* genes, and/or second-line injectable drugs (SLID), aminoglycosides/cyclic peptides by mutations in the *rrs* gene (injectable antibiotics as capreomycin, viomycin/kanamycin, amikacin), and/or low-level kanamycin resistance by mutations in the *eis* gene (MTBDRsl kit insert 2018) (Fig. 23.5b).

Although endorsed by WHO and other international recommendation agencies, these tests have some limitations. A major shortcoming of the DNA probe technology is its incapacity in identifying mixed cultures. However, additional probes might yield a result in case of any mixed cultures. Additionally, a little number of species only are differentiated by these additional probes. Moreover, these MTBDRplus and MTBDRsl assays are not validated on smear-negative specimens and extrapulmonary specimens, confirming to the kit limitations. Vanquishing such limitations, specific gene probes coupled with hybridization, polymerase chain reaction-based assays, and DNA sequencing platforms have been adopted nowadays.

(c) *Cartridge-based nucleic acid amplification test (CBNAAT)*: Majority of real-time PCR assays have now been implemented in rapid and specific detection of

M. tuberculosis in the clinical specimens. CBNAAT assays have high sensitivity unlike smear microscopy particularly in HIV-positive people and less time-consuming ability (less than two hours) comparing to sputum culture (several weeks). Along with the rapid detection of drug resistance, it shows the uniqueness and easy implementation of this technology.

The Xpert MTB/RIF from Cepheid A quantum leap in TB diagnostics is invention of a nucleic acid amplification test that is fully automated, as sputum smear microscopy turns out to be a century-old technique. Xpert after endorsement from WHO has been rolled out in many countries including India for the rapid screening of rifampicin drug resistance and also adopted by technical and operational guidelines in India for tuberculosis control program. Xpert MTB/RIF is also recommended as a stand-alone diagnostic screening test in individuals at risk of MDR-TB. The use of GeneXpert has not only diminished the TTD for rifampicin drug resistance significantly from days to only two hours, but also the *near-patient device* can be operated with minimal laboratory infrastructure requirement in a mobile van to reach outskirt areas. However, GeneXpert does not obliterate the need of conventional microscopy, culture, and anti-tuberculosis drug sensitivity that are required to monitor the progression of treatment and for drug resistance determination to drugs other than rifampicin (ISTC 3rd edition 2014). Recent developments from Cepheid show promise and commitment to develop and implement other ATT drugs cartridges, which are used in MDRTB treatment including second-line drugs also (Fig. 23.6a, b).

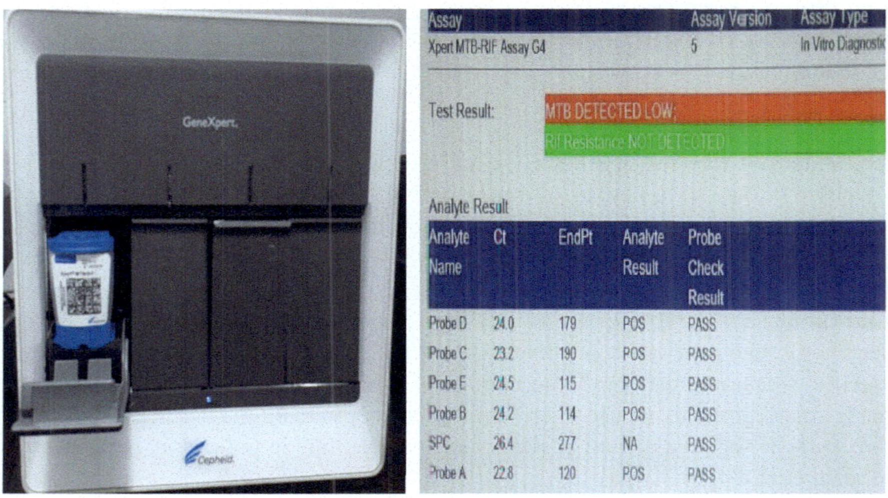

Fig. 23.6 Above figure (**a**) showing GeneXpert system with a cartridge and figure (**b**) showing GeneXpert result in a patient sensitive to rifampicin

(d) *TrueNat*: The TrueNat TB test (in-house product from India) has shown promising results as a new molecular test that can diagnose TB in one hour and simultaneously detect drug resistance to rifampicin. The TrueNat machine is designed more likely as a point-of-care machine, is semiautomated, and can handle outdoor situations without electricity and fulfills the objective with maximum conclusion on minimum prerequisites. Sample testing can be done as soon as a suspected TB patient with symptoms is seen. The system is battery operated, and with the help of a handheld device, it can be used in far-flung areas in the periphery (Chaitali et al. 2014). The procedure involves DNA extraction which spends 25 minutes initially and another 35 min to diagnose TB, which is an additional one hour for testing of rifampicin drug resistance (The Hindu Webpage 2017).

(e) *Loop-mediated isothermal amplification (LAMP)*: TB LAMP is a manual nucleic acid amplification test comprising of DNA amplification along with gene detection in a single step. This test can be used at outreach centers as a near-patient assay and does not require sophisticated instruments and laboratory infrastructure (manufactured by Eiken Chemical Company Ltd (Tokyo, Japan) (Eiken Chemicals Webpage 2017)).

Advantages LAMP is rapid with turnaround time (TAT) of 40 min; it gives result usually observed through the naked eye under ultraviolet light. Additional 40% more patients are detected due to increased sensitivity than smear microscopy as in smear-negative samples where bacillary load is less. Once smear microscopy results are known and documented, it can also be implemented as an add-on test (WHO Policy Guidance 2016).

Limitations It does not differentiate among drug-resistant TB and drug-sensitive TB, and it is having less specificity as compared to smear microscopy, thereby producing increment in false-positive results.

(f) *Genedrive*: Recently TB detection and rifampin-resistant TB detection are now possible through a new assay based on NAAT platform. Genedrive Company (formerly called Epistem) developed Genedrive, but its low accuracy is not supporting the WHO requirements.

Limitation Sensitivity was found to be zero percent in smear-negative samples and specificity 45.6% in smear-positive samples which is considered suboptimal compared to smear microscopy. More number of false-negative result, failing to detect many cases, provides no use over smear microscopy. Failing TB detection even in direct smear-negative culture-positive samples, resulting in poor assay performance (Shenai et al. 2015).

By and large, quality enhancement techniques for gene amplification are built up to be exceedingly sensitive and explicit for analysis of tuberculosis (specificity) straightforwardly from clinical examples. Contingent on the bacteriological status and copy number of target sequence, affectability (sensitivity) has extended from 70% to 100% though explicitness (specificity) between 80% and 100% has been accounted for different modalities.

23.7 Typing the Bug

23.7.1 Molecular Typing of *Mycobacterium tuberculosis*

Molecular typing involves DNA fingerprinting of *M. tuberculosis* strains, thereby improving the understanding in complexity of TB transmission. Moreover, the recognition of genotype families has facilitated many research groups working on the population structure of the *M. tuberculosis* complex and exploring its transmission dynamics. Several molecular typing assays for *Mycobacterium tuberculosis* are now available, with disparate levels of reproducibility, discriminative power, and technical competence requirement.

Spoligotyping (Spacer Oligonucleotide Typing)
Spoligotyping is a direct, less expensive, quick, and reproducible assay to contemplate the phylogeny of *M. tuberculosis* complex strains and to connect phenotypic highlights of disengages with the genotype family the *Mycobacterium* speak to. Clinical specimens of *M. tuberculosis* complex are recognized by the presence or absence of at least one spacer.

Polymorphism in the direct repeat (DR) locus is detected in the mycobacterial chromosome. Direct repeats conserved regions (36 bp) are interspersed with unique spacer sequences (from 35 bp to 41 bp) in size. 43 spacer sequences are used in spoligotyping for *M. tuberculosis* complex strains. Multiple copies of variable DNA spacers in the genomic DR region of *M. tuberculosis* complex isolates are produced through PCR. Amplified PCR products are further hybridized on a membrane having covalently bound spacer oligonucleotides, deduced from DR region sequences (Brudey et al. 2006). The SpolDB4, a publicly available database on the *M. tuberculosis* complex, is one of the largest and contains spoligo patterns from approximately 40,000 clinical isolates representing 122 countries. Spoligotyping is more commonly applied to identify the genotype family; and this technique is less fit for strain typing.

MIRU-VNTR
Mycobacterium Interspersed Repetitive Units - Variable Number Tandem Repeats typing (MIRU-VNTR) assay is developed to identify the genetic polymorphisms within bacterial species. MIRU-VNTR works on the basis of the number of tandem repeats differences in the 24 stretches of the genome of *M. tuberculosis* complex strains. Using amplification, up to 24 loci are amplified with the help of primers

specific for the flanking regions of each repeat locus; later on to deduce the number of tandem repeats present, sizes of the amplified stretches of tandem repeats are determined (Kremer et al. 2005). A numerical code is generated by the detection of number of tandem repeats at the different loci. That code serves as a DNA fingerprint of the respective tuberculosis bacteria. VNTR typing has multiple applications, most commonly contact tracing, source-case finding, and ruling out transmission reliably.

Advantage Its results are easy to compare and is user-friendly than IS6110 RFLP typing. It is mostly suitable as a high-discrimination typing method for highly conserved genotypes.

RFLP Typing
Restriction fragment length polymorphism (RFLP typing) assay detects the number of IS6110 insertion sequences present in the genome difference in strains (Kremer et al. 1999). Due to the presence of highly variable genomic insertion sites in *M. tuberculosis* complex strains, highly variable banding patterns can be obtained. Since it uses its high level of discrimination power and reproducibility, RFLP assay proves to be the most extensively used method for typing bacterial strains over the last few years.

Advantage Although requiring highly experienced manpower and sophistication, RFLP is still superior because of the level of discrimination. Therefore it is also used for strain typing.

23.8 Sequencing the Bug

23.8.1 DNA Sequencing

DNA sequencing involves the amplification and identification of genetic mutations in different genomic regions. It also offers a fast and specific identification of mycobacterial species and drug resistance simultaneously. The principle used in DNA sequencing methods is the determination of species-specific nucleotide sequences, its comparison to quality-controlled sequences known widely available from MTB databases. Although DNA sequencing was invented by F. Sanger in 1977 known as chain termination sequencing, latest developments in the NGS technology were only made after the complete sequencing of *Mycobacterium tuberculosis* genome in 1998 (Cole et al. 1998). There is a rapid surge in development of diagnostic assays using DNA sequencing platform mostly involving pyrosequencing, next-generation sequencing (NGS), and/or whole genome sequencing (WGS).

Pyrosequencing

Pyrosequencing strategy depends on the ongoing real-time sequencing of short stretches at the time of DNA synthesis (sequencing by synthesis) and has demonstrated as a promising new instrument for the fast distinguishing proof of mycobacteria with positive cultures as well as utilizing processed smear-positive sputum examples. This strategy is less demanding to perform and is more affordable than conventional sequencing, yet the shorter DNA stretches are not as discriminating. Automated sequencers can provide results within 1 to 3 days (Pai et al. 2009). Routinely DNA sequencing is labor-intensive and can cost a lot in countries with less infrastructure and poor health settings.

Whole Genome Sequencing (WGS)

Whole genome sequencing provides comprehensive data on resistance mutations and strain typing for monitoring transmission, but unlike for conventional molecular tests, this has previously been achievable only from cultures of *M. tuberculosis*. Detection of known resistance mutations provides a personalized treatment of drug-resistant tuberculosis as compared to empirical treatment. This personalized therapy not only helps improved outcomes but also spares the use of resistant antimicrobials providing a basis of universal DST-guided treatment. Studying whole genomes also allows simultaneous identification of all known resistance mutations as well as specific markers with which transmission can be monitored (Brown et al. 2015a, b).

Small fractions of *M. tuberculosis* genome are interrogated with the help of classic genotyping methods that target highly variable genetic elements. Hence, small evolutions occurring in other genomic regions are not captured. However, NGS assay gives access to nearly full genome sequences. Frequent advancements and increase in affordability have extended the use of NGS technology for research and epidemiological studies. Various genetic markers are investigated with the help of microbial genomics, hampering treatment and infection prognosis. Entire genome sequencing (WGS) is turning into a moderate and available strategy that can recognize microevolution inside MTB ancestries as they are transmitted between hosts (Kwong et al. 2015).

The WGS results in better detection of types of various mutations than other genotypic assays. Moreover, WGS could avoid false positives occurring due to polymorphism present within the specific gene regions. Multiple individual DNA strands can be independently sequenced, providing a true discrimination of heterogenetic variation (e.g., heteroresistance) and quantitative analysis.

Nowadays, a variety of NGS platforms are available, for example, the most rapid Oxford Nanopore (minION, PromethION, GridION), Bio-Rad Laboratories (GnuBio), PacBio (Sequel and RS2), Thermo Fisher (SOLiD, S5, personal genome machine [PGM], Proton), Qiagen (GeneReader), Vela Diagnostics NGS platform (Singapore), and the most common Illumina (MiSeq, HiSeq, and NextSeq).

WGS needs culturing of *M. tuberculosis,* a slow-growing microorganism, for obtaining affluent amount of genomic DNA, making it a time-consuming assay.

However, recently invented WGS assay can be done without prior specimen culture. This method is based on specific biotinylated RNA baits which are specific to MTB and can sequence full MTB genomes directly from non-cultured sputum samples (Brown et al. 2015a, b). In routine, WGS used for clinical specimens is most common in high-income countries because of its high cost, dedicated infrastructure, and requirement of skilled manpower with analysis expertise.

As MTB requires early detection of drug resistance, WGS holds great benefit as a handshake of diagnostic and epidemiological tool combination in a single assay. Compared to conventional molecular tests which focus on mutations in hotspot regions of genes involved in resistance to first- and second-line anti-TB drugs, data generated from WGS analysis applies holistic approach of creating a genome-based personalized medicine in healthcare. Sequencing data generated by investigators and researchers are not yet directly comparable, and universal databases access is limited for surveillance and studies of epidemiological importance.

TBDReaMDB is a database available for optimal prediction of drug resistance mutations; the need of a comprehensive and well-curated database is a must for better correlating genotypic data with phenotypic drug susceptibility testing data. These challenges are being addressed by large international consortia (e.g., http://patho-ngentrace.eu/) in order to accelerate the implementation of NGS-based epidemiology and diagnostics (Sandgren et al. 2009).

The advent of novel anti-TB drugs or modification in regimens requires alignment with novel diagnostic platforms, thereby helping scaled use of new drugs and in developing appropriate genotypic assays to identify drug-resistant alleles to the new drugs.

Crude information produced by NGS and WGS is colossal and presents a huge test to the scientists. Mapping or the genome assembly, base variant calling, and near phylogenetic investigations are a few NGS information data management techniques. Aptitude in bioinformatics abilities are required for handling and breaking down the information utilizing programming and calculations in algorithms which are very perplexing and regularly unlinked to one another.

23.9 What We Can Do with the Bug

23.9.1 Future Prospects in TB Diagnostics

Evolution in laboratory diagnosis of *Mycobacterium tuberculosis* has diminished time lapsed in identification and drug susceptibility results. Continuous efforts have been made for increasing reproducibility, improvement of performance, and minimizing cost and infrastructure requirement. Some of the future projections that are going to be implemented in TB diagnostics are discussed here:

23.9.1.1 Alere Determine™ LAM Assay

It is a lateral flow (LF)-based immunodiagnostic strip test for the detection of lipoarabinomannan (LAM) antigen in urine. LF-LAM support the diagnosis of TB in HIV-positive adult patients. Patients unable to expectorate sputum, with other immunocompromised condition, and patients unable to move out will definitely be benefitted with the use of LF-LAM assay, especially in ill diagnostic geographic conditions (WHO Policy Update 2015).

23.9.1.2 Cepheid GeneXpert Ultra

An updated version of GeneXpert has already undergone multicentric trials in 2017. Ultra works on the same principle of CBNAAT as used in the earlier GeneXpert, but with greater sensitivity is comparatively high in direct smear-negative rather culture-positive specimens, pediatric specimens, extrapulmonary specimens (notably cerebrospinal fluid), and especially for paucibacillary specimens mostly from HIV-positive individuals. CBNAAT Ultra cartridge is renovated with improved assay performance, resulting in a limit of detection (LOD) of 16 bacterial colony-forming units (cfu) per ml compared to previously 114 cfu/ml in Xpert® MTB/RIF (WHO Meeting Report of Technical Expert Consultation 2017).

In addition, XDR TB assay has also incorporated probes for detecting mutations responsible for resistance toward second-line anti-TB drug groups, namely, INH, FLQ, and AMG, using asymmetric PCR. Sloppy molecular beacons and fluorescent melt curve analysis is supporting the PCR phenomenon to identify genotype drug resistance (Roh et al. 2015). Currently under pilot run, the XDR TB assay release date and price are not known till date.

23.9.1.3 TRCReady® 80

TRCReady® 80 is a molecular diagnostic tool, developed by Tosoh Bioscience (Japan) as a stand-alone assay with a capacity of processing up to eight samples, and requires a computer platform for operation. TRCRapid® M.TB kit is marketed by Tosoh Company for use on this platform. Fully automated sample purification, amplification, and detection are incorporated in this assay (Drouillon et al. 2009). Amplification of MTBC specific target is done by using transcription reverse-transcription concerted reaction (TRCR) amplification technology. Total fluorescence is measured in real time as increase in amplification reactions and takes only 30 minutes to complete.

23.9.1.4 Volatile Organic Compounds (VOCs)

The VOCs here are mycobacterial metabolites derived from the host's digestion, incorporating microorganisms in the gut or the host's reaction to the mycobacterial disease. Companies have used various names for already developed VOC devices for, e.g., gas chromatography (Menssana Research Inc., USA), metal-oxide sensors (Nanosynth Materials and Sensors, USA, and The eNose Company, Netherlands), and metabolite detection by chemical reaction (Metabolomx, USA). The development pipeline still focuses on products that are at varying trial stages (Bruins et al. 2013; Lim et al. 2016).

These TB screening devices are portable and handheld. At the point when a patient associated with experiencing TB inhales into the device tube, VOCs created by MTB in the lungs tie with titanium oxide nanotubes immobilized in the gadget. An electrical impulse is being produced upon that binding and is captured and read by applications in smartphones. Clinically suspected patients with a positive result are then recommended to undergo mycobacterial confirmation testing. Some of the proposed VOC platforms are Breathscanner™ and Aeonose™.

23.9.1.5 Assays for Diagnosing Latent TB Infection (LTBI)

C-Tb, a novel skin test is built up and named by Statens Serum Institut in Denmark, for distinguishing LTBI. The test estimates the body's resistant reaction to two explicit MTB antigens that are not decoded in the BCG immunization: ESAT-6 and CFP10. Improved TSTs assays in pipeline include Generium Pharmaceutical (the Diaskintest, Russian Federation), and china-based skin test known as ESAT-6 has already undergone phase II trials in China (Sun et al. 2013).

23.9.1.6 Automated Microscopy Platforms

With the advent of automated imaging, the conventional visual examination for acid-fast bacilli (AFB) by direct sputum smear microscopy is also advanced. Some examples are:

- The *TBDx system*: It is a mechanized computerized automated microscopy platform that is accessible from Signature Mapping Medical Sciences Inc., USA. This stage comprises of a great magnifying lens (higher resolution power) in the microscope and imaging framework that related to a slide holding adjustment that can peruse up to 200 arranged smears utilizing fluorescent microscopy in a solitary run (Ismail et al. 2015).
- *Capture-XT*™: An MTB cell advancement gadget is in development stage named as QuantuMDx (UK), to enhance the affectability (sensitivity) of SSM past current fundamental concentration techniques, for example, sputum sedimentation. The organization has begun to examine the utilization of an MTB cell concentration innovation as a stand-alone assay, namely, Capture-XT™. The gadget will be battery worked and will most likely work for 8 h on a solitary charge. Pilot testing of the innovation has already initiated.

23.9.1.7 Surface Plasmon Resonance (SPR) Sensing

SPR-based biosensors systems having several ultrasensitive detectors will be expected in the next 5 years, as a fully integrated point-of-care SPR imaging-based diagnostic system (FU et al. 2008).

23.9.1.8 Use of Nano-particles

Quantum dots 1–6 nm nanoparticles can be functionalized with capture probe or antibodies (Resch-Genger et al. 2008) coated with capture probe. Mirkin and colleagues built up a promising nanotechnology-based stage for multiplex identification of protein and nucleic acids utilizing a sandwich of standardized magnetic and bar-code probes tag tests (Cheng et al. 2006).

23.9.1.9 Lab on a Chip

Single molecule detection will become possible by the detection of minute amounts of genetic materials. Assays capable of detecting even single molecules are now being developed, e.g., *cantilever beams and nanotube paddle resonators* (Witkamp et al. 2008). These new detection frameworks could be pioneered to deliver "lab on chip" frameworks for the identification of pertinent biomolecules.

23.9.1.10 Future NGS Platforms

Smaller, robust, and user-friendly stages for use in comparatively smaller research facilities are at present being developed. These include Genalysis® from DNAe, Gene Electronic Nano-Integrated Ultra-Sensitive (GENIUS) from GenapSys, and the Genia sequencer from the Genia Corporation (all USA). These innovative devices are proposed to have little cartridge-based frameworks with incorporated sample amplification and sequencing at a cost focus of US$ 100 or less; be that as it may, further subtleties are restricted (Pankhurst et al. 2016).

Barriers and challenges are there always, but present achievements and future probabilities in diagnostics offer tremendous potential and hope. The amalgamation of interdisciplinary sciences like molecular biophysics, nanobiotechnology, immunology, biochemistry, genomics, biomedical sciences, and bioinformatics gives ever-upgrading and revolutionizing scientific and logical research and relocating diagnostic development from the biological to the molecular level as an indicative improvement from the natural to the atomic dimension. These advancements promise of future TB diagnostics that are rapid, accurate, and cost-effective and help address multiple drug-resistant TB strains in high-burden, low-resource countries.

However, stand-alone diagnostic test currently cannot assure of all three "quick," "cheap," and "easy" demands. The gaps in the diagnostic services must be addressed urgently for eradicating tuberculosis. Clinical diagnosis must be supported with the early adherence to anti-TB drug therapy for the benefit of public health in management of tuberculosis. Selection bias compromised interpretation of assay results, and partial selection is also involved in failure of many diagnostic assays which has shown optimistic results initially. Even after decades of research, focus is still on finding new biomarkers required for better detection that is desperately needed along with all available diagnostic methods. With the recent enthusiasm of various stakeholders, policy makers, funding agencies, and grants, the need is now to foster innovations that deliver state-of-the-art tools to diagnose TB with confidence using affordable approaches. High-cost techniques from the developed nations need to be subsidized to support the high-TB-burden countries with resource-limited settings. It is to be trusted that ongoing advances will prompt the improvement of novel demonstrative techniques and diagnostic strategies that are relevant to use in developing and creating countries, to reduce the morbidity and mortality with effective intervention most urgently required (Table 23.1).

Table 23.1 Summary of newer prospects in TB diagnostics pipeline at various stages (detailed table is freely available at https://www.finddx.org/dx-pipeline-status/) and cited with permission from FIND India

In feasibility stage (prototype but not design lock)	In development stage (design locked/pilot manufacturing)	In validation stage (manufacturing in place)	In regulatory stage (CE-IVD, FDA, CFDA, or similar regulatory approval in domestic country)
TB flow (Salus Discovery, LLC) *Biomarker type*: antigens	True array MDRTB (Akonni Biosystems) *Biomarker type*: nucleic acid (RNA, DNA)	Sensitive LAM (Fujifilm) *Biomarker type*: antigens	Sandwich vessel concentration diagnostic product (Hunan-Tech New Medicine) *Biomarker type*: whole bug
MTB antigens POC assays (multiple) Biomarker type: antigens	True array XDR-TB (Akonni Biosystems) *Biomarker type*: Nucleic acid (RNA, DNA)	Aeonose for TB (The eNose Company) *Biomarker type*: metabolite	TB resistance module series (autoimmune diagnostics) *Biomarker type*: nucleic acid (RNA, DNA)
Blood/urine cfDNA POC assays (multiple) *Biomarker type*: nucleic acid (RNA, DNA)	PoC (Bioneer) *Biomarker type*: nucleic acid (RNA, DNA)	ID-FISH assay (ID-FISH Technology, Inc.) *Biomarker type*: nucleic acid (RNA, DNA)	Accupower TB and MDR real-time PCR (Bioneer) *Biomarker type*: nucleic acid (RNA, DNA)
High-sensitivity TB rapid diagnostic (global good) *Biomarker type*: antigens	Xpert XDR (Cepheid) *Biomarker type*: nucleic acid (RNA, DNA)	INFINITY MDR-TB (AutoGenomics) *Biomarker type*: nucleic acid (RNA, DNA)	Tuberculosis drug resistance array kit (Capital Bio-Technologies) *Biomarker type*: nucleic acid (RNA, DNA)
TB Dx (not defined) (Unima) *Biomarker type*: antibody	Omni (Cepheid) *Biomarker type*: nucleic acid (RNA, DNA)	Accupower XDR-TB real-time PCR (Bioneer) *Biomarker type*: nucleic acid (RNA, DNA)	MTB drug-resistant mutation test kits (QuanDx) *Biomarker type*: nucleic acid (RNA, DNA)
Erythra TB (Erythra Inc.) *Biomarker type*: antigens	Q-POC TB/MDR TB (QuantuMDx) *Biomarker type*: nucleic acid (RNA, DNA)	INNO-LiPA Rif TB (Fuji-Rebio Europe) *Biomarker type*: nucleic acid (RNA, DNA)	Mycolor TK platform (Salubris) *Biomarker type*: whole bug
SemanticMD AI for TB (SemanticMD)	TB MultiTest (Selfdiagnostics	AdvanSure MDR-TB	Anyplex assays for MDR/XDR series

(continued)

Table 23.1 (continued)

In feasibility stage (prototype but not design lock)	In development stage (design locked/pilot manufacturing)	In validation stage (manufacturing in place)	In regulatory stage (CE-IVD, FDA, CFDA, or similar regulatory approval in domestic country)
Biomarker type: others, imaging	Deutschland GmbH) *Biomarker type*: nucleic acid (RNA, DNA)	GenoBlot assay (LG Life Sciences) *Biomarker type*: nucleic acid (RNA, DNA)	(Seegene, Inc.) *Biomarker type*: nucleic acid (RNA, DNA)
ReView-TB (ChironX) *Biomarker type*: others, imaging	Molecular stool (multiple) *Biomarker type*: nucleic acid (RNA, DNA)	PrimeSuite TB (Longhorn Vaccines and Diagnostics LLC) *Biomarker type*: nucleic acid (RNA, DNA)	VersaTrek (Thermo Fisher) *Biomarker type*: whole bug
Breath test (Avisa Pharma Inc.) *Biomarker type*: metabolite	FluoroType MTBXDR Ver 1.0 (Hain Lifesciences) *Biomarker type*: nucleic acid (RNA, DNA)	QMAC DST (Quanta Matrix, Inc.) *Biomarker type*: whole bug	Sensititre MYCOTB MIC Plate (Thermo Fisher) *Biomarker type*: whole bug
NS-POC TB detector (Nanosynth) *Biomarker type*: metabolite	LabChip-based rapid POCT (MicoBiomed) *Biomarker type*: nucleic acid (RNA, DNA)	VereMTB detection (Veredus Laboratories Pte Ltd.) *Biomarker type*: nucleic acid (RNA, DNA)	Molecu Tech REBA MDR/XDR assays series (YD Diagnostics) *Biomarker type*: nucleic acid (RNA, DNA)
TB Breathalyser (Rapid Biosensor System) *Biomarker type*: antigens	UltraFast LabChip real-time PCR-MDR-TB kit (MicoBiomed) *Biomarker type*: nucleic acid (RNA, DNA)		MeltPro MTB (MDR-TB, XDR-TB kits) (Zeesan Biotech) *Biomarker type*: nucleic acid (RNA, DNA)
NA-NOSE (Breathtec Biomedical, Inc.) *Biomarker type*: antigens	RTT TB Lophius Biosciences *Biomarker type*: NA		LAM-ELISA (Otsuka Pharmaceuticals Co. Ltd.) *Biomarker type*: antigens
Blood host marker POC tests (Multiple) *Biomarker type*: protein			OMNIgene SPUTUM (DNA Genotek, Inc.) *Biomarker type*: NA

(continued)

Table 23.1 (continued)

In feasibility stage (prototype but not design lock)	In development stage (design locked/pilot manufacturing)	In validation stage (manufacturing in place)	In regulatory stage (CE-IVD, FDA, CFDA, or similar regulatory approval in domestic country)
CYCLE MDR-TB (FRIZ Biochem GmBH) *Biomarker type*: nucleic acid (RNA, DNA)			PrimeStore MTM (Longhorn Vaccines and Diagnostics LLC) *Biomarker type*: NA
GeneDrive MTB (Genedrive PLC) *Biomarker type*: nucleic acid (RNA, DNA)			FluoroType MTB VER 1.0 (Hain Lifesciences) *Biomarker type*: nucleic acid (RNA, DNA)
InSilixa HYDRA-1K (InSilixa, Inc.) *Biomarker type*: nucleic acid (RNA, DNA)			
Scanogen (Scanogen Inc.) *Biomarker type*: nucleic acid (RNA, DNA)			
MtB Drug Resistance Dx (Omniome, Inc.) *Biomarker type*: nucleic acid (RNA, DNA)			
Low-cost easy-to-use NGS (multiple) *Biomarker type*: nucleic acid (RNA, DNA)			
BioNanoPore (NanoLogix, Inc.) *Biomarker type*: whole bug			
mfloDx MDR/XDR-TB (EMPE Diagnostics) *Biomarker type*: nucleic acid (RNA, DNA)			
Not defined (Mobidiag) *Biomarker type*: nucleic acid (RNA, DNA)			
Incipient TB assay (Abbott Laboratories) *Biomarker type*: NA			
T-Cell Immune Prof (Becton-Dickinson) *Biomarker type*: NA			
QFT-Predict (Qiagen) *Biomarker type*: NA			
QIA-TB Signature (Qiagen) *Biomarker type*: NA			

(continued)

Table 23.1 (continued)

In feasibility stage (prototype but not design lock)	In development stage (design locked/pilot manufacturing)	In validation stage (manufacturing in place)	In regulatory stage (CE-IVD, FDA, CFDA, or similar regulatory approval in domestic country)
Not defined (Biomerieux/ Bioaster) *Biomarker type*: nucleic acid (RNA, DNA)			
Molecular bacterial load assay (MBLA) (LifeArc/ University of St Andrews Infection Group) *Biomarker type*: nucleic acid (RNA, DNA)			

Acknowledgment Foundation for Innovative Newer Diagnostics (FIND) India deserves a special acknowledgment for providing permission to cite a table (annexure) summarizing the TB diagnostic pipeline showing progress at different development stages.

References

Abe C, Hirano K, Tomiyama T (1999) Simple and rapid identification of the Mycobacterium tuberculosis complex by immunochromatographic assay using anti-MPB64 monoclonal antibodies. J Clin Microbiol 37(11):3693–3697

Brock TD (1999) The etiology of tuberculosis Koch R (1882).Trans. In: Brock TD (ed) Milestones in microbiology: 1546 to 1940. American Society for Microbiology, Washington

Brown AC, Bryant JM, Einer-Jensen K, Holdstock J, Houniet DT, Chan JZ et al (2015a) Rapid whole-genome sequencing of mycobacterium tuberculosis isolates directly from clinical samples. J Clin Microbiol 53:2230–2237

Brown CA et al (2015b) Rapid whole-genome sequencing of Mycobacterium tuberculosis Isolates directly from clinical samples. J Clin Microbiol 53(7):2230–2237. https://doi.org/10.1128/JCM.00486-15

Brudey K, Driscoll JR, Rigouts L, Prodinger WM, Gori A, Al-Hajoj SA et al (2006) Mycobacterium tuberculosis complex genetic diversity: mining the fourth international spoligotyping database (SpolDB4) for classification, population genetics and epidemiology. BMC Microbiol 6(6):23

Bruins M, Rahim Z, Bos A, van de Sande WW, Endtz HP, van Belkum A (2013) Diagnosis of active tuberculosis by e-nose analysis of exhaled air. Tuberculosis (Edinb) 93(2):232–238

Centers for Disease Control (2000) [homepage on the internet]. Atlanta, Georgia, USA: CDC. Acid-fast direct smear microscopy training package. Available at: http://stacks.cdc.gov/view/cdc/31282. Cited 14 August 2018

Chaitali N et al (2014) Evaluation of the Indian TrueNAT micro RT-PCR device with GeneXpert for case detection of pulmonary tuberculosis. www.researchgate.net/publication/8233_Evaluation_of_the_Indian_TrueNAT_micro_RT-PCR_device_with_GeneXpert_for_case_detection_of_pulmonary_tuberculosis

Cheaper, Indigenous TB test enters final validation phase, The Hindu (2017). www.thehindu.com/sci-tech/health/cheaper-indigenous-tb-test-enters-final-validationphase/article19193085.ece
Cheng MM-C et al (2006) Current opinion in chemical biology. NanoSci Technol Inst 10:11–19
Cole ST, Brosch R, Parkhill J, Garnier T, Churcher C et al (1998) Deciphering the biology of Mycobacterium tuberculosis from the complete genome sequence. Nature 393:537–544
Cruz-Knight W, Blake-Gumbs L (2013) Tuberculosis: an overview. Prim Care 40(3):743e56. Available from: https://www.sciencedirect.com/science/article/abs/pii/S0095454313000651?via%3Dihub
De Kantor IN, Kim SJ, Frieden T, Laszlo A, Luelmo F, Norval PY et al (1998) Laboratory services in tuberculosis control. WHO Global Tuberculosis Programme. WHO/TB/98.258. WHO, Geneva
Diel R, Loddenkemper R, Nienhaus A (2012) Predictive value of interferon-g release assays and tuberculin skin testing for progression from latent TB infection to disease state: a meta-analysis. Chest 142(1):63e75. Available from: http://journal.publications.chestnet.org/article.aspx?articleid¼1206608
Drobniewski FA, Caws M, Gibson A, Young D et al (2003) Modern laboratory diagnosis of tuberculosis. Lancet Infect Dis 3(3):141–147
Drouillon V, Delogu G, Dettori G, Lagrange PH, Benecchi M, Houriez F et al (2009) Multicenter evaluation of a transcription-reverse transcription concerted assay for rapid detection of Mycobacterium tuberculosis complex in clinical specimens. J Clin Microbiol 47(11):3461–3465
Eiken Chemical Co. Ltd (2017) The principle of LAMP method [Homepage on the internet]. Available from http://loopamp.eiken.co.jp/e/lamp/. Cited 12 August 2018
Fu E et al (2008) Chapter 10. In: Schasfoort BM, Tudos AJ (eds) Handbook of surface Plasmon resonance. Royal Society of Chemistry, Cambridge
Goldman L, Schafer AI (2011) Tuberculosis: disease overview. In: Goldman L, Schafer AI (eds) Goldman's Cecil medicine: expert consult premium edition, 24th edn. Saunders Elsevier, St. Louis
Hillemann D, Rüsch-Gerdes S, Richter E (2005) Application of the Capilia TB assay for culture confirmation of Mycobacterium tuberculosis complex isolates. Int J Tuberc Lung Dis 9 (12):1409–1411
Hirano K, Aono A, Takahashi M, Abe C (2004) Mutations including IS6110 insertion in the gene encoding the MPB64 protein of Capilia TB-negative Mycobacterium tuberculosis isolates. J Clin Microbiol 42(1):390–392
International Standard for Tuberculosis Care, 3rd edition (2014). www.who.int/tb/publications/standards-tb-care-2014/
Iseman MD (2000) A clinical guide to tuberculosis. Lippincott Williams & Wilkins, Philadelphia
Ismail NA, Omar SV, Lewis JJ, Dowdy DW, Dreyer AW, van der MH et al (2015) Performance of a novel algorithm using automated digital microscopy for diagnosing tuberculosis. Am J Respir Crit Care Med 191(12):1443–1449
Kremer K, van Soolingen D, Frothingham R, Haas WH, Hermans PW, Martín C (1999) Comparison of methods based on different molecular epidemiological markers for typing of Mycobacterium tuberculosis complex strains: interlaboratory study of discriminatory power and reproducibility. J Clin Microbiol 37(8):2607–2618
Kremer K, Au BK, Yip PC, Skuce R, Supply P, Kam KM et al (2005) Use of variable-number tandem-repeat typing to differentiate Mycobacterium tuberculosis Beijing family isolates from Hong Kong and comparison with IS6110 restriction fragment length polymorphism typing and spoligotyping. J Clin Microbiol 43(1):314–320
Kwong JC, McCallum N, Sintchenko V, Howden BP (2015) Whole genome sequencing in clinical and public health microbiology. Pathology 47:199–210
Leão SC, Martin A, Mejia GI, Palomino JC, Robledo J, Telles MAS et al (2004) Practical handbook for the phenotypic and genotypic identification of mycobacteria. TropMed Central Antwerp, Antwerp. Introduction available at: http://www.esmycobacteriology.eu/PDF%20files/foreword.pdf

Lim SH, Martino R, Anikst V, Xu Z, Mix S, Benjamin R et al (2016) Rapid diagnosis of tuberculosis from analysis of urine volatile organic compounds. ACS Sens 1(7):852–856
Mase SR, Ramsay A, Ng V, Henry M, Hopewell PC, Cunningham J et al (2007) Yield of serial sputum specimen examinations in the diagnosis of pulmonary tuberculosis: a systematic review. Int J Tuberc Lung Dis 11(5):485–495
Migliori GB, Zellweger JP, Abubakar I, Ibraim E, Caminero JA, De Vries G et al (2012) European Union standards for tuberculosis care. Eur Respir J 39(4):807–819
Minion J, Pai M, Ramsay A, Menzies D, Greenaway C (2011) Comparison of LED and conventional fluorescence microscopy for detection of acid-fast bacilli in a low-incidence setting. PLoS One 6:e22495
MTBDRsl kit insert (2018) [Homepage on the internet]. https://www.immunodiagnostic.fi/wp-content/uploads/MTBDRsl-V2_kit-insert.pdf. Cited 19 August 2018
Mudur GS "One-hour test for TB" (2017) [Homepage on the internet]. https://www.telegraphindia.com/india/one-hour-test-for-tb-183556. Cited 21 August 2018
Nikolayevskyy V, Balabanova Y, Simak T, Malomanova N, Fedorin I, Drobniewski F (2009) Performance of the GenoType MTBDRplus assay in the diagnosis of tuberculosis and drug resistance in Samara, Russian Federation. BMC Clin Pathol 10(9):2
Pai M, Minion J, Sohn H, Zwerling A, Perkins MD (2009) Novel and improved technologies for tuberculosis diagnosis: progress and challenges. Clin Chest Med 30:701–716
Palomino JC (2012) Current developments and future perspectives 127. for TB diagnostics. Future Microbiol 7:59–71
Pankhurst LJ, Del Ojo EC, Votintseva AA, Walker TM, Cole K, Davies J et al (2016) Rapid, comprehensive, and affordable mycobacterial diagnosis with whole-genome sequencing: a prospective study. Lancet Respir Med 4(1):49–58
Resch-Genger U et al (2008) Quantum dots versus dyes as fluorescent labels. Nature Methods 5:763–775
Rieder HL, Van Deun A, Kam KM, Kim SJ, Chonde TM, Trébucq A et al (2007) Priorities for tuberculosis bacteriology services in low-income countries, 2nd edn. International Union against Tuberculosis and Lung Disease, Paris
Rodrigues C, Shenai S, Sadani M, Sukhadia N, Jani M, Ajbani K et al (2009) Evaluation of the Bactec MGIT 960 TB system for recovery and identification of Mycobacterium tuberculosis complex in a high through put tertiary care centre. Indian J Med Microbiol 27:217–221
Roh SS, Smith LE, Lee JS, Via LE, Barry CE III, Alland D et al (2015) Comparative evaluation of sloppy molecular beacon and dual-labeled probe melting temperature assays to identify mutations in Mycobacterium tuberculosis resulting in rifampin, fluoroquinolone and aminoglycoside resistance. PLoS One. 10(5):e0126257
Sandgren A, Strong M, Muthukrishnan P, Weiner BK, Church GM, Murray MB (2009) Tuberculosis drug resistance mutation database. PLoS Med 6(2):e1000002
Scarparo C, Piccoli P, Rigon A, Ruggiero G, Nista D, Piersimoni C (2001) Direct identification of mycobacteria fromMB/BacT alert 3D bottles: comparative evaluation of two commercial probe assays. J Clin Microbiol 39(9):3222–3227
Shen GH, Chen CH, Hung CH, Wu KM, Lin CF, Sun YW et al (2009) Combining the Capilia TB assay with smear morphology for the identification of Mycobacterium tuberculosis complex. Int J Tuberc Lung Dis 13:371–376
Shenai S, Armstrong DT, Valli E et al (2015) Analytical and clinical evaluation of the epistem genedrive assay for detection of mycobacterium tuberculosis. J Clin Microbiol 54 (4):1051–1057. https://doi.org/10.1128/JCM.02847-15
Stop TB Partnership and World Health Organization (2008) New Laboratory Diagnostic Tools for TB Control. World Health Organization, Geneva. Available at: http://apps.who.int/tdr/publications/non-tdrpublications/diagnostic-tool-tb/pdf/diagnostic-tool-tb.pdf
Sun QF, Xu M, Wu JG, Chen BW, Du WX, Ding JG et al (2013) Efficacy and safety of recombinant Mycobacterium tuberculosis ESAT-6 protein for diagnosis of pulmonary tuberculosis: a phase II trial. Med Sci Monit 19:969–977

Thillai M, Pollock K, Pareek M, Lalvani A (2014) Interferon-gamma release assays for tuberculosis: current and future applications. Expert Rev Respir Med 8(1):67e78. Available from: http://informahealthcare.com/doi/abs/10.1586/17476348.2014.852471

Tortoli E, Mariottini A, Mazzarelli G (2003) Evaluation of INNO-LiPA MYCOBACTERIA v2: improved reverse hybridisation multiple DNA probe assay for mycobacterial identification. J Clin Microbiol 41(9):4418–4420

WHO Global Tuberculosis Report (2017). http://apps.who.int/iris/bitstream/handle/10665/259366/9789241565516eng.pdf;jsessionid=50ED827765AA9D9A35DB7D0127F3D574?sequence=1

WHO Meeting Report of a Technical Expert Consultation (2017) Non-inferiority analysis of Xpert MTB/RIF Ultra compared to Xpert MTB/RIF. World Health Organization, Geneva

Witkamp B et al (2008) Self-detecting gate tunable nanotube paddle resonators. Appl Phys Lett 93:111909–1–3

World Health Organization (2008) Policy guidance on Drug Susceptibility Testing (DST) of second-line anti-tuberculosis drugs. WHO, Geneva. WHO/HTM/TB/2008.392. Available at: http://www.who.int/tb/dots/laboratory/policy/en/print.html

World Health Organization (2010a) A roadmap for ensuring quality tuberculosis diagnostics services within national laboratory strategic plans. The Global Laboratory Initiative, Advancing TB Diagnosis. Available at: http://www.tbevidence.org/documents/rescentre/books/GLI_Roadmap_2010.pdf

World Health Organization (2010b) Policy Statement on Non-commercial culture and DST Methods for rapid screening of patients at risk of multidrug-resistant Tuberculosis. WHO, Geneva. Available at: http://www.who.int/tb/dots/laboratory/policy/en/print.html

World Health Organization (2015) The use of lateral flow urine lipoarabinomannan assay (LF-LAM) for the diagnosis and screening of active tuberculosis in people living with HIV: WHO policy update, Geneva

World Health Organization (2016) The use of loop mediated isothermal amplification (TB LAMP) for the diagnosis of TB: policy guidance. World Health Organization, Geneva. Available from http://www.who.int/tb/publications/lamp-diagnosis-molecular/en/

Breaking the Transmission of TB: A Roadmap to Bridge the Gaps in Controlling TB in Endemic Settings

24

Pooja Singh, Jasmine Samal, Sheeba Zarin, Ravikrishnan Elangovan, Seyed E. Hasnain, and Nasreen Z. Ehtesham

Abstract

Tuberculosis (TB), caused by *Mycobacterium tuberculosis (M.tb)*, remains an enormous health burden with nearly 2 billion people worldwide being affected by it, though only 10% progress to active disease. Currently around 58% of people infected with *M.tb* are being diagnosed and treated. The lack of a standard research setting and limited resource in terms of early diagnosis of TB lead to high incidence of transmission of TB infection in high-TB-burden countries. Therefore, early diagnosis of TB and an effective vaccination can primarily break the cycle of transmission of TB. Recommended diagnostic tests, currently available, have several limitations, making them unsuitable for resource-limited settings and remote areas. The healthcare settings are the highest risk zones for transmission of drug-resistant *M.tb* strains. Multidrug-resistant (MDR) TB is resilient to diagnosis and pose major complications to patient's health during treatment. In order to limit the

Pooja Singh, Jasmine Samal and Sheeba Zarin contributed equally with all other contributors.

P. Singh
JH Institute of Molecular Medicine, Jamia Hamdard, New Delhi, Delhi, India

J. Samal · S. Zarin · N. Z. Ehtesham (✉)
Inflammation Biology and Cell Signaling Laboratory, ICMR-National Institute of Pathology, Safdarjung Hospital Campus, New Delhi, Delhi, India
e-mail: nasreenzehtesham.nip@gov.in

R. Elangovan
Department of Biochemical Engineering and Biotechnology, Indian Institute of Technology-Delhi, New Delhi, Delhi, India

S. E. Hasnain
JH Institute of Molecular Medicine, Jamia Hamdard, New Delhi, Delhi, India

Dr Reddy's Institute of Life Sciences, University of Hyderabad Campus, Hyderabad, India
e-mail: vc@jamiahamdard.ac.in

S. E. Hasnain et al. (eds.), Mycobacterium tuberculosis: *Molecular Infection Biology, Pathogenesis, Diagnostics and New Interventions*,
https://doi.org/10.1007/978-981-32-9413-4_24

spread of MDR TB, we need to implement better diagnostic tools and health measures which may eventually interrupt transmission of *M.tb* from TB patients to uninfected individuals. Complete and correct execution of regime of anti-TB therapies at both the individual and community level could help in minimizing the transmission of TB. This chapter gives an insight on strategies that aid in interruption of transmission of TB, especially in high-TB-burden areas across the globe.

Keywords
Tuberculosis · Transmission · *Mycobacterium tuberculosis* · Diagnosis · Vaccine

Abbreviations

BCG	Bacillus Calmette-Guerin
DNA	Deoxyribonucleic acid
GoI	Government of India
LPA	Line probe assays
LAMP	Loop-mediated isothermal amplification
MTB	*Mycobacterium tuberculosis*
MDR	Multidrug resistant
NSP	National Strategic Plan
NHP	Nonhuman primate
NAATs	Nucleic acid amplification tests
RNTCP	Revised National Tuberculosis Control Program
RIF	Rifampin
TTD	Time to detection
TB	Tuberculosis
WHO	World Health Organization
XDR	Extensively drug resistant

24.1 Introduction

Tuberculosis continues to be one of the leading causes of deaths in spite of being declared as a global emergency by World Health Organization (WHO), way back in 1993. The year 2017 recorded 10 million people getting infected with *Mycobacterium tuberculosis* (*M.tb)* and the death toll due to TB rose to 1.6 million (WHO 2018). The present rate of decline in TB cases is recorded as 2% per year; however in order to achieve global END TB target, it is imperative to achieve a decline in TB cases to as much as 5% annually. An increase in development of drug resistance in various *M.tb* strains poses a serious public health threat. In spite of efforts to create awareness for TB eradication and huge investment in terms of providing access to primary healthcare and anti-TB drugs, transmission of *M.tb* remains nearly uninterrupted. It is therefore necessary that we reformulate our strategies for tackling TB cases and derive novel approaches in terms of deciphering novel drug targets, designing new drugs, and generating vaccines. A better understanding of the ways of transmission and preventing the dissemination of disease from the source can reduce the overall incidence of TB dramatically (Nardell 2004).

It is estimated that almost three million TB cases worldwide remain undiagnosed annually (WHO 2014). These undiagnosed cases serve as source for transmission of *M.tb* from TB patients to uninfected individuals in the community. TB control program strategies can be effective in strict sense only if the true burden of TB is considered, taking into account the underreporting of TB cases. Underreporting of TB cases could be due to multiple reasons like inaccessibility to TB services, financial instability of patients, use of obsolete techniques that fail to detect *M.tb* in smears, noncompliance of patients to follow up treatment, and poor data recording (Tollefson et al. 2016). From the patient point of view, a delay or error in diagnosing TB diagnosis at an early stage of infection not only results in loss of critical time for timely management of TB but also poses risk in terms of inadvertent transmission of *M.tb* to patient contacts. As a result of delay in treatment of TB, patients acquire full-blown symptoms of the disease and may or may not respond to medicaments, with greater chances of acquiring multidrug resistance that may increase the duration of infectivity and may also lead to death. In 2015, only 40% of multidrug-resistant TB (MDR-TB) cases were estimated out of total 340,000 notified TB cases, globally (WHO 2017). India poses 27% of the global burden of TB, of which 2.5% are new TB cases and 16% previously treated TB cases (WHO 2017).

The goal of the National Strategic Plan (NSP) 2017–2025 by the Government of India (GoI) is to achieve a rapid decline in TB morbidity and mortality that may pave way for elimination of TB by 2025. The requirements for moving toward TB elimination have been integrated into four strategic pillars of "Detect-Treat-Prevent-Build" (RNTCP 2017). India's TB control program to break the TB transmission cycle has come up with enhanced interventions and strategies such as contact screening of family members, preventive treatment of all children (<5 years) who have not been diagnosed with TB, and mandatory compliance to undertake the drug susceptibility test. To achieve END TB target, there should be universal access to diagnostic facilities for early identification of presumptive TB cases at the first point-of-care site, private or public sectors, and prompt diagnosis using diagnostic tests with sensitivity greater than 90% (Fig. 24.1).

24.2 Approaches to Interrupt the Transmission of TB

(1) *Effective TB Vaccination*

Vaccines are one of the most effective public health interventions in improving immunity against infectious pathogens. An effective vaccine is needed to interrupt transmission of *M.tb* from TB patients to uninfected individuals, thereby leading to elimination of TB. Currently, the only vaccine available against TB is the *Mycobacterium bovis*-derived Bacillus Calmette-Guerin (BCG), discovered in 1921 (Calmette 1927). While BCG has shown to protect against severe forms of TB such as TB meningitis and miliary TB, in infants and children (Rodrigues et al.

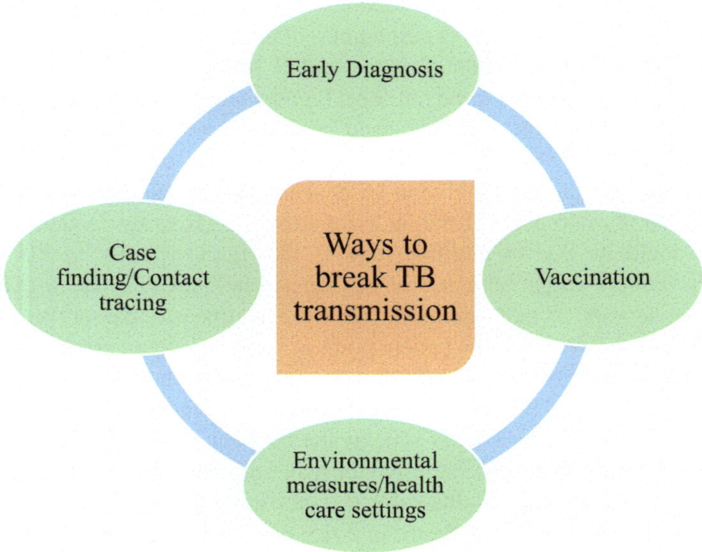

Fig. 24.1 A schematic overview of ways to break TB transmission: One of the four essential and inter-dependable steps for controlling the transmission of TB is to have a vaccination that exhibits long-term memory response. Early detection of TB is important not only to implement medicament at an early stage of infection but also in segregating patients from the uninfected individuals. Healthcare settings can provide easy and correct access to TB treatment regime to the patients. Preventive healthcare steps such as use of face masks by healthy contacts may prevent inadvertent acquisition of *M.tb* from aerosol generated during coughing or sneezing by the patients

1993; Trunz B Bourdin, Fine PEM 2006), however its protective efficacy wanes in 10–20 years. BCG also exhibits limited protection against adult pulmonary TB (Fine 1995). Such variations in efficacy have been attributed to several factors such as interaction with environmental mycobacteria, genetic difference of immunized populations and their immune response to various BCG sub-strains, and geographic diversity in clinical isolates of *M.tb* along with some subsidiary factors (Andersen and Doherty 2005; Kernodle S. Douglas 2010; Springett and Sutherland 1994; Wilson et al. 1995).

Current vaccination strategies against TB aim to intervene in transmission of *M.tb* at three stages: (a) pre-exposure vaccines, aim at prevention of getting infected; (b) postexposure vaccines, aim at elimination/containment of latent TB and prevention of reactivation; and (c) therapeutic vaccines, aim as an adjunct to chemotherapy. Several ongoing efforts to achieve promising vaccine candidates are aimed at improving the immunogenicity of the current BCG vaccine by generating recombinant BCG or by replacing BCG with new platforms such as whole cell lysate or viral vectors. Table 24.1 summarizes the list of vaccine candidates that are being tested in various stages of clinical trials. Recently, vaccine candidate M72/AS01 (subunit vaccine containing mycobacterial antigens 32A and 39A with adjuvant AS01) has been successful in preventing active TB disease, with an efficacy of 54% in phase 2b

Table 24.1 TB vaccine candidates currently in clinical trials

Preclinical	Phase 1	Phase 2a	Phase 2b	Phase 3
BCG-ZMPI		MTBVAC	M72 + ASO1	VPM1002
H64:CFA01	Ad5 Ag85A	TB/ Flu04L	DAR-901	*MIP*
PPE15-85A	GamTBVac	ID93/ GLA-SE	H56:IC31	*M. vaccae*
CMV6Ag	H56:IC31	RUTI		
CysVac2/Ad	ID93/GLA-SE	H4:IC31		
BCG-ZMP1	ChadOx1.85A MVA 85A Aerosol			
MVA multiphasic vac				
BCG, ChadOx/MVA PPE15-85A				

clinical trial with participants from Kenya, South Africa, and Zambia (Van Der Meeren et al. 2018). Another vaccine candidate VPM1002, a recombinant BCG vaccine in which the BCG urease C gene has been replaced with a gene encoding for listeriolysin from *Listeria monocytogenes*, is undergoing phase 3 clinical trial in India (Nieuwenhuizen et al. 2017). In order to check the efficacy of this vaccine in reducing the rate of relapse and resurgence of TB, this vaccine candidate is being tested in patients with relapse TB cases. The use of a whole cell bacteria, known as *Mycobacterium indicus pranii* (MIP) as an adjunct with the first-line antimicrobial treatment for Category II TB patients, has shown promising results in the phase 2b trial and is currently being evaluated in phase 3 trial (Sharma et al. 2017).

The major focus now should be to have an increasing number of vaccine candidates in the pipeline and to move the current ones from one clinical phase to another at a faster rate, since vaccine candidate such as MVA85A, in spite of showing encouraging results (in phase 2b trial), failed to demonstrate additional protection in infants as compared to BCG (Tameris et al. 2013). There are several parameters which need to be understood deeply while designing and screening different vaccine candidates. Firstly, we need to better understand the molecular pathogenesis and immunogenicity of mycobacterial antigens at the laboratory level before advancing to clinical stages. Secondly, a systematic and efficient screening of vaccines in animal models is needed to assess their efficacy in providing protection and sustained memory response, thereby preventing progression to active disease. We need multiple assays to be performed using in vivo models for validation and reliability of our data. Macaques, the nonhuman primate (NHP) model that closely resembles the complete spectrum of TB disease and pathogenesis, could serve as excellent model for evaluating vaccine efficacy before entering the clinical testing (Dockrell 2016). Thirdly, identification of ideal biomarkers and immune correlates can be a big boost in assessing vaccine candidates for their potential in inducing -

long-term protection and memory response. Thus, an effective vaccine to impede TB transmission can significantly contribute in TB control and eradication in areas where TB is highly endemic.

(2) *Early Diagnosis*

Early and accurate diagnosis of TB is one of the major tools for controlling transmission of *M.tb*, especially the drug-resistant strains to uninfected healthy individuals. A novel and fast diagnostic tool for identification of TB-infected cases would decrease the rate of transmission in the following three ways:

(a) *To have potency in treatment regimen and prevent long-term complications arising due to no diagnosis or inappropriate diagnosis*

Direct smear microscopy and culture of *M.tb* are standard procedures for TB diagnosis. While both these techniques are relatively inexpensive, they do have certain limitations. Direct smear microscopy test is simple and rapid but comes along with low sensitivity, ranging from 20% to 60% (Steingart et al. 2006). Culturing of *M.tb*, which is widely accepted as the gold standard for TB detection, is highly sensitive but unacceptably slow to provide results, as it takes a minimum of 2 weeks or as long as 6–8 weeks in case of solid culture techniques (UNITAID 2017). Other nucleic acid amplification tests (NAAT)-based diagnostic methods have gained attention in TB diagnosis due to reduced turnaround time with acceptable sensitivity and specificity. In recent times, commercially available NAATs are being used globally for TB diagnosis, despite of not being endorsed by the WHO (UNITAID 2017). WHO-endorsed tests include Xpert® MTB/RIF (Xpert), various line probe assays (LPA), and loop-mediated isothermal amplification (TB-LAMP) (UNITAID 2016 and Nathavitharana et al. 2016). The TB-LAMP test based on isothermal amplification of *M.tb* DNA is a low-complexity NAAT which was conditionally recommended by WHO in 2016 for use in microscopy centers and in high-tiered test facilities (WHO 2016), but its low-throughput nature limits its application in high-burden countries where a high-throughput assay is required. It also cannot help in substituting the rapid NAAT-based tests to detect TB and rifampicin (RIF) resistance among MDR-TB risk groups. In 2016, an Indian company, Molbio Diagnostics, introduced Truenat™ for the detection of TB (Truenat™ MTB) as well as for RIF resistance (Truenat™ MTB-RIF). These are currently under evaluation phase in India by FIND (UNITAID 2017). A major challenge for delayed TB diagnosis is nonadherence to treatment regime. Noncompliance to adhere to TB treatment is mainly attributed to ignorance of the patient in understanding the duration of TB medicament and significance of completing the entire treatment cycle, nonavailability of drugs at the healthcare facility, side effects of TB medications, and the lack of social support.

(b) *Correct and timely diagnosis of TB cases for preventing TB transmission*

Delays and inefficiency in diagnosis and treatment of TB can increase the risk of transmission. GeneXpert is a diagnostic test for TB and RIF resistance that is simple to perform and provides results within 2 h. However, GeneXpert is not a high-throughput method, has high maintenance cost, and requires electric supply and temperature control for operation. These factors pose major challenge to scale up its use, in resource-limited settings and less developed geographically remote locations (UNITAID 2016). Light microscopy has a threshold limit of 5000–10,000 bacilli/ml for detection of *M.tb* bacilli. However, infectivity dose of *M.tb* could be as low as less than 10 bacilli. So, smear-negative patients can also be potent transmitters of the infection and thus should not be ignored. Although liquid culture is highly sensitive with a limit of detection as low as 10–100 cfu/ml, it has an extensive detection time of around 2–6 weeks. Time to detection (TTD) is inversely proportional to bacillary load and thus a lower TTD means a higher transmission risk (O'Shea et al. 2014). In the time gap between culturing the bacteria and obtaining the culture positivity result, the patient is inadvertently transmitting the disease to others. There is an urgent need to decrease this window period for an overall better control of the disease.

(c) *Prevention in transmission of drug-resistant TB strains*

The major reason in emergence of MDR and XDR-TB cases in high-TB-burden countries is failure of early diagnosis of TB infection. Involvement of private health organizations such as NGOs to scale up TB control, can raise awareness about drug resistance and improve access to TB care and treatment. Although assays like MTBDRplus line probe assay (LPA) provide information on most MDR and XDR-TB markers, its high sensitivity is impaired due to diligent processing steps and open hybridization format, thereby leading to amplicon contamination (UNITAID 2017).

The aforementioned tests suffer from one or the other limitations, including but not limited to high cost, requirement for maintenance and trained manpower, dependence on proprietary reagents, supply of constant electricity, and ambient temperature control, thereby making it unsuitable for resource-limited settings and remote areas (UNITAID 2017; Trebucq et al. 2011). Therefore, a cost-effective, rapid, portable, sensitive, and specific diagnostic test with the high-throughput ability for TB remains an unmet need till this day.

(d) *Case Finding/Contact Tracing*

The first step in breaking the transmission cycle of TB is early identification of patients with active TB. Successful TB control depends majorly on case findings and treatment provision. Routine screening for cases of active TB or latent infection should be promoted in high-risk populations (active case finding) instead of passive case finding, i.e., waiting for patients with symptoms to turn up for medical assistance. Investigation on contacts of TB patients helps in limiting transmission. However, a delay in initiating notable action against TB transmission is always observed. Earlier studies from countries with high TB incidence have shown that sociodemographic characteristics, financial status of patients, and fear of stigma of disease are important determinants that attribute to delay in seeking care from specialist (American Thoracic Society). Ignorance and superstition about the disease, its spread and causation, etc. are still prevalent in many areas. The early stage symptoms in TB are mostly nonspecific. Misinterpretation of active TB symptoms with other infections such as viral cough and cold has been associated with patient delay in seeking medical advice. In most of the cases, self-treatment with traditional home remedies and intake of drugs without supervision of medical practitioner is usually the first behavioral process in patients. The delay in seeking health supervision from medical practitioners increases the likelihood of transmission of TB to contacts. Many TB patients have no knowledge or understanding about TB. Different notions and myths are associated with TB infection, most of which are inconsistent with the fact that a bacteria causes TB. Older unwarranted beliefs which still are prevalent in society include sharing of utensils, food and water transmitting TB, TB due to hard physical work or exposure to cold or smoke (Ailinger and Dear 1997; Wandwalo and Mørkve 2000; Lambert and Van der Stuyft 2005; Ayisi et al. 2011), and TB inherited from parents to offspring (Liefooghe et al. 1997; Lin et al. 2007). Misconceptions regarding the knowledge, perception, and practices of TB exist both at individual and community level (Sharma et al. 2007), which lead to discrimination and stigmatization of the illness. TB diagnosis is often associated with social isolation and different psychosocial consequences, which is dreaded more than the disease itself. These stigmas limit prevention of TB because at-risk individuals tend to avoid screening and seeking healthcare facilities. This highlights the need for counseling, creating awareness and motivation, which is missing in the present-day TB control programs.

(e) *Environmental Control/Healthcare Settings*

Several environmental factors such as closed indoors, minimal light exposure, and high humidity favor *M.tb* to stay viable and infectious in patient's air droplets, accelerating transmission to uninfected individuals. Enclosed spaces with poor ventilation and positive air pressure increase the rate of TB transmission. Other parameters such as physical proximity and longer and more exposure to infectious person increases the risk of getting infected. A good quality healthcare service is an important component at all stages of TB treatment right from diagnosing to drug resistance screening and to treatment response monitoring. Still, the biggest challenge is to provide access to quality TB diagnostic services. Available lab services suffer from different shortcomings like poor infrastructure, limited human resource capacity, and a weak underlying healthcare system. In many countries, healthcare facilities are not easily accessible and care is mostly provided by nonspecialized individuals. Deficiency in specimen collection at district level along with lack of proper transportation adds on to the challenges. Also, the microscopy centers in different high-burden countries lack skilled personnel and equipments (Denkinger et al. 2013). As a consequence, repeated visit at the healthcare facilities seems to be the common cause of delay in diagnosis and treatment leading toward the nonspecific anti-tubercular.

24.3 Mathematical Modeling

Another approach to break the transmission cycle of TB is mathematical modeling. This provides an insight to improve the understanding of TB epidemiology and also helps to strategize TB prevention and control. Various mathematical models were made by assuming response to TB control, risk of transmission, the incidence of active cases, and source of infection. A detailed mathematical modeling study of diagnostic pathway was conducted to estimate the impact of new diagnostic tools (Trauer et al. 2014). The reduction in occurrence and mortality of TB was observed by new diagnostic tools against sputum smear microscopy. These models can be useful for guiding decisions for the development of novel techniques and the effect of different diagnostic pathways. Another study demonstrated that the change in rate of contact will help in better estimation of effect of novel diagnostic tests. Expanded collaboration between policy makers and epidemiologists could prove to be beneficial for future attempts to control TB. Apart from physical and environment approaches, mathematical modeling has gained considerable interest and serves as better prediction tool to stop transmission of TB infection, especially in high-burden countries.

In summary, it is important to determine the major factors responsible for transmission of TB, as breaking the transmission chain can profoundly decrease the incidence of TB in high endemic countries. In addition, identifying potential cases, effective diagnosis, and treatment measures can significantly contribute to eradicate TB worldwide. The major gaps in our knowledge of understanding the drivers of transmission of TB need to be addressed effectively in order to curb the prevalence of TB in high-TB-burden areas.

References

Ailinger RL, Dear MR (1997) Latino immigrants' explanatory models of tuberculosis infection. Qual Health Res 7:521–531

Andersen P, Doherty TM (2005) The success and failure of BCG – implications for a novel tuberculosis vaccine. Nat Rev Microbiol 3:656–662

Ayisi JG, van't Hoog AH, Agaya JA, Mchembere W, Nyamthimba PO, Muhenje O, Marston BJ (2011) Care seeking and attitudes towards treatment compliance by newly enrolled tuberculosis patients in the district treatment programme in rural western Kenya: a qualitative study. BMC Public Health 11:515

Calmette A (1927) La Vaccination préventive contre la tuberculose par le "BCG". Masson et Cie, Paris

Denkinger CM, Nicolau I, Ramsay A, Chedore P, Pai M (2013) Are peripheral microscopy centres ready for next generation molecular tuberculosis diagnostics? Eur Respir J 42:544–547

Dockrell HM (2016) Towards new TB vaccines: What are the challenges? Pathog Dis 74:ftw016

Douglas KS (2010) Decrease in the effectiveness of Bacille Calmette-Guérin Vaccine against pulmonary tuberculosis: a consequence of increased immune suppression by microbial antioxidants, not overattenuation. Clin Infect Dis 51:177–184

Fine PEM (1995) Variation in protection by BCG: implications of and for heterologous immunity. Lancet 346:1339–1345

Lambert ML, Van der Stuyft P (2005) Editorial: delays to tuberculosis treatment: shall we continue to blame the victim? Trop Med Int Heal 10:945–946

Liefooghe R, Baliddawa JB, Kipruto EM, Vermeire C, De Munynck AO (1997) From their own perspective. A Kenyan community's perception of tuberculosis

Lin H-H, Ezzati M, Murray M (2007) Tobacco smoke, indoor air pollution and tuberculosis: a systematic review and meta-analysis. PLoS Med 4:e20

Nardell EA (2004) Catching droplet nuclei. Am J Respir Crit Care Med 169:553–554. https://doi.org/10.1164/rccm.2401003

Nathavitharana RR, Hillemann D, Schumacher SG, Schlueter B, Ismail N, Omar SV, Sikhondze W, Havumaki J, Valli E, Boehme C, Denkinger CM (2016) Multicenter noninferiority evaluation of hain genotype MTBDRplus version 2 and Nipro NTM+MDRTB Line probe assays for detection of rifampin and isoniazid resistance. J Clin Microbiol 54:1624–1630

Nieuwenhuizen NE, Kulkarni PS, Shaligram U, Cotton MF, Rentsch CA, Eisele B, Grode L, Kaufmann SHE (2017) The recombinant Bacille Calmette-Guérin Vaccine VPM1002: ready for clinical efficacy testing. Front Immunol 8:1147

O'Shea MK, Koh GCKW, Munang M, Smith G, Banerjee A, Dedicoat M (2014) Time-to-detection in culture predicts risk of mycobacterium tuberculosis transmission: a cohort study. Clin Infect Dis 59:177–185

RNTCP (2017) World TB day slogans

Rodrigues LC, Diwan VK, Wheeler JG (1993) Protective effect of BCG against Tuberculous meningitis and Miliary Tuberculosis: a meta-analysis. Int J Epidemiol 22:1154–1158

Sharma N, Malhotra R, Taneja DK, Saha R, Ingle GK (2007) Awareness and perception about tuberculosis in the general population of Delhi. Asia Pacific J Public Health 19:10–15

Sharma KS, Katoch K, Sarin R, Balambal R, Kumar Jain N, Patel N, Murthy KJR, Singla N, Saha PK, Khanna A, Singh U, Kumar S, Sengupta A, Banavaliker JN, Chauhan DS, Sachan S, Wasim M, Tripathi S, Dutt N, Jain N, Joshi N, Penmesta SRR, Gaddam S, Gupta S, Khamar B, Dey B, Mitra DK, Arora SK, Bhaskar S, Rani R (2017) Efficacy and safety of Mycobacterium indicus pranii as an adjunct therapy in category II pulmonary tuberculosis in a randomized trial. Sci Rep 7:3354

Springett VH, Sutherland I (1994) A re-examination of the variations in the efficacy of BCG vaccination against tuberculosis in clinical trials. Tuber Lung Dis 75:227–233

Steingart KR, Henry M, Ng V, Hopewell PC, Ramsay A, Cunningham J, Urbanczik R, Perkins M, Aziz MA, Pai M (2006) Fluorescence versus conventional sputum smear microscopy for tuberculosis: a systematic review. Lancet Infect Dis 6(9):570–581

Tameris MD, Hatherill M, Landry BS, Scriba TJ, Snowden MA, Lockhart S, Shea JE, McClain JB, Hussey GD, Hanekom WA, Mahomed H, McShane H, MVA85A 020 Trial Study Team the M 020 TS (2013) Safety and efficacy of MVA85A, a new tuberculosis vaccine, in infants previously vaccinated with BCG: a randomised, placebo-controlled phase 2b trial. Lancet 381:1021–1028

Tollefson D, Ngari F, Mwakala M, Gethi D, Kipruto H, Cain K, Bloss E (2016) Under-reporting of sputum smear-positive tuberculosis cases in Kenya. Int J Tuberc Lung Dis 20:1334–1341

Trauer JM, Denholm JT, McBryde ES (2014) Construction of a mathematical model for tuberculosis transmission in highly endemic regions of the Asia-pacific. J Theor Biol 358:74–84

Trébucq A, Enarson DA, Chiang CY, Van Deun A, Harries AD, Boillot F, Detjen A, Fujiwara PI, Graham SM, Monedero I, Rusen ID, Rieder HL (2011) Xpert ® MTB/RIF for national tuberculosis programmes in low-income countries: when, where and how? Int J Tuberc Lung Dis 15:1567–1572

Trunz B Bourdin, Fine PEM DC (2006) Articles. Lancet 367:1173–1180
UNITAID (2016) UNITAID
UNITAID (2017) Tuberc diagnostics technol landscape
Van Der Meeren O, Hatherill M, Nduba V, Wilkinson RJ, Muyoyeta M, Van Brakel E, Ayles HM, Henostroza G, Thienemann F, Scriba TJ, Diacon A, Blatner GL, Demoitié M-A, Tameris M, Malahleha M, Innes JC, Hellström E, Martinson N, Singh T, Akite EJ, Khatoon Azam A, Bollaerts A, Ginsberg AM, Evans TG, Gillard P, Tait DR (2018) Phase 2b controlled trial of M72/AS01E vaccine to prevent Tuberculosis. N Engl J Med 379:1621–1634
Wandwalo ER, Mørkve O (2000) Delay in tuberculosis case-finding and treatment in Mwanza, Tanzania. Int J Tuberc Lung Dis 4(2):133–138
WHO (2014) TB Global 2014
WHO (2016) World Health Organization Geneva
WHO (2017) Global Tuberculosis report 2017-main text
WHO (2018) WHO Global Tuberculosis report
Wilson ME, Fineberg HV, Colditz GA (1995) Geographic latitude and the efficacy of bacillus Calmette-Guérin vaccine. Clin Infect Dis 20:982–991

Challenges and Advances in TB Drug Discovery

25

Garima Khare, Prachi Nangpal, and Anil K. Tyagi

Abstract

In this chapter, we provide a comprehensive review of the recent developments and challenges associated with tuberculosis drug discovery. The chapter begins with an overview of the global TB burden with an emphasis on the high-burden countries such as India and the probable reasons associated with high disease burden. We have discussed the targets for the WHO End TB Strategy along with the requirements to achieve them. The chapter further provides an insight into the major obstacles of TB control, the problems associated with the current chemotherapy, the need for new anti-TB drugs and expectations from an ideal TB therapy. The chapter also provides a comprehensive review of the candidate drugs in the TB drug clinical pipeline with description of their identification, mechanistic action and in vitro and in vivo efficacy data along with clinical trial progress. We then provide details about the commonly employed approaches like whole cell phenotypic approach, target-based virtual screening and repurposing of drugs for TB drug discovery along with the advantages and major challenges associated with these approaches. In this regard, the success of whole cell-based phenotypic screening has been highlighted in view of discovery of the two recently FDA-approved anti-TB drugs, namely, bedaquiline and delamanid. The chapter also deals with another promising strategy for TB drug discovery based on rational drug design with a focus on some of the leads identified by this

Garima Khare and Prachi Nangpal contributed equally with all other contributors.

G. Khare · P. Nangpal
Department of Biochemistry, University of Delhi South Campus, New Delhi, Delhi, India

A. K. Tyagi (✉)
Department of Biochemistry, University of Delhi South Campus, New Delhi, Delhi, India

Guru Gobind Singh Indraprastha University, New Delhi, Delhi, India
e-mail: vc@ipu.ac.in

S. E. Hasnain et al. (eds.), Mycobacterium tuberculosis: *Molecular Infection Biology, Pathogenesis, Diagnostics and New Interventions*,
https://doi.org/10.1007/978-981-32-9413-4_25

approach. We have also emphasized the recent advancements towards newer approaches like antisense RNA-based therapeutics, use of natural products, gene-editing tools such as CRISPR-CAS system and immunotherapy for the development of anti-TB molecules. Besides, the chapter also describes the development of methods to enhance the bioavailability of drugs such as novel delivery systems like nanoparticles/liposomes and devices for sustained release.

Keywords
Tuberculosis · *Mycobacterium tuberculosis* · TB drugs · Target-based virtual screening · Whole cell phenotypic screening · Repurposing approach

Abbreviations

TB	Tuberculosis
M. tuberculosis	*Mycobacterium tuberculosis*
WHO	World Health Organization
HIV	Human immunodeficiency virus
AIDS	Acquired immunodeficiency syndrome
RR-TB	Rifampicin-resistant tuberculosis
MDR-TB	Multidrug-resistant tuberculosis
XDR-TB	Extremely drug-resistant tuberculosis
FDA	Food and Drug Administration
BCG	Bacillus Calmette-Guerin
MIC	Minimum inhibitory concentration
IC	Inhibitory concentration
EBA	Early bactericidal activity
PK	Pharmacokinetic
NTZ	Nitazoxanide
LZD	Linezolid
NCI	National Cancer Institute
AES	Allelic exchange substrate
SAR	Structure activity relationship
GFP	Green fluorescent protein
PS-ODN	Phosphorothioate oligodeoxynucleotide
PLG	Poly(lactide-co-glycolide)
CRISPR	Clustered regularly interspaced short palindromic repeats
Cas	CRISPR-associated system
PAM	Protospacer adjacent motif
NHEJ	Non-homologous end joining

25.1 Global TB Scenario

Tuberculosis (TB) is a major global threat to public health. In spite of the presence of various intervention strategies against TB, the disease continues to persist and leads to loss of millions of human lives each year. TB is a complex disease due to multiple outcomes that can be manifested upon infection with the causative agent, *Mycobacterium tuberculosis*. The disease is caused by the inhalation of aerosol droplets containing the pathogen expelled from a diseased individual by coughing/sneezing.

TB is the ninth leading cause of human deaths worldwide and the leading cause of deaths from a single infectious agent, ranking above HIV/AIDS (WHO 2017). The epidemiology of the disease is indeed alarming and requires attention towards its urgent control. According to the WHO report on tuberculosis, 10.4 million people developed TB in 2016 worldwide, of which ~1.0 million were HIV positive (WHO 2017). 65% of these total TB incident cases were estimated to be in males and children accounted for 6.9% TB cases in the year 2016 (WHO 2017). Globally, 1.3 million HIV-negative people died of TB (down from 1.7 million in 2000) with an additional 0.37 million TB deaths observed in HIV-positive individuals in the year 2016 (WHO 2017). Drug-resistant TB continues to be a major threat with an estimated 0.6 million new cases resistant to rifampicin (RR-TB) in 2016 at a global level, of which 0.49 million had multidrug-resistant TB (MDR-TB). There were about 0.24 million deaths globally from MDR/RR-TB in 2016 (WHO 2017).

TB Burden is not Uniform Across the Globe

India, Indonesia, China, the Philippines and Pakistan (in descending order) accounted for 56% of the total estimated TB cases with India representing the highest TB burden globally (25%, 2.6 million TB cases) (WHO, 2017).

In 2016, ~82% of TB deaths among HIV-negative people occurred in the WHO African Region and the WHO South-East Asia Region. India accounted for 33% of TB deaths among HIV-negative people and for 26% of the combined total TB deaths in HIV-negative and HIV-positive people globally (WHO, 2017).

India, China and the Russian Federation accounted for almost half (47%) of the global MDR-TB cases (WHO, 2017).

It has been observed over the years that India has been one of the highest TB-burden countries with high rate of incidence, mortality and resistant cases (WHO 2017). The reason for this can be largely attributed to the nature of the disease transmission, high population and overcrowding. Moreover, India also has major problems of malnutrition, poor hygienic conditions, poor supply of drugs and unregulated use of medicines, which have contributed in a big way to the high disease burden.

WHO End TB Strategy has proposed a target, with reference to the estimates of 2015, to achieve a 95% reduction in TB deaths and a 90% reduction in TB incidence (new cases per year) by 2035 (Uplekar et al. 2015). However, fulfilment of this goal demands providing TB care, preventive methods and awareness of health coverage at global level along with a deliberate collaboration among various stakeholders to tackle the socio-economic factors related to TB (WHO 2017). Moreover, it is of utmost importance to develop groundbreaking technological advancements in the next 5–7 years, whose implementation ought to result in reduction in the TB incidence rate at a level faster than in the past (WHO 2017).

25.2 Major Obstacles to TB Control

Since the discovery of *M. tuberculosis* by Robert Koch in 1882, as the causative agent of TB, the global TB epidemic till date remains persistent, highlighting the shortcomings of the current control measures available for combating tuberculosis. The bacteria spreads very easily through aerosols (a few droplet nuclei of 1–5 microns in diameter) which are coughed by an active TB patient and are inhaled by an uninfected person, and this easy transmission mode poses a big challenge to curtail the spread of this disease. BCG, the only vaccine available against TB, is highly effective in preventing childhood tuberculosis, but it is unsatisfactory in preventing pulmonary tuberculosis in adults showing a variable protective efficacy ranging from 0% to 80% (Colditz et al. 1994). The current diagnostic tests such as X-ray, Mantoux test, culture-based test and GeneXpert suffer from several limitations of being cost ineffective and time consuming and having shortcomings of sensitivity or specificity. Current efforts are being made towards the development of a better TB vaccine as well as an easy, rapid and effective diagnostic test.

The current chemotherapeutic regimen for the treatment of drug-susceptible tuberculosis consists of four first-line drugs, namely, rifampicin, isoniazid, pyrazinamide and ethambutol, administered for a span of 6 months (Guidelines for treatment of drug-susceptible tuberculosis and patient care (2017 update). http://apps.who.int/iris/bitstream/handle/10665/255052/9789241550000-eng.pdf;jsessionid=86C860DB7117D77D4A072A39ABCD6429?sequence=1). The standard treatment regimen for drug-susceptible TB comprises of 2 months of initiation phase consisting of all the four drugs (termed as 2HRZE), followed by a 4-month-long continuation phase consisting of rifampicin and isoniazid (termed as 4HR) (Guidelines for treatment of drug-susceptible tuberculosis and patient care (2017 update). http://apps.who.int/iris/bitstream/handle/10665/255052/9789241550000-eng.pdf;jsessionid=86C860DB7117D77D4A072A39ABCD6429?sequence=1). These drugs are administered in a daily dosing frequency, and the use of fixed-drug combination (FDC) has been recommended by WHO over separate drugs (Guidelines for treatment of drug-susceptible tuberculosis and patient care (2017 update). http://apps.who.int/iris/bitstream/handle/10665/255052/9789241550000-eng.pdf;jsessionid=86C860DB7117D77D4A072A39ABCD6429?sequence=1). This protracted therapy is one of the major reasons for the TB patients to default on the therapy, which leads to non-compliance and non-adherence. Thus, in spite of having an extremely effective treatment for treating active TB, the non-compliance to the therapy has led to an inexorable increase in the emergence of drug-resistant strains of the pathogen leading to drug-resistant TB cases (WHO guidelines for the programmatic management of drug-resistant tuberculosis. http://apps.who.int/iris/bitstream/handle/10665/130918/9789241548809_eng.pdf;jsessionid=BBA844744A619EF170C00783C655E148?sequence=1; Chiang et al. 2010). This kind of resistance can be classified as an acquired (secondary) drug resistance; however, primary drug resistance can also occur by infection with a drug-resistant strain of the pathogen (WHO guidelines for the programmatic management of drug-resistant tuberculosis. http://apps.who.int/iris/bitstream/handle/10665/130918/9789241548809_eng.pdf;jsessionid=BBA844744A619EF170C00783C655E148?sequence=1; Chiang et al. 2010). Resistant cases of TB can be divided into various categories depending on the kind of resistance to isoniazid (isoniazid-resistant TB), to rifampicin (RR-TB), to both rifampicin and isoniazid (MDR-TB), to any fluoroquinolone and to at least one of the injectable second-line drugs (amikacin, kanamycin, capreomycin) in addition to both rifampicin and isoniazid (XDR-TB) (WHO guidelines for the programmatic management of drug-resistant tuberculosis. http://apps.who.int/iris/bitstream/handle/10665/130918/9789241548809_eng.pdf;jsessionid=BBA844744A619EF170C00783C655E148?sequence=1; Chiang et al. 2010). Treatment of the drug-resistant strains requires the use of various combinations of the second-line drugs such as levofloxacin, moxifloxacin, gatifloxacin, amikacin, kanamycin, capreomycin, streptomycin, ethionamide, cycloserine, linezolid and clofazimine. The therapy for resistant cases may last anywhere from 9 to 24 months. The emergence of these multidrug-resistant (MDR) strains poses a serious challenge to the world's health and towards the global control of this disease. It has been estimated that an active TB patient undergoing the treatment may transmit the infection to at least ten uninfected people suggesting the requirement of stringent control measures (www.who.int/mediacentre/factsheets/fs104/en/). Besides, the unpleasant side effects and a high pill burden of these drugs make the treatment of drug-resistant TB and compliance to the therapy an extremely daunting task. The prevalence of HIV makes the situation even more precarious due to an enhanced susceptibility of the HIV-infected immunocompromised people to TB infection. Additionally, the situation is complicated by the difficulty faced by using these anti-TB drugs along with antiretroviral therapy, which shows negative drug-drug interactions and, hence, precludes the use of these anti-TB drugs in HIV-positive patients (López-Cortés et al. 2002). Further, the key challenge is also to treat individuals who are subclinically infected with *M. tuberculosis* and are at a lifetime risk of reactivation TB due to various reasons such as HIV infection, anti-TNF therapy, diabetes, malnutrition or lowering of immunity (Narayanan et al. 2010; Gardam et al. 2003; Stevenson et al. 2007). The complexity also arises due to the fact that the pathogen has the ability to modulate the immune system thereby evading the immune surveillance. Its ability to persist in a latent and low metabolic state for years poses significant challenge like drug tolerance. Hence, latent TB disease requires several months of treatment for complete sterilization.

Thus, in view of the above challenges in the TB treatment, it is of utmost importance to develop novel chemotherapeutic regimens that are able to (i) reduce the duration of this long drawn therapy, (ii) reduce the pill burden, (iii) target the latent pathogen, (iv) target drug-resistant strains of the pathogen and (v) can be easily co-administered with HIV medication.

Ideal Drug Regimen

An ideal anti-TB drug regimen should be able to achieve the following:

- Shorter duration of the therapy
- Lower pill burden along with less toxicity of the drugs
- A cost effective therapy
- Ability to target latent and resistant forms of *M. tuberculosis*
- Better co-administration properties with other drugs
- Lower dosing frequency

Although considerable efforts have been made in the field of TB drug discovery as is evident from a relatively filled drug pipeline now as compared to a few decades earlier, the progress of TB drug discovery program has been extremely slow with only two new anti-TB drugs, bedaquiline (marketed as Sirturo) and delamanid, receiving the FDA approval in the last 50 years that too with limited access and reserved only for the treatment of MDR-TB. Hence, looking at the high attrition rate in the TB drug discovery, it is important to develop robust approaches for identifying better candidate drug molecules that can be channelled for clinical assessment so that more anti-TB drugs can reach the market.

This chapter provides a comprehensive review of the recent developments in the field of TB drug discovery and addresses the major challenges associated with the current approaches being employed for the identification of new drugs.

25.3 TB Drug Clinical Pipeline: The Prospective Future Anti-TB Drugs

Although the TB drug pipeline remained almost empty for a long period of time till the 1990s, it is indeed optimistic to see that it is currently filled with more than 20 molecules, which are being assessed under different clinical phases (https://www.newtbdrugs.org/pipeline/clinical) (Fig. 25.1). Moreover, apart from the molecules in clinical trials, many more molecules are in the lead optimization phase or in the early preclinical stages and have the promise of entering the clinical pipeline (Fig. 25.1) (https://www.newtbdrugs.org/pipeline/clinical).

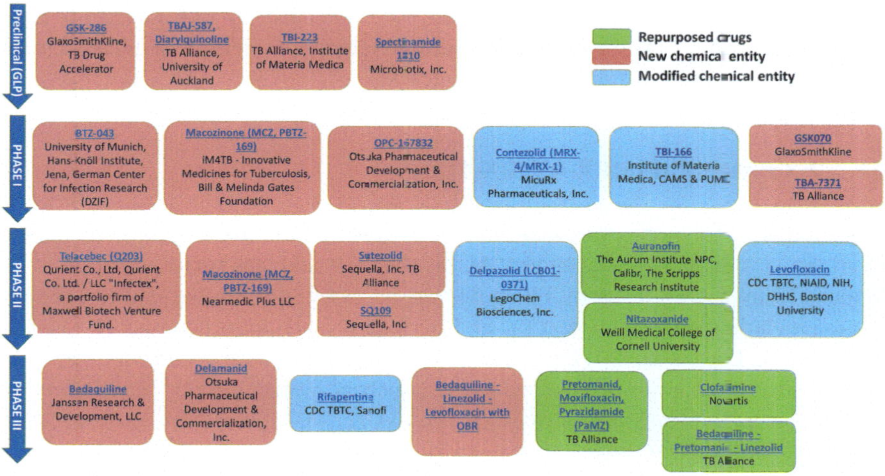

Fig. 25.1 Candidate anti-TB drugs in various stages of clinical trials

The number of candidate molecules in the TB drug pipeline provides no assurance that these molecules may reach the advance stages of clinical development as is evident from the fact that only few molecules are presently being evaluated in phase III and almost 50% are being evaluated as combination regimens of these drugs. This highlights a high attrition rate and the necessity of more molecules to fill the pipeline. Besides, the failure of the current candidate molecules reflects the limitations and caveats of the existing drug discovery approaches and emphasizes the importance of devising innovative strategies and tools to develop new molecules or increase the efficacy of molecules that qualify for clinical trials.

The anti-TB molecules currently in the clinical pipeline belong to various classes like fluoroquinolones, diarylquinolines, nitroimidazoles, benzothiazinones, etc., targeting various proteins/enzymes/pathways of *M. tuberculosis* including cell wall biosynthesis enzymes, energy metabolism and protein synthesis. Most of these molecules have been identified either through whole cell phenotypic screening or by repurposing of the drugs already in use for other diseases. Candidate drugs in TB pipeline are mentioned below.

25.3.1 BTZ043

The antitubercular agent BTZ043 (belonging to the nitrobenzothiazinone (BTZ) class) specifically blocks the mycobacterial enzyme decaprenyl-phosphoribose-2-′-epimerase (DprE1), responsible for the synthesis of a cell wall component D-arabinofuranose, and shows MIC in nanomolar range against the members of the *M. tuberculosis* complex (Makarov et al. 2009). BTZ043 showed superior antitubercular activity in comparison to isoniazid in vivo in mouse model with a low toxicological potential and was also well tolerated in rats and mini pigs (Kloss et al. 2017). The molecule is currently being evaluated in phase I trial (https://www.newtbdrugs.org/pipeline/clinical).

25.3.2 Contezolid (MRX-4/MRX-1)

Linezolid (LZD) belongs to oxazolidinone class of antibiotics that showed potent in vitro and in vivo activities against *M. tuberculosis* and was subsequently used in humans to treat drug-resistant TB; however, its use was restricted due to toxicity issues, including myelosuppression and peripheral and optic neuropathy (Mehta et al. 2016; Lee et al. 2012). Contezolid (MRX-1), a new oxazolidinone, was developed to treat gram-positive infections, such as methicillin-resistant *Staphylococcus aureus* and vancomycin-resistant *enterococci*, and showed decreased toxicity as compared to LZD (Gordeev and Yuan 2014; Shoen et al. 2018). MRX-1 showed promising in vitro as well as in vivo activity against both drug-susceptible and drug-resistant *M. tuberculosis* in mice model and is currently being evaluated in phase III trials (https://www.newtbdrugs.org/pipeline/clinical, Gordeev and Yuan 2014).

25.3.3 OPC-167832

OPC-167832, a newly synthesized carbostyril derivative discovered by Otsuka healthcare, inhibits decaprenylphosphoryl-β-D-ribose 2′-oxidase (DprE1), essentially involved in cell wall biosynthesis of *M. tuberculosis* (http://www.cptrinitiative.org/wp-content/uploads/2017/05/Jeffrey_Hafkin_CPTR2017_JH.pdf). This anti-TB compound shows in vitro as well as in vivo efficacy against both laboratory and clinically isolated strains including multidrug-resistant and extensively drug-resistant *M. tuberculosis* (http://www.cptrinitiative.org/wp-content/uploads/2017/05/Jeffrey_Hafkin_CPTR2017_JH.pdf). Furthermore, OPC-167832, when administered along with delamanid, showed superior efficacy to the standard regimen RHZE (rifampicin + isoniazid + pyrazinamide + ethambutol) in mice (http://www.cptrinitiative.org/wp-content/uploads/2017/05/Jeffrey_Hafkin_CPTR2017_JH.pdf). OPC-167832 is presently being evaluated in phase I trial (https://www.newtbdrugs.org/pipeline/clinical).

25.3.4 GSK070

GSK070 is an oxaborole derivative that targets the leucyl-tRNA synthetase (LeuRS) required for charging the tRNALue with leucine, thereby inhibiting protein synthesis (Palencia et al. 2016; Rock et al. 2007). GSK70 was observed to demonstrate in vitro and in vivo efficacy against *M. tuberculosis* with potent enzyme inhibition of *M. tuberculosis* LeuRS ($IC_{50} = 0.216$ μM) (Palencia et al. 2016). GSK70 recently qualified for phase I clinical assessment (https://www.newtbdrugs.org/pipeline/clinical).

25.3.5 Macozinone (PBTZ169)

Lead optimization studies of the compound BTZ043 by using medicinal chemistry resulted in PBTZ169, which is a piperazinobenzothiazinone derivative. PBTZ169 covalently binds to DprE1, thereby inhibiting cell wall biosynthesis (Trefzer et al. 2010; Makarov et al. 2014). PBTZ169 has an MIC_{99} against *M. tuberculosis* in nanomolar range (0.3 ng/ml) and has shown additive effects with many TB therapeutic agents and has also demonstrated synergistic effects with bedaquiline and clofazimine in preclinical models (Makarov et al. 2014). Innovative Medicines for Tuberculosis (iM4TB) foundation (Lausanne, Switzerland) has initiated the phase I clinical study of PBTZ169 to investigate dose-related safety (https://www.newtbdrugs.org/pipeline/compound/macozinone-mcz-pbtz-169).

25.3.6 TBI-166

Clofazimine, which is a very potent anti-TB compound belonging to riminophenazine class of drugs, has shown extremely impressive bactericidal and sterilizing efficacy against TB both in vitro and in mouse models of the disease (Reddy et al. 1996). However, it has poor solubility and a long half-life which results in side effects including pronounced skin discoloration (Job et al. 1990; Levy and Randall 1970). Lead optimization of clofazimine led to the identification of TBI-166, which has shown improved physicochemical and pharmacokinetic properties with similar efficacy as the parent compound (Lu et al. 2011). Based on these preliminary, preclinical and toxicological studies, TBI-166 was approved for phase I clinical trials (https://www.newtbdrugs.org/pipeline/clinical).

25.3.7 TBA-7371

TBA-7371 is a novel molecule belonging to pyrazolopyridone class of inhibitors (1,4-azaindole series) identified by performing lead optimization studies of an imidazopyridine compound. This molecule inhibits DprE1 (decaprenylphosphoryl-β-D-ribose 2′-epimerase) by binding to it non-covalently and shows an IC_{50} value of 10 nM with an MIC range of 0.78–3.12 μM and demonstrates efficacy in a rodent model of tuberculosis (Shirude et al. 2013, 2014; Yuan and Sampson 2018). The TB Alliance has initiated its phase I trial to evaluate its safety, tolerability, pharmacokinetics and the pharmacokinetic interactions (https://www.newtbdrugs.org/pipeline/clinical).

25.3.8 Telacebec (Q203)

Q203 resulted from the lead optimization of imidazo[1,2-a]pyridine amides that target the respiratory cytochrome *bc* complex, which in turn disturbs the electron motive force (Pethe et al. 2013; Lu et al. 2018; Kang et al. 2014). The small molecule Q203 inhibits the growth of MDR and XDR *M. tuberculosis* clinical isolates in culture broth medium with MIC in the low nanomolar range as well as shows therapeutic efficacy in mice (Pethe et al. 2013; Lu et al. 2018; Kang et al. 2014). In addition, Q203 displayed good pharmacokinetic and safety profiles; hence, it was qualified for the phase I trial for a dose escalation study which revealed encouraging results. Phase II trial for early bactericidal activity (EBA) evaluation of Q203 will be soon initiated in South Africa (https://www.newtbdrugs.org/pipeline/compound/telacebec-q203).

25.3.9 Sutezolid (Previously Known as PNU-100480)

Sutezolid exhibited increased in vitro as well as in vivo antimycobacterial activity, when compared with linezolid against both drug-susceptible and drug-resistant TB and showed improved safety profile highlighting its potential as an anti-TB agent (Barbachyn et al. 1996; Cynamon et al. 1999; Shaw and Barbachyn 2011; Wallis et al. 2010, 2011; Alffenaar et al. 2011). In fact, recent studies also showed that the use of sutezolid along with standard therapy was able to shorten the treatment duration by preventing relapse, thus, suggesting that sutezolid may have sterilizing activity against drug-susceptible TB and MDR-TB (Barbachyn et al. 1996; Shaw and Barbachyn 2011). Sutezolid was well tolerated and safe up to a daily dose of 1200 mg up to 14 days and phase II trials showed early bactericidal activity. Hence, it is considered that sutezolid may show clinical efficacy in a larger phase II trial (https://www.newtbdrugs.org/pipeline/clinical, Wallis et al. 2010).

25.3.10 Delpazolid (LCB01-0371)

LCB01-0371 is also a new oxazolidinone compound similar to linezolid with cyclic amidrazone. In vitro activity of LCB01-0371 was found to be similar to linezolid with an improved safety profile (Zong et al. 2018; Kim et al. 2017; Jeong et al. 2010). In vivo activity of LCB01-0371 against systemic infections in mice was also evaluated, and it was found to be more active than linezolid against these systemic infections (Zong et al. 2018; Kim et al. 2017; Jeong et al. 2010). It is now in phase II trial for the evaluation of EBA studies (https://www.newtbdrugs.org/pipeline/clinical).

25.3.11 SQ109

SQ109 is a novel 1,2-ethylenediamine molecule having a novel mechanism of action targeting MmpL3, which is a mycolic acid transporter required for mycolic acid incorporation into the *M. tuberculosis* cell wall (Tahlan et al. 2012; Grzegorzewicz et al. 2012). SQ109 inhibited the growth of both drug-susceptible and multidrug-resistant *M. tuberculosis* strains, including extensively drug-resistant *M. tuberculosis* strains (Sacksteder et al. 2012; Protopopova et al. 2005). It also exhibited synergistic effect with no adverse pharmacokinetic (PK) parameters and also improved the overall efficacy of the regimen when given along with standard treatment in mice (Sacksteder et al. 2012; Chen et al. 2006; Nikonenko et al. 2007). Three phase I studies are completed for SQ109 in the USA along with two phase II studies in Africa in drug-sensitive TB patients (https://www.newtbdrugs.org/pipeline/compound/sq109).

25.3.12 Auranofin (Brand Name: Ridaura)

Auranofin is a gold complex FDA-approved drug to treat rheumatoid arthritis (Suarez-Almazor et al. 2000). The drug is able to reduce and improve arthritis-related symptoms like pain, tender and swollen joints and morning stiffness. Auranofin was identified by employing a cell-based screen under nutrient-deprivation conditions against *M. tuberculosis* (Harbut et al. 2015). It exhibits potent inhibition of the growth of both replicating and nonreplicating *M. tuberculosis*, which was found to be bactericidal in nature (Harbut et al. 2015). Phase II trial is initiated to study the efficacy of the auranofin against *M. tuberculosis* (https://www.newtbdrugs.org/pipeline/clinical).

25.3.13 Levofloxacin

Levofloxacin is a second-generation fluoroquinolone, which shows enhanced activity against gram-positive pathogens, including *S. pneumoniae* and *S. aureus*, and was shown to be effective in the treatment of upper and lower respiratory tract infections in adults (Peterson et al. 2009; Alsultan et al. 2015). Levofloxacin is one of the essential medicines listed by World Health Organization and may be used for the treatment of tuberculosis (https://www.newtbdrugs.org/pipeline/compound/levofloxacin). TBTC 32/NIAID OPTI-Q phase II studies will determine the levofloxacin dose and exposure required to achieve the maximal reduction in *M. tuberculosis* burden in Peru and South Africa (https://www.newtbdrugs.org/pipeline/compound/levofloxacin).

25.3.14 Nitazoxanide (NTZ)

NTZ is a synthetic nitrothiazolyl salicylamide prodrug that is deacetylated in the gastrointestinal tract to the active metabolite tizoxanide and is approved for the treatment of giardiasis and cryptosporidiosis (Aslam and Musher 2007) NTZ showed activity against other protozoa, helminths, rotavirus and hepatitis C (Aslam and Musher 2007; Stachulski et al. 2011; Adagu et al. 2002; Theodos et al. 1998; Darling and Fried 2009; Korba et al. 2008; Rossignol et al. 2008, 2010). It was also shown to kill both replicating and nonreplicating *M. tuberculosis* with a MIC of 16 μg/ml (de Carvalho et al. 2009). No resistant mutants were found on treatment of

M. tuberculosis with various concentrations of NTZ suggesting that NTZ may have multiple targets (de Carvalho et al. 2009). It is currently undergoing phase II efficacy trial (https://www.newtbdrugs.org/pipeline/clinical).

25.3.15 Combination Regimens

The drugs that have successfully completed the phase II efficacy trials are now being evaluated in combination regimens in phase III trials to evaluate their therapeutic efficacy in shortening the treatment regimen.

25.3.16 Bedaquiline

Bedaquiline belongs to the class of diarylquinolines with a novel mechanism of action targeting the mycobacterial ATP synthase (Andries et al. 2005; Koul et al. 2007). It was found to exhibit activity against both drug-susceptible and drug-resistant strains of the pathogen having a strong bactericidal and sterilizing properties (Andries et al. 2005; Diacon et al. 2009). Bedaquiline got US FDA approval in 2012 as an anti-TB drug; however, its usage is reserved for the treatment of MDR-TB cases only (https://www.newtbdrugs.org/pipeline/compound/bedaquiline-0). It is now being evaluated in various combination regimens in which bedaquiline and pretomanid will be administered along with existing and repurposed anti-TB drugs for the treatment of biologically confirmed pulmonary multidrug-resistant TB (MDR-TB) (https://www.newtbdrugs.org/pipeline/compound/bedaquiline-0). Besides, another trial is scheduled to be carried out to assess its efficacy, when administered with delamanid (https://www.newtbdrugs.org/pipeline/compound/bedaquiline-0).

25.3.17 Rifapentine

Rifapentine is a semisynthetic derivative of rifamycin family with MIC_{99} value of 0.25 μg/ml in liquid medium (Sensi et al. 1959; Bemer-Melchior et al. 2000). It is currently approved for intermittent dosing in the treatment of TB. Also, it is currently included in combination regimen for phase III trials to test whether these regimens can shorten the treatment duration (https://www.newtbdrugs.org/pipeline/compound/rifapentine). The regimens being evaluated are (i) a single replacement of rifampin with rifapentine, initial 2 months of isoniazid, rifapentine, ethambutol and pyrazinamide, followed by another 2 months of isoniazid and rifapentine, and (ii) a double replacement of rifampin with rifapentine and ethambutol with moxifloxacin – initial 2 months of isoniazid, rifapentine, moxifloxacin and pyrazinamide, followed by another 2 months of isoniazid, rifapentine and moxifloxacin (https://www.newtbdrugs.org/pipeline/compound/rifapentine).

25.3.18 Delamanid

Delamanid (OPC-67683, Deltyba®) belongs to bicyclic nitroimidazole class of compounds, which showed a marked antituberculosis activity in vitro and a superior therapeutic efficacy in the chronic mouse model (Matsumoto et al. 2006; Tsubouchi et al. 2016). In 2014, European Medicines Agency (EMA) approved delamanid for the treatment of adult pulmonary MDR-TB (Yuan and Sampson 2018). Delamanid-resistant mutants revealed that it inhibits genes involved in F420-dependent deazaflavin nitroreductase bioactivation pathway (Fujiwara et al. 2018). In addition, in a phase IIb global trial, delamanid was found to increase the rate of 2-month sputum culture conversion, when it was added to an already optimized background regimen for the treatment of MDR-TB patients (Diacon et al. 2011; Gler et al. 2012). Clinical studies also revealed the efficacy of delamanid containing regimens in highly resistant TB patients that included cases with extensively drug-resistant TB (Skripconoka et al. 2013). Currently, for the evaluation of safety and efficacy of delamanid at a 200 mg oral daily dose, a phase III trial is being conducted (https://www.newtbdrugs.org/pipeline/compound/delamanid-0). Moreover, combined usage of delamanid and bedaquiline is also in progress to evaluate whether their combination can enhance the efficacy against MDR-TB (https://www.newtbdrugs.org/pipeline/compound/delamanid-0).

25.3.19 Clofazimine

Clofazimine, as described above, shows potent antitubercular activity (Reddy et al. 1996; Xu et al. 2012) and is now being evaluated in phase III trials for its efficacy, safety and tolerability, when it is administered in various regimens such as (i) TMC207 plus PA-824 plus pyrazinamide plus clofazimine, (ii) TMC207 plus PA-824 plus clofazimine, (iii) TMC207 plus pyrazinamide plus clofazimine and (iv) clofazimine alone, in adult patients with newly diagnosed, smear-positive pulmonary tuberculosis (https://www.newtbdrugs.org/pipeline/compound/clofazimine).

25.3.20 Pretomanid-Moxifloxacin-Pyrazinamide Regimen

Pretomanid (PA-824, Pa) belongs to nitroimidazo-oxazine class of compounds, which shows a very potent MIC of 0.125 μg/ml against *M. tuberculosis* and shows bactericidal activity during the initial and continuation phases of treatment in murine model with no issues of cross-resistance with other existing TB drugs (Stover et al. 2000; Tyagi et al. 2005). In addition, combination of PA-824, moxifloxacin and pyrazinamide (PaMZ) cured mice more rapidly than the first-line regimen of rifampin, isoniazid and pyrazinamide (Nuermberger et al. 2008; Tasneen et al. 2011).

PaMZ represents the first regimen to undergo clinical evaluation for multidrug TB treatment and has shown very positive results in phase II trial to conclude that it has potential of curing both the susceptible and resistant TB (https://www.newtbdrugs.org/pipeline/regimen/pretomanid-moxifloxacin-pyrazinamide-regimen). Besides, PaMZ also had better co-administration ability with the antiretrovirals, which is useful as an improved treatment option for HIV-TB co-infected patients (https://www.newtbdrugs.org/pipeline/regimen/pretomanid-moxifloxacin-pyrazinamide-regimen). TB Alliance is now focusing on advancing the BPaMZ regimen consisting of bedaquiline, PA-824, moxifloxacin and pyrazinamide (https://www.newtbdrugs.org/pipeline/regimen/pretomanid-moxifloxacin-pyrazinamide-regimen).

25.4 Advantages and Pitfalls of Various TB Drug Discovery Approaches

25.4.1 Target-Based Virtual Screening

The growing burden of antibiotic resistance propelled research towards the development of new antibiotics resulting in various drug discovery approaches. The genome sequence of *M. tuberculosis* was elucidated in the year 1998 (Cole et al. 1998), which gave a tremendous boost to the approaches such as target-based virtual screening, generation of gene knockout strains for the identification and validation of drug targets and system biology, etc. Target-/structure-based virtual screening relies on the use of computer-based software to identify potential inhibitors against the target structure by docking a library of small molecules into the active site and shortlisting the best binding molecules. An important prerequisite for this approach is the availability of a three-dimensional structure of the target protein, which gained its pace after the determination of genome sequence of *M. tuberculosis*. Another crucial step involved in virtual screening is the identification of an important drug target, which plays a key role in the pathogenesis of *M. tuberculosis*. This became possible as the knowledge about *M. tuberculosis* genes increased with the availability of its genome sequence, which also aided in developing gene deletion mutants. These knockout strains are then evaluated for their ability to grow in vitro, inside macrophages and subsequently in animal models for final validation; however, the major limitation of this technique, for a long time, has been a low frequency of site-specific recombination making it difficult to obtain the mutants and the inherent challenge of slow growth rate of *M. tuberculosis*. With newer methods of recombineering that were subsequently developed based on linear allelic exchange substrate, transposon mutagenesis and mycobacteriophage systems, generating knockout strains in *M. tuberculosis* became easier and the validation of essential and important drug targets gained a much faster pace (Van Kessel and Hatfull 2007; Sassetti et al. 2001). A study by Sassetti et al. revolutionized the drug discovery efforts by identifying several *M. tuberculosis* genes required for the in vitro and in vivo growth of the pathogen by using transposon mutagenesis, which led to a huge list of important drug targets along with the information on their essentiality for the growth of the pathogen (Sassetti et al. 2001; Sassetti and Rubin 2003). For instance, enzymes belonging to energy metabolism were shown to be essential for the growth of *M. tuberculosis* by Sassetti et al., and this information led to the identification of Q203 as an important small molecule inhibiting the cytochrome bc1 complex, which is currently being evaluated in clinical trials (https://www.newtbdrugs.org/pipeline/clinical, Kang et al. 2014). Additionally, many drug targets including UDP-galactopyranose mutase, 2C-methyl-D-erythritol 4-phosphate pathway and enzymes belonging to purine nucleotide biosynthetic pathway (GuaB2) were also shown to be essential by transposon mutagenesis by Sassetti et al. and have been employed as drug targets (Kincaid et al. 2015; Eoh et al. 2009; Singh et al. 2017). In addition, many other deletion mutants have been developed by using the above-mentioned methods, which have helped in the identification of several drug targets including MbtE, SapM, BioA, PptT, DrpE1, Rv3484, etc. (Reddy et al. 2013; Puri et al. 2013; Kar et al. 2017; Leblanc et al. 2012; Crellin et al. 2011; Malm et al. 2018). For instance, the *mbtE* mutant was developed by using linear AES method of homologous recombination, and MbtE was shown to be essential for the virulence and survival of the pathogen in broth culture and in macrophages. Besides, the mutant was also shown to be attenuated for its growth in guinea pigs as infection with it exhibited significantly reduced bacillary load and histopathological damage in the organs, in comparison to *M. tuberculosis*-infected animals (Reddy et al. 2013).

Target-based virtual screening has been employed extensively since the last decade, and till date many compounds have been screened against many important drug targets. A screening effort that evaluated 20,000 compounds against FtsZ, which plays an essential role in cell division, led to the identification of the inhibitor 297F, which showed inhibitory activity against *M. tuberculosis* in vitro growth (Lin et al. 2014). In another study, inhibitors were identified against *M. tuberculosis* thiamine phosphate synthase, an enzyme involved in the biosynthesis of thiamine by employing virtual screening resulting in promising inhibitors with IC_{50} of 34 μg/ml and MIC_{99}-6 μg/ml (Khare et al. 2011). In a recent study by Singh et al., structure-based virtual screening was employed against the active site of BioA, an enzyme involved in biotin biosynthetic pathway, which is essential for *M. tuberculosis* in vivo survival. It resulted in a few hits with the most potent hit displaying an MIC_{90} of 20 μg/ml (Singh et al. 2018). In another study, virtual screening was carried out by employing NCI library at the active site of *M. tuberculosis* 4′-phosphopantetheinyl transferase (PptT), which is involved in phosphopantetheinylation of several important proteins in a post-translational manner (Rohilla et al. 2018). This study led to the identification of a number of molecules with potent inhibition of the PptT enzymatic activity ($IC_{50} \leq 10$ μg/ml). Further, by employing a structure similarity approach based on chemoinformatics, a potent analogous molecule was identified with IC_{50} of 0.25 μg/ml, MIC_{99} of 10 μg/ml and negligible cytotoxicity (Rohilla et al. 2018). Recently, a study by Rohilla et al. showed that virtual screening of NCI library against IdeR, an essential iron regulatory transcriptional factor of *M. tuberculosis*, identified potent hits exhibiting IC_{50} values of 1–2 μg/ml (Rohilla et al. 2017). Major advantage of target-based screening is that there is a prior knowledge about the structure of the compound in question and the drug target, which makes it easier to elucidate the mechanism of action of the inhibition observed. Besides, the structural as well as the chemical knowledge about the compound helps in performing better lead optimization studies to obtain an improved antitubercular compound.

One of the major limitations of this strategy is the determination of the crystal structure of many *M. tuberculosis* proteins as they usually do not express well in *E. coli* and remain insoluble when expressed. Methods such as comparative homology modelling provide an alternative for the absence of crystal structures; however, they may not always result in potent inhibitors. Additionally, although this strategy holds promise in identifying a potent inhibitor against the target, many resulting enzyme inhibitors are unable to show a good MIC value in vitro (Singh et al. 2018; Rohilla et al. 2017, 2018; Kumar et al. 2017). *M. tuberculosis* is a complex pathogen comprising of a very strong cell wall which hampers the entry of most of the small inhibitor molecules, which are screened and shortlisted by employing target-based virtual screening. Even though the shortlisted molecules that are identified by screening chemical libraries against a validated target show high binding affinity for the target, their inability to enter the cell due to highly impermeable and hydrophobic cell wall of the pathogen might prevent them from arresting the cell growth. Besides, efflux of the compounds by the mycobacterial cells may also contribute to their poor MIC values (Kumar et al. 2017; Rodrigues et al. 2012; Zuniga et al. 2015; Kanji et al. 2016; Zhang et al. 2017; Parthasarathy et al. 2016). For instance, target-based virtual screening against many targets including PimA and PanC, both shown to be essential drug targets, resulted in the identification of several hits by screening many chemical libraries; however, none of the shortlisted molecules showed cellular activity emphasizing the possibility of their lack of entry into the cell (Kumar et al. 2017). Moreover, since *M. tuberculosis* is highly lipid rich, it is expected that a more lipophilic molecule will have a better permeability inside the mycobacterial cells and thus may be more potent as it has also been seen in the case of current anti-TB drugs, which are peculiarly more lipophilic than inhibitors against other bacteria (Kumar et al. 2017; Piccaro et al. 2015; Machado et al. 2018). However, in medicinal chemistry, the physicochemical characteristics such as high lipophilicity are not considered favourable for the development of a molecule into drug. This may explain the high attrition rate in the virtual screening approach with no identified compound reaching the TB drug pipeline, since most of the libraries selected for the screening are chosen as per the existing drug-likeness rules and exclude the lipophilic molecules. Hence, it seems more logical to reconsider the drug-likeness rules in the case of TB drug discovery, and new thinking beyond the existing dogma of Lipinski's rule of five is needed to succeed (Piccaro et al. 2015; Machado et al. 2018).

25.4.2 Whole Cell Phenotype Screening

The fact that none of the molecules that are being evaluated in clinical pipeline are derived from structure-based screening, there was a paradigm shift from computational methods to whole cell phenotypic screening approach for the identification of new anti-TB agents having novel mechanism of action. The method involves direct screening of small molecule libraries against the growth of *M. tuberculosis* in broth culture. The use of this approach gained pace after the discovery of bedaquiline (TMC207) and the successful identification of its target ATP synthase through isolation of resistant colonies and whole genome sequencing (Andries et al. 2005). Moreover, the success of this strategy is evident from the number of drugs in the clinical pipeline, which have been identified by whole cell phenotypic screening. Apart from bedaquiline (TMC207), various molecules such as SQ109, OPC-67683, PBTZ-169 and Q203 have been identified by employing this approach exhibiting potent inhibition of *M. tuberculosis* growth and are in various clinical, preclinical or early drug discovery stages (Makarov et al. 2014; Pethe et al. 2013; Lu et al. 2018; Sacksteder et al. 2012; Protopopova et al. 2005; Matsumoto et al. 2006; Tsubouchi et al. 2016). The power of this approach is that it circumvents the problems associated with target-based screens like compound penetration, target redundancy, etc.; however, a major drawback of this approach is the lack of knowledge about the mode of action of the drug, and hence, no inputs can be obtained via structure activity relationship studies and medicinal chemistry for better and improved drug designing (Zuniga et al. 2015). In fact, many times, for example, in the case of multiple targets, it is not possible to isolate resistant colonies which makes target identification a hard task. However, genomic tools for the identification of mutations in resistant colonies and transcriptional profiling studies in the presence and absence of drug along with the advent of proteomics and metabolomics have been shown to be promising and valuable tools for finding the drug targets. In addition if one is able to identify a potent inhibitor by using whole cell approach, it is sometimes difficult to understand its inhibitory activity in vivo due to a number of microenvironments the bacteria faces including hypoxic conditions, oxidative stress, acidic stress as well as the various metabolic states that the pathogen can acquire (Kumar et al. 2017; Koul et al. 2011). Hence, the development of better screening methods like carbon starvation model, hypoxic model and replicating and nonreplicating *M. tuberculosis* model systems that can mimic and represent the in vivo situation that *M. tuberculosis* encounters in human host will provide hope for improved drug designing (Kumar et al. 2017; Sala et al. 2010; Koul et al. 2008). More recently, TB investigators are focusing towards developing rapid and faster methods for screening inhibitors directly against the intraphagosomal growth of *M. tuberculosis* inside macrophages, the host niche where the bacteria resides, by employing GFP expressing strains of *M. tuberculosis* (Khare et al. 2013).

Discovery of Bedaquiline and Delamanid

Since the discovery of rifampicin in 1963, only two drugs with novel mechanisms of action, i.e. bedaquiline and delamanid, have been discovered and approved for treatment of tuberculosis.

Bedaquiline, marketed as Sirturo, was identified as a diarylquinoline, R207910 by whole cell screening approach that inhibits ATP synthase of *M. tuberculosis*. It inhibits both drug-sensitive and drug-resistant *M. tuberculosis in vitro* and *in vivo* and showed better bactericidal activity in comparison to isoniazid and rifampin (Andries *et al.*, 2005; Koul *et al.*, 2007; Diacon *et al.*, 2009).

Delamanid (OPC-67683, Deltyba), was identified by using the phenotypic approach that targets F420-dependent deazaflavin nitroreductase bio-activation pathway. It showed potent inhibition of replicating, dormant, and intracellular *M. tuberculosis*. Moreover, it exhibited bactericidal activity against *M. tuberculosis* in mouse model of TB (Matsumoto *et al.*, 2006).

25.4.3 Screening of Natural Products

The limited chemical space and diversity provided by screening libraries of small molecules result in the identification of hits with limited target diversity, which also highlights why certain pathways are always the frequent hits (Kumar et al. 2017). Hence, considerable attention has been drawn towards the use of natural products as starting point for TB drug discovery that can provide diverse chemical space with novel scaffolds, which are not present in the pharmaceutical libraries used for target-based as well as whole cell screening methods (Kumar et al. 2017; Cragg and Newman 2013). Natural products from medicinal plants or antibacterial bioactive compounds have shown promising results in terms of growth inhibitory potential against *M. tuberculosis* (Rodrigues Felix et al. 2017; Hartkoorn et al. 2012; Pruksakorn et al. 2010; Steinmetz et al. 2007). These include secondary metabolites isolated from plants and other microbial sources such as bacteria, fungi, marine organisms and algae, belonging to various chemical types such as terpenes (sesquiterpenes, diterpenes, sesterterpenes, triterpenes), steroids (sterols), alkaloids (indole, quinoline, pyridoacridone, manzamine alkaloids, etc.) and aromatics (flavonoids, chalcones, coumarins, lignans, xanthones, anthracenes, anthraquinones, naphthalenes, chromones, etc.). Apart from the secondary metabolites, polyketides (acetylenic fatty acids, polycyclic esters, quinones, etc.) and peptides have also shown growth inhibitory potential. Moreover, these metabolites are bioactive in nature; their relative bioavailability is quite high which thereby increases their access to the site of action (Quan et al. 2017). One of the naturally occurring classes of compounds is phenazines that are biosynthetically produced by *Actinobacteria* phylum, which are used against bacteria and fungi as broad-spectrum antibiotics (Quan et al. 2017; Laursen and Nielsen 2004). There is a renewed interest in considering riminophenazines as lead compounds for TB drug discovery, especially after clofazimine belonging to the same class was shown to work very effectively against MDR-TB, when administered in combination with gatifloxacin, ethambutol, pyrazinamide, prothionamide, kanamycin and high-dose isoniazid for 9 months (Reddy et al. 1999; Van Deun et al. 2010). However, because of the toxicity observed with the use of clofazimine, analogues of this natural product were developed, and as mentioned above, TBI-166 has been identified as a promising compound (https://www.newtbdrugs.org/pipeline/clinical, Reddy et al. 1996; Job et al. 1990; Levy and Randall 1970; Lu et al. 2011). Likewise, piperidines is another class of naturally available molecules isolated from black pepper, which have been used as wide range of drugs such as neuroleptics, vasodilators, antipsychotics and opioids (Quan et al. 2017). SQ109, the drug that is currently in the clinical pipeline, is an adamantine-containing hydroxydipiperidine that exhibits potent growth inhibitory property against *M. tuberculosis* in vitro and in vivo (Sacksteder et al. 2012; Protopopova et al. 2005). Similarly, BTZ043 also belongs to piperidine-containing benzothiazinone displaying inhibitory activity against the clinical isolates of *M. tuberculosis* and drug-resistant strains (Makarov et al. 2009; Kloss et al. 2017; Quan et al. 2017). Notably, both SQ109 and BTZ043 were identified through screening of dipiperidines and sulphur containing heterocycle libraries, respectively, emphasizing the strength of using natural products for the identification of antitubercular compounds given that these metabolites/peptides comprise of broad plethora of diverse scaffolds and pharmacophores. Moreover, other molecules in the clinical pipeline also belong to various classes of secondary metabolites like mycins and quinolones. However, the reasons for apprehension in employing this approach and a lag observed in the success of using these natural products are (i) difficulties in extraction of high yields of purified compounds, (ii) complexity in determining the structure of the compound, (iii) accessibility of the source material reproducibly, (iv) lack of safety studies and (v) lack of information about mechanism of action. Hence, innovative research is required to overcome these shortcomings.

25.4.4 Repurposing of Drugs

Another important strategy that has a lot of translational scope, which has gained focus and has resulted in promising molecules currently in the clinical pipeline, is the repurposing of the existing drugs also known as therapeutic switching or repositioning approach (Fig. 25.1) (Maitra et al. 2015). Repurposing drug approach employs the use of already existing drugs against various other diseases. Some of the drugs that are used to treat a particular condition can also interact with some other important target(s) and show its effect, which provides a window to analyse its therapeutic efficacy. Repurposing of the known drugs is promising as it is less time consuming in terms of the translation from preclinical work to the market, is less risky and is also less costly (Fig. 25.2).

This approach benefits from the fact that these molecules are already well characterized in the context of its target validation, hit-to-lead optimization, in vivo metabolic studies and their safety and toxicity profiling (Maitra et al. 2015). Only 1 in

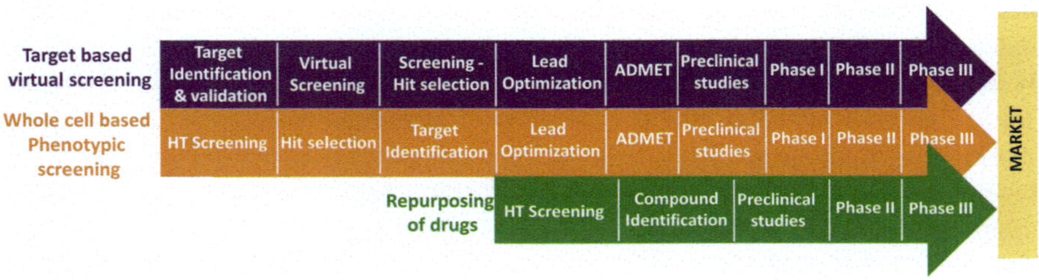

Fig. 25.2 The figure depicts various steps involved in the most commonly used drug discovery approaches for the identification of new anti-TB drugs

10,000 new chemical entities entering into pharmaceutical research actually makes it to the market; hence, the compounds already found to be safe in early-stage trials are less risky to begin with. It takes minimum a decade for a non-repurposed drug to reach the market as compared to only ~4 years for a repurposed drug. The most significant example of this approach is the use of sildenafil, which was used as an antihypertensive drug, also been shown to shorten the TB treatment in mouse model studies, when used along with standard drug regimen (Maiga et al. 2012). Moreover, three drugs, namely, clofazimine, linezolid and moxifloxacin, currently present in the TB drug clinical pipeline have resulted from repurposing approach (Van Deun et al. 2010; Till et al. 2002; Yanagihara et al. 2002; Alvirez-Freites et al. 2002). Clofazimine was initially used to treat leprosy and was shown to be successful in treating MDR- and XDR-TB; however, due to its side effects, it is being evaluated in combination with other TB drugs and further analogues are being prepared and tested such as TBI-166 (https://www.newtbdrugs.org/pipeline/clinical, Reddy et al. 1996; Job et al. 1990; Levy and Randall 1970; Lu et al. 2011; Xu et al. 2012; Van Deun et al. 2010; Garrelts 1991). Fluoroquinolones are potent broad-spectrum antibiotics that inhibit topoisomerases II and IV, in turn inhibiting DNA replication (Drlica and Zhao 1997). Moxifloxacin and gatifloxacin, which are the new generation fluoroquinolones, have shown sterilizing properties against *M. tuberculosis* in both in vitro and in vivo studies and are currently being used as second-line treatment for TB (Alvirez-Freites et al. 2002). Moxifloxacin is now being tested in phase III trials in combination with other drugs to evaluate its efficacy in shortening the treatment regimen (https://www.newtbdrugs.org/pipeline/clinical). Other classes of drugs such as members of the avermectin family, which are used for treating helminthic infections, are being tested for their activity against *M. tuberculosis* (Lim et al. 2013).

Efforts have also been made by the Indian scientists towards development of new drugs by using the repurposing approach. For example, Singhal et al. showed that the FDA-approved drug metformin used for treating diabetes was able to inhibit the intracellular growth of *M. tuberculosis* and enhance efficacy of the existing first-line TB drugs, thereby suggesting its use as part of adjunctive TB therapy (Singhal et al. 2014). Brindha et al. carried out virtual screening of FDA-approved drugs against the potential targets of *M. tuberculosis*, namely, TrpD and CoaA, and further screened the top ranking molecules for their ability to inhibit the in vitro growth of

M. tuberculosis resulting into the identification of two potential inhibitors of susceptible as well as resistant strains of *M. tuberculosis*, namely, lymecycline and cefpodoxime (Brindha et al. 2017). Moreover, lymecycline and cefpodoxime exhibited synergistic activity with rifampin and isoniazid against *M tuberculosis*, which suggests the potential of these drugs for the treatment of tuberculosis (Brindha et al. 2017). In another study, it was shown that administration of verapamil, an efflux pump inhibitor, as part of adjunctive therapy to infected mice was able to shorten the duration of standard TB regimen by accelerating the bacterial clearance and lowering down the relapse rates in comparison to mice that received only standard chemotherapy (Gupta et al. 2013). In addition, when verapamil was administered along with bedaquiline and clofazimine, it was able to sharply decrease the MIC of these drugs by 8- to 16-fold (Gupta et al. 2014). Another drug, namely, statin, which is used for lowering down of blood cholesterol, when given along with the first-line antitubercular drugs reduces the lung bacillary load in chronically infected mice (Dutta et al. 2016). Thus, drug repurposing approach is a very attractive strategy and provides alternatives for the treatment of drug-resistant cases.

Rational Drug Design: Another Promising Strategy

Topographical features near the active site of an important drug target are exploited to rationally design novel inhibitors/substrate mimics/transition state analogs and develop inhibitors with better binding affinity and more potent inhibitory activity.

A research group led by Courtney Aldrich at the University of Minnesota, Minneapolis, USA, has identified several potent rationally designed inhibitors against various drug targets such as MbtA that is involved in mycobactin synthesis, Biotin Protein Ligase, which biotinylates various important enzymes of lipid metabolism and BioA, which is involved in biotin synthesis (Qiao *et al.*, 2007; Duckworth *et al.*, 2011; Shi *et al.*, 2011; Shi and Aldrich 2012).

More examples of rationally designed inhibitors against TB targets have emerged against dihydrofolate reductase involved in the folate biosynthesis and Lumazine synthase involved in the riboflavin biosynthesis (Akthar 2016; Morgunova *et al.*, 2005).

25.4.5 New Approaches

a. *RNA-Based Therapeutics*

Apart from the major challenges associated with the TB chemotherapy, the biggest concern that arises in developing new anti-TB agent is the emergence of drug resistance. Even if a new molecule is identified as a potent anti-TB agent targeting a novel protein or metabolic pathway, the problem of developing drug resistance against the newly identified drug still prevails, and thus, cutting-edge research and innovative technologies are required, which can circumvent the problem of emergence of resistant strains. One such strategy is the use of RNA-based therapeutics such as antisense oligonucleotides, which may represent antitubercular compounds of the future. Antisense oligonucleotides complementary to the target mRNA sequence are designed which lead to the formation of target mRNA/antisense oligo duplex (Bai and Luo 2012). This duplex recruits RNaseH and cleaves the target mRNA molecule resulting in silencing of the target gene (Bai and Luo 2012). One of the limitations, however, associated with the use of oligonucleotides can be their unstable nature and inefficient delivery. Recent studies have shown that modified oligonucleotides such as phosphorothioate oligodeoxynucleotide (PS-ODNs) have enhanced cellular stability, and the use of nanoparticles or liposomes results in better delivery, stability and increased bioavailability. In a study by Meng et al., a new formulation, which involved the use of anionic liposome for encapsulation and delivery of mecA-specific PS-ODNs to target methicillin-resistant *Staphylococcus aureus*, was employed (Meng et al. 2009). Infected mice when treated with the encapsulated PS-ODN833 downregulated mecA and rescued the animals from MRSA-caused septic death (Meng et al. 2009). Harth et al. investigated the effect of hairpin loop extensions by using a sequence-specific PS-ODNs having 3′ and 5′ hairpin loop extensions targeting the 30–32 kDa protein complex (antigen 85 complex) that is involved in the mycolic acid synthesis of *M. tuberculosis* (Harth et al. 2007). These PS-ODNs with hairpin loop extensions inhibited bacterial growth in broth culture, inside human macrophages, and also reduced target gene transcription by >90% and showed increased bacterial sensitivity to isoniazid (Harth et al. 2007). Although very preliminary studies have been conducted by employing antisense RNA as a therapeutic tool against *M. tuberculosis*, due to promising results witnessed in these studies, further research is needed to take this approach forward.

b. *Delivery Systems*

b1. *Nanoparticles/Liposomes*

Recent studies are also being focused on developing better delivery strategies to increase the bioavailability of the anti-TB agents. A drug either administered via intravenous route or given orally gets distributed throughout the body, and hence, there are a limited number of molecules that reach the target site (Nasiruddin et al. 2017). Moreover, there can be non-specific or adverse side effects as well. In the case of mycobacterial infection, the drugs need to further reach the bacteria residing in macrophages and inside granulomas. Besides, the short half-life and rapid clearance of the drug also limit its effectiveness (Greenblatt 1985). Hence, to overcome this challenge, either we need very effective drugs or better delivery methods that can enhance the effectiveness of the drug. To this end, nanoparticles or liposomal preparations encapsulating the TB drugs are being developed. Nanoparticles are taken up very efficiently by cells and have the property of controlled, slow and persistent drug release making them an attractive and promising tool for drug delivery. Many modified nanoparticles have been prepared, for example, PEGylated nanoparticles to increase the bioavailability of drugs (Pandey et al. 2003; Sharma et al. 2004). In a study, a single subcutaneous injection of PLG nanoparticles encapsulated with the first-line TB drugs, rifampicin, isoniazid and pyrazinamide resulted in sustained plasma drug levels for 32 days and in lungs and spleen for 36 days (Pandey and Khuller 2004). This led to complete sterilization of the organs of *M. tuberculosis*-infected mice and demonstrated better therapeutic efficacy as compared with daily oral intake of free drugs (Pandey and Khuller 2004). Liposomes offer an inherent advantage that they have fusogenic abilities to fuse with the macrophages and can efficiently release the drugs into the macrophage cells, which is the primary niche of mycobacteria. In fact, target-based delivery systems are also being developed to avoid non-specific interactions of the drug-encapsulated nanoparticles or liposomes by decorating them with molecules such as mannose residues or O-SAP (O-stearyl amylopectin), which can specifically target the macrophages (Mahajan et al. 2010; Vyas et al. 2004). Thus, the superior drug bioavailability can help enhance its therapeutic usefulness even at low doses of the formulation, which may further help in reducing the period of chemotherapy and patient's compliance. In a study by Deol et al., it was demonstrated that liposome-encapsulated anti-TB drugs, isoniazid and rifampicin, showed better efficacy than free drugs against tuberculosis in mice model (Deol et al. 1997). Thus, advances should also be made towards improving various delivery systems, which can help in slow and sustained release of the drugs to finally increase the effectiveness and the associated therapeutic efficacy.

b2. *Devices for Sustained Release*

One of the problems associated with the control of TB relates to poor adherence of the patients to the lengthy chemotherapy, which requires daily administration, high pill burden and frequent dosing. Hence, current research is also being focused on the development of orally administered devices, which can help in increasing the bioavailability of the drug along with sustained release, with holding capacities for close to a month's pill dosage (Caffarel-Salvador et al. 2017). Such devices would make it feasible to provide the patient with the one time-large dose along with controlled release systems, which would help in getting rid of the needle injections and its associated complications. A similar device having these properties and ability to go and reside inside the GI tract is currently under preclinical evaluation. Such devices are advantageous over the existing injectable or oral drugs because of the ease of administration, a low immunological response and a greater accommodation of the drug inside the GI tract (Caffarel-Salvador et al. 2017). These devices would lead to drastic increase in the patient's adherence to the therapy, which would prove a boon for the prevention of drug-resistant TB cases.

c. *Gene-Editing Tools*

Recombineering techniques based on homologous recombination for the generation of deletion mutants are not a very efficient way to validate important drug targets due to high frequency of illegitimate recombinations and time-consuming procedure. Besides, multiple steps and specialized reagents are required, which make the procedure cost ineffective. As identification and validation of essential genes is a crucial requirement for the discovery of novel molecules, newer tools are needed that can speed up this process. CRISPR/CAS technology provides an alternative to the conventional method of gene silencing and has proven to be very useful in *M. tuberculosis* (Choudhary et al. 2015; Singh et al. 2016). The clustered regularly interspaced short palindromic repeats (CRISPR)/CRISPR-associated (Cas) system is a novel genome-editing tool found in bacteria and archaea responsible for the adaptive immune system of prokaryotes providing them with resistance to invading foreign viruses or plasmids (Makarova et al. 2011). The type II CRISPR/Cas system comprises of two short RNA and the DNA endonuclease Cas9. The short RNAs direct Cas9 DNA endonuclease to the target DNA sequence called the protospacer on the target DNA next to the protospacer adjacent motif (PAM) for site-specific cleavage resulting in double-stranded breaks, which can be repaired either by (i) the efficient, however, error-prone non-homologous end joining (NHEJ) pathway or (ii) the high-fidelity but less efficient homology-directed repair (HDR) pathway (Jiang and Doudna 2017). The double-stranded DNA breaks that are repaired by NHEJ pathway result in premature stop codon within the ORF of the target gene by creating either deletions or insertions or frameshift mutations (Jiang and Doudna 2017). The desired result is a loss-of-function mutation within the target gene (Jiang and Doudna 2017). More recently, CRISPR/CAS interference method is developed, which utilizes a small guide RNA that directs the enzymatically inactive CAS endonuclease to specific gene target resulting in interference in the transcription (Marraffini and Sontheimer 2010). Recently, Choudhary et al. showed that by employing an optimized CRISPR/CAS interference system, complete repression of individual or multiple target genes in mycobacteria could be achieved, thus providing a simple, rapid and cost-effective tool for the selective loss of gene expression in mycobacteria (Choudhary et al. 2015). In another such study, prevention of expression of several essential *M. tuberculosis* genes including *pknB* was reported emphasizing the ability of the system to modulate the extent of transcription inhibition (Singh et al. 2016).

d. *Immunotherapeutic Approach*

With an aim to shorten the treatment duration, newer strategies are being employed like the use of vaccines as an adjunct to standard chemotherapy. The basic rationale behind this approach is that vaccines may have the ability to alter the immune responses from unprotective Th2 type to protective Th1 type, which are typically needed for *M. tuberculosis* control. Hence, by administering the vaccine along with chemotherapy, it is believed that the combined effect exerted by both together might help in faster clearance of the bacteria from the host, which can have implications in reducing the duration of standard chemotherapy. For instance, immunotherapy with DNA expressing α-crystallin, an antigen associated with latency, was able to significantly reduce the chemotherapy period when compared with the chemotherapy alone (Chauhan et al. 2013). In another study, therapeutic vaccination of ID93/GLA-SE as an adjunct to chemotherapy decreased the bacillary load as well as improved the survival time of mice, when compared with mice that were given chemotherapy alone suggesting the possible benefits of adjunctive immunotherapy in shortening the treatment time (Coler et al. 2013). Vaccination with multivalent DNA vaccine encoding Ag85B, MPT-64 and MPT-83 in combination with isoniazid and pyrazinamide was effective in prevention of TB reactivation (Yu et al. 2008). Silva et al. demonstrated that immunotherapy with plasmid DNA encoding *Mycobacterium leprae* 65 kDa heat-shock protein (hsp65) in association with chemotherapy shortens the duration of treatment, improves the treatment of latent TB infection and is also effective against MDR-TB (Silva et al. 2005).

Concluding remarks

The efforts made by TB investigators in the last 50 years have succeeded in the identification of two new drugs, bedaquiline and delamanid, that are now approved for the treatment of multidrug resistant tuberculosis. Besides, it is optimistic to see many (~20) molecules being evaluated in the various phases of clinical trials. However, due to a high attrition rate, more efforts are needed to fill the pipeline with a higher number of molecules which can ultimately succeed in developing a very efficient regimen for TB cure that can shorten the therapy, can target all forms of the pathogen, can reduce the burden of resistant TB cases and has negligible toxicity to improve patient's compliance. Several conventional approaches such as target based virtual screening, whole cell phenotypic screening, rational drug design and repurposing of drugs have led to the identification of a number of potent inhibitors. Infact, most of the drugs such as SQ109, Q203, PA-824 etc., in the clinical pipeline have been identified through whole cell approach including bedaquiline and delamanid, highlighting the importance and success of this approach. However, there are certain scientific challenges that impede the progress of TB drug discovery such as low lipophilicity of the molecules under screening, low permeability of the enzyme inhibitors into the mycobacterial cell wall and insufficient target/pathway knowledge in the case of molecules that are identified via whole cell approach. Hence, considering the highly lipid rich cell wall of *M. tuberculosis*, new thinking is required beyond the dogma of fulfilling the Lipinski rule of drug likeliness, especially in the light of the fact that more lipophilic drugs will have a better chance to succeed and progress. Newer technologies like CRISPR/CAS system will be helpful in providing better knowledge about the drug target/pathway, which would help in accelerating the pace of identification of better drug targets. Additionally, innovative research is required to develop better delivery systems to enhance the bioavailability of drugs which can reduce the pill burden and frequency of dosing. Besides, there are a number of other non-scientific glitches, which are holding back the success of research in TB drug discovery. It is evident that the increased regulatory stringencies have slowed down the translation of potent inhibitors from bench to the market. The time between 1940s-1960s, the golden era, witnessed the discovery of maximum number of antibiotics including the currently available TB drugs and the translation of these antibiotics took only a couple of years to reach the market. However, the current regulatory concerns such as requirement of the ADMET compliance and rigorous clinical assessment of the candidate drugs, have increased the stringency to a level, which eliminates even the potent molecules that could have a likelihood of reaching the market. Moreover, a regulated marketing of the anti-TB drugs including the second-line drugs will help in curtailing the growing pool of resistant TB cases, which has impeded the control of the disease. Lastly, since drug discovery requires a huge expenditure, an increase in funding will have a huge impact on channelling more number of molecules to the pipeline especially with WHO's goal of 95% reduction in TB deaths and 90% reduction in TB incidence by 2035. Importantly, bridging academic interface and the industry sector would greatly enhance the translation of important research leads, which would considerably accelerate the drug discovery development leading to channelling of several novel lead molecules into clinical trials.

Figure 25.3 shows the problems and possible solutions involved in the TB drug discovery program.

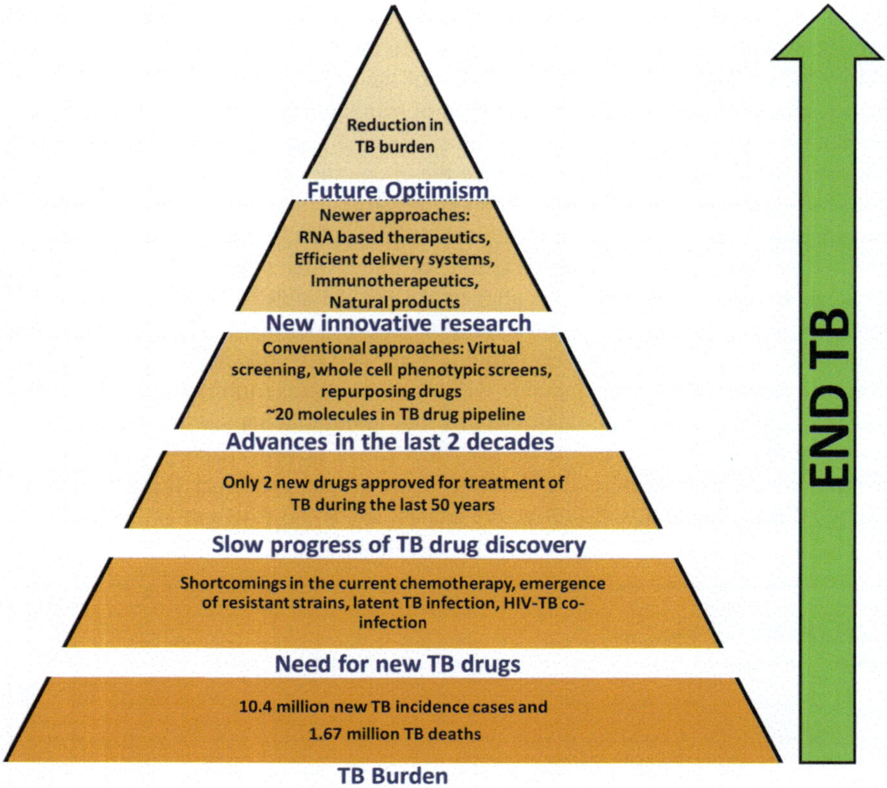

Fig. 25.3 TB drug discovery program: problems and possible solutions

References

Adagu IS, Nolder D, Warhurst DC et al (2002) In vitro activity of nitazoxanide and related compounds against isolates of *Giardia intestinalis*, *Entamoeba histolytica* and *Trichomonas vaginalis*. J Antimicrob Chemother 49:103–111

Akhter M (2016) Identification of novel *Mycobacterium tuberculosis* dihydrofolate reductase inhibitors through rational drug design. Int J Mycobacteriol 5(Suppl 1):S96

Alffenaar JW, van der Laan T, Simons S et al (2011) Susceptibility of clinical *Mycobacterium tuberculosis* isolates to a potentially less toxic derivate of linezolid, PNU-100480. Antimicrob Agents Chemother 55:1287–1289

Alsultan A, An G, Peloquin CA (2015) Limited sampling strategy and target attainment analysis for levofloxacin in patients with tuberculosis. Antimicrob Agents Chemother 59:3800–3807

Alvirez-Freites EJ, Carter JL, Cynamon MH (2002) In vitro and in vivo activities of gatifloxacin against *Mycobacterium tuberculosis*. Antimicrob Agents Chemother 46:1022–1025

Andries K, Verhasselt P, Guillemont J et al (2005) A diarylquinoline drug active on the ATP synthase of *Mycobacterium tuberculosis*. Science 307:223–227

Aslam S, Musher DM (2007) Nitazoxanide: clinical studies of a broad-spectrum anti-infective agent. Future Microbiol 2:583–590

Bai H, Luo X (2012) Antisense antibacterials: from proof-of-concept to therapeutic perspectives. In: Bobbarala V (ed) A search for antibacterial agents. IntechOpen, London, pp 319–344
Barbachyn MR, Hutchinson DK, Brickner SJ et al (1996) Identification of a novel oxazolidinone (U-100480) with potent antimycobacterial activity. J Med Chem 39:680–685
Bemer-Melchior P, Bryskier A, Drugeon HB (2000) Comparison of the in vitro activities of rifapentine and rifampicin against *Mycobacterium tuberculosis* complex. J Antimicrob Chemother 46:571–576
Brindha S, Sundaramurthi JC, Vincent S et al (2017) In silico and in vitro screening of FDA-approved drugs for potential repurposing against tuberculosis. bioRxiv: 228171
Caffarel-Salvador E, Abramson A, Langer R et al (2017) Oral delivery of biologics using drug-device combinations. Curr Opin Pharmacol 36:8–13
Chauhan P, Jain R, Dey B et al (2013) Adjunctive immunotherapy with α-crystallin based DNA vaccination reduces Tuberculosis chemotherapy period in chronically infected mice. Sci Rep 3:1821
Chen P, Gearhart J, Protopopova M et al (2006) Synergistic interactions of SQ109, a new ethylene diamine, with front-line antitubercular drugs in vitro. J Antimicrob Chemother 58:332–337
Chiang CY, Centis R, Migiori GB (2010) Drug-resistant tuberculosis: past, present, future. Respirology 15:413–432
Choudhary E, Thakur P, Pareek M et al (2015) Gene silencing by CRISPR interference in mycobacteria. Nat Commun 6:6267
Colditz GA, Brewer TF, Berkey CS et al (1994) Efficacy of BCG vaccine in the prevention of tuberculosis: meta-analysis of the published literature. JAMA 271:698–702
Cole ST, Brosch R, Parkhill J et al (1998) Deciphering the biology of *Mycobacterium tuberculosis* from the complete genome sequence. Nature 393:537–544
Coler RN, Bertholet S, Pine SO et al (2013) Therapeutic immunization against *Mycobacterium tuberculosis* is an effective adjunct to antibiotic treatment. J Infect Dis 207:1242–1252
Cragg GM, Newman DJ (2013) Natural products: a continuing source of novel drug leads. Biochim Biophys Acta 1830:3670–3695
Crellin PK, Brammananth R, Coppel RL (2011) Decaprenylphosphoryl-β-D-ribose 2′-epimerase, the target of benzothiazinones and dinitrobenzamides, is an essential enzyme in *Mycobacterium smegmatis*. PLoS One 6:e16869
Cynamon MH, Klemens SP, Sharpe CA et al (1999) Activities of several novel oxazolidinones against *Mycobacterium tuberculosis* in a murine model. Antimicrob Agents Chemother 43:1189–1191
Darling JM, Fried MW (2009) Nitazoxanide: beyond parasites toward a novel agent for hepatitis C. Gastroenterology 136:760–763
de Carvalho LP, Lin G, Jiang X et al (2009) Nitazoxanide kills replicating and nonreplicating *Mycobacterium tuberculosis* and evades resistance. J Med Chem 52:5789–5792
Deol P, Khuller GK, Joshi K (1997) Therapeutic efficacies of isoniazid and rifampin encapsulated in lung-specific stealth liposomes against *Mycobacterium tuberculosis* infection induced in mice. Antimicrob Agents Chemother 41:1211–1214
Diacon AH, Pym A, Grobusch M et al (2009) The diarylquinoline TMC207 for multidrug-resistant tuberculosis. N Engl J Med 360:2397–2405
Diacon AH, Dawson R, Hanekom M et al (2011) Early bactericidal activity of delamanid (OPC-67683) in smear-positive pulmonary tuberculosis patients. Int J Tuberc Lung Dis 15:949–954
Drlica K, Zhao X (1997) DNA gyrase, topoisomerase IV, and the 4-quinolones. Microbiol Mol Biol Rev 61:377–392
Duckworth BP, Geders TW, Tiwari D et al (2011) Bisubstrate adenylation inhibitors of biotin protein ligase from *Mycobacterium tuberculosis*. Chem Biol 18:1432–1441
Dutta NK, Bruiners N, Pinn ML et al (2016) Statin adjunctive therapy shortens the duration of TB treatment in mice. J Antimicrob Chemother 71:1570–1577
Eoh H, Brennan PJ, Crick DC (2009) The *Mycobacterium tuberculosis* MEP (2C-methyl-d-erythritol 4-phosphate) pathway as a new drug target. Tuberculosis (Edinb) 89:1–11

Fujiwara M, Kawasaki M, Hariguchi N et al (2018) Mechanisms of resistance to delamanid, a drug for *Mycobacterium tuberculosis*. Tuberculosis (Edinb) 108:186–194

Gardam MA, Keystone EC, Menzies R et al (2003) Anti-tumour necrosis factor agents and tuberculosis risk: mechanisms of action and clinical management. Lancet Infect Dis 3:148–155

Garrelts JC (1991) Clofazimine: a review of its use in leprosy and *Mycobacterium avium* complex infection. DICP 25:525–531

Gler MT, Skripconoka V, Sanchez-Garavito E et al (2012) Delamanid for multidrug-resistant pulmonary tuberculosis. N Engl J Med 366:2151–2160

Gordeev MF, Yuan ZY (2014) New potent antibacterial oxazolidinone (MRX-I) with an improved class safety profile. J Med Chem 57:4487–4497

Greenblatt DJ (1985) Elimination half-life of drugs: value and limitations. Annu Rev Med 36:421–427

Grzegorzewicz AE, Pham H, Gundi VA et al (2012) Inhibition of mycolic acid transport across the *Mycobacterium tuberculosis* plasma membrane. Nat Chem Biol 8:334–341

Guidelines for treatment of drug-susceptible tuberculosis and patient care (2017 update). http://apps.who.int/iris/bitstream/handle/10665/255052/9789241550000-eng.pdf;jsessionid=86C860DB7117D77D4A072A39ABCD6429?sequence=1. Last assessed on 15 March 2019

Gupta S, Tyagi S, Almeida DV et al (2013) Acceleration of tuberculosis treatment by adjunctive therapy with verapamil as an efflux inhibitor. Am J Respir Crit Care Med 188:600–607

Gupta S, Cohen KA, Winglee K et al (2014) Efflux inhibition with verapamil potentiates bedaquiline in Mycobacterium tuberculosis. Antimicrob Agents Chemother 58:574–576

Harbut MB, Vilchèze C, Luo X et al (2015) Auranofin exerts broad-spectrum bactericidal activities by targeting thiol-redox homeostasis. Proc Natl Acad Sci USA 112:4453–4458

Harth G, Zamecnik PC, Tabatadze D et al (2007) Hairpin extensions enhance the efficacy of mycolyl transferase-specific antisense oligonucleotides targeting *Mycobacterium* tuberculosis. Proc Natl Acad Sci USA 104:7199–7204

Hartkoorn RC, Sala C, Neres J et al (2012) Towards a new tuberculosis drug: pyridomycin–nature's isoniazid. EMBO Mol Med 4:1032–1042

http://www.cptrinitiative.org/wp-content/uploads/2017/05/Jeffrey_Hafkin_CPTR2017_JH.pdf. Last accessed on 15 March 2019.

https://www.newtbdrugs.org/pipeline/clinical. Last accessed on 15 March 2019

https://www.newtbdrugs.org/pipeline/compound/bedaquiline-0. Last assessed on 15 March 2019

https://www.newtbdrugs.org/pipeline/compound/clofazimine. Last assessed on 15 March 2019

https://www.newtbdrugs.org/pipeline/compound/delamanid-0. Last assessed on 15 March 2019

https://www.newtbdrugs.org/pipeline/compound/levofloxacin. Last assessed on 15 March 2019

https://www.newtbdrugs.org/pipeline/compound/macozinone-mcz-pbtz-169. Last assessed on 15 March 2019

https://www.newtbdrugs.org/pipeline/compound/rifapentine. Last assessed on 15 March 2019

https://www.newtbdrugs.org/pipeline/compound/sq109. Last assessed on 15 March 2019

https://www.newtbdrugs.org/pipeline/compound/telacebec-q203. Last assessed on 15 March 2019

https://www.newtbdrugs.org/pipeline/regimen/pretomanid-moxifloxacin-pyrazinamide-regimen. Last assessed on 15 March 2019

Jeong JW, Jung SJ, Lee HH et al (2010) In vitro and in vivo activities of LCB01-0371, a new oxazolidinone. Antimicrob Agents Chemother 54:5359–5362

Jiang F, Doudna JA (2017) CRISPR–Cas9 structures and mechanisms. Annu Rev Biophys 46:505–529

Job CK, Yoder L, Jacobson RR et al (1990) Skin pigmentation from clofazimine therapy in leprosy patients: a reappraisal. J Am Acad Dermatol 23:236–241

Kang S, Kim RY, Seo MJ et al (2014) Lead optimization of a novel series of imidazo [1, 2-a] pyridine amides leading to a clinical candidate (Q203) as a multi-and extensively-drug-resistant anti-tuberculosis agent. J Med Chem 57:5293–5305

Kanji A, Hasan R, Zaver A et al (2016) Alternate efflux pump mechanism may contribute to drug resistance in extensively drug-resistant isolates of *Mycobacterium tuberculosis*. Int J Mycobacteriol 5(Suppl 1):S97–S98

Kar R, Nangpal P, Mathur S et al (2017) bioA mutant of *Mycobacterium tuberculosis* shows severe growth defect and imparts protection against tuberculosis in guinea pigs. PLoS One 12: e0179513

Khare G, Kar R, Tyagi AK (2011) Identification of inhibitors *against Mycobacterium tuberculosis* thiamin phosphate synthase, an important target for the development of anti-TB drugs. PLoS One 6:e22441

Khare G, Kumar P, Tyag AK (2013) Whole cell screening based identification of inhibitors against the intraphagosomal survival of *Mycobacterium tuberculosis*. Antimicrob Agents Chemother 57:6372–6377

Kim TS, Choe JH, Kim YJ et al (2017) Activities of LCB01-0371, a novel oxazolidinone, against *Mycobacterium abscessus*. Antimicrob Agents Chemother 61:pii:e02752-16

Kincaid VA, London N, Wangkanont K et al (2015) Virtual screening for UDP-galactopyranose mutase ligands identifies a new class of antimycobacterial agents. ACS Chem Biol 10:2209–2218

Kloss F, Krchnak V, Krchnakova A et al (2017) In Vivo dearomatization of the potent Antituberculosis agent BTZ043 via Meisenheimer complex formation. Angew Chem Int Ed Engl 56:2187–2191

Korba BE, Montero AB, Farrar K et al (2008) Nitazoxanide, tizoxanide and other thiazolides are potent inhibitors of hepatitis B virus and hepatitis C virus replication. Antiviral Res 77:56–63

Koul A, Dendouga N, Vergauwen K et al (2007) Diarylquinolines target subunit c of mycobacterial ATP synthase. Nat Chem Biol 3:323–324

Koul A, Vranckx L, Dendouga N et al (2008) Diarylquinolines are bactericidal for dormant mycobacteria as a result of disturbed ATP homeostasis. J Biol Chem 283:25273–25280

Koul A, Arnoult E, Lounis N et al (2011) The challenge of new drug discovery for tuberculosis. Nature 469:483–490

Kumar A, Chettiar S, Parish T (2017) Current challenges in drug discovery for tuberculosis. Expert Opin Drug Discov 12:1–4

Laursen JB, Nielsen J (2004) Phenazine natural products: biosynthesis, synthetic analogues, and biological activity. Chem Rev 104:1663–1686

Leblanc C, Prudhomme T, Tabouret G et al (2012) 4′-Phosphopantetheinyl transferase PptT, a new drug target required for *Mycobacterium tuberculosis* growth and persistence in vivo. PLoS Pathog 8:e1003097

Lee M, Lee J, Carroll MW et al (2012) Linezolid for treatment of chronic extensively drug-resistant tuberculosis. N Engl J Med 367:1508–1518

Levy L, Randall HP (1970) A study of skin pigmentation by clofazimine. Int J Lepr Other Mycobact Dis 38:404–416

Lim LE, Vilchèze C, Ng C et al (2013) Anthelmintic avermectins kill *M. tuberculosis*, including multidrug resistant clinical strains. Antimicrob Agents Chemother 57:1040–1046

Lin Y, Zhu N, Han Y et al (2014) Identification of anti-tuberculosis agents that target the cell-division protein FtsZ. J Antibiot (Tokyo) 67:671–676

López-Cortés LF, Ruiz-Valderas R, Viciana P et al (2002) Pharmacokinetic interactions between efavirenz and rifampicin in HIV-infected patients with tuberculosis. Clin Pharmokinet 41:681–690

Lu Y, Zheng M, Wang B et al (2011) Clofazimine analogs with efficacy against experimental tuberculosis and reduced potential for accumulation. Antimicrob Agents Chemother 55:5185–5193

Lu P, Asseri AH, Kremer M et al (2018) The anti-mycobacterial activity of the cytochrome bcc inhibitor Q203 can be enhanced by small-molecule inhibition of cytochrome bd. Sci Rep 8:2625

Machado D, Girardini M, Viveiros M et al (2018) Challenging the "drug-likeness" dogma for new drug discovery in tuberculosis. Front Microbiol 9:1367

Mahajan S, Prashant CK, Koul V et al (2010) Receptor specific macrophage targeting by mannose-conjugated gelatin nanoparticles-an in vitro and in vivo study. Curr Nanosci 6:413–421

Maiga M, Agarwal N, Ammerman NC et al (2012) Successful shortening of tuberculosis treatment using adjuvant host-directed therapy with FDA-approved phosphodiesterase inhibitors in the mouse model. PLoS One 7:e30749

Maitra A, Bates S, Kolvekar T et al (2015) Repurposing—a ray of hope in tackling extensively drug resistance in tuberculosis. Int J Infect Dis 32:50–55

Makarov V, Manina G, Mikusova K et al (2009) Benzothiazinones kill *Mycobacterium tuberculosis* by blocking arabinan synthesis. Science 324:801–804

Makarov V, Lechartier B, Zhang M et al (2014) Towards a new combination therapy for tuberculosis with next generation benzothiazinones. EMBO Mol Med 6:372–383

Makarova KS, Haft DH, Barrangou R et al (2011) Evolution and classification of the CRISPR–Cas systems. Nat Rev Microbiol 9:467–477

Malm S, Maaß S, Schaible UE et al (2018) In vivo virulence of *Mycobacterium tuberculosis* depends on a single homologue of the LytR-CpsA-Psr proteins. Sci Rep 8:3936

Marraffini LA, Sontheimer EJ (2010) CRISPR interference: RNA-directed adaptive immunity in bacteria and archaea. Nat Rev Genet 11:181–190

Matsumoto M, Hashizume H, Tomishige T et al (2006) OPC-67683, a nitro-dihydro-imidazooxazole derivative with promising action against tuberculosis in vitro and in mice. PLoS Med 3:e466

Mehta S, Das M, Laxmeshwar C et al (2016) Linezolid-associated optic neuropathy in drug-resistant tuberculosis patients in Mumbai, India. PLoS One 11:e0162138

Meng J, Wang H, Hou Z et al (2009) Novel anion liposome-encapsulated antisense oligonucleotide restores susceptibility of methicillin-resistant *Staphylococcus aureus* and rescues mice from lethal sepsis by targeting mecA. Antimicrob Agents Chemother 53:2871–2878

Morgunova E, Meining W, Illarionov B et al (2005) Crystal structure of lumazine synthase from *Mycobacterium tuberculosis* as a target for rational drug design: binding mode of a new class of purinetrione inhibitors. Biochemistry 44:2746–2758

Narayanan S, Swaminathan S, Supply P et al (2010) Impact of HIV infection on the recurrence of tuberculosis in South India. J Infect Dis 201:691–703

Nasiruddin M, Neyaz M, Das S (2017) Nanotechnology-based approach in tuberculosis treatment. Tuberc Res Treat 2017:4920209

Nikonenko BV, Protopopova M, Samala R et al (2007) Drug therapy of experimental tuberculosis (TB): improved outcome by combining SQ109, a new diamine antibiotic, with existing TB drugs. Antimicrob Agents Chemother 51:1563–1565

Nuermberger E, Tyagi S, Tasneen R et al (2008) Powerful bactericidal and sterilizing activity of a regimen containing PA-824, moxifloxacin, and pyrazinamide in a murine model of tuberculosis. Antimicrob Agents Chemother 52:1522–1524

Palencia A, Li X, Bu W et al (2016) Discovery of novel oral protein synthesis inhibitors of *Mycobacterium tuberculosis* that target leucyl-tRNA synthetase. Antimicrob Agents Chemother 60:6271–6280

Pandey R, Khuller GK (2004) Subcutaneous nanoparticle-based antitubercular chemotherapy in an experimental model. J Antimicrob Chemother 54:266–268

Pandey R, Zahoor A, Sharma S et al (2003) Nanoparticle encapsulated antitubercular drugs as a potential oral drug delivery system against murine tuberculosis. Tuberculosis (Edinb) 83:373–378

Parthasarathy K, Sivaraj A, Sundar R (2016) Identification of new efflux pump proteins from multidrug resistant *Mycobacterium tuberculosis* and screening for peptide based efflux pump inhibitors. Int J Mycobacteriol 45:404–405

Peterson J, Yektashenas B, Fisher AC (2009) Levofloxacin for the treatment of pneumonia caused by *Streptococcus pneumoniae* including multidrug-resistant strains: pooled analysis. Curr Med Res Opin 25:559–568

Pethe K, Bifani P, Jang J et al (2013) Discovery of Q203, a potent clinical candidate for the treatment of tuberculosis. Nat Med 19:1157–1160

Piccaro G, Poce G, Biava M (2015) Activity of lipophilic and hydrophilic drugs against dormant and replicating *Mycobacterium tuberculosis*. J Antibiot (Tokyo) 68:711–714

Protopopova M, Hanrahan C, Nikonenko B et al (2005) Identification of a new antitubercular drug candidate, SQ109, from a combinatorial library of 1, 2-ethylenediamines. J Antimicrob Chemother 56:968–974

Pruksakorn P, Arai M, Kotoku N et al (2010) Trichoderins, novel aminolipopeptides from a marine sponge-derived Trichoderma sp., are active against dormant mycobacteria. Bioorg Med Chem Lett 20:3658–3663

Puri RV, Reddy PV, Tyagi AK (2013) Secreted acid phosphatase (SapM) of Mycobacterium tuberculosis is indispensable for arresting phagosomal maturation and growth of the pathogen in guinea pig tissues. PLoS One 8:e70514

Qiao C, Gupte A, Boshoff HI et al (2007) 5′-O-[(N-acyl) sulfamoyl] adenosines as antitubercular agents that inhibit MbtA: an adenylation enzyme required for siderophore biosynthesis of the mycobactins. J Med Chem 50:6080–6094

Quan D, Nagalingam G, Payne R et al (2017) New tuberculosis drug leads from naturally occurring compounds. Int J Infect Dis 56:212–220

Reddy VM, Nadadhur G, Daneluzzi D et al (1996) Antituberculosis activities of clofazimine and its new analogs B4154 and B4157. Antimicrob Agents Chemother 40:633–636

Reddy VM, O'Sullivan JF, Gangadharam PR (1999) Antimycobacterial activities of riminophenazines. J Antimicrob Chemother 43:615–623

Reddy PV, Puri RV, Chauhan P et al (2013) Disruption of mycobactin biosynthesis leads to attenuation of *Mycobacterium tuberculosis* for growth and virulence. J Infect Dis 208:1255–1265

Rock FL, Mao W, Yaremchuk A et al (2007) An antifungal agent inhibits an aminoacyl-tRNA synthetase by trapping tRNA in the editing site. Science 316:1759–1761

Rodrigues Felix C, Gupta R, Geden S et al (2017) Selective killing of dormant *Mycobacterium tuberculosis* by marine natural products. Antimicrob Agents Chemother 61: pii: e00743-17

Rodrigues L, Machado D, Couto I et al (2012) Contribution of efflux activity to isoniazid resistance in the *Mycobacterium tuberculosis* complex. Infect Genet Evol 12:695–700

Rohilla A, Khare G, Tyagi AK (2017) Virtual Screening, pharmacophore development and structure based similarity search to identify inhibitors against IdeR, a transcription factor of *Mycobacterium tuberculosis*. Sci Rep 7:4653

Rohilla A, Khare G, Tyagi AK (2018) A combination of docking and cheminformatics approaches for the identification of inhibitors against 4′ phosphopantetheinyl transferase of Mycobacterium tuberculosis. RSC Adv 8:328–341

Rossignol JF, Kabil SM, El-Gohary Y et al (2008) Clinical trial: randomized, double-blind, placebo-controlled study of nitazoxanide monotherapy for the treatment of patients with chronic hepatitis C genotype 4. Aliment Pharmacol Ther 28:574–580

Rossignol JF, Elfert A, Keeffe EB (2010) Treatment of chronic hepatitis C using a 4-week lead-in with nitazoxanide before peginterferon plus nitazoxanide. J Clin Gastroenterol 44:504–509

Sacksteder KA, Protopopova M, Barry CE 3rd et al (2012) Discovery and development of SQ109: a new antitubercular drug with a novel mechanism of action. Future Microbiol 7:823–837

Sala C, Dhar N, Hartkoorn RC et al (2010) Simple model for testing drugs against nonreplicating *Mycobacterium tuberculosis*. Antimicrob Agents Chemother 54:4150–4158

Sassetti CM, Rubin EJ (2003) Genetic requirements for mycobacterial survival during infection. Proc Natl Acad Sci USA 100:12989–12994

Sassetti CM, Boyd DH, Rubin EJ (2001) Comprehensive identification of conditionally essential genes in mycobacteria. Proc Natl Acad Sci USA 98:12712–12717

Sensi P, Margalith P, Timbal MT (1959) Rifomycin, a new antibiotic; preliminary report. Farmaco Sci 14:146–147

Sharma A, Pandey R, Sharma S et al (2004) Chemotherapeutic efficacy of poly (DL-lactide-co-glycolide) nanoparticle encapsulated antitubercular drugs at sub-therapeutic dose against experimental tuberculosis. Int J Antimicrob Agents 24:599–604

Shaw KJ, Barbachyn MR (2011) The oxazolidinones: past, present, and future. Ann NY Acad Sci 1241:48–70

Shi C, Aldrich CC (2012) Design and synthesis of potential mechanism-based inhibitors of the aminotransferase BioA involved in biotin biosynthesis. J Org Chem 77:6051–6058

Shi C, Geders TW, Park SW et al (2011) Mechanism-based inactivation by aromatization of the transaminase BioA involved in biotin biosynthesis in *Mycobacterium tuberculosis*. J Am Chem Soc 133:18194–18201

Shirude PS, Shandil R, Sadler C et al (2013) Azaindoles: noncovalent DprE1 inhibitors from scaffold morphing efforts, kill *Mycobacterium tuberculosis* and are efficacious in vivo. J Med Chem 56:9701–9708
Shirude PS, Shandil RK, Manjunatha MR et al (2014) Lead optimization of 1, 4-azaindoles as antimycobacterial agents. J Med Chem 57:5728–5737
Shoen C, DeStefano M, Hafkin B et al (2018) In vitro and in vivo activity of contezolid (MRX-I) against *Mycobacterium tuberculosis*. Antimicrob Agents Chemother 62 pii: e00493-18
Silva CL, Bonato VL, Coelho-Castelo AA et al (2005) Immunotherapy with plasmid DNA encoding mycobacterial hsp65 in association with chemotherapy is a more rapid and efficient form of treatment for tuberculosis in mice. Gene Ther 12:281–287
Singh AK, Carette X, Potluri LP et al (2016) Investigating essential gene function in *Mycobacterium tuberculosis* using an efficient CRISPR interference system. Nucleic Acids Res 44:e143
Singh V, Donini S, Pacitto A et al (2017) The inosine monophosphate dehydrogenase, GuaB2, is a vulnerable new bactericidal drug target for tuberculosis. ACS Infect Dis 3:5–17
Singh S, Khare G, Bahal RK et al (2018) Identification of *Mycobacterium tuberculosis* BioA inhibitors by using structure-based virtual screening. Drug Des Devel Ther 12:1065–1079
Singhal A, Jie L, Kumar P et al (2014) Metformin as adjunct antituberculosis therapy. Sci Transl Med 6:263ra159
Skripconoka V, Danilovits M, Pehme L et al (2013) Delamanid improves outcomes and reduces mortality for multidrug-resistant tuberculosis. Eur Respir J 41:1393–1400
Stachulski AV, Pidathala C, Row EC et al (2011) Thiazolides as novel antiviral agents. 1. Inhibition of hepatitis B virus replication. J Med Chem 54:4119–4132
Steinmetz H, Irschik H, Kunze B et al (2007) Thuggacins, macrolide antibiotics active against *Mycobacterium tuberculosis*: isolation from myxobacteria, structure elucidation, conformation analysis and biosynthesis. Chemistry 13:5822–5832
Stevenson CR, Critchley JA, Forouhi NG et al (2007) Diabetes and the risk of tuberculosis: a neglected threat to public health? Chronic Illn 3:228–245
Stover CK, Warrener P, Van Devanter DR et al (2000) A small-molecule nitroimidazopyran drug candidate for the treatment of tuberculosis. Nature 405:962–966
Suarez-Almazor ME, Spooner CH, Belseck E et al (2000) Auranofin versus placebo in rheumatoid arthritis. Cochrane Database Syst Rev 2:CD002048
Tahlan K, Wilson R, Kastrinsky DB et al (2012) SQ109 targets MmpL3, a membrane transporter of trehalose monomycolate involved in mycolic acid donation to the cell wall core of *Mycobacterium tuberculosis*. Antimicrob Agents Chemother 56:1797–1809
Tasneen R, Li SY, Peloquin CA et al (2011) Sterilizing activity of novel TMC207-and PA-824-containing regimens in a murine model of tuberculosis. Antimicrob Agents Chemother 55:5485–5492
Theodos CM, Griffiths JK, D'Onfro J et al (1998) Efficacy of nitazoxanide against cryptosporidium parvum in cell culture and in animal models. Antimicrob Agents Chemother 42:1959–1965
Till M, Wixson RL, Pertel PE (2002) Linezolid treatment for osteomyelitis due to vancomycin-resistant *Enterococcus faecium*. Clin Infect Dis 34:1412–1414
Trefzer C, Rengifo-Gonzalez M, Hinner MJ et al (2010) Benzothiazinones: prodrugs that covalently modify the decaprenylphosphoryl-β-D-ribose 2′-epimerase DprE1 of *Mycobacterium tuberculosis*. J Am Chem Soc 132:13663–13665
Tsubouchi H, Sasaki H, Ishikawa H et al (2016) Discovery of delamanid for the treatment of multidrug-resistant pulmonary Tuberculosis. In: Fischer J, Childers WE (eds) Successful drug discovery, vol 2. Wiley-VCH Verlag GmbH & Co, KGaA, Weinheim, pp 137–161
Tyagi S, Nuermberger E, Yoshimatsu T et al (2005) Bactericidal activity of the nitroimidazopyran PA-824 in a murine model of tuberculosis. Antimicrob Agents Chemother 49:2289–2293
Uplekar M, Weil D, Lonnroth K et al (2015) WHO's new end TB strategy. Lancet 385:1799–1801
Van Deun A, Maug AK, Salim MA et al (2010) Short, highly effective, and inexpensive standardized treatment of multidrug-resistant tuberculosis. Am J Respir Crit Care Med 182:684–692

van Kessel JC, Hatfull GF (2007) Recombineering in *Mycobacterium tuberculosis*. Nat Methods 4:147–152

Vyas SP, Kannan ME, Jain S et al (2004) Design of liposomal aerosols for improved delivery of rifampicin to alveolar macrophages. Int J Pharm 269:37–49

Wallis RS, Jakubiec WM, Kumar V et al (2010) Pharmacokinetics and whole-blood bactericidal activity against *Mycobacterium tuberculosis* of single doses of PNU-100480 in healthy volunteers. J Infect Dis 202:745–751

Wallis RS, Jakubiec W, Kumar V et al (2011) Biomarker-assisted dose selection for safety and efficacy in early development of PNU-100480 for tuberculosis. Antimicrob Agents Chemother 55:567–574

WHO guidelines for the programmatic management of drug-resistant tuberculosis.. http://apps.who.int/iris/bitstream/handle/10665/130918/9789241548809_eng.pdf;jsessionid=EBA844744A619EF170C00783C655E148?sequence=1 (Last assessed on 6.09.2018)

WHO: Global Tuberculosis Report 2017. http://www.who.int/tb/publications/global_report/gtbr2017_main_text.pdf

www.who.int/mediacentre/factsheets/fs104/en/. Last accessed on 22 August 2018

Xu HB, Jiang RH, Xiao HP (2012) Clofazimine in the treatment of multidrug-resistant tuberculosis. Clin Microbiol Infect 18:1104–1110

Yanagihara K, Kaneko Y, Sawai T et al (2002) Efficacy of linezolid against methicillin-resistant or vancomycin-insensitive *Staphylococcus aureus* in a model of hematogenous pulmonary infection. Antimicrob Agents Chemother 46:3288–3291

Yu DH, Hu XD, Cai H (2008) Efficient tuberculosis treatment in mice using chemotherapy and immunotherapy with the combined DNA vaccine encoding Ag85B, MPT-64 and MPT-83. Gene Ther 15:652–659

Yuan T, Sampson NS (2018) Hit generation in TB drug discovery: from genome to granuloma. Chem Rev 118:1887–1916

Zhang Y, Zhang J, Cui P et al (2017) Identification of novel efflux proteins Rv0191, Rv3756c, Rv3008 and Rv1667c involved in Pyrazinamide resistance in Mycobacterium tuberculosis. Antimicrob Agents Chemother 61:pii:e00940-17

Zong Z, Jing W, Shi J et al (2018) Comparison of in vitro activity and MIC distributions between the novel oxazolidinone delpazolid and linezolid against multidrug-resistant and extensively drug-resistant *Mycobacterium tuberculosis* in China. Antimicrob Agents Chemother 62:pii: e00165-18

Zuniga ES, Early J, Parish T (2015) The future for early-stage tuberculosis drug discovery. Future Microbiol 10:217–229

Repurposing of Carbapenems for the Treatment of Drug-Resistant Tuberculosis

26

Pankaj Kumar, Urvashi B. Singh, Gyanu Lamichhane, and Elizabeth Story-Roller

Abstract

The World Health Organization (WHO) estimated that there were 1.6 million deaths worldwide in 2017 from tuberculosis (TB). Today, TB kills more humans than any other infectious agent, even surpassing HIV/AIDS or malaria (WHO TB report 2018). Prolonged treatment with poorly tolerated drugs to treat drug-resistant tuberculosis (DR-TB) has further exacerbated this global health crisis. Overall success rate of treating multidrug-resistant (MDR) and extensively drug-resistant (XDR) TB is only 54%, which means for a significant number of patients, especially those with XDR-TB, prognosis is death. Factors that make existing treatment options of MDR/XDR-TB inadequate are (1) high incidence of undesirable side effects from the only drugs that are available today to treat these infections, (2) the protracted treatment duration (more than 18 months) and (3) the high cost of the treatment. The WHO has approved several new and repurposing of existing antibiotics for treatment of DR-TB aimed at improving outcomes of MDR/XDR-TB treatment. These recommendations have focused on optimizing all aspects of TB treatment including shortening treatment duration, optimizing dosages and developing new combination regimens. Although new anti-TB

P. Kumar (✉)
Department of Biochemistry, School of Chemical and Life Sciences, Jamia Hamdard, New Delhi, Delhi, India
e-mail: pankajkumar@jamiahamdard.ac.in

U. B. Singh
Tuberculosis Section, Department of Microbiology, A.I.I.M.S., New Delhi, Delhi, India

G. Lamichhane · E. Story-Roller
Division of Infectious Diseases, School of Medicine, Johns Hopkins University, Baltimore, MD, USA

S. E. Hasnain et al. (eds.), Mycobacterium tuberculosis: *Molecular Infection Biology, Pathogenesis, Diagnostics and New Interventions*,
https://doi.org/10.1007/978-981-32-9413-4_26

drugs, like bedaquiline, delamanid and pretomanid, have been successfully developed, their slow implementation, suboptimal availability and unanticipated cardiac toxicities mean availability of additional antibiotics would provide choices and improve our ability to treat DR-TB. Repurposing of already available antibiotics that are safe and readily accessible can be a practical solution for treating DR-TB, especially in resource-poor settings where majority of TB patients live. One class of antibiotics that matches this profile are the powerful carbapenems of the β-lactam group. There is a long history of their use in treating bacterial infections, they are safe and widely available, and emerging evidence shows that they can be immediately repurposed to treat DR-TB. This book chapter describes recent important developments in the repurposing of carbapenems for MDR and XDR-TB treatment.

Keywords
Tuberculosis · Drug resistance · Drug repurposing · Carbapenems

Abbreviations

WHO	World Health Organization
TB	Tuberculosis
MDR	Multidrug resistant
XDR	Extensively drug resistant
CrfA	Carbapenem-resistant factor A
PBP	Penicillin-binding protein
DHP-1	Dehydropeptidase 1

26.1 Introduction

According to the most recent WHO report, global tuberculosis (TB) is one of the top ten causes of death worldwide (Organization 2018). In 2017 alone, 10 million people fell ill with TB and 1.6 million succumbed to death. Two-thirds of TB cases were in ten countries alone: India (27%), China (9%), Indonesia (8%), the Philippines (6%), Pakistan (5%), Nigeria (4%), Bangladesh (4%) and South Africa (3%). About 1.7 billion people, 23% of the world's population, are estimated to harbour latent TB infection and are thus at risk of developing active TB disease during their lifetime (Organization 2018). Global prevalence of multidrug-resistant (MDR) TB and emergence of extensively drug-resistant (XDR) TB has further elevated the severity of this public health crisis. Three countries India (24%), China (13%) and Russian Federation (10%) account for almost half of the world's cases of MDR/rifampin-resistant TB. Amongst the cases of MDR-TB in 2017, 8.5% were estimated to have XDR-TB. The WHO has classified two additional classes of emerging drug-resistant

TB, extremely drug-resistant TB (XXDR-TB) and totally drug-resistant TB (TDR-TB) (Organization 2013). In 2016, the WHO reported that 58 countries and territories were treating 7234 XDR-TB patients with the majority of cases notified from four countries: India (2130), Ukraine (1206), the Russian Federation (1205) and South Africa (719) (Organization 2016; Prasad et al. 2017). A cross-sectional survey conducted amongst adults and children receiving antiretroviral therapy (ART) in Mumbai, India, revealed that DR-TB was diagnosed in 68 (34%) of 202 culture-positive cases. Of these 68 cases, 38% were MDR-TB, 21% were pre-XDR (MDR-TB plus resistance to either a fluoroquinolone or a second-line injectable drug), 6% XDR-TB and 2% XXDR-TB (Isaakidis et al. 2014). Therefore, there is an urgent need to develop new drugs or drug regimens to address the challenges associated with DR-TB.

Major efforts made in the past decade to develop new therapeutics against TB have delivered three new antibiotics, bedaquiline, delamanid and pretomanid, for TB treatment (Zumla et al. 2015; Wong et al. 2013; Andries et al. 2005; Diacon et al. 2009; Matsumoto et al. 2006; Singh et al. 2008). However, it is still unclear if these drugs will be safe for long-term use as is required for treating DR-TB and whether they would be affordable to TB patients, the majority of whom are financially destitute. An appealing solution could be repurposing of existing antibiotics that have not been used to treat TB, but exhibit potency against it. In pursuit of this approach, several existing drugs (sulfamethoxazole, sulfadiazine, linezolid, minocycline, amoxicillin/clavulanate and carbapenems such as meropenem, etc.) are currently being investigated for their efficacies to treat DR-TB (Sharma et al. 2017; Silva et al. 2018). Below we summarize recent developments in repurposing of carbapenems to treat DR-TB.

26.2 Oldie but Goldie: ß-Lactam Antibiotic for TB Treatment

One of the antibiotic classes for which there is empirical evidence of successful use in treating DR-TB is the β-lactams (Keener 2014). There are five major subclasses of β-lactams, penicillins, cephalosporins, monobactams, carbapenems and penems, that are available today for treating bacterial infections (Fig. 26.1). The order of these β-lactam subclasses as written above also represents the chronology of their development and introduction to the pharmaceutical market. The discovery and commercial development of penicillins was a major event in the modern antibiotic era. Beginning in the 1940s, penicillin was widely adopted to treat bacterial infections. However, penicillins exhibited poor activity against *M. tuberculosis*, the bacterium that causes TB. The existence of native β-lactamase that hydrolyses penicillins made them ineffective against *M. tuberculosis*. Later when cephalosporins were developed, similar lack of activity against *M. tuberculosis* and their hydrolysis by β-lactamase was documented. Based on these observations, it was considered that antibiotics bearing the β-lactam core would be susceptible to *M. tuberculosis* β-lactamase and therefore inefficacious against TB. While this concept is generally true with penicillin and cephalosporin subclasses, emerging evidence shows that this

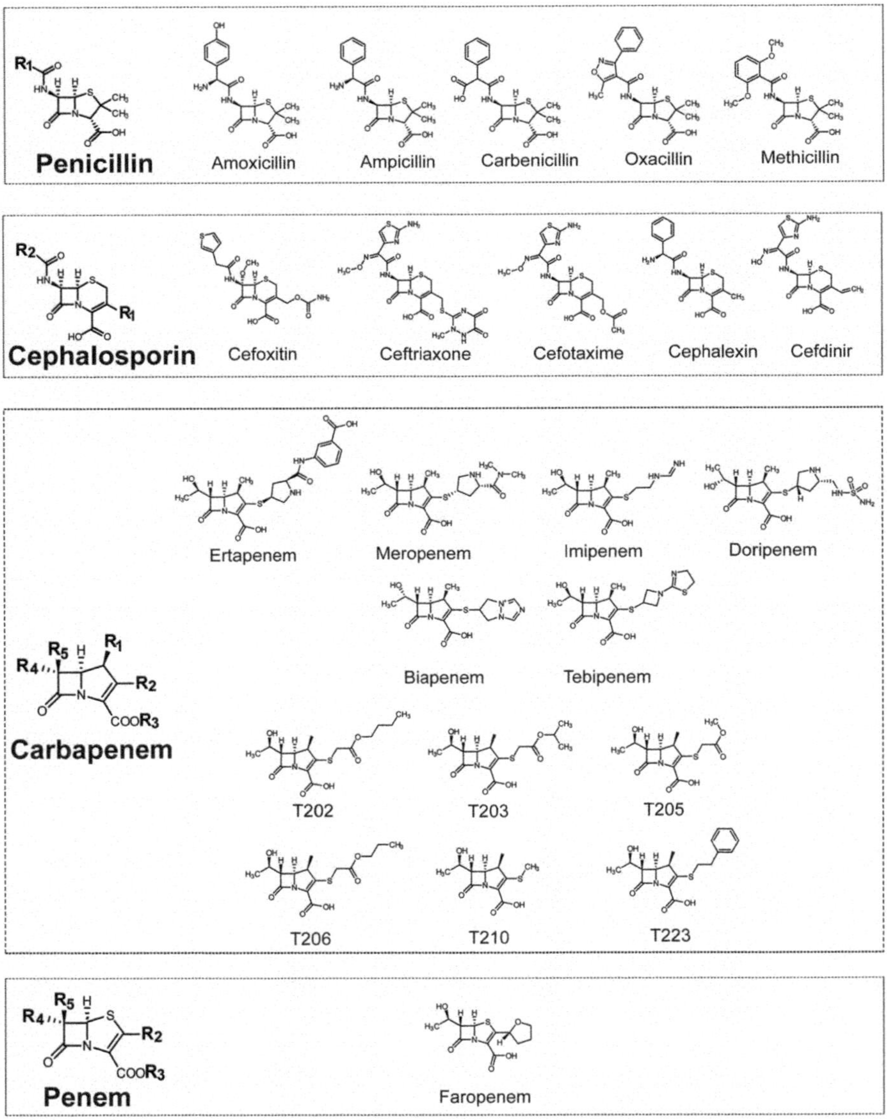

Fig. 26.1 Chemical structure of ß-lactam antibiotic: Penicillin, Cephalosporin, Carbapenem and Penem

may not be applicable to newer subclasses of β-lactams, namely, carbapenems and penems. Recent biochemical studies of *M. tuberculosis* and β-lactams have shown that carbapenems and penems are not readily hydrolysed by its β-lactamase. In addition, and perhaps more importantly, these β-lactams are uniquely effective at inhibiting an alternate enzyme class with which *M. tuberculosis* generates its cell wall peptidoglycan (Kumar et al. 2017b; Correale et al. 2013), a structural molecule that is required for survival, growth and cell division.

β-Lactams exert their activity by inhibiting synthesis of peptidoglycan. The final step of peptidoglycan synthesis involves polymerization of stem peptides to generate a single three-dimensional molecule that becomes the peptidoglycan. According to the historical model, only one enzyme class, namely, D,D-transpeptidases (also known as penicillin-binding proteins), catalyses this reaction (Walsh 2003)). This model of peptidoglycan was developed using rapidly growing model bacteria such as *B. subtilis* and *E. coli*. While this model is representative of many bacteria, recent discoveries demonstrate that peptidoglycan of *M. tuberculosis* is unique, and the final step of peptidoglycan synthesis is undertaken by two distinct enzyme classes: the D,D-transpeptidases and the newly identified L,D-transpeptidases (Erdemli et al. 2012; Gupta et al. 2010; Brammer Basta et al. 2015; Schoonmaker et al. 2014; Walsh 2003). In *M. tuberculosis,* the L,D-transpeptidases catalyse higher proportions of the reaction and therefore are potentially likely to be at least as important as the D, D-transpeptidases (Gupta et al. 2010). Recent studies have shown that, unlike penicillins and cephalosporins, carbapenems and penems are effective at inhibiting L,D-transpeptidases of *M. tuberculosis* (Kumar et al. 2017b; Brammer Basta et al. 2015; Erdemli et al. 2012; Gupta et al. 2010; Moraes et al. 2015; Correale et al. 2013; Kim et al. 2013). These studies suggest that higher potencies of carbapenems and penems against *M. tuberculosis* are likely based not only on their ability to inhibit D, D-transpeptidases and resist hydrolysis by β-lactamase (Hugonnet et al. 2009) but more critically because of their unique ability to inhibit the L,D-transpeptidases (Fig. 26.2).

Assessment of minimum inhibitory concentrations confirmed that carbapenems and penems are more potent than penicillins and cephalosporins against

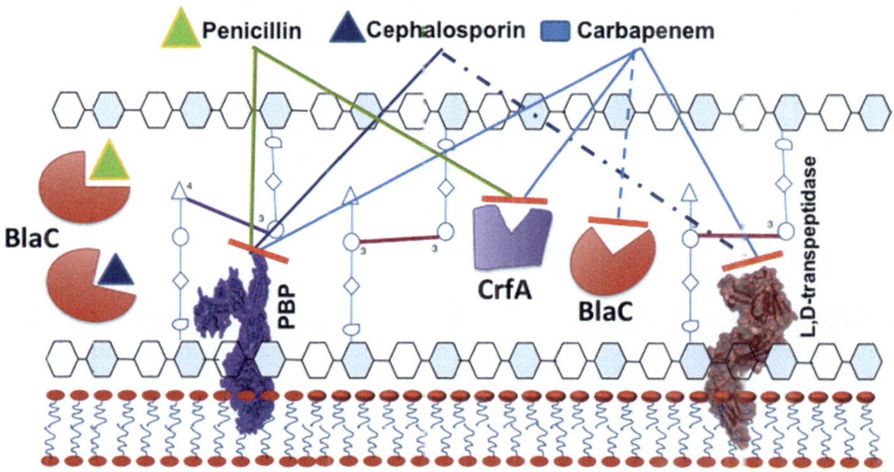

Fig. 26.2 Carbapenem inhibits multiple targets in Mycobacterium tuberculosis. *BlaC* ß-lactamase, *PBP* Penicillin Binding Protein, *CrfA* Carbapenem Resistance Factor A

M. tuberculosis (Hugonnet et al. 2009; Kaushik et al. 2015; Solapure et al. 2013; Horita et al. 2014). Carbapenems also inhibit a new class of ß-lactam hydrolysing enzymes in *M. tuberculosis*, carbapenem-resistant factor A (CrfA), which can confer *M. tuberculosis* resistance to carbapenems (Kumar et al. 2017b).

To date, meropenem, imipenem and ertapenem have been tested for synergy with clavulanate in XDR-TB patients through intravenous route of administration. Faropenem, a penem, has been clinically tested in a limited number of patients in combination with other TB drugs (Silva et al. 2018; Arbex et al. 2016; Tiberi et al. 2016a, b, c; van Rijn et al. 2016; Zuur et al. 2018; Payen et al. 2018; Diacon et al. 2016; Deshpande et al. 2016). There is no published data on the clinical use of doripenem, panipenem or biapenem (Zhang et al. 2016), but their efficacies have been studied in vitro or in vivo in preclinical models of TB (Kaushik et al. 2015; Kumar et al. 2017a, b; Kaushik et al. 2017a, b; Story-Roller and Lamichhane 2018; Walsh 2003).

26.3 Empirical Evidence of Efficacy of Carbapenems Against TB

Clinicians have evaluated the potential utility of ß-lactams for treating MDR and XDR-TB. Anecdotally, in select MDR-TB patients, regimens containing amoxicillin/clavulanate and other second-line agents have been effective against MDR-TB, though this approach has met considerable scepticism (Yew and Chau 1996). Thereafter, different subclasses of ß-lactams, including various penicillins, cephalosporins and carbapenems, have been anecdotally tried for the treatment of MDR and XDR-TB (Ramon-Garcia et al. 2016; Jaganath et al. 2016; Chambers et al. 2005). Amongst the different subclasses of ß-lactam drugs, only carbapenems have emerged as potent inhibitors of MDR-TB and XDR-TB. In 2005, clinical efficacy of imipenem was tested in ten patients in conjunction with an aminoglycoside or capreomycin, which indicated that imipenem was therapeutically useful for TB treatment (Chambers et al. 2005). Thereafter, several carbapenems have been tested for their clinical efficacy in MDR-TB and XDR-TB settings (Sotgiu et al. 2015a, b, Sotgiu et al. 2016; Jaganath et al. 2016) and are currently in various stages of clinical trials in combination with other anti-TB drugs as seen in Table 26.1.

26.4 Repurposing of Meropenem

After Hugonnet et al. (Hugonnet et al. 2009) demonstrated in a laboratory setting that the carbapenem drug meropenem, combined with clavulanate, is consistently effective against XDR-TB, there has been tremendous progress in the repurposing of carbapenems for MDR/XDR-TB treatment. Soon after this seminal publication, an in vivo study evaluated the efficacy of meropenem/clavulanate in a chronically infected mouse model of TB (England et al. 2012). In this study, one group of mice was treated with meropenem alone at 300 mg/kg body weight by subcutaneous injection twice daily, and a second group was given the same dosage of meropenem

Table 26.1 Different clinical trials of carbapenems for treatment of drug-resistant TB

Drug	Clinical trial identifier	Sponsor institute	Clinical study phase status	Title of study
Rifampin + *Meropenem* + Amoxicillin/ Clavulanate	NCT03174184	Johns Hopkins University	2 (recruiting)	A Phase 2a study of the early bactericidal activity of Rifampin (RIF) in combination with Meropenem Plus Amoxicillin/ Clavulanate among adults with Rifampin-resistant or Rifampin-susceptible pulmonary Tuberculosis
Meropenem + Amoxiycillin/ clavulanic *Faropenem*, Amoxiycillin/ clavulanic acid	NCT02349841	Task Foundation NPC	2 (completed)	A Phase 2 trial to evaluate the early bactericidal activity, safety and tolerability of Meropenem, administered intravenously, Plus Amoxiycillin/CA and Faropenem, administered orally, Plus Amoxiycillin/ CA in adult patients with newly diagnosed, smear-positive pulmonary Tuberculosis.
Rifampicin, rifabutin, isoniazid, high dose isoniazid, pyrazinamide, ethambutol, levofloxacin, moxifloxacin, ofloxacin, gatifloxacin, amikacin, capreomycin, kanamycin, streptomycin, ethionamide, prothionamide, cycloserine, terizidone, pretomanid,	NCT03237182	Centre for the AIDS Programme of Research in South Africa	4 (recruiting)	The individualized M(X) drug-resistant TB treatment strategy study A strategy to improve treatment outcomes in patients with drug-resistant TB

(continued)

Table 26.1 (continued)

Drug	Clinical trial identifier	Sponsor institute	Clinical study phase status	Title of study
linezolid, sutezolid, clofazimine, bedaquiline, delamainid, para-aminosalicylic acid, *imipenem*/cilastatin, *Meropenem*, amoxicillin/ clavulanate, clarithromycin, azithromycin and thioacetazone				
Isoniazid, Rifampicin, Pyrazinamide Ethambutol, Levofloxacin, Moxifloxacin, Gatifloxacin, Amikacin, Capreomycin, Kanamycin (Streptomycin), Ethionamide, Cycloserine, terizidone, Clofazimine, Bedaquiline, Delamanid, p-aminosalicylic acid, Imipenem-cilastatin, Amoxicillin-clavulanat e, Thioacetazone	NCT03625739	Beijing Children's Hospital	(recruiting)	Population Pharmacokinetics of anti- tuberculosis drugs in children with Tuberculosis
Ertapenem	NCT01730664	University Medical Center Groningen	2 (completed)	Pharmacokinetics and Pharmacodynamics of *Ertapenem* in patients with Tuberculosis
Faropenem, Amoxicillin/ clavulanic acid 500 mg/125 mg Rifampicin 10 mg/ kg	NCT02393586	National University Hospital, Singapore	1 (completed)	Evaluating pharmacokinetics and whole blood bactericidal activity against Mycobacterium Tuberculosis of single doses of

(continued)

Table 26.1 (continued)

Drug	Clinical trial identifier	Sponsor institute	Clinical study phase status	Title of study
				Faropenem Plus Amoxicillin/ Clavulanic acid in healthy volunteers
Faropenem, Amoxicillin/ clavulanic acid 500 mg/125 mg cCefadroxil 1 g Rifampicin 10 mg/ kg	NCT02381470	National University Hospital, Singapore	2 (not yet recruiting)	Trial of *Faropenem* and Cefadroxil (in combination with Amoxicillin/ Clavulanic Acid and standard TB drugs) in patients with pulmonary Tuberculosis : measurement of early bactericidal activity and effects on novel biomarkers

in combination with 50 mg/kg of clavulanate given orally. Both meropenem alone and the combination of meropenem with clavulanate significantly reduced the bacterial burden in both the spleen and lungs. These drugs were also tested in New Zealand white rabbits, which are capable of developing hypoxic pulmonary lesions. To prevent hydrolysis of meropenem by renal dihydropeptidase-1 (DHP-1), which is much more active in rabbits compared to humans, the DHP-1 inhibitor cilastatin was co-administered with meropenem. Cilastatin increased drug exposure by 4-fold, but resulted in significant adverse effects leading to early cessation of the experiment (England et al. 2012). In summary, meropenem in synergy with clavulanate exhibits bactericidal activity in macrophage and murine models of TB, but assessment in higher animals such as rabbits is complicated by the increased activity of renal DHP-1 that hydrolyses carbapenems. Since the human isoform of DHP-1 is less active than in mice or rabbits, prospective clinical evaluation of meropenem/clavulanate in humans with chronic tuberculosis was absolutely warranted.

Following the discovery of potency of meropenem/clavulanate against TB in in vitro and in vivo models, a clinical trial (n = 96) evaluating the effectiveness and safety of meropenem-/clavulanate-containing regimens against MDR-TB and XDR-TB was performed (Tiberi et al. 2016b). In this study, similar sputum culture conversion rates were observed between meropenem-/clavulanate-containing and meropenem-/clavulanate-sparing regimens against XDR-TB; however, treatment success in the XDR-TB cohort with meropenem/clavulanate was higher than previously achieved in the largest observational cohort study of MDR/XDR-TB treatment to date (46.8% vs. 40%) (Falzon et al. 2013). In another study that evaluated early bactericidal activity of meropenem in combination with amoxicillin and clavulanate in TB patients, this combination exhibited desirable bactericidal activity (NCCT02349841) (Diacon et al. 2016).

Another phase 2 clinical trial is underway for evaluating the early bactericidal efficacy of rifampin in combination with meropenem plus amoxicillin/clavulanate amongst adults with rifampin-resistant TB (NCT03174184). A phase 4 clinical trial, sponsored by the Centre for the AIDS Programme of Research in South Africa (NCT03237182), is currently in progress to develop an individualized treatment strategy for drug-resistant tuberculosis. This trial aims to use *M. tuberculosis* whole genome sequencing prior to selection of a drug-resistant tuberculosis treatment regimen that may include various drugs, including meropenem. Results from all three clinical trial studies on meropenem are currently pending (Table 26.1). Recently, a retrospective observational case series study was published assessing long-term follow-up of XDR-TB patients treated with meropenem/clavulanate (Payen et al. 2018). A total of 18 patients were followed for a median of four years after TB treatment. TB isolates from these patients were resistant to a median of 10 anti-tubercular drugs; 15 isolates were XDR-TB and three were pre-XDR. Out of 18, 15 patients achieved cure, and no relapses occurred in these patients on a median follow-up of 4 years (Payen et al. 2018). No adverse events were observed with meropenem/clavulanate, but the use of implantable intravenous device leads to catheter infection resulting in septicaemia in six patients.

26.5 Repurposing of Imipenem

Imipenem and meropenem were initially assessed for their in vitro antibacterial activity against *M. tuberculosis* by Watt et al. in 1992 (Watt et al. 1992). The first clinical evaluation of imipenem efficacy against TB included ten patients with TB isolates resistant to both isoniazid and rifampin (Chambers et al. 2005). In this study, patients were treated with imipenem in addition to standard first- or second-line anti-tubercular drugs. Eight of 10 patients achieved sputum culture conversion and seven patients remained culture negative off treatment, suggesting imipenem as a potential adjunctive treatment option for MDR-TB.

One clinical trial sponsored by Beijing Children's Hospital (NCT03625739, Table 26.1) is currently recruiting patients for the study of pharmacokinetics of anti-TB drugs, including imipenem, in children with TB.

A comparative study of imipenem/clavulanate and meropenem/clavulanate has also been performed to compare their relative efficacies against TB (Tiberi et al. 2016c). Eighty-four patients with MDR/XDR-TB were treated with imipenem/clavulanate, and 96 patients were treated with meropenem/clavulanate in conjunction with other TB drugs. The cohort of patients treated with a meropenem-/clavulanate-containing TB regimen achieved better results compared to imipenem/clavulanate. Specifically, culture conversion rates and treatment success rates were significantly higher with meropenem/clavulanate (Tiberi et al. 2016b). Both meropenem/clavulanate and imipenem/clavulanate were well tolerated with minimal adverse effects. Imipenem, however, is cheaper than meropenem and is more economically viable in developing countries. Imipenem/cilastatin is now categorized by WHO as a group 5 medication for the treatment of MDR/XDR-TB.

26.6 Repurposing of Ertapenem

The first clinical research of ertapenem for TB treatment was published in 2016 (Tiberi et al. 2016a). In vitro and pharmacological data suggested that ertapenem has a longer half-life than meropenem, allowing a more simplified single parenteral administration of ertapenem per day compared to thrice daily administration of meropenem or imipenem (Tremblay et al. 2010; Tiberi et al. 2016a) In this small retrospective cohort study, five patients with pulmonary TB (two patients with XDR and three with pre-XDR) had been treated with regimens containing ertapenem. Of these, four patients achieved cure and ertapenem was generally well tolerated. A phase 2 clinical trial studying pharmacokinetics and pharmacodynamics of ertapenem in patients with TB has recently been completed by the University Medical Center Groningen (NCT01730664, Table 26.1).

26.7 Repurposing of Faropenem

Faropenem is an orally bioavailable penem that has shown efficacy against TB both in vitro and in vivo, with superior bactericidal activity compared to meropenem (Dhar et al. 2015). Additional in vitro efficacy studies of several carbapenems, in addition to faropenem, against TB showed that faropenem was the most potent single agent after tebipenem against *M. tuberculosis* H37Rv strain (Kumar et al. 2017b). Faropenem was also noted to exhibit significant synergy when combined with rifampin against H37Rv as well as several clinical TB isolates (Kaushik et al. 2015). Of note, this synergistic effect was also observed to a lesser extent between rifampin and carbapenems doripenem, biapenem and meropenem.

Two clinical trials studying faropenem have been completed: a phase 1 trial by the National University Hospital, Singapore (NCT02393586), and a phase 2 trial by TASK Foundation NPC (NCT02349841). Both studies evaluated the anti-tubercular activity and pharmacokinetics of faropenem in synergy with amoxicillin/clavulanate. Additionally, the National University Hospital, Singapore, is conducting a phase 2 trial of faropenem/cefadroxil in combination with amoxicillin/clavulanate (NCT02381470). Results of the first trial have been published by Gurumurthy et al., in which blood samples from healthy volunteers were inoculated with *Mtb* H37Rv shortly after treatment with either rifampin alone, faropenem plus amoxicillin/clavulanate or a combination of all three drugs. They found that while faropenem plus amoxicillin/clavulanate exhibited no bactericidal activity, the three-drug combination was effective (Gurumurthy et al. 2017). Of note, this is in contrast to the previously cited study by Dhar et al. that showed efficacy of faropenem alone against an intracellular macrophage model of TB (Dhar et al. 2015).

Most recently, in 2018, 20 clinical isolates, including MDR and XDR, were tested for susceptibility against ertapenem and faropenem. Eighteen out of twenty isolates were resistant to ertapenem (in combination with amoxicillin/clavulanate). 50% of the isolates showed some level of susceptibility against faropenem that was further augmented in combination with amoxicillin/clavulanate (Giovanni Satta 2018).

26.8 Repurposing of Biapenem

Biapenem has been shown to exhibit bactericidal activity against *M. tuberculosis* in vitro and in vivo (Kaushik et al. 2017a, b, 2015; Kumar et al. 2017a, b; Zhang et al. 2016), and an in vivo study in mice directly comparing biapenem to faropenem against TB showed both agents to reduce bacterial loads in the lungs compared to untreated controls, with biapenem exhibiting superior killing. In this study, biapenem was found to be effective against both H37Rv and rifampin-resistant strains. The addition of rifampin resulted in further synergistic activity against H37Rv, although this was not observed amongst resistant strains (Kumar et al. 2017b).

To date, there is not a single clinical study on biapenem for the treatment of TB, but preclinical studies suggest this drug warrants further investigation. This is especially true in light of the fact that biapenem is much more resistant to hydrolysis by renal dehydropeptidase 1 (DHP-1) than meropenem, which may make it a more viable agent in vivo (Hikida et al. 1992).

26.9 Repurposing of Tebipenem

Although few studies evaluating tebipenem against TB have been performed, it does appear to exhibit anti-tubercular activity in vitro, with increased efficacy noted with the addition of the β-lactamase inhibitor, clavulanate (Horita et al. 2014; Kaushik et al. 2015).

26.10 Development of Evolved Carbapenems Against Tuberculosis

The ß-lactam antibiotics, especially carbapenems, are being repurposed for the treatment of drug-resistant tuberculosis. However, these antibiotics were developed to treat other non-tubercular infections. There is a need to understand the molecular mechanism of inhibition of *M. tuberculosis* by ß-lactams and develop new ß-lactams that exhibit improved efficacy against *M. tuberculosis.* An initiative of such kind was taken where new evolved carbapenems were developed based on structure-based drug designing by targeting a newly discovered class of L,D-transpeptidase enzymes that are involved in synthesis of 3 $\rightarrow$3 trans-peptide linkages in the peptidoglycan of *M. tuberculosis* (Kumar et al. 2017b).

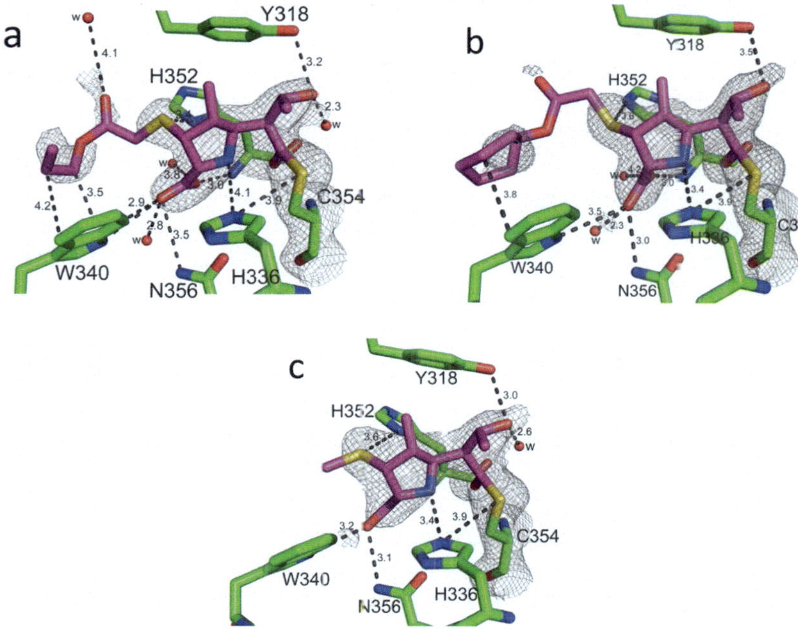

Fig. 26.3 Crystal structure of L,D-transpeptidase 2 (LdtMt2) in covalent complex with newly evolved anti-TB carbapenem drugs. (**a**) Crystal structure with T206; (**b**) with T208 and (**c**) with T210. (Kumar et al. 2017a, b)

Based on structure-activity relationship data (SAR), 12 evolved carbapenems were reported to show in vitro antibacterial activity with minimum inhibitory concentration (MIC_{90}) lower than that of meropenem, doripenem, biapenem, faropenem and tebipenem against *M. tuberculosis*. Four of synthesized evolved carbapenems, T205, T206, T208 and T210, were further evaluated for molecular interaction studies with L,D-transpeptidases Ldt_{Mt1} and Ldt_{Mt2}. These compounds exhibited irreversible inhibition of the L,D-transpeptidases. Crystal structures of T206, T208 and T210 were solved in complex with L,D-transpeptidase Ldt_{Mt2} (Fig. 26.3) (Kumar et al. 2017b). Crystal structures of evolved carbapenems with L,D-transpeptidase will be valuable for further development of more therapeutically potent carbapenems that are also specific against *M. tuberculosis*.

26.11 Future Prospects in the Use of Carbapenems for TB Treatment

ß-Lactam antibiotics are the most commonly used antibiotics worldwide, yet only recently have we developed an appreciation for their anti-tubercular activity (especially carbapenems) and the potentiating effects of ß-lactamase inhibitors like clavulanate. This class of antibiotics has been re-evaluated in recent years due to the shortage of new drugs for the treatment of DR-TB. There is limited preclinical and clinical data on the use of carbapenems in combination with conventional anti-TB drugs. More rigorous in vivo evaluations should be undertaken to identify optimized dosing strategies to treat DR-TB. Most of the existing carbapenems are administered intravenously, which is impractical in the clinical setting, especially in resource-limited areas. Particular emphasis should be placed on development of oral carbapenems, as both of the currently available oral carbapenem and penem formulations, tebipenem and faropenem, respectively, have shown potency against *M. tuberculosis*.

A coordinated effort is necessary in Asian, Southeast Asian and African countries that harbour the highest number of TB cases worldwide. These countries are in desperate need of novel drugs and combinations capable of effectively treating drug-resistant TB if we hope to eradicate this pathogen. Clinical trials assessing combinations of carbapenems with conventional anti-TB drugs for treatment of MDR-TB and XDR-TB have been undertaken in the USA and Europe, where carbapenem cost is high and TB prevalence is low. Countries with high TB burdens such as India and China have the potential to develop carbapenems at low cost and can easily enrol patients in clinical trials and consequently accelerate research and development of potent carbapenem containing regimens to treat DR-TB.

References

Andries K, Verhasselt P, Guillemont J, Gohlmann HW, Neefs JM, Winkler H, Van Gestel J, Timmerman P, Zhu M, Lee E, Williams P, de Chaffoy D, Huitric E, Hoffner S, Cambau E, Truffot-Pernot C, Lounis N, Jarlier V (2005) A diarylquinoline drug active on the ATP synthase of Mycobacterium tuberculosis. Science 307(5707):223–227. https://doi.org/10.1126/science.1106753

Arbex MA, Bonini EH, Kawakame Pirolla G, D'Ambrosio L, Centis R, Migliori GB (2016) Effectiveness and safety of imipenem/clavulanate and linezolid to treat multidrug and extensively drug-resistant tuberculosis at a referral hospital in Brazil. Rev Port Pneumol (2006) 22 (6):337–341. https://doi.org/10.1016/j.rppnen.2016.06.006

Brammer Basta LA, Ghosh A, Pan Y, Jakoncic J, Lloyd EP, Townsend CA, Lamichhane G, Bianchet MA (2015) Loss of a functionally and structurally distinct ld-Transpeptidase, LdtMt5, compromises cell wall integrity in Mycobacterium tuberculosis. J Biol Chem 290 (42):25670–25685. https://doi.org/10.1074/jbc.M115.660753

Chambers HF, Turner J, Schecter GF, Kawamura M, Hopewell PC (2005) Imipenem for treatment of tuberculosis in mice and humans. Antimicrob Agents Chemother 49(7):2816–2821. https://doi.org/10.1128/AAC.49.7.2816-2821.2005

Correale S, Ruggiero A, Capparelli R, Pedone E, Berisio R (2013) Structures of free and inhibited forms of the L,D-transpeptidase LdtMt1 from Mycobacterium tuberculosis. Acta Crystallogr D Biol Crystallogr 69(Pt 9):1697–1706. https://doi.org/10.1107/S0907444913013085

Deshpande D, Srivastava S, Nuermberger E, Pasipanodya JG, Swaminathan S, Gumbo T (2016) A Faropenem, Linezolid, and Moxifloxacin Regimen for both drug-susceptible and multidrug-resistant Tuberculosis in children: FLAME path on the Milky Way. Clin Infect Dis 63(Suppl 3): S95–S101. https://doi.org/10.1093/cid/ciw474

Dhar N, Dubee V, Ballell L, Cuinet G, Hugonnet JE, Signorino-Gelo F, Barros D, Arthur M, McKinney JD (2015) Rapid cytolysis of Mycobacterium tuberculosis by faropenem, an orally bioavailable beta-lactam antibiotic. Antimicrob Agents Chemother 59(2):1308–1319. https://doi.org/10.1128/AAC.03461-14

Diacon AH, Pym A, Grobusch M, Patientia R, Rustomjee R, Page-Shipp L, Pistorius C, Krause R, Bogoshi M, Churchyard G, Venter A, Allen J, Palomino JC, De Marez T, van Heeswijk RP, Lounis N, Meyvisch P, Verbeeck J, Parys W, de Beule K, Andries K, Mc Neeley DF (2009) The diarylquinoline TMC207 for multidrug-resistant tuberculosis. N Engl J Med 360 (23):2397–2405. https://doi.org/10.1056/NEJMoa0808427

Diacon AH, van der Merwe L, Barnard M, von Groote-Bidlingmaier F, Lange C, Garcia-Basteiro AL, Sevene E, Ballell L, Barros-Aguirre D (2016) Beta-Lactams against Tuberculosis–new trick for an old dog? N Engl J Med 375(4):393–394. https://doi.org/10.1056/NEJMc1513236

England K, Boshoff HI, Arora K, Weiner D, Dayao E, Schimel D, Via LE, Barry CE 3rd (2012) Meropenem-clavulanic acid shows activity against Mycobacterium tuberculosis in vivo. Antimicrob Agents Chemother 56(6):3384–3387. https://doi.org/10.1128/AAC.05690-11

Erdemli SB, Gupta R, Bishai WR, Lamichhane G, Amzel LM, Bianchet MA (2012) Targeting the cell wall of Mycobacterium tuberculosis: structure and mechanism of L,D-transpeptidase 2. Structure 20(12):2103–2115. https://doi.org/10.1016/j.str.2012.09.016

Falzon D, Gandhi N, Migliori GB, Sotgiu G, Cox HS, Holtz TH, Hollm-Delgado MG, Keshavjee S, DeRiemer K, Centis R, D'Ambrosio L, Lange CG, Bauer M, Menzies D, Collaborative Group for meta-analysis of individual patient data in M-T (2013) Resistance to fluoroquinolones and second-line injectable drugs: impact on multidrug-resistant TB outcomes. Eur Respir J 42 (1):156–168. https://doi.org/10.1183/09031936.00134712

Giovanni Satta XG, Canseco JO, Wey E, Drobniewski F, McHugh TD (2018) 2447. Ertapenem and Faropenem for the Treatment of drug-resistant Tuberculosis. Open Forum Infect Dis 5:1

Gupta R, Lavollay M, Mainardi JL, Arthur M, Bishai WR, Lamichhane G (2010) The Mycobacterium tuberculosis protein LdtMt2 is a nonclassical transpeptidase required for virulence and resistance to amoxicillin. Nat Med 16(4):466–469. https://doi.org/10.1038/nm.2120

Gurumurthy M, Verma R, Naftalin CM, Hee KH, Lu Q, Tan KH, Issac S, Lin W, Tan A, Seng KY, Lee LS, Paton NI (2017) Activity of faropenem with and without rifampicin against Mycobacterium tuberculosis: evaluation in a whole-blood bactericidal activity trial. J Antimicrob Chemother 72(7):2012–2019. https://doi.org/10.1093/jac/dkx081

Hikida M, Kawashima K, Nishiki K, Furukawa Y, Nishizawa K, Saito I, Kuwao S (1992) Renal dehydropeptidase-I stability of LJC 10,627, a new carbapenem antibiotic. Antimicrob Agents Chemother 36(2):481–483

Horita Y, Maeda S, Kazumi Y, Doi N (2014) In vitro susceptibility of Mycobacterium tuberculosis isolates to an oral carbapenem alone or in combination with beta-lactamase inhibitors. Antimicrob Agents Chemother 58(11):7010–7014. https://doi.org/10.1128/AAC.03539-14

Hugonnet JE, Tremblay LW, Boshoff HI, Barry CE 3rd, Blanchard JS (2009) Meropenem-clavulanate is effective against extensively drug-resistant Mycobacterium tuberculosis. Science 323(5918):1215–1218. https://doi.org/10.1126/science.1167498

Isaakidis P, Das M, Kumar AM, Peskett C, Khetarpal M, Bamne A, Adsul B, Manglani M, Sachdeva KS, Parmar M, Kanchar A, Rewari BB, Deshpande A, Rodrigues C, Shetty A, Rebello L, Saranchuk P (2014) Alarming levels of drug-resistant tuberculosis in HIV-infected patients in metropolitan Mumbai, India. PLoS One 9(10):e110461. https://doi.org/10.1371/journal.pone.0110461

Jaganath D, Lamichhane G, Shah M (2016) Carbapenems against Mycobacterium tuberculosis: a review of the evidence. Int J Tuberc Lung Dis 20(11):1436–1447. https://doi.org/10.5588/ijtld.16.0498

Kaushik A, Makkar N, Pandey P, Parrish N, Singh U, Lamichhane G (2015) Carbapenems and Rifampin exhibit synergy against Mycobacterium tuberculosis and Mycobacterium abscessus. Antimicrob Agents Chemother 59(10):6561–6567. https://doi.org/10.1128/AAC.01158-15

Kaushik A, Ammerman NC, Tasneen R, Story-Roller E, Dooley KE, Dorman SE, Nuermberger EL, Lamichhane G (2017a) In vitro and in vivo activity of biapenem against drug-susceptible and rifampicin-resistant Mycobacterium tuberculosis. J Antimicrob Chemother 72(8):2320–2325. https://doi.org/10.1093/jac/dkx152

Kaushik A, Gupta C, Fisher S, Story-Roller E, Galanis C, Parrish N, Lamichhane G (2017b) Combinations of avibactam and carbapenems exhibit enhanced potencies against drug-resistant Mycobacterium abscessus. Future Microbiol 12:473–480. https://doi.org/10.2217/fmb-2016-0234

Keener AB (2014) Oldie but goodie: Repurposing penicillin for tuberculosis. Nat Med 20 (9):976–978. https://doi.org/10.1038/nm0914-976

Kim HS, Kim J, Im HN, Yoon JY, An DR, Yoon HJ, Kim JY, Min HK, Kim SJ, Lee JY, Han BW, Suh SW (2013) Structural basis for the inhibition of Mycobacterium tuberculosis L, D-transpeptidase by meropenem, a drug effective against extensively drug-resistant strains. Acta Crystallogr D Biol Crystallogr 69 (Pt 3):420-431. doi:https://doi.org/10.1107/S0907444912048998

Kumar P, Kaushik A, Bell DT, Chauhan V, Xia F, Stevens RL, Lamichhane G (2017a) Mutation in an unannotated protein confers Carbapenem resistance in Mycobacterium tuberculosis. Antimicrob Agents Chemother 61(3). https://doi.org/10.1128/AAC.02234-16

Kumar P, Kaushik A, Lloyd EP, Li SG, Mattoo R, Ammerman NC, Bell DT, Perryman AL, Zandi TA, Ekins S, Ginell SL, Townsend CA, Freundlich JS, Lamichhane G (2017b) Non-classical transpeptidases yield insight into new antibacterials. Nat Chem Biol 13(1):54–61. https://doi.org/10.1038/nchembio.2237

Matsumoto M, Hashizume H, Tomishige T, Kawasaki M, Tsubouchi H, Sasaki H, Shimokawa Y, Komatsu M (2006) OPC-67683, a nitro-dihydro-imidazooxazole derivative with promising action against tuberculosis in vitro and in mice. PLoS Med 3(11):e466. https://doi.org/10.1371/journal.pmed.0030466

Moraes GL, Gomes GC, Monteiro de Sousa PR, Alves CN, Govender T, Kruger HG, Maguire GE, Lamichhane G, Lameira J (2015) Structural and functional features of enzymes of Mycobacterium tuberculosis peptidoglycan biosynthesis as targets for drug development. Tuberculosis (Edinb) 95(2):95–111. https://doi.org/10.1016/j.tube.2015.01.006

Organization WH (2016) Global Tuberculosis Report 2016

Organization WH (2018) Global Tuberculosis Report 2018

Payen MC, Muylle I, Vandenberg O, Mathys V, Delforge M, Van den Wijngaert S, Clumeck N, De Wit S (2018) Meropenem-clavulanate for drug-resistant tuberculosis: a follow-up of relapse-free cases. Int J Tuberc Lung Dis 22(1):34–39. https://doi.org/10.5588/ijtld.17.0352

Prasad R, Singh A, Balasubramanian V, Gupta N (2017) Extensively drug-resistant tuberculosis in India: current evidence on diagnosis & management. Indian J Med Res 145(3):271–293. https://doi.org/10.4103/ijmr.IJMR_177_16

Ramon-Garcia S, Gonzalez Del Rio R, Villarejo AS, Sweet GD, Cunningham F, Barros D, Ballell L, Mendoza-Losana A, Ferrer-Bazaga S, Thompson CJ (2016) Repurposing clinically approved cephalosporins for tuberculosis therapy. Sci Rep 6:34293. https://doi.org/10.1038/srep34293

Schoonmaker MK, Bishai WR, Lamichhane G (2014) Nonclassical transpeptidases of Mycobacterium tuberculosis alter cell size, morphology, the cytosolic matrix, protein localization, virulence, and resistance to beta-lactams. J Bacteriol 196(7):1394–1402. https://doi.org/10.1128/JB.01396-13

Sharma D, Dhuriya YK, Deo N, Bisht D (2017) Repurposing and revival of the drugs: a new approach to combat the drug resistant Tuberculosis. Front Microbiol 8:2452. https://doi.org/10.3389/fmicb.2017.02452

Silva DR, Dalcolmo M, Tiberi S, Arbex MA, Munoz-Torrico M, Duarte R, D'Ambrosio L, Visca D, Rendon A, Gaga M, Zumla A, Migliori GB (2018) New and repurposed drugs to treat multidrug- and extensively drug-resistant tuberculosis. J Bras Pneumol 44(2):153–160. https://doi.org/10.1590/s1806-37562017000000436

Singh R, Manjunatha U, Boshoff HI, Ha YH, Niyomrattanakit P, Ledwidge R, Dowd CS, Lee IY, Kim P, Zhang L, Kang S, Keller TH, Jiricek J, Barry CE 3rd (2008) PA-824 kills nonreplicating

Mycobacterium tuberculosis by intracellular NO release. Science 322(5906):1392–1395. https://doi.org/10.1126/science.1164571

Solapure S, Dinesh N, Shandil R, Ramachandran V, Sharma S, Bhattacharjee D, Ganguly S, Reddy J, Ahuja V, Panduga V, Parab M, Vishwas KG, Kumar N, Balganesh M, Balasubramanian V (2013) In vitro and in vivo efficacy of beta-lactams against replicating and slowly growing/nonreplicating Mycobacterium tuberculosis. Antimicrob Agents Chemother 57(6):2506–2510. https://doi.org/10.1128/AAC.00023-13

Sotgiu G, Alffenaar JW, Centis R, D'Ambrosio L, Spanevello A, Piana A, Migliori GB (2015a) Therapeutic drug monitoring: how to improve drug dosage and patient safety in tuberculosis treatment. Int J Infect Dis 32:101–104. https://doi.org/10.1016/j.ijid.2014.12.001

Sotgiu G, Centis R, D'Ambrosio L, Migliori GB (2015b) Tuberculosis treatment and drug regimens. Cold Spring Harb Perspect Med 5(5):a017822. https://doi.org/10.1101/cshperspect.a017822

Sotgiu G, D'Ambrosio L, Centis R, Tiberi S, Esposito S, Dore S, Spanevello A, Migliori GB (2016) Carbapenems to treat multidrug and extensively drug-resistant Tuberculosis: a systematic review. Int J Mol Sci 17(3):373. https://doi.org/10.3390/ijms17030373

Story-Roller E, Lamichhane G (2018) Have we realized the full potential of beta-lactams for treating drug-resistant TB? IUBMB Life 70(9):881–888. https://doi.org/10.1002/iub.1875

Tiberi S, D'Ambrosio L, De Lorenzo S, Viggiani P, Centis R, Sotgiu G, Alffenaar JW, Migliori GB (2016a) Ertapenem in the treatment of multidrug-resistant tuberculosis: first clinical experience. Eur Respir J 47(1):333–336. https://doi.org/10.1183/13993003.01278-2015

Tiberi S, Payen MC, Sotgiu G, D'Ambrosio L, Alarcon Guizado V, Alffenaar JW, Abdo Arbex M, Caminero JA, Centis R, De Lorenzo S, Gaga M, Gualano G, Roby Arias AJ, Scardigli A, Skrahina A, Solovic I, Sulis G, Tadolini M, Akkerman OW, Alarcon Arrascue E, Aleska A, Avchinko V, Bonini EH, Chong Marin FA, Collahuazo Lopez L, de Vries G, Dore S, Kunst H, Matteelli A, Moschos C, Palmieri F, Papavasileiou A, Spanevello A, Vargas Vasquez D, Viggiani P, White V, Zumla A, Migliori GB (2016b) Effectiveness and safety of meropenem/clavulanate-containing regimens in the treatment of MDR- and XDR-TB. Eur Respir J 47(4):1235–1243. https://doi.org/10.1183/13993003.02146-2015

Tiberi S, Sotgiu G, D'Ambrosio L, Centis R, Abdo Arbex M, Alarcon Arrascue E, Alffenaar JW, Caminero JA, Gaga M, Gualano G, Skrahina A, Solovic I, Sulis G, Tadolini M, Alarcon Guizado V, De Lorenzo S, Roby Arias AJ, Scardigli A, Akkerman OW, Aleksa A, Artsukevich J, Auchynka V, Bonini EH, Chong Marin FA, Collahuazo Lopez L, de Vries G, Dore S, Kunst H, Matteelli A, Moschos C, Palmieri F, Papavasileiou A, Payen MC, Piana A, Spanevello A, Vargas Vasquez D, Viggiani P, White V, Zumla A, Migliori GB (2016c) Comparison of effectiveness and safety of imipenem/clavulanate- versus meropenem/clavulanate-containing regimens in the treatment of MDR- and XDR-TB. Eur Respir J 47(6):1758–1766. https://doi.org/10.1183/13993003.00214-2016

Tremblay LW, Fan F, Blanchard JS (2010) Biochemical and structural characterization of Mycobacterium tuberculosis beta-lactamase with the carbapenems ertapenem and doripenem. Biochemistry 49(17):3766–3773. https://doi.org/10.1021/bi100232q

van Rijn SP, van Altena R, Akkerman OW, van Soolingen D, van der Laan T, de Lange WC, Kosterink JG, van der Werf TS, Alffenaar JW (2016) Pharmacokinetics of ertapenem in patients with multidrug-resistant tuberculosis. Eur Respir J 47(4):1229–1234. https://doi.org/10.1183/13993003.01654-2015

Walsh C (2003) Where will new antibiotics come from? Nat Rev Microbiol 1(1):65–70. https://doi.org/10.1038/nrmicro727

Watt B, Edwards JR, Rayner A, Grindey AJ, Harris G (1992) In vitro activity of meropenem and imipenem against mycobacteria: development of a daily antibiotic dosing schedule. Tuber Lung Dis 73(3):134–136. https://doi.org/10.1016/0962-8479(92)90145-A

Wong EB, Cohen KA, Bishai WR (2013) Rising to the challenge: new therapies for tuberculosis. Trends Microbiol 21(9):493–501. https://doi.org/10.1016/j.tim.2013.05.002

Yew WW, Chau CH (1996) New antimycobacterial agents. Monaldi Arch Chest Dis 51(5):394–404

Zhang D, Wang Y, Lu J, Pang Y (2016) In Vitro activity of beta-Lactams in combination with beta-Lactamase inhibitors against multidrug-resistant Mycobacterium tuberculosis Isolates. Antimicrob Agents Chemother 60(1):393–399. https://doi.org/10.1128/AAC.01035-15

Zumla A, Chakaya J, Centis R, D'Ambrosio L, Mwaba P, Bates M, Kapata N, Nyirenda T, Chanda D, Mfinanga S, Hoelscher M, Maeurer M, Migliori GB (2015) Tuberculosis treatment and management–an update on treatment regimens, trials, new drugs, and adjunct therapies. Lancet Respir Med 3(3):220–234. https://doi.org/10.1016/S2213-2600(15)00063-6

Zuur MA, Ghimire S, Bolhuis MS, Wessels AMA, van Altena R, de Lange WCM, Kosterink JGW, Touw DJ, van der Werf TS, Akkerman OW, Alffenaar JWC (2018) Pharmacokinetics of 2,000 milligram ertapenem in Tuberculosis patients. Antimicrob Agents Chemother 62(5). https://doi.org/10.1128/AAC.02250-17

The manufacturer's authorised representative in the EU is Springer Nature Customer Service Centre GmbH, Europaplatz 3, 69115 Heidelberg, Germany. If you have any concerns regarding our products, please contact ProductSafety@springernature.com

Printed and bound by CPI Group (UK) Ltd, Croydon, CR0 4YY
30/09/2026
02225304-0019